Silicon Nanocrystals

Edited by
Lorenzo Pavesi and
Rasit Turan

Related Titles

Saile, V., Wallrabe, U., Tabata, O. (eds.)

LIGA and its Applications

2009

ISBN: 978-3-527-31698-4

Hierold, C. (ed.)

Carbon Nanotube Devices

Properties, Modeling, Integration and Applications

2008

ISBN: 978-3-527-31720-2

Rao, C. N. R., Müller, A., Cheetham, A. K. (eds.)

Nanomaterials Chemistry

Recent Developments and New Directions

2007

ISBN: 978-3-527-31664-9

Misawa, H., Juodkazis, S. (eds.)

3D Laser Microfabrication

Principles and Applications

2006

ISBN: 978-3-527-31055-5

Klauk, H. (ed.)

Organic Electronics

Materials, Manufacturing and Applications

2006

ISBN: 978-3-527-31264-1

Müllen, K., Scherf, U. (eds.)

Organic Light Emitting Devices

Synthesis, Properties and Applications

2006

IISBN: 978-3-527-31218-4

Silicon Nanocrystals

Fundamentals, Synthesis and Applications

Edited by
Lorenzo Pavesi and Rasit Turan

WILEY-VCH Verlag GmbH & Co. KGaA

The Editors

Prof. Lorenzo Pavesi
Physics Department/Science Faculty
University of Trento
Via Sommarive 14
38100 Povo (Trento)
Italy

Prof. Dr. Rasit Turan
Department of Physics
Middle East Technical University
06531 Ankara
Turkey

Library of Congress Card No.: applied for

British Library Cataloguing-in-Publication Data
A catalogue record for this book is available from the British Library.

Bibliographic information published by the Deutsche Nationalbibliothek
The Deutsche Nationalbibliothek lists this publication in the Deutsche Nationalbibliografie; detailed bibliographic data are available on the Internet at http://dnb.d-nb.de.

© 2010 WILEY-VCH Verlag GmbH & Co. KGaA, Weinheim

Typesetting Thomson Digital, Noida, India
Printing and Binding Strauss GmbH, Mörlenbach
Cover Design Grafik-Design Schulz, Fußgönheim

Printed in the Federal Republic of Germany
Printed on acid-free paper

ISBN: 978-3-527-32160-5

Contents

Silicon Nanocrystals: Fundamentals, Synthesis and Applications. Edited by Lorenzo Pavesi and Rasit Turan
Copyright © 2010 WILEY-VCH Verlag GmbH & Co. KGaA, Weinheim
ISBN: 978-3-527-32160-5

List of Contributors

Aleksei Anopchenko
University of Trento
Physics Department
Nanoscience Laboratory
via Sommarive 14
38050 Povo (Trento)
Italy

Isaac Balberg
The Hebrew University
The Racah Institute of Physics
Jerusalem 91904
Israel

Simona Boninelli
Università di Catania
MATIS CNR-INFM and Dipartimento
di Fisica e Astronomia
via S. Sofia 64
95123 Catania
Italy

Elisabeeta Borsella
ENEA
ENEA – Frascati Research Centre
Via Enrico Fermi 45
00045 Frascati (Rome)
Italy

Ceyhun Bulutay
Bilkent University
Department of Physics
Bilkent, Ankara 06800
Turkey

Iain Calder
Group IV Semiconductor Inc.
400 March Road
Ottawa, ON
Ontario, K2K 3H4
Canada

Han-Yun Chang
New Jersey Institute of Technology
Department of Electrical and Computer
Engineering
Newark, NJ
USA

Dominik Clément
Diehl BGT Defence GmbH & Co. KG
Fischbachstr. 16
D-90552 Roethenbach
Germany

Gavin Conibeer
University of New South Wales
ARC Photovoltaics Centre of Excellence
Sydney
Australia

Silicon Nanocrystals: Fundamentals, Synthesis and Applications. Edited by Lorenzo Pavesi and Rasit Turan
Copyright © 2010 WILEY-VCH Verlag GmbH & Co. KGaA, Weinheim
ISBN: 978-3-527-32160-5

Nicola Daldosso
University of Trento
Physics Department
Nanoscience Laboratory
via Sommarive 14
38050 Povo (Trento)
Italy

Ilker Dogan
Middle East Technical University
Physics Department
Inonu Blvrd
06531 Ankara
Turkey

Christian Dufour
Université de Caen
CIMAP, UMR CEA/CNRS/ENSICAEN
6 Boulevard Maréchal Juin
14050 Caen Cedex
France

Robert Elliman
Australian National University
Research School of Physical Sciences
and Engineering
Department of Electronic Materials
Engineering
Canberra, ACT 0200
Australia

Mauro Falconieri
ENEA – Casaccia Research Centre
Via Anguillarese, 301
00123 S. Maria di Galeria (Rome)
Italy

Giorgia Franzò
Università di Catania
MATIS CNR-INFM and Dipartimento
di Fisica e Astronomia
via S. Sofia 64
95123 Catania
Italy

Minoru Fujii
Kobe University
Graduate School of Engineering
Department of Electrical and Electronic
Engineering
Rokkodai, Nada
Kobe 657-8501
Japan

Bernard Gelloz
Tokyo University of Agriculture and
Technology
2-24-16 Nakacho, Koganei-shi
184-8588 Tokyo
Japan

Arife Gencer
Middle East Technical University
Physics Department
Inonu Blvrd
06531 Ankara
Turkey

Fabrice Gourbilleau
Université de Caen
CIMAP, UMR CEA/CNRS/ENSICAEN
6 Boulevard Maréchal Juin
14050 Caen Cedex
France

Romain Guider
University of Trento
Physics Department
Nanoscience Laboratory
via Sommarive 14
38050 Povo (Trento)
Italy

Nathalie Herlin
Centre CEA de Saclay
Gif-sur-Yvette
91191 Cedex
France

Shaoyun Huang
Tokyo Institute of Technology
Quantum Nanoelectronics Research
Center and SORST-JST
O-okayama 2-12-1, Meguro
Tokyo 152-8552
Japan

Fabio Iacona
Università di Catania
MATIS CNR-INFM and Dipartimento
di Fisica e Astronomia
via S. Sofia 64
95123 Catania
Italy

Alessia Irrera
Università di Catania
MATIS CNR-INFM and Dipartimento
di Fisica e Astronomia
via S. Sofia 64
95123 Catania
Italy

Aylin Karakuscu
Università di Trento
Dipartimento di Ingegneria dei
Materiali e Tecnologie Industriali
Via Mesiano 77
38050 Trento
Italy

Uwe Kortshagen
University of Minnesota
Department of Mechanical Engineering
111 Church Street SE
Minnesota, MN 55455
USA

Dmitry Kovalev
University of Bath
Department of Physics
Bath BA2 7AY
UK

Jing Li
McMaster University
Department of Engineering Physics and
Centre for Emerging Device
Technologies
Hamilton, ON
Canada

Victor Loschenov
Natural Sciences Centre
General Physics Institute
38 Vavilov Str.
119997, Moscow
Russia

Lorenzo Mangolini
University of Minnesota
Department of Mechanical Engineering
111 Church Street SE
Minnesota, MN 55455
USA

Ingrid Mann
Kini University
School of Science and Engineering
Osaka
Japan

Peter Mascher
McMaster University
Department of Engineering Physics
and Centre for Emerging Device
Technologies
Hamilton, ON
Canada

Giuseppe Miserocchi
Università di Milano Bicocca
DIMS
Via Cadore, 48
20025 Monza
Italy

Daniel Navarro-Urrios
University of Trento
Physics Department
Nanoscience Laboratory
via Sommarive 14
38050 Povo (Trento)
Italy

Shunri Oda
Tokyo Institute of Technology
Quantum Nanoelectronics Research
Center and SORST-JST
O-okayama 2-12-1, Meguro
Tokyo 152-8552
Japan

Stefano Ossicini
Università di Modena e Reggio Emilia
CNR-INFM-S3 and Dipartimento di
Scienze e Metodi dell'Ingegneria
via Amendola 2 Pad. Morselli
I-42100 Reggio Emilia
Italy

Lorenzo Pavesi
University of Trento
Physics Department
Nanoscience Laboratory
via Sommarive 14
38050 Povo (Trento)
Italy

Alessandro Pitanti
University of Trento
Physics Department
Nanoscience Laboratory
via Sommarive 14
38050 Povo (Trento)
Italy

Xavier Portier
Université de Caen
CIMAP, UMR CEA/CNRS/ENSICAEN
6 Boulevard Maréchal Juin
14050 Caen Cedex
France

Francesco Priolo
Università di Catania
MATIS CNR-INFM and Dipartimento
di Fisica e Astronomia
via S. Sofia 64
95123 Catania
Italy

Richard Rizk
Université de Caen
CIMAP, UMR CEA/CNRS/ENSICAEN
6 Boulevard Maréchal Juin
14050 Caen Cedex
France

Ilarai Rivolta
Università di Milano Bicocca
DIMS
Via Cadore, 48
20025 Monza
Italy

Tyler Roschuk
McMaster University
Department of Engineering Physics and
Centre for Emerging Device
Technologies
Hamilton, ON
Canada

Anastasiya Ryabova
Natural Sciences Centre
General Physics Institute
38 Vavilov Str.
119997, Moscow
Russia

Ayse Seyhan
Middle East Technical University
Physics Department
Inonu Blvrd
06531 Ankara
Turkey

Gian Domenico Soraru
Università di Trento
Dipartimento di Ingegneria dei
Materiali e Tecnologie Industriali
Via Mesiano 77
38050 Trento
Italy

Rita Spano
University of Trento
Physics Department
Nanoscience Laboratory
via Sommarive 14
38050 Povo (Trento)
Italy

Celine Ternon
Université de Caen,
CIMAP, UMR CEA/CNRS/ENSICAEN
6 Boulevard Maréchal Juin
14050 Caen Cedex
France

Leonid Tsybeskov
New Jersey Institute of Technology
Department of Electrical and Computer
Engineering
Newark, NJ 07102-1982
USA

Rasit Turan
Middle East Technical University
Department of Physics
Inonu Blvrd
06531 Ankara
Turkey

Jonathan Veinot
University of Alberta
Department of Chemistry
Edmonton, Alberta T6G 2G2
Canada

Dayang Wang
Max Planck Institute of Colloids and
Interfaces
Am Mühlenberg
14476 Potsdam
Germany

Jacek Wojcik
McMaster University
Department of Engineering Physics
and Centre for Emerging Device
Technologies
Hamilton, ON L8S4L7
Canada

Selcuk Yerci
Boston University
Department of Electrical and
Computer Engineering
Boston, MA 02215
USA

Zhizhong Yuan
University of Trento
Physics Department
Nanoscience Laboratory
via Sommarive 14
38050 Povo (Trento)
Italy

Margit Zacharias
Albert-Ludwigs-University Freiburg
Institute of Microsystem Technology
(IMTEK)
Georges-Köhler-Allee 103
79110 Freiburg
Germany

1
Introduction

Silicon Nanocrystals; Fundamentals, Synthesis, and Applications

Lorenzo Pavesi and Rasit Turan

Bulk Si crystal is the main material of today's microelectronic, photovoltaic, and MEMS technologies. Reducing the size of Si crystal to the nanoscale level brings about new properties and functionalities. Si nanocrystals (Si-nc) are expected to pave way for the new and exciting applications in microelectronic, photonic, photovoltaic, and nanobiotech industries. Being fully compatible with the existing technologies makes the use of Si nanocrystals easier and more attractive than other kind of nanoparticles. Thus, it is of great technological and scientific interest to know the physical and chemical properties of Si nanocrystals, its production methods, applications, and the way of their characterization.

This book presents fundamentals of Si nanocrystals and their applications in microelectronics, photonics, photovoltaics, and nanobiotechnology. Methods of preparation, growth kinetics, basic physical, chemical and surface properties, various applications such as flash memories, photonic and photovoltaic components, biosensing are extensively reviewed.

Figure 1.1 reports the number of results one gets when one looks for Si-nc in Google™ Scholar. A steady rise in the number of publications witnessing the rising interest in this material is observed. The initial search results refer to four papers reporting about different properties of Si-nc: the first is the paper by Pavesi *et al.* on the observation of optical gain [1], the second is the paper by Tiwari *et al.* on the use of Si-nc for memories [2], the third is the paper by Wilson *et al.* on the demonstration of quantum size effects in Si-nc [3], and the fourth is the paper by Mutti *et al.* on the observation of room temperature luminescence in Si-nc [4]. When one makes the same search on Google™ the first result concerns the use of Si-nc for photovoltaics [5]. All the key ingredients that make Si-nc appealing for photonics and microelectronics are discussed in these papers: quantum size effects make new phenomena appear in silicon, such as room temperature visible photoluminescence, optical gain, coulomb blockade and multiexciton generation.

However, Si-nc has a much wider application spectrum than bare silicon photonics or microelectronics. The goal of this book is to present the various applications of

Silicon Nanocrystals: Fundamentals, Synthesis and Applications. Edited by Lorenzo Pavesi and Rasit Turan
Copyright © 2010 Wiley-VCH Verlag GmbH & Co. KGaA, Weinheim
ISBN: 978-3-527-32160-5

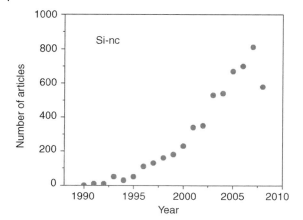

Figure 1.1 Number of articles versus year as reported by Google™ Scholar in a search performed on December 31, 2008. The search keys were "low-dimensional silicon" or "silicon nanocrystal(s) or "silicon nanocluster(s)." The total number of articles integrated over the years is 6220, while the results for "silicon photonics" is 1740 and for "porous silicon" is 28 800. The single point referred to year 2008 is low due to the time delay between the search date and the actual database construction.

Si-nc and introduce the first comprehensive collection of chapters on the technology, the synthesis and the applications of this new material.

The book is introduced with a chapter on fundamentals of Si nanocrystals, where a comprehensive theoretical overview of the electronic structure and optical properties of Si nanocrystals is given. The theoretical calculations are compared with experimental and first principles data. Chapter 3 discusses the main issues on optical properties of intrinsic, compensated and doped nanocrystals. Extensive experimental data and their analyses from a fundamental point of view can be found in this chapter. In Chapter 4, electrical transport through SiO_2 containing Si nanocrystals is presented. Transport mechanisms in low and high density regimes are shown to be different in nature. Hoping conduction, tunneling, and percolation mechanisms are discussed in detail. Chapter 5 deals with thermal properties of nanostructured Si. Lattice thermal conductivity in semiconductors, and specifically phonon boundary scattering, which is primarily responsible for the observed thermal conductivity reduction in different forms of Si nanostructures are discussed. A brief overview of thermal conductivity measurement is also provided in this chapter. Nanoparticles exhibit enormous surface area to volume ratios. It is expected that surface properties, including chemistry, of these versatile materials will strongly influence their material characteristics. Chapter 6 presents key studies of foundational silicon surface chemistry as it pertains to methods for controlling silicon nanocrystal surfaces for their functionalization. One of the extraordinary applications of Si nanocrystals can be found in Chapter 7 where the exciting story of chase after Si nanocrystals in cosmic dusts is presented. In Chapter 8, the control of the size of nanocrystals is discussed where it is shown that this can be achieved by using a Si/SiO_2 superlattice structure.

A comprehensive review of investigation of defect structure with ESR studies is also presented in Chapter 8.

Chapters 9–11 summarize the main three fabrication techniques for Si-nc: namely ion implantation (Chapter 9), PECVD (Chapter 10), and sputtering (Chapter 11). In order to form nanocrystals, the most common approach is to form a substoichiometric glass (i.e., rich in Si) and then by thermal annealing to cause a partial phase separation that yields a nucleation of small Si aggregates (Si-nc) in a dielectric matrix. The three fabrication techniques differ in their methods of producing the substoichiometric glass. Excess Si ions can be inserted into the glass by ion implantation, which is compatible with the existing Si technology. The number of ions embedded into the matrix can be determined precisely and the nanocrystal size can thus be controlled. As an alternative method, PECVD has been applied to form Si nanocrystals in matrices such as SiO_2, Si_3N_4, and SiC. PECVD offers a low temperature process that is fully compatible with the mass production of integrated electronic circuits. PECVD is widely used for thin film deposition in the semiconductor industry. Magnetron sputtering has also been used to fabricate single and multilayer structures successfully. As a low temperature and nontoxic method, sputtering is an attractive technique for applications such as solar cells, flash memories. In Chapter 12, an alternative technique to form the substoichiometric glass is presented: the sol–gel technique, which offers a low cost alternative to vacuum base technologies. After a brief description of the technology and an introduction on the polymer pyrolysis, the synthesis of Si-nc/SiO_2 materials, either as bulk or thin films is described, followed by the synthesis of Si-nc and SiC-nc containing SiO_2. In Chapter 13, synthesis of Si nanocrystals by means of nonthermal techniques is presented. Free-standing crystalline silicon particles of about 10 nm can be prepared by nonthermal plasmas at gas temperatures as low as room temperature. Use of several plasma processes, both inductively and capacitively coupled plasmas are shown to yield silicon nanocrystals several tens of nanometers in size.

Formation of nanocrystals in porous Si has been of interest since its invention in the beginning of 1990s. In spite of the fact that commercially viable light-emitting devices have not been achieved, the interest for porous Si has kept on growing due to its wide range of applications such as photonics, sensing, biotechnology, and acoustic. These interesting features of porous Si can be found in Chapter 14.

Chapter 15–20 presents an extensive review of various applications of Si-nc with updated research results. Chapter 15 describes how Si nanocrystals are applied to flash memory cells. The high dense integration capability, low voltage operation and low power consumption allow them to be one of pioneering candidates for the next generation flash memory. Chapter 15 attends to address the key subjects of nc-Si flash memory, encompassing fabrication methodologies, operation mechanisms, tolerance, present advances, and future prospects. Photonic applications are given in Chapter 16, where the exploitation of Si nanocrystals in various photonic building blocks for an all Si nanophotonics is reviewed. Applications of Si nanocrystals for guiding, modulating and generating, and/or amplifying light are reviewed. Integration of Si nanocrystal-based light sources and/or amplifiers within CMOS photonics platform is particularly focused in this chapter. Si-based lightening is a revolutionary development in Si technology. One of the expected gains of Si at nanometer dimension is the light generation based on quantum size effect. This promising feature of Si nanocrystals is extensively discussed in Chapter 17. A new generation of solar cell devices toward record high converging efficiencies are expected to utilize nanostructured materials. All Si quantum dot solar cell devices have attracted quite much attention in recent years. These new approaches are described in Chapter 18. Silicon nanoparticles have the potential to be useful in the biomedical applications since silicon is inert, nontoxic, abundant, and economical. Moreover, the silicon surface is apt for chemical functionalization, thus allowing for numerous stabilization and bioconjugation steps. Chapter 19 summarizes the use of Si nanoparticles in these biomedical applications. An interesting application of Si nanocrystals is found in Chapter 20 where highly explosive binary system based on porous silicon layers is described. The porous layers exhibit morphological properties that are different from most other energetic materials. Their production is completely compatible with the standard silicon technology and a large number of small explosive elements can be produced simultaneously on a single bulk silicon wafer

Finally characterization issues are discussed in Chapter 21 where the most important characterization and diagnostic techniques such as TEM, Raman, and XRD are presented from a metrology point of view.

References

1 Pavesi, L., Negro, L.D., Mazzoleni, C., Franzo, G., and Priolo, F. (2000) Optical gain in Si nanocrystals. *Nature*, **408**, 440.

2 Tiwari, S., Rana, F., Hanafi, H., Hartstein, A., Crabbe, E.F., and Chan, K. (1996) A silicon nanocrystals based memory. *Appl. Phys. Lett.*, **68**, 1377.

3 Wilson, W.L., Szajowski, P.F., and Brus, L.E. (1993) Quantum confinement in size-selected, surface-oxidized silicon nanocrystals. *Science*, **262**, 1242.

4 Mutti, P., Ghislotti, G., Bertoni, S., Bonoldi, L., Cerofolini, G.F., Meda, L., Grilli, E., and Guzzi, M. (1995) Room temperature visible luminescence from silicon nanocrystals in silicon implanted SiO layers. *Appl. Phys. Lett.*, **66**, 851.

5 Beard, M.C., Knutsen, K.P., Yu, P., Luther, J.M., Song, Q., Metzger, W.K., Ellingson, R.J., and Nozik, A.J. (2007) Multiple exciton generation in colloidal silicon nanocrystals. *Nano Lett.*, **7**, 2506.

2
Electronic and Optical Properties of Silicon Nanocrystals

Ceyhun Bulutay and Stefano Ossicini

2.1
Introduction

Several strategies have been researched over the last few years for light generation and amplification in silicon. One of the most promising one is based on silicon nanocrystals (Si NCs) with the aim of taking advantage of the reduced dimensionality of the nanocrystalline phase (1–5 nm in size) where quantum confinement, band folding, and surface effects play a crucial role [1, 2]. Indeed, it has been found that the Si NC bandgap increases with decrease in size and a photoluminescence (PL) external efficiency in excess of 23% is obtained [3]. Si NC-based LED with high efficiency have been obtained by using Si NC active layers [4] and achieving separate injection of electrons and holes [5]. Moreover, optical gain under optical pumping has been already demonstrated in a large variety of experimental conditions [6–11].

After the initial impulse given by the pioneering work of Canham on photoluminescence from porous Si [12], nanostructured silicon has received extensive attention. This activity is mainly centered on the possibility of getting relevant optoelectronic properties from nanocrystalline Si. The huge efforts made toward matter manipulation at the nanometer scale have been motivated by the fact that desirable properties can be generated by just changing the system dimension and shape. Investigation of phenomena such as the Stokes shift (difference between absorption and emission energies), the PL emission energy versus nanocrystals size, the doping properties, the radiative lifetimes, the nonlinear optical properties, the quantum-confined Stark effect (QCSE), and so on can give a fundamental contribution to the understanding of how the optical response of such systems can be tuned. A considerable amount of work has been done on excited Si NCs [2, 13–23], but a clear understanding of some aspects is still lacking. The question of surface effects, in particular oxidation, has been addressed in the last few years. Both theoretical calculations and experimental observations have been applied to investigate the possible active role of the interface on the optoelectronic properties of Si NCs. Different models have been proposed: Baierle *et al.* [24] have considered the role of the surface geometry distortion of small hydrogenated Si clusters in the excited state.

Silicon Nanocrystals: Fundamentals, Synthesis and Applications. Edited by Lorenzo Pavesi and Rasit Turan
Copyright © 2010 Wiley-VCH Verlag GmbH & Co. KGaA, Weinheim
ISBN: 978-3-527-32160-5

Wolkin *et al.* [25] have observed that oxidation introduces defects in the Si NC bandgap, which pin the transition energy. They claimed the formation of a Si=O double bond as the pinning state. The same conclusion has been recently reached by other authors [26–29], whereas Vasiliev *et al.* [30] have pointed out that similar results can also be obtained for O connecting two Si atoms (single bond) at the Si NC surface. The optical gain observed in Si NC embedded in SiO_2 has added impetus to these studies. Interface radiative states have been suggested to play a key role in the mechanism of population inversion at the origin of the gain [6, 8, 31]. Thus, the study of the nature and the properties of the interface between the Si NC and the SiO_2 host matrix has become crucial.

The calculation of the electronic and optical properties of nanostructures is a difficult task. First-principle studies are very demanding and in order to investigate very large systems, empirical methods are needed. In this paper we present and resume a comprehensive study of the structural, electronic, and optical properties of undoped and doped Si nanostructures terminated by different interfaces and, in particular, embedded in silicon dioxide matrix. For smaller nanocrystals we will present *ab initio* results, in particular the absorption and emission spectra and the effects induced by the creation of an electron–hole pair are calculated and discussed in detail, including the many-body effects. The aim is to investigate in a systematic way the structural, electronic, and stability properties of silicon nanostructures as a function of size and capping species, as well as pointing out the main changes induced by the nanostructure excitation.

The indisputable superiority of the first-principles approaches is gloomed by their applicability in systems of less than a thousand atoms with the current computer power. On the other hand, fabricated NCs of sizes 2–5 nm embedded in an insulating host matrix require computationally more feasible techniques that can handle more than 10 000 atoms including the surrounding matrix atoms. There exist several viable computational approaches for low-dimensional structures with a modest computational budget. The most common ones are the envelope function **k·p** [32], semiempirical tight binding [33], and semiempirical pseudopotential techniques [34]. The choice of a computational approach should be made according to the accuracy demands, but a subjective dimension is also introduced by the established biases of the particular practitioners. As the simplest of all, the envelope function **k·p** approach lacks the atomistic touch and more importantly, both qualitative and significant quantitative errors were identified, mainly derived from the states that were not accounted by the multiband **k·p** Hamiltonian [35]. As an atomistic alternative, the semiempirical tight binding approach has been successfully employed by several groups; see, Ref. [33] and references therein. Its traditional rival has always been the semiempirical pseudopotential approach. About a decade ago, Wang and Zunger proposed a more powerful technique that solves the pseudopotential-based Hamiltonian using a basis set formed by the linear combination of bulk bands (LCBBs) of the constituents of the nanostructure [36, 37]. Its main virtue is that it enables an insightful choice of a basis set with moderate number of elements. It should be mentioned that the idea of using bulk Bloch states in confined systems goes back to earlier times; one of its first implementations being the studies of Ninno *et al.* [38, 39]. Up to now, it has been tested on self-assembled quantum dots [36, 37], superlattices [40, 41], and high-electron mobility transistors [42]. In the context of Si NCs, very recently it has been used for studying the effects of NC aggregation [43], the linear [44], and third-order nonlinear optical properties [45], and also for characterizing the Auger recombination and carrier multiplication [46]. The fact that it is a pseudopotential-based method makes it more preferable over the empirical tight binding technique for the study of *optical* properties. For these reasons, in the case of large NCs, we shall make use of the linear combination of bulk bands technique. For both small NCs analyzed with *ab initio* techniques and larger NCs dealt with atomistic semiempirical approaches, a comparison with the experimental outcomes will be presented and discussed, whenever possible.

The organization of this review is based on two main sections for discussing the methodology and results of small and large NCs. A description of the theoretical methods used in the *ab initio* calculations is given in Section 2.2. The first-principle study of the physical systems is then presented starting from the analysis of hydrogenated Si NC (Section 2.2.1). We then consider the effect of oxidation (Section 2.2.2), of doping (Section 2.2.3) and finally of an embedding matrix (Section 2.2.4). As for the larger NCs embedded in a wide bandgap matrix, in Section 2.3, first a theoretical framework of the semiempirical approach is presented. Next, the comparisons of our results for the case of effective optical gap (Section 2.3.1), and radiative lifetime (Section 2.3.2) are presented. This is followed by the linear optical absorption properties (Section 2.3.3), where our interband, intraband, and excited state absorption results are summarized. The important subject of third-order nonlinear optical susceptibility is discussed in Section 2.3.4. Finally, a theoretical analysis of the quantum-confined Stark effect in Si NCs is presented in Section 2.3.5.

2.2
Ab Initio Calculation for Small Nanocrystals

The results have been obtained through plane-wave, pseudopotential density functional (DFT) calculations. In the case of the H-Si NC (see Section 2.2.1) all the calculations have been performed with the ABINIT code [47]. Norm-conserving, nonlocal Hamann-type pseudopotentials have been used. The Kohn–Sham wave functions have been expanded within a plane-wave basis set, choosing an energy cutoff of 32 Ry. The calculations performed are not spin-polarized. Each H-Si NC has been embedded within a large cubic supercell, containing vacuum in order to make nanocrystal–nanocrystal interactions negligible. The calculations for each cluster have been performed both in ground and in excited state. Full relaxation with respect to the atomic positions is performed for all the considered systems. We have addressed the issue of excited state configurations, which is mostly relevant for Si NC with a high surface-to-volume ratio using the so called Δ-SCF method [48–51], where total energies are calculated both in the ground state (GS) and in the excited state (ES). Here the ES corresponds to the electronic configuration in which the highest occupied single-particle state (highest occupied molecular orbital (HOMO)) contains a hole, while the lowest unoccupied single-particle state (lowest unoccupied molecular orbital (LUMO)) contains the corresponding electron. Thus, one can extract the absorption and emission energies and through their difference calculate the Stokes or Frank–Condon shift due to the lattice relaxation induced by the electronic excitation.

For the oxidized and doped Si NCs (see Sections 2.2.2 and 2.2.3) all the DFT calculations have been performed using the ESPRESSO package [52], within the generalized gradient approximation (GGA) using Vanderbilt ultrasoft [53] pseudopotentials for the determination of the structural properties and norm-conserving pseudopotential within the local density approximation (LDA) at the optimized geometry to evaluate the electronic and optical properties. This choice is due to the fact that Vanderbilt ultrasoft pseudopotentials allow the treatment of several hundreds of atoms per unit cell in the atomic relaxation process but the lift of the norm-conservation condition is a crucial and well-known problem for the calculating the optical transition matrix elements. The Si NC have been embedded in large supercells in order to prevent interactions between the periodic replicas (about 6 Å of vacuum separates neighbor clusters in all the considered systems). A careful analysis has been performed on the convergence of the structural and electronic properties with respect to both the supercell side and the plane-wave basis set cutoff. Since our aim is to allow a direct comparison between the experimental data and the theoretical results, we have calculated the transition energies within the Δ-SCF approach as well as the absorption and emission optical spectra. The excited state corresponds to the electronic configuration in which the highest occupied single-particle state (HOMO) contains a hole, while the lowest unoccupied single-particle state (LUMO) contains the corresponding electron. The optical response of the nanocrystal has been evaluated for both the ground and the excited state optimized geometries through the imaginary part of the dielectric function $\varepsilon_2(\omega)$

$$\varepsilon_2^\alpha(\omega) = \frac{4\pi^2 e^2}{m^2 \omega^2} \sum_{v,c,k} \frac{2}{V} |\langle \psi_{c,k}|p_\alpha|\psi_{v,k}\rangle|^2 \delta[E_c(k) - E_v(k) - \hbar\omega], \tag{2.1}$$

where $\alpha = (x, y, z)$, E_v and E_c denote the energies of the valence $\psi_{v,k}$ and conduction $\psi_{c,k}$ band states at a k point (Γ in our case), and V is the supercell volume. Owing to the strong confinement effects, only transitions at the Γ point have been considered. In order to perform emission spectra calculations, we have used the excited state geometry and the ground state electronic configuration. Thus, strictly speaking, $\varepsilon_2(\omega)$ corresponds to an absorption spectrum in the new structural geometry of the excited state: in this way we are simply considering the emission, in a first approximation, as the time reversal of the absorption [54] and therefore as a sort of photoluminescence spectra of the nanocrystals. Moreover in several cases we go beyond the one-particle approach including the self-energy corrections, by means of the GW approximation [55], and the excitonic effects, through the solution of the Bethe–Salpeter equation (BSE) [56]. To take into account the inhomogeneity of the system local field effects (LFEs) have been considered, too. This new approach where many-body effects, such as the self-energy corrections and the hole–electron interaction, are combined with a study of the structural distortion due to the impurity atoms in the excited state, allows calculating accurately the Stokes shift between the absorption and photoluminescence spectra.

2.2.1
Hydrogenated Silicon Nanocrystals

In this section we present an analysis of the structural, electronic, and optical properties of hydrogenated silicon nanocrystals (H-Si NCs) as a function of size and symmetry and, in particular, we will point out the main changes induced by the nanocrystal excitation. The calculations for each cluster have been performed both in ground and in excited state. The starting configuration for each cluster has been fixed with all Si atoms occupying the same position as in the bulk crystal, and passivating the surface with H atoms placed along the bulk crystal directions, at a distance determined by studying the SiH_4 molecule. It is worth pointing out that the starting H–Si NC has T_d symmetry, which is kept during relaxation in the ground state configuration. Nevertheless for excited state configurations such symmetry is generally lost, due to the occupation of excited energy levels. We have first of all investigated the structural distortions caused by the relaxation of these structures in different electronic configurations. The analysis of the structural properties reveals that the average Si–Si bond approaches the bulk bond length as the cluster dimension increases. In particular, on moving from the center of the cluster toward the surface, a contraction of the outer Si shells is observed. The presence of an electron–hole pair in the clusters causes a strong deformation of the structures with respect to their ground state configuration, and this is more evident for smaller systems and at the surface of the H-Si NC. This is expected, since for large clusters the charge density perturbation is distributed throughout all the structures, and the effect it locally induces becomes less evident. Baierle *et al.* [24] and Allan *et al.* [57] stressed the importance of bond distortion at the Si NC surface in the excited state (ES) in creating an intrinsic localized state responsible for the PL emission. The structural analysis is immediately reflected in the electronic structure. It can be noted from Table 2.1 that the expected decrease in the energy gap with increase in the cluster dimension and also that the excitation of the electron–hole pair causes a reduction in the energy gap as much significant as smaller is the cluster. For small excited clusters, the HOMO and LUMO become strongly localized in correspondence of the distortion, giving rise to defect-like states that reduce the gap. The distortion induced by the nanocluster excitation can give a possible explanation of the observed Stokes shift in these systems. The radiation absorption of the cluster in its ground state configuration induces a transition between the HOMO and LUMO levels, which is optically allowed for all these clusters. Such a transition is followed by a cluster relaxation in the excited state configuration giving rise to distorted

Table 2.1 Calculated values for the ground and excited state HOMO–LUMO energy gaps and for the absorption and emission energies calculated within the Δ-SCF approach for the considered H-Si NC.

	Absorption	GS HOMO–LUMO gap	Emission	EXC HOMO–LUMO gap
Si_1H_4	8.76	7.93	0.38	1.84
Si_5H_{12}	6.09	5.75	0.42	0.46
$Si_{10}H_{16}$	4.81	4.71	0.41	0.55
$Si_{29}H_{36}$	3.65	3.58	2.29	2.44
$Si_{35}H_{36}$	3.56	3.50	2.64	2.74

All values are in eV.

geometries (as previously shown) and to new LUMO and HOMO, whose energy difference is smaller than that in the ground state geometry. It is between these two last states that emission occurs, thus explaining the Stokes shift. It is also worth pointing out how such a shift changes as a function of the dimension. As the distortion is smaller for larger clusters, it is expected that the Stokes shift decreases with increase in the dimension. This is shown in Table 2.1 where the absorption and emission energies together with the Stokes shift calculated as described in Section 2.2 are reported.

A number of papers present in literature consider the HOMO–LUMO gaps of the ground and excited state as the proper absorption and emission energies; this leads to wrong results, mostly for the smaller clusters. In fact, from Table 2.1 it is clearly seen that the smaller the H-Si NC, the larger is the difference between the absorption and HOMO–LUMO ground state gap and between emission and HOMO–LUMO excited state gap. In our calculations, on going from smaller to larger clusters the difference between the HOMO–LUMO gap in the ground state and the absorption gap becomes smaller. In particular, the GS HOMO–LUMO gap tends to be smaller than the absorption energy while the ES HOMO–LUMO gap tends to be larger than the emission energy. In conclusion, trying to deduce the Stokes shift simply from the HOMO–LUMO gaps leads to errors, especially for small clusters. In particular the GS HOMO–LUMO gap tends to be smaller than the absorption energy while the ES HOMO–LUMO gap tends to be larger than the emission energy. In conclusion, trying to deduce the Stokes shift simply from the HOMO–LUMO gaps leads to not negligible errors that are more significant for smaller clusters. When comparing our results for the ground state with other DFT calculations we note that there is in general a good agreement between them. It is worth mentioning that our results for the absorption gaps of the Si_1H_4 (8.76 eV) and Si_5H_{12} (6.09 eV) clusters agree quite well with the experimental results of Itoh *et al.* [58]. They have found excitation energies of 8.8 and 6.5 eV, respectively. Regarding the Stokes shift, really few data exist in literature, and, in particular for really small H-Si NC (from Si_1H_4 to $Si_{10}H_{16}$), no data exist. The dependence of the Stokes shift from the H-Si NC size qualitatively agrees with the calculations of Puzder *et al.* [48] and Franceschetti and Pantelides [49].

2.2.2
Oxidized Silicon Nanocrystals

Recent experimental data have shown strong evidence that the surface changes of silicon nanocrystals exposed to oxygen produce substantial impact on their opto-electronics properties; thus oxidation at the surface has to be taken into account. In this section, first we will analyze the optical properties of oxidized Si NCs, using the Δ-SCF method [48–51]. Each Si NC has been embedded within a large cubic supercell, containing vacuum in order to make nanocrystal–nanocrystal interactions negligible. The starting configuration for each cluster has been fixed with all Si atoms occupying the same position as in the bulk crystal, and passivating the surface with H atoms placed along the bulk crystal directions, at a distance determined by studying the SiH_4 molecule. Two classes of systems have been studied, the Si_{10} and the Si_{29} core-based nanoclusters, and three types of Si/O bonds have been introduced at the cluster surface: the Si−O back bond, the Si>O bridge, and the Si=O double bond. Through formation energy calculation we have found that the configuration with the back-bonded oxygen is not favored with respect to the other two, and moreover the bridge-bonded configuration has been demonstrated to lead to the stablest isomer configuration [59, 60], too. Full relaxation with respect to the atomic positions is performed within DFT limit for all systems, both in the ground and in the excited configurations, using norm-conserving LDA pseudopotential with an energy cutoff of 60 Ry [52]. The ionic relaxation has produced structural changes with respect to the initial geometry, which strongly depend on the type of surface termination. In the case of $Si_{10}H_{14}{=}O$, the changes are mainly localized near the O atom, in particular the angle between the double-bonded O and its linked Si atom is modified. In the bridge structure, instead, the deformation is localized around the Si−O−Si bond determining a considerable strain in the Si−Si dimer distances [60]. Similar results are obtained for the larger Si_{29}-based clusters. The only difference is that now the distortion induced by the promotion of an electron is smaller, as expected, since for larger clusters the charge density perturbation is distributed throughout the structure, and the locally induced effect becomes less evident. These structural changes are reflected in the electronic and optical properties.

In Table 2.2 absorption and emission gaps are reported: the redshift of the emission gap with respect to the absorption is less evident for the case of the cluster

Table 2.2 Absorption and emission energy gaps and Stokes shift calculated as total energy differences within the Δ-SCF approach.

	Absorption	Emission	Stokes shift
$Si_{10}H_{14}{=}O$	2.79	1.09	1.70
$Si_{10}H_{14}{>}O$	4.03	0.13	3.90
$Si_{29}H_{34}{=}O$	2.82	1.17	1.65
$Si_{29}H_{34}{>}O$	3.29	3.01	0.28

All values are in eV.

with the double-bonded oxygen (see the Stokes shift values); the same can be observed for the double-bonded $Si_{29}H_{34}O$.

The double-bonded oxygen seems, hence, almost size independent: actually, the presence of this kind of bond creates localized states within the gap that are not affected by quantum confinement as previously predicted [61].

As already stated in Section 2.2 we have calculated not only the transition energies within the Δ-SCF approach but also the absorption and emission optical spectra. Actually, for both the calculated GS and the ES optimized geometry, we have evaluated the optical response through first-principles calculations, also beyond the one-particle approach. We have considered the self-energy corrections by means of the GW method and the excitonic effects through the solution of the Bethe–Salpeter equation. The effect of the local fields is also included. In Table 2.3, the calculated gaps at different levels of approximation (DFT-LDA, GW, and BS-LF approaches) are reported for both the Si_{10} and Si_{29}-based nanocrystals.

The main result common to absorption and emission is the opening of the LDA bandgap with the GW corrections by amounts weakly dependent on the surface termination but much larger than the corresponding 0.6 eV of the Si bulk case. Looking at the BS-LF calculations, we note a sort of compensation (more evident in the GS than in the ES) of the self-energy and excitonic contributions: the BS-LF values return similar to the LDA ones. The only exception are the BS-LF calculations for the excited state geometries of the clusters with Si−O−Si bridge bonds at the surface. Concerning the differences between the values of the Stokes shifts calculated through the Δ-SCF approach in Table 2.2 or through the MBPT in Table 2.3 they are essentially due to the ability or not of the two methods of distinguish dark transitions. In the MBPT the oscillator strengths of each transition are known, while the Δ-SCF approach only gives the possibility of finding the energy of the first excitation: if this transition is dark (and the Δ-SCF approach do not give this information), the associated energy is not the real optical gap. A clearer insight on the MBPT results is offered by Figure 2.1 (left panel), where the calculated absorption and emission spectra for all the oxidized Si_{10}-based clusters are depicted and

Table 2.3 Absorption and emission gaps calculated as HOMO–LUMO differences within DFT-LDA and GW approaches and as the lowest excitation energy when excitonic and local field effects (BS-LF) are included.

	Absorption			Emission			Stokes shift
	LDA	GW	BS-LF	LDA	GW	BS-LF	
$Si_{10}H_{14}{=}O$	3.3 (2.5)	7.3 (6.5)	3.7 (2.7)	0.8	4.6	1.0	2.7
$Si_{10}H_{14}{>}O$	3.4	7.6	4.0	0.1	3.5	1.5	2.5
$Si_{29}H_{34}{=}O$	2.5	6.0	3.7 (3.1)	0.9	4.1	1.2	2.5
$Si_{29}H_{34}{>}O$	2.3	4.8	2.3	0.4	3.0	2.2 (0.3)	0.1

In the last column the Stokes shift calculated in the BS-LF approximation is reported. In parenthesis the lowest dark transitions (when present) are also given. All values are in eV.

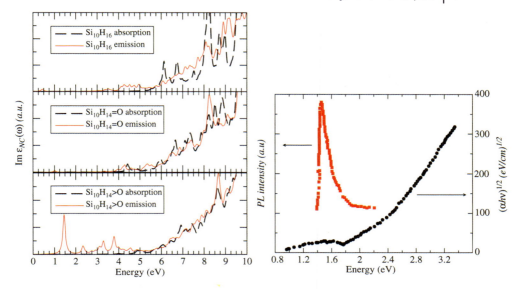

Figure 2.1 Emission (solid line) and absorption (dashed line) spectra: imaginary part of the dielectric function for the three considered Si NC. On the left: $Si_{10}H_{16}$ (top panel), $Si_{10}H_{14}$=O (central panel), and $Si_{10}H_{14}$ > O (bottom panel). On the right: experimental results for emission (red curve on the left) and absorption (on the right) by Ref. [25].

compared with the fully hydrogenated cluster. Self-energy, local-field, and excitonic effects (BS-LF) are fully taken into account.

Concerning the absorption spectra (Figure 2.1, dashed lines), all the three cases show a similar smooth increase in the absorption features. The situation for the emission related spectra is different (Figure 2.1, solid lines). Here, whereas the situation remains similar for the fully hydrogenated $Si_{10}H_{16}$ (top panel) cluster and for the $Si_{10}H_{14}$=O (central panel) cluster, in the case of a Si−O−Si bridge bond (Figure 2.1 (bottom panel)) an important excitonic peak, separated from the rest of the spectrum, is evident at 1.5 eV. Actually bound excitons are also present in the fully hydrogenated (at 0.4 eV) and in the $Si_{10}H_{14}$=O (at 1.0 eV) clusters, nevertheless, the related transitions are almost dark and the emission intensity is very low. Only in the case of the Si−O−Si bridge bond a clear PL peak appears thanks to the strong oscillator strength of the related transition. The right panel of Figure 2.1 shows the experimental absorption and emission spectra measured by Ma *et al.* [62] for Si-nanodots embedded in SiO_2 matrix. A strong photoluminescence peak appears around 1.5 eV. Comparison of the experimental spectra with our results suggest that the presence of a Si−O−Si bridge bond at the surface of Si NC and the relative deformation localized around the Si−O−Si bond can explain the nature of luminescence in Si nanocrystallites: only in this case the presence of an excitonic peak in the emission related spectra, redshifted with respect to the absorption onset, provides an explanation for both the observed SS and the near-visible PL in Si NC. Similar results have been obtained in the case of Si_{29}-based clusters. Only in the case of O in bridge position, there is a cage distortion at the interface that allows the presence of significant emission features in the optical region.

2.2.3
Doped Silicon Nanocrystals

2.2.3.1 Single-Doped Silicon Nanocrystals

We resume, here, the effects of size and shape of Si NCs on the incorporation of group III (B and Al), group IV (C and Ge), and group V (N and P) impurities. Single-doping has been investigated both in spherical and in faceted-like Si NCs [63, 64]. The spherical Si NC are built taking all the bulk Si atoms contained within a sphere of a given radius and terminating the surface dangling bonds with H; whereas the faceted Si NCs are resulting from a shell-by-shell construction procedure that starts from a central atom and adds shells of atoms successively. Spherical-like Si NCs are the $Si_{29}H_{36}$, $Si_{87}H_{76}$, $Si_{147}H_{100}$, and $Si_{293}H_{172}$ clusters and the faceted Si NCs are the Si_5H_{12}, $Si_{17}H_{36}$, $Si_{41}H_{60}$, and $Si_{147}H_{148}$ clusters. The average diameter of the doped and undoped Si NCs after relaxation is about 2.3 nm for the largest crystal considered. The substitutional impurity site is the one of the Si atom at the center of the NC. As for impurities in bulk Si, Jahn–Teller distortions occur in the neighborhood of the impurity sites and the bond lengths show a dependence with respect to the size and shape of the Si NCs. After ionic relaxation the $Si-X$ bond lengths (X = B, Al, C, Ge, N, and P) tend to be longer for faceted than for spherical-like Si NCs. A little variation of the impurity levels with respect to the shape of the Si NCs is observed. Boron and aluminum give rise to shallow acceptor levels, whereas phosphorus gives rise to shallow donor level and nitrogen to a deep donor level. The energetic positions of the impurity levels become deeper as the size of the Si NC decreases and tend nearly to the position of the corresponding impurity levels of Si bulk as the size of the Si NC increases. For all the impurities considered the lowest-energy transitions occur at lower energies than the ones in the corresponding undoped Si NCs. The optical absorption spectra of medium and large doped Si NCs exhibit an onset of absorption characteristic of indirect-gap materials. High peaks in the spectra can be found in the Si NCs whose diameters are smaller than 1.0 nm. Moreover the radiative lifetimes are sensibly influenced by the shape, especially for the small Si NCs, whereas these influences disappear when the size of the nanoparticles increase.

Starting from the Si_nH_m nanocluster [51], the formation energy (FE) for the neutral X impurity can be defined as the energy needed to insert the X atom with chemical potential μ_X within the cluster after removing a Si atom (transferred to the chemical reservoir, assumed to be bulk Si)

$$E_f = E(Si_{n-1}XH_m) - E(Si_nH_m) + \mu_{Si} - \mu_X. \tag{2.2}$$

Here E is the total energy of the system, μ_{Si} is the total energy per atom of bulk Si, and μ_X is the total energy per atom of the impurity. The results show that for smaller Si-MCs a larger energy is needed for the formation of the impurity. There is a slight tendency in the formation energy that suggests that the incorporation of the impurities is more favored in spherical than in faceted Si NCs. This tendency is not valid for the neutral Ge and P impurities, which present a formation energy nearly independent of the shape, and by the Al impurity, for which the incorporation is slightly favored for faceted Si NCs. We have also calculated how the FE changes as a function of the impurity position within the Si NC [63]. For the B neutral impurity in the large $Si_{146}BH_{100}$ cluster we have moved the impurity from the cluster center toward the surface along different paths still considering substitutional sites. It comes out that as far as the internal core is concerned, variations not higher than 0.06 eV are found. On the contrary, an energy drop between 0.25 and 0.35 eV is found as the B impurity is moved to the Si layer just below the surface. This is explained by considering that such positions are the only ones, which allow a significant atomic relaxation around the impurity, because in the other cases the surrounding Si cage is quite stable. Thus, as the B atom is moved toward the surface, the FE decreases, making the subsurface positions more stable.

2.2.3.2 Codoped Silicon Nanocrystals

As already mentioned, simultaneous doping with n- and p-type impurities represent a way to overcome the low radiative recombination efficiency in our systems so, starting from the already described hydrogenated Si NCs and following the work of Fujii *et al.* [65], we have doped the $Si_{35}H_{36}$ cluster locating the B and P impurities in substitutional positions just below the nanocrystal surface. It is worth mentioning that this arrangement represents the most stable configuration, as confirmed by theoretical and experimental works [66–68]. Full relaxation with respect to the atomic positions has been allowed and electronic properties have been computed through DFT calculations. We have found that in all the cases of codoping, the formation energy is strongly reduced, favoring this process with respect to the single doping. The choice of studying the small $Si_{33}BPH_{36}$ cluster (see Figure 2.2) (diameter around 1 nm) is due to the fact that the GW-BSE calculation [69], necessary for obtaining the

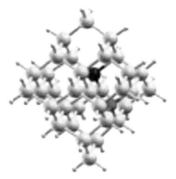

Figure 2.2 Relaxed structure of the $Si_{33}BPH_{36}$ codoped nanocrystal (diameter = 1.10 nm). Gray balls represent Si atoms, while the light gray balls are the hydrogens used to saturate the dangling bonds. B (dark gray) and P (black) impurities have been located at subsurface position in substitutional sites on opposite sides of the nanocrystals. The relaxed impurity distance is DBP = 3.64 Å.

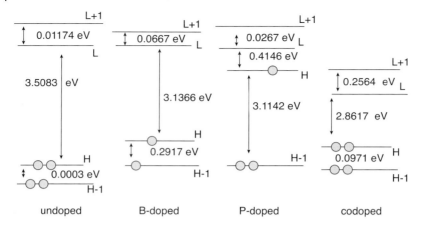

Figure 2.3 Calculated energy levels at Γ point for the $Si_{33}BPH_{36}$-NC. Alignment has been performed locating at the same energy the fully occupied levels with the same type of localization.

optical spectra, are very computing demanding. The energy levels of this system at the Γ point calculated at the optimized geometries are shown in Figure 2.3 where only the levels corresponding to the HOMO, LUMO, HOMO-1, and LUMO + 1 states are depicted. The calculated square modulus contour plots related to HOMO and LUMO states, have shown their localization within the Si NC, in particular the HOMO state is localized on the B impurity while the LUMO is localized on the P one. The presence of these donor and acceptor states lowers the energy gap from 3.51 eV for the pure cluster to 2.86 eV for the doped one. In principle, starting with a bigger cluster, for which the energy gap is smaller than in this case, it is possible through codoping to tune the gap also below the bulk Si bandgap as experimentally observed by Fuji *et al.* [70].

In order to give a complete description, within the many-body framework, of the codoped Si NC response to an optical excitation, we consider both the self-energy corrections by means of the GW method [71] to obtain the quasiparticle energies and the excitonic effects through the solution of the Bethe–Salpeter equation. The effect of local fields is also included, to take into account the inhomogeneity of the systems.

To carry out emission spectra calculations, we have used the excited state geometry and the ground state electronic configuration as already described in Section 2.2. Thus, the electron–hole interaction is also considered here in the emission geometry. Figure 2.4 (right panel) shows the calculated absorption and emission spectra fully including the many-body effects. The electron–hole interaction yields significant variations with respect to the single-particle spectra (shown in the left panel), with an important transfer of the oscillator strength to the low energy side. Moreover, in the emission spectrum the rich structure of states characterized, in the low energy side, by the presence of excitons with largely different oscillator strengths, determines excitonic gaps well below the optical absorption onset. Thus, the calculated emission spectrum results to be redshifted to lower energy with respect to the absorption one. This energy difference between emission and absorption, the Stokes shift, can be lead back to the relaxation of the Si NCs after the excitation process.

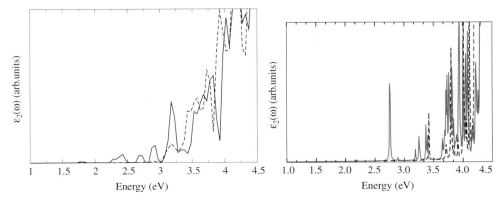

Figure 2.4 Left panel: single-particle imaginary part of the dielectric function for the codoped $Si_{33}BPH_{36}$ nanocrystal in the ground (dashed line) and excited (solid line) geometries. Right panel: absorption (dashed line) and emission (solid line) many-body spectra of $Si_{33}BPH_{36}$.

The new important features that appear in the emission of many-body spectra are related to the presence of both B and P impurities as shown in Figure 2.5, which gives the real-space probability distribution $|\psi_{exc}(r_e, r_h)|^2$ for the bound exciton as a function of the electron position r_e when the hole is fixed in a given r_h position. In this case the hole is fixed on the boron atom and we see that the bound exciton is mainly localized around the phosphorus atom. From Table 2.4, it can be seen that the single-particle DFT results strongly underestimate the absorption and emission edge with respect to the GW + BSE calculation, in which the excitonic effect are taken into account. This means that, in this case, the cancellation between GW gap opening (which gives the electronic gap) and BSE gap shrinking (which originates the excitonic gap) is only partial [72]. It is also interesting to note that the calculated Stokes shift are almost independent on the level of the computation.

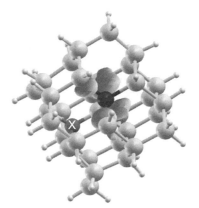

Figure 2.5 Excitonic wave function of $Si_{33}BPH_{36}$ (atom colors as in Figure 2.3). The gray isosurface represents the probability distribution of the electron, with the hole fixed on the B impurity.

Table 2.4 Absorption and Emission gaps calculated as HOMO–LUMO difference through a DFT, the many-body GW and a GW-BSE approach.

$Si_{33}BPH_{36}$	DFT	GW	GW + BSE
Absorption (eV)	2.80	5.52	3.35
Emission (eV)	1.79	4.37	2.20
Δ (eV)	1.01	1.15	1.15

The difference between the GW electronic gap and the GW + BSE optical excitonic gap gives the exciton binding energy E_b. We note the presence of exciton binding energies as big as 2.2 eV, which are very large if compared with bulk Si ($\sim$15 meV) or with carbon nanotubes [73, 74], where $E_b \sim 1$ eV, but similar to those calculated for undoped Si NCs [75] of similar size and for Si and Ge small nanowires [76, 77].

It is interesting to note that the HOMO–LUMO transition in the emission spectrum at 2.20 eV is almost dark while an important excitonic peak is evident at about 2.75 eV (see Figure 2.4), redshifted with respect to the first absorption peak.

2.2.4
Silicon Nanocrystals Embedded in a SiO₂ Matrix

In this section our goal is to build a simple model to study the properties of Si nanocrystals embedded in SiO_2 matrix from a theoretical point of view [78, 79]. *Ab initio* calculations of the structural, electronic, and optical properties of Si NCs embedded in a crystalline SiO_2 matrix have been carried out starting with a cubic cell ($l = 14.32$ Å) of SiO_2 β-cristobalite (BC) [80]. The crystalline cluster has been obtained by a spherical cutoff of 12 O atoms at the center of a $Si_{64}O_{128}$ BC geometry, as shown in Figure 2.6 (top panels). In this way, we have built an initial supercell of 64 Si and 116 O atoms with 10 Si bonded together to form a small crystalline skeleton. No defects (dangling bonds) are present at the interface and all the O atoms at the Si NC surface are single bonded with the Si atoms of the cluster.

The amorphous silica model has been generated using classical molecular dynamics simulations of quenching from a melt; the simulations have been done using semiempirical ionic potentials [81], following a quench procedure [82]. The amorphous dot has been obtained by a spherical cutoff of 10 O atoms from a $Si_{64}O_{128}$ amorphous silica cell, as shown in Figure 2.6 (bottom panels). For both the crystalline and the amorphous cases we have performed successive *ab initio* dynamics relaxations with the SIESTA code [83] using Troullier–Martins pseudopotentials with nonlocal corrections, and a mesh cutoff of 150 Ry. No external pressure or stress was applied, and we left all the atom positions and the cell dimensions totally free to move.

Electronic and optical properties of the relaxed structures have been calculated in the framework of DFT, using the ESPRESSO package [52]. Modification in the band structure and in the absorption spectrum due to the many-body effects has been also computed.

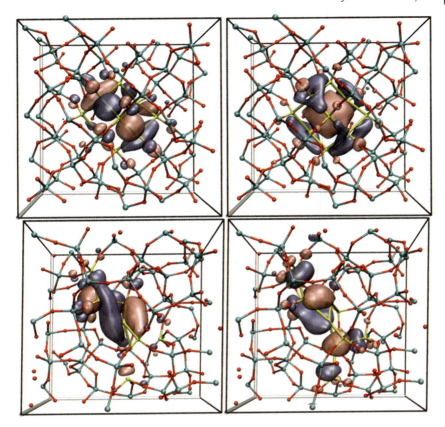

Figure 2.6 Stick and ball pictures of the final optimized structure of Si_{10} in a β-cristobalite matrix (top panels) or in a SiO_2 glass (bottom panes). Dark gray spheres (red) represent O atoms, light gray (cyan) represents Si of the matrix, and white (yellow) represents Si atoms of the nanocrystal. Also the isosurfaces at fixed value (7% of max. amplitude) of the square modulus $|\Psi|^2$ of HOMO (left) and LUMO (right) Kohn–Sham orbitals for the Si_{10} crystalline (top) and amorphous (bottom) clusters in the silica matrix are shown. Pink (blue) represents the positive (negative) sign of Ψ.

Parallel to the design of the Si_{10}/SiO_2 system, both crystalline and amorphous, we have studied three other systems: (i) the pure matrix (SiO_2) (the BC for the crystalline case and the glass for the amorphous one), (ii) the isolated Si NC extracted by the nc-silica complex (both crystalline and amorphous) relaxed structure and capped by hydrogen atoms (Si_{10}-H), and (iii) the Si NC and the first interface oxygens extracted as in point ii), passivated by H atoms (Si_{10}-OH) [84]. The goal is to separate the properties related to the Si NC from those related to the matrix in order to check the possible role of the Si NC/matrix interface; in fact the comparison of the results relative to different passivation regimes (H or OH groups) could give some insight on the role played by the interface region.

Table 2.5 Energy gap values in eV for crystalline and amorphous silica, embedded Si NC, OH-terminated Si NC, H-terminated Si NC.

System	SiO_2	Si_{10}/SiO_2	Si_{10}-OH	Si_{10}-H
Crystalline	5.44	1.77	1.60	4.66
Amorphous	5.40	1.41	1.55	1.87

From the analysis of the relaxed crystalline Si_{10}/SiO_2 supercell, it is found that the Si NC has a strained structure with respect to the bulk Si [85], while the BC matrix around is strongly distorted near the Si NC and progressively reduces its stress going from the NC to the external region [86, 87]. Concerning the structure of the cluster in the amorphous glass, the system (both the cluster and the matrix) completely loses memory of the starting tetrahedral symmetry configuration. There are not dangling bonds at the cluster surface while some bridge-bonded oxygens appear that were not present in the crystalline case [88].

Table 2.5 shows the comparison between the E_G of all the considered systems. For the crystalline case we initially observe a strong reduction in the Si_{10}/SiO_2 gap with respect to both the bulk BC SiO_2 and the isolated Si NC passivated by H atoms; finally, as reported in Ref. [25, 27–29, 89, 90], the passivation by OH groups tends to redshift the energy spectrum, resulting in a gap similar to that of the embedded Si NC.

The smaller gap of the Si_{10}/SiO_2 system with respect to the SiO_2 BC bulk case is clearly due to the formation, at the conduction and valence band edges, of confined, flat states. These new states are not simply due to the Si NC; actually they are not present in the case of the isolated Si NC capped by H atoms, but are instead visible for the isolated Si NC passivated by OH groups. This means that a strong influence on the electronic properties of the host matrix is played not only by the presence of the Si NC but, in particular, by the interface region where O atoms play a crucial role. Deep inside the bands, the typical behavior of the bulk matrix is still recognizable.

Concerning the electronic properties of the amorphous systems, the E_G of the embedded Si NC and of the isolated Si_{10}a-OH-NC are similar to that of the crystalline case and, as in the crystalline case, one can find band edges states due to the interface region. However, contrary to the crystalline case, the Si_{10}a-H-NC shows a strongly reduced bandgap. This difference can be addressed to the amorphization effects, indicating that in the amorphous case the localization process at the origin of the E_G lowering is mainly driven by the disorder [88].

In Figure 2.6 the square modulus contour plot of the HOMO (top left) and of the LUMO (top right) for the Si_{10} crystalline cluster in the BC matrix are reported. In both the cases the spatial distribution is mainly localized in the Si NC region and in the O atoms immediately around the cluster, that is, at the Si NC/SiO_2 interface. Also the HOMO and LUMO states for the amorphous case (reported in Figure 2.6, bottom left and right) follow the shape of the Si NC, strongly deformed with respect to the ordered case. Thus they are still localized on the Si NC and on the vicinal O atoms. In

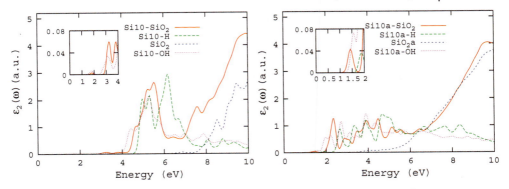

Figure 2.7 Imaginary part of the dielectric function for the Si_{10} crystalline (top) and amorphous (bottom) embedded clusters compared with that of H-terminated clusters, OH-terminated clusters, and of the silica matrices. The insets show an enlargement of the spectra at low energies.

our opinion, this supports the important role played by the interface between the Si NC and the matrix.

The electronic properties are reflected in the optical ones. Figure 2.7 shows the calculated imaginary part of the dielectric function for all the structures described above. As one can see for both the crystalline and the amorphous case, three regions due to the hosting matrix, the Si NC, and the interface are clearly distinguishable. At high energy (>7 eV) the spectra of the embedded Si NC resemble that of silica bulks while in the energy region between ~4 and ~6 eV for the crystalline and between ~2 and ~6 eV for the amorphous case is clearly the contribution of the Si NC itself. Below these intervals, anyway, new transitions exist that are not due to the matrix alone or to the isolated, hydrogenated cluster. These low energy transitions originate from the interplay between the Si NC and the embedding matrix and in particular are also due to the O atoms present at the interface region. Indeed, comparing the imaginary part of the dielectric function (see the insets) for the embedded cluster (solid line) with that of the OH-terminated cluster (dotted line), one can see that the presence of OH groups at the surface of the cluster induces low energy transitions in fair agreement with the new peaks observed for the Si NC in the matrix. This result is in agreement with other works [44], sustaining the idea that the deformation of the nanoclusters does not seem to be the determinant for the absorption onset at low energies.

The main differences between crystalline and amorphous case are related to the intensity of the peaks in the visible region (<3 eV). Here the localization process due to the disorder enhances the intensities of the optical transitions, which in the crystalline case are very low. Many-body effects have been considered in both the crystalline and the amorphous embedded Si NC. We have included quasiparticle effects within the GW approach [55] and excitonic effects within the Bethe–Salpeter equation [56, 69].

Table 2.6 shows the calculated E_G within DFT, GW, GW + BSE approximations. In the ordered (amorphous) case, the inclusion of GW corrections spreads up the gap of about 1.9 (1.7) eV, while the excitonic and local fields correction reduces it of about 1.5 (1.6) eV. Thus, the total correction to the LDA results to be in the order of 0.4 (0.1) eV.

Table 2.6 Many-body effects on the gap values (in eV) for the crystalline and amorphous embedded dots.

Si$_{10}$/SiO$_2$	DFT	GW	GW + BSE
Crystalline	1.77	3.67	2.17
Amorphous	1.41	3.11	1.51

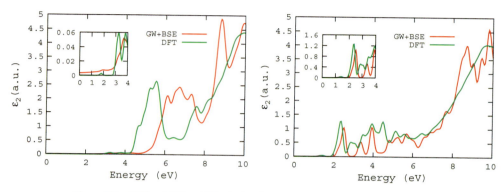

Figure 2.8 DFT and GW + BSE calculated imaginary part of the dielectric function for crystalline (top) and amorphous (bottom) case.

Now finally the E_G of the crystalline and amorphous embedded Si NC are quite different, that is, 2.17 and 1.51 eV, respectively.

This difference is more evident in Figure 2.8 that shows a comparison of the absorption spectra for the embedded Si NC for both crystalline and amorphous case. We see that even if the inclusion of many-body effects does not much change the situation at the onset of the absorption (see insets), it strongly depletes the intensity of the transitions in the 4–6 eV region for the crystalline case, whereas for the amorphous one the situation in the 2–6 energy region remains more or less the same. This is a further strong indication of the importance of localization processes for the optical properties of embedded Si NC.

Until now, no experimental measurements on Si NC with diameter in the order of 1 nm have been performed. Thus, a straightforward comparison with experimental data is not possible but the fitting of some recent measurements [91–97] of Si NC of different sizes shows a fair agreement with our results.

2.3
Pseudopotential Calculations for Large Nanocrystals

As mentioned in Section 2.1, for large NCs, we employ the linear combination of bulk bands technique, which makes use of the semiempirical pseudopotentials for describing the atomistic environment. Note that we shall ignore many body effects, unlike our treatment in the previous section, where the excitonic effects were incorporated within the BSE. In this technique, the NC wave function with a state index j is expanded in terms of the bulk Bloch bands of the constituent core and embedding medium (matrix) materials

$$\psi_j(\vec{r}) = \frac{1}{\sqrt{N}} \sum_{n,\vec{k},\sigma} C^\sigma_{n,\vec{k},j} \, e^{i\vec{k}\cdot\vec{r}} u^\sigma_{n,\vec{k}}(\vec{r}), \tag{2.3}$$

where N is the number of primitive cells within the computational supercell, $C^\sigma_{n,\vec{k},j}$ is the expansion coefficient set to be determined, and σ is the constituent bulk material label pointing to the NC core and embedding medium. $u^\sigma_{n,\vec{k}}(\vec{r})$ is the cell-periodic part of the Bloch states that can be expanded in terms of the reciprocal lattice vectors $\{\vec{G}\}$ as

$$u^\sigma_{n,\vec{k}}(\vec{r}) = \frac{1}{\sqrt{\Omega_0}} \sum_{\vec{G}} B^\sigma_{n\vec{k}}(\vec{G}) \, e^{i\vec{G}\cdot\vec{r}}, \tag{2.4}$$

where Ω_0 is the volume of the primitive cell. The atomistic Hamiltonian for the system is given by

$$\hat{H} = -\frac{\hbar^2 \nabla^2}{2m_0} + \sum_{\sigma,\vec{R}_j,\alpha} W^\sigma_\alpha(\vec{R}_j) v^\sigma_\alpha(\vec{r}-\vec{R}_j-\vec{d}^\sigma_\alpha), \tag{2.5}$$

where m_0 is the free electron mass, $W^\sigma_\alpha(\vec{R}_j)$ is the weight function that takes values 0 or 1 depending on the type of atom at the position $\vec{R}_j - \vec{d}^\sigma_\alpha$; an intermediate value between 0 and 1 can be used for the alloys or modeling the interface region. v^σ_α is the screened spherical pseudopotential of atom α of the material σ. We use semiempirical pseudopotentials for Si developed particularly for strained Si/Ge superlattices that reproduces a large variety of measured physical data such as bulk band structures, deformation potentials, electron–phonon matrix elements, and heterostructure valence band offsets [98]. With such a choice, this approach benefits from the empirical pseudopotential method, which in addition to its simplicity has another advantage over the more accurate density functional *ab initio* techniques that run into well-known bandgap problem [99], which is a disadvantage for the correct prediction of the excitation energies.

The formulation can be cast into the following generalized eigenvalue equation [37, 42]:

$$\sum_{n,\vec{k},\sigma} H_{n'\vec{k}'\sigma',n\vec{k}\sigma} \, C^\sigma_{n,\vec{k}} = E \sum_{n,\vec{k},\sigma} S_{n'\vec{k}'\sigma',n\vec{k}\sigma} \, C^\sigma_{n,\vec{k}}, \tag{2.6}$$

where

$$H_{n'\vec{k}'\sigma',n\vec{k}\sigma} \equiv \left\langle n'\vec{k}'\sigma' | \hat{T} + \hat{V}_{\text{xtal}} | n\vec{k}\sigma \right\rangle,$$

$$\left\langle n'\vec{k}'\sigma' | \hat{T} | n\vec{k}\sigma \right\rangle = \delta_{\vec{k}',\vec{k}} \sum_{\vec{G}} \frac{\hbar^2}{2m} |\vec{G}+\vec{k}|^2 B^{\sigma'}_{n'\vec{k}'}(\vec{G})^* B^\sigma_{n\vec{k}}(\vec{G}),$$

$$\left\langle n'\vec{k}'\sigma'|\hat{V}_{\text{xtal}}|n\vec{k}\sigma\right\rangle = \sum_{\vec{G},\vec{G}'} B^{\sigma'}_{n'\vec{k}'}(\vec{G}')^* B^{\sigma}_{n\vec{k}}(\vec{G}) \sum_{\sigma'',\alpha} V^{\sigma''}_{\alpha}\left(|\vec{G}+\vec{k}-\vec{G}'-\vec{k}'|^2\right)$$

$$\times W^{\sigma''}_{\alpha}(\vec{k}-\vec{k}')\, e^{i(\vec{G}+\vec{k}-\vec{G}'-\vec{k}')\cdot\vec{d}^{\sigma''}_{\alpha}},$$

$$S_{n'\vec{k}'\sigma',n\vec{k}\sigma} \equiv \left\langle n'\vec{k}'\sigma'|n\vec{k}\sigma\right\rangle.$$

Here, the atoms are on the regular sites of the underlying Bravais lattice: $\vec{R}_{n_1,n_2,n_3} = n_1\vec{a}_1 + n_2\vec{a}_2 + n_3\vec{a}_3$, where $\{\vec{a}_i\}$ are its direct lattice vectors of the Bravais lattice. Both the NC and the host matrix are assumed to possess the same lattice constant and the whole structure is within a supercell that imposes the periodicity condition $W(\vec{R}_{n_1,n_2,n_3} + N_i\vec{a}_i) = W(\vec{R}_{n_1,n_2,n_3})$, recalling its Fourier representation $W(\vec{R}_{n_1,n_2,n_3}) \rightarrow \sum \tilde{W}(q)\, e^{i\vec{q}\cdot\vec{R}_{n_1,n_2,n_3}}$, implies $e^{i\vec{q}\cdot N_i\vec{a}_i}=1$, so that $\vec{q} \rightarrow \vec{q}_{m_1,m_2,m_3} = \vec{b}_1\frac{m_1}{N_1} + \vec{b}_2\frac{m_2}{N_2} + \vec{b}_3\frac{m_3}{N_3}$, where $\{\vec{b}_i\}$ are the reciprocal lattice vectors of the *bulk* material. Thus the reciprocal space of the supercell arrangement is not a continuum but is in the grid form composed of points $\{\vec{q}_{m_1,m_2,m_3}\}$, where $m_i = 0, 1, \ldots, N_i-1$.

2.3.1
Effective Optical Gap

The hallmark of quantum size effect in NCs has been the widening of the optical gap, as demonstrated by quite a number of theoretical and experimental studies performed within the last decade. Figure 2.9 contains a compilation of some

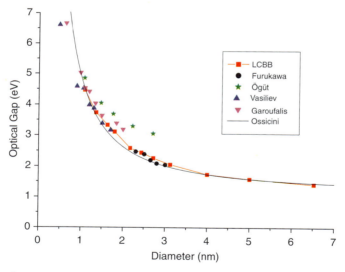

Figure 2.9 Comparison of optical gap as a function of Si NC diameter of LCBB results with previous experimental and theoretical data: Furukawa and Miyasato [100], Öğüt *et al.* [101], Vasiliev *et al.* [102], Garoufalis *et al.* [103], and the fitting by Ossicini *et al.* [1].

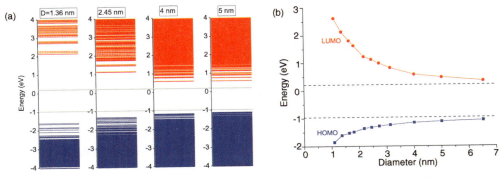

Figure 2.10 (a) The evolution of states for four increasing diameters of Si NCs and (b) HOMO and LUMO variation with respect to diameter. The bulk band edges of Si are marked with a dashed line for comparison.

representative results. For Si NCs, it can be observed that there is a good agreement among the existing data, including LCBB results. In the latter approach, the optical gap directly corresponds to the LUMO–HOMO energy difference, as calculated by the single-particle Hamiltonian in Eq. (2.5). Such a simplicity relies on the finding of Delerue and coworkers that the self-energy and Coulomb corrections almost exactly cancel each other for Si NCs larger than a diameter of 1.2 nm [72]. The evolution of the individual states with increasing size is shown in Figure 2.10a. In particular, the HOMO and LUMO variation is plotted in Figure 2.10b. It can be clearly observed that for a diameter larger than about 6 nm the quantum size effect essentially disappears.

2.3.2
Radiative Lifetime

An excellent test for the validity of the electronic structure is through the computation of the direct photon emission. The associated radiative lifetime for the transition between HOMO and LUMO is obtained via time-dependent perturbation theory utilizing the momentum matrix element, which was first undertaken by Dexter [104]. Here, we use the expression offered by Califano et al. [105] that differs somewhat in taking into account the local field effects:

$$\frac{1}{\tau_{\mathrm{fi}}} = \frac{4}{3}\frac{n}{c^2}F^2\alpha\omega_{\mathrm{fi}}^3|r_{\mathrm{fi}}|^2, \qquad (2.7)$$

where $\alpha = e^2/\hbar c$ is the fine structure constant, $n = \sqrt{\varepsilon_{\mathrm{matrix}}}$ is the refractive index of the host matrix, $F = 3\varepsilon_{\mathrm{matrix}}/(\varepsilon_{\mathrm{NC}} + 2\varepsilon_{\mathrm{matrix}})$ is the screening factor within the real-cavity model [106], ω_i is the angular frequency of the emitted photon, and c is the speed of light. Using the dipole length element between the initial (i) and final states (f), the expression $r_{\mathrm{fi}} = \langle f|p|i\rangle/(im_0\omega_{\mathrm{fi}})$, we can rewrite the Eq. (2.7) as,

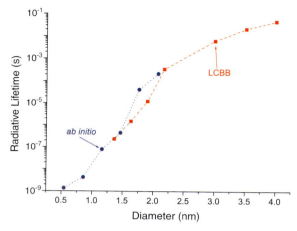

Figure 2.11 The variation in the radiative lifetime with respect to diameter for Si NCs.

$$\frac{1}{\tau_{fi}} = \frac{16\pi^2}{3} n F^2 \frac{e^2}{h^2 m_0^2 c^3} (E_f - E_i) |\langle i|p|f \rangle|^2. \tag{2.8}$$

To obtain the thermally averaged radiative lifetime, the Boltzmann average is performed over the states close to the HOMO and LUMO as

$$\frac{1}{\langle \tau_r \rangle} = \frac{\sum\limits_{fi} \frac{1}{\tau_{fi}} e^{-\beta(E_{fi} - E_G)}}{\sum\limits_{fi} e^{-\beta(E_{fi} - E_G)}}, \tag{2.9}$$

where E_G is the HOMO–LUMO gap, $\beta = 1/(k_B T)$, and k_B is the Boltzmann constant.

The results [107] for the radiative lifetime in Si NCs as a function of diameter are shown in Figure 2.11. The LCBB values for large NCs merge very well with the *ab initio* calculations for small NCs [108]. It should be noted that the radiative lifetime is reduced exponentially, as the NC size is reduced, turning the indirect bandgap bulk materials into efficient light emitters. It needs to be remembered that the non-radiative processes in Si NCs, Auger recombination and carrier multiplication, are still much more efficient than the radiative process [46]. As another remark, for larger NCs the level spacings become comparable to phonon energies. Therefore, the direct recombination as considered here, needs to be complemented by the phonon-assisted recombination beyond approximately 3–4 nm-diameters.

2.3.3
Linear Optical Absorption

Once the electronic wave functions of the NCs are available, their linear optical properties can be readily computed. The three different types of direct photon (zero-phonon) absorption processes are illustrated in Figure 2.12. These are interband, intraband, and excited state absorptions. In the latter, the blue (dark-colored) arrow

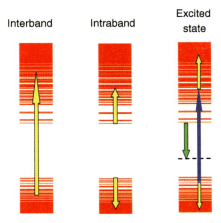

Figure 2.12 Illustration for the three different absorption processes in NCs considered here: interband, intraband, and excited state absorption. The yellow (light-colored) arrows indicate the direct photon absorption transitions, the blue (dark-colored) arrow represents optical pumping, and the downward green arrow corresponds to luminescence, which can be to a interface state (dashed line).

represents optical pumping and following carrier relaxation, the downward green arrow corresponds to luminescence that can be up to a final interface state (dashed line) [6]. For all these processes, the relevant quantity is the imaginary part of the dielectric function (which was denoted by ε_2 in Section 2.2). Within the independent-particle approximation and the artificial supercell framework, it can be put to an alternative form [109],

$$\text{Im}\{\varepsilon_{aa}(\omega)\} = \frac{(2\pi e\hbar)^2}{m_0 V_{\text{SC}}} \sum_{c,v} \frac{f_{cv}^{aa}}{E_c - E_v} \frac{\Gamma/(2\pi)}{\left[E_c - E_v - \hbar\omega\right]^2 + (\Gamma/2)^2}, \qquad (2.10)$$

where $a = x, y, z$ denotes the Cartesian components of the dielectric tensor and

$$f_{cv}^{aa} = \frac{2|\langle c|p_a|v\rangle|^2}{m_0(E_c - E_v)}, \qquad (2.11)$$

is the oscillator strength of the transition. In these expressions, e is the magnitude of the electronic charge, and Γ is the full-width at half maximum value of the Lorentzian broadening. The label v (c) corresponds to occupied (empty) valence (conduction) states, referring only to their orbital parts in the absence of spin-orbit coupling; the spin summation term is already accounted in the prefactor of Eq. (2.10). Finally, V_{SC} is the volume of the supercell that is a fixed value, chosen conveniently large to accommodate the NCs of varying diameters. However, if one replaces it with that of the NC, V_{NC}, this corresponds to the calculation of $\text{Im}\{\varepsilon_{aa}\}/f_v$, where $f_v = V_{\text{NC}}/V_{\text{SC}}$ is the volume filling ratio of the NC. For the sake of generality, this is the form in which the results will be presented here. The electromagnetic *intensity* absorption coefficient

$\alpha(\omega)$ is related to the imaginary part of the dielectric function through [110]

$$\text{Im}\{\varepsilon_{aa}(\omega)\} = \frac{n_r c}{\omega}\alpha_{aa}(\omega), \tag{2.12}$$

where n_r is the index of refraction and c is the speed of light.

In the case of intraband absorption, its rate depends on the amount of excited carriers. Therefore, we consider the absorption rate for *one* excited electron or hole that lies at an initial state i with energy E_i. As there are a number of closely spaced such states, we perform a Boltzmann averaging over these states as $e^{-\beta E_i}/\sum_j e^{-\beta E_j}$. We further assume that the final states have no occupancy restriction, which can easily be relaxed if needed. The expression for absorption rate per an excited carrier in each NC becomes

$$\frac{\alpha_{aa}}{f_v} = \frac{\pi e^2}{2m_0 c n_r \omega V_{NC}} \sum_{i,f} \frac{e^{-\beta E_i}}{\sum_j e^{-\beta E_j}} f_{fi}^{aa}[E_f - E_i]\frac{\Gamma/(2\pi)}{[E_f - E_i - \hbar\omega]^2 + (\Gamma/2)^2}, \tag{2.13}$$

where again a is the light polarization direction.

Finally, we include the surface polarization effects, also called local field effects using a simple semiclassical model, which agrees remarkably well with more rigorous treatments [111]. We give a brief description of its implementation. First, using the expression

$$\varepsilon_{SC} = f_v\varepsilon_{NC} + (1-f_v)\varepsilon_{matrix}, \tag{2.14}$$

we extract (i.e., de-embed) the size-dependent NC dielectric function, ε_{NC}, where ε_{SC} corresponds to Eq. (2.10), suppressing the Cartesian indices. ε_{matrix} is the dielectric function of the host matrix; for simplicity, we set it to the permittivity value of SiO_2, that is, $\varepsilon_{matrix} = 4$. Since the wide bandgap matrix introduces no absorption up to an energy of about 9 eV, we can approximate $\text{Im}\{\varepsilon_{NC}\} = \text{Im}\{\varepsilon_{SC}\}/f_v$. One can similarly obtain the $\text{Re}\{\varepsilon_{NC}\}$ within the random-phase approximation [112], hence get the full complex dielectric function ε_{NC}. According to the classical Clausius–Mossotti approach, which is shown to work also for NCs [113], the dielectric function of the NC is modified as

$$\varepsilon_{NC,LFE} = \varepsilon_{matrix}\left[\frac{4\varepsilon_{NC} - \varepsilon_{matrix}}{\varepsilon_{NC} + 2\varepsilon_{matrix}}\right], \tag{2.15}$$

to account for LFE. The corresponding supercell dielectric function, $\varepsilon_{SC,LFE}$ follows using Eq. (2.14). Similarly, the intensity absorption coefficients are also modified due to surface polarization effects, cf. Eq. (2.12).

2.3.3.1 Interband Absorption

For applications such as in solar cells, an efficient interband absorption over a certain spectrum is highly desirable. Depending on the diameter, Si NCs possess strong absorption toward the UV region. In Figure 2.13 we compare LCBB results [44] with the experimental data of Wilcoxon *et al.* for Si NCs [114]. Overall, there is a good agreement, especially with LFE. The major discrepancies can be attributed to

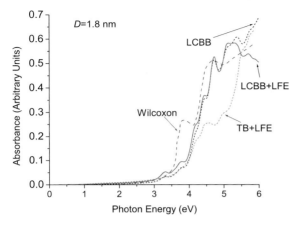

Figure 2.13 Comparison of LCBB absorbance results with the available data: experimental work of Wilcoxon *et al.* [114] and the theoretical tight binding results of Trani *et al.* [111]. For LCBB spectra, a Lorentzian broadening energy full width of 200 meV is used.

excitonic effects that are not included in the LCBB results. In Figure 2.12 we also display the tight binding result of Trani *et al.* that also includes LFE [111].

2.3.3.2 Intraband Absorption

The electrical injection of carriers or doping opens the channel for intraband absorption, also termed as intersubband absorption, which has practical importance for mid- and near-infrared photodetectors [115]. We assume that these introduced carriers eventually relax towards their respective band edges and attain a thermal distribution. Therefore, we perform a Boltzmann averaging at room temperature (300 K) over the initial states around LUMO (HOMO) for electrons (holes). The absorption coefficients to be presented are for unity volume filling factors and for one carrier per NC; they can easily be scaled to a different average number of injected carriers and volume filling factors. In Figure 2.14 the Si NCs of different diameters are compared [44]. The intraband absorption is observed to be enhanced as the NC size grows up to about 3 nm followed by a drastic fall for larger sizes. For both holes and electrons very large number of absorption peaks are observed from 0.5 to 2 eV. Recently, de Sousa *et al.* have also considered the intraband absorption in Si NCs using the effective mass approximation and taking into account the multivalley anisotropic band structure of Si [116]. However, their absorption spectra lack much of the features seen in Figure 2.13. Mimura *et al.* have measured the optical absorption in heavily phosphorus doped Si NCs of a diameter of 4.7 nm [117]. This provides us an opportunity to compare the LCBB results on the intraconduction band absorption in Si NCs. There is a good order-of-magnitude agreement. However, in contrast to LCBB spectra in Figure 2.14 that contains well-resolved peaks, they have registered a smooth spectrum that has been attributed by the authors to the smearing out due to size and shape distribution within their NC ensemble [117].

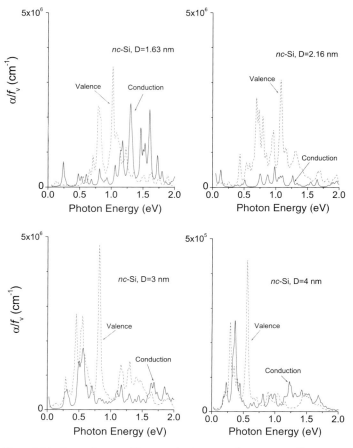

Figure 2.14 Intravalence and intraconduction state absorption coefficients in Si NCs of different diameters per excited carrier and at unity filling factor. A Lorentzian broadening energy full width of 30 meV is used. Mind the change in the vertical scale for 4 nm-diameter case.

2.3.3.3 Excited State Absorption

An optical pumping well above the effective bandgap leads to excited state absorption, also termed as photoinduced absorption, which is an undesired effect that can inhibit the development of optical gain [118]. Recent experiments on excited state absorption concluded that more attention should be devoted to the role of the excitation conditions in the quest for the silicon laser [119–121]. For this reason, we consider another intraband absorption process where the system is under a continuous interband optical pumping that creates electrons and holes with excess energy. We consider three different excitation wavelengths: 532, 355, and 266 nm that, respectively, correspond to the second-, third-, and fourth-harmonic of the Nd-YAG laser at 1064 nm. The initial states of the carriers after optical pumping are chosen to be at the pair of states with the maximum oscillator strength among interband transitions under the chosen excitation. As a general trend, it is observed that the excess energy is

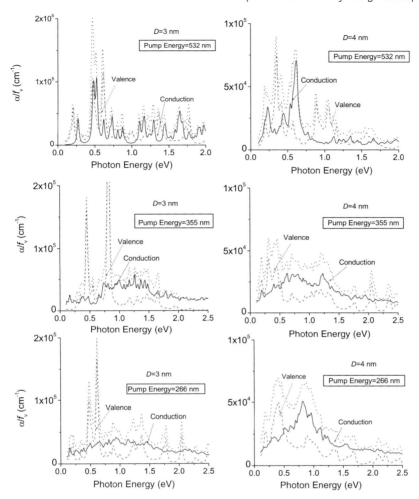

Figure 2.15 Excited state absorption within valence and conduction states of Si NCs per excited carrier and at unity filling factor under three different optical pumping wavelengths of 532, 355, and 266 nm. Dotted lines in black color refer to the total absorption coefficients. Two different diameters are considered, 3 and 4 nm. A Lorentzian broadening energy full width of 30 meV is used.

unevenly partitioned, mainly in favor of the conduction states [44, 46]. Once again a Boltzmann averaging is used to get the contribution of states within the thermal energy neighborhood.

Considering 3 and 4 nm diameters, the results [44] are shown in Figure 2.15. Note that the 532 nm excitation results are qualitatively similar to those in intraband absorption, cf. Figure 2.14. This is expected on the grounds of small excess energy for this case. Some general trends can be extracted from these results. First of all, the conduction band absorption is in general smooth over a wide energy range. On the other hand, the valence band absorption contains pronounced absorption at several narrow energy windows mainly below 1 eV and they get much weaker than the conduction band absorption in the remaining energies. As the excitation energy increases, the absorption coefficient per excited carrier in general decreases. In connection to silicon photonics, we should point out that the excited state absorption is substantial including the important 1.55 μm fiber optics communication wavelength. These results provide a more comprehensive picture than the reported experimental measurements [119–121] that are usually obtained at a single energy of the probe beam. Finally, it needs to be mentioned that for both intraband and excited state absorptions displayed in Figures 2.14 and 2.15, the high energy parts will be masked by the interband transition whenever it becomes energetically possible.

2.3.4
Third-Order Nonlinear Optical Properties

Recent experimental reports show that Si NCs have promising nonlinear optical properties and device applications [122–124]. One group of very important optical nonlinearities is the third-order nonlinearities, which involve nonlinear refraction coefficient or optical Kerr index n_2 and two-photon absorption coefficient β. These nonlinearities are crucial in all-optical switching and sensor protection applications [125] as well as in the up-conversion of the subbandgap light for the possible solar cell applications [126]. The third-order nonlinear optical susceptibility expression is obtained through perturbation solution of the density matrix equation of motion [125]. To simplify the notation, in this section we denote the quantities that refer to unity volume filling factor by an overbar. The final expression is given by [125]

$$
\begin{aligned}
\bar{\chi}^{(3)}_{dcba}(-\omega_3;\omega_\gamma,\omega_\beta,\omega_\alpha) &\equiv \frac{\chi^{(3)}_{dcba}(-\omega_3;\omega_\gamma,\omega_\beta,\omega_\alpha)}{f_v} \\
&= \frac{e^4}{V_{NC}\hbar^3}\mathbf{S}\sum_{lmnp}\frac{r^d_{mn}}{\omega_{nm}-\omega_3}\left[\frac{r^c_{nl}}{\omega_{lm}-\omega_2}\left(\frac{r^b_{lp}r^a_{pm}f_{mp}}{\omega_{pm}-\omega_1}\right.\right. \\
&\left.\left. -\frac{r^a_{lp}r^b_{pm}f_{pl}}{\omega_{lp}-\omega_1}\right)-\frac{r^c_{pm}}{\omega_{np}-\omega_2}\left(\frac{r^b_{nl}r^a_{lp}f_{pl}}{\omega_{lp}-\omega_1}-\frac{r^a_{nl}r^b_{lp}f_{ln}}{\omega_{nl}-\omega_1}\right)\right],
\end{aligned}
$$

$$(2.16)$$

where the subscripts $\{a, b, c, d\}$ refer to Cartesian indices, $\omega_3 \equiv \omega_\gamma + \omega_\beta + \omega_\alpha$, $\omega_2 \equiv \omega_\beta + \omega_\alpha$, $\omega_1 \equiv \omega_\alpha$ are the input frequencies, r_{nm} is the matrix element of the position operator between the states n and m, $\hbar\omega_{nm}$ is the difference between energies of these states, $\mathbf{S}$ is the symmetrization operator [125], indicating that the following expression should be averaged over the all possible permutations of the pairs (c, ω_γ), (b, ω_β), and (a, ω_α), and finally $f_{nm} \equiv f_n - f_m$, where f_n is the occupancy of the state n. The r_{nm} is calculated for $m \neq n$ through $r_{nm} = \frac{p_{nm}}{im_0\omega_{nm}}$, where p_{nm} is the momentum matrix element. Hence, after the solution of the electronic structure, the computational machinery is based on the matrix elements of the standard momentum operator, p, the calculation of which trivially reduces to simple summations.

The above susceptibility expression is evaluated without any approximation, taking into account all transitions within the 7 eV range. This enables a converged spectrum up to the ultraviolet spectrum. In the case of relatively large NCs the number of states falling in this range becomes excessive making the computation quite demanding. For instance, for the 3 nm NC the number of valence and conduction states (without the spin degeneracy) exceeds 3000. As another technical detail, the perfect C_{3v} symmetry of the spherical NCs results in an energy spectrum with a large number of degenerate states [44]. However, this causes numerical problems in computing the susceptibility expression given in Eq. (2.16). This high symmetry problem can be practically removed by introducing two widely separated vacancy sites deep inside the matrix. Their sole effect is to introduce a splitting of the degenerate states by less than 1 meV.

The refractive index and the absorption, in the presence of the nonlinear optical effects become, respectively, $n = n_0 + n_2 I$, $\alpha = \alpha_0 + \beta I$, where n_0 is the linear refractive index, α_0 is the linear absorption coefficient, and I is the intensity of the light. $\bar{n}_2$ is proportional to $\mathrm{Re}\{\bar{\chi}^{(3)}\}$, and is given by [127]

$$\bar{n}_2(\omega) = \frac{\mathrm{Re}\{\bar{\chi}^{(3)}(-\omega;\omega,-\omega,\omega)\}}{2n_0^2\varepsilon_0 c}, \tag{2.17}$$

where c is the speed of light. Similarly, $\bar{\beta}$ is given by [127]

$$\bar{\beta}(\omega) = \frac{\omega \mathrm{Im}\{\bar{\chi}^{(3)}(-\omega;\omega,-\omega,\omega)\}}{n_0^2\varepsilon_0 c^2}, \tag{2.18}$$

where ω is the angular frequency of the light. Note that Eqs. (2.17) and (2.18) are valid only in the case of negligible absorption. The degenerate two-photon absorption cross section $\bar{\sigma}^{(2)}(\omega)$ is given by [125]

$$\bar{\sigma}^{(2)}(\omega) \equiv \frac{\sigma^{(2)}(\omega)}{f_v} = \frac{8\hbar^2\pi^3 e^4}{n_0^2 c^2}\sum_{i,f}\left|\sum_m \frac{r_{fm}r_{mi}}{\hbar\omega_{mi}-\hbar\omega-i\hbar\Gamma}\right|^2 \delta(\hbar\omega_{fi}-2\hbar\omega), \tag{2.19}$$

where Γ is the inverse of the lifetime; the corresponding full width energy broadening of 100 meV is used throughout. The sum over the intermediate states, m, requires all the interband and intraband transitions. As we have mentioned previously, we compute such expressions without any approximation by including all the states that contribute to the chosen energy window. Finally, $\bar{\sigma}^{(2)}(\omega)$ and $\bar{\beta}$ are related to each other through $\bar{\beta} = 2\hbar\omega\bar{\sigma}^{(2)}(\omega)$.

The LFEs lead to a correction factor in the third-order nonlinear optical expressions given by [128], $L = \left(\frac{3\varepsilon_{\mathrm{matrix}}}{\varepsilon_{\mathrm{NC}}+2\varepsilon_{\mathrm{matrix}}}\right)^2 \left|\frac{3\varepsilon_{\mathrm{matrix}}}{\varepsilon_{\mathrm{NC}}+2\varepsilon_{\mathrm{matrix}}}\right|^2$ where $\varepsilon_{\mathrm{matrix}}$ and $\varepsilon_{\mathrm{NC}}$ are the dielectric functions of the host matrix and the NC, respectively. We fix the local field correction at its *static* value, since when the correction factor is a function of the wavelength it brings about unphysical negative absorption regions at high energies.

We consider four different diameters, $D = 1.41$, 1.64, 2.16, and 3 nm. Their energy gap, E_G as determined by the separation between the LUMO and the HOMO energies

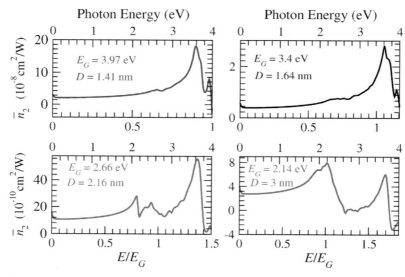

Figure 2.16 $\bar{n}_2$ (i.e., at unity filling) as a function of the photon energy for different NC sizes.

show the expected quantum size effect [44]. The $\bar{n}_2$ is plotted in Figure 2.16, which increases with the decrease in NC size for all frequencies [45]. The smallest diameter gives us the largest $\bar{n}_2$. When compared to the n_2 of bulk Si in this energy interval ($\sim 10^{-14}$ cm^2/W) [129–131], calculated NC n_2 is enhanced by as much as ($\sim 10^4 f_v$) for the largest NC. For Si NCs having a diameter of a few nanometers, Prakash *et al.* [122] have obtained n_2 in the order of $\sim 10^{-11}$ cm^2/W, which, in order of magnitude, agrees with LCBB results when a typical f_v is assumed for their samples.

In Figure 2.17 we have plotted $\bar{\beta}$ against the photon energy [45]. Peaks at high energies are dominant in the spectrum and $\bar{\beta}$ decreases with the growing NC volume. The obtained β is about $10^5 f_v$ cm/GW for the largest NC at around 1 eV. When compared to the experimental bulk value (1.5–2.0 cm/GW measured at around 1 eV) [129], calculated NC β is scaled by a factor of $300-400 f_v$ times. Prakash *et al.* [122] have observed β to be between ($10^1 - 10^2$ cm/GW) at 1.53 eV, which is close to LCBB values provided that f_v is taken into account.

We should note that $\bar{\beta}$ is nonzero down to static values due to band tailing. Another interesting observation is that the two-photon absorption threshold is distinctly beyond the half bandgap value, which becomes more prominent as the NC size increases. This can be explained mainly as the legacy of the NC core medium, silicon that is an indirect bandgap semiconductor. Hence, the HOMO–LUMO dipole transition is very weak, especially for relatively large NCs. We think that this is the essence of what is observed also for the two-photon absorption. As the NC size gets smaller, the HOMO–LUMO energy gap approaches to the direct bandgap of bulk silicon, while the HOMO–LUMO dipole transition becomes more effective.

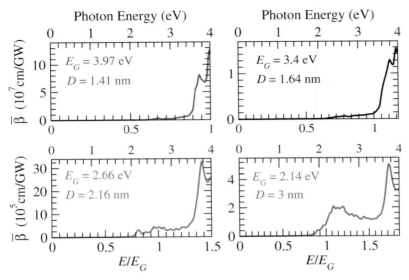

Figure 2.17 $\overline{\beta}$ (i.e., at unity filling) as a function of the photon energy for different NC sizes.

2.3.5
Quantum-Confined Stark Effect in Si Nanocrystals

In 1984, Miller *et al.* discovered a variant of the Stark effect in quantum wells, which was much more robust due to the confinement of the carriers, hence named as the quantum-confined Stark effect [132]. It took more than a decade to observe the same effect in NCs, which were chemically synthesized CdSe colloids [133]. Interestingly, it has taken again more than a decade to realize it with Si NCs [134]. Based on the importance of QCSE for silicon-based photonics, in this section we provide a theoretical analysis of QCSE in embedded Si NCs. The basic electrostatic construction of the problem is presented in Figure 2.18 with the assumption that the NCs are well separated. If we denote the uniform applied electric field in the matrix region far away from the NC as F_0, then the solution for electrostatic potential is given in spherical coordinates by [110]

$$
\Phi(r, \theta) = \begin{cases} -\dfrac{3}{\varepsilon + 2} F_0 r \cos \theta, & r \leq a \\[2mm] -F_0 r \cos \theta + \left(\dfrac{\varepsilon - 1}{\varepsilon + 2} \right) F_0 \dfrac{a^3}{r^2} \cos \theta, & r > a \end{cases}
\tag{2.20}
$$

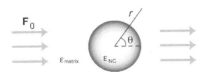

Figure 2.18 An illustration for the QCSE showing the geometry and the variables.

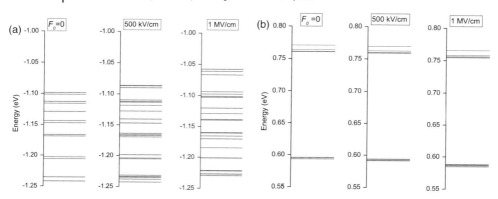

Figure 2.19 (a) The Stark shift of the (a) valence and (b) conduction states of a 5 nm-diameter Si NC. The electric field values quoted refer to F_0, that in the matrix region far away from the NC.

where $\varepsilon \equiv \varepsilon_{NC}/\varepsilon_{matrix}$ is the ratio of the permittivities of the inside and outside of the NCs. The effect of this external field can be incorporated in the framework of LCBB by adding the $V_{ext} = e\Phi$ term to the potential energy matrix elements. A computationally convenient approach is to assume that the external potential is relatively smooth so that its Fourier transform can be taken to be band-limited to the first Brillouin zone of the underlying unit cell, which leads to [42],

$$\left\langle n'\vec{k}'\sigma'|\hat{V}_{ext}|n\vec{k}\sigma\right\rangle = \sum_{\vec{G},\vec{G}'}\mathrm{Rect}_{\vec{b}_1,\vec{b}_2,\vec{b}_3}(\vec{G}+\vec{k}-\vec{G}'-\vec{k}')B^{\sigma'}_{n'\vec{k}'}(\vec{G}')^* B^{\sigma}_{n\vec{k}}(\vec{G}),\qquad (2.21)$$

here $\mathrm{Rect}_{\vec{b}_1,\vec{b}_2,\vec{b}_3}$ is the rectangular pulse function, which yields unity when its argument is within the first Brillouin zone defined by the reciprocal lattice vectors, $\{\vec{b}_1,\vec{b}_2,\vec{b}_3\}$, and zero otherwise.

In Figure 2.19 we show the evolution of the valence and conduction states of a 5 nm-diameter Si NC under increasing electric fields. This clearly indicates that valence states are more prone to Stark shift. The LUMO–HOMO bandgap Stark shift is plotted in Figure 2.20. As in any electronic structure calculation, this is done at 0 K. However, a shift of more than 80 meV points out that the QCSE can be easily measured at room temperature. The dotted line in this figure is a quadratic fit. The deviation for large electric fields indicates the change in charge displacement regime.

Finally, in Figure 2.21 the intravalence and intraconduction state electroabsorption curves of a 5 nm-diameter Si NC are shown, under zero and 1 MV/cm electric fields. As expected from the rigidity of the conduction states under the electric field from Figure 2.19, the electroabsorption effect can be best utilized in p-doped Si NCs. These results can be important for the assessment of the prospects of nanocrystalline Si-based infrared electroabsorption modulators.

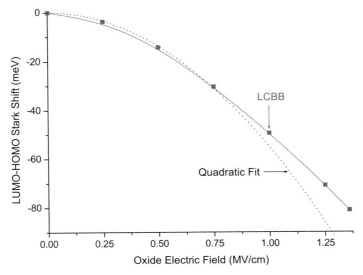

Figure 2.20 The LUMO–HOMO bandgap Stark shift of a 5 nm-diameter Si NC. The solid line is to guide the eyes.

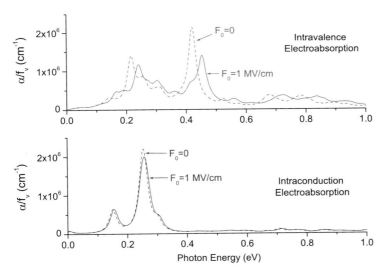

Figure 2.21 The intravalence and intraconduction state electroabsorption curves of a 5 nm-diameter Si NC at $T = 300$ K.

Acknowledgments

The C.B. would like to thank Hasan Yildirim and Cem Sevik for their valuable contributions. S.O. thanks all the contributors to the publications, who have been important for this work. This work has been supported by the European FP6 Project

SEMINANO with the contract number NMP4 CT2004 505 285, by the Turkish Scientific and Technical Council TÜBĪTAK with the project number 106T048, by MIUR PRIN2007 Italy, and Progetto Italia-Turchia CNR.

References

1 Ossicini, S., Pavesi, L., and Priolo, F. (2003) *Light Emitting Silicon for Microphotonics*, vol. 194, Springer Tracts on Modern Physics, Springer-Verlag, Berlin.

2 Bisi, O., Ossicini, S., and Pavesi, L. (2000) *Surf. Sci. Rep.*, **38**, 5.

3 Gelloz, B., Kojima, A., and Koshida, N. (2005) *Appl. Phys. Lett.*, **87**, 031107.

4 Franz, G., Irrera, A., Moreira, E.C., Miritello, M., Iacona, F., Sanfilippo, D., Di Stefano, G., Falica, P.G., and Priolo, F. (2003) *Appl. Phys.*, **A77**, 57.

5 Walters, R.J., Bourianoff, G.I., and Atwater, H.A. (2005) *Nature Mater.*, **4**, 143.

6 Pavesi, L., Dal Negro, L., Mazzoleni, C., Franzó, G., and Priolo, F. (2000) *Nature (London)*, **408**, 440.

7 Khriachtchev, L., Rasanen, M., Novikov, S., and Sinkkonen, J. (2001) *Appl. Phys. Lett.*, **79**, 1249.

8 Dal Negro, L., Cazzanelli, M., Pavesi, L., Ossicini, S., Pacifici, D., Franzó, G., Priolo, F., and Iacona, F. (2003) *Appl. Phys. Lett.*, **82**, 4636.

9 Ruan, J., Fauchet, P.M., Dal Negro, L., Cazzanelli, M., and Pavesi, L. (2003) *Appl. Phys. Lett.*, **83**, 5479.

10 Cazzanelli, M., Kovalev, D., Dal Negro, L., Gaburro, Z., and Pavesi, L. (2004) *Phys. Rev. Lett.*, **93**, 207042.

11 Luterova, K., Dohnalova, K., Servcek, V., Pelant, I., Likforman, J.-P., Crgut, O., Gilliot, P., and Hönerlage, B. (2004) *Appl. Phys. Lett.*, **87**, 3280.

12 Canham, L.T. (1990) *Appl. Phys. Lett.*, **57**, 1046.

13 Bensahel, D., Canham, L.T., and Ossicini, S. (eds) (1993) *Optical Properties of Low Dimensional Silicon Structures*, Kluwer, Dordrecht, Amsterdam.

14 Hamilton, B. (1995) *Semicond. Sci. Technol.*, **10**, 1187.

15 Kanemitsu, Y. (1995) *Phys. Rep.*, **263**, 1.

16 John, G.S. and Singh, V.A. (1995) *Phys. Rep.*, **263**, 93–151.

17 Canham, L.T. (ed.) (1997) *Properties of Porous Silicon*, INSPEC: The Institution of Electrical Engineers, London.

18 Cullis, A.G., Canham, L.T., and Calcott, P.D.J. (1997) *J. Appl. Phys.*, **82**, 909.

19 Takahashi, Y., Furuta, T., Ono, Y., Tsishima, T., and Tabe, M. (1995) *Jpn. J. Appl. Phys.*, **34**, 950.

20 Lockwood, D.J., Lu, Z.H., and Baribeau, J.-M. (1996) *Phys. Rev. Lett.*, **76**, 539.

21 Novikov, S.V., Sinkkonen, J., Kilpela, O., and Gastev, S.V. (1997) *J. Vac. Sci. Technol. B*, **15**, 1471.

22 Kanemitsu, Y. and Okamoto, S. (1997) *Phys. Rev. B*, **56**, R15561.

23 Mulloni, V., Chierchia, R., Mazzoleni, C., Pucker, G., Pavesi, L., and Bellutti, P. (2000) *Phil. Mag. B*, **80**, 705–718.

24 Baierle, R.J., Caldas, M.J., Molinari, E., and Ossicini, S. (1997) *Solid State Commun.*, **102**, 545.

25 Wolkin, M.V., Jorne, J., Fauchet, P.M., Allan, G., and Delerue, C. (1999) *Phys. Rev. Lett.*, **82**, 197.

26 Puzder, A., Williamson, A.J., Grossman, J.C., and Galli, G. (2002) *Phys. Rev. Lett.*, **88**, 097401.

27 Luppi, M. and Ossicini, S. (2003) *J. Appl. Phys.*, **94**, 2130.

28 Luppi, M. and Ossicini, S. (2005) *Phys. Rev. B*, **71**, 035340.

29 Luppi, M. and Ossicini, S. (2003) *Phys. Stat. Sol. (a)*, **197**, 251.

30 Vasiliev, I., Chelikowsky, J.R., and Martin, R.M. (2002) *Phys. Rev. B*, **65**, 121302(R).

31 Ossicini, S., Arcangeli, C., Bisi, O., Degoli, E., Luppi, M., Magri, R., Dal Negro, L., and Pavesi, L. (2003) *Towards the First Silicon Laser*, vol. 93 (eds L. Dal Negro, S. Gaponenko, and L. Pavesi), Nato Science Series, Kluwer Academic Publishers, Dordrecht, p. 271.

32 Di Carlo, A. (2003) *Semicond. Sci. Technol.*, **18**, R1.

33 Delerue, C. and Lannoo, M. (2004) *Nanostructures: Theory and Modelling*, Springer-Verlag, Berlin.

34 Harrison, P. (2005) *Quantum Wells, Wires and Dots*, 2nd edn, John Wiley & Sons, Inc., New York.

35 Wood, D.M. and Zunger, A. (1996) *Phys. Rev. B*, **53**, 7949.

36 Wang, L.-W., Franceschetti, A., and Zunger, A. (1997) *Phys. Rev. Lett.*, **78**, 2819.

37 Wang, L.-W. and Zunger, A. (1999) *Phys. Rev. B*, **59**, 15806.

38 Ninno, D., Wong, K.B., Gell, M.A., and Jaros, M. (1985) *Phys. Rev. B*, **32**, 2700.

39 Ninno, D., Gell, M.A., and Jaros, M. (1986) *J. Phys. C: Solid State Phys.*, **19**, 3845.

40 Botti, S. and Andreani, L.C. (2001) *Phys. Rev. B*, **63**, 235313.

41 Botti, S., Vast, N., Reining, L., Olevano, V., and Andreani, L.C. (2004) *Phys. Rev. B*, **70**, 045301.

42 Chirico, F., Di Carlo, A., and Lugli, P. (2001) *Phys. Rev. B*, **70**, 045314.

43 Bulutay, C. (2007) *Physica E*, **38**, 112.

44 Bulutay, C. (2007) *Phys. Rev. B*, **76**, 205321.

45 (a) Yildrm, H. and Bulutay, C. (2008) *Opt. Commun.*, **281**, 4118; (b) Yildrm, H. and Bulutay, C. (2008) *Opt. Commun.*, **281**, 6146.

46 Sevik, C. and Bulutay, C. (2008) *Phys. Rev. B*, **77**, 125414.

47 First-Principles computation of material properties: The ABINIT software project (URL http://www.abinit.org).

48 Puzder, A., Williamson, A.J., Grossman, J.C., and Galli, G. (2003) *J. Am. Chem. Soc.*, **125**, 2786.

49 Franceschetti, A. and Pantelides, S.T. (2003) *Phys. Rev. B*, **68**, 033313.

50 Luppi, E., Degoli, E., Cantele, G., Ossicini, S., Magri, R., Ninno, D., Bisi, O., Pulci, O., Onida, G., Gatti, M., Incze, A., and Del Sole, R. (2005) *Opt. Mater.*, **27**, 1008.

51 Degoli, E., Cantele, G., Luppi, E., Magri, R., Ninno, D., Bisi, O., and Ossicini, S. (2004) *Phys. Rev. B*, **69**, 155411.

52 Baroni, S., Dal Corso, A., de Gironcoli, S., Giannozzi, P., Cavazzoni, C., Ballabio, G., Scandolo, S., Chiarotti, G., Focher, P., Pasquarello, A., Laasonen, K., Trave, A.,

Car, R., Marzari, N., and Kokalj, A. (2009) *J. Phys. Condens: Matter*, **21**, 395502, http://www.pwscf.org/.

53 Vanderbilt, D. (1990) *Phys. Rev. B*, **41**, R7892.

54 Bassani, F. and Pastori Parravicini, G. (1975) *Electronic States and Optical Transitions in Solids*, Pergamon Press, New York.

55 Hedin, L. (1965) *Phys. Rev. A*, **139**, 796.

56 Onida, G., Reining, L., and Rubio, A. (2002) *Rev. Mod. Phys.*, **74**, 601, and references therein.

57 Allan, G., Delerue, C., and Lannoo, M. (1996) *Phys. Rev. Lett.*, **76**, 2961.

58 Itoh, U., Toyoshima, Y., Onuki, H., Washida, N., and Ibuki, T. (1986) *J. Chem. Phys.*, **85**, 4867.

59 Gatti, M. and Onida, G. (2005) *Phys. Rev. B*, **72**, 045442.

60 The $Si_{10}H_{14}>O$ cluster we considered here corresponds to the $Si_{10}H_{14}O$-sym of Ref. [11], where O make a bridge between "second neighbors" Si atoms. We have obtained similar results have for the $Si_{10}H_{14}O$-asym case, where O is in between two "first neighbors" Si atoms.

61 Luppi, M. and Ossicini, S. (2003) *J. Appl. Phys.*, **94**, 2130.

62 Ma, Z., Liao, X., Kong, G., and Chu, J. (1999) *Appl. Phys. Lett.*, **75**, 1857.

63 Cantele, G., Degoli, E., Luppi, E., Magri, R., Ninno, D., Iadonisi, G., and Ossicini, S. (2005) *Phys. Rev. B*, **72**, 113303.

64 Ramos, L.E., Degoli, E., Cantele, G., Ossicini, S., Ninno, D., Furthmüller, J., and Bechstedt, F. (2007) *J. Phys.: Condens. Matter*, **19**, 466211.

65 Fujii, M., Yamaguchi, Y., Takase, Y., Ninomiya, K., and Hayashi, S. (2005) *Appl. Phys. Lett.*, **87**, 211919.

66 Ossicini, S., Degoli, E., Iori, F., Luppi, E., Magri, R., Cantele, G., Trani, F., and Ninno, D. (2005) *Appl. Phys. Lett.*, **87**, 173120.

67 Ciacchi, L.C. and Payne, M.C. (2005) *Phys. Rev. Lett.*, **95**, 196101.

68 Garrone, E., Geobaldo, F., Rivolo, P., Amato, G., Boarino, L., Chiesa, M., Giamello, E., Gobetto, R., Ugliengo, P., and Vitale, A. (2005) *Adv. Mater.*, **17**, 528.

69 Code, E.X.C. and Olevano, V. http://www.bethe-salpeter.org.

70 Fujii, M., Yamaguchi, Y., Takase, Y., Ninomiya, K., and Hayashi, S. (2004) *Appl. Phys. Lett.*, **85**, 1158.

71 We have used the nonselfconsistent G_0W_0 approach within the RPA plasmon pole approximation. We use a planewave-frequency space code.

72 Delerue, C., Lannoo, M., and Allan, G. (2000) *Phys. Rev. Lett.*, **84**, 2457.

73 Spataru, C.D., Ismail-Beigi, S., Benedict, L.X., and Louie, S.G. (2004) *Phys. Rev. Lett.*, **92**, 077402.

74 Chang, E., Bussi, G., Ruini, A., and Molinari, E. (2004) *Phys. Rev. Lett.*, **92**, 196401.

75 Luppi, E., Iori, F., Magri, R., Pulci, O., Degoli, E., Ossicini, S., and Olevano, V. (2007) *Phys. Rev. B*, **75**, 033303.

76 Bruno, M., Palummo, M., Marini, A., Del Sole, R., Olevano, V., Kholod, A.N., and Ossicini, S. (2005) *Phys. Rev. B*, **72**, 153310.

77 Bruno, M., Palummo, M., Marini, A., Del Sole, R., and Ossicini, S. (2007) *Phys. Rev. Lett.*, **98**, 036807.

78 Luppi, M. and Ossicini, S. (2005) *PRB*, **71**, 035340.

79 Guerra, R., Magri, R., Martin-Samos, L., Pulci, O., Degoli, E., and Ossicini, S. (2009) *Superlatt. Microstruct.*, **46**, 246.

80 Kageshima, H. and Shiraishi, K. (1996) *Proceedings of the 23rd International Conference on Phys. Semicon* (eds M. Scheffler and R. Zimmermann), World Scientific, Singapore, p. 903.

81 van Beest, B.W.H., Kramer, G.J., and van Santen, R.A. (1990) *Phys. Rev. Lett.*, **64**, 1955.

82 Martin-Samos, L., Limoge, Y., Crocombette, J.-P., Roma, G., Anglada, E., and Artacho, E. (2005) *Phys. Rev. B*, **71**, 014116.

83 Ordejón, P., Artacho, E., and Soler, J.M. (1996) *Phys. Rev. B (Rapid Commun.)*, **53**, R10441.

84 In the last two cases a relaxation run only for the H atoms has been performed before the calculation of the optoelectronic properties.

85 Yilmaz, D.E., Bulutay, C., and Çağin, T. (2008) *Phys. Rev. B*, **77**, 155306.

86 Daldosso, N., Luppi, M., Ossicini, S., Degoli, E., Magri, R., Dalba, G., Fornasini, P., Grisenti, R., Rocca, F., Pavesi, L., Boninelli, S., Priolo, F., Spinella, C., and Iacona, F. (2003) *Phys. Rev. B*, **68**, 085327.

87 Watanabe, T. *et al.* (2004) *Appl. Surf. Sci.*, **237**, 125–133.

88 Hadjisavvas, G. and Kelires, P.C. (2007) *Physica E*, **38**, 99–105.

89 Ramos, L. *et al.* (2005) *Appl. Phys. Lett.*, **87**, 143113.

90 Ramos, L. *et al.* (2005) *Phys. Rev. B.*, **71**, 035328.

91 De la Torre, J., Souifi, A., Poncet, A., Bremond, G., Guillot, G., Garrido, B., and Morante, J.R. (2005) *Solid State Electron.*, **49**, 1114.

92 Duan, X. and Lieber, C.M. (2000) *Adv. Mater.*, **12**, 298.

93 Kanzawa, Y., Kageyama, T., Takeda, S., Fujii, M., Hayashi, S., and Yamamoto, K. (1997) *Solid State Commun.*, **102**, 553.

94 Guha, S., Qadri, B., Musket, R.G., Wall, M.A., and Shimizu-Iwayama, T. (2000) *J. Appl. Phys.*, **88**, 3954.

95 Takeoka, S., Fujii, M., and Hayashi, S. (2000) *Phys. Rev. B*, **62**, 16820.

96 Watanabe, K., Fujii, M., and Hayashi, S. (2001) *J. Appl. Phys.*, **90**, 4761.

97 Lioudakis, E., Antoniou, A., Othonos, A., Christofides, C., and Nassiopoulou, A.G., Lioutas, Ch.B., Frangis, N. (2007) *J. Appl. Phys.*, **102**, 083534.

98 Friedel, P., Hybertsen, M.S., and Schlüter, M. (1989) *Phys. Rev. B*, **39**, 7974.

99 Martin, R.M. (2004) *Electronic Structure*, Cambridge University Press, Cambridge.

100 Furukawa, S. and Miyasato, T. (1988) *Phys. Rev. B*, **38**, 5726.

101 Öğüt, S., Chelikowsky, J.R., and Louie, S.G. (1997) *Phys. Rev. Lett.*, **79**, 1770.

102 Vasiliev, I., Öğüt, S., and Chelikowsky, J.R. (2001) *Phys. Rev. Lett.*, **86**, 1813.

103 Garoufalis, C.S., Zdetsis, A.D., and Grimme, S. (2001) *Phys. Rev. Lett.*, **87**, 276402.

104 Dexter, D.L. (1958) *Solid State Physics*, vol. 6, Academic Press, New York, p. 358.

105 Califano, M., Franceshetti, A., and Zunger, A. (2005) *Nano Lett.*, **5**, 2360.

106 Glauber, R.J. and Lewenstein, M. (1991) *Phys. Rev. A*, **43**, 467.

107 Sevik, C. (2008) PhD Thesis, Bilkent University.

108 Weissker, H.-Ch., Furthmüller, J., and Bechstedt, F. (2004) *Phys. Rev. B*, **69**, 115310.

109 Weissker, H.-Ch., Furthmüller, J., and Bechstedt, F. (2002) *Phys. Rev. B*, **65**, 155327.

110 Jackson, J.D. (1975) *Classical Electrodynamics*, 2nd edn, John Wiley & Sons, Inc., New York.

111 Trani, F., Ninno, D., and Iadonisi, G. (2007) *Phys. Rev. B*, **75**, 033312.

112 Trani, F., Cantele, G., Ninno, D., and Iadonisi, G. (2005) *Phys. Rev. B*, **72**, 075423.

113 Mahan, G.D. (2006) *Phys. Rev. B*, **74**, 033407.

114 Wilcoxon, J.P., Samara, G.A., and Provencio, P.N. (1999) *Phys. Rev. B*, **60**, 2704.

115 Ryzhii, V., Khmyrova, I., Mitin, V., Stroscio, M., and Willander, M. (2001) *Appl. Phys. Lett.*, **78**, 3523.

116 de Sousa, J.S., Leburton, J.-P., Freire, V.N., and da Silva, E.F., Jr. (2005) *Appl. Phys. Lett.*, **87**, 031913.

117 Mimura, A., Fujii, M., Hayashi, S., Kovalev, D., and Koch, F. (2000) *Phys. Rev. B*, **62**, 12625.

118 Malko, A.V., Mikhailovsky, A.A., Petruska, M.A., Hollingsworth, J.A., and Klimov, V.I. (2004) *J. Phys. Chem. B*, **108**, 5250.

119 Elliman, R.G., Lederer, M.J., Smith, N., and Luther-Davies, B. (2003) *Nuclear Instrum. Methods Phys. Res., Sect. B*, **206**, 427.

120 Trojánek, F., Neudert, K., Bittner, M., and Malý, P. (2005) *Phys. Rev. B*, **72**, 075365.

121 Forcales, M., Smith, N.J., and Elliman, R.G. (2006) *J. Appl. Phys. B*, **100**, 014902.

122 Prakash, G.V., Cazzanelli, M., Gaburro, Z., Pavesi, L., Iacona, F., Franzo, F., and Priolo, J.G. (2002) *J. Appl. Phys.*, **91**, 4607.

123 King, S.M., Chaure, S., Doyle, J., Colli, A., Ferrari, A.C., and Blau, W.J. (2007) *Opt. Commun.*, **276**, 305.

124 He, G.S., Zheng, Q., Yong, K.-T., Erogbogbo, F., Swihart, M.T., and Prasad, P.N. (2008) *Nano Lett.*, **8**, 2688.

125 Boyd, R.W. (2003) *Nonlinear Optics*, Academic Press, San Diego.

126 Thurpke, T., Green, M.A., and Würfel, P. (2002) *J. Appl. Phys.*, **92**, 4117.

127 Sheik-Bahae, M., Said, A.A., Wei, T.H., Hagan, D.J., and Van Styrland, E.W. (1990) *IEEE J. Quantum. Electron.*, **26**, 760.

128 Sipe, J.E. and Boyd, R.W. (2002) *Optical Properties of Nanostructured Random Media* (ed. V.M. Shalev), Topics in Applied Physics, vol. 82, Springer, Berlin-Heidelberg.

129 Bristow, A.D., Rotenberg, N., and van Driel, H.M. (2007) *Appl. Phys. Lett.*, **90**, 191104.

130 Lin, Q., Zhang, J., Piredda, G., Boyd, R.W., Fauchet, P.M., and Agrawal, G.P. (2007) *Appl. Phys. Lett.*, **91**, 021111.

131 Dinu, M., Quochi, F., and Garcia, H. (2003) *Appl. Phys. Lett.*, **82**, 2954.

132 Miller, D.A.B., Chemla, D.S., Damen, T.C., Gossard, A.C., Wiegmann, W., Wood, T.H., and Burrus, C.A. (1984) *Phys. Rev. Lett.*, **53**, 2173.

133 Empedocles, S.A. and Bawendi, M.G. (1997) *Science*, **278**, 2114.

134 Kulakci, M., Serincan, U., Turan, R., and Finstad, T.G. (2008) *Nanotechnology*, **19**, 455403, and references therein.

3
Optical Properties of Intrinsic and Shallow Impurity-Doped Silicon Nanocrystals

Minoru Fujii

3.1
Introduction

The electronic band structure of silicon (Si) crystals is significantly modified when the size is reduced to below the exciton Bohr radius ($\sim$4.9 nm) of bulk Si crystals due to the quantum size effect. The quantum size effect manifests itself as a high-energy shift of the luminescence band and an enhancement of the spontaneous emission rate. Furthermore, enlargement of the singlet–triplet splitting energy of exciton states with decreasing size has been demonstrated [1–4]. The modified band structure opens up new application of Si nanocrystals in the fields where bulk Si crystal has not been involved. Besides well-known high-efficiency visible photoluminescence (PL), one of the most exotic new feature of Si nanocrystals is their ability to generate a kind of active oxygen species called singlet oxygen by energy transfer from excitons to oxygen molecules (O_2) adsorbed onto the surface of Si nanocrystals [5]. This feature is due to the molecule-like energy structure of Si nanocrystals and is a direct consequence of the quantum size effects. Similarly, efficient photosensitization of rare earth ions by Si nanocrystals has been reported and this phenomenon is expected to lead to the development of an efficient planar waveguide-type optical amplifier operating at the optical telecommunication wavelength. Since almost all new features of Si nanocrystals arise from the modified electronic band structure, its deep understanding is indispensable to explore new nano-Si-based research fields and devices. Fortunately, the energy structure of Si nanocrystals is almost clarified, at least qualitatively, by detailed optical spectroscopy. In Section 3.2, we briefly summarize fundamental optical properties of intrinsic Si nanocrystals [1–8].

The properties of Si nanocrystals can be controlled by the size, shape, surface termination, and so on. An additional freedom of material design can be introduced by impurity doping. The introduction of shallow impurities in a semiconductor significantly modifies its optical and electrical transport properties. Actually, a precise control of an impurity profile is key to achieve desirable functions in almost all kinds

Silicon Nanocrystals: Fundamentals, Synthesis and Applications. Edited by Lorenzo Pavesi and Rasit Turan
Copyright © 2010 WILEY-VCH Verlag GmbH & Co. KGaA, Weinheim
ISBN: 978-3-527-32160-5

of semiconductor devices. Therefore, impurity-doped semiconductor nanostructures have been a subject of intensive research [9–13]. Unfortunately, contrary to many theoretical studies [14–25], experimental work on shallow impurity-doped Si nanocrystals [26–47] makes a poor progress because of difficulties in their preparation and characterization. The main difficulty arises from the fluctuation of impurity number per nanocrystal in a nanocrystal assembly. For Si nanocrystals as small as a few nanometers in diameter, the expression of the doping level in the form of "impurity concentration" is not suitable and it should be expressed as "impurity numbers" because it changes digitally. For example, doping of one impurity atom into a nanocrystal 3 nm in diameter ($1.4 \times 10^{-20}\,\mathrm{cm}^3$, about 700 atoms) corresponds to an impurity concentration of $7.0 \times 10^{19}\,\mathrm{atoms/cm}^3$. At this doping level, bulk Si is a degenerate semiconductor and exhibits metallic behavior. Therefore, in nanometer-sized crystals, addition or subtraction of a single impurity atom drastically changes the electronic structure and the resultant optical and electrical transport properties. To explore research fields of doped Si nanocrystals and to realize devices based on them, development of a technique to control the "impurity number" with extremely high accuracy is indispensable.

Although accurate control of the "impurity number" in a Si nanocrystal has not been achieved, recent research partly reveals the properties of doped Si nanocrystals. In Section 3.3, we present experimental results on shallow impurity-doped Si nanocrystals. In Section 3.4, we discuss the concept of compensation and counterdoping in Si nanocrystals. We show that compensated Si nanocrystals exhibit different properties from those of intrinsic Si nanocrystals and demonstrate that shallow impurity control in combination with geometrical confinement of carriers extends the controllable range of the properties of Si nanocrystals.

Finally, before starting the main part, we would like to clarify here the type of Si nanocrystals considered in this chapter. Si nanocrystals can be prepared by various methods and the properties are slightly different depending on the preparation procedure because of different defect density, different degree of interaction between nanocrystals, different surface termination, and so on. In this chapter, we will consider the following nanocrystals:

1) The surface is oxygen (O)-terminated for majority of Si nanocrystal samples discussed. The properties of O-terminated and hydrogen (H)-terminated Si nanocrystals are qualitatively similar, but quantitatively there are some differences. The advantage of H-terminated Si nanocrystals is that almost all theoretical calculations are based on them and thus understanding the experimentally obtained phenomena is easier. On the other hand, they are not very stable and H-terminated surface is slowly oxidized even at room temperature. Therefore, for device application of Si nanocrystals, O-terminated Si nanocrystals are probably more realistic.

2) Oxide barrier between nanocrystals is thick enough to prevent carrier transport between nanocrystals. Understanding the mechanism of carrier transport between Si nanocrystals is very important for device applications and many groups are working on the subject [33, 43, 46, 48]. However, in the case of

conductive nanocrystal assemblies, the properties are mainly determined by interfaces and thus they are not very suitable for the study of the properties of Si nanocrystals themselves.

3) The diameter of nanocrystals is limited to be less than 10 nm because quantum size effects start to play a role below this size. Above this size, Si nanocrystals can be regarded as bulk.

3.2
PL Properties of Intrinsic Silicon Nanocrystals

3.2.1
Fundamental Properties

The luminescence properties of Si nanocrystals have been intensively studied for many years. Size-dependent high-energy shift of the luminescence maximum and shortening of the lifetime have commonly been observed for both H- and O-terminated Si nanocrystals prepared by a variety of methods. Furthermore, a very large temperature dependence of luminescence lifetime, that is, approximately two orders of magnitude different from liquid He temperature to room temperature, has been reported. All these phenomena could at least be qualitatively explained by quantum confinement of excitons in the space smaller than the Bohr radius of the bulk exciton, although quantitative differences are reported between H- and O-terminated Si nanocrystals, especially when the size is very small. Nitrogen (N)-terminated Si nanocrystals or Si nanocrystals in SiN_x exhibit quite different properties from those of H- or O-terminated ones, and the PL mechanism is still under debate. In this section, we will briefly summarize fundamental PL properties of O-terminated Si nanocrystals.

Figure 3.1a shows PL spectra of Si nanocrystals embedded in SiO_2 prepared by cosputtering of Si and SiO_2 and annealing at temperatures higher than 1100 °C [3, 49]. By controlling the size, the luminescence maximum shifts from very close to the bulk Si bandgap to 1.6 eV. In O-terminated Si nanocrystals, the high-energy shift is almost always limited to 1.6 eV. This is considered to be due to the formation of O-related states in the bandgap of Si nanocrystals [50]. On the other hand, in H-terminated Si nanocrystals, the range can be extended to ~2.3 eV [2]. Figure 3.1b shows PL spectra at 4 K. In addition to the main peak, a peak appears at a lower energy. This peak is assigned to the recombination of a conduction band electron with a hole in a deep surface trap at Si–SiO_2 interfaces [51]. Similar to the main peak, the low-energy peak exhibits a size-dependent shift. However, the amount of the shift is about half that of the main peak [3, 51–54]. The size-dependent shift of the low-energy peak is considered to reflect that of the conduction band edge.

Figure 3.2 shows PL lifetime as a function of temperature for two samples with different size distributions [3]. The temperature dependence is larger for smaller Si nanocrystals. The temperature dependence can be well explained by the model proposed by Calcott *et al.* [55]; the exciton state is split into a singlet state and a triplet

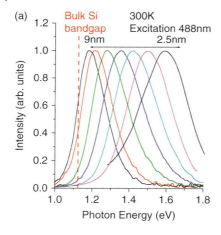
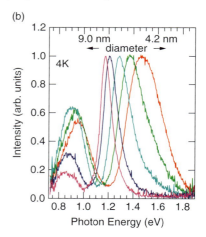

Figure 3.1 PL spectra of Si nanocrystals embedded in SiO_2 thin films (a) at room temperature and (b) at 4 K. The average diameters are changed from about 9 to 2.5 nm in (a) and from 9 to 4.2 nm in (b). Excitation wavelength is 488 nm. (b) Reproduced with permission from Ref. [3]. Copyright (2000) by the American Physical Society. http://link.aps.org/abstract/PRB/v62/ p16820.

state due to the electron–hole exchange interaction. By fitting the temperature dependence of the lifetime, one can estimate both the lifetimes of the triplet and singlet states and the splitting energy. Figure 3.3a shows temperature dependence of the lifetimes as a function of the PL detection energy, that is, the size. The lifetime of the triplet state is almost independent of the size because the transition is forbidden, while that of the singlet state depends strongly on the size and becomes short with

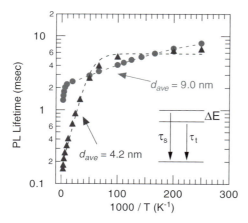

Figure 3.2 Temperature dependence of PL lifetime for two Si nanocrystal samples with different size distributions. Inset is a schematic illustration of singlet and triplet splitting of an exciton state. τ_s and τ_t represent singlet and triplet lifetimes, respectively, and ΔE the splitting energy. Reproduced with permission from Ref. [3]. Copyright (2000) by the American Physical Society. http://link.aps.org/abstract/ PRB/v62/p16820.

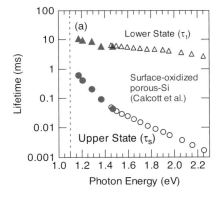
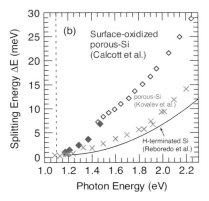

Figure 3.3 (a) Lifetimes of the upper singlet (τ_s) and lower triplet (τ_t) excitonic states and (b) exchange splitting energy (ΔE) as a function of the monitored PL energy. The data of the solid circles, solid triangles, and solid diamonds are obtained for Si nanocrystals in SiO_2 prepared by sputtering [3]. The open circles, open triangles, and open diamonds represent the data for surface-oxidized porous Si [55]. The crosses in (b) are for fresh porous Si [2]. The solid curve represents the calculated ΔE for H-terminated Si nanocrystals [56]. Reproduced with permission from Ref. [3]. Copyright (2000) by the American Physical Society. http://link.aps.org/abstract/PRB/v62/ p16820.

decreasing size. In indirect bandgap semiconductors, geometrical confinement of carriers increases the overlap of electron and hole wavefunctions in momentum space and thus enhances the oscillator strength. The shortening of the singlet lifetime is due to this effect. From this effect, one can expect Si nanocrystals to behave as direct bandgap semiconductors. However, there is some evidence suggesting that the momentum conservation rule is only partially broken and Si nanocrystal strongly preserves the indirect bandgap nature of bulk Si crystals [2]. In fact, even for very small Si nanocrystals with PL in the visible range, the PL radiative lifetime is longer than several microseconds. The indirect bandgap nature of Si nanocrystals is discussed in detail in Sections 3.2.2 and 3.2.3.

It is worth noting that for large Si nanocrystals with the PL maximum near the bandgap energy of bulk Si crystal, the singlet lifetime is close to 1 ms. Excitons must stay for such a long period in very small space without recombining nonradiatively at room temperature. For this, Si nanocrystals should be defect-free and the surface should be perfectly terminated. Because of the difficulty in preparing such Si nanocrystal samples, reports on Si nanocrystals with the PL maximum below 1.2 eV at room temperature are very few.

In Figure 3.3b, the singlet–triplet splitting energy is plotted as a function of PL detection energy. It increases monotonously with decreasing the size. This is because of better overlap of electron and hole wavefunctions by confinement. The splitting energy is larger in the case of O-termination than that of H-termination. The splitting energy of H-terminated Si nanocrystals agrees well with that obtained by calculation [56]. In Figure 3.3a and b, all the data are changed continuously in the very wide energy range starting from the bulk Si bandgap. This continuity is one of the evidence that the PL arises from quantum-confined excitons.

3.2.2
Effect of Size and Shape Distribution on the PL Bandwidth

In general, optical transitions between quantized electronic states result in atomic-like very sharp absorption and emission peaks. However, as can be seen in Figure 3.1, the PL band of Si nanocrystals is very broad (full width at half maximum, FWHM: 200–300 meV). Even at very low temperatures, there is no significant narrowing of the bandwidth [3]. This suggests that the large width is due to inhomogeneous broadening. The group of Linnros was first to measure PL from a single Si nanocrystal at low temperatures [57]. They demonstrated without any doubt that each Si nanocrystal exhibits atomic-like very narrow PL peaks (FWHM of a few meV at 35 K), and the broadening observed for nanocrystal ensembles is just due to size and shape distributions. In their measurements, a pair of peaks is observed for each Si nanocrystal [57]. The high-energy one corresponds to a zero-phonon line, while the other one is assigned to a transition accompanied by the emission of a momentum conservation phonon, that is, an optical phonon at Δ minima or a confined acoustic phonon [58]. The observation of the peak pair evidences that the momentum conservation is only partially broken in Si nanocrystals, and they still strongly preserve the indirect bandgap character. Note that, at room temperature, the PL peak of a single Si nanocrystal is rather broad (FWHM of 120–150 meV) due to participation of different phonons in optical transitions [59].

The evidence of indirect nature of Si nanocrystals is also obtained by hole-burning spectroscopy [60] and resonant PL spectroscopy [55, 61]. In resonant PL spectroscopy, when PL is excited by photons with the energy within the broad PL band, only nanocrystals with the bandgap energy smaller than the excitation energy are excited. This results in significant narrowing of the PL band and the appearance of features corresponding to momentum conservation phonons.

3.2.3
Resonant Quenching of PL Band Due to Energy Transfer

Besides several techniques described above, the information from specific size of Si nanocrystals from inhomogeneously broadened PL bands can be extracted by introducing an "energy acceptor" that selectively kills luminescence from nanocrystals with a specific size by nonradiative energy transfer. This phenomenon can be used to investigate fundamental physics of Si nanocrystals when the state to which energy transfer is made is narrow. Here, we demonstrate two examples that introduce "holes" into the PL band of Si nanocrystal assemblies.

The first one is rare earth ions. Rare earth ions exhibit sharp atomic-like absorption and emission spectra due to the intra-4f shell transitions. If a kind of rare earth, for example, Er, is doped into SiO_2 films containing Si nanocrystals, there is strong interaction between Si nanocrystals and Er^{3+}, that is, excitation energy of Si nanocrystals is effectively transferred to Er^{3+} nonradiatively and the 4f shell is excited [62–67]. For Si nanocrystals, this is an introduction of a nonradiative recombination pass and thus the PL is quenched. For the energy transfer, energy conservation rule should be satisfied. Therefore, only Si nanocrystals with the bandgap energy corresponding to the energy difference of the discrete electronic states of Er^{3+} can transfer energy resonantly, while emission or absorption of phonons is required for others. Since the energy transfer rate is different between the resonant and nonresonant energy transfer, the PL quenching does not occur uniformly, but dips (holes) are observed in inhomogeneously broadened PL bands [64, 65]. Figure 3.4a shows PL spectra of Er and Si nanocrystals doped SiO_2 at low temperature [65]. The spectra of two samples with different Si nanocrystal size distributions are shown. We can clearly see features around 1.55 and 1.28 eV. These energies correspond to the transitions between the $^4I_{9/2}$ and $^4I_{15/2}$ states and the $^4I_{11/2}$ and $^4I_{15/2}$ states of Er^{3+}, respectively, and are the signature of resonant energy transfer (see Figure 3.4b). In Figure 3.4a, we notice that the dips appear at exactly the resonant energy (1.55 eV) and about 56 meV below the energy. The mechanism for the appearance of the low energy dip is schematically shown in Figure 3.4b. For the energy transfer to the $^4I_{9/2}$ state of Er^{3+}, the bandgap energy of Si nanocrystals should be 1.55 eV. Si nanocrystals with the bandgap energy of 1.55 eV exhibit PL either at 1.55 eV or at "1.55 eV minus the energy of a momentum conservation phonon." The most probable phonon emission process is the emission of a TO

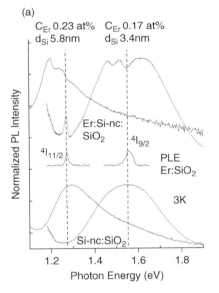

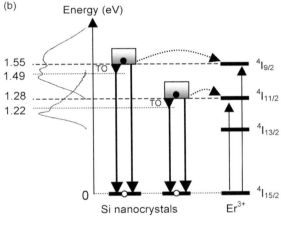

Figure 3.4 (a) (Upper) PL spectra of Si nanocrystals and Er-doped SiO_2 at 3 K. Spectra of two samples with different Er concentration and Si nanocrystal sizes are shown. Dips due to resonant quenching are clearly seen. (Middle) PLE spectra of Er-doped SiO_2 detected at 0.81 eV. (Bottom) PL spectra of Si nanocrystals at 3 K for two samples with different size distributions. (b) Schematic illustration to explain the origin of dips in PL spectra of Si nanocrystals. Reproduced with permission from Ref. [65]. Copyright (2005) by the American Physical Society.

phonon at the Δ minima (56 meV). Therefore, if PL from Si nanocrystals with the bandgap energy of 1.55 eV is lost by energy transfer to the $^4I_{9/2}$ state, dips appear both at 1.55 eV and at 1.55 eV–56 meV. Therefore, the observation of the dip pair is an evidence that the momentum conservation is only partially broken and zero-phonon and phonon-assisted transitions simultaneously occur.

Similar features can be seen in a more exotic system. When O_2 molecules are adsorbed onto the surface of H-terminated Si nanocrystals, the excitation energy is transferred to the adsorbed O_2 molecule in the ground triplet state and it is excited to the singlet state, that is, singlet oxygen is generated [5, 68]. This energy transfer results in resonant quenching of PL from Si nanocrystals. Figure 3.5 shows time-resolved PL spectra of O_2 molecule-adsorbed Si nanocrystal at low temperature [69]. The spectra are divided by those of Si nanocrystals in vacuum. Inset in Figure 3.5 is an expansion of the region around 1.6 eV after 80 ns from the excitation. We can see a pair of dips as the case of Er doping. Again, the higher energy dip exactly coincides with the energy difference of the $^1\Sigma$ the $^3\Sigma$ states of O_2 molecules [70] and the lower energy dip stays at about 56 meV below the resonant energy (see Figure 3.5b). In Figure 3.5, a pair of dips can also be seen at 1.96 and 1.9 eV, indicating that very small

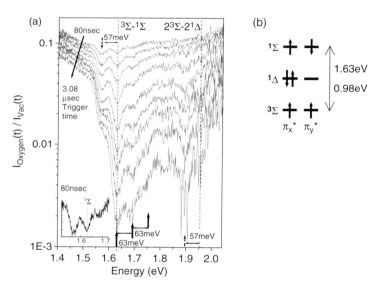

Figure 3.5 (a) Time-resolved PL spectra of a porous Si layer having O_2 molecules physisorbed on the surface divided by those of porous Si in vacuum ($I_{\text{Oxygen}}(t)/I_{\text{Vac}}(t)$). The measurement gate width is 100 ns. The data for the measurement delay time with respect to the excitation pulse of 008, 0.18, 0.28, 0.48, 0.78, 1.08, 1.48, 2.08, and 3.08 μs are shown. The energy positions of the $^3\Sigma$ to $^1\Sigma$ transition of O_2 monomers and that of the $2^3\Sigma$ to $2^1\Delta$ transition of $(O_2)_2$ dimers are indicated by dashed lines. Inset is the expansion of $^3\Sigma$ to $^1\Sigma$ transition at 80 ns. (b) Schematic illustration of energy diagram of O_2. Arrows represent spin-up and spin-down. Reproduced with permission from Ref. [69]. Copyright (2005) by the American Physical Society. http://link.aps.org/abstract/PRB/v72/e165321.

Si nanocrystals with the bandgap energy close to 2 eV still strongly preserve the indirect bandgap character.

3.2.4
PL Quantum Efficiency of Intrinsic Si Nanocrystals

As discussed above, the radiative decay rate of Si nanocrystals is enhanced due to better overlap of electron and hole wavefunctions by confinement. However, high PL quantum efficiency of Si nanocrystals [71–76] cannot always be explained by the enhanced radiative rate. It arises mainly from restriction of carrier transport within nanocrystals. If carrier transport is limited within a nanocrystal and there is no nonradiative recombination processes in the nanocrystal, the PL quantum efficiency should be one. On the other hand, if there is at least one nonradiative recombination process, the microsecond to millisecond radiative lifetime cannot compete with the nonradiative process and the quantum efficiency is close to zero. This means that in Si nanocrystal assemblies, the PL quantum efficiency of nanocrystals constituting the assembly is either one or zero, and the ratio of the "bright" and "dark" nanocrystals determines the total quantum efficiency of the assembly. The model roughly explains how the total quantum efficiency of Si nanocrystal assemblies is determined. However, it is too crude for quantitative discussion, and in actual Si nanocrystal assemblies, the PL quantum efficiency of "bright" Si nanocrystals is not always 100%. We will show experimental results that assess this point.

PL quantum efficiency of Si nanocrystals contributing to PL in an assembly can be extracted by the modulation of the radiative decay rate [77–79]. PL decay rate is modified when Si nanocrystals are placed close to an interface due to the change in the photonic mode density. If the distance between the nanocrystal and the interface is changed, both the mode density and the radiative rate oscillate. The effect appears as an oscillation of the PL decay rate. By comparing experimentally obtained oscillation of PL decay rates with calculated oscillation of the radiative rate, PL quantum efficiency as well as radiative and nonradiative decay rates of Si nanocrystals contributing PL are obtained. Figure 3.6a and b shows radiative (w_r) and nonradiative (w_{nr}) decay rates, respectively, as a function of wavelength obtained in this procedure for four samples prepared under different conditions. PL spectra of these samples are shown in Figure 3.6d [79]. Radiative decay rates for four samples are on a single curve and increase to shorter wavelength. The increase in the radiative decay rate is due to better overlap of electron and hole wavefunctions in momentum space by confinement. The fact that all the data obtained for different samples fit on a single curve indicates that the radiative recombination rate is determined only by the size of Si nanocrystals. On the other hand, the nonradiative rate in Figure 3.6b depends strongly on samples and therefore on sample preparation procedures.

Figure 3.6c shows PL quantum efficiency obtained by the ratio of the radiative rate and the sum of the radiative and nonradiative rates [$w_r/(w_r + w_{nr})$]. Note that the quantum efficiency obtained in this procedure reflects that of nanocrystals participating in the PL process and contribution from completely dead Si nanocrystals is not involved. Therefore, the quantum efficiency should be different from (higher

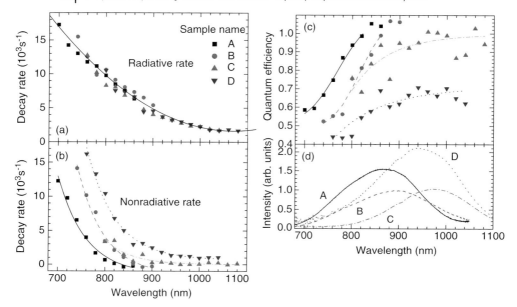

Figure 3.6 (a) Radiative decay rates, (b) nonradiative decay rates, and (c) PL quantum efficiencies, and (d) PL spectra of four Si nanocrystal samples prepared under different conditions; samples A (closed squares), B (closed circles), C (triangles), and D (inverted triangles), as a function of wavelength. (d) PL spectra of samples A (solid curve), B (dashed curve), C (dashed-dotted curve), and D (dotted curve). Reproduced with permission from Ref. [79]. Copyright (2006) by the American Physical Society. http://link.aps.org/abstract/PRB/v73/e245333.

than) that obtained by usual procedures by using integral spheres. In Figure 3.6c, the quantum efficiency is very high. In particular, it is almost one at a longer wavelength side for some samples. Therefore, for Si nanocrystal samples in Figure 3.6 at a specific wavelength range, the "bright" and "dark" model is satisfactory. On the other hand, in some samples and in some wavelengths, the quantum efficiency is not one. This means that even in "bright" Si nanocrystals, there is a competition between radiative and nonradiative processes. As shown in Figure 3.6c, the quantum efficiency is always lower at shorter wavelengths where the radiative recombination rate is high. A possible explanation of this wavelength dependence is as follows. For large Si nanocrystals with very small radiative rates, the radiative process cannot compete with nonradiative processes. As a result, each nanocrystal becomes either perfectly "bright," that is, quantum efficiency of 100%, or completely dead, that is, PL signal cannot be detected. On the other hand, the increased radiative recombination rate of smaller particles by stronger confinement makes it possible to compete with nonradiative processes. This results in the quantum efficiency in between zero and one. Therefore, low quantum efficiency in Figure 3.6c does not always mean that the total PL quantum efficiency of a whole sample is low. Instead, participation of more nanocrystals in the PL process due to higher radiative rates increases the total quantum efficiency.

3.3
Shallow Impurity-Doped Si Nanocrystals

One of the largest problems for the growth of doped Si nanocrystals is that impurity atoms are pushed out of nanocrystals to surrounding matrices by the so-called self-purification effect. This effect can be understood by considering very high formation energy of doped Si nanocrystals [21]. Impurity concentration in nanocrystals is thus always different from average concentration in a whole system. In the worst case, the number of impurity in a nanocrystal becomes zero even when average concentration is rather high. Therefore, development of a technique to characterize impurities, especially "active" impurities doped into nanocrystals, is crucial. A conductivity measurement is impossible or extremely difficult. Electron spin resonance (ESR) spectroscopy has been used to characterize n-type Si and is also useful for studying n-type Si nanocrystals [31, 34, 45]. In Section 3.3.4, we present ESR spectra of phosphorus (P)-doped Si nanocrystals. Unfortunately, this technique is not very sensitive to p-type Si nanocrystals. Infrared absorption due to intraconduction or intravalence band transitions is also used [31, 39], but it is not very sensitive. As an alternative tool to characterize doped Si nanocrystals, we propose PL spectroscopy. In Sections 3.3.2 and 3.3.3, we discuss PL properties of boron (B)- and P-doped Si nanocrystals, respectively, and show how the PL properties are modified by the doping and that PL spectra are very sensitive to doping.

3.3.1
Preparation of Impurity-Doped Si Nanocrystals

Various methods have been reported for the growth of intrinsic Si nanocrystals. On the other hand, a limited number of studies are published concerning the growth of shallow impurity-doped Si nanocrystals with a diameter below 10 nm. Shallow impurity-doped Si nanocrystals are usually grown by plasma decomposition of silane (SiH_4) by adding dopant precursors (diborane (B_2H_6) and phosphine (PH_3)) [33, 44, 46]. With this method, a variety of morphologies from densely packed nanocrystalline films to nanoparticle powder can be produced by controlling process parameters. Another method to grow doped Si nanocrystals is to use phase separation of silicon suboxide (SiO_x) into Si and SiO_2 at high-temperature annealing. With this method, Si nanocrystals embedded in glass matrices are produced. Thin films of SiO_x have been prepared by various methods, for example, cosputtering of Si and SiO_2, ion implantation of Si into SiO_2, vacuum evaporation of SiO, chemical vapor deposition by using SiH_4 and N_2O precursors, and so on. If dopant atoms are incorporated in SiO_x, some of them are doped into Si nanocrystals during phase separation by annealing.

The following is the procedure we use to grow impurity-doped Si nanocrystals. Many of the data shown in this chapter are obtained for samples prepared by this method. We grow Si nanocrystals by a cosputtering method [31]. In the case of P- (B-) doped Si nanocrystals, Si, SiO_2, and P_2O_5 (B_2O_3) are simultaneously sputter-deposited and they are annealed in N_2 gas atmosphere at 1100–1250 °C. During the annealing, Si nanocrystals are grown in phosphosilicate (PSG) (borosilicate (BSG)) thin films. The P (B) concentration in PSG (BSG) thin films is at maximum several percentages. Similarly, if Si, SiO_2, B_2O_3, and P_2O_5 are simultaneously sputter-deposited, B and P codoped Si nanocrystals are grown in borophosphosilicate glass (BPSG) matrices. Note that impurity concentration in nanocrystals is different from that of matrices because the segregation coefficient, that is, the ratio of equilibrium concentration of impurity atoms in Si and SiO_2, strongly depends on the kind of impurities.

3.3.2
PL from B-Doped Si Nanocrystals

The introduction of extra carriers by impurity doping makes the three-body Auger process possible [8]. The Auger rate is calculated as a function of the size of Si nanocrystals and is estimated to be of the order of nanoseconds [16]. This value is four to five orders of magnitude larger than the radiative rate of excitons, and thus one shallow impurity can almost completely kill PL from a Si nanocrystal. In other words, PL quantum efficiency of impurity-doped Si nanocrystals is considered to be much lower than that of intrinsic Si nanocrystals. Since we measure ensembles of many Si nanocrystals with different impurity concentrations (numbers), with increasing average impurity concentration in a nanocrystal assembly, the PL intensity is expected to continuously decrease. This effect is really observed in p-type Si nanocrystals. Figure 3.7a shows the examples. With increasing B concentration, the PL intensity decreases monotonously [27, 29]. The decrease in the PL intensity is accompanied by the shortening of the lifetime. Another possible explanation for PL

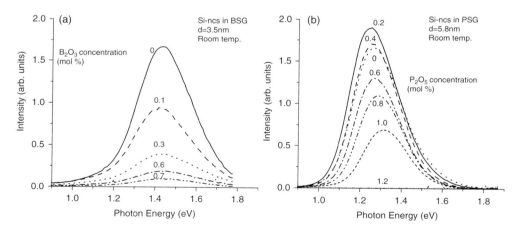

Figure 3.7 PL spectra of (a) B-doped and (b) P-doped Si nanocrystals at room temperature. B and P concentrations are changed. (a) Reproduced with permission from Ref. [29]. Copyright (1999) by Elsevier Science Ltd. (b) Reproduced with permission from Ref. [34]. Copyright (2002) by the American Physical Society. http://link.aps.org/abstract/PRL/v89/e206805.

quenching by B doping is defect formation due to a strain induced by doping because of the small size of B atoms (21% smaller than a Si atom) [20, 44, 80].

3.3.3
PL from P-Doped Si Nanocrystals

In P-doped Si nanocrystals, the situation is different from B-doped Si nanocrystals. When the P concentration is relatively low, the PL intensity increases slightly compared to that of the undoped Si nanocrystals (Figure 3.7b) [30, 31, 34, 38]. This effect can be seen more clearly when the size of Si nanocrystals is smaller. The increase in the PL intensity indicates that nonradiative recombination processes are quenched by P doping. In fact, defect-related infrared emission band observed at low temperatures is quenched by P doping [30], and the defect-related ESR signal almost disappears as we will see in Section 3.3.4 [34]. Two scenarios are considered as the mechanism of quenching nonradiative recombination processes. The first one is that the number of dangling bonds on the surface of Si nanocrystals decreases because of higher flexibility of bond angles and lengths in PSG than SiO_2. The second scenario is that electrons supplied by P doping are captured by the dangling bonds, which inactivate the nonradiative recombination centers and compensate donors [46, 81]. Although the definite evidence of these mechanisms is not yet obtained, the second one seems to be more plausible. It is worth noting that the impurity concentration dependence of the PL intensity depends strongly both on the size and the concentration of Si nanocrystals and on annealing parameters such as temperature, duration, atmosphere, cooling speed, and so on [29]. Therefore, the P concentration that shows the maximum PL intensity is different depending on preparation parameters.

3.3.4
Electron Spin Resonance Studies of Shallow Impurity-Doped Si Nanocrystals

As discussed in detail in Chapter 2, the electronic states of shallow impurities in Si nanocrystals are expected to be strongly modified from those of the bulk ones if the size is close to the effective Bohr radius of impurities. In particular, the binding (ionization) energy should be enhanced because donors (acceptors) are squeezed three-dimensionally [9, 10, 16]. The hyperfine structure (hfs) in ESR of shallow donors is a sensitive sensor of the modification of their electronic states because the strength of the hyperfine interactions is directly related to the localization of the dopant state electrons [82, 83]. If donor wavefunctions are squeezed by spatial confinement, the hyperfine splitting should be enhanced.

Figure 3.8 shows X-band ESR spectra obtained at 40 K. The signal with $g = 2.006$ is due to dangling bond defects at Si–SiO_2 interfaces [30]. Although the intensity of this signal is very large, it quickly vanishes by P doping by the mechanism discussed in Section 3.3.3. The signal with $g = 2.002$ can be assigned to an EX center characterized by the involved hyperfine structure of 16 G splitting [84]. The EX center signal is also observed for sputter-deposited pure SiO_2. It also becomes weaker with increasing

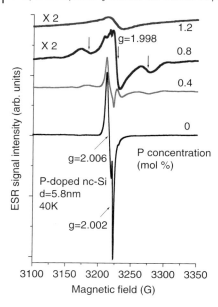

Figure 3.8 ESR spectra of Si nanocrystals dispersed in PSG matrices. P concentration is changed from 0 to 1.2 mol%. Arrows indicate the position of the features at 3180, 3225, and 3270 G. Reproduced with permission from Ref. [34]. Copyright (2002) by the American Physical Society. http://link.aps.org/abstract/PRL/v89/e206805.

P concentration. At high P concentration, a broad signal with $g = 1.998$ emerges. This signal is assigned to conduction electrons in Si nanocrystals [28, 82, 85, 86]. The intensity of this signal is nearly inversely proportional to the temperature except for a very low temperature range below 15 K even if P concentration is very high [45, 47]. This Curie-like paramagnetism indicates that Si nanocrystals do not become metallic even under heavily doped conditions at least when they are grown under thermo-dynamic equilibrium conditions.

On both sides of the conduction electron signal with $g = 1.988$, two broad bands are observed at 3180 and 3270 G. These features are seen only at a moderate P concentration range. At the highest P concentration (1.2 mol%), the features completely disappear and only the broad conduction electron signal remains. The most probable origin of these features is the P donor hfs; a donor electron localizes at P nucleus and interacts with the nuclear spin leading to the split of a single resonance line into a line doublet [82]. In the present samples, the size of Si nanocrystals is close to the effective Bohr radius of P donors in bulk Si crystal. This means that if more than two P donors exist in one nanocrystal, their wavefunctions are overlapped, resulting in the delocalization of electrons even at low temperatures. Therefore, Si nanocrystals containing more than two P donors will show only a conduction electron signal. With this simple model, P concentration dependence of the ESR signal in Figure 3.8 and that of the PL spectra in Figure 3.7b can be well explained. For the sample with the P concentration of 0.4 mol%, PL intensity starts to decline. This is accompanied by the appearance of broad infrared absorption due to intraconduction band transitions of confined electrons [31, 34]. In this P concentration, a part of Si nanocrystals has P donors and others are considered to be still intrinsic. As a result, we can see weak hfs and the conduction electron signal. With further increasing P concentration (0.8 mol%), the ratio of doped Si nanocrystals increases, resulting in the increase in both the hfs and the conduction electron signal. In this P concentration, the conduction electron signal is much stronger than the hfs, indicating that the number of Si nanocrystals having more than two P donors is rather large. At the P concentration of 1.2 mol%, only the conduction electron signal is observed in ESR. At this concentration, PL is almost completely quenched. This suggests that almost all nanocrystals contain more than two P donors.

In Figure 3.8, the separation of the hyperfine splitting (ΔH(hfs)) is about 90 G, which is more than twice larger than that of the bulk value (42 G). This enhancement of the hyperfine splitting can be caused by the quantum confinement of P donors because P donors are confined in a space close to the effective Bohr radius of P donors in bulk Si crystals (1.67 nm). This model can be directly proved by studying the size dependence of ΔH(hfs). In fact, clear size dependence of the ΔH(hfs) is observed [34]. With decreasing size of Si nanocrystals, ΔH(hfs) drastically increases. This is the evidence that the observed large hyperfine splitting arises from quantum confinement of P donors in Si nanocrystals. The observed enhancement of ΔH(hfs) is successfully reproduced by first-principles calculations [18, 25].

3.3.5
Location of Dopant Atoms

There are many theoretical studies on preferential location of impurities in Si nanocrystals. On the other hand, it is almost impossible to experimentally control the location of impurities in nanocrystals. However, information on the location can be experimentally obtained. The group of Kortshagen [44] assigned the location of impurities by investigating the effects of oxidization and etching of Si nanocrystals on PL spectra. The Si nanocrystals were grown by plasma decomposition of SiH_4 and the doping was achieved by introducing B_2H_6 or PH_3 in plasma. They showed that the effect of oxidation is quite different between P and B doping. In the case of P doping, exposure of the sample to air for 5 days at room temperature results in the recovery of the PL intensity to the level of intrinsic Si nanocrystals. Furthermore, wet chemical etching of oxide layer results in approximately 80% decrease in P concentration in the sample. These are the strong indication that P dopant is located at or close to the surface; after oxidation, P dopant is embedded in surface Si oxide and no longer exhibits an effect on the PL from Si nanocrystals.

On the contrary, in B-doped Si nanocrystals, recovery of PL intensity by oxidation is limited. Furthermore, B concentration increases after removing the oxide layer. These results suggest that B is primarily incorporated into the Si nanocrystal core. It is worth noting that doping efficiency of B is much smaller than that of P probably due to larger formation energy of B-doped Si nanocrystals than P-doped ones [21, 22].

We have to be aware that the above result is not necessarily universal. The preferential location of impurities may depend on nanocrystal growth processes and the surface termination. At high-temperature growth, the incorporation of impurities in nanocrystals is determined by thermodynamic equilibrium solubility. On the other hand, if thermodynamic equilibrium is not established, kinetic factors such as activation barriers will control doping [13, 44]. Therefore, properties of dopant may be quite different between Si nanocrystals grown by chemical route, decomposition of SiH_4, phase separation of SiO_x, and so on. In fact, one recent calculation suggests that preferable location of B is on the surface and that of P is in the core in H-terminated Si nanocrystals [24]. A preferential doping of B in the subsurface is also shown by another group [20, 22]. These predictions are not consistent with experimental results.

3.4
P and B Codoped Si Nanocrystals

For the fabrication of Si-based devices, a counterdoping process is routinely used to make an n-type region in a p-type region or vice versa. The question then arises whether counterdoping is possible in Si nanocrystals or the concept of counterdoping is the same as that of the bulk. In this section, we will discuss this issue. Especially, we will focus on carrier compensation in Si nanocrystals and the properties of compensated Si nanocrystals.

One of the motivations to study compensated Si nanocrystals is that they may have higher radiative rate than that of intrinsic Si nanocrystals. As discussed in Section 3.3, because of the very efficient Auger process, heavily shallow impurity-doped Si nanocrystals are very bad candidates for light-emitting applications. However, the additional localization of carriers by impurities is a possible approach to enhance oscillator strength of excitons in Si nanocrystals. The Auger process can be avoided if isoelectronic impurities are used. Unfortunately, Si does not have proper isoelectronic impurities that can strongly localize excitons at room temperature and enhance the PL intensity. An alternative approach is to simultaneously dope the same number of p- and n-type impurities in Si nanocrystals (compensated Si nanocrystals). The codoping and compensation may result in further localization of excitons without being overshadowed by the Auger process.

To study compensated Si nanocrystals, a major technological problem is how to dope exactly the same number of B and P atoms into Si nanocrystals. If the number is not exactly equal, excitons will recombine nonradiatively via the Auger process. Although a precise control of the number of p- and n-type impurities in a Si nanocrystal seems to be impossible, recent calculations demonstrated preferential formation of nanocrystals with equal number of p- and n-type impurities. The group of Ossicini [20, 21, 23] demonstrated by first-principles calculations that the formation energy of Si nanocrystals drastically decreases when pairs of B and P are doped into Si nanocrystals; the formation energy of a pair of B- and P-doped Si nanocrystals ($Si_{147}H_{100}$ clusters) is about 1 eV smaller than that of B-doped ones and about 0.7 eV smaller than that of P-doped ones. This indicates that if Si nanocrystals are grown by the procedure described in Section 3.3.1, that is, the phase separation of P- and B-doped Si-rich SiO_2 (Si-rich BPSG) by annealing, nanocrystals with equal number of B and P are preferentially grown because they are energetically favorable.

3.4.1
PL Properties of P and B Codoped Si Nanocrystals

Figure 3.9 shows PL spectra of B- and P-doped Si nanocrystals [35, 37, 39]. B concentration in a sample is almost fixed while P concentration is changed. Without P doping, that is, when only B is doped, the PL intensity is very weak due to the efficient Auger process. By doping P simultaneously, the intensity recovers and the peak shifts to lower energy. The recovery of PL intensity strongly suggests that carriers are compensated in majority of nanocrystals in the sample. Therefore, simultaneous B- and P-doped and compensated Si nanocrystals can be really prepared by the process shown in Section 3.3.1 and they have rather high PL quantum efficiency [35, 37, 39, 87].

It is interesting to note that the luminescence energy of the codoped sample in Figure 3.9 is far below the bandgap energy of the bulk Si crystals. Although PL peak energy shifts slightly to lower energy by doping either P or B, below bulk bandgap PL can be realized only when both kinds of impurities are doped [37, 39]. This implies that both donor and acceptor states are involved in the recombination process. Similar low-energy PL has been observed for heavily doped and compensated bulk Si crystals (Si:P,B) due to the transition from the conduction band tail (donor band) to the valence band tail (acceptor band) [88]. The mechanism of the low-energy PL

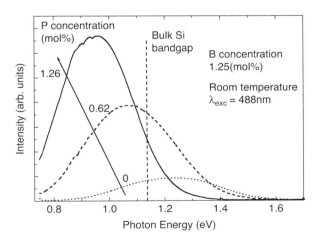

Figure 3.9 PL spectra of B-doped (dotted curve) and B- and P-doped (dashed and solid curves) Si nanocrystals at room temperature. B concentration is fixed, while P concentration is changed. Reproduced with permission from Ref. [35]. Copyright by the American Institute of Physics. http://link.aip.org/link/?JAPIAU/94/1990/1.

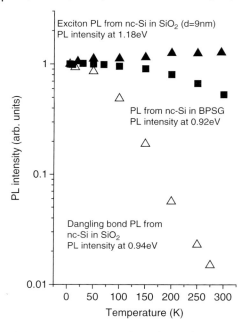

Figure 3.10 Temperature dependence of PL peak intensities for B and P codoped Si nanocrystals in BPSG (■) and exciton PL of intrinsic Si nanocrystals in pure SiO$_2$ (▲). Intensity of dangling bond-related PL (△) is also plotted. Reproduced with permission from Ref. [35]. Copyright by the American Institute of Physics. http://link.aip.org/link/?JAPIAU/94/1990/1.

observed in Si nanocrystals is considered to be essentially the same. The very large spectral width could be an inhomogeneous broadening caused by the distributions of the size and the impurity concentrations. However, there are some crucial differences between compensated bulk Si and compensated Si nanocrystals. The largest difference appears on the temperature dependence of the PL intensity. Figure 3.10 shows PL peak intensities of intrinsic (▲) and codoped (■) Si nanocrystals detected at 1.18 and 0.94 eV, respectively. There is almost no temperature quenching for the intrinsic Si nanocrystals. On the other hand, small temperature quenching, that is, PL intensity at room temperature is about half that at 5 K, can be observed in codoped Si nanocrystals. This temperature quenching is significantly smaller than that of the heavily doped bulk Si crystals. In heavily doped bulk Si crystals, the low-energy PL can be detected only at very low temperatures because excited electrons and holes are easily thermalized or migrated in impurity bands and recombine nonradiatively [88]. Probably, large ionization energy of impurities and restriction of carrier transport within nanocrystals result in the very small temperature dependence of the PL.

The observed small temperature quenching is also completely different from that of dangling bond-related PL that appears at almost the same energy as the main PL band of codoped samples (see Figure 3.1b). The open triangles in Figure 3.10 represent PL intensity of the dangling bond-related PL detected at 0.92 eV. The PL intensity decreases very rapidly with increasing temperature and that at room temperature is about 1% of that at low temperatures. The completely different temperature dependence suggests that the origin of low-energy PL of codoped samples is different from dangling bond-related PL observed for intrinsic Si nanocrystals.

The significantly different PL spectra between intrinsic and compensated Si nanocrystals suggest that the electronic band structure is quite different. This is consistent with theoretical calculations [21]. Therefore, although both intrinsic and compensated Si nanocrystals have no extra carriers, we can consider them as completely different materials. In Figure 3.11, PL spectra of intrinsic (a) and codoped (b) Si nanocrystals are compared. The tunable range of PL energy of intrinsic Si nanocrystals is limited above the bulk Si bandgap because the PL shift is caused only by the quantum size effects. The tunable range is shifted to lower energy for impurity codoped Si nanocrystals (Figure 3.11b) without significant reduction in the PL intensity.

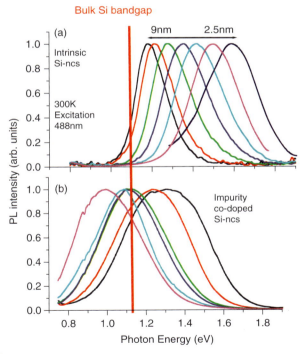

Figure 3.11 Normalized PL spectra of (a) intrinsic and (b) p- and n-type impurities codoped Si nanocrystals at room temperature. The lowest possible PL energy of intrinsic Si nanocrystals is the bulk Si bandgap, while that of codoped Si nanocrystals is extended to 0.9 eV.

3.4.2
PL Lifetime of P and B Codoped Si Nanocrystals

The codoping and compensation of impurities in Si nanocrystals may result in further localization of excitons. This effect is expected to be observed as shortening of PL lifetime compared to that of intrinsic Si nanocrystals. Figure 3.12a shows PL decay curves of intrinsic and P and B codoped Si nanocrystals at 5 K. The energy where the PL signal is detected is indicated by arrows in the inset. The exciton PL of intrinsic Si nanocrystals shows nearly a single-exponential decaying behavior. The lifetime is very long (about 8 ms) due to nearly 100% occupation of the triplet state at 5 K. Compared to this, the lifetime of codoped Si nanocrystals is much shorter and the decay curve deviates from a single-exponential function. In a broad PL band of codoped Si nanocrystals, there is a distribution of the lifetime in the range between 100 and 500 µs. The nonexponential decay curve and the wide distribution of lifetime suggest that the samples are very inhomogeneous probably due to the distributions of impurity numbers and sites as well as the size and shape of Si nanocrystals. Note that the lifetime of dangling bond-related PL is shorter than 1 µs [66]. This again excludes the possibility that the low-energy PL of codoped Si nanocrystals is related to dangling bond-related PL of intrinsic Si nanocrystals. Figure 3.12b shows temperature dependence of PL lifetime of a codoped sample. The lifetime becomes shorter from about 450 to 150 µs with increasing temperature.

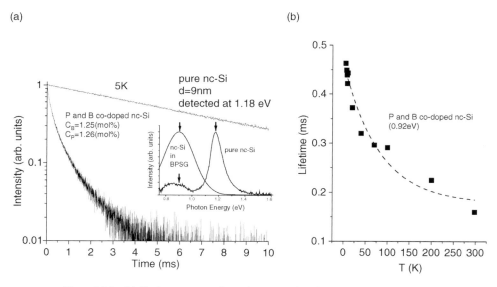

Figure 3.12 (a) PL decay curves of P and B codoped and intrinsic Si nanocrystals at 5 K. In the inset, energy positions for the PL detection are indicated by arrows. (b) Temperature dependence of PL lifetime for P and B codoped Si nanocrystals. Reproduced with permission from Ref. [35]. Copyright by the American Institute of Physics. http://link.aip.org/link/?JAPIAU/94/1990/1.

The size of codoped Si nanocrystals in Figure 3.12 is rather large ($\sim$10 nm in diameter). Compared to PL lifetime of intrinsic Si nanocrystals with almost the same size ($\sim$1 ms at room temperature, see Figure 3.3), that of the codoped Si nanocrystals is short. However, the difference is not very large (at most of the order of magnitude). Therefore, unfortunately, strong localization of carriers by codoping does not significantly enhance the radiative recombination rate. Probably, the effect will be enhanced for smaller Si nanocrystals, and by simultaneously controlling the size and impurity concentration, we could achieve strong enhancement of the radiative recombination rate.

3.4.3
Codoped But Not Compensated Si Nanocrystals

Let us consider Si nanocrystals with unequal number of n- and p-type impurities. In bulk Si crystals, the electrical properties are determined by the difference in the concentration of n- and p-type impurities and thus n- (p-) type regions are produced in p- (n-) type regions by counterdoping. On the other hand, in Si nanocrystals, the properties are not expected to be simply determined by the difference in the number of n- and p-type impurities. For example, a Si nanocrystal with one B atom is considered to be not the same material as that with two B atoms and one P atom. However, experimental demonstration of the difference is very difficult. There is one preliminary experimental result that shows the difference between singly B-doped Si nanocrystals and B and P codoped (B concentration > P concentration) Si nanocrystals [47]. In B-doped Si nanocrystals, ESR signals from confined holes are not detected. This is due to very fast spin relaxation time by the degeneracy of the valence band maximum. On the other hand, in codoped samples with higher B concentration, a broad signal that might be assigned to confined holes is observed. If this is the case, the valence band structure of codoped Si nanocrystals is modified from that of B-doped Si nanocrystals. However, at present, the results are very primitive and no definite conclusion can be achieved.

3.5
Summary

More than 15 years have passed since the discovery of visible PL by Canham [89]. So far, the electronic band structure of Si nanocrystals has almost been clarified. Unfortunately, it became clear that Si strongly preserves the indirect bandgap nature even if the size is a few nanometers and that strong PL is mainly due to the restriction of carrier transport in nanocrystals. This suggests that Si nanocrystals are very suitable for PL materials but not ideal for electroluminescence (EL) devices because high conductivity between nanocrystals is not compatible with high luminescence efficiency. On the other hand, if one understands fundamental properties of Si nanocrystals, new application fields emerge. Some of them are discussed in detail in this book. Very recently, the possibility of multiexciton generation in Si nanocrystal assemblies is reported [90, 91]. This effect has potential to significantly enhance conversion efficiency of Si-based solar cells.

In intrinsic Si nanocrystals, the properties can be controlled only by size, shape, and surface termination. Shallow impurity doping adds new freedom to control the properties of Si nanocrystals. For example, the size is the only parameter to control PL energy of intrinsic Si nanocrystals. On the other hand, there are some combinations of size and impurity concentrations in the case of codoped Si nanocrystals. The additional freedom of material design may extend application fields of Si nanocrystals. Furthermore, different optical properties of intrinsic and impurity-doped Si nanocrystals suggest that doped Si nanocrystals can be regarded as new materials that may be a new building block to realize Si-based electronic and optoelectronic devices.

References

1 Cullis, A.G., Canham, L.T., and Calcott, P.D.J. (1997) The structural and luminescence properties of porous silicon. *J. Appl. Phys.*, **82**, 909–965.

2 Kovalev, D., Heckler, H., Polisski, G., and Koch, F. (1999) Optical properties of Si nanocrystals. *Phys. Status Solidi B*, **215**, 871–932.

3 Takeoka, S., Fujii, M., and Hayashi, S. (2000) Size-dependent photo-luminescence from surface-oxidized Si nanocrystals in a weak confinement regime. *Phys. Rev. B*, **62**, 16820–16825.

4 Heitmann, J., Muller, F., Zacharias, M., and Gosele, U. (2005) Silicon nano-crystals: size matters. *Adv. Mater.*, **17**, 795–803.

5 Kovalev, D. and Fujii, M. (2005) Silicon nanocrystals: photosensitizers for oxygen molecules. *Adv. Mater.*, **17**, 2531–2544.

6 Lockwood, D.J. (1998) *Light Emission in Silicon: From Physics to Devices*, Academic Press, San Diego, CA.

7 Kovalev, D. and Fujii, M. (2008) Chapter 4, in *Annual Review of Nano Research*, vol. 2 (eds G. Cao and J. Brinker), World Scientific, Singapore.

8 Kovalev, D. and Fujii, M. (2008) Chapter 15, in *Silicon Nanophotonics: Basic Principles, Present Status and Perspectives* (ed. L. Khriachtchev), World Scientific, Singapore.

9 Tsu, R. and Babić, D. (1994) Doping of a quantum dot. *Appl. Phys. Lett.*, **64**, 1806–1808.

10 Ferreyra, J.M. and Proetto, C.R. (1995) Strong-confinement approach for

impurities in quantum dots. *Phys. Rev. B*, **52**, R2309–R2312.

11 Holtz, P.O. and Zhao, Q.X. (2004) *Impurities Confined in Quantum Structures*, Springer-Verlag, Berlin.

12 Erwin, S.C., Zu, L.J., Haftel, M.I., Efros, A.L., Kennedy, T.A., and Norris, D.J. (2005) Doping semiconductor nanocrystals. *Nature*, **436**, 91–94.

13 Norris, D.J., Efros, A.L., and Erwin, S.C. (2008) Doped nanocrystals. *Science*, **319**, 1776–1779.

14 Delerue, C., Lannoo, M., Allan, G., and Martin, E. (1995) Theoretical descriptions of porous silicon. *Thin Solid Films*, **255**, 27–34.

15 Allan, G., Delerue, C., Lannoo, M., and Martin, E. (1995) Hydrogenic impurity levels, dielectric constant, and Coulomb charging effects in silicon crystallites. *Phys. Rev. B*, **52**, 11982–11988.

16 Delerue, C., Lannoo, M., Allan, G., Martin, E., Mihalcescu, I., Vial, J.C., Romestain, R., Muller, F., and Bsiesy, A. (1995) Auger and Coulomb charging effects in semiconductor nanocrystallites. *Phys. Rev. Lett.*, **75**, 2228–2231.

17 Zhou, Z.Y., Friesner, R.A., and Brus, L. (2003) Electronic structure of 1 to 2 nm diameter silicon core/shell nanocrystals: surface chemistry, optical spectra, charge transfer, and doping. *J. Am. Chem. Soc.*, **125**, 15599–15607.

18 Melnikov, D.V. and Chelikowsky, J.R. (2004) Quantum confinement in phosphorus-doped silicon nanocrystals. *Phys. Rev. Lett.*, **92**, 046802.

19 Zhou, Z.Y., Steigerwald, M.L., Friesner, R.A., Brus, L., and Hybertsen, M.S. (2005) Structural and chemical trends in doped silicon nanocrystals: first-principles calculations. *Phys. Rev. B*, **71**, 245308.

20 Cantele, G., Degoli, E., Luppi, E., Magri, R., Ninno, D., Iadonisi, G., and Ossicini, S. (2005) First-principles study of n- and p-doped silicon nanoclusters. *Phys. Rev. B*, **72**, 113303.

21 Ossicini, S., Degoli, E., Iori, F., Luppi, E., Magri, R., Cantele, G., Trani, F., and Ninno, D. (2005) Simultaneously B- and P-doped silicon nanoclusters: formation energies and electronic properties. *Appl. Phys. Lett.*, **87**, 173120.

22 Ossicini, S., Iori, F., Degoli, E., Luppi, E., Magri, R., Poli, R., Cantele, G., Trani, F., and Ninno, D. (2006) Understanding doping in silicon nanostructures. *IEEE J. Sel. Top. Quant.*, **12**, 1585–1591.

23 Iori, F., Degoli, E., Magri, R., Marri, I., Cantele, G., Ninno, D., Trani, F., Pulci, O., and Ossicini, S. (2007) Engineering silicon nanocrystals: theoretical study of the effect of codoping with boron and phosphorus. *Phys. Rev. B*, **76**, 085302.

24 Xu, Q., Luo, J.W., Li, S.S., Xia, J.B., Li, J.B., and Wei, S.H. (2007) Chemical trends of defect formation in Si quantum dots: the case of group-III and group-V dopants. *Phys. Rev. B*, **75**, 235304.

25 Chan, T.L., Tiago, M.L., Kaxiras, E., and Cheikowsky, J.R. (2008) Size limits on doping phosphorus into silicon nanocrystals. *Nano Lett.*, **8**, 596–600.

26 Kanzawa, Y., Fujii, M., Hayashi, S., and Yamamoto, K. (1996) Doping of B atoms into Si nanocrystals prepared by rf cosputtering. *Solid State Commun.*, **100**, 227–230.

27 Fujii, M., Hayashi, S., and Yamamoto, K. (1998) Photoluminescence from B-doped Si nanocrystals. *J. Appl. Phys.*, **83**, 7953–7957.

28 Müller, J., Finger, F., Carius, R., and Wagner, H. (1999) Electron spin resonance investigation of electronic states in hydrogenated microcrystalline silicon. *Phys. Rev. B*, **60**, 11666.

29 Mimura, A., Fujii, M., Hayashi, S., and Yamamoto, K. (1999) Quenching of photoluminescence from Si nanocrystals caused by boron doping. *Solid State Commun.*, **109**, 561–565.

30 Fujii, M., Mimura, A., Hayashi, S., Yamamoto, K., Urakawa, C., and Ohta, H. (2000) Improvement in photoluminescence efficiency of SiO_2 films containing Si nanocrystals by P doping: an electron spin resonance study. *J. Appl. Phys.*, **87**, 1855–1857.

31 Mimura, A., Fujii, M., Hayashi, S., Kovalev, D., and Koch, F. (2000) Photoluminescence and free-electron absorption in heavily phosphorus-doped Si nanocrystals. *Phys. Rev. B*, **62**, 12625–12627.

32 Pawlak, B.J., Gregorkiewicz, T., Ammerlaan, C.A.J., Takkenberg, W., Tichelaar, F.D., and Alkemade, P.F.A. (2001) Experimental investigation of band structure modification in silicon nanocrystals. *Phys. Rev. B*, **64**, 115308.

33 Liu, X.N., Xu, G.Y., Sui, Y.X., He, Y.L., and Bao, X.M. (2001) Electron spin resonance in doped nanocrystalline silicon films. *Solid State Commun.*, **119**, 397–401.

34 Fujii, M., Mimura, A., Hayashi, S., Yamamoto, Y., and Murakami, K. (2002) Hyperfine structure of the electron spin resonance of phosphorus-doped Si nanocrystals. *Phys. Rev. Lett.*, **89**, 206805.

35 Fujii, M., Toshikiyo, K., Takase, Y., Yamaguchi, Y., and Hayashi, S. (2003) Below bulk-band-gap photoluminescence at room temperature from heavily P- and B-doped Si nanocrystals. *J. Appl. Phys.*, **94**, 1990–1995.

36 Švrček, V., Slaoui, A., and Muller, J.C. (2004) *Ex situ* prepared Si nanocrystals embedded in silica glass: formation and characterization. *J. Appl. Phys.*, **95**, 3158–3163.

37 Fujii, M., Yamaguchi, Y., Takase, Y., Ninomiya, K., and Hayashi, S. (2004) Control of photoluminescence properties of Si nanocrystals by simultaneously doping n- and p-type impurities. *Appl. Phys. Lett.*, **85**, 1158–1160.

38 Tchebotareva, A.L., Dooda, M.J.A.d., Biteenb, J.S., Atwaterb, H.A., Polman, A. (2005) Quenching of Si nanocrystal photoluminescence by doping with gold or phosphorous. *J. Lumin.*, **114**, 137–144.

39 Fujii, M., Yamaguchi, Y., Takase, Y., Ninomiya, K., and Hayashi, S. (2005) Photoluminescence from impurity codoped and compensated Si nanocrystals. *Appl. Phys. Lett.*, **87**, 1158–1160.

40 Chiesa, M., Amato, G., Boarino, L., Garrone, E., Geobaldo, F., and Giamello, E. (2005) ESR study of conduction electrons in B-doped porous silicon generated by the adsorption of Lewis bases. *J. Electrochem. Soc.*, **152**, G329–G333.

41 Garrone, E. *et al.* (2005) A nanostructured porous silicon near insulator becomes either a p- or an n-type semiconductor upon gas adsorption. *Adv. Mater.*, **17**, 528–531.

42 Baldwin, R.K., Zou, J., Pettigrew, K.A., Yeagle, G.J., Britt, R.D., and Kauzlarich, S.M. (2006) The preparation of a phosphorus doped silicon film from phosphorus containing silicon nanoparticles. *Chem. Commun.*, 658–660.

43 Lechner, R., Wiggers, H., Ebbers, A., Steiger, J., Brandt, M.S., and Stutzmann, M. (2007) Thermoelectric effect in laser annealed printed nanocrystalline silicon layers. *Phys. Status Solidi (RRL)*, **1**, 262–264.

44 Pi, X.D., Gresback, R., Liptak, R.W., Campbell, S.A., and Kortshagen, U. (2008) Doping efficiency, dopant location, and oxidation of Si nanocrystals. *Appl. Phys. Lett.*, **92**, 123102.

45 Sumida, K., Ninomiya, K., Fujii, M., Fujio, K., Hayashi, S., Kodama, M., and Ohta, H. (2007) Electron spin-resonance studies of conduction electrons in phosphorus-doped silicon nanocrystals. *J. Appl. Phys.*, **101**, 033504.

46 Stegner, A.R., Pereira, R.N., Klein, K., Lechner, R., Dietmueller, R., Brandt, M.S., Stutzmann, M., and Wiggers, H. (2008) Electronic transport in phosphorus-doped silicon nanocrystal networks. *Phys. Rev. Lett.*, **1**, 026803.

47 Fujio, K., Fujii, M., Sumida, K., Hayashi, S., Fujisawa, M., and Ohta, H. (2008) Electron spin resonance studies of P and B codoped Si nanocrystals. *Appl. Phys. Lett.*, **93**, 021920.

48 Schierning, G., Theissmann, R., Wiggers, H., Sudfeld, D., Ebbers, A., Franke, D., Witusiewicz, V.T., and Apel, M. (2008) Microcrystalline silicon formation by silicon nanoparticles. *J. Appl. Phys.*, **103**, 084305.

49 Kanzawa, Y., Kageyama, T., Takeoka, S., Fujii, M., Hayashi, S., and Yamamoto, K. (1997) Size-dependent near-infrared photoluminescence spectra of Si nanocrystals embedded in SiO_2 matrices. *Solid State Commun.*, **102**, 533–537.

50 Wolkin, M.V., Jorne, J., Fauchet, P.M., Allan, G., and Delerue, C. (1999) Electronic states and luminescence in porous silicon quantum dots: the role of oxygen. *Phys. Rev. Lett.*, **82**, 197–200.

51 Hill, N.A. and Whaley, K.B. (1995) A theoretical study of light emission from nanoscale silicon. *J. Electron. Mater.*, **25**, 269–285.

52 Meyer, B.K., Hofmann, D.M., Stadler, W., Petrova-Koch, V., Koch, F., Emanuelsson, P., and Omling, P. (1993) Photo-luminescence and optically detected magnetic resonance investigations on porous silicon. *J. Lumin.*, **57**, 137–140.

53 Meyer, B.K., Hofmann, D.M., Stadler, W., Petrova-Koch, V., Koch, F., Omling, P., and Emanuelsson, P. (1993) Defects in porous silicon investigated by optically detected and by electron paramagnetic resonance techniques. *Appl. Phys. Lett.*, **63**, 2120–2122.

54 Mochizuki, Y., Mizuta, M., Ochiai, Y., Matsui, S., and Ohkubo, N. (1992) Luminescent properties of visible and near-infrared emissions from porous silicon prepared by the anodization method. *Phys. Rev. B*, **46**, 12353–12357.

55 Calcott, P.D.J., Nash, K.J., Canham, L.T., Kane, M.J., and Brumhead, D. (1993) Identification of radiative transitions in highly porous silicon. *J. Phys.: Condens. Matter*, **5**, L91–L98.

56 Reboredo, F.A., Franceschetti, A., and Zunger, A. (1999) Excitonic transitions and exchange splitting in Si quantum dots. *Appl. Phys. Lett.*, **75**, 2972–2974.

57 Sychugov, I., Juhasz, R., Valenta, J., and Linnros, J. (2005) Narrow luminescence linewidth of a silicon quantum dot. *Phys. Rev. Lett.*, **94**, 087405.

58 Fujii, M., Kanzawa, Y., Hayashi, S., and Yamamoto, K. (1996) Raman scattering from acoustic phonons confined in Si nanocrystals. *Phys. Rev. B*, **54**, R8373–R8376.

59 Valenta, J., Juhasz, R., and Linnros, J. (2002) Photoluminescence spectroscopy of single silicon quantum dots. *Appl. Phys. Lett.*, **80**, 1070–1072.

60 Kovalev, D., Heckler, H., Averboukh, B., Ben-Chorin, M., Schwartzkopff, M., and Koch, F. (1998) Hole burning spectroscopy of porous silicon. *Phys. Rev. B*, **57**, 3741–3744.

61 Kovalev, D., Heckler, H., Ben-Chorin, M., Polisski, G., Schwartzkopff, M., and Koch, F. (1998) Breakdown of the k-conservation rule in Si nanocrystals. *Phys. Rev. Lett.*, **81**, 2803–2806.

62 Fujii, M., Yoshida, M., Kanzawa, Y., Hayashi, S., and Yamamoto, K. (1997) 1.54 μm photoluminescence of Er^{3+} doped into SiO_2 films containing Si nanocrystals: evidence for energy transfer from Si nanocrystals to Er^{3+}. *Appl. Phys. Lett.*, **71**, 1198–1200.

63 Fujii, M., Yoshida, M., Hayashi, S., and Yamamoto, K. (1998) Photoluminescence from SiO_2 films containing Si nanocrystals and Er: effects of nanocrystalline size on the photoluminescence efficiency of Er^{3+}. *J. Appl. Phys.*, **84**, 4525–4531.

64 Watanabe, K., Fujii, M., and Hayashi, S. (2001) Resonant excitation of Er^{3+} by the energy transfer from Si nanocrystals. *J. Appl. Phys.*, **90**, 4761–4767.

65 Imakita, K., Fujii, M., and Hayashi, S. (2005) Spectrally resolved energy transfer from excitons in Si nanocrystals to Er ions. *Phys. Rev. B*, **71**, 193301.

66 Izeddin, I., Timmerman, D., Gregorkiewicz, T., Moskalenko, A.S., Prokofiev, A.A., Yassievich, I.N., and Fujii, M. (2008) Energy transfer in Er-doped SiO_2 sensitized with Si nanocrystals. *Phys. Rev. B*, **78**, 035327.

67 Fujii, M., Hayashi, S., and Yamamoto, K. (1998) Excitation of intra-4f shell luminescence of Yb^{3+} by energy transfer from Si nanocrystals. *Appl. Phys. Lett.*, **73**, 3108–3110.

68 Gross, E., Kovalev, D., Kunzner, N., Diener, J., Koch, F., Timoshenko, V.Y., and Fujii, M. (2003) Spectrally resolved electronic energy transfer from silicon nanocrystals to molecular oxygen mediated by direct electron exchange. *Phys. Rev. B*, **68**, 115405.

69 Fujii, M., Kovalev, D., Goller, B., Minobe, S., Hayashi, S., and Timoshenko, V.Y. (2005) Time-resolved photoluminescence studies of the energy transfer from excitons confined in Si nanocrystals to oxygen molecules. *Phys. Rev. B*, **72**, 165321.

70 Turro, N.J. (1991) *Modern Molecular Photochemistry*, University Science Books, Sausalito, CA.

71 Vial, J.C., Bsiesy, A., Gaspard, F., Hérino, R., Ligeon, M., Muller, F., Romestain, R., and MacFarlane, R.M. (1992) Mechanisms of visible-light emission from electro-oxidized porous silicon. *Phys. Rev. B*, **45**, 14171–14176.

72 Wilson, W.L., Szajowski, P.F., and Brus, L.E. (1993) Quantum confinement in size-selected, surface-oxidized silicon nanocrystals. *Science*, **262**, 1242–1244.

73 Skryshevsky, V.A., Laugier, A., Strikha, V.I., and Vikulov, V.A. (1996) Evaluation of quantum efficiency of porous silicon photoluminescence. *Mater. Sci. Eng. B*, **40**, 54–57.

74 Ledoux, G., Guillois, O., Porterat, D., Reynaud, C., Huisken, F., Kohn, B., and Paillard, V. (2000) Photoluminescence properties of silicon nanocrystals as a function of their size. *Phys. Rev. B*, **62**, 15942–15951.

75 Jurbergs, D., Rogojina, E., Mangolini, L., and Kortshagen, U. (2006) Silicon nanocrystals with ensemble quantum yields exceeding 60%. *Appl. Phys. Lett.*, **88**, 233116.

76 Mangolini, L. and Kortshagen, U. (2007) Plasma-assisted synthesis of silicon nanocrystal inks. *Adv. Mater.*, **19**, 2513–2519.

77 Kalkman, J., Gersen, H., Kuipers, L., and Polman, A. (2006) Excitation of surface plasmons at a SiO_2/Ag interface by silicon quantum dots: experiment and theory. *Phys. Rev. B*, **73**, 075317.

78 Walters, R.J., Kalkman, J., Polman, A., Atwater, H.A., and de Dood, M.J.A. (2006) Photoluminescence quantum efficiency of dense silicon nanocrystal ensembles in SiO₂. *Phys. Rev. B*, **73**, 132302.

79 Miura, S., Nakamura, T., Fujii, M., Inui, M., and Hayashi, S. (2006) Size dependence of photoluminescence quantum efficiency of Si nanocrystals. *Phys. Rev. B*, **73**, 245333.

80 Rai-Choudhury, P. and Salkovitz, E.I. (1970) Doping of epitaxial silicon: effect of dopant partial pressure. *J. Cryst. Growth*, **7**, 361–367.

81 Lenahan, P.M. and Conley, J.F. (1998) What can electron paramagnetic resonance tell us about the Si/SiO₂ system? *J. Vac. Sci. Technol. B*, **16**, 2134–2153.

82 Feher, G. (1959) Electron spin resonance experiments on donors in silicon. I. Electronic structure of donors by the electron nuclear double resonance technique. *Phys. Rev.*, **114**, 1219–1244.

83 Stutzmann, M., Biegelsen, D.K., and Street, R.A. (1987) Detailed investigation of doping in hydrogenated amorphous silicon and germanium. *Phys. Rev. B*, **35**, 5666–5701.

84 Stesmans, A. and Scheerlinck, F. (1995) Electron-spin-resonance analysis of the natural intrinsic EX center in thermal SiO₂ on Si. *Phys. Rev. B*, **51**, 4987–4997.

85 Bardelebena, H.J.v., Ortega, C., Grosmana, A., Morazzania, V., Siejkaa, J., and Stievenard, D. (1993) Defect and structure analysis of n⁺-, p⁺- and p-type porous silicon by the electron paramagnetic resonance technique. *J. Lumin.*, **57**, 301–313.

86 Murakami, K., Masuda, K., Gamo, K., and Namba, S. (1977) Effects of ion-implanted atoms upon conduction electron spin resonance (CESR) in a Si:P system. *Appl. Phys. Lett.*, **30**, 300–302.

87 Imakita, K., Fujii, M., Yamaguchi, Y., and Hayashi, S. (2005) Interaction between Er ions and shallow impurities in Si nanocrystals within SiO₂. *Phys. Rev. B*, **71**, 115440.

88 Levy, M., Yu, P.Y., Zhang, Y., and Sarachik, M.P. (1994) Photoluminescence of heavily doped, compensated Si:P,B. *Phys. Rev. B*, **49**, 1677–1684.

89 Canham, L.T. (1990) Silicon quantum wire array fabrication by electrochemical and chemical dissolution of wafers. *Appl. Phys. Lett.*, **57**, 1046–1048.

90 Beard, M.C., Knutsen, K.P., Yu, P.R., Luther, J.M., Song, Q., Metzger, W.K., Ellingson, R.J., and Nozik, A.J. (2007) Multiple exciton generation in colloidal silicon nanocrystals. *Nano Lett.*, **7**, 2506–2512.

91 Timmerman, D., Izeddin, I., Stallinga, P., Yassievich, I.N., and Gregorkiewicz, T. (2008) Space-separated quantum cutting with silicon nanocrystals for photovoltaic applications. *Nat. Photon.*, **2**, 105–109.

4
Electrical Transport Mechanisms in Ensembles of Silicon Nanocystallites

Isaac Balberg

4.1
Introduction

While the optical [1–3] properties of various ensembles of individual Si nanocrystallites (NCs) and the memory-charge storage (CS) characteristics of two-dimensional
(2D) arrays of Si NCs [4] have been investigated by many researchers, relatively little
attention was paid to the transport properties of 3D ensembles of such quantum dots
(QDs) [5]. The interest in the last systems, however, is expected to follow the new basic
physics that it reveals and the potential applications of such systems. Following the
reasonable (though still controversial [6]) understanding of granular metals [7, 8],
where the electrical conduction takes place via metallic particles, additional significant insights into the transport mechanism [5, 9] and new phenomena are expected
in Si NCs due to the presence of confined levels [10]. In particular, the combination of
Coulomb blockade (CB) effects and the quantum confinement (QC) restrictions can
yield various resonant tunneling effects [11] as well as various quantum phase
transformations [12] that are beyond the scope of this chapter. From the application
point of view, one realizes that significant electroluminescence (EL) can come about
only as a result of efficient transport in 3D ensembles of Si NCs [4, 13, 14] that are
dense enough to yield strong light emission. Finding the conditions for the
optimization of the luminescence and the transport is then the route to achieve
efficient Si-based photoelectronic devices [15].

In this chapter, we present a review on the electrical transport in the ensembles of
Si NCs. Since we are concerned with the transport in 3D ensembles of quantum dots,
we will only briefly review the main results obtained from lower dimensional
ensembles (i.e., 1D-like [16, 17] or perpendicular to 2D arrays [18–21]) of Si QDs
in order to provide a reference for the effects of the small size of the single Si NCs on
the transport in ensembles of larger dimensions. We will see that in spite of many
studies of these 0D-like and 2D systems, which are essentially associated with single
isolated NCs, and their potential use as nonvolatile memories, only the basics of the

Silicon Nanocrystals: Fundamentals, Synthesis and Applications. Edited by Lorenzo Pavesi and Rasit Turan
Copyright © 2010 WILEY-VCH Verlag GmbH & Co. KGaA, Weinheim
ISBN: 978-3-527-32160-5

corresponding transport mechanisms in them can be considered as properly founded and generally accepted.

There is even much less understanding of the transport mechanism in ensembles of 3D systems of Si NCs as there have been no comprehensive studies of the transport mechanisms in them [5]. In this chapter, we present probably the first comprehensive report of the transport mechanisms in ensembles of Si NCs embedded in insulating continuous matrices, adding (to the very common studies of the temperature dependence of the transport properties that is commonly applied for the determination of transport mechanisms) the very important parameter of the density of the NCs, N, the role of which has been overlooked in previous studies. As we will see, this parameter provides a very useful scale along which one can follow the basic transport mechanisms in 3D ensembles. In particular, the corresponding consideration also adds the important system connectivity aspect that has hardly been discussed in the past. This is in contrast with the interdot tunneling aspect that was emphasized in almost all previous studies. We note, however, that the need to consider these two aspects was first suggested by Burr *et al.* in 1997 [13] and, although no specific application was made there, their work, as well as the density of the Ge NCs, first considered by Fujii *et al.* [9] within the context of transport, may be viewed as the major pioneering works on the transport in 3D ensembles of column IV semiconductor NCs.

We limit ourselves to the ensembles of Si NCs that are smaller than 10 nm since for these NCs QC and CB effects become significant even at room temperature [1]. We note in passing that in the literature the term Si NCs has been used in many works for much larger particles [22] and these will not be considered here. Also, we do not discuss here the ensembles of Si NCs that are embedded in the Si matrix of porous silicon. The transport mechanism in these systems that has been considered previously [3, 5, 23–25] is quite different from that of Si NCs embedded in an insulating matrix since the Si matrix in porous silicon provides current routes that are parallel to those controlled solely by the Si NCs. In addition, in view of the limited scope of this chapter, we will not provide here an extended review of the theories and previous experimental results, but we will refer to those that we found essential for this chapter. We will try, however, to give the physical essence of the transport mechanisms that appear to be involved in the 3D systems, leaving it to the reader to consider more detailed and quantitative accounts of them in the cited literature.

This chapter is structured as follows. In Section 4.2, we provide the reader with a brief background, giving first the basic concepts to be used in this review (Section 4.2.1), and then we give (in Section 4.2.2) a short critical review of the achievements made, and the questions that arise, following the previous studies of transport in low-dimensional and 3D systems of Si NCs. Considering the above, we turn to the results of our experimental work presenting them in a systematic manner. We believe that these results and their interpretation give an initial comprehensive framework for the discussion of the transport mechanisms in 3D ensembles of Si NCs. Correspondingly, first we will present in Section 4.3 our method of sample fabrication (that enables us to carry out a systematic and comprehensive study) and this is followed by the description of the measurements that were carried out in order to appreciate the combined effects of inter-NCs conduction and the connectivity of the system on the transport mechanisms. Then, in Section 4.4, we present our experimental observations that reflect these two ingredients emphasizing the great importance of knowing the structure of the 3D system as a first step for interpreting their transport properties. Following the conclusions of our experimental results in a broader context, we show in Section 4.5 that while we were able to obtain a comprehensive picture of the transport mechanisms in 3D ensembles of Si NCs, we are still far from understanding the transport mechanism in detail and/or account for most of ours and others experimental observations quantitatively. The corresponding progress that was made, the general picture that emerges, and the major questions that remain unanswered will be discussed. We will conclude that section by mentioning some of our recent experimental findings in order to demonstrate the richness of transport-based new phenomena that are associated with the nanosize of the Si NCs in their ensembles and that, as stated above, may be of considerable interest from both the basic physics and the application points of view.

4.2
Background

4.2.1
Basic Concepts Associated with Transport and Quantum Dots

Percolation theory describes the effect of a system's connectivity on its geometrical and physical properties [26, 27]. Considering a system of N metallic spheres embedded randomly in an insulating matrix of a unit volume, there will be a critical density of spheres N_c such that a continuous path of touching spheres that will be associated with the onset of a finite macroscopic conductivity σ of the system will form for $N \geq N_c$. If these metallic spheres have a radius b, the corresponding volume fractions that they occupy will be given by $x = (4\pi/3)b^3N$ and $x_c = (4\pi/3)b^3N_c$, respectively. These quantities, to be expressed in terms of volume % (vol%), will be our fundamental density characterization parameters throughout this study. It can be shown that the dependence of the global conductivity on the above quantities is given by [26–28]

$$\sigma \propto (x-x_c)^t, \tag{4.1}$$

where t is a constant that (for simple systems in the continuum [26–29]) has (in 3D) a universal dimensional value of $t_0 = 2$. In the case where there is a tunneling conduction between the spheres, the "touching" of the conducting particles is not well defined (as, say, in porous media [27]) since there will be some tunneling probability between any two spheres that are at any distance r [27, 29, 30]. Considering the exponential decay of the local tunneling between two spheres that is given by [31]

$$G = G_0 \exp(-2\chi(r-b)), \tag{4.2}$$

it is expected, however, that the contribution of the farther neighbors of a given particle will diminish with the increase in their inter-(surface)-distance $r - 2b$. Here, G_0 is a constant that accounts for the charge carrier concentration and the local particle environment around $r = 0$, and $1/2\chi$ is the tunneling decay (or sometimes referred to as the localization) length of the corresponding charge carrier [28, 32]. It turns out that this behavior can still yield a percolation-like behavior, but t is no more universal and its value can be approximated by $t_0 + u$, where $u \approx 2\chi(a - b)$ and a is some average of the intersphere distance [27, 29, 30, 33]. In this study, for reasons to be clarified below, we will apply the concept of "touching" to two particles the surfaces of which are separated by about a lattice constant (≈ 0.5 nm in the Si lattice) of the conducting particle. Tentatively then, for the description of the systems to be considered below, we assume that two Si NCs "touch" if they are no more than 0.5 nm apart. Since the features of the transport between "touching" NCs are new, we will discuss them in some detail in Sections 4.4 and 4.5.

The transfer of an electron from a given neutral particle to an adjacent neutral particle charges this particle by one (positive) elementary charge (q) and that of the adjacent particle by one (negative) elementary charge [7, 8]. If the capacitance of the individual particle in its corresponding environment is C_0, the energy needed to be supplied for the above "electron–hole" transfer by tunneling is then [5, 7, 8, 11]

$$E_{CB} = q^2/C_0. \tag{4.3}$$

This energy, which is a hindrance to the transfer of charge carriers, is known as the Coulomb blockade energy [5, 11, 34]. In passing, we note that in the simplest case of isolated spheres we have $C_0 = 4\pi\varepsilon\varepsilon_0 b$, where $\varepsilon\varepsilon_0$ is the dielectric constant of the matrix.

In general, we can say that a tunneling process is thermally activated when it requires a supply of energy. A series of such events that yields a global conduction is defined as a hopping conduction mechanism [28, 32]. The more known of this family of such conduction mechanisms is the variable range hopping (VRH) process in which the preferred optimized conduction path is determined by the competing requirements of the availability of the minimum energy supply and the minimum tunneling distance. In the case where the required energy is due to the disorder in the system, the temperature dependence of the conductivity in 3D is expected to behave as [28, 32]

$$\sigma = \sigma_A \exp[-(T_A/T)^\gamma], \tag{4.4}$$

where σ_A and T_A are constants of the system and $\gamma = 1/4$. In the case where electrostatic interactions are involved, the energy needed to overcome the corresponding Coulomb blockade [7, 8] or Coulomb gap [5, 28, 34] was initially suggested [7, 8] to yield a $\gamma = 1/2$ value. Refinement of this theory later [35] has shown that γ can vary in the interval $1/4 < \gamma < 1$, according to the energetic distribution of the CB energies [33]. The first attempt to derive a $\gamma = 1/2$ value from a combined effect of the distribution of the QC levels in different NCs (that enhances the effect of the particles' size distribution) and the CB was made by Simanek [36]. The choice of the optimized percolation path when both contributions are considered leads to $\gamma = 1/2$ value, on the one hand, and to a simple expression for T_A (in Eq. (4.4)) on the other hand. In what follows, we will refer to the general behavior described by Eq. (4.4), for any value of γ, as an Arrhenius-like dependence. At high temperatures, or for small energies involved in all the above processes, one expects, following the scenario of a near-neighbor hopping [28, 32], a $\gamma = 1$ value. This follows the fact that the interparticle tunneling route will prefer the nearest neighbors that correspond to the average lowest tunneling resistance that enables a continuous percolation path. In that case, one can further show that [28, 30]

$$\sigma \propto \exp(-\alpha x^{-1/3}), \tag{4.5}$$

where α is a known constant. This dependence applies to the dilute (low-x) limit and can be easily shown [30] to be $\sigma \propto \exp(-\alpha x^{-1})$ in the denser particles case. We will refer to the general behavior described by Eq. (4.4), for any value of γ, as an Arrhenius plot dependence.

Another type of a thermally activated processes that will be important in our discussion may be described by the older, yet less known, Berthelot dependence [37] that (when generalizing again for any γ) is given by

$$\sigma = \sigma_B \exp[(T/T_B)^{\gamma}], \tag{4.6}$$

where σ_B and T_B are appropriate parameters of the system. One notes that it is easy to distinguish between the behaviors of Eqs. (4.4) and (4.6) by plotting the log σ versus log T dependence. In the first case (Eq. (4.4)), this dependence yields a convex curve while in the latter case (Eq. (4.6)) this dependence yields a concave curve. The latter, much less abundant and much less understood, dependence has been found in some semiconductor [38, 39] and nanosemiconductor [40–42] systems and was interpreted as associated with a tunneling barrier effect. In some cases [38, 40, 41], it has been interpreted as due to thermally activated barrier vibration process, while in others [39, 42], it has been attributed to the narrowing, with increasing energy, of the interparticle barrier such that there is a competition, for the thermally activated tunneling probability, between the thermal excitation from the conduction level and the width of the barrier. The higher the temperature, the more carriers there are that can tunnel through the narrower part of the barrier, and the maximum tunneling rate behaves as given in Eq. (4.6). This simple model is very appealing since all it requires is the solution of the Schrödinger equation [42]. However, it appears that this behavior is applicable to a rather special barrier shape, but it does not explain why a barrier that forms between two particles will have the particular shape that was suggested [42]. On the other hand, the fact that the transport between the NCs is controlled in this model by potential barriers that are very different from the "conventional" (polycrystalline-like) semiconductor parabolic barriers [39] is not too surprising since no space charge region can be developed within the nanometer size NCs [42]. Considering the other explanation [40, 41], based on the enhancement of the tunneling rate through a vibrating tunneling barrier, we note that no independent proof for the existence of such a vibrating barrier was given. In passing, we further note that a Berthelot-type behavior has also been shown to be consistent with thermionic emission above the narrow barrier [41], and this, as well as the vibrating barrier models, shows that the *I–V* characteristics will depend on temperature. For completeness, we also note the fact that there are models that consider a related mechanism but yield the Arrhenius-like dependence as given by Eq. (4.4) (with $\gamma = 2/3$ [43]). For our purpose, it is important to note that the above two types of mechanisms apply to very narrow potential barriers such as we encountered for high densities of Si NCs in our samples. Following all the above, we will refer, regardless of the particular value of γ, to an Eq. (4.4)-like behavior as an Arrhenius-like and to an Eq. (4.6)-like behavior as a Berthelot-like behavior.

Another, even much less known type of a concave log σ versus log *T* behavior is that of the type [44, 45]

$$\sigma = \sigma_0 T^\delta, \tag{4.7}$$

where σ_0 is a constant and δ increases with increasing temperature. The corresponding model that was suggested for this is associated with the subject of our concern here as follows. We have pointed out above that the Arrhenius behavior is known for the various hopping mechanisms and that γ can take values between $\gamma = 1/4$ and $\gamma = 1$ such that, with increasing temperature, γ-values also increase [28, 32]. The behavior described by Eq. (4.7) was suggested [44, 45] for a hopping conduction that essentially parallels the same process as the $\gamma = 1/4$ to $\gamma = 1$ transition (i.e., from VRH to near-neighbor hopping) with increasing *T*. In this model, $\delta = N_e - 2/(N_e + 1)$, where N_e is the number of hops in the transport path. The basic physics of this process is that, as in VRH, the hopping range increases with decreasing temperature and thus N_e, and δ, increases with increasing temperature. In contrast to the above local barrier models, we have a network effect, as in the VRH mechanism, but it appears to yield a 1D-like conduction path rather than a 3D-like network.

Turning to the concepts associated with the nanometer size regime, we realize that they are essentially 3D quantum wells in which the energy level separations are given by [1–3]

$$E_{QC} \propto 1/m^* b^2, \tag{4.8}$$

where m^* is the effective "effective mass," that is, the effective mass [46] that can be deduced from the calculated or measured electronic structure of the quantum dot for the corresponding carrier in that quantum well configuration [47, 48]. As the first approximation, it is common to use [49, 50] the bulk "effective mass," which is 0.3 for the electron and 0.5 for the hole in Si [50] in units of the free electron mass. For the semiconductor QDs, the density of the levels (unlike in the one-dimensional quantum well case) increases with energy [1, 34, 48, 50]. The filling of the levels with available electrons makes the corresponding level separation (above 1 K) unimportant for metals [34] but very important for semiconductors in general [1] and Si [48, 50] in particular, even at room temperature. This is obvious since even the ground-states that provide the levels for the conduction of the electrons and the holes, and are responsible for the optical properties, are not filled (as at the bottom of the corresponding bands in bulk semiconductors [46]). In order to get a carrier transfer by tunneling between two NCs, we need the matching of the two ground (or upper lying) states for tunneling between them [9, 51] or we "need to provide" the energy differences between them, E_{QC} (by field or thermal excitation) as in the hopping process [52, 53]. This effect can be termed as resonant tunneling [54] between two NCs, and the conduction process in the entire macroscopic system can be described as sequential resonance tunneling [5]. Following the above discussion (Eqs. (4.3) and (4.8)), we can say that the energy needed for the "hopping" between two NCs can be expressed to first order by [5, 9, 49, 55–57]

$$E = E_{CB} + \Delta E_{QC} = A/b + B/m^*b^2, \qquad (4.9)$$

where ΔE_{QC} is the corresponding energy difference between the two NCs (see Eq. (4.8)) and A and B are system-dependent parameters.

4.2.2
Previous Studies of Transport in Systems of Si

As pointed out in Section 4.1, the understanding of the electronic properties of the single NC is (unlike the case in granular metals) quite necessary in order to interpret the transport data in ensembles [9, 56] of semiconductor NCs. We start our review of the previous studies by considering the information available on the single or single-like NCs. It turns out that the knowledge we have on the level separation in Si quantum dots came almost exclusively from optical measurements [1–3] and these results are reasonably accounted for by relatively simple theoretical models [1–3, 48, 50]. On the other hand, the determination of the electronic structure by electrical measurements, which is more relevant for studying the transport properties via semiconductor quantum dots [51], is much less conclusive in the case of Si NCs. In fact, following the success of the local electrical spectroscopy of single quantum dots that are given between two metal electrodes (in particular in II–VI and III–V semiconductor NCs [51, 55]), trials were conducted to carry out scanning tunneling and conductive atomic force spectroscopies (STS [58] and C-AFM [59]) on Si NCs, indicating that the current is via the crystallites. On the other hand, with many difficulties associated with the interpretation of those results the best information that could have been deduced from them was that the confined level separations and the CB energy are of the order of 0.1–0.3 eV for NCs in the 3–5 nm size regime. Also, charges induced by electrical force microscopy (EFM) have been used by a few groups [60–62] to demonstrate that charge storage can be induced in individual Si NCs when they are removed well from each other. Another (more complicated to interpret but simpler to fabricate) structure that was studied by many authors [63–73] is the 2D array configuration that is diluted enough so that the interaction between the NCs is negligible. In this rather popular configuration [4], the current is perpendicular to the 2D arrays so that the measured current amounts to the sum of the currents through the individual NCs. The separation between the metal electrodes and the quantum dot is tunable, thus yielding the so-called double-barrier tunnel junction (DBTJ) configuration [5, 11, 51]. These studies of many parallel DBTJ-like configurations involved essentially a two-terminal MOS device where M is the top metal, separated or not, by a thin SiO_2 layer from the O layer, that is, the SiO_2 matrix layer contains the Si NCs, and S, that is the Si wafer (separated or not separated from the O layer) that is also the substrate of the structure. The conventional MOS current–voltage (I–V) characteristics of these systems disclosed the typical electrical rectification of such a structure and also included a quasiregular collection of plateaus, jumps, peaks, and switching events. In many of these studies, ad hoc models were proposed to associate the experimentally found features with the energies E of the form given by Eq. (4.9). Similar attempts were made to interpret peaks observed in the capacitance–voltage (C–V) characteristics that were measured on such 2D, MOS-like, array structures [66–68, 71, 72]. In view of the fact that the results obtained are very different in different works and the comparison with theories was done by ad hoc models, it seems at present that while these studies definitely indicate tunneling via the NCs, the results do not add up to an electronic level structure or even to a universal clear separation between the CB and QC effects as was obtained in corresponding studies of the isolated NCs of II–VI or III–V semiconductors [51, 55, 74], which were mentioned above. In particular, the various levels detected may or may not be associated with the quantum confinement in the NCs as other states can be present in the system [58, 75]. On the other hand, the various measurements have definitely revealed the presence of charge stored and CB effects [4, 67, 71].

The better defined systems for the electrical spectroscopy of isolated Si NCs seem to emerge from structures where only a couple of Si NCs were present between the two contacts [65, 76–78], so as to yield a well-defined DBTJ-like configurations, which are reminiscent of the DBTJs studied in the II–VI and III–V NCs [51, 54, 74]. For the Si NCs systems, the room-temperature I–V characteristics did not reveal many features, but they were nearly symmetric and nonlinear as to be expected from tunnel junctions. On the other hand, at low temperatures clear reproducible structures of peaks and plateaus were observed in the I–V characteristics. Although these have been convincingly associated with sequential charging of the NCs, no clear information regarding quantum confinement levels has been revealed. While, again, these studies [76–78] still do not provide a comprehensive, unified model, for E_{CB} and E_{QC} for single NCs, we will assume that as in the III–V and II–VI semiconductors, the two effects play a role in the transport between the NCs. We will see below that, similarly, at present, while for the 3D ensembles the role of the CB is apparent, the effects of the QC on the transport are as yet suggestive but not conclusively proven from the available transport data.

Following the above facts of the electronic structure, from the optical and the transport via single NCs data, we turn now to the available transport data on 3D ensembles of Si NCs when these are embedded in an insulating (generally SiO_2) matrix. As far as we know, all the previous studies on such 3D systems were carried out in a two-probe (or "sandwich") configuration that always involves the possible contribution of contact effects that are not always easy to separate from the bulk effects that one would like to study [9, 13, 56, 79–81]. As usual, the two main experimental methods for the evaluation of transport mechanisms in solids are the measurement of the temperature dependence of the conductivity $\sigma(T)$ and the measurement of the current–voltage (I–V) characteristics. Starting with the $\sigma(T)$ results that were obtained from 3D ensembles [9, 56, 79–84], it appears that, considering the uncertainties in the expression of the experimental data in terms of an exact γ (Eq. (4.4)-like) dependence [43], the first attempt to suggest a rather well-defined transport mechanism in 3D systems based on $\sigma(T)$ data was that of Fujii *et al.* [79], who measured the temperature dependence of the conductivity $\sigma(T)$ of a cosputtered Si-SiO$_2$ film between 20 and 300 K. Their results yielded a $\gamma = 1/4$ behavior that was interpreted to be due to VRH. However, in this case the samples were not annealed and the Si content was not specified, so that it is not clear whether it is the Si NCs network or the amorphous Si (a-Si) tissue in the sample that is responsible for the observations. Similar considerations apply to other studies, in particular, those that involved hydrogenated a-Si [80, 82, 84] fabrication techniques. On the other hand, in a latter study by Fujii *et al.* [9] on well-characterized cosputtered samples of Ge NCs embedded in SiO$_2$, a $\gamma = 1/2$ behavior was found for x-values smaller than $x = 15.3$ vol%. This dependence was interpreted to be due to the distribution of the quantized conduction band edges and the presence of a CB, according to the Simaneck model [36]. The authors also claimed that they obtained similar results for a similar system of Si NCs. We note in passing that due to the higher intrinsic carrier concentration in Ge, in comparison with that of Si, the measurements on Ge NCs seem to yield a more reliable determination of the $\sigma(T)$ dependence than in the studies of this dependence in ensembles of Si NCs. A later study of Si NCs (that were "capped" by an SiO$_2$ shell that is of the order of 2–3 nm thick) by Rafiq *et al.* [56] yielded a γ-value of $1/2$. For these data, the x-value of the corresponding samples can be estimated to be in the range of $30 \leq x \leq 45$ vol% and the preferred interpretation was also along the Simanek model [36]. While this interpretation accounts for the specific data, it leaves quite a few questions as to the type of system on which these findings were observed. We assume, however, that both the data of Fujii *et al.* [9] and the data of Rafiq *et al.* [56] suggest some kind of a "resonant"-hopping under CB in ensembles of Si NCs. We will see later that while, basically, their interpretation seems to be justified, the range of x-values for which this suggestion is applicable is limited, and thus this mechanism does not represent a general conduction mechanism in ensembles of Si NCs. It is to be stressed that the uncertainty in the exact determination of the $\sigma(T)$ dependence in other works did not provide the desired wider basis for the establishment of their conclusions. For example, a similar study of Banjeree [81] on the ensembles of Ge NCs embedded in a SiO$_2$ matrix was claimed to yield a $\gamma = 1/4$ value. However, a more careful analysis of his data reveals more difficulties [43] in deriving the reliable γ-values from the limited T range available than yielding the actual γ-values. In fact, one can fit his data almost equally well to a linear log $\sigma \propto$ log T or a $\gamma = 1/4$ behavior. The corresponding measurements that were carried out in perpendicular to the very thin layer of Ge NCs and on an unspecified density of the Ge NCs can at best represent the inter-NCs conduction rather than a 3D VRH behavior as claimed by the author. We are left essentially, thus far, only with results of Fujii *et al.* [9] and Rafiq *et al.* [56] to assume, with a high degree of reliability, that a Simaneck-like hopping under CB conditions takes place in dilute 3D systems of Si NCs. We note, however, in passing, that if the $\sigma(x)$ dependence of Eq. (4.4) would have been checked, a more specific interpretation could have been achieved.

Turning to the I–V or σ–V characteristics reported by various authors, the picture that emerges from the reported nonlinear characteristics appears to be quite blurred. The corresponding interpretations were quite different in different studies and involved or did not involve NC size effects. This include rectification in the corresponding tested device [56], simple inter-NCs tunneling [81, 85, 86], Fowler–Nordheim tunneling [13, 20], Poole–Frekel detrapping mechanism [85], space charge-limited currents [13], resonant tunneling [19], and Coulomb blockade [20] mechanisms.

4.3
Experimental Details

As was pointed out in the introduction, the limited scope of this chapter does not allow us to review all the works on the transport mechanism in the nanocomposites of ensembles of Si NCs that are embedded in an insulating (usually SiO_2) matrix. In particular, we will not mention the various methods used for their fabrication [4, 13, 20]. Since our focus is on the transport in relatively thick films of these nanocomposites and the effect of their density on the transport properties, we will concentrate on, probably, the most suitable method for such studies [7, 8]. Indeed, our samples were fabricated mostly by the cosputtering technique [87]. In particular, we emphasize that this method provides an extra handle for studying the transport in the nanocomposites that are of interest here. We will give some essential details on the typical cosputtering system and the deposition conditions that have been used by our group, noting that they are not too different from those used by other groups for the deposition of granular metals [7, 8] and Si-SiO_2 [79] films. On the other hand, our emphasis here is on the flexibility of this method to provide a very wide scale of nanoparticles densities, under exactly the same conditions [88].

The radio frequency (rf) sputtering technique is based on ion (usually and in our studies [89–97], Ar^+) bombardment of a material (the target) from which, as a result, groups of atoms are ejected and then reach a surface on which they are deposited (the substrate) [7, 8]. The role of the rf is to enable the discharge of the target surface when it is made of an insulating substance such as SiO_2. One can use various target types [7–9] or a few target [87, 89] configurations. In particular, for immiscible compounds one can produce, using this technique, a multiphase film [7, 8, 87, 89, 90]. Then, the relative content of each phase can be determined by the thickness profile of the deposited single-phase film in comparison with the thickness profile of the multiphase film [88].

For the typical samples in our various studies of ensembles of Si NCs, we have used the following system and deposition conditions. In general, we used two separate targets, 5 cm in diameter each, such that their centers were 15 cm apart. One target was of high purity (99.999%) sintered silicon pellets or, electronic quality silicon wafers, and the other consisted of pure (99.995%) fused quartz. The substrates that we used were typically, 13 cm long, 1 cm wide, and 0.7 mm thick quartz slides, or 0.3 mm thick, Si wafers that were cut into such slides. For the sputtering process, the slides were laid 6 cm above the line connecting the centers of the two targets and parallel to it [89].

The application of the above mentioned method [88] of thickness analysis of the films has shown that they contain between 5 vol% Si (at the substrate end adjacent to the SiO_2 target) and 95 vol% Si (at the substrate end adjacent to the Si target). The variation in the Si volume content, x (given in vol%) as a function of the position along the slide was weak for $x < 20$ and $x > 65$ regimes while in the range $20 < x < 65$ it varied roughly linearly with the distance along the substrate. The latter regime was along the central 8 cm range of the substrate. In passing, we point out that one can get similar results by using two, half-circle targets [90], or a single-target configuration, where small Si pieces are placed nonuniformly on the fused quartz target [79]. The deposition conditions in our studies were not too different from those used in the deposition of granular metal systems [7, 8] and in other studies of Si-SiO$_2$ composites [79]. For example, the base pressure in the vacuum chamber was typically 5×10^{-6} Torr, the radio frequency was 12.56 MHz, and the rf power was 60 W for each target. The voltages measured on the targets during sputtering were approximately 250 V on the Si target and approximately 400 V on the SiO_2 target. For the films used in our many studies [92–97], the typical thickness was about 1 μm and this thickness was obtained after 3 h of cosputtering.

The films so produced were found, by Raman scattering, infrared (IR) spectroscopy, and X-ray diffraction (XRD), to consist of amorphous SiO_2 and amorphous silicon (a-Si). After annealing of these films (at 1150–1200 °C for a typical duration of 40 min, under a flow of 4 l/min of pure N_2), we reapplied the above structural characterization methods and used transmission electron microscopy (TEM). Since the structural data obtained by applying this latter method are of fundamental importance for understanding the transport mechanisms, we carried out, in particular, a comprehensive study of the Si NCs/SiO$_2$ ensembles by cross-sectional high-resolution TEM (200 kV Technai F20, FEI) [93]. It is important to note that while some small amounts of silicon NCs were found (by Raman scattering) in the nonannealed a-Si/SiO$_2$ samples, no a-Si was detected in the samples that were annealed under the above-mentioned conditions. All the above-mentioned structural determination methods revealed [93, 94] that the films consist of amorphous SiO_2 and of Si NCs that have typical diameters, around 3 nm, at the Si-poor end and diameters around 10 nm at the Si-rich end. As described in more detail below, we have used these structurally well-characterized samples to evaluate the transport [91, 92, 95] and the optical [96, 97] mechanisms in our 3D ensembles of Si NCs, using their densities N (i.e., x, see Section 4.2.1) as a guiding parameter.

For the electrical measurements on the quartz-substrated films, we used sputtered aluminum or silver, 1 mm wide, electrodes (that were 0.2–1 μm thick) with 1 mm separation between them [91, 92, 95], while for the Si-substrated films we used evaporated aluminum (0.2 μm thick) or Hg point contacts [89, 92]. The areas of these contacts were on the order of 10^{-3} cm^2. With the first type of substrate, we carried standard four probe conductivity measurements [91, 92] and conductive AFM (C-AFM) [95] under various conditions, such as temperature variation, current–voltage (*I*–*V*) characteristics under various scanning rates and temperatures, and current–time (*I*–*t*) characteristics under various applied voltages and temperatures. The above contact separation enabled us, in the first configuration, to map the dependence of the sample conductivity on *x*, σ(*x*), with a resolution of $\Delta x \approx 2\,\text{vol\%}$, and thus it was possible to check their dependence with a resolution good enough to compare the σ(*x*) profiles with the theoretical predictions that were mentioned in Section 4.2.1. The importance of this first configuration is that it enables to take the precautions necessary to avoid contact effects (e.g., four probe measurements [91]), while the importance of the second configuration [92], which was used mainly for the measurement of the capacitance–voltage, *C*–*V*, characteristics, is that we could indirectly extract some information on the transport mechanism from it. This was important in particular in the low-*x* regime, where the resistance is too high to be measured in the coplanar [91] and even in the sandwich (vertical) configurations [13, 89]. Since the *C*–*V* measurement and the parameters that can be extracted from it are quite standard [46, 98], we would not elaborate on them here but would describe how the stored charge was extracted in our work [92] from the corresponding *C*–*V* characteristics.

Our rather standard *C*–*V* measurements were all taken at a frequency of 1 MHz, a voltage sweep rate of 0.5 V/s, and in the sweep direction from $V \leq 0$ to $V > 0$. In all measurements, the sign of the voltage was that of the Hg or the Al "top" electrode. For the derivation of the stored charge from the *C*–*V* characteristics, we had initially employed the maximum–minimum procedure [98, 99] that yields the properties of the corresponding ideal system, that is, the dopant density in the Si substrate and the film's (the "oxide's") capacitance C_{ox} from the shape of the *C*–*V* curve. The charge stored in the film (the "oxide fixed charge") Q_f and the charge stored in the Si/SiO$_2$ interface (the "surface states") Q_{ss} were derived from the voltage deviations ΔV_{FB} and ΔV_{mg} (toward the accumulation end and the inversion end, respectively) of the *C*–*V* characteristic from the above ideal behavior [46, 98]. The parameters of interest are then the flat band charge $Q_{\text{FB}} = -C_{\text{ox}}\,\Delta V_{\text{FB}}$ and the midgap charge $Q_{\text{mg}} = -C_{\text{ox}}\,\Delta V_{\text{mg}}$ [46, 98, 100] that we utilize below. Noting that $Q_f \approx Q_{\text{mg}}$ and that $Q_{\text{ss}} \approx Q_{\text{FB}} - Q_{\text{mg}}$, we derived both the quantities from the *C*–*V* characteristics and presented them in terms of $N_i = Q_i/q$, where N_i is the density of those charges per cm^2, *i* is FB or mg, and *q* is the (positive) elementary charge.

To confirm that the current routes in the samples are via the NCs, we applied our conductive AFM technique that enables to record the current between a sample-contacting conducting tip and a side (a "back") electrode of the sample, both in the dark and under illumination [95]. This is rather important in studies such as ours since it enables to indicate if the current takes place via the crystallites or IT is in the matrix that surrounds them. In turn this enables to evaluate whether a suggested transport mechanism (that is based on other measurements) is consistent with the observed current paths [101]. We note that this precaution has not been taken simultaneously earlier with the studies of global transport of the 3D systems of Si NCs thus leaving some doubt as to the certainty of a suggested transport mechanism.

Our *I–V* measurements were carried out by applying a voltage (from a high-voltage source Kiethley-240) to the sample and determining the current by the voltage drop across a small in-series connected resistor. This voltage was measured using a (Systron-Donner-7205) multimeter. For the measurement of the photoconductivity we applied a He–Ne laser (633 nm with a flux of 70 mW/cm^2) [91, 92] or a solid-state laser (473 nm with a flux of 250 mW/cm^2) [95]. The difference between the current, under illumination and without illumination, was considered to be the photocurrent through the sample. As we show below, the comparison between the dark conductivity σ and the so deduced photoconductivity σ_{ph} has provides useful information regarding the transport mechanism. All macroscopic transport measurements were carried out in the room-temperature–liquid nitrogen temperature range by using a cryostat (Janis, VPF-475) with a temperature controller (Lakeshore-321).

In view of the wide interest in the optical properties of our Si NCs systems, our standard characterization tool of the samples was the *x* dependence of their spectral photoluminescence (PL) characteristics [91, 92, 96, 97]. The PL was measured under the 488 nm excitation of an air-cooled (Spectra Physics model 269) Ar ion laser with a variable power of 6–40 mW or by a water-cooled Ar laser with a power of 1.5 kW. In the first configuration [90–92, 102], the detection was done with an optical fiber that led the luminescence directly to the spectrometer (Control Development Inc.), and in the other setup, the PL signal was dispersed by a 1/4 m spectrometer onto a photomultiplier that fed a photon counting system [96, 97].

4.4
Experimental Results and Their Interpretation

In the introduction, we have emphasized the importance of having the parameter *x* as an extra handle for the evaluation of the transport mechanisms in ensembles of Si NCs. We start with the over-all picture of the dependence of the electrical conductivity on the Si-phase content in our samples and we underline the experimental difficulties in studying the problem at hand. In Figure 4.1, we show the typical *x* dependencies of the conductivity $\sigma(x)$ and the photoconductivity $\sigma_{ph}(x)$ throughout a wide portion of the entire *x* range. We see that up to about *x* = 38 vol%, we measure the background conductivity of the experimental system and thus, from this single measurement, we cannot draw any conclusions regarding the transport mechanism in the *x* < 38 vol% regime. In the 38 < *x* < 60 vol% regime, we see a strong rise in the conductivities and then a saturation (with a slight decrease) of both $\sigma(x)$ and $\sigma_{ph}(x)$. We already note at this stage that both $\sigma(x)$ and $\sigma_{ph}(x)$ have the same dependence on *x* showing that since the number of charge carriers contributing to each is different, their *x* dependence represents well their "common mobility." Noting that these results [91] can be fitted to the predictions of percolation theory, as given in Eq. (4.1), it is apparent that this dependence reflects the connectivity of the system. This fit yields the percolation threshold x_c and the critical exponent t of the system (that in the present case are

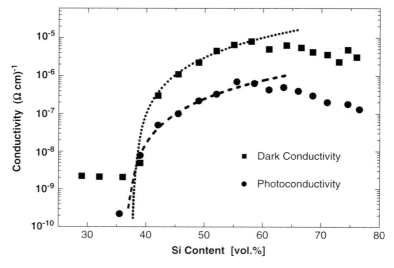

Figure 4.1 Typical dependencies of the dark and photoconductivity on the volume content of the Si NCs phase in our typical cosputtered film. The data at the regime of the strong variation of $\sigma(x)$ can be described by a percolation-like dependence. The experimental points for $x < x_c = 37$ vol% indicate the experimental limit for the derivation of the conductivity in this case of 1 μm thick films. (From Ref. [91].)

about 37 vol% and 2, respectively [91]). We found that within our deposition and postdeposition treatments of many samples the typical values of x_c were in the $25 < x < 50$ vol% range and the typical t values were in the $2 < t < 5$ range. We will use the corresponding x_c value as a prime reference point for x in the discussion below. In what follows, x and x_c will always be given in units of vol%.

After introducing the basic features of the macroscopic conductivity scene in Figure 4.1 and in order to be able to consider the expectations regarding the transport in the $0 < x < 38$ range as well as the other distinguishable regimes of $38 < x < 60$ and $x > 60$, we turn to the structural information that we deduced from our comprehensive HRTEM study [93, 94] of that sample. In Figure 4.2, we show images that are typical of the three significantly distinguishable x regimes of the sample. The low-x $(0 < x < 18)$ regime consists of well-dispersed (geometrically isolated) spherical NCs. The intermediate x $(18 < x < x_c)$ regime clearly shows a small distortion of the crystallites to ellipses for the geometrically isolated NCs, the creation of "touching" pairs, and the further distortion of the NCs in the regions of relatively high NC densities, so as to accommodate the crystallites that are grown in their proximity. Also, note that the Si crystallographic lattice planes in the images of the NCs are not parallel in the various crystallites suggesting the independent nucleation of the crystallites during the annealing process. For a more precise definition of "touching," we consider two NCs as "touching" if the distance between their surfaces does not exceed the order of a lattice constant (≈ 0.5 nm in Si). Finally, at the high-x $(x > x_c)$ regime we see further growth of the individual NCs, further distortion of the

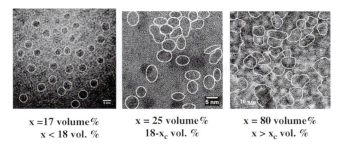

x = 17 volume%
x < 18 vol. %

x = 25 volume%
18-x_c vol. %

x = 80 volume%
x > x_c vol. %

Figure 4.2 HRTEM micrographs of our annealed cosputtered films for three Si phase contents: 17, 25, and 80 vol%. The crystallites, detected by the Si lattice fringes, are marked by their border. One notes that with the increase of x the crystallites grow and change from a spherical shape to shapes that could accommodate their high density. This increase causes "more and more" of them to "touch" their neighbors. (From Ref. [92].)

crystallites, and, most important, the apparent presence of very large clusters of "touching" NCs. Note that a priori it is not obvious from the 2D HRTEM images that in the latter regime we have a 3D network of connected "touching" crystallites. To confirm that the latter is the case, and to show that in our study the macroscopically measured currents pass through the NCs and not via possible parallel paths (such as a possible residual a-Si network that may still be present in the sample), we have carried out (as far as we know the first and the only) conductance AFM measurements on samples for which the macroscopic conductivity was simultaneously measured. The images shown in Figure 4.3 for the high-*x* regime confirm the existence of such a network, enabling us to interpret our data accordingly. Also note that this finding is a significant improvement over the topographic and electric field microscopy (EFM) images, where only the presence and the charging of the NCs are revealed.

Following the general features of the $\sigma(x)$ and $\sigma_{ph}(x)$ dependencies and the structural information described above, we can clearly conclude that we have five

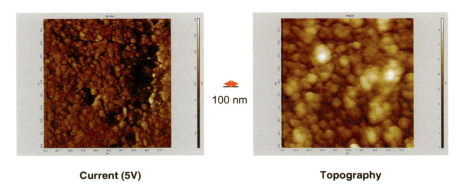

100 nm

Current (5V)   **Topography**

Figure 4.3 Images derived from topographic AFM (right) and a C-AFM (left) scans of our samples in the high-*x* ($x = 65$, and $x > x_c = 45$) regime, showing that the current takes place through the 3D network of the Si NCs.

different structural–electrical regimes, such that in each of them we may expect a different transport mechanism to dominate. These regimes are (a) the low-x regime, of isolated uniformly dispersed spherical NCs; (b) the transition between the low-x regime and the intermediate x regime where some of the NCs already "touch" their neighbors; (c) the intermediate x regime where clusters of "touching" NCs are formed; (d) the percolation transition regime ($x \approx x_c$) where the above clusters form a global continuous network; and (e) the high-x regime where the percolation cluster of "touching" NCs is well formed and geometrically non-"touching" NCs are rarely found. In what follows, we will describe our principal experimental data by dividing the descriptions of our observations, and the relevant expectations from the models and data mentioned in Section 4.2, into each of these regimes. One should note that for the above regimes, only in the last two, direct experimental measurements of the transport could have been carried out because of the high planar resistance of samples in the first three regimes. For the latter regimes, we have carried out some indirect measurements, which combined with the structural data presented in Figure 4.2 and the above-reviewed (Section 4.2.2) information on the single NCs can shed light on the transport mechanisms in them. The picture that emerges following all that and the data given in the literature will be discussed in Section 4.5. Owing to the limited scope of this chapter and the fact that we have previously mentioned some specific experimental difficulties associated with a particular measurement [91] and some possible alternative explanations to some of our particular results [103], we do not discuss them here, bringing in only our current view of the transport mechanisms as suggested by the available data.

4.4.1
The Low-x Regime

The low-x system shown in Figure 4.2 consists of a geometry that is expected to yield hopping conduction, that is, tunneling between adjacent particles, such that if there is a difference between the energies of two particles it is provided thermally or by an applied electric field. This scenario is similar to that encountered in the transport between impurities in semiconductors [28, 32] and in granular metals [7, 8] in the so-called dielectric regime. As we mentioned in Section 4.2.1, in the former case the distribution in the particles' energy is attributed to the disorder in the energy levels in the network and the corresponding conduction is the well-known variable range hopping, while in the granular metals the energy distribution is attributed primarily to the variation in the local charging energies associated with the size and/or environment of the individual particles. In semiconductor NCs, both ingredients are also included except that the distribution of the electronic "conducting" levels is more emphasized, as manifested by the effects of resonant tunneling mechanism (see Eq. (4.9)) [9, 36].

Following the fact that unlike the situation in granular metals and ensembles of Ge NCs embedded in SiO_2, the simple $\sigma(x)$ dependence could not be measured reliably in this high-resistance regime of Si NCs, we turned to other, though indirect, experimental approaches that can shed light on the conduction mechanism in that low NCs

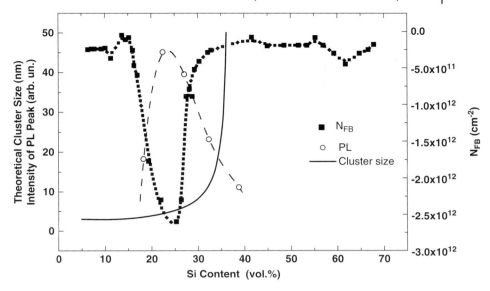

Figure 4.4 Stored negative charge N_{FB} (in units of elementary charges per cm^2), as well as the PL spectral peak intensity, as a function of the Si-phase content for the entire x range of the sample. Also indicated in this figure is the percolation theory prediction for the average cluster size of the Si NCs, assuming that the NCs diameter (at $x \rightarrow 0$ limit) is 3 nm and that the percolation threshold is at $x_c = 36$ vol%. (From Ref. [92].)

density regime. To consider this regime, we examined the charge storage and the photoluminescence in this and the other two subsequent regimes for which the electrical conductivity could not have been measured. The corresponding results of the PL intensity (at its spectral peak [91, 96, 97]) as a function of x and the CS as determined from the estimate of Q_{FB} (i.e., the charge stored in the film, see Section 4.3 [89, 92]) are shown for the entire x range in Figure 4.4. As is clearly shown in this figure, we found an excellent correlation between the negative CS (given by the areal density of elementary charges) and the PL. Combining those results with the structural data of Figure 4.2, for the present $x < 18$ regime, where the concentration of NCs, N, grows with x, we can clearly conclude [103] that the increase in the PL and the CS is associated with the increase in the concentration of geometrically isolated NCs. We must assume that the crystallites get charged by a resonant tunneling process that is combined with a hopping charging scenario as in granular metals.

As we found here and previously, from the spectral shift of the PL peak with the variation in the size of the NCs, and as is well confirmed for our ensembles of Si NCs, the presence of quantum confinement within our isolated NCs is well established [93, 94, 96, 97]. On the other hand, the results shown in Figure 4.4 confirm the presence of NCs charging [92, 103]. We also note that the optical bandgap opening observed in our samples [96, 97] is of the order expected from recent theoretical estimations of the band shifts (0.1–0.3 eV [50]), and that the size of the NCs in our samples suggests CB energies ($\approx$0.1 eV) that are of the order estimated in other studies [49] and in granular metals [7, 8] for particles of similar sizes. We have then an indirect evidence for the possible interplay between resonant tunneling and charging effects that, at least qualitatively, suggest that the physical nature of the inter-NC transport in the low-x regime must consist of resonant tunneling under CB. In passing, we remark that for similar scenarios (i.e., when the grains do not touch) in granular metals, for which we could measure $\sigma(x)$ [33], we have recently found [30] that the hopping-like behavior is also confirmed by the $\sigma(x)$ dependence. It seems then that our indirect evidence supports the basic processes suggested from the direct (but "vertical" transport) measurements in Refs [9, 56].

The route to get more conclusive data, beyond these basic conclusions, is to try to fabricate samples with relatively high N but with essentially no "touching" NCs, so that the $\sigma(T)$ dependence could be measured as a function of the NC size. Our preliminary results indicate that this may be feasible [104]. We should point out, however, that in order to confirm, conclusively, that the Simaneck model [36] suggested in Refs [9, 56] is applicable to the dilute ensembles of Si NCs, a simultaneous structural study of the NCs sizes and the inter-NCs distance distributions will be needed.

4.4.2
The Low-x to Intermediate-x Transition Regime

Above, we followed the increase in the CS and the PL with increasing x and this was well accounted for by the increase in the concentration of the "nontouching" NCs in the low-x regime. While this concentration keeps increasing with x (as seen in Figure 4.2), we see in Figure 4.4 that a further increase in x causes the CS and the PL to decrease, yielding a peak in their x-dependence at some Si phase content, x_d, which is around $x = 24$ in this particular case. Following the fact that for $x > 17$ we see, from our HRTEM images (Figure 4.2), that while there is an increase in the total concentration of the NCs, N, there are less and less "nontouching" individual NCs, with increasing x. We explain the decrease in both the CS and the PL intensity to be due to the decrease in the content of the latter NCs, N_1 [103]. In particular, we expect that the "touching" NCs do not contribute to PL since the exciton's Bohr diameter in Si is 8.6 nm [1] and the NCs here have a diameter of the order of 5 nm [93, 94] and "doubling" the size of the NCs will cause an effective deconfinement of the electron–hole pairs. Hence, the only contributing NCs are the N_1 "isolated" ones. This suggestion is strongly supported when we consider the average cluster size dependence on x as predicted by percolation theory [26]. In this theory the increase of the average cluster size with x is accompanied by the disappearance of most of the geometrically isolated NCs in favor of the growth of the larger clusters [26, 92]. Here, we see that at the experimental x_d the average cluster size is indeed about twice the size of the individual NC. The above conclusion is even more strongly supported by the fact that, as shown in Figure 4.4 for the present sample, for which $x_c = 36$, we found that $x_d \approx 2x_c/3$. This observation is in excellent agreement with the findings on the cluster statistics in percolation theory [26] that suggest that the peak in N_1 (the concentration of single NCs, i.e., "nontouching" NCs here) will be below the percolation with increasing x, below the percolation threshold and at $2x_c/3$.

Similarly, this deconfinement (hence the notation x_d) is expected to affect the CS in the NCs since a significant increase in N, in general, and the facilitated transport between "touching" NCs (that consist of very narrow tunnel barriers as shown in Figure 4.2), in particular, will provide more routes for the discharging of the NCs. Indeed, the latter has been shown previously to weaken the retention of charges in their sites [105, 106]. Hence, as in the PL peak case, we expect, in view of the competition between the increase in N (as charging sites) and the shorter average "stay time" with the increase $N - N_1$, at the expense of N_1, to have a peak in the CS. While the presence of the above PL and CS peaks is not surprising, the fact that they are at exactly the same x (i.e., x_d) is. This is in particular so since while the two deconfinements are expected to be effective, their consequences on the measured properties are not (e.g., in the PL, we have a competition between radiative and nonradiative processes [96, 97] that are not relevant to the CS). Although this question is still unresolved, it seems that, as we saw in Figure 4.4, the clue lies with the doubling of the average cluster size at x_d, which is expected to yield the largest relative change in the effect of both the QC and the CB energies.

From the transport point of view, the dominant contribution to conduction in this regime is expected to be the tunneling between the NCs. The hopping conduction will be manifested, however, in comparison with that in the low-x regime, by having smaller confinement and average CS energies. On the other hand, these energies will have a broader span in their values, so that VRH-like behavior under a "weaker" CB is expected to replace the resonant tunneling-like behavior as x is increased. This can be a possible explanation for the VRH ($\gamma = 1/4$) behaviors that were reported in some works [79–81]. However, the lack of even an approximate x determination in those works does not enable us to confirm this possibility.

We then relate the above collective transition at x_d to the formation of "enough" finite clusters that consist of "touching" NCs. While, as we saw above, this has a profound effect on the CS and the PL, one does not expect that it will have a significant effect on the transport since the "bottleneck" for transport will still be the tunneling between the clusters of various sizes (usually few and small in that x regime). On the other hand, the nature of the transport is expected to change with the increase in the size of the clusters, from resonant tunneling under Coulomb blockade to VRH type.

4.4.3
The Intermediate-x Regime

As in the previous regimes, we could not directly measure the conductivity in this regime, and we have to consider the transport in it by using our above indirect data and the data available for the related regime in granular metals [7, 107]. A priori, one would expect that this regime will simply enhance the features that were suggested in the previous regime. This is because, from the results of Figures 4.2 and 4.4, it appears that the main effect of the increase in x is the general increase in N, on the one hand, and the growth of clusters of "touching" NCs, at the expense of the "nontouching" individual NCs, on the other hand. The conduction mechanism that dominates the macroscopic conductance here is expected to be tunneling between all these entities as in the previous x regime. This is obviously the case in the low-x end of this regime. However, with the further increase in x in this regime, one would expect (from Figure 4.2) that arrays of low resistance "blobs" that are connected by tunneling junctions will form [30]. The blobs grow to essentially become "shorts," while the junctions become narrower and narrower to yield a higher tunneling conductivity between various (single or multiparticle) clusters. In the low-x end of this regime, we will have a VRH-like transport while at the high-x end of this regime, the role of the cluster "shorts" will become more and more pronounced. This scenario is very much like the situation encountered in granular metals [107], except that here (as we see below) the "touching" within the clusters is associated with a different conduction mechanism (see Section 4.4.5) from the one involved in the coalescence [108] that dominates the conduction there [8, 107]. As in that case, the outcome of the competition between the intercluster and intracluster tunneling is sensitive to the structure that is imposed by the sample preparation conditions, in general, and the sample annealing conditions, in particular. Following the considerations outlined in the previous section, the lack of information regarding the x values in the samples for which $\sigma(T)$ dependencies were reported does not allow us to attribute the observed dependencies to a particular transport mechanism. Following this we can only base our anticipations on the images of Figure 4.2 and the results of Figure 4.4, expecting that in the present regime there will be a mixture of mechanisms that will vary from a Simanek-like [36] inter-NCs hopping to some intercluster hopping [107, 109] as x increases through this regime.

In reality, the situation appears to be even more complicated and this has not been resolved yet, even in the granular metals [7]. A priori one can expect that, for the same x, a high density of small NCs will yield a larger conductance than a lower density of larger NCs [8], and if we assume that in such a system there is a critical conductance in the system we would expect that it will increase with increasing x. However, this critical conductance can also decrease with x as the cluster separation will increase. This situation was found in granular metals [107] where, by annealing, one can affect (for the same x) the competition between the formation of a large number of individual NCs and the formation of larger clusters. Following the fact that, for the same x, the smaller the grains the smaller their separation and therefore the larger the tunneling between them [7], the annealing causes a diminishing conductance in the intermediate x regime. Our preliminary manipulation of the fabrication conditions to achieve such scenarios [104] has indicated that while the derivation of transport data in that regime is possible, the combination of quite a few types of conduction mechanisms in this regime and the additional (in comparison with granular metals) contribution of a parallel a-Si matrix in cosputtered Si-SiO$_2$ systems may further complicate the evaluation of the transport mechanism. Since, as we saw in Figure 4.1, in the well-annealed samples this regime does not reveal conductance, we cannot even guess, at the present stage, which is the dominant conduction mechanism there. On the other hand, it may be that the measured conductance in the presence of both, green PL and conductance, as we found [104] in this regime for samples prepared at lower annealing temperatures, is a combination of the conduction via these NCs and the a-Si network that still exists there, in a situation that is reminiscent of the conduction in porous silicon [25].

Clearly, more studies of how the annealing affects the conduction are needed in order to form a NCs network with a measurable conductance but with no parallel competing amorphous network. Judging from granular metals [30, 33] this may be achieved by increasing the conductance of the Si NCs, say, by doping (see Section 4.5). Also for this intermediate-x regime, it seems to be very useful to apply local probe microcopies and/or nanolithography, to determine the local current path such as shown above in Figure 4.3, for the high-x end of this regime. We note, however, in passing that the latter two regimes are crucial for optoelectronic applications since simultaneously achieving efficient transport and efficient PL are the two essential ingredients needed for their development.

4.4.4
The Percolation Threshold Regime

In this regime, the conductivity and the photoconductivity can be directly measured and thus experimental data that could be analyzed in view of the possible models are finally possible. The transition in the conductivity in the $x_d < x \approx x_c$ regime of Figures 4.1 and 4.4 can be well accounted for as a percolation transition. Indeed the $\sigma(x)$ dependence in this regime is well characterized by the percolation power law that is given by Eq. (4.1). In previous works, we noted that in granular metals, there are two transitions, one that takes place for noncoalescing grains at $x_c = 18$ [33] and the other that takes place for coalescing grains [7, 33, 111] at around $x_c = 50$ vol%. The first is associated with the onset of hopping under CB conditions [6–8, 30, 33] and the second is associated with the formation of a continuous metallic network [7, 107, 110, 111]. In the former case, a t value of 3.3 was observed while in the latter case the universal $t = 2$ value was observed [7, 107, 110].

Returning to the system of interest here, we note that, as we saw in Figure 4.2 (at variance with the well-known percolation transition in granular metals), the Si NCs do not coalesce, geometrically. The latter was manifested by the different orientations of the crystallographic plains in the "touching" crystallites in spite of their close proximity. While there is a very narrow separation between the "touching" NCs that is on the order of the lattice constant, some type of barrier that is reminiscent, but very different from the barriers between two semiconductors [39] or a semiconductor–oxide–semiconductor junction [46], must form. In this case, we emphasize that the 2D image of the high-x regime in Figure 4.2 represents a 3D system that may be fully connected electrically as suggested by Figure 4.3. In that case, however, we may have single isolated NCs or clusters, the transport between which is by regular tunneling and that these are the "bottlenecks" of the percolation cluster. A hint that this may happen is provided by the nonuniversal percolation exponent ($t > 2$) that we observed, as shown by the best fit of the data in Figure 4.5 to the predicted critical behavior (Eq. (4.1)). We have found in particular that the t values are all in the range of

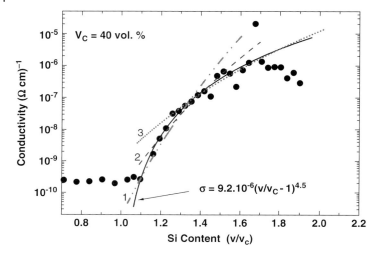

Figure 4.5 A typical $\sigma(x)$ characteristic in the vicinity of the percolation threshold. Fittings to percolation theory predications (Eq. (4.1) and the solid curve) and to hopping behavior (Eq. (4.5) and the other curves) are shown in order to indicate that for this regime only the first possibility applies.

$5 \geq t \geq 2$, as to be expected from the apparent spatial distribution of the various resistances of the conducting elements (that is likely to be due to the distance distribution between the NCs [30, 36]) in the system. This is also in agreement with the above-mentioned findings on granular metals when x increases from the dielectric regime [30, 33] to the coalescing particle [7, 110] regime.

In order to check, as we have done previously for granular metals [30], to which end of this range of behaviors do our samples (around x_c) belong, we have also attempted to fit the data with the full possible range of the exponential $\alpha = a v_c^{1/3}$ parameters (see Eq. (4.5)). As seen in Figure 4.5, the trials with various α parameters, which correspond to the best fits at the high, the intermediate, and the lower sets of the data points, show that the hopping scenario cannot account for them. This consequence is important since it indicates that hopping-like behavior does not dominate the present and the high-x regimes, where the conductance is conveniently measurable. In passing, we note that the consequence of the presence of simultaneous conduction via coalescing particles and tunneling between noncoalescing particles have hardly been discussed in the past, even for granular metals [107, 109–111]. Here, in ensembles of Si NCs, the consequences of this competition makes the understanding of the transport more difficult since, as we show in Section 4.4.5, the interparticle transport is more complicated for "touching" particles than for coalescing particles (that one encounters in granular metals [7, 8, 108].

So far, we have seen that the above percolation scenario describes well the system around x_c while we still do not know the interparticle tunneling mechanism. However, we can expect that the closer we approach the percolation threshold from above, by lowering x, we may find more (though still few) geometrically isolated

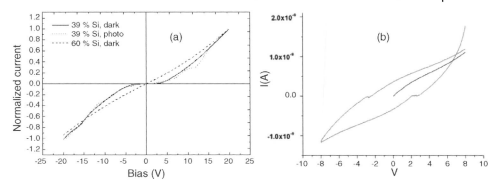

Figure 4.6 (a) The *I–V* characteristics as obtained in the close vicinity (*x* = 39) and far above (*x* = 60) the percolation threshold. It is clearly seen that a "voltage gap" is opened only for the lower *x* (≈ x_c = 37 vol%) sample. Such characteristics were obtained under slow (typically <0.02 V/s) scan rates (a). Note also the very similar *V* dependence of the dark and photocurrents. The presence of charging effects around x_c is manifested by the difference between these characteristics and the characteristics that were obtained under scans that are taken at much faster (typically >0.1 V/s) scan rates (b): the latter behavior indicates (after the initial current increase, solid line) a well-defined "voltage gap," as well as a hysteresis (shaded line), that must be associated with the variations in the stored charge in the system. (From Ref. [91].)

crystallites that are critical for the electrical network. Such single-particle "bottlenecks" are expected to show a DBTJ behavior [91]. In particular, these are expected to show local charging effects [11]. The fulfillment of this expectation is exhibited by both the current and the photocurrent *I–V* characteristics that are shown in Figure 4.6a. We see there that a "voltage gap" develops around *V* = 0 for *x* values in the close vicinity of x_c, but it does not exist away (*x* > x_c) from this vicinity. We also see, in Figure 4.6b, the consequences of the formation of such a gap and its coexistence with a hysteresis at *x* ≈ x_c. This indicates that charge is stored in the continuous network. All this well suggests that, indeed, DBTJ conditions do exist in the vicinity of x_c. Another manifestation of the same phenomenon is exhibited in Figure 4.7, where we show the *I–t* characteristics in the close vicinity and far above the percolation threshold. We see that in the former case the charging phenomena are apparent by the current decay while for larger *x*-values they are absent. The charge stored was calculated here, as for Figure 4.6, by using the Wagner circuit [112] (that is also illustrated in Figure 4.7) as a model. Both the charge derived from the *I–t* characteristics and those derived from the hysteresis of Figure 4.6b were found to be in agreement with those derived for the corresponding *x* from *C–V* data such as those shown in Figure 4.4. All these measurements indicated that the CS is on the order of 3 nC, which corresponds to about 5×10^{15} cm^{-3} elementary charges in our samples. Note that this means that less than 1% of the crystallites in the system, at that *x* regime, are charged, supporting our conjecture that in practice only the single individual NCs retain their charge, on the one hand, and the existence of DBTJs in this regime, on the other hand. All these data suggest that the DBTJ-like behavior, as

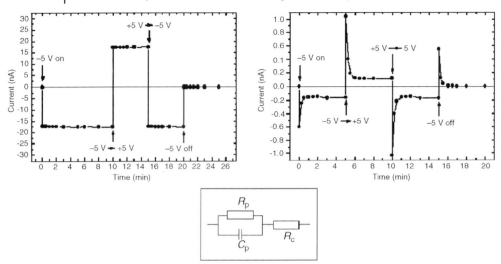

Figure 4.7 Typical *I–t* characteristics as obtained in the close vicinity of x_c ($x = 39$, right) and far from it ($x = 60$, left). Note that there is evidence (by the current decay and response to switching) for substantial charge storage only for the close vicinity of x_c. The stored charge was analyzed by applying a Wagner-circuit model that is illustrated (by two resistors, R_p and R_c, and a capacitor C_p) in this figure.

proposed by the illustration (of an isolated DBTJ (a) and a DBTJ between two large clusters (b)) presented in Figure 4.8, for this regime, is valid and consistent with the concept of the critical resistor (or resistors) in the system [30]. These clusters serve as leads to the "isolated" NC and are thus expected to bring about a DBTJ-like structure. For a more apparent possible scenario that may account for such a model, we show in Figure 4.8 also a part of the image in Figure 4.2 that may be associated with such a configuration. The corresponding possible tunneling current tracks via such a possible DBTJ are indicated. What we do not know though is whether the voltage gaps in Figure 4.6 are due to the level splitting associated with the QC and/or the CB of the individual DBTJ crystallites. In particular, the scaling of the measured voltage in terms of the DBTJ, as shown in Figure 4.6, will require a specific model. We note though that this voltage is of the order reported for single NCs [13, 21, 56, 86] as well as on 2D systems where the in-plane conduction (i.e., via an ensemble of the NCs) has been measured [113], thus supporting this suggestion [91].

4.4.5
The High-x Regime

From the experimental point of view, the $x > x_c$ regime is the most convenient to study since it is associated with a relatively high conductance. In particular, this enables a systematic study of $\sigma(T)$ that, as noted in the introduction, is the most abundant tool for studying transport mechanisms in solids. The results we found

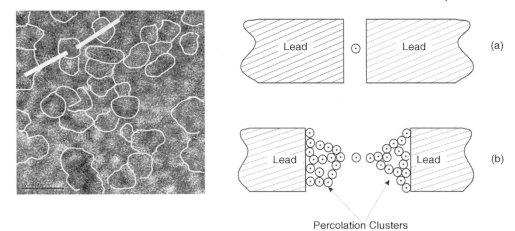

Figure 4.8 Model of a single DBTJ (a) and as suggested for the local environment in the percolation transition regime (b), as well as the possible presence of a DBTJ in the present network of the Si NCs as expected from our HRTEM images. The bright segments represent the possible NC separations via which the tunneling current in the DBTJ takes place. (From Ref. [91].)

using this experimental tool were quite surprising as they did not conform to any of the dependencies mentioned in Section 4.2.1 over the entire temperature range. Rather, we found that, with increasing temperature, there is a general transition from dependencies that are similar to those of Eqs. (4.6) and (4.7) to a dependence that is in the form given by Eq. (4.4). Following this, we turned to a more basic analysis of the data, that is, their presentation in terms of a log σ–log T dependence, noting that Eq. (4.4) corresponds to a convex log σ–log T dependence and Eqs. (4.6) and (4.7) to a concave log σ–log T dependence.

Our experimental results observed in this regime can be summarized as follows. We found a concave dependence that turned to a convex dependence in the log σ–log T plots, both in the conductivity and in the photoconductivity. With the initial increase in the temperature, the corresponding slope was found to increase with temperature as shown in Figure 4.9. When a fit to Eq. (4.6) was attempted (in the "concave" regime), we found that the γ-values and the T_B values were not constant in any portion of the curve but changed smoothly and simultaneously from high to low values as the temperature increased. In fact, a linear dependence of T_B on γ was revealed in the annealed samples. Correspondingly, when a fit to the behavior of Eq. (4.7) was attempted we saw an increase in the δ values with increasing temperature. While the deviation from a linear log σ–log T behavior (over the entire temperature rang) was different in different samples, as a rule, we found that the log σ–log T behavior of the photoconductivity is much milder (lower γ and T_B values) than in the dark conductivity, when a fit to Eq. (4.6) is attempted. On the other hand, the annealed (Si NCs/SiO$_2$ ensembles) yielded a more pronounced (larger γ and T_B values) behavior than the nonannealed (a-Si/SiO$_2$) samples for

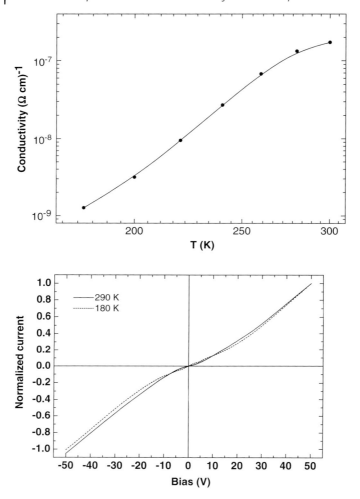

Figure 4.9 Temperature dependence of the conductivity (measured at $V = 100$ V, in this high-x (≈ 50) $> x_c$ ($=37$) regime) and the corresponding normalized *I–V* characteristics. Note in particular the typical transition from a concave to a convex curve dependence with increasing temperature in the $\sigma(T)$ dependence, and there is no change in the shape of the *I–V* characteristics with increasing temperature.

both the conductivity and the photoconductivity. Correspondingly, the attempted fittings to Eq. (4.7) have shown an increase in δ that turned to a sharp decrease in δ as room temperature was approached. Also, the effect of doping appeared to be similar to the effect of illumination. All the above behaviors became milder with the increase in *x*.

It is obvious that the immediate consequence of these experimental results is that we have a competition between two mechanisms such that one is responsible for the concave log σ–log *T* behavior and the other is responsible for the convex log σ–log *T* behavior. At this stage we have only these features for suggesting a possible transport mechanism, and we do not try to derive particular γ-values as these depend on temperature.

In trying to evaluate the data, we found that every one of the mechanisms that we mentioned in Section 4.2.1 for the behavior of σ(T) is qualitatively consistent with the structure that is suggested by Figures 4.2 and 4.8 for the high-x regime. We have narrow barriers (between "touching" NCs), on the one hand, and "bottlenecks" for the conduction, on the other hand. We must consider further details of the data to reasonably narrow down the possible mechanism scenarios. For this purpose, we also checked the temperature dependence of the *I–V* characteristics finding that, as shown in Figure 4.9, their shapes are independent of temperature. This enables us to tentatively eliminate both the vibrating barrier model scenario [40, 41] and the charging effect granular metal-like scenario [7] that we proposed previously [25] for porous silicon. On the other hand, while the indications from the results given in Section 4.4.4 and Figure 4.8 are that bottlenecks do exist, the fact that Eq. (4.7)-like behavior has been shown to be found in a-Si microstructures [44] does not lend support to the basic view of that kind of model.

The fact that the above-described behavior is weakened in the photoconductivity and doped samples can be explained by the narrow high barrier [42] and the "bottleneck" [44] models since in both models the carriers are photoexcited to relatively higher energy levels. However, both models do not explain the turnover to the convex log σ–log T (i.e., to an Arrhenius-like) behavior at the higher temperatures. Since the hopping-like aspect is provided by the Eq. (4.7)-like behavior, this Arrhenius-like behavior is very likely (in view of the above-mentioned effect of illumination and doping) to reflect a thermal excitation of the charge carriers to the conducting levels, which may or may not be associated with QC effects [9, 56, 114]. We are thus left, at present, with a scenario that includes one or a combination of the single barrier [42] or the collective [44] hopping processes, as well as a carrier concentration excitation process, thus yielding the apparent monotonic change of γ and/or δ as described above. This process appears to be of tunneling via "narrow barriers" between semiconductor NCs with some scattering along the conduction path. The transport is such that at low temperatures only few NCs participate in the tunneling process while at the high temperatures the hindrance to transport is in the matching between the quantum confined levels and/or carrier excitation.

As is obvious from the discussion, while at this stage the above scenario seems to be the most likely one (the contribution of a parallel a-Si tissue is negligible as we saw in Figure 4.3 and the CS is diminished as we saw in Figures 4.4, 4.6, and 4.7), further research will be needed to establish the exact type of processes that are responsible for the above, rather new, observations. In particular, local probe microscopy as a function of temperature is expected to be a very useful tool in order to find whether the observations have to do with single tunneling barriers or with larger scale contributions of the network. At this stage, we consider the mixture of the above three models as the most likely, but still tentative, transport scenario in this high-x regime.

4.5
Discussion and Overview

Although not a topic of this chapter, the understanding of the electronic structure of the single NC when it is a quantum dot is of crucial importance. This topic was studied theoretically [1–3, 47, 48, 50] and by many experimental optical (mainly PL) studies [1–3, 96, 97]. It is quite generally agreed that the broadening of the bandgap with the reduction of the NCs size is on the order of up to 0.8 eV as the NC diameter is reduced to 3 nm [94]. The variations in the conduction band edge, and the valence band edges [47, 48, 50] and the PL mechanism [2, 97] are still under debate. In fact, details of the level structure (such as the energy of critical points in the Brillouin zone, as achieved in Ge-SiO$_2$ structures [115], have not been determined conclusively yet. Even more important from the point of view of this chapter is the fact that, unlike the situation in III–V and II–VI single NCs [51, 57, 74], the various microscopic or microscopic-like electrical measurements that were reviewed in Section 4.2.2 did not add much information to this picture. It also seems, at present, that the many results of the *I–V* and *C–V* measurements on small clusters of Si NCs of 2D systems (in the vertical configuration) do not add up to a convincing picture that can separate the contributions of the resonant tunneling and CB effects. This is in particular so since no two studies yielded the same set of the single NC energies and since there were no studies that were combined with optical or other characterizing measurements (as done previously with single NCs of the II–V or the II–VI compounds [51, 57, 74]) that can confirm the ad hoc models that were proposed to explain particular features in the experimental data. In fact, as described in Section 4.2.2, the clear presence of the CB effects seems to overshadow the QC effect in spite of the expectation that the energies involved are of the same order for the NCs sizes that were studied [49]. Under these conditions, the basis on which one can interpret the available 3D transport data is still quite narrow, in particular when comparison is made with transport in granular metals [7, 8] in the temperature range where only the charging energies are significant [34]. The transport via the (essentially) single isolated Si NCs has been discussed in terms of DBTJs where the carriers tunnel from one electrode to the other via the quantum confined or other levels in which they are, or they are not, trapped [14, 49, 67]. Considering this major limitation of our knowledge, we will try, using this rather narrow basis and the available data reported here and in the literature (see Section 4.2.2), to provide an initial basis for the discussion of the corresponding electrical transport mechanisms in our 3D systems.

In this section, we will make an attempt to integrate the findings reported in the literature and the results presented in Section 4.4 in order to come up with a better picture of the transport mechanisms in 3D ensembles of Si quantum dots. Because of the limited scope of this chapter, we will not discuss particular experimental data [103] and we will critically review only those data that seem reliable and/or crucial enough to be considered as the cornerstones of the basis that we are trying to lay down. As will be shown below, while it seems that we are able to present here a basis for the evaluation of these mechanisms, the details, in general, and their quantification, in particular, are far from being accounted for.

Turning to the subject of interest, we consider, systematically, the transport in ensembles of NCs, with the increase in their density. Starting with the very dilute (low-x) ensembles of Si NCs, the structurally found presence of only individual spherical NCs suggests that only a hopping-like transport mechanism can take place between them. There are, however, very few direct data that confirm this expectation in ensembles of Si NCs and the more convincing data that are available come from ensembles of Ge NCs that are embedded in a SiO_2 matrix [9]. The more detailed nature of this hopping process is much more questionable. In particular, it is not clear between which levels does the tunneling take place. On the other hand, it is clear that in this case the carrier injection from the contact adjusts itself to the concentration of carriers in the NCs as is manifested by the larger conductivities in granular metals [7, 8] and in ensembles of Ge NCs embedded in SiO_2 [9]. Noting that there are at least five models available for this regime [6, 7, 28, 35, 36], on the one hand, and that we have rather few reliable data, on the other hand, it seems that it is too early to suggest the particular hopping-Coulomb blockade effect that is responsible for the observations in this regime. However, at present, judging from the temperature dependence of the conductivity that yielded a $\gamma = 1/2$ value and the "right order of magnitude" of T_A values that were reported for both Ge [9] and Si [56] NCs, the results as suggested by the authors [9, 56] can be interpreted on the basis of "resonant tunneling" under the presence of a Coulomb blockade in systems with corresponding distributions of the NC sizes and inter-NC distances between them. Our work provides further support to their interpretation. First, while a conclusive comparison with their data with all possible models (Section 4.2.1 has not been made, our optical data [94, 97, 98] for such dilute ensembles show that QC exists in the NCs. Second, the nonuniversal behavior [27, 30] that we have found (although only at the $x \approx x_c$ regime) indicates the presence of a wide distribution of these distances. We can assume then that at present the Simanek [36] hopping mechanism as suggested in Refs [9, 56] is the most likely to represent correctly the low density system.

It is rather unfortunate that for the dilute limit (the low-x) regime, for which the theoretical background is the most developed, the acquisition of the experimental data is the most difficult one, in particular due to the relatively high resistivity of 3D systems where "nontouching" Si NCs are embedded in an insulating matrix. It seems that three experimental approaches are called for in order to enable a more detailed picture of the transport mechanisms in that regime. The first is to try local electrical probe microscopy, combined with optical spectroscopy and global conductivity measurements on the same ensembles of Si NCs, in order to overcome the above difficulty. As an initial attempt to circumvent the above difficulty one may, however, try the use of Ge [9] and/or doped Si NCs. Our recent experimental results (see below) suggest that such doping which enables to increase the carrier concentration is possible.

The next two regimes are those that have to do with the clustering of the Si NCs. These regimes have not been studied at all in the case of semiconductor NCs embedded in insulators, and even not in any detail in granular metals [107]. The reason for this seems to be the expectation that the "bottleneck" for transport in these regimes is still the tunneling under Coulomb blockade, whether it is between individual grains (where the blockade is significant) or between larger clusters (where the blockade is relatively small). Although this model is certainly applicable in the dilute end of our intermediate x regime, this assumption becomes much less reliable as one considers the dense end of these regimes, that is, as one approaches the percolation threshold [107]. Thus, in principle, we may have a competition between critical resistors that originate from cluster overlaps [111] or intercluster tunneling [109], such that each can provide the critical (average) resistor in the system. Another complication in this regime is the need to understand the particular transport mechanism between two "touching" Si NCs. All the background of granular metals is mentioned here since the situation there is expected to be a good starting point for the discussion of semiconductor nanocomposites, in the similar x regimes, in general, and for ensembles of Si NCs in particular. However, in the latter and the present case the system provides much richer physical phenomena and their applications. These originate, of course, from the relatively "well separated" levels of the quantum dots, which unlike the case in metals enables controllable carrier population (electrically or optically) of the relevant levels [9, 97]. This brings about the PL and the charge storage with "trapping," effects [4, 14] that have no counterparts in granular metals. We note in passing that this difference enables the most significant applications of (dense enough) Si NCs systems, that is, the light emitting diodes) (LEDs) [15] and the nonvolatile memories [4]. What appears to be clear, at present, for these intermediate density regimes is that the dominant transport mechanism is intercluster tunneling and that for photogenerated carriers there is a competition between the radiative recombination that provides the PL and the nonradiative recombination that provides the photoconductivity [91, 92, 96]. The larger the Si NCs clusters and the larger the concentration of these clusters, at the expense of the single crystallite "clusters," the more dominant the latter effect.

The consideration of the cluster formation in terms of percolation theory revealed to us, in the intermediate-x regimes, what we called the delocalization threshold x_d for carrier confinement in general and for the electron–hole pairs in particular. The latter is well recognized by the common peak of the PL and the CS dependencies on the density of the Si NCs. Hence, for those regimes we can summarize the transport mechanism as being determined by a combination of the hopping (that we concluded in the dilute case) and the transport mechanism between two "touching" NCs (to be discussed below for the high-density regime). It is clear that further understanding of the intermediate x regimes will have to wait until a better understanding of the (easier to understand, in principle) two extreme regimes will be provided. On the experimental end, it appears that an efficient tool to study those regimes is the utilization of the electron beam-induced deposition (EBID) technique, by which one can fabricate contacts that are tens of nanometers apart, enabling to follow the above conduction mechanisms in local environments of variable size.

Considering the x regime in the close vicinity of the percolation threshold, the situation is similar to the one described above with, however, the significant addition of a percolation cluster that connects in parallel to the above-described networks [107, 111]. This situation is, in principle, even more complicated than the intermediate density regimes. On the other hand, the dominant mechanism is easier to follow in this case [91]. This is because here the percolation cluster may contain a few DBTJs that dominate the conduction path and thus may simplify the understanding of the transport mechanism in this threshold regime, on the one hand, and enable to extract the information, so difficult to obtain, on the isolated Si NCs (see above) from macroscopic measurements [91], on the other hand. In this regime, we clearly observed the development of a "voltage gap" [91] and its association with charging effects that have to do with a Coulomb blockade effect. On the other hand, at present we do have only an indirect evidence [95] for the possible contribution of the quantum confinement to the transport. For this regime, two possible routes of experimental approaches appear promising. The first is to prepare and test samples of variable width of the NC size distribution and the second is to try a photon-energy phototransport spectroscopy that may indicate (at least at low temperatures) the importance of the confined levels in the transport. However, this may be the case only at the "higher x end" of the percolation transition, while at the "lower x end" (where the conductivity is controlled by both the percolation cluster and its combination with clusters that are "connected" to it by tunneling) the situation may be much more complicated. Such a scenario that yields a nonmonotonic $\sigma(x)$ dependence, as the percolation threshold approaches, was found originally in a system of the granular metal of tungsten grains embedded in an Al_2O_3 matrix [107]. While no quantitative analysis was reported for that system, it was apparent that the formation of larger single grains, at the expense of cluster formation (that may be controlled by proper annealing), can considerably reduce (for x just below x_c). The contribution of these networks in comparison with that of the continuous metallic network of the percolation cluster then yields a "sharper" percolation transition to the degree that it can be well fitted to continuum percolation theory (where one assumes infinitely small partially overlapping spheres that cover a part of the space [27]). In fact, we have observed [104] in our study of Si NCs ensembles all the $\sigma(x)$ behaviors that were reported previously for the granular $W-Al_2O_3$ system. However, as in the case of granular metals, since this regime involves a combination of structure and transport mechanism variations, its understanding beyond the DBTJ process [91] will also wait till the simpler extreme regimes are understood in more details, as mentioned above.

Turning to the high-density regime, we encounter a situation that is adjacent to Si NCs "touch" and we can safely assume that the current paths are via connected chains of "touching" NCs. We found in our study that this "touching" is quite different from the situation in granular metals where when particles coalesce [108] no potential barrier is formed between them. In Si NCs, this "touching," while taking place over a distance that is of the order of a lattice constant (0.5 nm), it involves, at least, a different crystallographic orientation of the "touching" crystallites. As far as we know, our study of the transport in this regime is the first to consider this inter-Si NC conduction mechanism.

This new mechanism is manifested by a temperature-activated process that can be described by a concave to a convex log σ–log T transition. This may be a local barrier effect that dominates the local critical resistor in the network, such that a very special type of tunneling barrier between two adjacent particles is involved [42], or a global effect that has to do with a disordered semiconductor-like conduction in the system [32]. In the first case we do not know how such a barrier comes about between two "touching" NCs but we may attribute their special shape to the nanosize with which we are dealing in the system mentioned here. We should emphasize that if this is indeed the observed effect, it is very different from the situation with macroscopic barriers such as the classical parabolic barriers that are encountered in polycrystalline Si [39], where the barrier is formed by the space charge that exists between the "touching" crystals [42]. On the other hand, the global mechanism that was suggested to explain such a $\sigma(T)$ dependence is simple, in principle, but the role of the "touch" between the NCs in such a model is not clear. This question is now studied intensively by our group. However, again, our possibility possibility seems to be the accumulated charge in parts of the network as we suggested previously [95]. However, again, more local probe spectroscopies will be needed in order to follow either of these suggestions. We also note that the turn to a convex log σ–log T dependence, at higher temperatures, appears to be a carrier excitation effect rather than a mobility excitation effect. This is based on its resemblance to findings from a-Si [44], a-Si:H [116], and μc-Si:H [117] structures. The most likely donor in samples that have not been intentionally doped appears to be oxygen [116, 118, 119]. The fact that in our samples of P-doped and B-doped Si NCs, the observed (at the high-temperature range) activation energies are lower than those of the undoped samples confirms these expectations.

Before concluding, let us point out that the contribution of the relatively large number of conduction mechanisms is expected to yield a very rich set of behaviors that have not been reported previously for this system. To illustrate that and to make the clear distinction between the present system and granular metals, we mention three of our corresponding observations. Very recently, we have found that the doping of the Si NCs system can be achieved and manifested not only in the PL as reported previously [120] but also in the contribution to the carrier concentration as is the case in bulk semiconductors. This result seems to be at variance with theoretical expectations [121] of doping only the NCs surface, such that, as in amorphous (nonhydrogenated) Si, the dopant is expected to saturate bonds rather than contribute to the conduction [122]. As mentioned above, our studies seem to indicate that the origin of the carriers that contribute to the conductivity of the nonintentionally doped NCs is as in other disordered Si systems [116–119] the oxygen-induced donors. Another phenomenon that we have observed recently is the long-time (hours) conductivity relaxation effects that are very similar to those found on some porous silicon structures [123]. So far, these effects have been well characterized, but no specific mechanism was suggested in order to account for the resemblance of the observed behaviors to those of an electronic glass out of equilibrium [124]. While the detailed out-of-equilibrium behavior in our Si NCs system has not been analyzed yet, it appears to us that, as in other nanostructures out of equilibrium, the hopping and charging mechanisms that we described above can be the basis for the interpretation of these results. Finally, our finding [95] of an anomalous photovoltaic effect in Si NCs ensembles in the proximity of x_c (just above it) was interpreted by us as a charge separation effect in parts of the percolation cluster where the interplay between the quantum confinement effect and the Coulomb blockade effect becomes important in the presence of a well-defined direction for the size variation of the NCs. The electron– hole charge separation and the range of this separation is expected to be determined by the difference between the energy level splitting in the conduction band and in the valence band (due to the smaller effective "effective mass" of the electrons [50]). Further local probe measurements may confirm or disprove this model as well as elucidate the inter-NCs conduction between crystallites of different sizes.

In conclusion, we attempted to provide a framework for the discussion of the transport mechanisms in 3D ensembles of Si NCs that are embedded in an insulating matrix. Although the rather few relevant data and complexity of the system do not allow, at present, to pinpoint or quantify the transport mechanisms in this system, we can categorize them and suggest gross models for them according to the density of the NCs. In general, while tunneling conduction and Coulomb blockade effects are apparent for all densities, the effect of quantum confinement (that is clearly evident in the optical data) is not as well established from the electrical data. It appears, however, that hopping conduction, under Coulomb blockade and resonant tunneling, dominates the global conductance in the low-density regime, while the interfaces between touching NCs control the global conductivity in the higher density regimes. The findings for the high NC density end, the hopping that exists at the low NC density end, and the DBTJ behavior, found at the percolation threshold, are the three fundamental transport mechanisms on which we suggest to base the understanding of the transport mechanism in the systems reviewed in this chapter.

Acknowledgments

This work was supported in part by the Israel Science Foundation (ISF) and in part by the Israeli Ministry of Science. The author would like to thank I.V. Antonova, Y. Goldstein, O. Millo, A.G. Nassiopoulou, and A. Sa'ar for their valuable contribution to the work reported in this chapter. The author is holding the Enrique Bergman chair in Solar Energy Research at the HU.

References

1 Lockwood, D.J. (ed.) (1998) *Light Emission in Silicon: From Physics to Devices*, Academic Press, New York.

2 Kovalev, D., Heckler, H., Polsski, G., and Koch, F. (1999) *Phys. Stat. Sol. (b)*, **215**, 871.

3 Bissi, O., Ossicini, S., and Pavesi, L. (2000) *Surf. Sci. Rep.*, **38**, 1.

4 Nassiopoulou, A.G. (2004) *Encycl. Nanosci. Nanotechnol.*, **9**, 793.

5 Balberg, I. (2008) *J. Nanosci. Nanotechnol.*, **8**, 745.

6 Kolek, A. (1994) *J. Phys.: Condens. Matter*, **6**, 469.

7 Abeles, B., Sheng, P., Coutts, M.D., and Arie, Y. (1975) *Adv. Phys.*, **24**, 407.

8 Abeles, B. (1976) *Appl. Solid State Sci.*, **6**, 1.

9 Fujii, M., Mamezaki, O., Hayashi, S., and Yamomoto, K. (1998) *J. Appl. Phys.*, **83**, 1507.

10 Gaponenko, S.V. (1998) *Optical Properties of Semiconductor Nanocrystals*, Cambridge University, Cambridge.

11 Grabert, H. and Devoret, M.H. (eds) (1991) *Single Electron Tunneling*, Plenum, New York.

12 Stafford, C.A. and Das Sarma, S. (1994) *Phys. Rev. Lett.*, **72**, 3590.

13 Burr, T.A., Seraphin, A.A., Werwa, E., and Kolenbraner, K.D. (1997) *Phys. Rev. B*, **56**, 4818.

14 Walters, R.J., Bourianoff, G.I., and Atwater, H.A. (2005) *Nat. Mater.*, **4**, 143.

15 Lockwood, D.J. and Pavesi, L. (2004) *Silicon Photon. Top. Appl. Phys.*, **94**, 1.

16 Nishiguchi, K. and Oda, S. (2000) *J. Appl. Phys.*, **88**, 4186.

17 Tan, T.Y., Kamiya, T., Durrani, Z.A.K., and Ahemed, H. (2003) *J. Appl. Phys.*, **94**, 633.

18 Khalafalla, M., Mizuta, H., Oda, S., and Durrani, Z.A.K. (2006) *Curr. Appl. Phys.*, **6**, 536.

19 Wan, A., Wang, T.H., Zhu, M., and Lin, C.L. (2002) *Appl. Phys. Lett.*, **81**, 538.

20 Kim, K. (1998) *Phys. Rev. B*, **57**, 13072.

21 Shalchian, M., Grisolia, J., BenAssayag, G., Coffin, H., Atarodi, S.M., and Claverie, A. (2005) *Sold State Electron.*, **49**, 1198.

22 Stielerl, D., Dalal, V.L., Muthukrishnan, K., Noack, M., and Schares, E. (2006) *J. Appl. Phys.*, **100**, 036106.

23 Lubianiker, Y. and Balberg, I. (1997) *Phys. Rev. Lett.*, **78**, 2433.

24 Reshotko, M.R., Sa'ar, A., and Balberg, I. (2003) *Phys. Stat. Sol. (a)*, **197**, 113.

25 Urbach, B., Axelrod, E., and Sa'ar, A. (2007) *Phys. Rev. B*, **75**, 205303.

26 Stauffer, D. and Aharony, A. (1994) *Introduction to Percolation Theory*, Taylor and Francis, London.

27 Balberg, I. (2009) *Springer Encyclopedia of Complexity* Vol. **2**, Percolation (ed. M. Sahimi), Springer, Berlin.

28 Shklovskii, B.I. and Efros, A.L. (1984) *Electronic Properties of Doped Semiconductors*, Springer-Verlag, Berlin.

29 Johner, N., Grimaldi, C., Balberg, I., and Ryser, P. (2008) *Phys. Rev. B*, **77**, 174204.

30 Balberg, I. (2008) *J. Phys. D.*, **42**, 064003.

31 Wiesendanger, R. (1994) *Scanning Probe Microscopy and Spectroscopy*, Cambridge University Press, Cambridge.

32 Mott nd, N.F. and Davis, E.A. (1979) *Electron Processes in Non-Crystalline Materials*, Clarendon, Oxford.

33 Balberg, I., Azulay, D., Toker, D., and Millo, O. (2004) *Int. J. Mod. Phys. B*, **18**, 2091.

34 Kastner, M.A. (1992) *Rev. Mod. Phys.*, **64**, 849.

35 Sheng, Ping and Klafter, J. (1983) *Phys. Rev. B*, **27**, 2583.

36 Simanek, E. (1981) *Solid State Commun.*, **40**, 1021.

37 Berthelot, M. (1862) *Ann. Chim. Phys.*, **66**, 110.

38 Mares, J.J., Kristofik, J., and Smid, V. (1992) *Semicond. Sci. Tecnol.*, **7**, 119.

39 Street, R.A. (1976) *Adv. Phys.*, **25**, 397.

40 Mares, J.J., Kristofik, J., Pangrac, J., and Haspodkova, A. (1993) *Appl. Phys. Lett.*, **63**, 180.

41 Stewart, D.R., Ohlberg, D.A.A., Beck, P.A., Lau, C.N., and Stanley Williams, R. (2005) *Appl. Phys. A*, **80**, 1379.

42 Kapoor, M., Singh, V.A., and Johri, G.K. (2000) *Phys. Rev. B*, **61**, 1941.

43 Houtepen, A.J., Kockmann, D., and Vanmaekelbergh, D. (2008) *Nanoletters*, **8**, 3516.

44 Yakimov, A.I., Stepina, N.P., and Dvurechenskii, A.V. (1994) *J. Phys.: Condens. Matter*, **6**, 2583.

45 Glazman, L.I. and Matveev, K.A. (1988) *Sov. Phys. JETP*, **67**, 1276.

46 Sze, S.M. and Ng, K.K. (2007) *Physics of Semiconductor Devices*, John Wiley & Sons, Inc., Hoboken, NJ.

47 Ren, S.Y. and Dow, J.D. (1992) *Phys. Rev. B*, **45**, 6492.

48 Hill, N.A. and Whaley, K.B. (1996) *J. Electron. Mater.*, **25**, 269.

49 Huang, S., Banerjee, S., Yung, T.T., and Oda, S. (2003) *Appl. Phys. Lett.*, **94**, 7261.

50 Mahdouani, M., Bourguiga, R., and Jaziri, S. (2008) *Physica E*, **42**, 228.

51 Banin, U. and Millo, O. (2003) *Annu. Rev. Phys. Chem.*, **54**, 465.

52 Chandller, R.E., Houtepen, A.J., Nelson, J., and Vanmeakelbergh, D. (2007) *Phys. Rev. B*, **75**, 085325.

53 Toker, D., Balberg, I., Zelaya-Angel, O., Savir, E., and Millo, O. (2006) *Phys. Rev. B*, **73**, 045317.

54 Raisky, O.Y., Wang, W.B., Alfano, R.R., Reynolds, C.L., Stampone, D.V., and Focht, M.W. (1999) *Appl. Phys. Lett.*, **74**, 129.

55 Niquet, Y.M., Delerue, C., Lannoo, M., and Alan, G. (2001) *Phys. Rev. B*, **64**, 113305.

56 Rafiq, M.A., Tsuchiya, Y., Mizuta, H., Oda, S., Uno, S., Durnani, Z.A.K., and Milne, W.I. (2006) *J. Appl. Phys.*, **100**, 014303.

57 Balberg, I., Savir, E., Dover, Y., Portillo Moreno, O., Lozada-Morales, R., and Zelaya-Angel, O. (2007) *Phys. Rev. B*, **75**, 153301.

58 Baron, T., Gentile, T., Magnea, N., and Mur, P. (2001) *Appl. Phys. Lett.*, **79**, 1175.

59 Salem, M.A., Mizuta, H., Oda, S., Fu, Y., and Willander, M. (2005) *Jap. J, Appl. Phys.*, **44**, L88.

60 Salem, M.A., Mizuta, H., and Oda, S. (2004) *Appl. Phys. Lett.*, **85**, 3262.

61 Feng, T., Yu, H., Dicken, M., Heath, J.R., and Atwater, H.A. (2005) *Appl. Phys. Lett.*, **86**, 033103.

62 Makihara, K., Xu, J., Ikeda, M., Murakami, H., Higashi, S., and Miyazaki, S. (2006) *Thin solid Films*, **508**, 186.

63 Dongs, E.M., Skowronski, C.S., and Farmer, K.R. (1998) *Appl. Phys. Lett.*, **73**, 3712.

64 Tanaka, Y.A., Fujii, M., Hayashi, S., and Yamamoto, K. (1999) *J. Appl. Phys.*, **86**, 3199.

65 Rokhinson, L.P., Guo, L.J., Chou, S.Y., and Tsui, D.C. (2000) *Appl. Phys. Lett.*, **76**, 1591.

66 Shi, Y., Yuan, X.L., Wu, J., Bu, H.M., Yang, H.G., Han, P., Zheng, Y.D., and Hiramoto, T. (2000) *Supperlatt. Microstruct.*, **28**, 387.

67 Shi, J., Wu, L., Huang, X., Liu, J., Ma, Z., Li, W., Li, X., Xu, J., Li, A., and Chen, K. (2002) *Solid State Commun.*, **123**, 437.

68 Huang, S., Banerjee, S., Tung, R.T., and Oda, S. (2003) *J. Appl. Phys.*, **94**, 7261.

69 Yu, Z., Aceves, M., Carrillo, J., and Flores, F. (2003) *Nanotechnology*, **14**, 959.

70 Puglisi, R.A., Lombardo, S., Ammendola, G., Nicota, G., and Gerardi, C. (2003) *Mater. Sci. Eng.*, **23**, 1047.

71 Yu, L.W., Chen, K.J., Wu, L.C., Dai, M., Li, W., and Huang, X.F. (2005) *Phys. Rev. B*, **71**, 245305.

72 Luna-Lopez, A., Aceves-Mijares, M., Malik, O., and Glaenzer, R. (2005) *J. Vac. Sci. Technol. A*, **3**, 534.

73 Grisolia, J., Shalchian, M., BenAssayage, G., Coffin, H., Bonafos, C., Schamm, S., Atarodi, S.M., and Calverie, A. (2005) *Mater. Sci. Eng. B*, **124–125**, 494.

74 Klein, D.L., Roth, R., Lim, A.K.L., Alvisatos, A.P., and McEuen, P.L. (1997) *Nature*, **389**, 699.

75 Kapaetanakis, E., Normand, P., Tsoukalas, D., Beltsios, K., Stoemenos, J., Zhang, S., and van der Berg, J. (2000) *Appl. Phys. Lett.*, **77**, 3450.

76 Nishiguchi, K. and Oda, S. (2000) *J. Appl. Phys.*, **88**, 4186.

77 Augke, R., Eberhardt, W., Single, C., Prins, F.E., Wharam, D.A., and Kern, D.P. (2000) *Appl. Phys. Lett.*, **76**, 2065.

78 Gunther, A., Khoury, M., Milicic, S., Vasileska, D., Thornton, T., and Goodnick, S.M. (2000) *Suprlatt. Microstruct.*, **27**, 373.

79 Fujii, M., Inoue, Y., Hayashi, S., and Yamamoto, K. (1996) *Appl. Phys. Latt.*, **68**, 3749.

80 Wang, T., Wei, W., Xu, G., and Zhang, C. (2002) *J. Mod. Phys. B*, **16**, 4289.

81 Banerjee, S. (2002) *Physica E*, **15**, 164.

82 Taguchi, M., Tsutsumi, Y., Bhatt, R.N., and Wagner, S. (1996) *J. Non-Cryst. Solids*, **198–200**, 899.

83 Lombardo, S., Coffa, S., Bongiorno, C., Spinella, C., Castagna, E., Sciuto, A., Gerardi, C., Ferrari, F., Fazio, B., and Privitera, S. (2000) *Mater. Sci. Eng. B*, **69–70**, 295.

84 Concari, S.B. and Buitrago, R.H. (2004) *J. Non-Cryst. Solids*, **338–340**, 331.

85 Busseret, C., Souifi, A., Baron, T., Monfray, S., Buffet, N., Gautier, E., and Semeria, M.N. (2002) *Mater. Sci. Eng. C*, **19**, 237.

86 De la Torre, J., Souifi, A., Lemiti, M., Poncet, A., Busseret, C., Guillot, G., Bremond, G., Gonzalez, O., Garrido, B., and Morante, J.R. (2003) *Physica E*, **17**, 604.

87 Miller, N.C. and Shivn, A.G. (1967) *Appl. Phys. Lett.*, **10**, 86.

88 Hanak, J.J. (1970) *J. Mater. Sci.*, **5**, 964.

89 Antonova, I.V., Gulyaev, M.B., Yanovitskaya, Z.Sh., Valodin, V.A., Marin, D.V., Efremov, M.D., Goldstein, Y., and Jedrzejewski, J. (2006) *Semiconductors*, **40**, 1198.

90 Posada, Y., Balberg, I., Fonseca, L.F., Resto, O., and Weisz, S.Z. (2001) *Mater. Res. Soc. Symp. Proc.*, **638**, F14.44.

91 Balberg, I., Savir, E., Jedrzejewski, J., Nassiopoulou, A.G., and Gardelis, S. (2007) *Phys. Rev. B*, **75**, 225329.

92 Antonova, I.V., Gulyaev, M., Savir, E., Jedrzejewski, J., and Balberg, I. (2008) *Phys. Rev. B*, **77**, 125318.

93 Dovrat, M., Goshen, Y., Popov, I., Jedrzejewski, J., Balberg, I., and Sa'ar, A. (2005) *Phys. Stat. Sol. (c)*, **2**, 3440.

94 Sa'ar, A., Dovrat, M., Jedrzejewski, J., and Balberg, I. (2007) *Physica E*, **38**, 122.

95 Levi Aharoni, H., Azulay, D., Millo, O., and Balberg, I. (2008) *Appl. Phys. Lett.*, **92**, 112109.

96 Dovrat, M., Oppenheim, Y., Jedrzejewski, J., Balberg, I., and Sa'ar, A. (2004) *Phys. Rev. B*, **69**, 155311.

97 Sa'ar, A., Reichman, Y., Dovrat, M., Kraf, D., Jedrzejewski, J., and Balberg, I. (2005) *Nanoletters*, **5**, 2443.

98 Nicolian, E.N. and Brews, J.R. (1981) *MOS (Metal Oxide Semiconductor) Physics and Technology*, John Wiley & Sons, Inc., New York.

99 Luna-Lopez, A., Aceves-Mijaes, M., Malik, O., and Glaenzer, R. (2005) *J. Vac. Sci. Technol. A*, **23**, 534.

100 Zambuto, M. (1989) *Semiconductor Devices*, McGraw-Hill, New York.

101 Azulay, D., Balberg, I., Chu, V., Conde, J.P., and Millo, O. (2005) *Phys. Rev. B*, **71**, 113304.

102 Reshotko, M.R. and Balberg, I. (2001) *Appl. Phys. Lett.*, **78**, 763.

103 Balberg, I. (2008) *Phys. Stat. Sol. (c)*, **5**, 3771.

104 Balberg, I. (2005) *Proceedings of the first International Workshop on Semiconductor Nanocrystals*, Hungarian Academy of Science, Budapest, p. 291.

105 Huang, S., Banerjee, S., Tung, R.T., and Oda, S. (2000) *J. Appl. Phys*, **93**, 578.

106 Han, K., Kim, I., and Shin, H. (2001) *IEEE Trans. Electr. Devices*, **48**, 874.

107 Abeles, B., Pinch, H.L., and Gittleman, J.I. (1975) *Phys. Rev. Lett.*, **35**, 247.

108 Feng, D.I., Kohn, T.J., Wakoh, K., Hirose, T., and Sumiyama, K. (2001) *Appl. Phys. Lett.*, **78**, 1535.

109 Mantese, J.V., Goldburg, W.I., Darling, D.H., Craighead, H.G., Gibson, U.J., Buhrman, R.A., and Webb, W.W. (1981) *Solid State Commun.*, **37**, 353.

110 Balberg, I., Wagner, N., Goldstein, Y., and Weisz, S.Z. (1990) *Mater. Res. Soc. Symp. Proc.*, **195**, 233.

111 Fonseca, L.F. and Balberg, I. (1993) *Phys. Rev. B*, **48**, 14915.

112 Gross, B. and de Figueireda, M.T. (1995) *J. Phys. D: Appl. Phys.*, **18**, 617.

113 Kamal, A.H.M., Lutzen, J., Sanborn, B.A., Sidorov, M.V., Kozicki, M.N., Smith, D.J., and Ferry, D.K. (1998) *Semicond. Sci. Technol.*, **13**, 1328.

114 Ciurea, M.L., Tedorescu, V.S., Iancu, V., and Balberg, I. (2006) *Chem. Phys. Lett.*, **423**, 225.

115 Alonso, M.I., Garriga, M., Bernardi, A., Goni, A.F., Lopeandia, A.F., Garcia, G., Rodriguez-Viejo, J., and Labar, J.L. (2008) *Phys. Stat. Sol. (a)*, **205**, 888.

116 Balberg, I., Naidis, R., Fonseca, L., Weisz, S.Z., Conde, J.P., Alpuim, P., and Chu, V. (2001) *Phys. Rev. B*, **63**, 113201.

117 Balberg, I., Dover, Y., Naides, R., Conde, J.P., and Chu, V. (2004) *Phys. Rev. B*, **69**, 035203.

118 Paesler, N.A., Anderson, D.A., Freeman, E.C., Moddel, G., and Paul, W. (1975) *Phys. Rev. Lett.*, **41**, 1492.

119 Kamei, T. and Wada, T. (2004) *J. Appl. Phys.*, **96**, 2087.

120 Fujii, M., Mimura, A., Hayashi, S., Yamamoto, K., Urakawa, C., and Ohta, H. (2000) *Appl. Phys. Lett.*, **87**, 1855.

121 Ossicini, S., Degoli, E., Iori, F., Magri, R., Canlele, G., Trani, F., and Ninno, D. (2004) *Appl. Phys. Lett.*, **87**, 173120.

122 Street, R.A. (1982) *Phys. Rev. Lett.*, **49**, 1187.

123 Borini, S., Boarino, L., and Amato, G. (2007) *Phys Rev. B*, **75**, 165205.

124 Grenet, T., Dehaye, J., Sabra, M., and Gay, F. (2007) *Eur. Phys. J. B*, **56**, 183.

5

Thermal Properties and Heat Transport in Silicon-Based Nanostructures

Han-Yun Chang and Leonid Tsybeskov

5.1
Introduction

For the past 40 years, crystalline Si (c-Si) continues to be the major material for microelectronics, and modern silicon technology is superior compared to other semiconductors (e.g., II–VI and III–V compounds). In addition to the unique electronic and structural properties of bulk c-Si, silicon dioxide (SiO_2) and Si/SiO_2 interfaces, single-crystal Si possesses one of the best known lattice thermal conductivity [1, 2]. This exceptional heat conductance is critically important for Si device heat management and circuit reliability. However, most of the modern complementary metal–oxide–semiconductor (CMOS) platforms are no longer single-crystal Si wafers but rather thin layers of Si-on-insulator (SOI), ultrathin strained Si and SiGe heterostructures that are the foundation of SiGe bipolar transistors (HBTs), and high-mobility metal-oxide–semiconductor field-effective transistors (MOSFETs). Major properties of these Si-based nanostructures are very different from those of bulk c-Si. For example, thermal conductivity in ultrathin SOI layers, SiGe alloys, and Si/SiGe nanostructures could be reduced by more than an order of magnitude compared to that in c-Si [3–6], and heat dissipation has become an important issue for modern nanoscale electronic devices and circuits. Thus, the understanding and improvement of heat management in Si-based nanostructures is critically important for the evolution of microelectronic industry.

On the other hand, many interesting applications of nanostructured Si (ns-Si) in photonic devices and CMOS-compatible light emitters were recently discussed [7–11]. These ns-Si materials and devices can be produced by electrochemical anodization (i.e., porous Si [12]), chemical vapor deposition (CVD) using thermal decomposition of SiH_4 [13–15], Si ion implantation into a SiO_2 matrix [16], and deposition of amorphous Si/SiO_2 layers followed by thermal crystallization [17–19]. These ns-Si materials and devices produce an efficient and tunable light emission in the near-infrared and visible spectral region [20, 21]. Also, it has been shown that under photoexcitation with energy density $>10\,mJ/cm^2$, optical gain is possibly

Silicon Nanocrystals: Fundamentals, Synthesis and Applications. Edited by Lorenzo Pavesi and Rasit Turan
Copyright © 2010 WILEY-VCH Verlag GmbH & Co. KGaA, Weinheim
ISBN: 978-3-527-32160-5

achievable [14, 22–25]. Such energy density absorbed by a submicrometer film generates a lot of heat. Thus, thermal properties and heat dissipation in ns-Si-based photonic devices need to be understood in detail.

Other ns-Si systems of interests are one-dimensional (1D) (e.g., Si and SiGe nanowires (NWs) [26–28]) and three-dimensional (3D) (e.g., cluster morphology Si/SiGe cluster multilayers separated by approximately 10 nm thick Si layers [29–31]) nanostructures. Many device applications were recently proposed (e.g., Si nanowire electrical interconnects [32, 33], Si nanowire thermoelectric devices [34, 35], Si nanowire sensors [36–38], SiGe cluster-based memory devices [39–41], SiGe cluster-based near-infrared light emitters [42–46], etc.), and detailed understanding of thermal processes in these nanostructures is absolutely necessary.

In general, thermal energy in a solid is transported by means of particles or quasiparticles moving to restore thermodynamic equilibrium in a system with a temperature gradient ∇T. In metals, heat is mainly conducted by mobile electrons (i.e., electronic thermal conductivity $\varkappa_e$), while in insulators and lightly doped semiconductors, heat is primarily transported by phonons. Phonons are quasiparticles representing lattice vibrations, and phonon processes are responsible for the lattice thermal conductivity $\varkappa_l$. An ideal pure single crystal is an excellent thermal conductor because phonon inharmonicity (i.e., inelastic phonon scattering) is the only phonon mean free path limitation. When electronically active dopants (donors and acceptors) and other impurities are introduced into a pure single crystal, phonon scattering associated with impurity atoms severely reduces the lattice conductivity. For crystallites with reduced dimensions, such as micro- or nanocrystals (nc), the conductivity decreases as the grain size of the crystallites decreases, mainly due to phonon scattering at the grain boundaries. This effect becomes dominant as the grain size approaches the phonon mean free path in a single crystal. Thermal conductivity in amorphous Si (a-Si) is drastically reduced, mainly because the long-range order in a nearly ideal Si crystal lattice is replaced with an amorphous network of Si atoms without pure phonon modes. Furthermore, amorphous materials might have a significant number of structural defects providing additional phonon scattering. Similarly, crystalline alloys (e.g., SiGe, SiCGe, etc.) have strongly reduced thermal conductivity due to compositional disorder. Figure 5.1 summarizes experimental results on thermal conductivity in different forms of c-Si, SiGe alloy-based nanostructures, a-Si and ns-Si-based materials [3, 47–49]. In these Si-based nanostructures and compounds, thermal conductivity is dramatically reduced compared to that in c-Si, from approximately 10^3 to less than 10^{-1} W/(m K) at $T = 100$ K. For example, thermal conductivity of nc-Si/SiO$_2$ multilayers is, probably, one of the lowest thermal conductivities in nonporous solids; it is comparable with thermal conductivity of foam insulation and air (Figure 5.1). In addition, thermal conductivity temperature dependence is found to be qualitatively different in c-Si, Si-based alloys, amorphous materials, and nanostructures, which will be explained later.

We begin this chapter with a very brief introduction of the classical kinetic theory that is the foundation of thermal physics. We mainly consider lattice thermal conductivity in semiconductors and specifically focus on phonon boundary scattering, which is primarily responsible for the observed thermal conductivity reduction

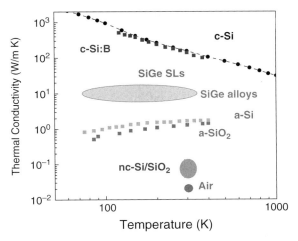

Figure 5.1 Temperature dependence of thermal conductivity of Si-based materials and nanostructures, including c-Si, a-Si, a-SiO$_2$, lightly B-doped c-Si (4×10^{14} cm^{-3}), Si/Si$_{0.7}$Ge$_{0.3}$ SL (200 Å/100 Å), Si$_{0.7}$Ge$_{0.3}$ alloy, and nc-Si/a-SiO$_2$ SL. Note the double logarithmic scale. Data compiled using Refs [3,47–49].

in different forms of Si nanostructures (Figure 5.1). A brief overview of thermal conductivity measurement techniques will be provided in the subsequent section. Finally, we will discuss and compare thermal properties of several major forms of Si-based nanostructures including:

ultrathin SOI and Si nanowires
epitaxially grown Si/SiGe cluster nanostructures
porous Si prepared by electrochemical anodization
nc-Si/SiO$_2$ multilayers

5.2
Thermal Conductivity in Bulk Solids and Nanostructures

5.2.1
Kinetic Theory: Thermal Properties and Heat Flow

It can be found in major textbooks that for a long rod with different temperatures at each end, thermal conductivity $\varkappa$ can be characterized by a steady state of heat flow, directly proportional to temperature gradient ∇T [50, 51]:

$$j_U = -\varkappa \nabla T, \tag{5.1}$$

where j_U is the flux of thermal energy or the energy transmitted across unit area per unit time. Heat capacity of one particle can be defined as $c \equiv \partial E/\partial T$, where E is the energy of the particle; $C = n \cdot c$ is the total heat capacity, where n is the concentration of the particles. The rate of energy change under the presence of a temperature gradient is

$$\frac{\partial E}{\partial t} = c \frac{\partial T}{\partial t} = c \vec{v} \cdot \nabla T, \tag{5.2}$$

where $\vec{v}$ is the velocity at which the particle travels. The mean free path, at which a particle can travel before being scattered, is $l \equiv v\tau$, where τ is the phonon relaxation time. Hence, the total thermal energy flux per unit area is

$$j_U = -n \cdot c \langle \vec{v} \cdot \vec{v} \rangle \tau \nabla T = -\frac{1}{3} n \cdot c \langle v^2 \rangle \tau \nabla T. \tag{5.3}$$

Comparing with Eq. (5.1), thermal conductivity can be written as

$$\varkappa = \frac{1}{3} n \cdot c \langle v^2 \rangle \tau = \frac{1}{3} Cvl. \tag{5.4}$$

This definition is true for all particles that contribute to heat conduction in solid, and the total thermal conductivity is

$$\varkappa = \frac{1}{3} \sum_i C_i v_i l_i. \tag{5.5}$$

5.2.2
Lattice Thermal Conductivity

Lattice thermal conductivity is provided by phonons and is the prominent heat conduction factor in insulators, intrinsic and lightly doped semiconductors. Phonon dispersion relationship $\omega(q)$, where ω is the phonon frequency and q is the wave vector, can be found in most solid-state physics books, and c-Si phonon dispersion is shown in Figure 5.2 [50, 52].

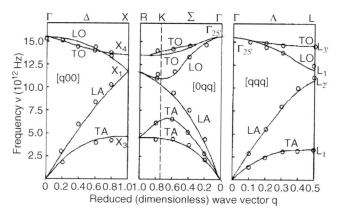

Figure 5.2 Phonon dispersion curves in Si along high-symmetry axes. The circles are experimental data points and the continuous curves are calculated with the adiabatic bond charge model. From Refs [50, 52].

Since phonon group velocity is $\vec{v}_{\mathrm{g}} = \partial \omega(q)/\partial q$, lattice heat conduction in solids is provided mainly through acoustic phonons. The heat flux of a phonon mode q is the product of the thermal energy of that mode and the group velocity $\vec{v}_{\mathrm{g}}$. For a phonon distribution N_q, the total thermal energy flux carried by all phonon modes is

$$j_{\mathrm{U}} = \sum_q N_q \hbar \omega \vec{v}_{\mathrm{g}}. \tag{5.6}$$

At thermal equilibrium, the average number of phonons with wave vector q is

$$N_0 = \frac{1}{\exp(\hbar \omega / k_{\mathrm{B}} T) - 1}. \tag{5.7}$$

In the presence of a temperature gradient, the Boltzmann statistics expects the phonon distribution N_q to undergo a scattering process in order to restore thermal equilibrium N_0 at a rate proportional to the differences from equilibrium:

$$\frac{N_q - N_0}{\tau} = -(\vec{v}_{\mathrm{g}} \cdot \nabla T) \frac{\partial N_0}{\partial T}. \tag{5.8}$$

Thus, Eq. (5.6) can be written as

$$j_{\mathrm{U}} = -\frac{1}{3} \sum_q \hbar \omega(q) v_{\mathrm{g}}^2 \tau \frac{\partial N_0}{\partial T} \nabla T. \tag{5.9}$$

And since $j_{\mathrm{U}} = -\varkappa \nabla T$, lattice thermal conductivity is found to be

$$\varkappa_{\mathrm{l}} = \frac{1}{3} \sum_q \hbar \omega(q) v_{\mathrm{g}}^2 \tau \frac{\partial N_0}{\partial T}. \tag{5.10}$$

Evaluation of Eq. (5.10) requires the precise phonon dispersion of a real crystal and the relaxation times for each phonon mode. The *Debye model* simplifies the calculation by replacing v_{g} with a constant phonon velocity v, which is roughly equal to the velocity of sound in solid and is the same in all phonon modes (i.e., spherically symmetric, dispersionless approximation) (Figure 5.3). The model is based on a linear dispersion

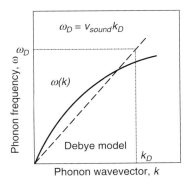

Figure 5.3 Schematic of phonon dispersion and the Debye model approximation.

relation, $\omega(q) = v \cdot q$, which is true for small q in crystals (i.e., near the center of the Brillouin zone). These approximations give the differential Debye heat capacity

$$C(x)dx = \frac{3k_B}{2\pi^2 v^3} \left(\frac{T}{\theta_D}\right)^3 \frac{x^4 e^x}{(e^x - 1)^2} dx, \tag{5.11}$$

where $x \equiv \hbar\omega/\tau = \hbar\omega/k_B T \equiv \theta_D/T$ and $\theta_D = \hbar\omega/k_B$ is the Debye temperature (e.g., for c-Si $\theta_D^{Si} = 645$ K [50]). The lattice thermal conductivity

$$\varkappa_l = \frac{k_B}{2\pi^2 v} \left(\frac{T}{\theta_D}\right)^3 \int\limits_0^x \tau(x) \frac{x^4 e^x}{(e^x - 1)^2} dx \tag{5.12}$$

stays in the basic form of $\varkappa_l = (1/3) \int_0^x C(x)vl(x)dx$ from simple kinetic equations. At low temperature ($T \ll \theta_D$), thermal conductivity follows $\varkappa_l \propto T^3$ dependence, which is similar to heat capacity temperature dependence at low temperature (Eq. (5.11)). At higher temperature, phonon collision significantly reduces the mean free path, and $\varkappa_l \propto 1/T^\alpha$ with $1 \leq \alpha \leq 2$. Thus, qualitatively, the thermal conductivity temperature dependence should be relatively the same for the majority of crystalline solids where lattice thermal conductivity is the dominant heat transport mechanism [53].

In an ideal single crystal at temperature $T > \theta_D$, lattice thermal conductivity limitations are primarily due to phonon–phonon interactions, that is, phonon collision. A phonon collision process has to satisfy the conservation of energy and momentum. In terms of phonon frequency and wave vector, it can be written as

$$\omega_1 + \omega_2 = \omega_3 \tag{5.13a}$$

$$\vec{q}_1 + \vec{q}_2 = \vec{q}_3 + \vec{g}. \tag{5.13b}$$

In Eq. (5.13b), if $\vec{g} = 0$, the process is called normal process or N-process, and the total initial and last crystal momenta are absolutely equal (i.e., the total momentum of the phonons is conserved). In contrast, in umklapp process (or U-process), crystal momenta differ by a nonzero reciprocal lattice vector $\vec{g} \neq 0$. Thus, N-process can be described as phonon scattering confined within the first Brillouin zone, while in U-process it is not. A schematic comparison of N- and U- processes is shown in Figure 5.4. Considering the process of phonon scattering within the ideal crystal lattice, we find that the modes of vibration are the same for wave vectors differing by one reciprocal lattice vector $\vec{g} = 2\pi/a$, where a is the lattice constant. In U-process, when two phonons scatter and the resultant $\vec{q}_3$ falls outside the first Brillouin zone, it must be brought back by adding or subtracting one phonon with wave vector equal to the reciprocal lattice vector $\vec{g}$. In a U-process, the thermal energy carried by the phonon group velocity of $\vec{q}_1$ and $\vec{q}_2$ is transferred to a very different (i.e., $\vec{q}_3$) direction in reciprocal space, and it quickly returns the phonon distribution to the equilibrium form. For N-process, however, $\vec{q}_3$ has an effective direction of energy flow carried by $\vec{q}_1$ and $\vec{q}_2$; thus, the process does not directly contribute to the system thermal resistance. In this approximation, without U-processes, thermal conductivity would be nearly infinite.

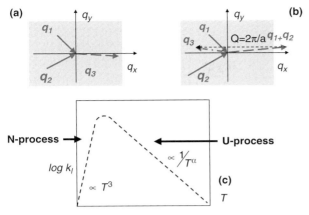

Figure 5.4 Schematic of (a) N- and (b) U-processes associated with (c) thermal conductivity limitations at low ($T \ll \theta_D$) and high ($T \gg \theta_D$) temperatures.

From Figure 5.4, both incident phonon wave vectors must be in the order of $1/2\vec{g}$ to possibly undergo U-process. At low temperature, the number of phonons qualified for U-process is limited, but it increases exponentially with temperature according to the Boltzmann statistics. Therefore, at sufficiently low temperature, the only scattering processes that occur at a reasonable rate are N-processes because U-processes are frozen out. In contrast, at high temperature ($T > \theta_D$), when all the phonon modes are excited, because $k_B T > \hbar\omega$, phonon scattering occurs mostly via U-processes.

The kinetic model explains reasonably well the thermal conductivity of bulk crystals and it immediately predicts thermal conductivity reduction for materials with reduced dimensions. Figure 5.5 compares thermal conductivity as a function of

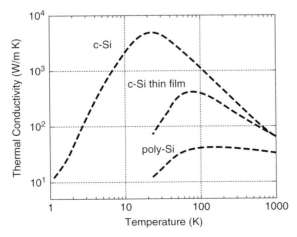

Figure 5.5 Temperature dependence of thermal conductivities of bulk c-Si, c-Si thin film, and poly-Si, demonstrating the effect of phonon boundary scattering in Si-based nanostructures. Note the double logarithmic scale. Data adopted from Refs [54–57].

temperature in bulk c-Si, c-Si thin films, and polycrystalline Si (poly-Si) films (data are compiled using Refs [54–57]). At low temperature, crystal lattice imperfections and grain boundaries in c-Si thin film and poly-Si limit the mean free path of the phonon and reduce the thermal conductivity dramatically, up to 100 times, compared to that in bulk c-Si. As temperature increases, the phonon mean free path eventually becomes comparable with the distance between lattice imperfections, grain size, and so on. Applying this observation to Figure 5.1, we conclude that in Si nanostructures, phonon collision at low and moderate temperatures no longer limits heat dissipation. Thus, lattice thermal conductivity as a function of temperature is mainly governed by grain boundaries, interface phonon scattering, and possibly other processes associated with changes in phonon dispersion.

5.2.3
Electronic Thermal Conductivity

The contribution of electronic thermal conductivity is significant for metals, heavily doped semiconductors, and devices operating with a high current density. A good example is carrier diffusion at high level of a nonuniform in-depth photoexcitation. In a crystal lattice, free electron has energy E_k as a function of the electron wave vector $\vec{k}$, which depends on the periodicity and lattice constant of the crystal [51]. The k-space is separated into Brillouin zones, and E_k is continuous within each zone and discontinuous at the zone boundaries, forming energy gaps. At equilibrium, the free electrons follow the Fermi–Dirac distribution and the average number of electron with energy E_k is

$$f_0 = \frac{1}{\exp(E_k - E_F/k_B T) + 1},\tag{5.14}$$

where E_F is the Fermi energy. In the presence of an electric field $\vec{E}$ and a temperature gradient ∇T, the electron distribution f_k undergoes scattering process in order to restore equilibrium:

$$\frac{f_k - f_0}{\tau(k)} = -\vec{v}_k \cdot \left(\frac{\partial f_0}{\partial T} \nabla T + e \frac{\partial f_0}{\partial E_k} \vec{E} \right),\tag{5.15}$$

where $\tau(k)$, $\vec{v}_k$, e are the relaxation time, the electron velocity, and the electron charge, respectively. The electrical current density and the flux of energy can be written as

$$\vec{J} = \int e v_k f_k d\vec{k}\tag{5.16}$$

and

$$\vec{Q} = \int (E_k - E_F) v_k f_k d\vec{k}.\tag{5.17}$$

In nondegenerated semiconductors, the Fermi level is inside the energy gap. The electron and hole distributions can be written, respectively, as

$$f_o^e \sim \exp\left(-\frac{E^e + E_F^e}{k_B T}\right) \tag{5.18a}$$

and

$$f_o^h \sim \exp\left(-\frac{E^h + E_F^h}{k_B T}\right). \tag{5.18b}$$

The electron and hole density and flux of energy can be written in similar form as Eqs (5.16) and (5.17). The electronic thermal conductivity can be found from the definition in Eq. (5.1) as

$$\varkappa_e = -\left[\frac{\vec{Q}}{\nabla T}\right]_{\vec{J}=0}. \tag{5.19}$$

If we define a set of general integrals

$$K_n = -\frac{1}{3}\int (\vec{v}_k)^2 \tau(k)(E_k - E_F)^n \frac{\partial f_0}{\partial E_k}\, dk, \tag{5.20}$$

then the electronic thermal conductivity and electrical conductivity can be found as

$$\varkappa_e = \frac{1}{T}\left[(K_2^e + K_2^h) - \frac{(K_1^e + K_1^h)^2}{K_0^e + K_0^h}\right] \tag{5.21}$$

and

$$\sigma = e^2(K_0^e + K_0^h), \tag{5.22}$$

where K_n^e, and K_n^h denote the integrals for electrons and holes.

To simplify the expression, we assume that the energy band is parabolic and the electron/hole relaxation times are

$$\tau_{e,g}(E) \propto E^a, \tag{5.23}$$

where a is a constant. The electronic thermal conductivity becomes

$$\varkappa_e = \sigma T\left(\frac{k_B}{e}\right)^2\left[\left(\frac{5}{2} + a\right) + \left(5 + 2a + \frac{E_G}{k_B T}\right)^2 \frac{n_e \mu_e n_h \mu_h}{(n_e \mu_e + n_h \mu_h)^2}\right]. \tag{5.24}$$

The second term in Eq. (5.24) is the bipolar diffusion term. In some intrinsic semiconductors, with electrons and holes having the same concentrations and similar mobilities, electron–hole pairs have enough time to recombine at the hot end and convert the bandgap energy E_G into thermal energy (assuming a low efficiency of radiative recombination). Thus, in narrow bandgap semiconductors, thermal conductivity may be governed by the bipolar diffusion. In heavily doped semiconductors or semiconductors with a large difference in electron and hole mobilities, the bipolar diffusion term becomes less important. However, it is important to note that under high level of photoexcitation that generates a large number of photo carriers, carrier diffusion might contribute to the heat dissipation process. Another interesting case is the situation when significant temperature gradient is responsible for a high current density, which is highly desirable in thermoelectric devices. However, since most of the semiconductor nanostructures have relatively low overall diffusion coefficient, electronic thermal conductivity is less important compared to lattice thermal conductivity.

5.3
Measurements of Thermal Conductivity in Nanostructures

In general, experimental investigation of thermal conductivity requires two steps: (i) heating, by introducing thermal energy into the system and (ii) sensing, by detecting the change in temperature or related physical properties due to an increase in thermal energy. Both heating and sensing can be done by two major techniques: electrical and optical. In electrical heating, thermal energy is provided by electrical heaters directly coupled to the specimen, while in optical heating, a focused laser beam is usually used as the energy source. Miniature electrical thermometers (preferably with heat capacity much smaller than that of the studied samples) are attached to the sample and used to measure the temperature gradient (Figure 5.6). Thus, one of the major challenges in measurements of nanostructure thermal properties is the problem with the thermometer dimension and its heat capacity. Owing to the extremely small volume (e.g., typical quantum dot might have volume as small as $\sim 10\,\mathrm{nm}^3$), the only possibility is to apply electrical measurements to a large ensemble of the nanoscale objects, for example, quantum dot films, quantum well multilayers, superlattices (SLs), and so on. Optical sensing methods, on the other hand, might be able to collect information more selectively; for example, optical signal (e.g., luminescence, Raman scattering, etc.) could be used to distinct between emission or scattering originated within a sample or its substrate, and to measure localized temperatures. In addition, it is possible to measure optical properties of individual nano-objects using, for example, near-field optics in fluorescence or Raman thermometry [58, 59].

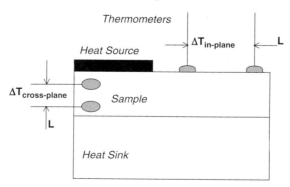

Figure 5.6 Schematic of a structure for steady-state measurements of thermal conductivity.

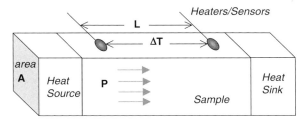

Figure 5.7 Schematic of heat flow in thermal conductivity measurements.

There are many variation of how the electrical heating and sensing elements can be assembled in the experiment. The most straightforward is the steady-state absolute method, which is illustrated in Figure 5.7. Thermal power P, generated by the heat source at the hot end provides heat that flows through a sample with the cross-sectional area A to the heat sink or the cold end; two thermometers are placed with the distance L apart and read the temperature difference ΔT. Thus, thermal conductivity can be calculated as [60]

$$\varkappa = P\frac{L}{A \cdot \Delta T}. \tag{5.25}$$

This setup, however, is limited in determining thermal conductivity of nano-structures. One of the major challenges is to find the exact $P = P_{in} - P_{loss}$ value: the input power can be determined accurately but the losses from heat radiation and thermal contacts between source, sample, and heat sink can be substantial. Moreover, this method measures only cross-sectional thermal conductivity and it cannot be used to study highly anisotropic nanostructures, for example, quantum wells and super-lattices. Also, as previously stated, most nanometer-dimension structures are fabricated on massive, micron-dimension substrates. The most accurate approach is to have the temperature sensor prepared directly beneath the nanostructure, which may not always be possible. Sometime, the substrate can be removed, but it is difficult to perform without damaging the thin film sample. To overcome this problem, several methods have been developed and will be discussed.

5.3.1
The 3ω Method

A thin strip of metal deposited onto a sample might work as both heaters and temperature sensors (Figure 5.8). In the 3ω method [61–65], an AC current $I(t) = I_0\cos(\omega t)$ with frequency ω and amplitude I_0 is passed through the deposited metal strip. Assuming that the resistance of the metal strip is R_h and it is linearly temperature dependent, we find that

$$R_h(t) = R_0[1 + C_{rt}T(t)], \tag{5.26}$$

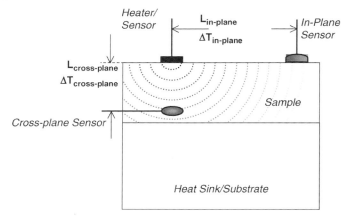

Figure 5.8 Schematic of an experimental setup with microfabricated heaters and sensors for 3ω and other electrical methods for in-plane or cross-plane thermal conductivity measurements.

where R_0 is the resistance without heating, C_{rt} is the resistance temperature coefficient, and $T(t)$ is the temperature of the metal strip. The power generated by the AC current can then be written as

$$P(t) = I^2R = I_0^2R_h\cos^2(\omega t) = \left(\frac{I_0^2R_h}{2}\right)_{\text{DC}} + \left(\frac{I_0^2R_h\cos(2\omega t)}{2}\right)_{2\omega} \tag{5.27}$$

and it contains a DC component and a 2ω modulated AC component. Similarly, the temperature increase in the metal strip and subsequently at the sample/strip contact will also contain (i) a DC with an amplitude T_{DC}, (ii) a 2ω AC part with an amplitude $T_{2\omega}$, and (iii) a phase shift φ:

$$T(t) = T_{\text{DC}} + T_{2\omega}\cos(2\omega + \varphi). \tag{5.28}$$

The resistance R_h can be rewritten as

$$R_h(t) = R_0\{1 + C_{rt}[T_{\text{DC}} + T_{2\omega}\cos(2\omega + \varphi)]\}$$
$$= [R_0(1 + C_{rt}T_{\text{DC}})]_{\text{DC}} + [R_0C_{rt}T_{2\omega}\cos(2\omega + \varphi)]_{2\omega}. \tag{5.29}$$

The voltage measured across the metal strip equals current times resistance:

$$V(t) = I(t)R_h(t) = V_{\text{source}} + V_{3\omega} + V_{1\omega}$$
$$= I_0R_0(1 + C_{rt}T_{\text{DC}})\cos(\omega t) + \frac{I_0R_0C_{rt}T_{2\omega}}{2}\cos(3\omega t + \varphi)$$
$$+ \frac{I_0R_0C_{rt}T_{2\omega}}{2}\cos(\omega t + \varphi). \tag{5.30}$$

The measurement includes voltages from the two heater DC resistance, modulated 1ω, and modulated 3ω components with a phase shift φ due to temperature rise. The 3ω voltage can be as small as three orders of magnitude less than the 1ω components but can be distinguished using a lock-in amplifier. The amplitude of the temperature increase $T_{2\omega}$ can be written as

$$T_{2\omega} = \frac{2V_{3\omega}}{I_0 R_0 C_{rt}} \approx \frac{2V_{3\omega}}{C_{rt} V_{1\omega}}. \tag{5.31}$$

To determine thermal conductivity of nanostructures such as thin films and superlattices, a simplified model is commonly used with the following approximations: linear source with heater width larger than the thickness of the cross section, one-dimensional frequency-dependent heat transport through the nanostructure, and two-dimensional (2D) heat transport in a semi-infinite substrate [61, 63, 66]. The temperature at the heater can be written as

$$T_{strip} = T_{subs} + \Delta T = T_{subs} + P\frac{t}{A}\frac{1}{\varkappa}, \tag{5.32}$$

where T_{subs} is temperature at the nanostructure/substrate interface, A is the cross-sectional area of the heater, and P, t, and $\varkappa$ are the power dissipated in, thickness, and thermal conductivity of the sample, respectively. T_{subs} can be calculated with a similar linear source and semi-infinite substrate model:

$$T_{subs} = \frac{P}{l\pi\varkappa_{subs}} \left[\frac{1}{2}\ln\left(\frac{\varkappa_{subs}}{C_{subs}(w/2)^2}\right) + \eta - \frac{1}{2}\ln(2\omega)\right], \tag{5.33}$$

where $\varkappa_{subs}$ and C_{subs} are thermal conductivity and heat capacity of the substrate, respectively.

The major advantage of the 3ω method is the temperature frequency dependence allows very high signal-to-noise ratio. Also, compared to the long time (minutes to even hours) to reach a constant heat flow in the steady-state method, the 3ω method provides a much more efficient measurement since the effect of a frequency-modulated temperature change can be detected after only a few periods of the input signal (as short as few seconds). By using a high enough frequency, the effect of AC heating can be confined to the area close to the heater, and it minimizes the effect of boundary conditions at the nanostructure/substrate interface. The AC signal is also less receptive to radiation loss, thus lowering the errors in the calculation of power. The limitation of the 3ω method is that the experiment only accounts for the cross-plane thermal conductivity. To avoid this limitation, the solution is to use a multiple strip 3ω setup for a complete profile of the device's thermal properties [67].

5.3.2
In-Plane Thermal Conductivity Measurements

In a nanoscale device, thermal conductivity varies by the geometry and the crystal orientation of the structure. The in-plane thermal conductivity is specifically important in a multilayered thin film structure where cross-plane thermal conductivity is usually very low. Although it can be determined by the aforementioned multiple striped 3ω method, the flow of the thermal energy in the cross-plane direction complicates the profiling. Several methods minimizing the effect of cross-plane heat conduction can be employed, especially when the in-plane thermal properties are of primary concern [68–71].

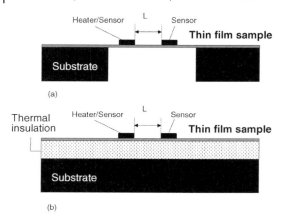

Figure 5.9 Schematic of structures for in-plane thermal conductivity measurements in thin films: (a) membranes and (b) thin films on thermal insulators.

The *membrane method* utilizes strips of electrical heaters and temperature sensors fabricated on top of the nanostructure [72–75]. The nanostructure is thermally isolated by removing a part of the substrate and leaving it suspended on a frame of the substrate material, which also works as a heat sink (Figure 5.9). Heat loss through radiation and convection can be minimized by conducting the experiment only with a small temperature increase and in a vacuum environment. Alternatively, to better ensure the thermal isolation and heat conduction in one direction, a cantilever beam setup can be used, rather than resting on an all-around frame. The nanostructure is suspended on the substrate at one end with a thermometer deposited on top while the heater is deposited on the opposite end.

The heaters and sensors are placed across the nanostructure, parallel to each other and perpendicular to the direction of the in-plane heat conduction under investigation. To reduce heat loss along the heaters and sensors, their cross-sectional dimensions are minimized. The heater can be used to convert steady-state, pulsed, or modulated electric signals to thermal energy. The sensors can be used to analyze the nanostructure's response to thermal gradient and its thermal conductivity and heat capacity. Thermal conductivity under steady-state heating can be calculated as

$$\varkappa = P\frac{L}{A \cdot \Delta T} = P\frac{L}{wd(T_\mathrm{h} - T_\mathrm{s})}, \tag{5.34}$$

where w is the length of the heater, d is the thickness of the sample membrane, P is power dissipated in the heater, L is the distance between heater and the heat sink, T_h is the heater temperature rise, and T_s is the temperature of the sink.

When removing the substrate is unfeasible, different methods can be used, although the model used to fit the result may become complex. One technique is to separate the nanostructure and the substrate with a low thermal conductivity layer (Figure 5.9b). A heater/sensor setup similar to the membrane method can be fabricated on top or in the separating layer. When the thermal energy enters the nanostructure, it is forced to move laterally, rather than into the substrate. Similar lateral conduction can be done by reducing the width of the heater [74].

5.3.3
Pump-Probe and Other Optical Measurements

Optical heating methods require precise determination of the exact amount of thermal energy absorption. Often, thermal diffusivity $D = \varkappa/C$ is obtained rather than direct measurement of thermal conductivity [76, 77]. One of the most commonly used techniques is the laser pump-probe method. A laser, focus to a very small point, is used to heat the surface. The duration of the pulses should be less than the time of the heat penetration into the nanostructure to avoid influence from substrate; thus, femtosecond, picosecond, and nanosecond pulsed lasers are used in such measurements [78–81]. Semiconductor nanostructures are usually coated with a thin layer of a film to better absorb the laser beam and reduce reflection.

Another laser beam is directed at the same spot and used as the probing signal. The increase in temperature due to the pumping laser pulses induces a proportional change in optical reflectivity due to the temperature dependence of the refractive index, which can be detected by collecting the reflected probing signal. For semiconductors, the change of reflectivity R is generally small, in the range of $(1/R)(\mathrm{d}R/\mathrm{d}T) \sim 10^{-3}\text{–}10^{-4}$.

Since the magnitude of the detected signal is small and might be comparable to the noise level, averaging of multiple laser pulses is needed to reduce the noise level. As many photodetectors are not fast enough for ultrashort (e.g., pico- and femtosecond) laser pulses, controllable time-delayed probes are used by varying the optical path [79].

In the pump-probe method, thermal energy is inserted into a very small spot, not only on the surface but also partially, depending on the penetration depth of the laser beam, into the nanostructure. Thus, the heat conduction is detected mostly along the optical axis of the pumping laser, which is the cross-plane and, in most cases, the growth direction. To measure in-plane thermal diffusivity, a transient grating technique is used. The pump laser is split into two beams to generate an interference pattern on the surface, which then create a transient grating due to the temperature dependence of the refractive index. The time-delayed probing signal is diffracted and thermal diffusivity of in the in-plane direction can be revealed [80, 82].

Similar, but much slower, pump-probe measurements can also be performed with a modulated continuous wave (cw) laser, a lock-in amplifier, and a probe laser from a different source to detect the reflectance intensity [83]. With a fixed probe beam position and a scanning pump beam, the distribution of thermal diffusivity across the sample or along the sample surface (lateral thermal conductivity) can be found [84]. Other than reflectance, photothermal emission can also be detected at the surface to attain thermal diffusivity. The laser pump beam can also stimulate thermal expansion at the heating site. Using a probe laser, the expansion can be detected as photothermal displacement due to change in volume (with interferometry) or as photothermal deflection due to change in refractive index [84].

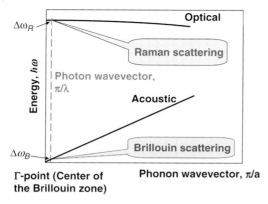

Figure 5.10 Schematic of light scattering processes involving optical phonons (Raman scattering) and acoustic phonons (Brillouin scattering).

5.3.4
Raman Scattering and Thermal Conductivity

Inelastic light scattering is a phenomenon associated with interaction between photons and phonons. Light scattering associated with acoustic phonons is known as Brillouin scattering, while scattering associated with optical phonons is named Raman scattering. Figure 5.10 illustrates both processes in reciprocal space: since the phonon wave vector π/a (where a is the lattice constant and in the order of Å) is significantly greater than photon wave vector π/λ (where the laser wavelength λ is in the order of 400–700 nm), light scattering occurs near the center of the Brillouin zone (Figure 5.10). Therefore, Brillouin scattering frequency (or Brillouin shift) is much smaller than Raman scattering frequency (or Raman shift). In general, the shift $\Delta\omega$ of the excitation photon frequency ν_{laser} (often measured in inverse wavelength $1/\Delta\lambda$ with typical units of reciprocal centimeters) could be associated with either phonon emission or phonon absorption (Figure 5.11). These processes are named Stokes shift ($\Delta h\nu = h\nu - \hbar\omega_{phonon}$) and anti-Stokes shift ($\Delta h\nu = h\nu + \hbar\omega_{phonon}$). Inelastic light scattering can be used to measure sample temperature because the intensities of Stokes and anti-Stokes components are proportional to phonon population, which is determined by Boltzmann statistics:

$$\frac{I_A}{I_S} \cong e^{-\frac{\hbar\omega_{phonon}}{k_B T}}, \tag{5.35}$$

where $\hbar\omega_{phonon}$ is the phonon energy, k_B is Boltzmann constant, and T is temperature. Figure 5.12 illustrates this statement, which is correct for any system with a nonresonant excitation, and shows Stokes/anti-Stokes Raman spectra in c-Si with the major optical phonon mode at $\pm 520\,\text{cm}^{-1}$ (first-order optical phonon mode) and much weaker feature near $\pm 300\,\text{cm}^{-1}$ (higher order acoustic phonons, overtones, etc. [85]). The anti-Stokes/Stokes components intensity ratio exponentially depends on the phonon frequency, and near the Brillouin zone center optical phonon energy is

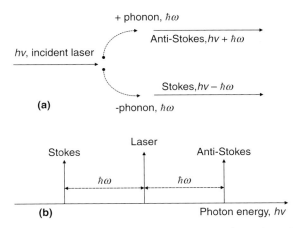

Figure 5.11 Schematic of Raman scattering: Stokes and anti-Stokes processes.

much greater compared to that in acoustic phonons. Therefore, it is more beneficial to use Raman rather than Brillouin scattering in determining specimen temperature, and "Raman thermometers" are successfully used in remote temperature sensing [86, 87].

In Raman scattering measurements, a laser beam by itself could be a significant source of thermal power delivered to a sample. Typically, a low-efficiency, nonresonant Raman scattering ($\sim 10^{-6}$) requires substantial, 10–100 mW, power of a cw laser beam focused onto an area of approximately 1–10 μm^2. For a Si-based sample, the penetration depth of visible laser is usually less than 1 μm. Such a laser beam could easily heat a c-Si sample surface up by approximately 10–100 K, and a sample

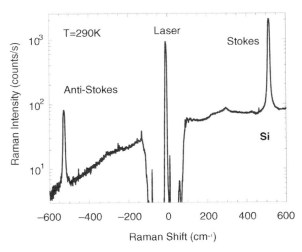

Figure 5.12 Room-temperature Raman spectrum in Si showing Stokes and anti-Stokes Raman peaks.

with approximately 10 times less thermal conductivity can easily be melted. Thus, reasonably short wavelength (<500 nm) laser beam can be used to increase the sample surface temperature substantially, and a significant temperature gradient between the surface and the bulk of the sample $\Delta T = T_{\text{surface}} - T_{\text{bulk}}$ can be easily established.

In addition to Stokes/anti-Stokes Raman peak ratio, Raman peak position and full width at half maximum (FWHM) are also sensitive to the sample temperature [88–90], mainly due to lattice thermal expansion and changes in the phonon dispersion. With increasing temperature, the Raman peaks are shifted to lower wavenumbers and broadened. The nonlinear dependence for Raman peak position $\Delta\omega_{\text{Raman}}$ and Raman spectrum FWHM $\Delta\Gamma_{\text{Raman}}$ as a function of temperature in Si can be simplified as $\Delta\omega_{\text{Raman}}(T)/\text{cm}^{-1} = -0.022(\Delta T/\text{K})$ and $\Delta\Gamma_{\text{Raman}}(T)/\text{cm}^{-1} = 0.011(\Delta T/\text{K})$. This method (i.e., sample temperature calculation based on Raman peak position and Raman spectrum FWHM) is less reliable due to other contributions to the Raman spectral line-shape, such as embedded strain, strain due to a mismatch in thermal expansion, and sample compositional/structural disorder [91–93].

Raman scattering can provide not only very accurate measurements of temperature but also temperature spatial distribution. The obvious choice is both to combine Raman spectroscopy with confocal microscopy for in-depth measurements and to use Raman mapping for a nonuniform in-plane sample. Another very interesting but technically challenging method is time-resolved Raman measurements that can be used to analyze fast dynamics of sample heating and cooling [94–99]. Another possibility is to use the analytical nature of Raman spectroscopy, which is sensitive to the presence of a relatively low concentration of impurities, contaminations, and so on. Thus, a different chemical composition of a thin film (e.g., SiGe) and a substrate (e.g., Si) can be used in monitoring temperature distribution in the direction of growth. Another important characteristic of Raman scattering is the Raman cross-sectional anisotropy in crystals, which can be used to determine sample crystallographic orientations and thermal conductivity in different crystallographic directions. We explain details of these measurements later, using examples of Si-based nanostructures including Si/SiGe and nc-Si/SiO$_2$ multilayers.

5.4
Thermal Properties of Si-Based Nanostructures

5.4.1
Two- and One-Dimensional Si Nanostructures: Si-on-Insulator and Si Nanowires

Silicon-on-insulator technology offers significant performance gain over bulk c-Si CMOS, and high-end microprocessors using SOI CMOS have been commercially available since 1997. As the technology moves beyond 0.1 μm, SOI application is spreading to lower end microprocessors and memory devices [100–102]. In general, SOI refers to a thin c-Si film on top of a SiO$_2$ film. Such 2D structures can be produced by oxygen ion implantation known as separation by implantation of oxygen

(SIMOX) [102] or by wafer bonding technology [103, 104]. The SiO_2 layer sandwiched between the c-Si thin film and a Si carrier wafer is known as the buried oxide (BOX) layer. A BOX layer has a low thermal conductivity and is responsible for a large thermal resistance between the SOI circuit and the carrier wafer. The strong reduction in the cross-plane thermal conductivity of SOI is accompanied by a reduced in-plane thermal conductivity of the thin c-Si layer. The reduction is found to be considerable for SOI layers with thicknesses $\leq 1\,\mu m$ [105]. The most probable sources for the in-plane thermal conductivity reduction in thin SOI layers are phonon scattering at the interfaces, interface roughness, interface-related nonuniform strain, and structural imperfections. Figure 5.13 ([6]) summarizes thermal conductivity as a function of temperature for SOI layers with thickness ranging from 1.6 to 0.42 μm. A similar trend has been found in ultrathin SOI with thickness approaching 0.1 μm [69]. The observed weak thermal conductivity temperature dependence in SOI materials is qualitatively similar to that in poly-Si (see Figure 5.5). In all these Si nanostructures, thermal conductivity at low temperatures is limited by the characteristic size (e.g., the thickness of the SOI layer, poly-Si and nc-Si grain diameter, etc.) because c-Si phonon mean free path can be as large as many microns. At room temperature, due to the increased contribution of U-processes, c-Si phonon mean free path is much smaller, only a fraction of a micron. We conclude that Si/SiO_2 interface-related structural defects are the most probable sources for the in-plane thermal conductivity reduction in SOI structures, and a submicron SOI layer with good quality Si/SiO_2 interfaces might conduct heat almost as efficiently as bulk c-Si (Figure 5.13).

A stronger thermal conductivity size dependence is found in one-dimensional Si nanostructures or nanowires with typical diameter <100 nm. The most common technique for NW fabrication is vapor–liquid–solid (VLS), and there are several in-depth reviews discussing VLS fabrication and major properties of semiconductor

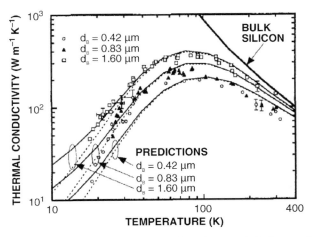

Figure 5.13 Thermal conductivities of SOI layers with thicknesses 0.42, 0.83, and 1.6 μm. The experimental data are compared with the calculations using the phonon-boundary scattering analysis. Adopted from Ref. [6].

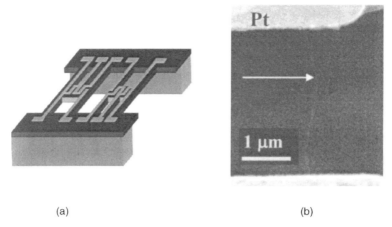

(a)   (b)

Figure 5.14 (a) Schematic and (b) SEM image of a microdevice containing two bridging membranes used for NW thermal conductivity measurements. From Ref. [111].

NWs [106–110]. One of the proposed applications for these nanostructures is thermoelectric devices. Thermal conductivity of individual Si NWs was studied using a microdevice containing two bridging membranes (Figure 5.14), a technique specifically developed for low-volume objects [35, 84, 111]. In individual Si NWs with diameters ranging between 115 and 22 nm, the measured thermal conductivity clearly exhibits a dependence on the NW diameter, which is summarized in Figure 5.15 (from Ref. [107]). Similar to SOI, it has been proposed that phonon boundary scattering is the dominant mechanism controlling thermal conductivity in Si NWs [107]. Thus, in 1D and 2D Si nanostructures, thermal conductivity is controlled by the nanostructure characteristic size (when it is comparable or smaller than phonon mean free path) and by quality of the interfaces.

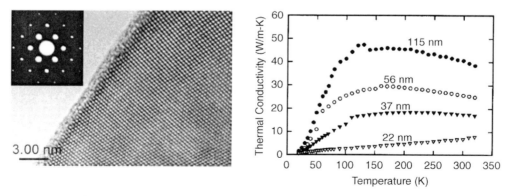

Figure 5.15 (a) TEM image and (b) thermal conductivity temperature dependence in controlled (indicated) diameter Si nanowires [107].

5.4.2
Epitaxially Grown Si/SiGe Nanostructures: Superlattices and Cluster Multilayers

Bulk crystalline SiGe alloys have quite low thermal conductivity, mainly due to their compositional disorder [112, 113]. It has been proposed that Si/SiGe superlattices with an enhanced carrier transport and an even lower thermal conductivity due to the nanometer-thick layers could be used in efficient thermoelectric devices [3, 4, 66, 71, 107]. Figure 5.16 summarizes some studies demonstrating that thermal conductivity in undoped Si/SiGe SLs mainly depends on interface roughness, density of interface defects, and nonuniform strain due to the 4.2% lattice mismatch between Si and Ge. In doped Si/SiGe SLs with a possibility of dopant segregation at the Si/SiGe heterointerfaces, thermal conductivity is found to be reduced by approximately 50–60% (Figure 5.16) [71]. It has also been proposed that in Si/SiGe 3D, that is, cluster morphology multilayers, thermal conductivity could be reduced even further [114]. We will briefly review Si/SiGe 3D nanostructure fabrication and will focus on their thermal properties measured by electrical and optical methods, especially by using Raman scattering.

The standard fabrication of 3D Si/SiGe multilayer nanostructures is based on the sequential physical sputtering of Si and Ge (SiGe) in molecular beam epitaxy (MBE) or the thermal decomposition of SiH_4 and GeH_4 in CVD at temperature approximately 550–650 °C. In this temperature range, both the high Si solid solubility in Ge and the strain-induced SiGe interdiffusion due to the 4.2% lattice mismatch are important. The MBE growth provides better control over the average SiGe cluster composition; however, because of interdiffusion during growth, the composition is not uniform within the cluster volume [115–117]. Figure 5.17 shows transmission electron microscopy (TEM) micrographs of 2D and 3D multilayer Si/SiGe nano-

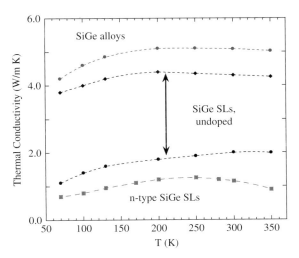

Figure 5.16 Thermal conductivity temperature dependence in SiGe alloys and SiGe superlattices. The data complied from Refs [3, 4, 66, 71].

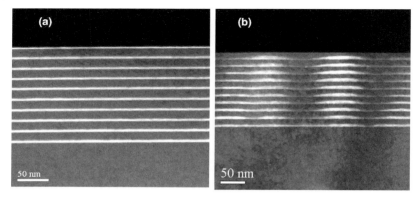

Figure 5.17 TEM micrograph of (a) Si/SiGe planar (2D) and (b) Si/SiGe cluster (3D) multilayers with Ge composition exceeding 30%. Courtesy D. J. Lockwood.

structures grown using MBE. In general, SiGe cluster growth commences with the spontaneous development of a $Si_{1-x}Ge_x$, planar, <1 nm thick wetting layer where x varies, mainly due to uncontrollable SiGe interdiffusion. With further influx of Ge and Si, the growth mode switches from 2D (layers) to 3D (clusters), which helps release some of the lattice mismatch-induced strain. The fully grown, 3–10 nm high, and initially near-pyramid-shaped SiGe clusters have a Ge-rich core (~50%, depending on the Ge flux), although the exact final cluster shape and composition strongly depend on the fabrication conditions. Detailed structural analysis also indicates that the Si matrix in the valleys between SiGe clusters is slightly compressed. To summarize, buried SiGe clusters with the highest Ge composition of close to 50% at the center of the clusters are surrounded by Si, which is tensile strained above each SiGe cluster and compressed laterally between the clusters to maintain a low overall strain. Each SiGe cluster consists of $Si_{1-x}Ge_x$ crystalline alloys with x increasing toward the cluster center [115–117].

In SiGe materials and systems, Raman spectroscopy is a unique characterization technique due to the multimodal nature of Raman scattering from optical phonons in SiGe [117–119]. Typical Raman spectra reveal three major vibrational modes known as the Si–Si vibration at approximately 520 cm^{-1}, the Si–Ge vibration at approximately 400 cm^{-1}, and the Ge–Ge vibration at approximately 300 cm^{-1} (Figure 5.18). In most cases, Si/SiGe 3D multilayers are grown on Si substrates. Thus, by choosing excitation wavelength with proper penetration depth, one can obtain the Raman signal associated with Si–Ge and Ge–Ge vibrations, collected entirely from the Si/ SiGe nanostructure with 10–20 periods and overall thickness <100 nm. At the same time, the Raman signal associated with the Si–Si vibration mode is primarily collected from the substrate, for example, by using excitation wavelength $\lambda = 514.5$ nm with a penetration depth of approximately 500 nm. Figure 5.19 supports this statement by comparing normalized Stokes and anti-Stokes Raman spectra in the vicinity of 500 cm^{-1}. The Raman peak at approximately 505 cm^{-1} corresponds to the Si–Si vibration in the presence of a Ge atom and/or associated with strained Si sandwiched between SiGe cluster multilayers [117]. In a sample without significant

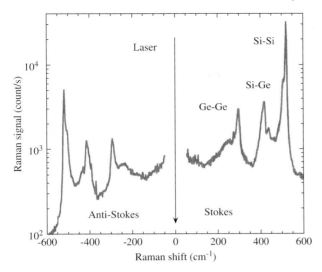

Figure 5.18 Raman spectra of a Si/SiGe multilayer sample (similar to those shown in Figure 5.17b) displaying Stokes/anti-Stokes Raman peaks. Major Raman features associated with Si–Si (520 cm^{-1}), Si–Ge (~400 cm^{-1}), and Ge–Ge (~300 cm^{-1}) vibrations are shown.

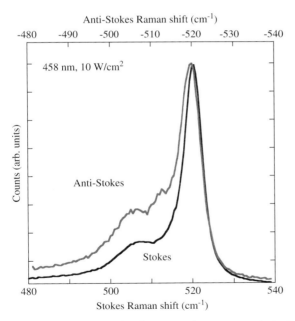

Figure 5.19 Normalized Stokes/anti-Stokes Raman peaks in the vicinity of Si–Si Raman mode showing relative enhancement of the Raman feature at approximately 505 cm^{-1}, presumably due to a higher temperature in the Si/SiGe multilayers compared to Si substrate temperature.

temperature gradient, these normalized Stokes/anti-Stokes Raman lines should be identical. The increase in anti-Stokes Raman signal at approximately 505 cm^{-1} indicates that under laser excitation with intensity >10 W/cm^2, the sample surface temperature (associated with Si/SiGe 3D nanostructures) is significantly higher than the Si substrate temperature. By knowing the laser light penetration depth and calculating the SiGe nanostructure and the Si substrate temperatures, thermal gradient and thermal conductivity of the Si/SiGe system can be estimated. Furthermore, by varying the laser wavelength (i.e., the penetration depth) and by monitoring temperature distribution in the cross-plane direction of the sample, it is possible to control the temperature gradient and verify these calculations (Figure 5.20). Another possibility to evaluate the overall Si/SiGe system thermal conductivity is to use Raman Stokes/anti-Stokes peak intensity ratio to calculate the sample surface temperature, which can be compared with the temperature of c-Si measured under identical conditions. Note that with a focused laser beam as the heat source, one should consider both in-plane and cross-plane heat dissipation. Thus, to produce an accurate result, this experimental method should be combined with a proper analysis [118]. In general, the Raman-scattering-based "thermometry" can be an accurate, contactless technique, especially useful in analyzing complex Si/SiGe nanostructures.

There are many reports on thermal conductivity in 2D and 3D Si/SiGe multilayers studied by 3ω or similar electrical methods [63, 71, 119–121], and fewer data collected by optical measurements [122–125]. The calculated thermal conductivity obtained from the optical measurements is found to be close to values obtained by the 3ω method. It has been noticed that in several samples with significant degree of SiGe

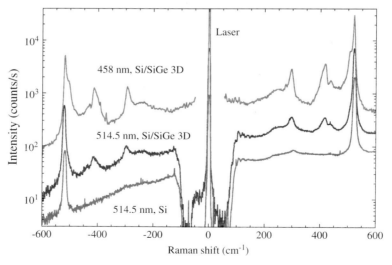

Figure 5.20 Stokes/anti-Stokes Raman spectra in a Si/SiGe multilayer sample on a Si substrate collected using different excitation wavelengths (indicated). Spectra are shifted vertically for clarity and c-Si Raman spectrum is shown for comparison.

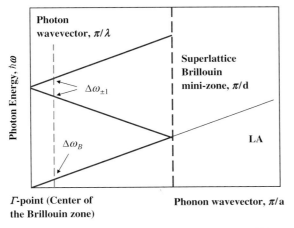

Figure 5.21 Schematic of acoustic phonon dispersion folding and light scattering in an ideal periodic structure consisting of alternating layers with different elastic properties.

cluster vertical self-ordering, thermal conductivity slightly increases [118]. Moreover, in quasiplanar Si/SiGe multilayers with Ge concentration approaching 40% (which is much higher compared to conventional Si/SiGe SLs with Ge concentration ~10–20%), thermal conductivity is found to be increased by approximately 100% compared to similar Si/SiGe composition in true 3D Si/SiGe cluster multilayers [118]. The increase in thermal conductivity in self-organized 3D Si/SiGe nanostructures is verified by Raman spectroscopy. In highly periodic Si/SiGe multilayers with reasonably abrupt Si/SiGe interfaces, folded longitudinal acoustic (FLA) phonon modes can be observed in the form of low-frequency peaks in Raman scattering [126–129]. The nature of this process is explained in Figure 5.21; it shows that in an ideal, periodic, multilayered heterostructure with abrupt and defect-free interfaces, acoustic phonon folding appears. In our experiments, an increase in thermal conductivity correlates with an increase in the FLA-phonon-related Raman signal [118]. The intensity of the FLA phonon Raman peaks also increases as the excitation wavelength decreases from 514.5 to 457.9 nm (Figures 5.22 and 5.23). It has been suggested that this dependence is associated with the Raman resonance in Si/SiGe 2D and 3D multilayer nanostructures at photon energy close to 2.7 eV [129, 130]. However, at least in part, FLA phonon peaks in Raman spectra might increase due to a significant increase in temperature in Si/SiGe multilayers (see Table 5.1). To conclude, these results suggest that in Si/SiGe 2D and 3D multilayer nanostructures, thermal conductivity is controlled by a number of parameters, including the multilayer period, the interface roughness, the presence of structural defects, strain, intermixing, and the degree of SiGe cluster vertical self-organization. The increase observed in thermal conductivity of ordered Si/SiGe multilayer nanostructures indicates that the previously reported large anisotropy in cross-plane and in-plane thermal conductivity in Si/SiGe multilayers [71, 114, 120, 131] can be further enhanced and employed in novel thermoelectric devices.

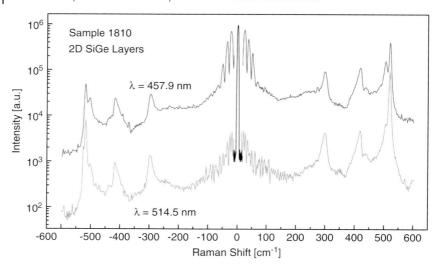

Figure 5.22 Stokes/anti-Stokes Raman spectra in a Si/SiGe planar (2D) multilayer sample (Figure 5.17a) collected using different excitation wavelengths (indicated). Spectra are shifted vertically for clarity. Note multiple Raman peaks at low wavenumbers associated with the FLA phonons.

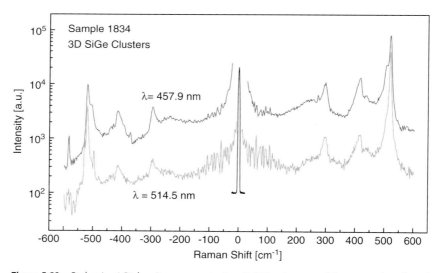

Figure 5.23 Stokes/anti-Stokes Raman spectra in a Si/SiGe cluster multilayer sample collected using different excitation wavelengths (indicated). Spectra are shifted vertically for clarity. Note the absent Raman features at low wavenumbers, presumably due to diffused interfaces and variations in SiGe cluster heights.

Table 5.1 Summary for the Raman spectroscopy-based temperature calculation with two different wavelengths, and the correlation between FLA intensities and temperature gradients.

Sample	x (nominal)	FLA	Temperature (K)					
			$\lambda = 457.9$ nm $P_{laser} \sim 100$ mW			$\lambda = 514.5$ nm $P_{laser} \sim 400$ mW		
			Si–Ge	Si–Si	ΔT	Si–Ge	Si–Si	ΔT
1810 (2D)	0.52	High	~380	~350	30	−370	~300	70
1814 (3D)	0.53	Lower	~440	~380	60	~420	~315	105
1831 (3D)	0.53	None	~440	~370	70	~460	~340	120
1834 (3D)	0.6	None	~410	~340	70	~450	~315	135

5.4.3
Electrochemically Etched Si (Porous Si)

Porous silicon (PSi) was first discovered in 1956 by researchers at Bell Laboratories interested in electropolishing of c-Si surface [132]. In the late 1980s, PSi was "rediscovered" as a nanostructured material by Canham [133] and Lehmann and Gösele [134]; they suggested that PSi exhibits quantum confinement effect due to reduced dimensions of Si crystallites. PSi has many other interesting properties including a highly nanoporous structure (typically >70% porosity, evaluated by the gravimetric method [135]) with extremely large surface-to-volume ratio and very low index of refraction (<1.2) [12]. Sample fabrication has been discussed in detail in many review studies, and it is typically based on anodic dissolution of c-Si in a hydrofluoric acid-based (HF-based) solution. PSi microstructure is usually dendrite-like with loosely connected, nanometer-sized Si crystallites (Figure 5.24) [12]. A large amount of hydrogen in electrochemically etched Si has been found; it has been proposed that hydrogen passivates PSi internal surfaces by forming Si−H bonds, which can be easily broken under intense radiation or by annealing at $T > 400\,^{\circ}$C [12].

Since PSi porosity is quite high, its thermal conductivity is anticipated to be low. This assumption has been confirmed by dynamic 3ω measurements performed at

(a)   (b)

Figure 5.24 (a) Cross-sectional SEM showing a dendrite-like structure of porous Si and (b) a higher resolution TEM image showing small-size Si nanocrystals. From Ref. [12].

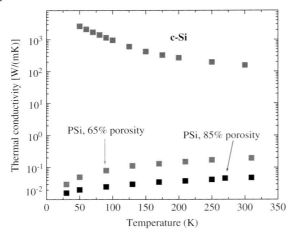

Figure 5.25 Comparative thermal conductivity temperature dependence in c-Si and PSi samples with different (indicated) porosity. Data compiled using Ref. [136].

temperatures 30–320 K in a low-pressure environment in order to suppress thermal convection and gaseous heat conduction [136]. In such measurements, the penetration of the thermal energy should be kept lower than the PSi layer thickness, while microtopology of PSi could be approximated by the effective medium model [12, 137]. Experimentally, PSi thermal conductivity $\varkappa_{PSi}$ is found to be $10^3–10^4$ times lower than that in c-Si (Figure 5.25). The thermal conductivity temperature dependence of PSi is similar to other Si-based nanosystems, where $\varkappa$ is limited by crystallite sizes, lattice imperfections, and grain boundaries rather than U-processes (see Figure 5.1); it slightly increases as temperature increases in the range of 10–400 K. In addition, $\varkappa_{PSi}$ exhibits dramatic dependence on the PSi porosity (Figure 5.26). Assuming that PSi

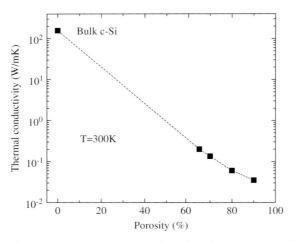

Figure 5.26 Room-temperature thermal conductivity in Si as a function of sample porosity. For zero porosity, c-Si thermal conductivity is used. PSi data from Ref. [136].

effective thermal conductivity can be estimated as $\varkappa_{eff} = f\eta\varkappa_{Si}$, where $\varkappa_{Si}$ is bulk c-Si thermal conductivity, $f = 1-p$ determines the fraction of solid volume, p is the sample porosity, and η is a percolation parameter [136]. Thus, it can be written as $\varkappa_{eff} = f\eta(1/3)\varrho c_l v d_c$, where ϱ is bulk Si density, c_l is lattice heat capacity, v is sound velocity, and d_c is the average crystallite diameter. In this equation, the phonon mean free path is replaced by the crystallite diameter, and the strong porosity volume dependence is taken into account as $f\eta \approx f^3$.

This approach explains reasonably well the experimentally observed very low PSi thermal conductivity. It also suggests that in this material, optical measurements (e.g., photoluminescence (PL), Raman scattering, etc.) should be performed with extreme care to avoid overheating and damage of sample by the laser beam. Simple calculations suggest that in a 70% porosity PSi layer, a laser beam with intensity of only 10 mW focused into 10 μm^2 and with the penetration depth of approximately 1 μm can increase the sample temperature by approximately 100 K. The same excitation would be enough to completely melt a 90% porosity PSi sample. Figure 5.27 (data taken from Ref. [138]) supports this estimation and shows a strong spectral shift of the Raman peak in micro-Raman measurements, which is mainly due to laser heating of the PSi sample. The PSi sample temperature increase is estimated to be approximately 400 K; thus, during this measurement, PSi sample temperature reaches 700 K. Another interesting observation is that PSi sample temperature detected by the Raman technique strongly depends on ambient gases, indicating that thermal convection in high-porosity Si nanostructures needs to be taken into account [139]. The data show that intrinsic PSi sample thermal conductivity with porosity approaching 90% is below thermal conductivity of air [136]. These experimental results immediately explain the observation of photoinduced hydrogen effusion [12, 140], photo-oxidation [141–144], continuous PL spectrum shift under moderate intensity cw laser excitation [145, 146], and many other photo/thermo-induced phenomena in PSi [12, 147–151]. These studies also suggest that electroluminescence (EL) measurements in PSi could be affected by significant sample

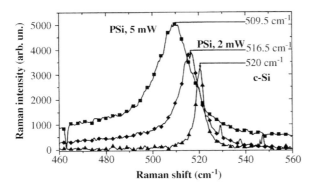

Figure 5.27 Raman spectra in PSi measured under different excitation intensities. The shift and broadening of the Raman spectrum are attributed to the sample temperature increase due to increasing laser intensity. From Ref. [138].

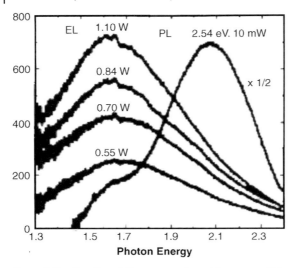

Figure 5.28 The PL and EL spectra collected from same PSi samples under different excitation conditions (indicated). The EL spectra redshift, in part, is attributed to a high sample temperature and blackbody emission. From Ref. [153].

heating by the electric current. In this case, blackbody radiation contributes to the EL spectra [152], and the phenomenon could be responsible for the consistently observed redshift of the EL spectra compared to the PL spectra (Figure 5.28) [153].

5.4.4
Nanocrystalline Si/SiO₂ Multilayers

In the mid-1990s, a search for nonporous, hydrogen-free, and stable light-emitting Si nanostructures has produced nc-Si/SiO₂ multilayers, one of the first Si nanostructures with relatively good control over Si nanocrystal sizes as small as <5 nm [18, 154–156]. Typically, samples of nc-Si/SiO₂ multilayers are grown on a standard ⟨100⟩ c-Si substrate using radio frequency (RF) or magnetron sputtering of a-Si/SiO₂ periodic multilayers. Sputtering has the clear advantage of a physical deposition process, that is, a linear dependence on the deposition rate as a function of time at a given RF or magnetron power. The process allows the simple cleaning of a target and substrate as a predeposition step and it can be performed in a moderate vacuum of approximately 10^{-7} Torr. A modern sputtering system also provides very good stabilization of the plasma and is able to bring the deposition rate down to 10 nm/min or less. For details on sample preparation and their structural properties, the reader may refer to Refs [18, 157, 158].

Thermal crystallization of a-Si/SiO₂ multilayers can be carried out in a conventional, hot-wall furnace with control over the ambient conditions to avoid unwanted oxidation. At moderate temperatures (200–400 °C), the interstitial oxygen is found to be quickly diffused into a-Si and initiate a-Si/SiO₂ intermixing [158]. It is also noticed that CVD or sputtered SiO₂ introduces less strain into layered a-Si/SiO₂ structures

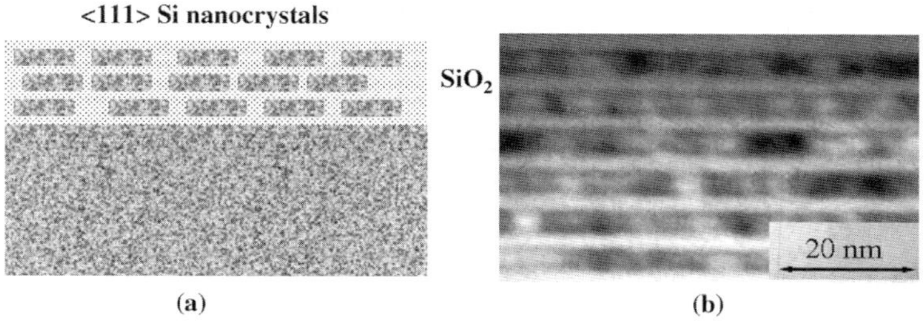

<111> Si nanocrystals

(a)   (b)

Figure 5.29   (a) Schematic and (b) cross-sectional TEM of an nc-Si/SiO$_2$ multilayer sample with clearly visible Si nanocrystals within nc-Si layers.

than SiO$_2$ fabricated by ozone or thermal oxidation, and this is most likely due to the lower density of the sputtered SiO$_2$. By considering these issues, it was proposed that the best growth choice would be sputtered or CVD-deposited a-Si/SiO$_2$ layers [158], but thermal crystallization must be performed using rapid thermal annealing (RTA), which was found to provide very efficient crystallization of nanometer-thick a-Si layers with their thickness down to approximately 2 nm [18]. It was also found that the combination of RTA with subsequent furnace annealing results in the segregation of crystallized nc-Si and a-SiO$_2$ layers with pronounced improvement in the nc-Si/SiO$_2$ interface abruptness and flatness. At the same time, thermal annealing at a temperature $\leq 1200\,°C$ improves the a-SiO$_2$ layer stoichiometry and density, and reduces the number of defects at the Si/SiO$_2$ interfaces. Thus, the standard sample preparation requires a cleaned $\langle 100 \rangle$ c-Si substrate, sputtering of a-Si/SiO$_2$ periodic multilayers, followed by RTA at $T \geq 750\,°C$ and furnace annealing at T approximately $1000\,°C$. Schematic and TEM micrograph of an nc-Si/SiO$_2$ multilayer sample are shown in Figure 5.29.

Several studies show that Si nanocrystals of different sizes in nc-Si/SiO$_2$ multilayers might have preferential shape and crystallographic orientations [159]. It has been found that Si nanocrystals smaller than 5 nm diameter usually have near-spherical shape and close to $\langle 100 \rangle$ crystallographic orientation in the direction of growth. At the same time, the larger (~ 7–10 nm) Si nanocrystals are rectangular (or "brick shaped") and have $\langle 111 \rangle$ crystallographic orientation [159]. It is known that the Raman intensity depends on the orientation of polarization vectors of the incident and scattered light relative to the crystallographic axes of the sample [160, 161]. To test the Raman polarization measurement setup, $\langle 100 \rangle$ and $\langle 111 \rangle$ single-crystal Si wafers are usually used because they have qualitatively different polarization Raman intensity dependence determined by the different crystal symmetry (Figure 5.30). Thus, using $\langle 100 \rangle$ c-Si substrate and Si nanocrystals with a direction different from $\langle 100 \rangle$ crystallographic direction, one can clearly distinguish between the Raman signal from Si nanocrystals and c-Si substrate. The simplest approach is to use photoexcitation with a short (<500 nm) wavelength and perform Raman scattering spectroscopy with the appropriate angles between the polarization of the incident and scattered light (see Figure 5.31). By performing the same measurements in Stokes/anti-Stokes spectral regions, one can find the nc-Si/SiO$_2$ multilayers and $\langle 100 \rangle$ Si

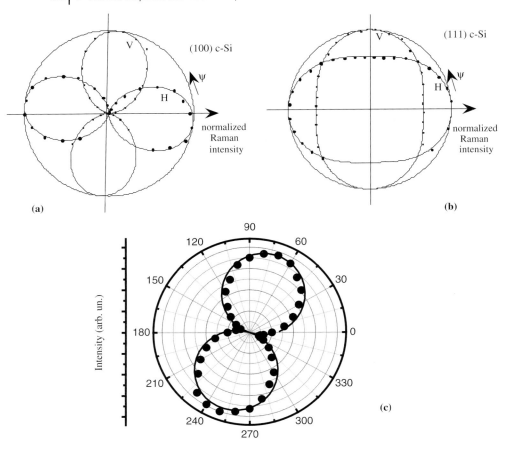

Figure 5.30 Intensity of polarized Raman spectra as a function of analyzer/polarizer angle for vertical (V) and horizontal (H) laser polarization in (a) $\langle 100 \rangle$ c-Si, (b) $\langle 111 \rangle$ c-Si, and (c) a thin-film nc-Si/SiO$_2$ multilayer sample grown on $\langle 100 \rangle$ Si substrate.

substrate temperatures and calculate the thermal conductivity of nc-Si/SiO$_2$ multilayers. The nc-Si/SiO$_2$ multilayer sample structure is found to be very convenient for such measurements because it has thermal conductivity several orders of magnitude lower than that in c-Si substrate, which is used as a heat sink.

Using the experimental setup described in Ref. [162] and 10–20 period nc-Si/SiO$_2$ multilayer samples, the polarized Raman signal recorded with approximately 70° (or ~250°) angle between polarization of the incident and scattered light is found to be mainly due to light scattering within the c-Si substrate, while at the approximately 160° (or ~340°) angle, the signal is mainly due to light scattering within the nc-Si/SiO$_2$ multilayers (Figure 5.30c). It is convincingly shown in Ref. [162] that the Raman signal related to <100> Si can be suppressed; thus, it is possible to separate the Raman signal from c-Si substrate and from Si nanocrystals. These measurements allow to accurately determine the nc-Si/SiO$_2$ multilayers and c-Si substrate tempera-

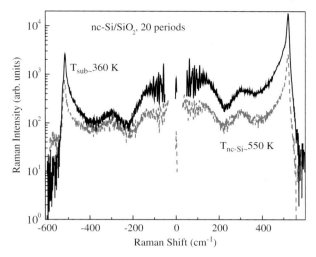

Figure 5.31 Polarized Stokes/anti-Stokes Raman spectra of nc-Si/SiO$_2$ multilayers on $\langle 100 \rangle$ Si indicating different (calculated and shown) temperatures of the nc-Si/SiO$_2$ multilayers and the c-Si substrate.

tures under cw laser irradiation (Figure 5.31). Under the described irradiation conditions, the nc-Si/SiO$_2$ multilayer temperature is approximately 200 K higher than the c-Si substrate temperature, and the thermal conductivity of nc-Si/SiO$_2$ multilayers was estimated to be $<10^{-1}$ W/(m K), which is an unusually low value for a nonporous solid.

For a sample with known thermal conductivity and melting temperature, laser damage threshold associated with sample surface melting (and other photo-thermo-induced phenomena) can be easily estimated. These calculations can be verified experimentally by using photo-induced reflection measurements in the pump-probe geometry. In Ref. [162], a mechanically chopped Ar$^+$ laser beam with a relatively low frequency (in order to establish thermal equilibrium after light is switched "off") has been used as the pump source. A much lower intensity, cw laser beam is used as the probe signal to measure sample reflectivity as a function of pump light intensity. Figure 5.32 shows the results of these measurements in nc-Si/SiO$_2$ multilayers. The threshold-like increase in sample reflectivity as a function of pump intensity indicates that the surface melting occurs at irradiation with intensities <10 kW/cm^2. This number is in good agreement with our estimation using thermal conductivity derived from polarized Raman Stokes/anti-Stokes measurements. Quantitatively similar results were obtained in experiments with nearly freestanding samples of nc-Si/SiO$_2$ multilayer films [163].

A strong interest in nonlinear optical properties of Si nanocrystals has been shown in a number of publications, in which nc-Si/SiO$_2$ multilayer films were subjected to pulsed laser irradiation with a high energy density [14, 22, 23, 164]. In these measurements, possible thermal effects should be carefully considered, especially for these low thermal conductivity nanostructures. Indeed, using thermal conduc-

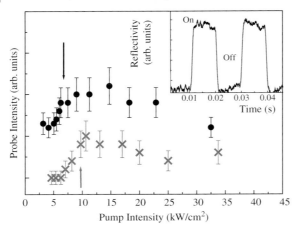

Figure 5.32 The intensity of probe light as a function of the intensity of pump light showing a sharp increase in the nc-Si/SiO₂ multilayer sample reflection associated with surface melting under approximately 10 kW/cm² pump intensity. The inset shows reflectivity as a function of time with pump light turned "on" and "off."

tivity and laser damage threshold derived from Raman and pump-probe measurements in nc-Si/SiO$_2$ multilayer films, the melting threshold under pulsed laser excitation is estimated to be approximately 10 mJ/cm^2, while visible destruction of the sample surface is found at significantly greater pulse energy density of approximately 100 mJ/cm^2 [162].

To verify this estimation and to better understand the nature of laser-induced structural transformations in Si nanocrystal thin films, a series of experiments involving excitations with a single pulse of a KrF laser with energy densities of 10–50 mJ/cm^2 were performed. The absorption coefficient in nc-Si/SiO$_2$ multilayer films with photoexcitation wavelength of 248 nm is in the range of 10^6 cm^{-1}, and light is fully absorbed by the top two to three layers of Si nanocrystals. The anticipated sample structural transformations are verified by polarized Raman spectroscopy with polarization forbidden for $\langle 100 \rangle$ c-Si substrate Raman signal. Figures 5.33 and 5.34 show modifications of the Raman signal in these studied Si nanostructures as the energy density of the single-pulse irradiation increases. All initial samples and samples irradiated by a low-intensity laser pulse show an asymmetric Raman peak centered at approximately 520 cm^{-1} related to Si nanocrystals, rather than c-Si substrate. All samples subjected to irradiation by *only one pulse* with an energy density >30 mJ/cm^2 show significant changes in the Raman spectra. These laser-induced phenomena in nc-Si/SiO$_2$ multilayer films strongly depend on the sample structure and can be divided into two different groups.

Under irradiation with pulse energy density $\geq$30 mJ/cm^2, nc-Si/SiO$_2$ multilayer films with SiO$_2$ thickness $\geq$5 nm show strong evidence of laser-induced amorphization. Figure 5.33 shows Raman spectra collected in polarization forbidden for $\langle 100 \rangle$ c-Si substrate. It is very clear that after laser irradiation, an additional contribution to the Raman signal appears at wavenumbers 460–500 cm^{-1}. After being irradiated by

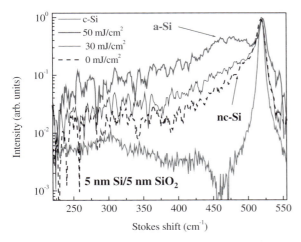

Figure 5.33 Raman spectra in an nc-Si/SiO$_2$ multilayer sample with thicker SiO$_2$ layers showing amorphization of small Si nanocrystals under pulsed laser irradiation (used laser energy densities are shown).

laser pulse with an energy density of 50 mJ/cm^2, a broad Raman feature centered around 480–490 cm^{-1}, which is associated with the amorphous phase of Si, becomes clearly pronounced.

In nc-Si/SiO$_2$ multilayer samples with SiO$_2$ thickness approximately 2 nm, pulsed laser irradiation decreases the Raman signal related to nc-Si phase and no a-Si phase is observed (Figure 5.34). Using an irradiation energy density of 50 mJ/cm^2, we observe total disappearance of the Raman signal associated with nc-Si phase. In all measured samples, we find a visible destruction of the sample surface after irradiation with much greater energy density of $\geq$100 mJ/cm^2.

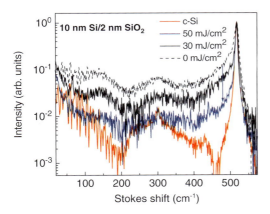

Figure 5.34 Raman spectra in an nc-Si/SiO$_2$ multilayer sample with thinner SiO$_2$ layers showing disappearance of nc-Si phase under pulsed laser irradiation, presumably due to sample melting and breakdown of the multilayer structure (used laser energy densities are shown).

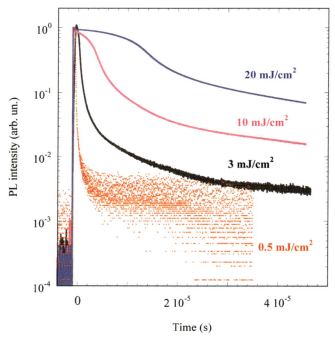

Figure 5.35 Integrated over 980–600 nm spectral region, PL decays in nc-Si/SiO$_2$ multilayers detected under different (indicated) pulsed excitation energy densities show the increase in the slow emission component as excitation increases. The pulse duration is approximately 10 ns.

Additional experimental data strongly supporting our conclusions are recently obtained using spectral-integrated, time-resolved measurements of light emission under pulsed laser irradiation with energy density close to the laser damage threshold in nc-Si/SiO$_2$ multilayer samples. Figure 5.35 shows the dynamics of the PL signal integrated over 950–400 nm spectral region. Using a nanosecond laser pulse with an energy density of 0.5 mJ/cm^2 (which should increase the sample temperature by less than 50 K), we observe optical emission (mostly PL from nc-Si) with a characteristic decay time $\leq 10^{-6}$ s. Using a higher level of optical excitation (pulse energy density >10 mJ/cm^2, which should generate an increase in sample temperature by 600–700 K), we find very different emission with a long decay time close to 10^{-3} s. It is unlikely that such a long process can be associated with room-temperature carrier recombination in nc-Si; it is much more likely that blackbody radiation generated by the hot sample surface strongly contributes to the detected slow emission. In order to prove this statement, the emission transient characteristics were measured using an infrared optical filter with cutoff at 800 nm (Figure 5.36). Under optical excitation with pulse energy density of 10 mJ/cm^2, we find that slow emission component is clearly associated with only long wavelengths (>800 nm), while a shorter wavelength emission is relatively fast even under a high level of laser excitation. Note that for sample surface temperature close to

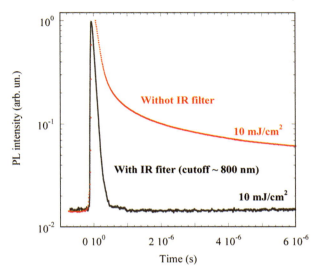

Figure 5.36 The PL decays in nc-Si/SiO$_2$ multilayers recorded with and without an IR 800 nm cutoff filter at the same excitation energy density and showing that the IR PL component is much slower compared to visible PL.

$T \leq 1000$ K, the corresponding blackbody radiation peak is at wavelength λ approximately 3 µm.

The observed laser-induced amorphization of nc-Si/SiO$_2$ multilayer samples is accompanied by a strong (approximately three times) increase in visible PL intensity (Figure 5.37) and a decrease in the sample optical transmission, that is, photodarkening (Figure 5.38). Compared to c-Si, a-Si has a greater absorption coefficient at photon

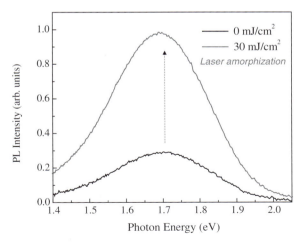

Figure 5.37 The PL spectra in nc-Si/SiO$_2$ multilayers before (0 mJ/cm^2) and after laser amorphization with laser energy density of 30 mJ/cm^2 and pulse duration of 10 ns.

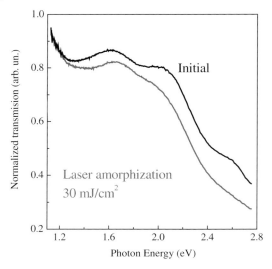

Figure 5.38 Normalized transmission spectra in nc-Si/SiO$_2$ multilayers before (0 mJ/cm^2) and after laser amorphization with laser energy density of 30 mJ/cm^2 and pulse duration of 10 ns.

energies below the mobility edge (which is equivalent to the energy gap in crystalline semiconductors) due to Urbach tail-related optical transitions [165]. Also, it is quite possible that the amorphization of Si nanocrystals relaxes carrier recombination selection rules, decreases carrier radiative lifetime, and produces more efficient PL. Finally, phase transitions are likely to be accompanied by strong nonlinear optical phenomena [166–168]; thus, nonlinear optical properties of Si nanocrystals [169, 170], at least in part, can be explained by amorphization and other heat-related effects.

The extremely low thermal conductivity $\varkappa_l$ in nc-Si/SiO$_2$ multilayers ($<10^{-1}$ W/(m K)) cannot be explained within the frame of kinetic theory by a simple reduction in the phonon mean free path l_{phonon} due to small Si crystallite size. Most likely, in small Si nanocrystals embedded into SiO$_2$, not only phonon mean free path but also phonon modes and phonon relaxation time change dramatically. For example, a significant modification in phonon modes is expected for heterosystems with a large difference in elastic properties of materials (e.g., Si/SiO$_2$) [75, 171–173]. The phonon relaxation time reduction is due to an increase in (i) electron–phonon scattering (which could be significant at high level of photoexcitation in Raman and other optical measurements); (ii) nc-Si/SiO$_2$ interface scattering; (iii) structural defect-induced scattering (which is inversely proportional to l_d^3, where l_d is the linear dimension of a defect); and (iv) phonon confinement. All these processes contribute to the experimentally observed extremely low thermal conductivity in nc-Si/SiO$_2$ multilayers. An attempt to include the discussed phenomena into calculations of nanostructure thermal conductivity has been proposed, and a good semiquantitative agreement is obtained [173].

5.5
Conclusions

This review confirms that thermal conductivity in different Si-based nanostructures is severely reduced compared to that in single c-Si. It also points out that the reduction of thermal conductivity is mainly due to increased phonon scattering at the nanostructure boundaries and interfaces. In particular, in Si nanoscale objects incorporated into different matrices (e.g., ultrathin SOI layers, nc-Si/SiO$_2$ multi-layers, Si/SiGe nanostructures, etc.), thermal conductivity is mainly controlled by boundary/interface-related structural defects, interface roughness, modification of acoustic phonon energy spectra, and decrease in phonon group velocity. The combined outcome is responsible for the Si nanostructure thermal conductivity reduction at room temperature 10^3–10^4 times less than c-Si. This fact is of great practical importance because it imposes many limitations on Si nanoscale systems and devices. The conclusion, however, does not imply that the many suggested applications of Si-based nanostructures in nonlinear optics, electronic, optoelectronic, and photonic devices are entirely impossible due to the projected extremely poor heat dissipation. It rather suggests that this problem needs to be seriously considered and, possibly, solved by using intelligent engineering solutions (e.g., an additional heat sink, etc.).

Specifically, we would like to emphasize possible applications of Si-based nanostructures in thermoelectric devices [174]. Thermoelectric figure of merit is proportional to electrical conductivity and inversely proportional to thermal conductivity. Thus, the reduction of thermal conductivity in Si-based nanostructures could be considered as "good news" for such devices. However, the same phenomena responsible for reducing a material's thermal conductivity, in general, are even more effective in reducing its electrical conductivity due to the strong reduction in carrier mobility, diffusion coefficient, and lifetime. For example, a-Si thermal conductivity is reduced by approximately 100 times compared to that in c-Si, while carrier mobility is reduced by a factor of 10^4–10^5 [175, 176]. Very similar trend is found in Si-based nanostructures including PSi [177, 178] and tunnel transparent nc-Si/SiO$_2$ multilayers [179]. Therefore, one can conclude that the effort in semiconductor nanostructure-based thermoelectric devices should not be entirely focused on the thermal conductivity reduction but rather on an engineering solution using the balance between thermal and electric transport in these truly fascinating systems.

Acknowledgment

We would like to thank many colleagues for their contribution to this work including Boris V. Kamenev (NJIT), Eun-Kyu Lee (NJIT), Volodimyr Duzhko (NJIT), Haim Grebel (NJIT), Andrei Sirenko (NJIT), Alex Bratkovsky (HP Labs), Ted Kamins (HP Labs), and David J. Lockwood (NRC Canada). Partial financial support for this work was provided by the US National Science Foundation and the foundation at NJIT.

List of symbols

A	cross-sectional area
a	constant (in "electronic thermal conductivity")
α	constant, $1 \leq \alpha \leq 2$ (in "$1/T$ law")
C	total heat capacity
C_{rt}	resistance temperature coefficient (3ω measurement)
C_{subs}	heat capacity of the substrate (3ω measurement)
c	heat capacity of a particle
c_l	lattice heat capacity (PSi)
d	thickness of the membrane (membrane method)
d_c	average crystallite diameter (PSi)
E	energy of a particle
E_F	Fermi energy
E_k	energy of a free electron
e	electron charge
η	percolation parameter (PSi effective thermal conductivity)
f	fraction of solid volume (PSi)
f_o^e	electron distributions
f_o^h	hole distributions
f_k	electron distribution
g	lattice wave vector
$\hbar$	Planck constant
$I(t)$	AC current (3ω measurement)
I_0	current amplitude
$\vec{J}$	electrical current density
$\vec{Q}$	flux of energy
j_U	flux of thermal energy
K_n	general integral
$\vec{k}$	electron wave vector
k_B	Boltzmann constant
$\varkappa$	thermal conductivity
$\varkappa_e$	electronic thermal conductivity
$\varkappa_{eff}$	effective thermal conductivity (PSi)
$\varkappa_l$	lattice thermal conductivity
$\varkappa_{PSi}$	PSi thermal conductivity
$\varkappa_{Si}$	bulk Si thermal conductivity
$\varkappa_{subs}$	thermal conductivity of the substrate (3ω measurement)
L	distance between sensors
l	mean free path
l_d	linear dimension of a defect
λ	laser wavelength
N_0	average number of phonons
N_q	phonon distribution
n	concentration of the particles

P	thermal power
p	sample porosity
φ	phase shift (3ω measurement)
q	phonon wave vector
R	reflectivity
R_0	resistance of the metal strip without heating (3ω measurement)
R_h	resistance of the metal strip (3ω measurement)
ϱ	bulk Si density (PSi)
σ	electrical conductivity
T	temperature
$T(t)$	temperature of the metal strip (3ω measurement)
ΔT	temperature difference between sensors
∇T	temperature gradient
T_h	heater temperature rise (membrane method)
T_s	temperature of the heat sink (membrane method)
T_{subs}	temperature at the nanostructure/substrate interface (3ω measurement)
t	thickness of the sample (3ω measurement)
τ	phonon relaxation time
$\tau(k)$	electron relaxation time
ν	sound velocity
$\vec{\nu}$	velocity of particle
$\vec{\nu}_g$	group velocity
$\vec{\nu}_k$	electron velocity
w	length of the heater (membrane method)
ω	phonon frequency
ω	current frequency (3ω measurement)
$\omega(q)$	phonon dispersion relationship
$\Delta\omega_{Raman}$	Raman peak position
$\Delta\Gamma_{Raman}$	Raman spectrum FWHM.

References

1 Ho, C.Y., Powell, R.W., and Liley, P.E. (1972) Thermal conductivity of the elements. *J. Phys. Chem. Ref. Data*, **1**, 279–421.

2 Srivastava, G.P. (2006) Lattice thermal conduction mechanism in solids, in *High Thermal Conductivity Materials* (eds S.L. Shindé and J.S. Goela), Springer, pp. 1–35.

3 Lee, S.M., David, G.C., and Rama, V. (1997) Thermal conductivity of Si–Ge superlattices. *Appl. Phys. Lett.*, **70**, 2957–2959.

4 Huxtable, S.T., Abramson, A.R., Tien, C.-L., Majumdar, A., LaBounty, C., Fan, X., Zeng, G., Bowers, J.E., Shakouri, A., and Croke, E.T. (2002) Thermal conductivity of Si/SiGe and SiGe/SiGe superlattices. *Appl. Phys. Lett.*, **80**, 1737–1739.

5 Vining, C.B., Laskow, W., Hanson, J.O., Van der Beck, R.R., and Gorsuch, P.D. (1991) Thermoelectric properties of pressure-sintered $Si_{0.8}Ge_{0.2}$ thermoelectric alloys. *J. Appl. Phys.*, **69**, 4333–4340.

6 Asheghi, M., Touzelbaev, M.N., Goodson, K.E., Leung, Y.K., and Wong, S.S. (1998) Temperature-dependent thermal conductivity of single-crystal silicon layers in SOI substrates. *J. Heat Transfer*, **120**, 30–36.

7 Soref, R. (2006) The past, present, and future of silicon photonics. *IEEE J. Sel. Top. Quantum Electron.*, **12**, 1678–1687.

8 Iyer, S.S. and Xie, Y.H. (1993) Light emission from silicon. *Science*, **260**, 40–46.

9 Rong, H., Liu, A., Jones, R., Cohen, O., Hak, D., Nicolaescu, R., Fang, A., and Paniccla, M. (2005) An all-silicon Raman laser. *Nature*, **433**, 292–294.

10 Jalali, B. and Fathpour, S. (2006) Silicon photonics. *J. Lightwave Technol.*, **24**, 4600–4615.

11 Pavesi, L., Gaponenko, S., and Dal Negro, L. (2003) *Towards the First Silicon Laser*, Springer.

12 Cullis, A.G., Canham, L.T., and Calcott, P.D.J. (1997) The structural and luminescence properties of porous silicon. *J. Appl. Phys.*, **82**, 909–965.

13 Park, N.M., Kim, T.S., and Park, S.J. (2001) Band gap engineering of amorphous silicon quantum dots for light-emitting diodes. *Appl. Phys. Lett.*, **78**, 2575–2577.

14 Pavesi, L., Dal Negro, L., Mazzoleni, C., Franzo, G., and Priolo, F. (2000) Optical gain in silicon nanocrystals. *Nature*, **408**, 440–444.

15 Baron, T., Mazen, F., Busseret, C., Souifi, A., Mur, P., Fournel, F., Séméria, M.N., Moriceau, H., Aspard, B., Gentile, P., and Magnea, N. (2002) Nucleation control of CVD growth silicon nanocrystals for quantum devices. *Microelectron. Eng.*, **61–62**, 511–515.

16 Lalic, N. and Linnros, J. (1998) Light emitting diode structure based on Si nanocrystals formed by implantation into thermal oxide. *J. Lumin.*, **80**, 263–267.

17 Lu, Z.H., Lockwood, D.J., and Baribeau, J.M. (1995) Quantum confinement and light emission in SiO_2/Si superlattices. *Nature*, **378**, 258–260.

18 Tsybeskov, L., Hirschman, K.D., Duttagupta, S.P., Zacharias, M., Fauchet, P.M., McCaffrey, J.P., and Lockwood, D.J. (1998) Nanocrystalline-silicon superlattice produced by controlled recrystallization. *Appl. Phys. Lett.*, **72**, 43–45.

19 Grom, G.F., Fauchet, P.M., Tsybeskov, L., McCaffrey, J.P., Labbé, H.J., Lockwood, D.J., and White, B.E. (2001) Raman spectroscopy of Si nanocrystals in nanocrystalline Si superlattices: size, shape and crystallographic orientation. *Mater. Res. Soc. Symp. Proc.*, **638**, F6.1.1–F6.1.6.

20 Orii, T., Hirasawa, M., and Seto, T. (2003) Tunable, narrow-band light emission from size-selected Si nanoparticles produced by pulsed-laser ablation. *Appl. Phys. Lett.*, **83**, 3395–3397.

21 Lockwood, D.J., Sullivan, B.T., and Labb, H.J. (1998) Visible light emission from Si/SiO_2 superlattices in optical microcavities. *J. Lumin.*, **80**, 75–79.

22 Dal Negro, L., Cazzanelli, M., Pavesi, L., Ossicini, S., Pacifici, D., Franz, G., Priolo, F., and Iacona, F. (2003) Dynamics of stimulated emission in silicon nanocrystals. *Appl. Phys. Lett.*, **82**, 4636–4638.

23 Cazzanelli, M., Navarro-Urriós, D., Riboli, F., Daldosso, N., Pavesi, L., Heitmann, J., Yi, L.X., Scholz, R., Zacharias, M., and Gösele, U. (2004) Optical gain in monodispersed silicon nanocrystals. *J. Appl. Phys.*, **96**, 3164–3171.

24 Cloutier, S.G., Kossyrev, P.A., and Xu, J. (2005) Optical gain and stimulated emission in periodic nanopatterned crystalline silicon. *Nat. Mater.*, **4**, 887–891.

25 Fauchet, P.M., Ruan, J., Chen, H., Pavesi, L., Dal Negro, L., Cazzaneli, M., Elliman, R.G., Smith, N., Samoc, M., and Luther-Davies, B. (2005) Optical gain in different silicon nanocrystal systems. *Opt. Mater.*, **27**, 745–749.

26 Cui, Y., Duan, X., Hu, J., and Lieber, C.M. (2000) Doping and electrical transport in silicon nanowires. *J. Phys. Chem. B*, **104**, 5215–5216.

27 Cui, Y. and Lieber, C.M. (2001) Functional nanoscale electronic devices assembled using silicon nanowire building blocks. *Science*, **291**, 851–853.

28 Wu, Y., Fan, R., and Yang, P. (2002) Block-by-block growth of single-crystalline Si/

SiGe superlattice nanowires. *Nano Lett.*, **2**, 83–86.

29 Deng, X. and Krishnamurthy, M. (1998) Self-assembly of quantum-dot molecules: heterogeneous nucleation of SiGe islands on Si(100). *Phys. Rev. Lett.*, **81**, 1473–1476.

30 Teichert, C., Lagally, M.G., Peticolas, L.J., Bean, J.C., and Tersoff, J. (1996) Stress-induced self-organization of nanoscale structures in SiGe/Si multilayer films. *Phys. Rev. B*, **53**, 16334–16337.

31 Floro, J.A., Sinclair, M.B., Chason, E., Freund, L.B., Twesten, R.D., Hwang, R.Q., and Lucadamo, G.A. (2000) Novel SiGe island coarsening kinetics: Ostwald ripening and elastic interactions. *Phys. Rev. Lett.*, **84**, 701–704.

32 Holmes, J.D., Johnston, K.P., Doty, R.C., and Korgel, B.A. (2000) Control of thickness and orientation of solution-grown silicon nanowires. *Science*, **287**, 1471–1473.

33 Wu, Y., Cui, Y., Huynh, L., Barrelet, C.J., Bell, D.C., and Lieber, C.M. (2004) Controlled growth and structures of molecular-scale silicon nanowires. *Nano Lett.*, **4**, 433–436.

34 Hu, J., Odom, T.W., and Lieber, C.M. (1999) Chemistry and physics in one dimension: synthesis and properties of nanowires and nanotubes. *Acc. Chem. Res.*, **32**, 435–445.

35 Hochbaum, A.I., Chen, R., Delgado, R.D., Liang, W., Garnett, E.C., Najarian, M., Majumdar, A., and Yang, P. (2008) Enhanced thermoelectric performance of rough silicon nanowires. *Nature*, **451**, 163–167.

36 Zhou, X.T., Hu, J.Q., Li, C.P., Ma, D.D.D., Lee, C.S., and Lee, S.T. (2003) Silicon nanowires as chemical sensors. *Chem. Phys. Lett.*, **369**, 220–224.

37 Zheng, G., Patolsky, F., Cui, Y., Wang, W.U., and Lieber, C.M. (2005) Multiplexed electrical detection of cancer markers with nanowire sensor arrays. *Nat. Biotechnol.*, **23**, 1294–1301.

38 Patolsky, F. and Lieber, C.M. (2005) Nanowire nanosensors. *Mater. Today*, **8**, 20–28.

39 Gupta, R., Yoo, W.J., Wang, Y., Tan, Z., Samudra, G., Lee, S., Chan, D.S.H., Loh, K.P., Bera, L.K., Balasubramanian, N., and Kwong, D.L. (2004) Formation of SiGe nanocrystals in HfO$_2$ using *in situ* chemical vapor deposition for memory applications. *Appl. Phys. Lett.*, **84**, 4331–4333.

40 Liu, Y., Tang, S., and Banerjee, S.K. (2006) Tunnel oxide thickness dependence of activation energy for retention time in SiGe quantum dot flash memory. *Appl. Phys. Lett.*, **88**, 213504.

41 Liu, Y., Dey, S., Tang, S., Kelly, D.Q., Sarkar, J., and Banerjee, S.K. (2006) Improved performance of SiGe nanocrystal memory with VARIOT tunnel barrier. *IEEE Trans. Electron Devices*, **53**, 2598–2602.

42 Tang, Y.S., Ni, W.X., Sotomayor Torres, C.M., and Hansson, G.V. (1995) Fabrication and characterisation of Si-Si$_{0.7}$Ge$_{0.3}$ quantum dot light emitting diodes. *Electron. Lett.*, **31**, 1385–1386.

43 Eberl, K., Schmidt, O.G., Duschl, R., Kienzle, O., Ernst, E., and Rau, Y. (2000) Self-assembling SiGe and SiGeC nanostructures for light emitters and tunneling diodes. *Thin Solid Films*, **369**, 33–38.

44 Zakharov, N.D., Talalaev, V.G., Werner, P., Tonkikh, A.A., and Cirlin, G.E. (2003) Room-temperature light emission from a highly strained Si/Ge superlattice. *Appl. Phys. Lett.*, **83**, 3084–3086.

45 Chang, W.H., Chou, A.T., Chen, W.Y., Chang, H.S., Hsu, T.M., Pei, Z., Chen, P.S., Lee, S.W., Lai, L.S., Lu, S.C., and Tsai, M.J. (2003) Room-temperature electroluminescence at 1.3 and 1.5 µm from Ge/Si self-assembled quantum dots. *Appl. Phys. Lett.*, **83**, 2958–2960.

46 Tsybeskov, L., Lee, E.-K., Chang, H.-Y., Kamenev, B.V., Lockwood, D.J., Baribeau, J.-M., and Kamins, T.I. (2008) Three-dimensional silicon-germanium nanostructures for CMOS compatible light emitters and optical interconnects. *Adv. Opt. Technol.*, 218032.

47 Glassbrenner, C.J. and Slack, G.A. (1964) Thermal conductivity of silicon and germanium from 3 °K to the melting point. *Phys. Rev. A*, **134**, A1058.

48 Cahill, D.G., Katiyar, M., and Abelson, J.R. (1994) Thermal conductivity of a-Si: H thin films. *Phys. Rev. B*, **50**, 6077.

49 Dismukes, J.P., Ekstrom, L., Steigmeier, E.F., Kudman, I., and Beers, D.S. (1964) Thermal and electrical properties of heavily doped Ge-Si alloys up to 1300 °K. *J. Appl. Phys.*, **35**, 2899–2907.

50 Kittel, C. (2005) *Introduction to Solid State Physics*, 8th edn, John Wiley & Sons, Inc., Hoboken, NJ.

51 Berman, R. (1976) *Thermal Conduction in Solids, Oxford Studies in Physics* (eds B. Bleaney, D.W. Sciama, and D.H. Wilkinson), Oxford University Press, Oxford.

52 Sze, S.M. (1981) *Physics of Semiconductor Devices*, 2nd edn, John Wiley & Sons, Inc., New York.

53 Ashcroft, N.W. and Mermin, N.D. (1976) *Solid State Physics*, Holt, Rinehart and Winston, New York.

54 Holland, M.G. (1963) Analysis of lattice thermal conductivity. *Phys. Rev. A*, **132**, 2461–2471.

55 Joshi, Y.P. and Verma, G.S. (1970) Analysis of phonon conductivity: application to Si. *Phys. Rev. B*, **1**, 750.

56 Narumanchi, S.V.J., Murthy, J.Y., and Amon, C.H. (2004) Submicron heat transport model in silicon accounting for phonon dispersion and polarization. *J. Heat Transfer*, **126**, 946–955.

57 Gomes, C.J., Madrid, M., Goicochea, J.V., and Amon, C.H. (2006) In-plane and out-of-plane thermal conductivity of silicon thin films predicted by molecular dynamics. *J. Heat Transfer*, **128**, 1114–1121.

58 Jigami, T., Kobayashi, M., Taguchi, Y., and Nagasaka, Y. (2007) Development of nanoscale temperature measurement technique using near-field fluorescence. *Int. J. Thermophys.*, **28**, 968–979.

59 Abel, M.R., Wright, T.L., King, W.P., and Graham, S. (2007) Thermal metrology of silicon microstructures using Raman spectroscopy. *IEEE Trans. Compon. Packag. Technol.*, **30**, 200–208.

60 Halliday, D., Resnick, R., and Walker, J. (1997) *Fundamentals of Physics*, 5th edn, John Wiley & Sons Inc., New York.

61 Lee, S.M. and Cahill, D.G. (1997) Heat transport in thin dielectric films. *J. Appl. Phys.*, **81**, 2590–2595.

62 Jacquot, A., Lenoir, B., Dauscher, A., Stölzer, M., and Meusel, J. (2002) Numerical simulation of the 3ω method for measuring the thermal conductivity. *J. Appl. Phys.*, **91**, 4733.

63 Borca-Tasciuc, T., Kumar, A.R., and Chen, G. (2001) Data reduction in 3ω method for thin-film thermal conductivity determination. *Rev. Sci. Instrum.*, **72**, 2139–2147.

64 Cahill, D.G. (1990) Thermal conductivity measurement from 30 to 750 K: the 3ω method. *Rev. Sci. Instrum.*, **61**, 802–808.

65 Cahill, D.G. and Pohl, R.O. (1987) Thermal conductivity of amorphous solids above the plateau. *Phys. Rev. B*, **35**, 4067–4073.

66 Yang, B., Liu, J.L., Wang, K.L., and Chen, G. (2002) Simultaneous measurements of Seebeck coefficient and thermal conductivity across superlattice. *Appl. Phys. Lett.*, **80**, 1758.

67 Chen, G., Zhou, S.Q., Yao, D.Y., Kim, C.J., Zheng, X.Y., Liu, Z.L., Wang, K.L., Sun, X., and Dresselhaus, M.S. (1998) Heat conduction in alloy-based superlattices. 17th International Conference on Thermoelectrics, Nagoya, Japan, May 24–28, 1998.

68 Chen, G. and Neagu, M. (1997) Thermal conductivity and heat transfer in superlattices. *Appl. Phys. Lett.*, **71**, 2761–2763.

69 Ju, Y.S. and Goodson, K.E. (1999) Phonon scattering in silicon films with thickness of order 100 nm. *Appl. Phys. Lett.*, **74**, 3005–3007.

70 Ju, Y.S., Kurabayashi, K., and Goodson, K.E. (1999) Thermal characterization of anisotropic thin dielectric films using harmonic Joule heating. *Thin Solid Films*, **339**, 160–164.

71 Borca-Tasciuc, T., Liu, W., Liu, J., Zeng, T., Song, D.W., Moore, C.D., Chen, G., Wang, K.L., Goorsky, M.S., Radetic, T., Gronsky, R., Koga, T., and Dresselhaus, M.S. (2000) Thermal conductivity of symmetrically strained Si/Ge superlattices. *Superlattices Microstruct.*, **28**, 199–206.

72 Espinosa, H.D., Prorok, B.C., and Fischer, M. (2003) A methodology for determining mechanical properties of freestanding thin films and MEMS materials. *J. Mech. Phys. Solids*, **51**, 47–67.

73 Chu, D., Touzelbaev, M., Goodson, K.E., Babin, S., and Pease, R.F. (2001) Thermal conductivity measurements of thin-film resist. *J. Vac. Sci. Technol. B*, **19**, 2874–2877.

74 Jacquot, A., Chen, G., Scherrer, H., Dauscher, A., and Lenoir, B. (2005) Improvements of on-membrane method for thin film thermal conductivity and emissivity measurements. *Sens. Actuators A*, **117**, 203–210.

75 Cahill, D.G., Ford, W.K., Goodson, K.E., Mahan, G.D., Majumdar, A., Maris, H.J., Merlin, R., and Phillpot, S.R. (2003) Nanoscale thermal transport. *J. Appl. Phys.*, **93**, 793–818.

76 Parker, W.J., Jenkins, R.J., Butler, C.P., and Abbott, G.L. (1961) Flash method of determining thermal diffusivity, heat capacity, and thermal conductivity. *J. Appl. Phys.*, **32**, 1679–1684.

77 Hatta, I. (1990) Thermal diffusivity measurements of thin films and multilayered composites. *Int. J. Thermophys.*, **11**, 293–303.

78 Capinski, W.S. and Maris, H.J. (1996) Improved apparatus for picosecond pump-and-probe optical measurements. *Rev. Sci. Instrum.*, **67**, 2720–2726.

79 Capinski, W.S., Maris, H.J., Ruf, T., Cardona, M., Ploog, K., and Katzer, D.S. (1999) Thermal-conductivity measurements of GaAs/AlAs superlattices using a picosecond optical pump-and-probe technique. *Phys. Rev. B*, **59**, 8105–8122.

80 Käding, O.W., Skurk, H., and Goodson, K.E. (1994) Thermal conduction in metallized silicon-dioxide layers on silicon. *Appl. Phys. Lett.*, **65**, 1629–1631.

81 Goodson, K.E., Käding, O.W., Rösler, M., and Zachai, R. (1995) Experimental investigation of thermal conduction normal to diamond–silicon boundaries. *J. Appl. Phys.*, **77**, 1385–1392.

82 Harata, A., Nishimura, H., and Sawada, T. (1990) Laser-induced surface acoustic waves and photothermal surface gratings generated by crossing two pulsed laser beams. *Appl. Phys. Lett.*, **57**, 132–134.

83 Stolk, P.A., Saris, F.W., Berntsen, A.J.M., van der Weg, W.F, Sealy, L.T., Barklie, R.C., Krotz, G., and Muller, G. (1994) Contribution of defects to electronic, structural, and thermodynamic properties of amorphous silicon. *J. Appl. Phys.*, **75**, 7266–7286.

84 Cahill, D.G., Goodson, K., and Majumdar, A. (2002) Thermometry and thermal transport in micro/nanoscale solid-state devices and structures. *J. Heat Transfer*, **124**, 223–241.

85 Temple, P.A. and Hathaway, C.E. (1973) Multiphonon Raman spectrum of silicon. *Phys. Rev. B*, **7**, 3685.

86 Arshinov, Y.F., Bobrovnikov, S.M., Zuev, V.E., and Mitev, V.M. (1983) Atmospheric temperature measurements using a pure rotational Raman lidar. *Appl. Opt.*, **22**, 2984–2990.

87 Michalski, L., Eckersdorf, K., and McGhee, J. (2001) *Temperature Measurement*, 2nd edn, John Wiley & Sons, Inc., New York, p. 501.

88 Hart, T.R., Aggarwal, R.L., and Lax, B. (1970) Temperature dependence of Raman scattering in silicon. *Phys. Rev. B*, **1**, 638.

89 Viera, G., Huet, S., and Boufendi, L. (2001) Crystal size and temperature measurements in nanostructured silicon using Raman spectroscopy. *J. Appl. Phys.*, **90**, 4175–4183.

90 Périchon, S., Lysenko, V., Remaki, B., Barbier, D., and Champagnon, B. (1999) Measurement of porous silicon thermal conductivity by micro-Raman scattering. *J. Appl. Phys.*, **86**, 4700–4702.

91 Camassel, J., Falkovsky, L.A., and Planes, N. (2001) Strain effect in silicon-on-insulator materials: investigation with optical phonons. *Phys. Rev. B*, **63**, 353091–3530911.

92 Alonso, M.I. and Winer, K. (1989) Raman spectra of c-$Si_{1-x}Ge_x$ alloys. *Phys. Rev. B*, **39**, 10056.

93 Mishra, P. and Jain, K.P. (2000) Temperature-dependent Raman scattering studies in nanocrystalline silicon and finite-size effects. *Phys. Rev. B*, **62**, 14790–14795.

94 Kash, J.A., Tsang, J.C., and Hvam, J.M. (1985) Subpicosecond time-resolved Raman spectroscopy of LO phonons in GaAs. *Phys. Rev. Lett.*, **54**, 2151–2154.

95 Pangilinan, G.I. and Gupta, Y.M. (1997) Use of time-resolved Raman scattering to determine temperatures in shocked carbon tetrachloride. *J. Appl. Phys.*, **81**, 6662–6669.

96 Tsen, K.T., Ferry, D.K., Botchkarev, A., Sverdlov, B., Salvador, A., and Morkoc, H. (1998) Time-resolved Raman studies of the decay of the longitudinal optical phonons in wurtzite GaN. *Appl. Phys. Lett.*, **72**, 2132–2134.

97 Von Der Linde, D., Kuhl, J., and Klingenberg, H. (1980) Raman scattering from nonequilibrium LO phonons with picosecond resolution. *Phys. Rev. Lett.*, **44**, 1505–1508.

98 Lowndes, D.H., Jellison, G.E., Jr., and Wood, R.F. (1982) Time-resolved optical studies of silicon during nanosecond pulsed-laser irradiation. *Phys. Rev. B*, **26**, 6747–6755.

99 Compaan, A., Lee, M.C., and Trott, G.J. (1985) Phonon populations by nanosecond-pulsed Raman scattering in Si. *Phys. Rev. B*, **32**, 6731–6741.

100 Trinh, P.D., Yegnanarayanan, S., Coppinger, F., and Jalali, B. (1997) Silicon-on-insulator (SOI) phased-array wavelength multi/demultiplexer with extremely low-polarization sensitivity. *IEEE Photonics Technol. Lett.*, **9**, 940–942.

101 Shahidi, G.G. (2001) SOI technology for the GHz era. Proceedings of the International Symposium on VLSI Technology, Systems, and Applications.

102 Alles, M. and Wilson, S. (1997) Thin film silicon on insulator: an enabling technology. *Semicond. Int.*, **20**, 67–74.

103 Lasky, J.B. (1986) Wafer bonding for silicon-on-insulator technologies. *Appl. Phys. Lett.*, **48**, 78–80.

104 Gösele, U. and Tong, Q.Y. (1998) Semiconductor wafer bonding. *Annu. Rev. Mater. Sci.*, **28**, 215–241.

105 Su, L.T., Chung, J.E., Antoniadis, D.A., Goodson, K.E., and Flik, M.I. (1994) Measurement and modeling of self-heating in SOI nMOSFET's. *IEEE Trans Electron Devices*, **41**, 69–75.

106 Zou, J. and Balandin, A. (2001) Phonon heat conduction in a semiconductor nanowire. *J. Appl. Phys.*, **89**, 2932–2938.

107 Li, D., Wu, Y., Kim, P., Shi, L., Yang, P., and Majumdar, A. (2003) Thermal conductivity of individual silicon nanowires. *Appl. Phys. Lett.*, **83**, 2934–2936.

108 Volz, S.G. and Chen, G. (1999) Molecular dynamics simulation of thermal conductivity of silicon nanowires. *Appl. Phys. Lett.*, **75**, 2056–2058.

109 Mingo, N. (2003) Calculation of Si nanowire thermal conductivity using complete phonon dispersion relations. *Phys. Rev. B*, **68**, 1133081–1133084.

110 Chen, Y., Li, D., Lukes, J.R., and Majumdar, A. (2005) Monte Carlo simulation of silicon nanowire thermal conductivity. *J. Heat Transfer*, **127**, 1129–1137.

111 Shi, L., Li, D., Yu, C., Jang, W., Kim, D., Yao, Z., Kim, P., and Majumdar, A. (2003) Measuring thermal and thermoelectric properties of one-dimensional nano-structures using a microfabricated device. *J. Heat Trans Transfer*, **125**, 881–888.

112 Rowe, D.M. and Shukla, V.S. (1981) The effect of phonon-grain boundary scattering on the lattice thermal conductivity and thermoelectric conversion efficiency of heavily doped fine-grained, hot-pressed silicon germanium alloy. *J. Appl. Phys.*, **52**, 7421–7426.

113 Scoville, N., Bajgar, C., Rolfe, J., Fleurial, J.P., and Vandersande, J. (1995) Thermal conductivity reduction in SiGe alloys by the addition of nanophase particles. *Nanostruct. Mater.*, **5**, 207–223.

114 Liu, W.L., Borca-Tasciuc, T., Chen, G., Liu, J.L., and Wang, K.L. (2001) Anisotropic thermal conductivity of Ge quantum-dot and symmetrically strained Si/Ge superlattices. *J. Nanosci. Nanotechnol.*, **1**, 39–42.

115 Baribeau, J.M., Rowell, N.L., and Lockwood, D.J. (2005) Advances in the growth and characterization of Ge quantum dots and islands. *J. Mater. Res.*, **20**, 3278–3293.

116 Tsybeskov, L., Kamenev, B.V., Baribeau, J.M., and Lockwood, D.J. (2006) Optical properties of composition-controlled three-dimensional Si/Si$_{1-x}$Ge$_x$

nanostructures. *IEEE J. Sel. Top. Quantum Electron.*, **12**, 1579–1584.

117 Lockwood, D.J., Baribeau, J.M., Kamenev, B.V., Lee, E.K., and Tsybeskov, L. (2008) Structural and optical properties of three-dimensional $Si_{1-x}Ge_x/Si$ nanostructures. *Semicond. Sci. Technol.*, **23**, 064003.

118 Chang, H.-Y., Tsybeskov, L., Sirenko, A.A., Lockwood, D.J., Baribeau, J.-M., Wu, X., and Dharma-Wardana, M.W.C. Raman measurements of surface temperature in three- and two-dimensional SiGe nanostructures with a high Ge concentration. Unpublished.

119 Fan, X., Zeng, G., LaBounty, C., Bowers, J.E., Croke, E., Ahn, C.C., Huxtable, S., Majumdar, A., and Shakouri, A. (2001) SiGeC/Si superlattice microcoolers. *Appl. Phys. Lett.*, **78**, 1580–1582.

120 Yang, B., Liu, W.L., Liu, J.L., Wang, K.L., and Chen, G. (2002) Measurements of anisotropic thermoelectric properties in superlattices. *Appl. Phys. Lett.*, **81**, 3588.

121 Chakraborty, S., Kleint, C.A., Heinrich, A., Schneider, C.M., Schumann, J., Falke, M., and Teichert, S. (2003) Thermal conductivity in strain symmetrized Si/Ge superlattices on Si(111). *Appl. Phys. Lett.*, **83**, 4184–4186.

122 Ezzahri, Y., Grauby, S., Dilhaire, S., Rampnoux, J.-M., Claeys, W., Zhang, Y., and Shakouri, A. (2006) Determination of thermophysical properties of Si/SiGe superlattices with a pump-probe technique. 2005 International Workshop on Thermal Investigation of ICs and Systems, pp. 235–243.

123 Christofferson, J., Maize, K., Ezzahri, Y., Shabani, J., Wang, X., and Shakouri, A. (2007) Microscale and nanoscale thermal characterization techniques. Proceedings of the 1st International Conference on Thermal Issues in Emerging Technologies, Theory and Applications (ThETA1).

124 Ezzahri, Y., Grauby, S., Dilhaire, S., Rampnoux, J.M., and Claeys, W. (2007) Cross-plan Si/SiGe superlattice acoustic and thermal properties measurement by picosecond ultrasonics. *J. Appl. Phys.*, **101**, 013705.

125 Khitun, A., Balandin, A., Liu, J.L., and Wang, K.L. (2000) In-plane lattice thermal

conductivity of a quantum-dot superlattice. *J. Appl. Phys.*, **88**, 696–699.

126 Dharma-Wardana, M.W.C., MacDonald, A.H., Lockwood, D.J., Baribeau, J.M., and Houghton, A.D.C. (1987) Raman scattering in Fibonacci superlattices. *Phys. Rev. Lett.*, **58**, 1761–1764.

127 Lockwood, D.J., Dharma-Wardana, M.W.C., Baribeau, J.M., and Houghton, D.C. (1987) Folded acoustic phonons in Si/Ge_xSi_{1-x} strained-layer superlattices. *Phys. Rev. B*, **35**, 2243.

128 Liu, J.L., Jin, G., Tang, Y.S., Luo, Y.H., Wang, K.L., and Yu, D.P. (2000) Optical and acoustic phonon modes in self-organized Ge quantum dot superlattices. *Appl. Phys. Lett.*, **76**, 586–588.

129 Tan, P.H., Bougeard, D., Abstreiter, G., and Brunner, K. (2004) Raman scattering of folded acoustic phonons in self-assembled Si/Ge dot superlattices. *Appl. Phys. Lett.*, **84**, 2632–2634.

130 Cazayous, M., Huntzinger, J.R., Groenen, J., Mlayah, A., Christiansen, S., Strunk, H.P., Schmidt, O.G., and Eberl, K. (2000) Resonant Raman scattering by acoustical phonons in Ge/Si self-assembled quantum dots: interferences and ordering effects. *Phys. Rev. B*, **62**, 7243–7248.

131 Venkatasubramanian, R. (2000) Lattice thermal conductivity reduction and phonon localizationlike behavior in superlattice structures. *Phys. Rev. B*, **61**, 3091–3097.

132 Uhlir, A., Jr. and Uhlir, I. (2005) Historical perspective on the discovery of porous silicon. *Phys. Status Solidi C*.

133 Canham, L.T. (1990) Silicon quantum wire array fabrication by electrochemical and chemical dissolution of wafers. *Appl. Phys. Lett.*, **57**, 1046–1048.

134 Lehmann, V. and Gösele, U. (1991) Porous silicon formation: a quantum wire effect. *Appl. Phys. Lett.*, **58**, 856–858.

135 Vial, J.C., Bsiesy, A., Gaspard, F., Hérino, R., Ligeon, M., Muller, F., Romestain, R., and Macfarlane, R.M. (1992) Mechanisms of visible-light emission from electro-oxidized porous silicon. *Phys. Rev. B*, **45**, 14171–14176.

136 Gesele, G., Linsmeier, J., Drach, V., Fricke, J., and Arens-Fischer, R. (1997)

Temperature-dependent thermal conductivity of porous silicon. *J. Phys. D: Appl. Phys.*, **30**, 2911–2916.

137 Spanier, J.E. and Herman, I.P. (2000) Use of hybrid phenomenological and statistical effective-medium theories of dielectric functions to model the infrared reflectance of porous SiC films. *Phys. Rev. B*, **61**, 10437–10450.

138 Périchon, S., Lysenko, V., Roussel, P., Remaki, B., Champagnon, B., Barbier, D., and Pinard, P. (2000) Technology and micro-Raman characterization of thick meso-porous silicon layers for thermal effect microsystems. *Sens. Actuators A*, **85**, 335–339.

139 Scheel, H., Reich, S., Ferrari, A.C., Cantoro, M., Colli, A., and Thomsen, C. (2006) Raman scattering on silicon nanowires: the thermal conductivity of the environment determines the optical phonon frequency. *Appl. Phys. Lett.*, **88**, 233114.

140 Koropecki, R.R., Arce, R.D., and Schmidt, J.A. (2004) Infrared studies combined with hydrogen effusion experiments on nanostructured porous silicon. *J. Non-Cryst. Solids*, **338–340**, 159–162.

141 Salonen, J., Lehto, V.P., and Laine, E. (1999) Photo-oxidation studies of porous silicon using a microcalorimetric method. *J. Appl. Phys.*, **86**, 5888–5893.

142 El Houichet, H., Oueslati, M., Bessaïs, B., and Ezzaouia, H. (1997) Photo-luminescence enhancement and degradation in porous silicon: evidence for nonconventional photoinduced defects. *J. Lumin.*, **71**, 77–82.

143 Czaputa, R., Fritzl, R., and Popitsch, A. (1995) Anomalous luminescence degradation behaviour of chemically oxidized porous silicon. *Thin Solid Films*, **255**, 212–215.

144 Juan, M., Bouillard, J.S., Plain, J., Bachelot, R., Adam, P.M., Lerondel, G., and Royer, P. (2007) Soft photo structuring of porous silicon in water. *Phys. Status Solidi A*, **204**, 1276–1280.

145 Elhouichet, H., Bessaïs, B., Younes, O.B., Ezzaouia, H., and Oueslati, M. (1998) Origin of the photoluminescence shifts in porous silicon. *EPJ Appl. Phys.*, **1**, 153–157.

146 Lockwood, D.J. (1993) Optical properties of porous silicon, in *Optical Phenomena in Semiconductor Structures of Reduced Dimensions* (eds D.J. Lockwood and A. Pinczuk), Springer, New York, pp. 409–426.

147 Kovalev, D., Heckler, H., Polisski, G., Diener, J., and Koch, F. (2001) Optical properties of silicon nanocrystals. *Opt. Mater.*, **17**, 35–40.

148 Tischler, M.A., Collins, R.T., Stathis, J.H., and Tsang, J.C. (1992) Luminescence degradation in porous silicon. *Appl. Phys. Lett.*, **60**, 639–641.

149 Chang, I.M., Pan, S.C., and Chen, Y.F. (1993) Light-induced degradation on porous silicon. *Phys. Rev. B*, **48**, 8747–8750.

150 Simons, A.J., Cox, T.I., Loni, A., Canham, L.T., and Blacker, R. (1997) Investigation of the mechanisms controlling the stability of a porous silicon electroluminescent device. *Thin Solid Films*, **297**, 281–284.

151 Koyama, H. and Fauchet, P.M. (2000) Laser-induced thermal effects on the optical properties of free-standing porous silicon films. *J. Appl. Phys.*, **87**, 1788–1794.

152 Balucani, M., Bondarenko, V., Franchina, L., Lamedica, G., Yakovtseva, V.A., and Ferrari, A. (1999) A model of radiative recombination in n-type porous silicon–aluminum Schottky junction. *Appl. Phys. Lett.*, **74**, 1960–1962.

153 Yoshida, T., Yamada, Y., and Orii, T. (1998) Electroluminescence of silicon nanocrystallites prepared by pulsed laser ablation in reduced pressure inert gas. *J. Appl. Phys.*, **83**, 5427–5432.

154 Zacharias, M., Bläsing, J., Veit, P., Tsybeskov, L., Hirschman, K., and Fauchet, P.M. (1999) Thermal crystallization of amorphous Si/SiO$_2$ superlattices. *Appl. Phys. Lett.*, **74**, 2614–2616.

155 Zacharias, M., Heitmann, J., Scholz, R., Kahler, U., Schmidt, M., and Bläsing, J. (2002) Size-controlled highly luminescent silicon nanocrystals: a SiO/SiO$_2$ superlattice approach. *Appl. Phys. Lett.*, **80**, 661.

156 Heitmann, J., Müller, F., Yi, L., Zacharias, M., Kovalev, D., and Eichhorn, F. (2004) Excitons in Si nanocrystals: Confinement and migration effects. *Phys. Rev. B*, **69**, 195309.

157 Tsybeskov, L., Grom, G.F., Jungo, M., Montes, L., Fauchet, P.M., McCaffrey, J.P., Baribeau, J.M., Sproule, G.I., and Lockwood, D.J. (2000) Nanocrystalline silicon superlattices: building blocks for quantum devices. *Mater. Sci. Eng. B*, **69**, 303–308.

158 Tsybeskov, L. and Lockwood, D.J. (2003) Nanocrystalline silicon/silicon dioxide superlattices: structural and optical properties, in *Semiconductor Nanocrystals: From Basic Principles to Applications* (eds A.L. Efros, D.J. Lockwood, and L. Tsybeskov), Springer, New York, pp. 209–238.

159 Grom, G.F., Lockwood, D.J., McCaffrey, J.P., Labb, H.J., Fauchet, P.M., White, B., Jr., Diener, J., Kovalev, D., Koch, F., and Tsybeskov, L. (2000) Ordering and self-organization in nanocrystalline silicon. *Nature*, **407**, 358–361.

160 Okada, T., Iwaki, T., Kasahara, H., and Yamamoto, K. (1985) Probing the crystallinity of evaporated silicon films by Raman scattering. *Jpn. J. Appl. Phys. 1*, **24**, 161–165.

161 Nakajima, A., Sugita, Y., Kawamura, K., Tomita, H., and Yokoyama, N. (1996) Microstructure and optical absorption properties of Si nanocrystals fabricated with low-pressure chemical-vapor deposition. *J. Appl. Phys.*, **80**, 4006–4011.

162 Kamenev, B.V., Grebel, H., and Tsybeskov, L. (2006) Laser-induced structural modifications in nanocrystalline silicon/amorphous silicon dioxide superlattices. *Appl. Phys. Lett.*, **88**, 143117.

163 Khriachtchev, L., Räsänen, M., and Novikov, S. (2006) Laser-controlled stress of Si nanocrystals in a free-standing Si/SiO$_2$ superlattice. *Appl. Phys. Lett.*, **88**, 013102.

164 Ruan, J., Fauchet, P.M., Dal Negro, L., Cazzanelli, M., and Pavesi, L. (2003) Stimulated emission in nanocrystalline silicon superlattices. *Appl. Phys. Lett.*, **83**, 5479–5481.

165 Cody, G.D., Tiedje, T., Abeles, B., Brooks, B., and Goldstein, Y. (1981) Disorder and the optical-absorption edge of hydrogenated amorphous silicon. *Phys. Rev. Lett.*, **47**, 1480–1483.

166 Hatano, M., Moon, S., Lee, M., Suzuki, K., and Grigoropoulos, C.P. (2000) Excimer laser-induced temperature field in melting and resolidification of silicon thin films. *J. Appl. Phys.*, **87**, 36–43.

167 Akhmanov, S.A., Vladimir, I.E.y., Nikolai, I.K., and Seminogov, V.N. (1985) Interaction of powerful laser radiation with the surfaces of semiconductors and metals: nonlinear optical effects and nonlinear optical diagnostics. *Soviet Physics Uspekhi*, **28**, 1084–1124.

168 Hess, P. (1996) Laser diagnostics of mechanical and elastic properties of silicon and carbon films. *Appl. Surf. Sci.*, **106**, 429–437.

169 Vijayalakshmi, S., Grebel, H., Yaglioglu, G., Pino, R., Dorsinville, R., and White, C.W. (2000) Nonlinear optical response of Si nanostructures in a silica matrix. *J. Appl. Phys.*, **88**, 6418–6422.

170 Prakash, G.V., Cazzanelli, M., Gaburro, Z., Pavesi, L., Iacona, F., Franz, G., and Priolo, F. (2002) Linear and nonlinear optical properties of plasma-enhanced chemical-vapour deposition grown silicon nanocrystals. *J. Mod. Opt.*, **49**, 719–730.

171 Balandin, A. and Wang, K.L. (1998) Significant decrease of the lattice thermal conductivity due to phonon confinement in a free-standing semiconductor quantum well. *Phys. Rev. B*, **58**, 1544–1549.

172 Chantrenne, P., Barrat, J.L., Blase, X., and Gale, J.D. (2005) An analytical model for the thermal conductivity of silicon nanostructures. *J. Appl. Phys.*, **97**, 104318.

173 Balandin, A. (2000) Thermal properties of semiconductor low-dimensional structures. *Phys. Low Dimens. Struct.*, **2000**, 1–28.

174 Chen, G. and Shakouri, A. (2002) Heat transfer in nanostructures for solid-state energy conversion. *J. Heat Transfer*, **124**, 242–252.

175 Tiedje, T., Cebulka, J.M., Morel, D.L., and Abeles, B. (1981) Evidence for exponential band tails in amorphous silicon hydride. *Phys. Rev. Lett.*, **46**, 1425–1428.

176 Street, R.A. (1991) *Hydrogenated Amorphous Silicon*, illustrated edn, Cambridge University Press.

177 Lebedev, E.A., Smorgonskaya, E.A., and Polisski, G. (1998) Drift mobility of excess carriers in porous silicon. *Phys. Rev. B*, **57**, 14607–14610.

178 Rao, P.N., Schiff, E.A., Tsybeskov, L., and Fauchet, P. (2002) Photocarrier drift-mobility measurements and electron localization in nanoporous silicon. *Chem. Phys.*, **284**, 129–138.

179 Duzhko, V. and Tsybeskov, L. (2003) Time-resolved carrier tunneling in nanocrystalline silicon/amorphous silicon dioxide superlattices. *Appl. Phys. Lett.*, **83**, 5229–5231.

6
Surface Passivation and Functionalization of Si Nanocrystals

Jonathan Veinot

6.1
Introduction

By their very nature, nanoparticles exhibit enormous surface area-to-volume ratios; according to one estimate, a 2.0 nm diameter icosahedral silicon particle possesses approximately 280 Si atoms with 120 (43%) residing at the surface of the particle [1]. Considering this point alone, it is reasonable that surface properties, including chemistry, of these versatile materials will strongly influence their material characteristics. Many examples exist for a variety of nanoparticle systems (e.g., Au and CdSe), where tailoring the surface of said particles offers the attractive possibility of controlling their interactions with their surroundings. A specific and dramatic example pertaining to Si nanomaterials is the influence of surface groups on the optical properties of porous silicon (p-Si). When a carbon–carbon multiple bond is in close bonding proximity to the p-Si surface, the photoluminescence intensity of p-Si is diminished, and if this double bond is in conjugation with an aromatic structure (e.g., styrenyl), the characteristic emission is completely quenched [2].

Unlike the widely studied chalcogenide-based methods for surface modification of gold and compound semiconductor (e.g., CdSe) particles that rely on exchangeable bonding interactions, much of the surface modification of silicon nanomaterials has employed the more robust polar covalent bonding of silicon–carbon and silicon–nitrogen and in some cases silicon–oxygen linkages. Numerous approaches to obtaining the first of these bonding configurations (e.g., hydrosilylation and halogen displacement) have been presented; still, questions and challenges regarding the effectiveness of Si-nc surface passivation and the associated particle stability remain. This is particularly evident when the effectiveness of surface coverage (i.e., about 50% for bulk Si(111) and 80% for bulk Si(100) surfaces) [3] and the bond strengths of various species present at particle surfaces are considered (see Table 6.1) [4].

The following discussion includes key studies of foundational silicon surface chemistry as it pertains to methods for controlling silicon nanocrystal (Si-nc) surfaces.

Silicon Nanocrystals: Fundamentals, Synthesis and Applications. Edited by Lorenzo Pavesi and Rasit Turan
Copyright © 2010 WILEY-VCH Verlag GmbH & Co. KGaA, Weinheim
ISBN: 978-3-527-32160-5

Table 6.1 Selected bond energies relevant when considering modification of silicon nanocrystal surfaces.

Bond	Bond energy (kJ mol^{-1})
Si—H	323
Si—C	369
Si—Cl	391
Si—O	368
Si—Si (bulk)	210–250
Si—Si (disilane)	310–340
Si—Si (disilene)	105–126

6.2
Functionalizing Freestanding Particles

Effective functionalization of Si-nc relies upon realizing a suitably reactive surface. Such surfaces may be prepared by numerous methods including formation of reactive surface during nanoparticle preparation (e.g., grinding of silicon wafers, reaction of Zintl salts), during liberation of particles from oxide matrices (i.e., treatment with HF), and finally treatment of freestanding Si-ncs with a variety of reagents (e.g., HF, LiAlH$_4$, halogens). In many cases, the resulting reactive surfaces are subsequently primed for further modification using a variety of procedures drawn from the vast chemistry of silicon molecular analogues. A review of the methods applied to silicon nanoparticles as well as an overview of how surface chemistry was evaluated for each material will be presented.

6.3
In Situ Surface Chemistry Tailoring

One of the earliest attempts to obtain surface-functionalized silicon nanoparticles was reported by Heath in 1992 using a heterogeneous reaction mixture containing sodium dispersion to simultaneously reduce SiCl$_4$ and RSiCl$_3$ (R = H and *n*-octyl) in a high-pressure bomb reactor (see Scheme 6.1) [5]. The size-polydispersed product obtained from this reaction was confirmed to contain diamond lattice Si. Unexpectedly, products obtained from this reaction did not show significant *n*-octyl surface functionalization evident by infrared spectroscopy that showed a preponderance of Si—O, Si—Cl, and Si—H surface termination. While better size control was achieved

$$SiCl_4 + RSiCl_3 + Na \rightarrow$$
$$Si \text{ (diamond lattice)} + NaCl$$

Scheme 6.1 Preparation of Si-ncs via simultaneous heterogeneous reduction of silicon tetrachloride and alkyl trichlorosilane.

with the *n*-octyl reagent, no tailoring of the surface chemistry was realized. The explanation provided by the authors for the absence of the alkyl surface highlights the complexity of Si-nc surface chemistry and the challenges associated with controlling these materials and their properties. It was proposed that the polar covalent nature of Si−C bond in which Si is donating electron density to α-C effectively reduces the α-C−β−C bond. This electronic interaction leads to activation of β-hydrogens and β-hydride elimination yielding an Si−H surface and elimination of the 1-alkene at high temperature. As a result, while the surface-bonded octyl groups passivate the particle to further growth leading to the observation of a narrower size distribution (i.e., R = H, 5–3000 nm; R = *n*-octyl, 5.5 ± 2.5 nm), they are ultimately liberated under reaction conditions.

In 2001, Holmes *et al.* prepared sterically stabilized Si-ncs by thermally decomposing diphenyl silane in supercritical octanol. The resulting particles ranged in diameter from 1.5 to 4.0 nm, and the FT-IR analysis confirmed that a surface organic layer was indeed bonded through an alkoxide linkage (Si−O−C) (see Figure 6.1). As it will become evident in the following discussion, FT-IR is an invaluable tool in evaluating Si-nc surface chemistry; however, it is not necessarily comprehensive. An XPS-derived Si:C ratio and a shell approximation that afforded an estimated number of surface Si atoms indicated 1.5 nm diameter particles have approximately

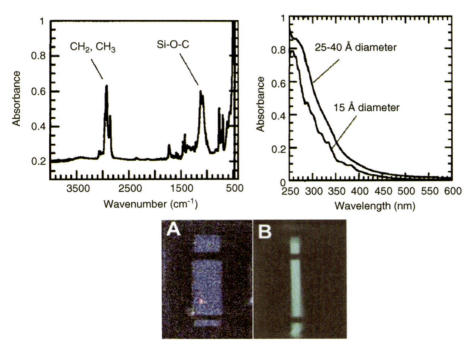

Figure 6.1 (Top left) FT-IR spectrum of alkoxy surface-terminated Si-ncs. (Top right) Absorption spectra of alkoxy-terminated Si-ncs of indicated dimensions. (Bottom) Photograph of luminescent Si-ncs in hexane upon excitation at 320 nm, A; *d* = 1.5 nm, B.

16 capping ligands that each occupy a surface area of 44 Å^2. These alcohol-soluble nanoparticles were highly photoluminescent and the smallest particles were stated to exhibit discrete energy transitions. It should be noted that despite the observation of these apparent discrete energy transitions, the authors state the absorption coefficient follows a quadratic relationship with the incident energy, indicating an indirect transition [6].

Fink and coworkers recently prepared freestanding Si-ncs via high-energy ball milling (HEBM) of high-purity silicon pieces in 1-octyne and inert atmosphere [7] (Figure 6.2). It is proposed that this method exploits fresh/clean surfaces formed during the milling process that subsequently react freely with the unsaturated carbon–carbon multiple bond in the reaction medium to yield highly crystalline, luminescent organic-soluble particles with organic surface passivation of varied surface bonding modalities as supported by FT-IR and ^{1}H NMR analysis (see Scheme 6.2). It should be noted that this contribution is one of the only few that claim to observe characteristic IR absorptions of the Si–C linkage at <1257, <806, and <796 cm^{-1} that are normally masked by oxide species that appear to be largely absent from these materials.

6.4
Aerosol-Based Functionalization

Si-ncs are readily prepared in nonthermal plasmas and further modified *in situ* to yield organic surface-stabilized materials. Liao and Roberts were the first to apply gas-phase reactions of this type to the surface modification of Si-ncs [8]. Particles were prepared from silane in a nonthermal reactor and extracted into a reactor chamber containing organic reagents of choice (e.g., amines, alkenes, and alkynes). Heating of the gas-phase mixture led to surface modification that was spectroscopically confirmed (i.e., FT-IR). Recently, Mangolini and Kortshagen modified this approach preparing Si-ncs with organic surfaces using plasma grafting [9]. The reactor shown schematically in Figure 6.3 allowed the attachment of a variety of organic molecules bearing a variety of functional groups. Particles prepared in this way are photoluminescent (QY up to 60%) [10], are crystalline, and show excellent solubility requiring no sonication. Still, FT-IR spectra of isolated particles are remarkably similar to those of solution-modified particles and clearly show the presence of characteristic absorptions that may be assigned to the organic surface and residual S–H. While the Si–C linkage is not conclusively identified, no C=C is evident consistent with a hydrosilylation reaction.

6.5
Solution-Based Postsynthetic Modification

The extensive body of literature outlining both the organometallic chemistry of silicon and the surface chemistry of bulk silicon offers a significant starting point to

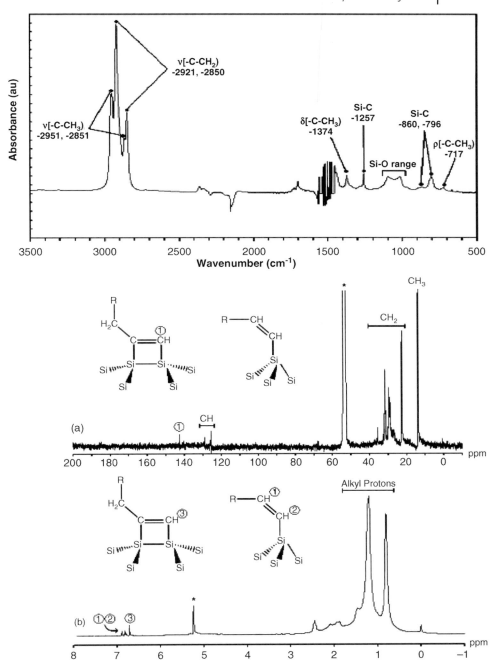

Figure 6.2 (Above) FT-IR spectrum of alkyl surface-functionalized Si-ncs prepared by high-energy ball milling. (Right) $^{13}C\{^{1}H\}$ and ^{1}H NMR spectra of Si-ncs. Reproduced with permission from Ref. [7]. Copyright Wiley-VCH Verlag GmbH.

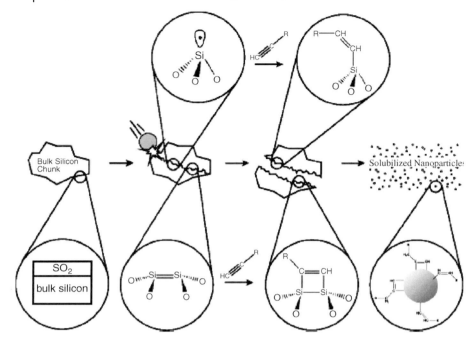

Scheme 6.2 Preparation of surface-functionalized Si-nc via high-energy ball milling. Open circles represent subsurface silicon atoms. Reproduced with permission from Ref. [7]. Copyright Wiley-VCH Verlag GmbH.

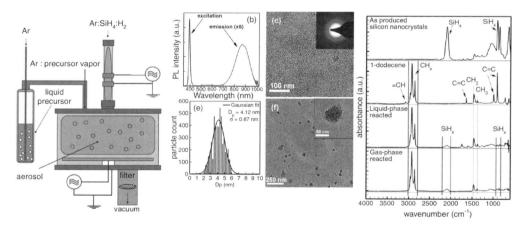

Figure 6.3 (Left) Schematic of the plasma/aerosol apparatus used to prepare surface-functionalized Si-ncs. (Middle) Optical response, size distribution, and TEM analysis of aerosol-functionalized Si-ncs. (Right) FT-IR spectra of Si–H, octadecene, Si–R (R = octadecyl) solution-functionalized Si-ncs, and Si–R (R = octadecyl) aerosol-functionalized Si-ncs. Reproduced with permission from Ref. [9]. Copyright Wiley-VCH Verlag GmbH.

tailoring Si-nc surface chemistry. Preparative methods exist that yield Si-ncs bearing a variety of surface functionalities (e.g., oxide, −H, −OH, and halogens). The following section outlines methods reported to date that offer tailoring of particle surface chemistry by directly bonding to the silicon surface and briefly discusses modification of a surface passivating oxide on Si-ncs.

6.6
Hydrosilylation

Without question, the most widely studied method for modifying the surface chemistry of Si-ncs is hydrosilylation of terminal alkenes – undoubtedly, a result of the vast body of studies pertaining to functionalization of porous-Si and bulk silicon that employs a similar methodology and the robust Si−C bond that is formed. In the light of this well-developed approach, Si-ncs functionalized using this general methodology have been reported to exhibit the highest photoluminescent quantum yields to date, which may result from particles not exposed to oxygen prior to their surface functionalization [10]. This observation further highlights the importance of surface chemistry and the effectiveness of surface-bonded alkyl groups to protect the underlying Si.

Prior to discussing the hydrosilylation reaction, and its affects on freestanding Si-ncs in detail, it is useful to note that the surface chemistry of the starting point (i.e., the Si−H-terminated surface) for this reaction exhibits subtleties that are not necessarily straightforward extensions from its bulk and porous-Si counterparts. For example, the rates of HF etching are significantly slower for nanoparticles (i.e., <1 nm/min for nanoparticles versus μm/min for silicon wafers) and there has been some suggestion that a fluorinated surface may play a role. Furthermore, it has also been reported that aqueous HF fails to remove all surface oxides [11]; however, the introduction of alcohol to the etching mixture to increase surface wettability produced oxide-free systems [12].

There are very few reported examples of detailed and systematic studies examining the aqueous HF etching process as it pertains to freestanding Si-ncs and their surface properties. Li et al. varied the concentration of aqueous HF used to etch the surface of particles recovered from their original reaction product via HF/HNO$_3$ liberation [11]. An FT-IR investigation of particles isolated from subsequent etching with 3% aqueous HF showed intense Si−H stretching modes at 2040–2231 cm^{-1} and relatively weak Si−O vibrations at about 1070 cm^{-1} that may result from postetching oxidation. Increasing the HF etching concentration to 5–15% resulted in increased complexity (i.e., broadening of structure) of the Si−H stretching frequencies that suggest the presence of SiH$_2$ and SiH$_3$ on the particle surfaces, and the presence of Si−O−H could not be precluded given the present structure in the Si−H stretching likely arising from backbonding as well as the significant Si−O absorption. In addition, higher HF concentration resulted in increased relative intensity of Si−O spectral features and the appearance of a broad absorption at about 3430 cm^{-1} assigned to silanol moieties. In this regard, the authors of this study suggest higher

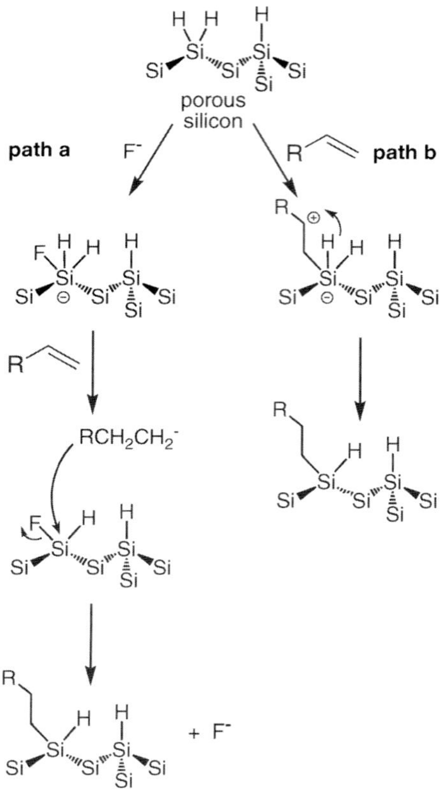

Scheme 6.3 Two independently proposed mechanisms for thermally initiated hydrosilylation reaction of Si—H-terminated surfaces with terminal alkenes. Reprinted in part with permission from Ref. [4]. Copyright 2002 the American Chemical Society.

HF concentrations catalyze the surface reaction with water to produce undesirable Si—OH functionalities.

As one may predict from the extensive literature pertaining to porous and bulk Si surface chemistry as well as the relative ease with which a hydride-terminated surface may be obtained, thermally (see Scheme 6.3) and photochemically (see Scheme 6.4) initiated addition of a Si—H bond across carbon–carbon multiple bonds has proven an effective method for controlling silicon nanoparticle surface chemistry. To a much lesser degree, transition metal-catalyzed hydrosilylation has been presented [13].

The hydrosilylation approach has led to a substantial control over Si-nc hydrophilicity ultimately rendering them soluble in a variety of organic (see Figure 6.4) [14] and in some cases aqueous solvents [13, 15]. FT-IR spectra of functionalized particles clearly show characteristic features of the surface-bonded organic moieties, residual hydride termination, and some surface oxide that admittedly masks the absorptions from the Si—C linkage at 1083 cm^{-1} (see Figure 6.4) [11, 16].

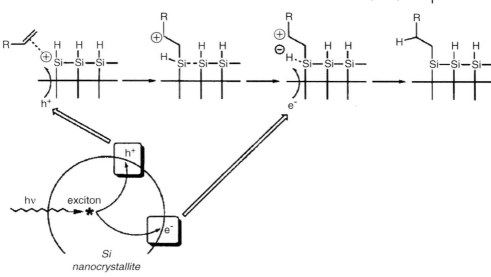

Scheme 6.4 Proposed mechanism for photochemically initiated hydrosilylation reaction of Si—H-terminated surfaces with terminal alkenes. Reprinted in part with permission from Ref. [4]. Copyright 2002 the American Chemical Society.

Consistent with the generally accepted radical mechanism for photochemically induced hydrosilylation discussed previously, spectroscopic analyses of functionalized Si-ncs obtained from this reaction indicate varied modes of attachment of the surface-bonded alkyl group (see Scheme 6.5). It has been determined by solution ^{1}H NMR that the organic surface cap attaches via the *alpha* or *beta* carbon depending upon the steric bulk of the capping ligand (see Figure 6.5) [16]. This reaction has been found to favor *alpha* and *beta* attachment for large and small alkenes, respectively. Similar observations were not reported for the thermally initiated reaction.

Aqueous compatibility of Si-ncs is a prerequisite of any biological application and is a focus of the efforts of many research groups. In this regard, studies focused on direct attachment of solubilizing surface groups such as propionic acid [15], hydrolyzable esters [17], or allyl amine [18], as well as polyacrylic acid [19] via hydrosilylation. Propionic acid-terminated surfaces were prepared via the photochemically initiated hydrosilylation reaction of hydride-terminated Si-ncs obtained from the HF/HNO_3 etched commercially available silicon-rich oxide with stabilized acrylic acid in the presence of HF. Intriguingly, the authors report the hydrosilylation reaction did not proceed efficiently in the absence of HF and indicate this provides a useful alternative to tedious solvent drying and degassing.

Solution samples prepared by transferring particles into water by dialysis remained suspended as nonopalescent, photoluminescent suspensions in water for extending periods. This is in stark contrast to straightforward solution mixtures of Si-ncs with acrylic acid that readily settle out (see Figure 6.6). It is important to note

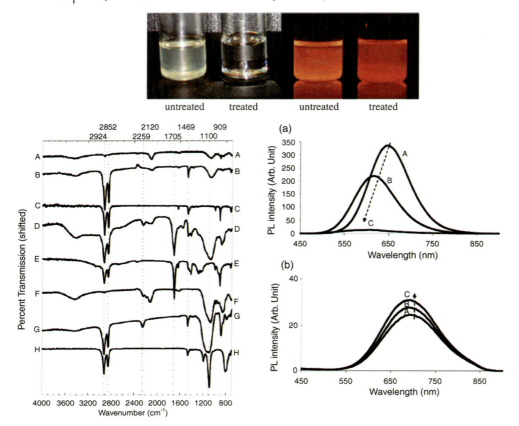

Figure 6.4 (Top) Solution and photoluminescent properties of alkyl-terminated Si-nc prepared via solution-phase photoinitiated hydrosilylation. (Bottom left) FT-IR spectra of Si—H-terminated Si-ncs (a), octadecyl-terminated Si-ncs (b), neat octadecene (c), undecylenic acid-terminated Si-ncs (d), neat undecylenic acid (e), Si—OH-terminated Si-ncs (f), octadecyl-trimethoxysilane-terminated Si-ncs (g), and neat octadecyltrimethoxysilane (h). (Bottom right) Photoluminescence stability of Si—H (top) and Si-alkyl-terminated particles. Reprinted in part with permission from Ref. [11]. Copyright 2004 the American Chemical Society.

that isolated Si-nc powders required significant ultrasonication to facilitate resuspension in water that resulted in particle surface oxidation. TEM analysis of these materials was complicated by particle agglomeration and small particle sizes. Still, HRTEM confirmed nanocrystals with 3.1 Å lattice spacing corresponding to (111) spacing in bulk silicon. As expected, FT-IR spectra of propionic acid-terminated Si-ncs again show characteristic absorptions of the acid surface group and no evidence of residual unsaturation from remaining acrylic acid. Unfortunately, given similar spectral signatures, it is impossible from this analysis to rule out the presence of oligomeric and polymeric impurities that may result from reaction of

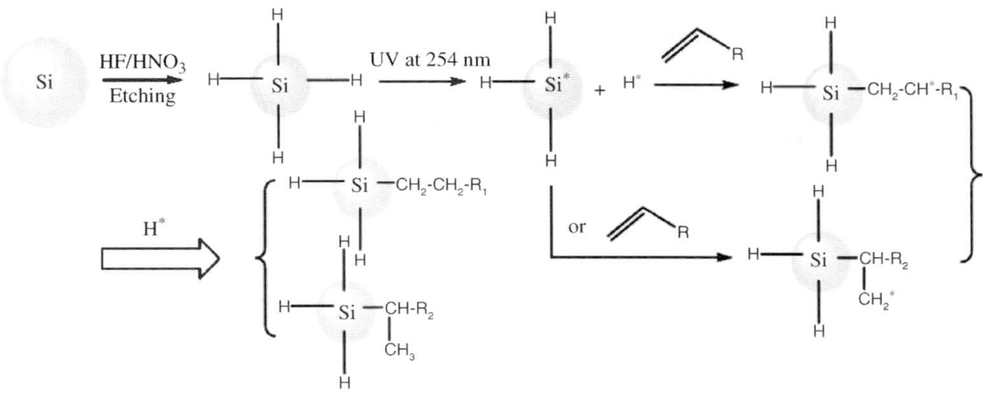

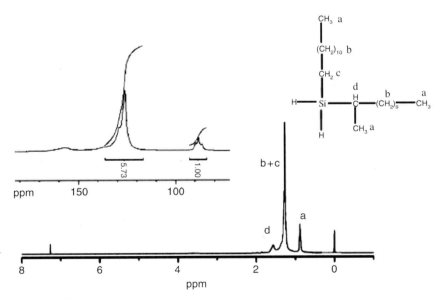

Scheme 6.5 Potential modes of attachment of alkyl surface termination obtained from photoinitiated hydrosilylation. Reprinted in part with permission from Ref. [16]. Copyright 2005 the American Chemical Society.

the acrylic acid with itself – a significant challenge when working with acrylates. Similar water solubility was observed for polyacrylic acid-modified Si-ncs by Li and Ruckenstein. [19].

Warner and coworkers approached the challenge of water solubility by adding acryl amine to Si-nc surfaces via platinum (i.e., platinic acid) catalyzed hydrosilylation

Figure 6.5 ^{1}H NMR spectra of alkyl-terminated Si-ncs highlighting the alpha and beta attachments of the surface group. Reprinted in part with permission from Ref. [16]. Copyright 2005 the American Chemical Society.

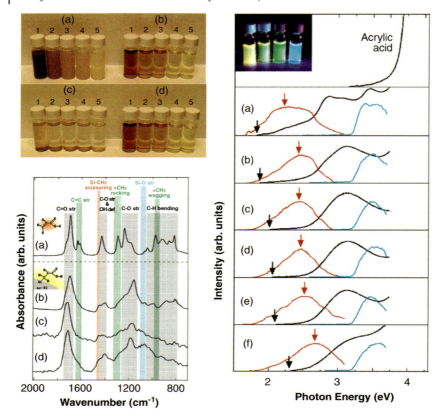

Figure 6.6 (Left top) Comparison of silicon nanoparticle in acrylic acid (a) without UV exposure and (b) after UV exposure. Parts (c) and (d) show the same samples after 1 week of storage. (Left bottom) FT-IR spectra of propionic acid (a), acid-functionalized Si-ncs (b), dialysis isolated acid-terminated Si-nc (c), and resuspended acid-terminated Si-ncs (d). (Right) Photoluminescence response of acid-terminated Si-ncs. Reprinted in part with permission from Ref. [15]. Copyright 2006 the American Chemical Society.

(see Scheme 6.6) [18b]. FT-IR again showed characteristic alkyl and amine absorptions of the surface groups as well as Si—CH$_2$ scissoring and symmetric bending at 1420 and 1260 cm^{-1}. Particles obtained in this way were photoluminescent ($\lambda_{EX} = 300$ nm; $\lambda_{EM} = 480$ nm) and were used to image HeLa cells (see Figure 6.7).

A third hydrosilylation-based approach to render Si-ncs soluble in water and THF reported by Rogozhina *et al.* saw the attachment of surface ester groups via the thermally activated reaction between methyl 4-pentenoate and the Si—H surface of Si-ncs obtained from etching of bulk silicon [17]. Following attachment, the ester surface was hydrolyzed with exposure to a methanol water solution containing NaOH. FT-IR analysis confirmed ester and acid surface groups; however, the Si—C linkage was not conclusively identified.

Hydrophobic   Hydrophilic

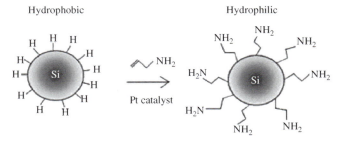

Scheme 6.6 Platinum-catalyzed alkyl surface termination of Si-ncs to yield amine surface functionality. Reproduced with permission from Ref. [18b]. Copyright 2005 Wiley-VCH Verlag GmbH.

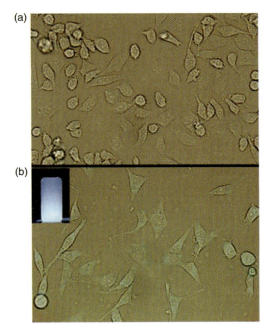

Figure 6.7 Photoluminescent imaging of HeLa cells with water-soluble amine surface-terminated Si-ncs. (Top) no Si-ncs, (bottom) Si-ncs present. Reproduced with permission from Ref. [18b]. Copyright 2005 Wiley-VCH Verlag GmbH.

6.7
Substitutional Approaches to Surface Functionalization

Yet another convenient, derivatizable Si-nc surface is the one bearing halogens (e.g., Cl, Br). Surfaces of this type may be directly obtained from the synthetic reaction, as is the case when particles are prepared from Zintl salt precursors, or it may be realized after postsynthetic modification upon reaction of Si$-$H-terminated particles with Cl$_2$.

The Si—Cl reactive surface opens the surface to modification using a wide variety of solution reagents. Bley and coworkers first demonstrated the reaction of a Si-nc chloride surface somewhat serendipitously and as a consequence of the alcoholic workup procedure of the reaction between excess silicon tetrachloride and a Zintl salt (i.e., ASi; A = Na, K, Mg). FT-IR analysis of the isolated pure nanoparticle product showed characteristic peaks for Si—O and C—H$_{sat}$ bending and stretching frequencies. It was proposed that the initial product from the redox reaction were, in fact, chloride surface-terminated particles. Subsequent reaction between these particles and the methanol used to remove salt impurities ultimately capped the particle surface with —OMe groups and rendered the product hydrophobic and soluble in organic solvents. Building upon this observed electrophilic behavior, the Kauzlarich team extended modification of the Si—Cl surface to include reactions with standard nucleophilic reagents such as alkyl lithium and Grignard reagents, which yielded alkyl surfaces [20]. In a similar fashion, Rogozhina *et al.* reacted a Si—Cl surface with butyl amine to obtain a Si—N-bonded organic surface layer [21].

Clearly, the Si—Cl surface offers significant chemical breadth as evident by its electrophilic reactivity and the demonstration that it may be readily converted to Si—H and Si—Br with exposure to lithium aluminum hydride and bromine, respectively (see Scheme 6.7). In the light of the vast chemistry of Si—Cl bonds, this mode of surface modification holds significant untapped potential in controlling Si-nc chemistry.

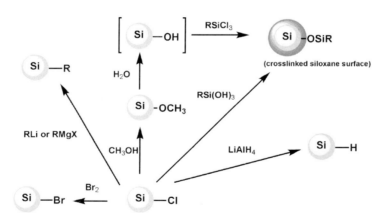

Scheme 6.7 Demonstrated reactivity of Si—Cl surface termination of Si-ncs.

6.8
Building on an Oxide Layer

Finally, a number of researchers have chosen to build upon a surface oxide on Si-nc.

Zou *et al.* modified chloride-terminated Si-ncs to yield an "ultrastable" siloxane-capped product. Two independent methods for the stepwise surface

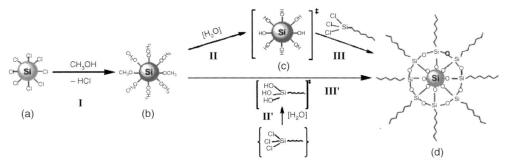

Scheme 6.8 Two independent routes for the formation of alkyl-terminated Si/SiO₂ core–shell ncs with stable optical response. Reprinted in part with permission from Ref. [22a]. Copyright 2005 the American Chemical Society.

functionalization were demonstrated. In one case, SI−Cl surfaces were converted to methoxy moieties upon exposure to methanol that were reacted with water and finally alkyl trichlorosilanes to yield particles with cross-linked siloxane surfaces. This same product was obtained via a second route in which the alkoxy groups were reacted directly with trihydroxysilanes (see Scheme 6.8) [22]. Modified Si-ns prepared in this way were isolated as a waxy, light-yellow solid that can be resuspended in nonpolar organic solvents (e.g., chloroform and hexane) to yield a yellow solution. Spectroscopic analysis (FT-IR and NMR) confirmed the surface structure included both oxide and organic moieties (see Figure 6.8).

The blue photoluminescent response of these siloxane-terminated surface particles shows excitation wavelength dependence and was found to be stable to photobleaching and ambient conditions for at least 60 days. This is in stark contrast to Si−Cl and Si−OR (OR = alkoxy) terminated particles whose luminescence was quenched within 15 days.

Swihart and coworkers also prepared oxide-terminated Si-ncs via the piranha (sulfuric acid:30% hydrogen peroxide, 7 : 3 by volume) or nitric acid treatment followed by reaction with alkyl trimethoxysilanes. These materials also exhibited solubility in organic solvents and stable photoluminescence [11].

6.9
How Many Surface Groups are on the Particle?

Despite all efforts to characterize the surface of Si-ncs, the degree of surface modification (i.e., efficiency of surface coverage) remains elusive. Numerous estimates have been put forth that offer approximate values of surface silicon atoms that serve as a starting point for estimating the surface coverage of alkyl-terminated Si-ncs [6, 23]. Holmes *et al.* used XPS integration data to estimate that 1.5 nm diameter particles obtained directly from the reaction of diphenylsilane in supercritical octanol bore approximately 16 surface groups – half the expected value of

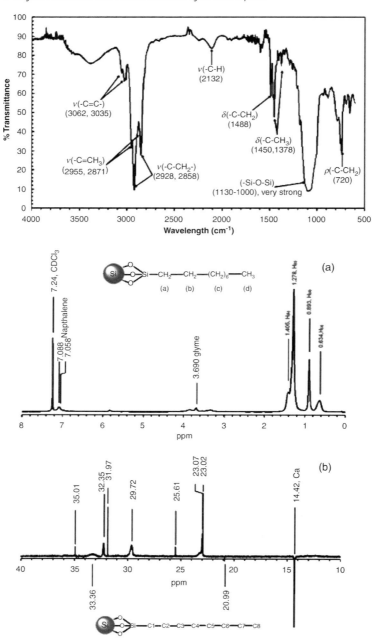

Figure 6.8 (Top) FT-IR spectrum of alkyl surface-terminated oxide-coated Si-ncs showing characteristic alkyl and Si—O—Si absorption frequencies. (Bottom) ^{1}H NMR (a) and ^{13}C NMR APT (b) spectra of alkyl surface-terminated oxide-coated Si-ncs. Reprinted in part with permission from Ref. [22a]. Copyright 2005 the American Chemical Society.

a close-packed monolayer. Yet another approach presented by Hua *et al.* employed thermogravimetric analysis (TGA) and standardized solution ^{1}H NMR of approximately 2 nm diameter particles led to the conclusion that particles obtained from photochemically initiated hydrosilylation possess a near-complete monolayer surface with 124 surface groups. This suggests postsynthetic modification of particle surfaces is more efficient than *in situ* functionalization. Still, the surfaces of functionalized Si-ncs isolated from hydrosilylation possess residual Si−H that likely result from sterically encumbered, incomplete functionalization of =SiH$_2$ on Si(100) surfaces of the particles and Si−O arising from surface oxidation.

6.10
Influence of Surface Chemistry

From the examples described in this chapter, it is obvious that the surface chemistry of Si-nc has developed dramatically – particles have been rendered soluble in aqueous and organic solvents. Likewise, material stability (i.e., prevention of oxidation) is strongly influenced by an organic surface. Few examples in the literature exist that definitively show the dependence of Si-nc optical response on surface chemistry – still, it is reasonable to expect surface chemistry to have a demonstrated impact on the optical properties of porous-Si [24]. It has also been predicted that surface chemistry will significantly influence the lowest unoccupied molecular orbital (LUMO) energy and hence the optical bandgap of very small FS-nc-Si [22a,25]. This may explain the unique as of yet unexplained discrepancies in the photoluminescent response of Si-ncs of seemingly identical dimension. In this regard, although significant advancements to tailor Si-nc surface chemistry have been made, methods affording well-defined and predictable surface structure are essential if these materials are to be fully understood and their applications realized.

6.11
Future Outlook and the Role of Surface Chemistry

Surface functionalization of silicon nanoparticles is in a relative state of infancy. To date, most surface modification methodologies have been confined to a narrow class of hydrosilyation reactions. While these methodologies have been extremely successful as evident by the preparation of materials of controlled properties (e.g., high-photoluminescence quantum yield, ambient stability, and tailored solubility), still further advances in functionalization procedures must be realized if these materials are to become practically relevant. One such important advancement aiming to prepare extremely stable Si nanoparticle with predictable properties uses an all-gas plasma procedure [26]. In parallel with the development of functionalization method, new applications must be explored. A recently reported surface functionalization-driven application involved biocompatible luminescent assemblies of silicon quantum dots formed through interaction between functionalized particles and polyethyleneglycol-grafted phospholipids that effectively image cancer cells [27]. Surface chemistry offers modes for interfacing silicon nanomaterials with their surroundings, and with the development of new methods the optical and chemical properties of these unique materials will be more fully appreciated and exploited.

References

1 Hua, F., Swihart, M.T., and Ruckenstein, E. (2005) *Langmuir*, **21**, 6054–6062.

2 Buriak, J.M., Stewart, M.P., Geders, T.W., Allen, M.J., Choi, H.C., Smith, J., Raftery, D., and Canham, L.T. (1999) *J. Am. Chem. Soc.*, **121**, 11491–11502.

3 Wayner, D.D.M. and Wolkow, R.A. (2002) *Perkin Trans. 2*, 23–34.

4 Buriak, J.M. (2002) *Chem. Rev.*, **102**, 1271–1308.

5 Heath, J.R. (1992) *Science*, **258**, 1131–1133.

6 Holmes, J.D., Ziegler, K.J., Doty, R.C., Pell, L.E., Johnston, K.P., and Korgel, B.A. (2001) *J. Am. Chem. Soc.*, **123**, 3743–3748.

7 Heintz, A.S., Fink, M.J., and Mitchell, B.S. (2007) *Adv. Mater.*, **19**, 3984–3988.

8 Liao, Y.-C. and Roberts, J.T. (2006) *J. Am. Chem. Soc.*, **128**, 9061–9065.

9 Mangolini, L. and Kortshagen, U. (2007) *Adv. Mater.*, **19**, 2513–2519.

10 Jurbergs, D., Rogojina, E., Mangolini, L., and Kortshagen, U. (2006) *Appl. Phys. Lett.*, **88**, 233116.

11 Li, X., He, Y., and Swihart, M.T. (2004) *Langmuir*, **20**, 4720–4727.

12 Hessel, C.M., Henderson, E.J., Cavell, R.G., Sham, T.-K., and Veinot, J.G.C. (2008) *J. Phys. Chem. C*, **111**, 6956–6961.

13 Tilley, R.D., Warner, J.H., Yamamoto, K., Matsui, I., and Fujimori, H. (2005) *Chem. Commun.*, 1833–1835.

14 Liu, S.-M., Sato, S., and Kimura, K. (2005) *Langmuir*, **21**, 6342–6329.

15 Sato, S. and Swihart, M.T. (2006) *Chem. Mater.*, **18**, 4083–4088.

16 Hua, F., Swihart, M.T., and Ruckenstein, E. (2005) *Langmuir*, **21**, 6054–6062.

17 Rogozhina, E.V., Eckhoff, D.A., Grattonb, E., and Braun, P.V. (2006) *J. Mater. Chem.*, **16**, 1421–1430.

18 (a) Tilley, R.D. and Yamamoto, K. (2006) *Adv. Mater.*, **18**, 2053–2056; (b) Warner, J.H., Hoshino, A., Yamamoto, K., and Tilley, R.D. (2005) *Angew. Chem., Int. Ed.*, **44**, 4550–4554.

19 Li, Z.F. and Ruckenstein, E. (2004) *Nano Lett.*, **4**, 1463–1467.

20 (a) Mayeri, D., Phillips, B.L., Augustine, M.P., and Kauzlarich, S.M. (2001) *Chem. Mater.*, **13**, 765–770; (b) Yang, C.-S., Bley, R.A., Kauzlarich, S.M., Lee, H.W.H., and Delgado, G.R. (1999) *J. Am. Chem. Soc.*, **121**, 5191–5195.

21 Rogoxhina, E., Belomoin, G., Smith, A., Barry, N., Akcakir, O., Braun, P.V., and Nayfeh, M.H. (2001) *Appl. Phys. Lett.*, **78**, 3711–3713.

22 (a) Zou, J., Baldwin, R.K., Pettigrew, K.A., and Kauzlarich, S.M. (2004) *Nano Lett.*, **4**, 1181–1186. (b) Zou, J. and Kauzlarich, S.M. (2008) *J Cluster Sci.*, **19**, 341–355.

23 Zhao, Y., Kim, Y.-H., Du, M.-H., and Zhang, S.B. (2004) *Phys. Rev. Lett.*, **93**, 015502.

24 (a) Laurerhaas, J.M. and Sailor, M.J. (1993) *Science*, **261**, 1567–1568. (b) Wolkin, M.V., Jorne, J., Fauchet, P.M., Allan, G., and Delerue, C. (1999) *Phys. Rev. Lett.*, **82**, 197–200.

25 Reboredo, F.A. and Galli, G. (2005) *J. Phys. Chem. B*, **109**, 1072.

26 Pi, X.D., Liptak, R.W., Nowak, J.D., Wells, N.P., Carter, C.B., Campbell, S.A., and Kortshagen, U. (2008) *Nanotechnology*, **19**, 245603.

27 Erogbogbo, F., Yong, K.T., Roy, I., Xu, G., Prasad, P.N., and Swihart, M.T. (2008) *ACS Nano*, **2**, 873–878.

7
Si-nc in Astrophysics

Ingrid Mann

7.1
Introduction

Nanoparticles are exotic and silicon nanoparticles are even flamboyant. This makes them an attractive target of investigation in astrophysics, a field of research largely based on observations. Even fickle fashion tastes can assume some deep meaning that becomes clear when its time has come. Recently, improved techniques have allowed to study the dust almost at the nanometer scales. Measurements disclose a new class of phenomena that arise as a result of the small sizes. Si is one of the most abundant elements and like most heavy elements it has the ability to condense into cosmic dust. It is therefore straightforward to consider that Si nanocrystals form in space.

Figure 7.1 shows abundances of different chemical elements in the matter that is contained in the solar system. These abundances are close to the current average cosmic abundances of the Galaxy. The light elements H and He and noble gases are in vast majority in the gas component. Elements heavier than He, in astrophysics generously denoted as "metals" depending on the temperature and conditions of the specific space environment, are partly contained in small solid dust particles. This chapter will follow the path of cosmic dust formation and then will present astronomical observations that possibly indicate the existence of silicon nanoparticles. Finally, the chapter presents laboratory measurements to form silicate nanostructures in cosmic analogue materials.

Dust particles evolve following three major stages in different environments where they are possibly observed: in circumstellar shells around evolved stars, in the interstellar medium (ISM) between stars, and in newly forming stars and their surrounding forming planetary systems. Dust particles initially form in the circumstellar shells that evolve around stars in a late stage of their evolution, as will be discussed in Section 7.2. The forming dust is expelled into the interstellar medium where dust destruction, dust alteration, and further dust growth occur (see Section 7.3). The ISM dust is incorporated into newly forming stars and the planetary systems that possibly surround them. Section 7.4 describes the path of dust

Silicon Nanocrystals: Fundamentals, Synthesis and Applications. Edited by Lorenzo Pavesi and Rasit Turan
Copyright © 2010 WILEY-VCH Verlag GmbH & Co. KGaA, Weinheim
ISBN: 978-3-527-32160-5

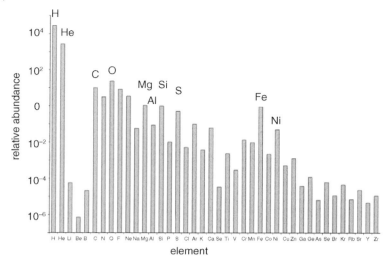

Figure 7.1 The solar system element abundances up to Zr normalized to Si. These are obtained from the abundances measured in the solar photosphere and in meteorites (from Ref. [1]). The solar element abundances are often used as a standard for the present average cosmic element abundances and these have evolved through nuclear processes of the past stellar evolutions. The indicated elements play a major role in dust formation discussed in the text.

during star formation and planetary system formation. Section 7.5 considers the dust in the solar system, which is the best-studied case of a planetary system.

In spite of many different types of observations, none of these three stages in the "cosmic dust life cycle" shows a clear evidence of the existence of silicon nanoparticles. As far as silicon nanoparticles are concerned, this seems to seal their fate. They simply do not form. Or, they do not form "simply?" Section 7.6 reports observational evidence that encouraged astrophysicist to follow up the possibilities of nanosilicates being formed. An enhanced extended red emission (ERE) is observed in many of the mentioned cosmic environments and silicon nanoparticles are discussed as one of the potential carriers of ERE. Section 7.7 finally discusses the possible formation of Si nanostructures in a solid, a path that may form silicon nanoparticles in different cosmic environments.

7.2
Late Stellar Evolution

The sources of dust in the interstellar medium of our Galaxy, and in a similar way in other galaxies, can be quantified with theoretical studies and observational results [2, 3]. The majority of dust production occurs in low and intermediate mass stars during their late stages of evolution when they form giant stars. Another source are supernovae that form out of stars that have greater mass. The stellar evolution can

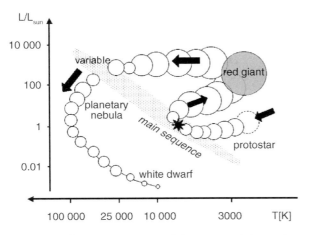

Figure 7.2 The Hertzsprung–Russell diagram indicating the change in luminosity, L_{star} and surface temperature T for a solar mass star during its evolution. The star spends the majority of its lifetime in the main sequence. Dust formation takes place mainly during the red giant and asymptotic giant phase, and ejected dust is also observed in the planetary nebulae. Planetary system formation takes place before and in the early main sequence stage. This sketch closely follows a presentation by Speck [4].

be followed when considering the Hertzsprung–Russell diagram that usually shows the numbers of observed stars as a function of their observed luminosity and temperature. The majority of observed stars are placed along a line in this diagram called main sequence, ranging from high temperature and high luminosity to low temperature and low luminosity. The luminosities and temperatures along the main sequence are those the stars display for the longest time during their lifetime. During its lifetime, a single star changes its luminosity and temperature and would follow a path within this diagram moving toward the main sequence as a young star and moving away from the main sequence in its final stage. This is sketched in Figure 7.2 for a star similar to the Sun and this is the typical evolution of stars with masses $M \leq 8M_{sun}$. The star is formed by the gravitational collapse of gas and dust in a cool interstellar cloud (i.e., molecular cloud). When the density in the interior of the newly forming star is sufficiently large, nuclear fusion reactions commence and a young stellar object evolves (shortly described in Section 7.4) until it reaches the main sequence that is characterized by stable hydrogen burning in the interior of the star and stable stellar brightness.

The main sequence ends when the majority of hydrogen in the interior of the star is exhausted: hydrogen-burning stops in the core and the core contracts. The gravitational contraction heats up the outer regions where hydrogen keeps burning to form helium. The high temperature accelerates the hydrogen burning and the surface brightness of the star increases. Subsequently, the outer layers vastly expand and the surface of the star cools down, a red giant star evolves. The core reaches sufficiently high temperatures so that helium-burning reactions ignite to form Be nuclei and then C nuclei. The star contracts, and depending on the mass of the star, further fusion reactions start and form heavier nuclei (chemical elements with higher mass number). The forming heavier elements are ejected to the outer layer (*stellar envelop*). This is associated with pulsation of the outer layers during which the entire envelope may be ejected. The ejected material forms a *circumstellar shell*. The described evolution refers to initially *red giant stars* and then to stars that evolve along the asymptotic giant branch (*AGB stars*). Red giant stars and AGB stars differ in their internal nuclear processes.

The stellar atmosphere of a giant star may extend to about a distance comparable to Earth orbit in the solar system. A solar mass AGB star ($M_{star} = M_{sun}$), for instance, has a stellar luminosity $L_{star} = 2 \times 10^4 L_{sun}$ (where L_{sun} is the luminosity of the Sun) while its surface temperature $T_{star} = 3000$ K is only half that of the Sun. The mass loss of AGB stars is to a large extent in the dust and amounts to $dM_{star}/dt = (10^{-6}–10^{-4})$ M_{sun}/year [5]. Dust particles condense from the heavy elements in the cooling gas. The dust around giant stars is further ejected by stellar radiation pressure and by interaction with the hot ionized gas ejected from the star. Observed AGB stars display different spectra revealing different element abundances and temperatures in their shells. The stars evolve from spectral M-type having oxygen-rich atmospheres to S-type where C and O are bound almost entirely in CO and finally C-type having carbon-rich atmospheres. This determines the condensation of the dust and it determines into what species Si condenses: silicon carbide SiC dust forms around cool carbon stars (C/O > 1), while silicate dust (SiO_2) forms in oxygen-rich atmospheres (C/O < 1).

The dust surrounding the star is observed by its thermal emission over a wide infrared spectral range. Emission bands that are characteristic of the emitting dust minerals (in particular, certain silicates, i.e., "silicate feature") are superimposed onto the broad thermal emission and allow to estimate the dust composition in the shells (an example will be shown in Section 7.4). An uncertainty arises, however, from ascribing the features to certain minerals, since the features vary with temperature, particle size, and other parameters. The SiC spectral features in carbon-rich AGB stars, for instance, change in a sequence that correlates with stellar evolution: shifts are due to optical depth and grain sizes in the system and in the case of an extremely optically thick shell, the features appear in absorption. On the basis of their observational and laboratory studies, Speck *et al.* [6] suggest that the size of grains that form decreases as the stars evolve.

Table 7.1 (dust condensates) lists the silicon-bearing species that are predicted to form as end products in M-stars and C-stars. The table also lists whether those species are detected by astronomical observations of the circumstellar shells or in presolar grains collected in the solar system. The presolar grains are isolated from meteorite samples, and measurements of certain isotope abundances allows to trace them back to a distinct nuclear process and hence distinct regions of formation (also mentioned in Section 7.4).

In a later stage, the shrinking star has a high temperature and pushes outward the dust and gas. The expanding shell is optically thin and intense stellar UV radiation also causes line emission of the newly formed elements. Dust emission

Table 7.1 Dust condensates in circumstellar shells.

M-stars			C-stars		
Mineral	**Formula**	**Observed**	**Mineral**	**Formula**	**Observed**
Gehlenit	$Ca_2Al_2SiO_7$	—	Silicon carbide	SiC	Shells Presolar Grains
Forsterite	Mg_2SiO_4	Shells Presolar Grains	Iron silicide	FeSi	—
Enstatite	$MgSiO_3$	Shell Presolar Grains	Forsterite	Mg_2SiO_4	Shells Presolar Grains
			Enstatite	$MgSiO_3$	Shells Presolar Grains

Condensation products containing Si predicted for the atmospheres of giant M-stars and C-stars. Observations of the predicted species by thermal emission features in the circumstellar shells or in laboratory analyses of presolar grains that were presumably formed in these stars are marked in the table. (The information given in this table is extracted from Ref. [7].)

is observed as well. Since its first observations resembled the visual brightness appearance of the giant outer planets of the solar system, this phenomenon is called *planetary nebula*.

The evolution of stars with larger masses ends as supernovae. The nuclear processes in supernovae generate the elements with masses beyond that of Fe. Though supernovae are essential for the formation of the heavy elements, and hence for the formation of dust, supernova dust is less frequently observed as presolar grains than dust from AGB stars. The generated dust is partly destroyed in the shock waves that are associated with the supernova [8]. Also, the size of the particles that are condensed in supernovae may play a role since particles with size below 10 nm have small probability to survive in the ISM.

7.3
Interstellar Medium and Dust Evolution

The ISM of our galaxy consists primarily of gas (order of 99% of the mass) and dust (order of 1% of the mass). The low gas densities in the ISM result in a very slow condensation rate, so that direct condensation to form dust nuclei directly from the gas phase is unlikely to occur in the ISM. The dust particles that have formed during late stellar evolution described above provide a condensation core for further condensation and possibly agglomeration in the ISM. This is counterbalanced by dust destruction; the most frequently occurring processes are given below.

The parameters of the different regions in the ISM are listed in Table 7.2 (ISM parameters). As initially suggested by McKee and Ostriker [9], the diffuse ISM is regarded as consisting of three distinct components: a hot interstellar cloud medium (HIM), a warm neutral or ionized medium (MNM, WIM), and a cold neutral medium

Table 7.2 Interstellar medium parameters.

	Temperature, T (K)	Gas number density, n (cm^{-3})	Filling factor, F
Major components			
Hot interstellar medium (HIM)	5×10^5	0.003	0.7–0.8
Warm neutral/ionized medium (WNM/WIM)	10^4	0.25	0.3
CNM	80	40	0.02–0.04
Other components			
Molecular clouds	20	100	($\approx$50% of ISM total mass)
HII regions (emission nebulae)	8000	10 (some $> 10^5$)	($\approx$1% of ISM total mass)
LIC	6300	0.25	—

The numbers for the major components are from Ref. [10], for the LIC from Ref. [11], and for molecular clouds and HII regions from Ref. [12].

(CNM). These components make up the vast majority of observable space between the stars (quantified by the filling factor) and they contain roughly half the mass that comprises the ISM. The other major components of the ISM are molecular clouds and HII regions. Molecular clouds contain roughly half the ISM mass, they are especially dense and cold regions, and due to low temperature and high density molecules can form. HII regions are characterized by a large fraction of ionized hydrogen. Emission lines arise from the interstellar gas being illuminated by the UV brightness of young forming stars.

The lifetime of dust in the ISM, considering the parameters of the three different regions of the ISM given above, is of the order of 10^8 years and considerable dust destruction occurs in its warm regions (see, for instance, Ref. [10]): fast supernova shock waves with propagation velocities of 50–200 km/s destroy dust particles by sputtering (mainly due to proton impact). The shock waves also induce turbulences, so that dust relative velocities and dust collision rates increase. The collisions generate smaller fragment particles and partly vaporize the dust. Growth of dust particles takes place by agglomeration during low velocity dust–dust collisions and by condensation of gaseous species onto existing core particles. In low-temperature regions the lighter elements (C, H, O, N) condense onto the cores and chemical reactions are induced by the particle and radiation environment. Complex chemical compounds form and also parts of the lighter elements are bound in species that survive at high temperature.

The ISM dust is studied by the polarization and the amount and spectral variation of the interstellar extinction. The interstellar extinction is the attenuation of the stellar brightness in the ISM caused by scattering and absorption of dust. The ISM dust is also observed in dark nebulae (due to dense dust cloud occulting light sources) and reflection nebulae (starlight reflected from clouds) and recently by the diffuse

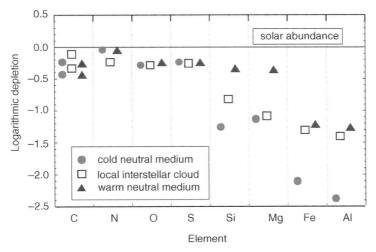

Figure 7.3 Depletion of the major dust forming elements in the gas phase of the cold neutral interstellar medium and in the warm neutral interstellar medium. The figure also shows the depletion of the local interstellar cloud surrounding the solar system, which is a warm cloud. The element depletions are shown from right to left according to the probable order of condensation of the elements. The logarithmic depletion shown is defined as the decadic logarithm of the ratio of gas-phase abundance to solar photospheric abundance. It is assumed that the observed regions contain the chemical elements with average cosmic abundances (i.e., solar photospheric abundances, indicated by the solid line) and that the missing amounts of the species are condensed into dust. The different values for Mg and Si abundances suggest that the dust in these regions has different compositions. (The figure is adapted from Ref. [13] and further references are given there.)

thermal emission of dust. The infrared brightness over parts of the sky forms a filamentary structure called *cirrus*, presumably resulting from thermal emission of dust being structured by dust, gas, magnetic fields, and/or interstellar radiation interactions.

Apart from direct observations, the dust component is estimated from the interstellar gas observations: by assuming the overall element abundance (being solar abundance) and comparing with the actually observed elements in the gas phase, one can attribute the depleted amount of the elements to the dust phase. Figure 7.3 shows the depletion of the major dust forming elements C, N, O, S, Si, Mg, Fe, Al in the cold neutral medium and in the warm neutral medium. The elements that are missing in the gas component compared to the cosmic abundances are associated with the dust. The variable Si content in the gas phase indicates that the Si abundance in the dust is variable and that the alteration of the dust component involves Si-bearing species.

Figure 7.3 also shows the depletion in the local interstellar cloud (LIC) surrounding our solar system. The cosmic abundances are here assumed to be those of the solar photosphere. Figure 7.4 shows the hence derived element abundances in the LIC dust. These element abundances were used to establish a model of the LIC dust that

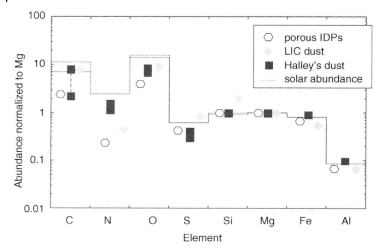

Figure 7.4 Elemental composition of dust in the local interstellar cloud (LIC dust, as derived from the element depletion in the gas phase, see Figure 7.3), of dust measured in the vicinity of comet Halley (Halley dust, *in situ* measurements from spacecraft), and of a class of interplanetary dust particles that were collected in the atmosphere (porous IDPs, laboratory analyses). The solid dashed lines denote the Mg-normalized composition of the solar photosphere describing the average cosmic abundances. The two lines indicate the uncertainty in the abundance, as do the two symbols that are given for some cases. The abundances of comet Halley dust are assumed to be uncertain by a factor of two. (This figure is adapted from Ref. [13] and further discussions of the uncertainties are given there.)

enters the solar system [13]. It is measured in-situ from spacecraft and for a discussion of LIC dust surrounding and entering the solar system the reader is referred to a recent review [14].

In order to explain direct observational results of dust in the ISM, Li and Greenberg [15] described both the interstellar extinction and the polarization with a dust model that consists of silicate core-organic refractory mantle particles, polycyclic aromatic hydrocarbons (PAHs), and carbonaceous particles. The carbonaceous particles consist of a refractory material of carbon and organic compounds. Other models to explain the observations of interstellar dust also consider silicates as major components. Following similar assumptions, the model to explain LIC dust that enters the solar system suggests that the larger dust with masses $>10^{-17}$ kg (approximately 100 nm) forms as agglomerates of core-mantle grains with the main compounds being Mg-rich pyroxene (i.e., enstatite, $MgSiO_3$) and Mg-rich olivine (i.e., forsterite, Mg_2SiO_4), further Fe-rich inclusions of troilite (FeS), kamacite (FeNi), and corundum (Al_2O_3), The mantle forms of organic refractory compounds of C, N, and O. Silicates and carbonaceous material may separately exist in the smaller grains (see [13]). It should be noted that assuming the core-mantle structure for the smaller grains that form the dust particles is motivated by the picture of the formation process, in most cases dust scattering models in which the materials are distributed within the grain in a different way, may give similar results.

7.4
Early Stellar Evolution and Planetary System Formation

Stars form through gravitational collapse in molecular clouds and the dust that is processed in the ISM is incorporated into the planetary systems that form around some of these stars. When the density in the interior of the collapsing cloud is sufficiently large, nuclear fusion reactions commence to form a star. The outer regions of the contracting cloud form a disk of gas and dust that partly accretes to the star and out of which a planetary system may form (protoplanetary disk, protoplanetary nebula). These young stellar objects consist of the disk brightness and the brightness of the forming star. The initially observed stellar brightness results from the gravitational energy of the contraction, the initial nuclear reactions, and the interaction with the accretion disk, and it is highly variable (T-Tauri stars for $M_{star} < 2M_{sun}$, Herbig Ae/Be stars for $M \approx 2-8M_{star}$). Many of the premain sequence stars have circumstellar disks that possibly host planet formation, so that the system transits to a planetary disk. The dust mass within the disk of classical T-Tauri stars ranges, for instance, between 3 and 300 Earthmasses. The total gas mass is hard to estimate, but dust and gas mass in the disks are sufficient so that planets could form. The disks are optically thick over a wide spectral range, which implies that they contain a wide spectrum of dust sizes. Dust evaporates in the inner zones of the protoplanetary nebula and later recondenses. Even if the detailed conditions are different and probably more complex than in stellar outflows, it is expected that similar minerals form [7]. Dust particles grow by coagulation, and it is assumed that a radial mixing transports dust particles from the inner regions outward. Agglomeration of form dust planetesimals, these are objects of meters to hundreds of kilometers in size that are similar to the asteroids, comets, and trans-Neptunian objects in the solar system. Further material alteration takes place within these parent bodies. The protoplanetary disk evolution ends when the gas component is removed from the system (by various processes) and when planetesimals and possibly planets have formed. Condensation of dust from the gas ceases to occur and new dust is formed by collisional fragmentation of the planetesimals. For a description of the transition from protoplanetary to planetary disk, the reader is refer to Takeuchi [16].

During the course of evolution of the disk, the central star evolves to its main sequence stage. This is characterized by stable hydrogen burning in the interior of the star, which serves as energy source for the stellar brightness. The stellar brightness during main sequence is comparably constant in time and approximately follows a blackbody spectrum determined by the surface temperature. The surface composition for the heavy elements has the element abundances of the interstellar cloud out of which the star was formed. This main sequence stage, depending on the star, lasts over hundreds of million years and the Sun is currently in its main sequence stage. In contrast to evolved stars, the mass loss of main sequence stars is small. The mass loss of the present Sun, for instance, is $dM_{star}/dt = 10^{-14}M_{sun}$ per year and this is in the solar wind, a hot ionized gas that streams outward from the solar corona. No dust condensation occurs in the solar wind and its interactions with the dust component are less important by far than the gas–dust interactions in the protoplanetary disks.

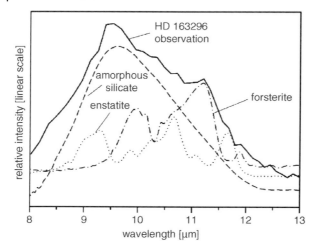

Figure 7.5 Thermal emission of dust in the debris disk of HD 163296 observed between 8 and 14 μm (from Ref. [17]). The continuum brightness is subtracted from the observations and the remaining brightness is explained by emission of silicates in their characteristic emission bands (silicate feature). The overall shape of the silicate feature is described by calculated thermal emission of amorphous silicate. The remaining features are explained with the shown emission of (crystalline) enstatite ($MgSiO_3$) and forsterite (Mg_2SiO_4) derived from laboratory measurements at meteorite material. (This figure is adapted from Köhler *et al.* [18].)

The protoplanetary disks are observed, among others, by their thermal emission brightness of the dust component. Recently, dust disks have also been observed around main sequence stars. They are considered to be planetary disks and are referred to as Vega-like systems or debris disks. Silicates are observed as a major component of dust in protoplanetary disks and in debris disks. For example, Figure 7.5 shows the observed thermal emission brightness of the debris disk of HD 163296 between 8 and 14 μm (from Ref. [17]). The brightness is compared to the characteristic emission of silicates in order to estimate the composition of the dust shell (from Ref. [18]): The continuum brightness is subtracted from the observations and the overall shape of the silicate feature is described by thermal emission of amorphous silicate. The remaining features agree well with features that were observed in laboratory measurements of crystalline silicates (enstatite, $MgSiO_3$, and forsterite, Mg_2SiO_4). The measurements were carried out at crystalline silicates isolated from meteorites: the enstatite was extracted from Kota-Kota, an enstatite chondrite of type (EH3), and forsterite was extracted from Ornans, a carbonaceous chondrite of type (CO 3.3). The petrological types (3 and 3.3) are ascribed as unequilibrated meteorites, meaning meteorite material that experiences little alteration while it was integrated into a larger parent body. The comparison of infrared spectra suggests that the dust and small bodies observed in the circumstellar system have an evolution history similar to that of solar system dust. The silicon is bound in silicates and the silicates are primarily Mg-rich. The observation of crystalline silicates (as opposed to amorphous silicates, observed in the ISM) suggests that some material alteration occurs in the silicates during their transition from the ISM, even if element abundances stay similar.

7.5
Dust in the Solar System

Most of the dust observed in the solar system is produced by comets and asteroids. The dust is distributed in the interplanetary medium between the planets and observed from Earth in the brightness of the Zodiacal light, sunlight scattered at the dust particles. The dust thermal emission brightness (Zodiacal emission) in the infrared is observed from satellite. Trans-Neptunian objects supposedly produce dust in the outer solar system (so that the solar system observed from outside would appear as a planetary debris disk), but until now this dust has evaded observations. Dust is frequently studied by observing comets and with *in situ* measurements from spacecraft. In addition, there is information from laboratory analyses of interplanetary dust particles (IDPs) collected in the Earth atmosphere (IDPs). Larger solid particles also partly survive when they enter the Earth atmosphere and their remnants collected on the ground, called meteorites, are analyzed in great detail.

Similar to models of interstellar dust and LIC dust, models of the optical properties of cometary dust suggest that the silicon is bound in Mg-rich silicates such as forsterite (Mg_2SiO_4) and enstatite ($MgSiO_3$). The results from laboratory measurements and *in situ* measurements in the solar system support these models. Figure 7.4, apart from the LIC dust, shows the element abundances of the cometary dust measured from spacecraft close to comet Halley and of a distinct type of IDPs. Cometary dust is believed to be relatively pristine and its thermal emission spectra are often studied in comparison to those of circumstellar dust (see e.g., Ref. [19]). The shown porous IDPs are believed to be the most pristine among the collected IDPs and they possibly originate from comets. The dust particles measured at comet Halley and the porous IDPs have similar element abundances and these are also similar to the LIC dust. Laboratory measurements show that the majority of Si in the IDPs is bound in silicates.

Nanodust particles most likely form in the solar system by fragmentation of larger interplanetary dust particles, but more likely than with astronomical observations they should be measured with *in situ* measurements from spacecraft [20]. Indeed, recent plasma measurements from spacecraft near Earth orbit reveal certain events that are suggested to result from high-velocity impacts of nanodust that is accelerated in the solar wind [21]. There is no direct information about the composition of nanodust in the solar system. Habbal *et al.* [22] explained an unidentified near-infrared emission that they observed in the solar corona with photoluminescence (PL) of silicon nanoparticles near the Sun. Mann and Murad [23] rejected this explanation of the observed coronal emission since the suggested photoluminescence does not occur at the high temperatures of dust in the inner solar system. They also studied the sublimation sequence of dust that approaches the Sun and on the basis of that suggest that metal-oxide particles form near the Sun, while SiO forms only as an intermediate phase at larger distance and this phase usually is not stable.

During solar system formation, a certain class of dust particles, called presolar dust grains (sometimes called "stardust"), has survived and these grains are separated from the meteorite material. The presolar grains can have sizes up to several micrometers and with the exception of the presolar diamonds they have sizes larger than several 10 nm. The presolar nature of the grains is evident from measured isotopic anomalies, and the isotopic abundances can be associated with stellar nuclear processes, in many cases of SiC and oxide grains with those in AGB stars (see [24], for a review of presolar grains). Some other types of presolar grains are linked by their derived isotopic composition to supernovae (a fraction of the SiC and all silicon nitride grains). The extracted diamonds have sizes of 10 nm or smaller, and the isotopic ratios that can confirm their presolar formation cannot be measured for single grains. It is suggested that at least a fraction of the detected diamond is formed after the solar system formation, possibly within the meteorite material. Si nano-crystals may form in a similar way within larger objects as will be discussed in Section 7.7.

7.6
Extended Red Emission and Si Nanoparticles as a Potential Carrier

While all the previously described observations indicate that silicon is bound in silicates or SiC, observation of the ERE stimulated investigations into silicon nanoparticles as a component of cosmic environments. The ERE is a broad excess brightness relative to the continuum that extends between 540 and 950 nm (see Figure 7.6). ERE was first noticed in the nebula surrounding HD 44179 (Red Rectangle, [25, 26]) and today is regarded a phenomenon that is common to many cosmic dust environments. The ERE is observed in reflection nebulae, dark nebulae, planetary nebulae, HII regions, halos of (other) galaxies, and in the diffuse ISM and cirrus (see [27], for a recent review). It peaks in most cases between 650 and 750 nm and has a width of 60–120 nm (exact numbers vary by authors and these are the number given in Ref. [27]).

Since ERE is observed in such a wide range of cosmic environments, luminescence, as opposed to thermal emission, is the probable process to generate ERE, and photoluminescence caused by the interstellar radiation field, in particular, is a process that can occur in all the environments considered. Also, the observational results concerning the correlation of the ERE with the local radiation field support photoluminescence being the mechanism [27]. Observations of the ERE in a HII region, because the region is optically thin, allowed to derive the efficiency of the process that generates the ERE and this efficiency is high (see e.g., [33]).

The photoluminescence process is sketched in Figure 7.7: a system (a semiconductor particle) by photon absorption proceeds from the ground state to a higher electronic state. It subsequently undergoes internal transitions (for instance, vibrational or rotational transitions) to an intermediate lower level from which an optical

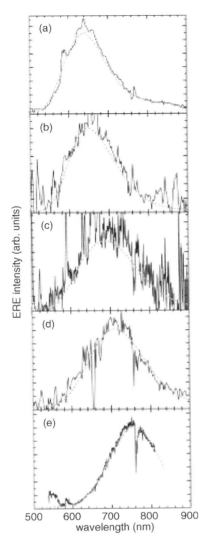

Figure 7.6 Comparison of ERE observations in different objects (solid lines) and the PL of Si nanocrystals (dashed line). The PL was determined either by experiment for (a) and (e) or by calculations for (b), (c), and (d). The observations are compared with synthesized spectra for a dust lognormal size distributions with $<d> = 2.85$, 3.3, and 3.6 nm and widths (full width half maximum, FWHM) of 1.08, 1.6, and 1.45 nm in (b), (c), and (d), respectively. The experimental PL spectra are for an average particle size of 2.71 nm in case (a) and 4.5 nm in case (e). The observational data show (a) and (e) the red rectangle, a post-AGB star/planetary nebula; (b) NGC 2023, a reflection nebula in the Orion constellation; (c) the halo of the galaxy M82; and (d) NGC 2327, a HII (ionized) region. (The observations shown are from (a) Ref. [28], (b) Ref. [28], (c) Ref. [29], (d) Ref. [30], and (e) Ref. [31] respectively. These figures were reproduced from Ref. [32], with the kind permission of *Astronomy & Astrophysics*).

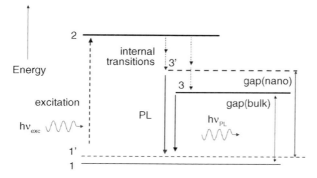

Figure 7.7 Principle of photoluminescence: UV photon absorption excites an electron from the ground state to a higher state (e.g., a high state in the conduction band of a semiconductor). The electron can reach a lower state (3) through a series of rotational or vibrational transitions. The electronic transition back to the ground state gives rise to the photoluminescence. In nanoparticles, the energy levels shift (to 1′ and 3′) and the bandgap increases with decreasing size of the particle. The presence of defects in solids increases the chance for internal transition so that de-excitation may occur without photoemission.

transition to the ground state is possible. The ground state is a state near the upper level of the valence band, while the excited state is a high state in the conduction band and the system relaxes to a lower level within the conduction band before transition across the bandgap occurs under photon emission.

The excitation can be considered as the generation of a free electron–hole pair, the electron being in the conduction band and the hole in the valence band.

The photon energy of the photoluminescence is close to the energy of the bandgap. Defects in the material (e.g., impurities, disorder sites in a crystal) facilitate non-radiative transitions to the ground state; hence, defects reduce the photoluminescence efficiency. Unlike the bulk material, the electrons and holes in a nanoparticle are spatially confined and this broadens the bandgap of the material. This is referred to as quantum confinement effect. Li [34] in a recent review of interstellar nanoparticles illustrates the broadening of the bandgap in case of quantum confinement in terms of the Heisenberg uncertainty principle of position and momentum. The broadening also depends on temperature. Reducing the size of particles increases the size of the bandgap and at the same time decreases the probability of an electron–hole pair combining without photon emission. Therefore, the photoluminescence for nanosized particles shifts to higher energies and to higher efficiencies. A requirement for the occurrence of the quantum confinement effect is the "surface passivation" of the silicon particle, meaning that the outer bonds that are not filled with silicon atoms form other chemical bonds, for instance, a hydrogen or oxygen bond. The lack of surface passivation ("dangling bonds") leads to nonradiative transitions. For a detailed discussion of optical properties of the nanoparticles and Si nanocrystallites in particular, the reader is referred to Chapter 2.

Note that a similar luminescence process can be observed for large molecules or molecular ions and that such systems are also discussed as potential carriers of the

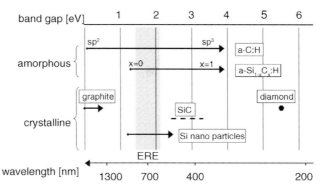

Figure 7.8 Bandgaps and related wavelengths of luminescence for different carbon- and silicon-bearing materials in comparison to the range of energies/wavelengths where maximum band positions of the ERE are observed (shaded area). The bandgap positions of the considered amorphous materials cover a wide range. The bandgap position of amorphous hydrogenated carbon (a-C:H) varies with the degree of sp^3 hybridization of the carbon atoms. For the amorphous hydrogenated silicon carbide (a-Si$_{1-x}$C$_x$:H), the bandgap varies with the concentration of C atoms. The bandgaps of the crystalline materials considered are more distinct. The range shown for SiC includes β-SiC with bandgap at 2.4 eV and α-SiC polytypes with bandgaps ranging from 3.0 to 3.3 eV. Bulk silicon has a bandgap of 1.17 eV and the arrow indicates the variation of bandgap due to quantum confinement for particles with decreasing size. The influence of quantum confinement for SiC is not shown in the figure since it has too large a bandgap compared to the range where ERE was observed. The crystalline carbon, where again the arrow indicates the influences of quantum confinement, has too a low bandgap compared to the observed ERE. (This figure was adapted from Ref. [32] and a detailed discussion is given there.)

ERE. Li [34], see references therein, summarizes the following suggested ERE carriers: submicron-sized carbonaceous materials, nm-sized carbon-based materials, nm-sized silicon-based materials, and particle-bombarded silicate grains. The variation in the photoluminescence with particle size and the high efficiency of their photoluminescence make nanoparticles good candidates for explaining ERE. Ledoux *et al.* [32, 35] have systematically studied the properties of nanoparticles of several carbon and silicon-bearing materials. Figure 7.8 shows the bandgaps of different carbon and silicon-bearing materials in comparison to the range of energies where maximum band positions of the ERE are observed (shaded area). Bulk silicon has a bandgap of 1.17 eV and this increases due to quantum confinement with decreasing particle size. The resulting wavelength range of the photoluminescence agrees well with the observed ERE, which provides the basis for suggesting silicon nanoparticles as carriers of the ERE, as was done in several works [32, 35, 36].

Ledoux *et al.* [32] also compare the photoluminescence yields as a function of the peak position of the photoluminescence band and its widths (as shown in Figure 7.9) with the observed positions and widths of the ERE. Among the considered materials, only the two considered types of silicon particles fall into the range of astronomical observations. The silicon hypothesis is further supported with a fit to observational

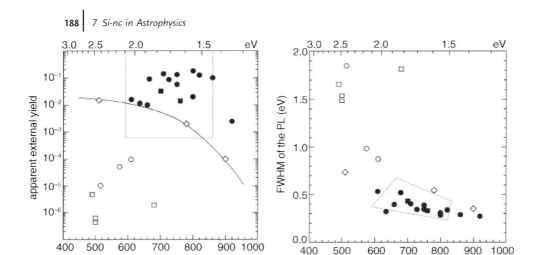

Figure 7.9 The left-hand side shows the measured photoluminescence yields as a function of the peak position of the PL band for different cosmic analogue materials. The measured PL widths (FWHM) for the same samples are plotted versus PL peak position in the figure on the right. The samples are natural coal (open squares), amorphous hydrogenated carbon (a-C:H, open circles), amorphous hydrogenated silicon carbide (a-SiC:H, open rhombi), porous silicon (p-Si, filled squares), and nanocrystalline silicon (nc-Si, filled circles), and the annotated symbols are valid for both figures. The gray-shaded area marks the range of peak wavelengths and PL yields derived from astronomical observations in the left figure and the respective range of peak width and peak position on the right. Only the two types of considered silicon particles fall into the range of astronomical observations. (These figures were reproduced from Ref. [32], with the kind permission of *Astronomy & Astrophysics*.)

data. Figure 7.6 shows observations of a planetary nebula, of a reflection nebula, of another galaxy, and of a HII region together with a suggested fit to the data assuming photoluminescence of Si nanocrystals [32].

Although this seems to strongly support the Si nanocrystals as carriers of the ERE, there are findings that support other models. Different from the bandgap value for diamond that Ledoux *et al.* used for their considerations (see Figure 7.8), Chang *et al.* [37] showed that after proton irradiation and thermal annealing diamond nanocrystals also generate photoluminescence in the spectral band of the ERE. Also, on the basis of laboratory measurements, Koike *et al.* [38] suggest thermoluminescence of forsterite and fused quartz as carriers of the ERE.

Results of other studies are even contrary to the silicon nanocrystal hypothesis. Li and Draine [39, 40] showed that, in order to agree with the observed infrared spectra of the ISM, Si nanoparticles should either form clusters or be attached to larger particles. And Witt and collaborators in 2006 suggest that observational findings about the radiation environment of the ERE seem to rule out Silicon nanoparticles as a carrier of the ERE [33]. Witt and Vijh [27] conclude their recent review by saying, "Yet, no current comprehensive model for interstellar dust explicitly accounts for the existence of the ERE. The identification of the ERE carrier remains a challenge for laboratory astrophysics, dust theory, and astronomical observations."

7.7
Formation of Si Nanoparticles under Nonequilibrium Conditions

The previously mentioned models of dust formation by condensation are not solely but partly based on equilibrium conditions. Deviations from the equilibrium are hard to describe and hence this is a suitable approach, which is also justified by the observation of the equilibrium products (listed in Table 7.2). Reality is likely to be more exciting than equilibrium. The timescales that the dust particles spend in their respective cosmic environments are long and surrounding particles (electrons, protons, high energy cosmic rays) and fields provide the energy to potentially change the internal structure of the solids. The small size of the particles may give a further boost to this.

It is worthwhile to consider laboratory measurements regarding these issues. Nuth and collaborators experimentally studied the formation of dust particles in a cooling gas flow to simulate the conditions in stellar outflows: the sizes of the forming condensates are of the order of 20–50 nm and their surfaces are highly reactive. The initial condensation temperature is of the order of 1500 K and the gas subsequently cools down. The detected forming condensates are not those expected for thermodynamic equilibrium conditions [41]. The initial stoichiometric composition being that of Mg and Fe silicates, Nuth and collaborators found that Mg oxides and Fe oxides (such as Fe_2O_3 and Fe_3O_4) condensed individually and that subsequently Mg silicate and Fe silicate form separately from the silicon oxide particles. Observations of the atomic ratio using an energy dispersion X-ray analysis system equipped with transmission electron microscope (TEM) showed that the forming SiO_x particles had x close to 1 (Y. Kimura, personal communication). It is suggested that the Mg and Fe silicates form by aggregation of the condensates and their subsequent processing, so that the minerals that finally form may be close to those that form under equilibrium conditions. Laboratory experiments showed, for instance, the formation of crystalline forsterite grains by coalescence and growth of Mg and SiO smoke particles [42].

The formation of silicon nanoparticles may occur as a result of internal alteration of other silicon-containing materials. Witt *et al.* [36] considered that since SiO is one of the most abundant and most strongly bonded molecules, it forms and condenses into oxygen-rich stellar outflows. They suggested that following the nucleation of SiO particles, the annealing and the separation into a Si core and a SiO_2 mantle would form the silicon nanoparticles and that the SiO_2 would facilitate the "passivation" of the Si crystal that is required for quantum confinement effects to occur for the Si nanoparticles.

The formation of silicon nanoparticles was also observed in laboratory experiments where evaporated SiO powder recondensed to Si and SiO_2 crystallites [43]. Moreover, experimental studies of SiO_x particles support the hypothesis that Si nanocrystallites form by annealing within a bulk sample. Kamitsuji *et al.* [44] reported the detection of Si nanocrystallites of about 10 nm diameters after heating a mixture of SiO_x particles with $x \approx 1$ (see Figure 7.10). These are detected in transmission electron microscope images by an enhanced contrast at the location of the Si

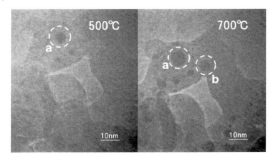

Figure 7.10 Transmission electron microscope image of a mixture of SiO_x after heating to 500 and 700 °C. Growth of Si nanocrystallites is seen in the black regions marked with white circles. (This figure is adapted from Ref. [44].)

nanocrystallites. Typical "stacking faults" of the Si nanocrystallites are observed at high-resolution images and the electron diffraction images show the characteristic rings of the silicon cube structure (see Figure 7.11). The Si nanocrystallites that are formed from SiO_x ($x \approx 1$) particles are still observed in the samples after heating to about 700 °C when subsequently cooled to room temperature again. Those Si nanocrystallites formed in SiO_2 particle samples survive heating to about 900 °C. The Si nanocrystallites have no influence on the measured IR spectra [44] and it is still

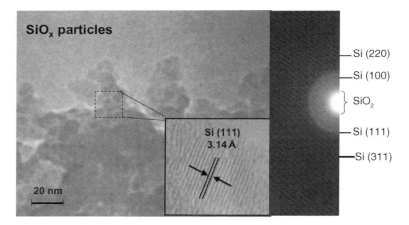

Figure 7.11 SiO_x particles at room temperature after heating to 900 °C. The left-hand side is a transmission electron microscope image and the right-hand side shows the electron diffraction pattern. Dark contrasts with approximately 10 nm diameter indicate the formation of Si crystallites. The high-resolution part of the photograph is taken in approximately the region that is indicated with the dashed square, it shows a structure forming along the (111) planes of cubic Si crystallites. The electron diffraction pattern shows the presence of silicon crystals. The given numbers denote the lattice planes for each crystal and the silicon patterns are for cubic silicon. The inner blurred diffraction rings are for β-cristobalite, a high-temperature phase of SiO_2. (This figure is by courtesy of Yuki Kimura, Tohoku University, Sendai.)

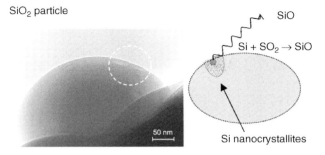

SiO$_2$ particle

SiO

Si + SO$_2 \rightarrow$ SiO

50 nm

Si nanocrystallites

Figure 7.12 Transmission electron microscope image (on the left) of SiO$_2$ particle at room temperature after annealing at 900 °C. Light contrast appeared as the result of partial evaporation and is possibly related to formation of Si crystallites. A possible mechanism of partial evaporation of the SiO$_2$ particle is sketched on the right. (This figure is by courtesy of Yuki Kimura, Tohoku University, Sendai.)

open whether photoluminescence can be observed. Luminescence caused by electron irradiation (cathodoluminescence) was observed at 400 nm though. The Si nanoparticles in this latter sample were produced by coevaporation of Ag and SiO (Y. Kimura, personal communication).

During the formation of Si particles, a significant amount of oxygen can dissolve into the Si particles while the structure of the silicon structure remains [45]. Formation of Si crystallites is also observed during partial sublimation of SiO$_2$ particles: the SiO$_2$ on the surface dissolves into SiO and Si, and while the SiO sublimates, the Si crystallites remain in the surface layer of the particle (Y. Kimura, personal communication). An image of a particle after heating and a sketch of the suggested process are shown in Figure 7.12. All these findings indicate that Si nanocrystallites most likely do form in cosmic environments and that they typically form within a larger bulk material. Moreover, their formation within the bulk material leaves the element abundances in the solid phase constant, so that it cannot be traced by gas-phase observations.

7.8
Conclusions

At this point, there is no clear evidence for the observation of silicon nanoparticles in cosmic environments. It is an open issue how the ERE is formed and silicon is only one potential carrier. The fact that ERE is not observed in the solar system, where *in situ* measurements provide additional information, further complicates the studies. Even if the observed ERE will turn out not to be generated by silicon nanocrystallites, its observations and thorough studies have initiated a new direction of astrophysical research, including that on the silicon nanoparticles. The evolution of silicon compounds in cosmic analogue materials on nanometer scale is complex and, among others, involves the formation of silicon nanocrystallites under certain conditions. It is up to future research to find out (a) how frequently these silicon nanocrystallites form and how stable they are, (b) whether they can form individually or can be released from the bulk material, and/or (c) whether they have noticeable influence on the optical properties of the bulk material.

References

1 Anders, E. and Grevesse, N. (1989) Abundances of the elements: meteoritic and solar. *Geochim. Cosmochim. Acta*, **53**, 197–214.

2 Dwek, E. (1998) The evolution of the elemental abundances in the gas and dust phases of the galaxy. *Astrophys. J.*, **501**, 643–665.

3 Calura, F., Pipino, A., and Matteucci, F. (2008) The cycle of interstellar dust in galaxies of different morphological types. *Astron. Astrophys.*, **479**, 669–685.

4 Speck, A.K. (1997) The mineralogy of dust around evolved stars, PhD Thesis, University College, The University of London.

5 Kozasa, T. and Sogawa, H. (1997) Formation of dust grains in circumstellar envelopes of oxygen-rich AGB stars. *Astrophys. Space Sci.*, **251**, 165–170.

6 Speck, A.K., Thompson, G.D., and Hofmeister, A.M. (2005) The effect of stellar evolution on SiC dust grain sizes. *Astrophys. J.*, **634**, 426–435.

7 Loddes, K. and Amari, S. (2005) Presolar grains from meteorites: remnants from the early times of the solar system, Chemie der Erde, **65(2)** 93–166.

8 Nozawa, T., Kozasa, T., Habe, A., Dwek, E., Umeda, H., Tominaga, N., Maeda, K., and Nomoto, K. (2007) Evolution of dust in primordial supernova remnants: can dust grains formed in the ejecta survive and be injected into the early interstellar medium? *Astrophys. J.*, **666**, 955–966.

9 McKee, C.F. and Ostriker, J.P. (1977) A theory of the interstellar medium: three components regulated by supernova explosions in an inhomogeneous substrate. *Astrophys. J.*, **218**, 148–169.

10 Jones, A.P., Tielens, A.G.G.M., Hollenbach, D.J., and McKee, C.F. (1994) Grain destruction in shocks in the interstellar medium. *Astrophys. J.*, **433**, 797–810.

11 Slavin, J.D. and Frisch, P.C. (2008) The boundary conditions of the heliosphere: photoionization models constrained by interstellar and *in situ* data. *Astron. Astrophys.*, **491**, 53–68.

12 Krügel, E. (2003) *The Physics of Interstellar Dust, IoP Series in Astronomy and Astrophysics*, The Institute of Physics, Bristol, UK.

13 Kimura, H., Mann, I., and Jessberger, E.K. (2003) Elemental abundances and mass densities of dust and gas in the local interstellar cloud. *Astrophys. J.*, **582**, 846–858.

14 Mann, I., (2010) Interstellar dust in the solar system. Annual Review of Astronomy and Astrophysics. vol. 48, Issue 1.

15 Li, A. and Greenberg, J.M. (1997) A unified model of interstellar dust. *Astron. Astrophys.*, **323**, 566–584.

16 Takeuchi, T. (2009) From protoplanetary disks to planetary disks: gas dispersal and dust growth, in *Small Bodies in Planetary Systems* (eds I. Mann, A.M. Nakamura, and T. Mukai), Lecture Notes in Physics, vol. 758, Springer-Verlag GmbH, Heidelberg, Germany, pp. 1–35.

17 Sitko, M.L., Grady, C.A., Lynch, D.K., Russell, R.W., and Hanner, M.S. (1999) Cometary dust in the debris disks of HD 31648 and HD 163296: two "baby" beta Pictoris stars. *Astrophys. J.*, **510**, 408–412.

18 Köhler, M. (2006) PhD Thesis, Münster University, Germany.

19 Li, A. and Greenberg, J.M. (1998) A comet dust model for the β Pictoris disk. *Astron. Astrophys.*, **331**, 291–313.

20 Mann, I., Murad, E., and Czechowski, A. (2007) Nanoparticles in the inner solar system. *Planet. Space Sci.*, **55**, 1000–1009.

21 Meyer-Vernet, N., Maksimovic, M., Czechowski, A., Mann, I., Zouganelis, I.,

Goetz, K., Kaiser, M.L., St. Cyr, O.C., Bougeret, J.L., and Bale, S.D. (2009) Dust detection by the wave instrument on STEREO: nanoparticles picked up by the solar wind? *Sol. Phys.*, **256**, 463–474.

22 Habbal, S.R., Arndt, M.B., Nayfeh, M.H. *et al.* (2003) On the detection of the signature of silicon nanoparticle dust grains in coronal holes. *Astrophys. J.*, **592**, L87–L90.

23 Mann, I. and Murad, E. (2005) On the existence of silicon nano-dust near the sun. *Astrophys. J.*, **624**, L125–L128.

24 Ott, U. (2003) The most primitive material in meteorites, in *Astromineralogy* (ed. T. Henning) Lecture Notes in Physics, vol. 609, Springer-Verlag GmbH, Heidelberg, Germany, pp. 236–265.

25 Cohen, M., Anderson, C.M., and Cowley, A. (1975) The peculiar object HD 44179 "The red rectangle." *Astrophys. J.*, **196**, 179–189.

26 Schmidt, G.D., Cohen, M., and Margon, B. (1980) Discovery of optical molecular emission from the bipolar nebula surrounding HD 44179. *Astrophys. J.*, **239**, L133–L138.

27 Witt, A.N. and Vijh, U.P. (2004) Extended red emission: photoluminescence by interstellar nanoparticles, astrophysics of dust. ASP Conference Series, vol. 309, Proceedings of the Conference, 26–30 May 2003, Estes Park, Colorado (eds Adolf N. Witt, Geoffrey C. Clayton, and Bruce T. Draine), pp. 115–139.

28 Witt, A.N. and Boroson, T.A. (1990) Spectroscopy of extended red emission in reflection nebulae. *Astrophys. J.*, **355**, 182–189.

29 Perrin, J.-M., Darbon, S., and Sivan, J.-P. (1995) Observation of extended red emission (ERE) in the halo of M82. *Astron. Astrophys.*, **304**, L21–L24.

30 Witt, A.N. (1988) *Proceedings of the Conference: Dust in the Universe*, Cambridge University Press, p. 1.

31 Rouan, D., Lecoupanec, P., and Leger, A. (1995) Proceedings of the 1st Franco-British Meeting on the Physics and Chemistry of the Interstellar Medium (ed. C.S. Jeffery), vol. 22, p. 37.

32 Ledoux, G., Guillois, O., Huisken, F., Kohn, B., Porterat, D., and Reynaud, C. (2001) Crystalline silicon nanoparticles as carriers for the extended red emission. *Astron. Astrophys.*, **377**, 707–720.

33 Witt, A.N., Gordon, K.D., Vijh, U.P. *et al.* (2006) The excitation of extended red emission: new constraints on its carrier from hubble space telescope observations of NGC 7023. *Astrophys. J.*, **636**, 303–315.

34 Li, A. (2004) Interaction of nanoparticles with radiation, astrophysics of dust. ASP Conference Series, vol. 309, Proceedings of the Conference, 26–30 May 2003, Estes Park, Colorado (eds Adolf N. Witt, Geoffrey C. Clayton, and Bruce T. Draine), pp. 417–453.

35 Ledoux, G., Ehbrecht, M., Guillois, O., Huisken, F., Kohn, B., Laguna, M.A., Nenner, I., Paillard, V., Papoular, R., Porterat, D., and Reynaud, C. (1998) Silicon as a candidate carrier for ERE. *Astron. Astrophys.*, **333**, L39–L42.

36 Witt, A.N., Gordon, K.D., and Furton, D.G. (1998) Silicon nanoparticles: source of extended red emission? *Astrophys. J. Lett.*, **501**, L111–L114.

37 Chang, H.-C., Chen, K., and Kwok, S. (2006) Nanodiamond as a possible carrier of extended red emission. *Astrophys. J.*, **639**, L63–L66.

38 Koike, C., Chihara, H., Koike, K., Nakagawa, M., Okada, M., Tsuchiyama, A., Aok, M., Awata, T., and Atobe, K. (2002) Thermoluminescence of forsterite and fused quartz as a candidate for the extended red emission. *Meteorit. Planet. Sci.*, **37**, 1591–1598.

39 Li, A. and Draine, B.T. (2001) On ultrasmall silicate grains in the diffuse interstellar medium. *Astrophys. J.*, **554**, 778–802.

40 Li, A. and Draine, B.T. (2002) Are silicon nanoparticles an interstellar dust component? *Astrophys. J.*, **564**, 803–812.

41 Nuth, J.A., Rietmeijer, F.J.M., and Hill, H.G.M. (2002) Condensation processes in astrophysical environments: The Composition and structure of cometary grains. *Meteorit. Planet. Sci.*, **37**, 1579–1590.

42 Kaito, C., Ojima, Y., Kamitsuji, K. *et al.* (2003) Demonstration of crystalline forsterite grain formation due to coalescence growth of Mg and SiO smoke particles. *Meteorit. Planet. Sci.*, **38**, 49–58.

43 Kaito, C. and Shimizu, T. (1984) High resolution electron microscopic studies of

amorphous SiO film. *Jpn. J. Appl. Phys.*, **23**, L7–L8.

44 Kamitsuji, K., Ueno, S., Suzuki, H. *et al.* (2004) Direct observation of the metamorphism of silicon oxide grains. *Astron. Astrophys.*, **422**, 975–979.

45 Kimura, Y., Ueno, H., Suzuki, H., Tanigaki, T., Sato, T., Saito, Y., and Kaito, C. (2003) Dynamic behavior of Au cluster on the surface of silicon nanoparticles. *Physica E*, **19**, 298–302.

8
Size-Controlled Si Nanocrystals using the SiO/SiO$_2$ Superlattice Approach: Crystallization, Defects, and Optical Properties

Margit Zacharias

8.1
Introduction

Silicon nanocrystals (Si NCs) are candidates for silicon-based light-emitting devices, increasing the efficiency of solar cells and nonvolatile memories if an appropriate size control can be realized. Synthesis of randomly distributed silicon nanocrystals can be realized by ion implantation of silicon into a SiO$_2$ matrix followed by thermally induced Ostwald ripening of silicon clusters and their crystallization or by deposition of substoichiometric oxide films by CVD or LPCVD, sputtering processes or reactive evaporation, and a thermally induced phase separation and silicon crystallization. All these methods result in a relatively broad size distribution of the synthesized silicon nanocrystals. Size control in these systems is normally realized by varying the silicon content within the SiO$_2$ matrix or by subsequent oxidation of nanocrystals. The crystal density cannot be controlled independently. A lower Si content results not only in smaller crystals but also in a drastic reduction in their number.

In recent years, various approaches for narrowing the size distribution of silicon nanocrystals have been developed including new techniques in the synthesis process as discussed in our review [1]. The first approach realizing an improved size control of silicon nanocrystals is the use of Si/SiO$_2$ superlattices introduced by Lockwood and coworkers [2, 3]. In this case, molecular beam epitaxy (MBE) combined with oxidization using UV ozone is used for a precise growth of nanometer-thick amorphous silicon layers in between SiO$_2$ layers. The presence of grain boundaries is a kind of "birth defect" of the amorphous silicon layer approach. Crystallization most likely results in grain boundaries between the nanocrystals that drastically degrade the optical and electronic properties. The use of such ultrathin amorphous Si layers between insulating materials was later adapted for a number of different materials deposited by various deposition methods. Si/SiN$_x$ superlattices were synthesized using electron beam silicon evaporation and periodic electron cyclotron resonance plasma nitridation [4]. Just to give two more examples here, Si/Si$_x$O$_y$N$_z$

Silicon Nanocrystals: Fundamentals, Synthesis and Applications. Edited by Lorenzo Pavesi and Rasit Turan
Copyright © 2010 WILEY-VCH Verlag GmbH & Co. KGaA, Weinheim
ISBN: 978-3-527-32160-5

superlattices [5] and Si/CaF$_2$ superlattices [6] were synthesized by LPCVD and room-temperature MBE, respectively. However, to observe quantum confinement properties of the Si nanocrystals after the thermal crystallization process, a dielectric (high bandgap) material should separate the nanocrystals within the layer, too.

8.2
Size Control of Si Nanocrystals by the SiO/SiO$_2$ Superlattice Approach

A real breakthrough was reported in 2002 by using a superlattice of SiO and SiO$_2$ that is nowadays a standard approach for fabricating ordered, layer arranged, size-controlled Si nanocrystals [7]. This method is based on the preparation of amorphous SiO$_x$/SiO$_2$ superlattices and thermal annealing for phase separation and crystallization. The preparation of SiO$_x$/SiO$_2$ superlattices ($1 \leq x \leq 2$) is a simple, elegant, and efficient method for the synthesis of size-controlled Si nanocrystals and could be used for a number of applications [8]. Nonstoichiometric oxides (SiO$_x$) are not stable at high temperatures and decompose by a phase separation into the two stable components (Si and SiO$_2$):

$$\text{SiO}_x \rightarrow \frac{x}{2}\text{SiO}_2 + \left(1 - \frac{x}{2}\right)\text{Si}. \tag{8.1}$$

Depending on temperature, amorphous Si clusters or Si nanocrystals will be observed. Figure 8.1 schematically demonstrates the desired control of size, separation, and density for applications that can be achieved using the SiO$_x$/SiO$_2$ superlattice approach in the following way: the size of the nanocrystals is controlled by the layer thickness of the SiO$_x$ layers (for more details, please see below). The stoichiometry of the SiO$_x$ influences the number of Si nanocrystals within the layers and their average density. An additional density control can be achieved by thicker barrier SiO$_2$ layers between adjacent SiO$_x$ layers. The preparation can be done by evaporation of SiO powder under high vacuum. Adding oxygen during the growth can be used for change in stoichiometry. Please note that already an oxygen partial pressure of 10^{-4} mbar is enough to completely oxidize the growing films into SiO$_2$ [9]. The base pressure in our evaporation chamber was 1×10^{-6} mbar for SiO layers. We observed that the particle size of the SiO powder and the chosen

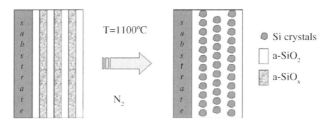

Figure 8.1 Model of the process for preparing layered arranged Si nanocrystals.

evaporation boat can also influence the deposition process. SiO pellets of 99.99% purity are used, which we grind to a powder. Tantalum boats are used for evaporation. A boat temperature exceeding 1000 °C (as used here for the evaporation of the SiO powder under oxygen partial pressure) results in an oxidation of the boat material. Only the vapor pressure of Ta_2O_5 is low enough to prevent a coevaporation of the metal oxide covering the boat. We observed metal contamination in the grown films that drastically deteriorate, especially the electrical properties if tungsten or molybdenum boats are used.

A detailed investigation of the phase separation of SiO_x resulting in Si nanocrystals after 1100 °C annealing under nitrogen atmosphere was reported by L.X. Yi [10]. The annealing under nitrogen atmosphere gives a better luminescence intensity than when Ar is used. We were not able to prove nitrogen in the films after the annealing process, thus only traces (of below 1%) might be included by diffusion during annealing. The evaporation process used for SiO/SiO₂ superlattices has a number of advantages: it is simple and the resulting layers do not contain any hydrogen and a very low level of nitrogen that is important for the crystallization process. Hydrogen stays in the films up to 450–500 °C and might hinder the phase separation that already starts at a lower temperature. The influence of nitrogen is not completely clear as yet, but traces supplied by diffusion during annealing seem to be of advantage. However, SiO_x films grown by CVD with N_2O as oxygen source normally contain up to 12% nitrogen, which stays in the films even at higher temperatures. Furthermore, CVD-grown SiO_x contains high concentrations of hydrogen that can be diffused out by annealing around 450–500 °C, but it might hinder the phase separation that starts at lower temperatures. Such films should be called a nonstoichiometric oxynitride and not a SiO_x film as often done in literature. Nitrogen obviously influences and even hinders the clustering of silicon in such oxynitride films, even from the beginning. We will only discuss films prepared by evaporation, that is, prepared without nitrogen.

In Figure 8.2, we compare cross-sectional bright field TEM images of (a) an as-prepared amorphous SiO/SiO₂ superlattice and (b) the same film after 1100 °C annealing under nitrogen. In (c), we demonstrate the effect of changing the thickness of the SiO layer. Figure 8.2d presents a bulk SiO film after annealing for comparison. The Si nanocrystals are observed as the darker contrast. Size control can be clearly seen here. The thick bulk film contains rather big crystals randomly distributed within the SiO₂ matrix whereas the nanocrystals are confined in the former SiO layers with uniform sizes controlled by the former SiO layer thickness. The size control works well for thickness of 2–6 nm. Figure 8.2b and d also gives the respective size distribution estimated from dark field images of samples (b) and (d). If the layer thickness is below 2 nm, then the resulting crystals are still in the range of around 2 nm but less in numbers. The reason for that is based on crystallization theory resulting in a critical crystallization diameter of the nanocrystals. Crystals below the critical size are not stable; they stay as amorphous clusters and contribute to the growth of the bigger ones by Ostwald ripening. If the SiO layer is thicker then 7 nm, then more than one starting nucleus will be established over the layer thickness and the size control is more and more lost. In bulk SiO_x films, there is a random

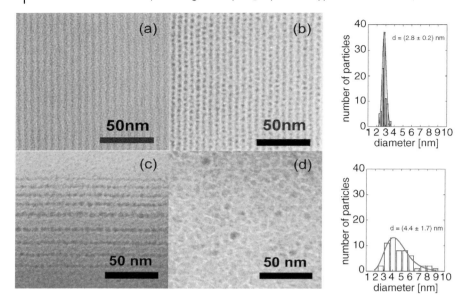

Figure 8.2 Cross-sectional bright field TEM images of (a) an as-prepared amorphous SiO/SiO$_2$ superlattice, (b) the same film after 1100 °C annealing under nitrogen, (c) the effect of changing the thickness of the SiO layer and the correspondent change in Si NCs size, and (d) a bulk SiO film after annealing for comparison.

nucleation within the matrix and an Ostwald ripening of the nuclei by diffusion and rearrangement of silicon and oxygen atoms.

The phase separation can be monitored by IR spectroscopy. In the range of 700–1500 cm^{-1}, various silicon–oxygen-related absorption bands can be seen. An example has been selected from Ref. 10 (Figure 8.3). The band around 810 cm^{-1} can be assigned to Si−O−Si bond bending motion in SiO$_2$. With higher annealing, a new band at 880 cm^{-1} appears and increases in intensity up to 400–500 °C. For annealing temperatures above 500 °C, this absorption band loses intensity and vanishes at 800 °C. We did semiempirical quantum mechanical calculations using a model as seen in Figure 8.4 simulating a ring of Si nanocrystals with an attached oxygen atom moving by thermal energy (vibrations) through the lattice. Such an oxygen atom is called nonbridging oxygen hole center (NBOHC). Our simulations gave two IR bands for the structure (880 and 1000 cm^{-1}) with intensity ratios similar to the ones seen in our IR spectra. So, we conclude that the 880 cm^{-1} IR band seen for intermediate annealing temperatures can be assigned to vibrations of Si rings with an attached O atom. And, of course, such rings are the first steps toward a nanocluster and will be definitely established at the beginning phase separation. For higher temperatures, the reordering is faster, the clusters grow larger, and the band vanishes again. However, such an NBOHC also represents a defect that might have a related signature in the PL spectra.

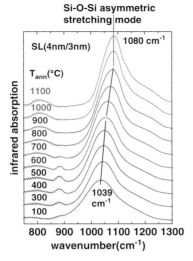

Figure 8.3 Evolution of oxygen-related IR bands with annealing observed on an example of a SiO/SiO₂ supperlattice film. Published with permission from Ref. [10].

The band above $1000\,\mathrm{cm}^{-1}$ is assigned to the asymmetric stretching motion of the oxygen atom of the Si—O—Si bridging configuration. Its peak position can be used for reasonable stoichiometry estimation in case of a homogeneous bulk SiO_x film. The as-prepared position for bulk SiO was measured to be at $980\,\mathrm{cm}^{-1}$ and that of the as-prepared bulk SiO_2 to be at $1060\,\mathrm{cm}^{-1}$. However, the IR absorption of the bulk SiO film is significantly broader than that of the SiO_2 film. A sharpening of the IR mode of the SiO film is observed for temperatures above $900\,^{\circ}\mathrm{C}$. Please note if the samples consist of a superlattice containing SiO and SiO_2 layers that both contribute to the absorption of the as-prepared sample, then a shift of the overall mode position to higher wavenumbers compared to a bulk SiO film will be observed ("effective medium" position). For the as-prepared superlattice ($100\,^{\circ}\mathrm{C}$), a position of $1039\,\mathrm{cm}^{-1}$

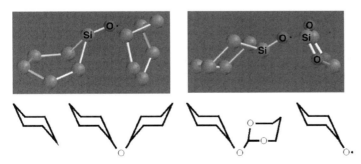

Figure 8.4 Model of Si rings used in our calculations containing a basic ring of six silicon atoms and the movement of a single O atom from cell to cell. Intermediate states represent what is called NBOHC.

was found. The observed IR band shifts from 1039 to 1052 cm^{-2} with annealing temperature in the range of 100–600 °C. A more pronounced shift is observed for annealing between 700 and 900 °C with a final position at 1080 cm^{-1}, representing the position observed for SiO_2. Annealing at higher temperatures does not result in any further change in the Si—O-related bands. Further significant questions related to annealing temperature and phase separation are as follows: Do the Si clusters remain in the amorphous state or do they transfer to Si nanocrystals, at which temperature will crystallization be complete, and does the size play a role in the process? section 8.3 will focus on crystallization and its size dependence.

8.3
Crystallization Behavior

The crystallization temperature T_c as a function of film thickness has been studied for amorphous semiconductors such as germanium or silicon embedded in oxides or nitrides. The investigation of X-ray diffraction of Raman scattering can be used for monitoring the transition from amorphous to crystalline state. In many cases, superlattice structures were used with thin amorphous semiconductor films sandwiched between oxides or nitrides [12–16]. It was found that the crystallization temperatures are strongly influenced by the material embedding the amorphous semiconductor like a sandwich [17, 18]. In many cases, when the semiconductor is sandwiched between an oxide and a nitride, a strong increase in the crystallization temperature T_c is observed when the film thickness of the semiconductor is reduced below a certain thickness, typically 20 nm. Please note that the further reported model was developed for amorphous Si or Ge layers embedded in amorphous SiO_2. Deviations will occur if a SiO_x layer is embedded instead of the a-Si or a-Ge. However, similar effects are expected. Modeling of the latter case is rather complicated because in addition to the change in the solid state (amorphous to crystalline) we have to consider the phase separation that results in embedded Si clusters within the SiO_2 matrix. Anyway, the compromise is that such films are annealed at rather high temperatures (1100 °C) to ensure a complete crystallization even for thin layers and hence small crystals.

The crystallization is evident by the occurrence of the (220) and (311) Bragg peaks in the XRD spectra. Decreasing the Si layer thickness leads to a correlated decrease in the average size of the nanocrystals as can be seen by the broadening of the Bragg peaks. Using a temperature greater than the crystallization temperature does not change the average size of the crystals within the error of measurement. Mainly, the number of crystals increases resulting in an increase in peak intensity. Figure 8.5 summarizes the crystallization temperature of an ensemble of superlattices based on different materials and interfaces as a function of layer thickness. An exponential increase in crystallization temperature was found when decreasing the layer thickness in all these different systems that were fitted by

$$T_c = T_{ac} + (T_{melt} - T_{ac}) \cdot e^{-d/C}. \tag{8.2}$$

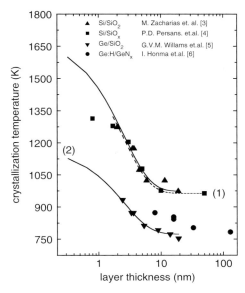

Figure 8.5 Dependence of the crystallization temperature on the layer thickness for amorphous Ge and Si sandwiched between oxide and nitride. The solid lines are based on a model fit following Eq. (8.2).

T_{melt} represents the melting temperature of bulk crystalline material, T_{ac} is the crystallization temperature of a thick bulk amorphous film, and d is the real thickness of the layer (Figure 8.5). The lines in Figure 8.5 represent a data fit using the above equation and parameters summarized in Table 8.1.

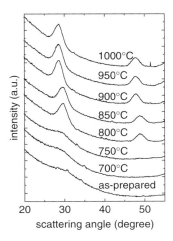

Figure 8.6 Example demonstration of the XRD spectra for Si/SiO$_2$ superlattice as a function of the annealing temperature. The crystallization is visible by the splitting of the (220) and (311) peaks.

Table 8.1 Fitting parameter used for the lines in Figure 8.5.

Superlattice material	T_{melt}	T_{ac}	c
Si/SiO$_2$ (curve 1)	1683 K	973 K	2.56 nm
Ge/SiO$_2$ (curve2)	1211 K	773 K	2.52 nm

This rather empirically developed equation fits amazingly well. Below please find a more detailed approach based on crystallization theory. The position of the Bragg peaks is affected by strain, too. The respective analysis is shown in Figure 8.7, which is the result of a detailed analysis of the inhomogeneous strain of Si/SiO$_2$ films with increasing temperature.

Please note the bulk melting point T_{melt} is used as a basic material property. Recently, however, there is some evidence suggesting that the melting behavior might also be affected by the nanostructure and the embedding material [19]. That is not included in our theory developed in 2000. However, very recently our crystallization model was used to also fit the behavior of phase change materials deposited in a superlattice structure that shows a similar increase in the crystallization behavior with decreasing layer thickness [20].

Amorphous SiO$_2$ interfaces embedding the thin amorphous Si layer (below 50 nm) will not result in a homogeneous and uninfluenced nucleation within the layers. We assume here that the crystallization nucleus is cylindrical in shape and symmetrically embedded in the amorphous material between the oxide interfaces. In contrast to previous models, an additional spacing l corresponding to a finite separation of the nucleus from the oxide boundaries is introduced as demonstrated schematically in Figure 8.8. In principle, for each combination of phases, we assume that a well-defined, that is, a sharp and perfect, interface bounded by bulk material can be formed, which is characterized by its specific free interface energy. We define γ_{ac}, γ_{oc}, and γ_{oa} as the specific interfacial free energies per unit area between the different material phases. However, for a sandwich structure as considered here, the interfaces between material o and material c are not well defined if the distance l between these materials is on the order of only a few lattice constants.

For $l = 0$, and hence $h = d$, the Persans model [21] should be rederived. However, for l very large ($l \rightarrow \infty$), the materials o and c are separated by two noninteracting, well-defined interfaces. We estimate now the interaction between these two interfaces for small l by defining an effective interface energy that interpolates between the well-defined limiting cases above and write

$$\gamma_{oc}^{eff} = \gamma_{ac} + (\gamma_{oc} - \gamma_{ac}) \cdot M \quad \text{with} \quad M = e^{-l/l_0}. \tag{8.3}$$

M is an effective order parameter that is normalized to unity for true oxide/crystalline interface and zero for true amorphous/crystalline interface. In view of the observed exponential dependence of the inhomogeneous strain on layer thickness, M is expected to be an exponentially decreasing function of the interface spacing l. If we now assume short-range interatomic forces, then l_0 can be interpreted as an average screening or bonding length that is related to the range of interatomic

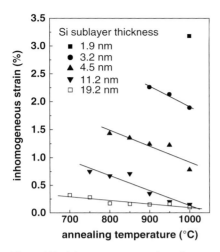

Figure 8.7 Inhomogeneous strains of nanocrystals as function of annealing temperature and Si layer thickness based on analysis of (220) and (311) Bragg peaks.

forces typical for the materials o and c. The nucleation energy barrier is given by the difference in the Gibbs free energies:

$$\Delta G = G_c - G_a = -\pi \cdot r^2 \cdot h \cdot \Delta G_v + 2 \cdot \pi \cdot r \cdot h \cdot \gamma_{ac} + 2 \cdot \pi \cdot r^2 \cdot \Delta \gamma_{eff}, \qquad (8.4)$$

with $\Delta G_v = G_{va} - G_{vc} > 0$ and $\Delta \gamma_{eff} = \gamma_{ac} + (\gamma_{oc} - \gamma_{ac} - \gamma_{oa}) \cdot e^{-l/l_0}$.

The terms in Eq. (8.4) describe the change in bulk free energy, the energy necessary for forming the new a/c interface, and the influence of the boundaries formed by the

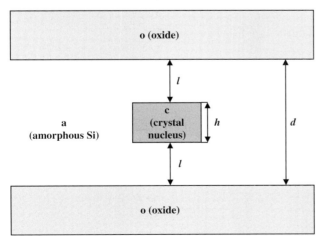

Figure 8.8 Model of the starting nanonucleus with a finite separation of a nucleus from oxide boundary.

oxide material o. The screening length l is given by $l = (d-h)/2$, with $d \geq h$. For $\gamma_{oc} - \gamma_{oa} > 0$, the crystal nucleation is inhibited, but $\gamma_{oc} - \gamma_{oa} < 0$ will enhance the formation of crystal nuclei. In our model, h and d are *geometrically independent parameters* that are *energetically* coupled by the above equation. This allows to consider the Gibbs free energy change in Eq (8.4) as a function of both the variables, r and h. The specific interface energies are not well known. At this point, we have to make an approximation: we ignore the fact that $\Delta\gamma_{eff}$ depends on the size h of the nucleus and replace l with an average value $\bar{l} = (l_{min} + l_{max})/2 = d/4$. With this approximation and assuming that the cylinder of radius r and height h in Figure 8.8 corresponds to the equilibrium shape of the critical nucleus, we can derive the radius-to-height ratio

$$\frac{r}{h} = \frac{\gamma_{ac}}{2\Delta\gamma_{eff}} \tag{8.5}$$

by minimizing the Gibbs free interface energy in Eq. (8.4) assuming a constant volume. Using this relation to eliminate one of the two variables in Eq. (8.4), that is, r, the nucleation barrier is given by the maximum of ΔG: $(\partial \Delta G / \partial h) = 0$, yielding

$$r^* = \frac{2\gamma_{ac}}{\Delta G_v} \quad \text{and} \quad h^* = \frac{4\Delta\gamma_{eff}}{\Delta G_v} \tag{8.6}$$

for the critical radius and the critical cylinder height of the nucleus. Inserting r^* into Eq. (8.4) yields for the nucleation barrier

$$\Delta G^* = \frac{8 \cdot \pi \cdot \gamma_{ac}^2 \cdot \Delta\gamma_{eff}}{\Delta G_v^2} \tag{8.7}$$

with $\Delta\gamma_{eff}$ given by $\Delta\gamma_{eff} = \gamma_{ac} + (\gamma_{oc} - \gamma_{ac} - \gamma_{oa}) \cdot e^{-d/4l_0}$.

The nucleation rate $N/t \sim \exp(-\Delta G^*/kT)$ is proportional to the Boltzmann factor; hence, the time t for the formation of a certain number N of nuclei is given by $t/N \sim \exp(\Delta G^*/kT)$. If we define the crystallization temperature T_c by the requirement that a certain fixed number N_c of nuclei are generated at a given fixed time t_c, then the crystallization temperature follows as

$$\Delta G^*/k \cdot T_c = \ln(t_c/N_c) + \text{const.} \quad \text{or} \quad k \cdot T_c \sim \Delta G^*. \tag{8.8}$$

If we now assume the bulk crystallization temperature T_{ac} by $k \cdot T_{ac} \sim \Delta G_{ac}^*$ with $\Delta G_{ac}^* = 8 \cdot \pi \cdot \gamma_{ac}^3/\Delta G_v^2$ (bulk nucleation barrier derived for $d \to \infty$), then we obtain

$$T_c/T_{ac} = \Delta G^*/\Delta G_{ac}^* \quad \text{or} \quad T_c = T_{ac}\frac{\Delta\gamma_{eff}}{\gamma_{ac}} = T_{ac}\left(1 + \frac{\gamma_{oc} - \gamma_{ac} - \gamma_{oa}}{\gamma_{ac}} \cdot e^{-d/4l_0}\right). \tag{8.9}$$

Please note this result has the functional form of our empirical Eq. (8.2). Comparison with Eq. (8.2) yields

$$T_{melt} = T_{ac}\frac{\gamma_{oc} - \gamma_{oa}}{\gamma_{ac}} \quad \text{and} \quad C = 4l_0. \tag{8.10}$$

The experimentally observed crystallization temperature T_c tends to be the temperature of the melting point of bulk crystalline silicon $T_{melt} = 1683$ K in the limit of zero Si layer thickness, whereas in the thick layer limit $T_c(d \rightarrow \infty) \approx T_{ac} = 973$ K. Similar results were obtained for germanium. According to our theoretical results, the enhancement of T_c in the limit $d \rightarrow 0$ is related to the difference in the specific energies of the interfaces between the involved crystalline and amorphous phases. However, taking into consideration that the melting of the crystalline phase is likewise associated with the nucleation of a crystalline–liquid interface, the above results at last give us a hint why the crystallization temperature T_c for $d \rightarrow 0$ and the melting point T_m of the crystalline phase are in such a good empirical correlation. The screening length l_0 should be related to the range of interatomic forces or to the length of elastic interactions between the respective interfaces. We found $C \approx 2.52$–2.56 nm independent of the material (Ge, Si). Hence, the screening length is given by $l_0 \approx 0.64$ nm, which would correspond to a range of 2–3 interatomic distances using the lattice parameter of Si and Ge.

The minimum lateral radius r^* of the cylindrical nucleus, which is determined only by γ_{ac} and the change in the bulk Gibbs free energy per unit volume ΔG_v, in the present approximation is the same as the radius of a free spherical crystalline nucleus in bulk amorphous silicon. With $\Delta G_v = \Delta g_v / a^3$, where $\Delta g_v = 0.100$ eV/atom is the free energy change associated with the crystallization of one atom and $\gamma_{ac} = \sigma_{ac} / a^2$ with the amorphous/crystalline interface energy per atom $\sigma_{ac} = 0.105$ eV/atom (see Ref. [22]), we derive the minimum lateral radius from the above equations as

$$r^* = \frac{2\sigma_{ca}}{\Delta g_v} a = 0.567 \text{ nm} \tag{8.11}$$

(four to five atoms in diameter). Hence, the base area of a critical cylinder consists of only 17 atoms and an average interatomic distance of $a = 0.27$ nm is used as for Si. The minimum height h^* of the cylindrical nucleus exponentially depends on the layer thickness via the effective specific interface energy. For thick layers, bulk behavior is restored for $d \rightarrow \infty$ yielding $h^* = h^*_{bulk} = 4\gamma_{ac}/\Delta G_v = 2r^* = 1.134$ nm, which corresponds to a fairly symmetrical nucleus with the same volume as a spherical nucleus with radius r^*. For very small layer thicknesses, the condition $l \geq 0$, that is, $h^* \leq d$, must be fulfilled in our model (see Figure 8.8), which gives the lower limit d_{min}. In order to determine d_{min}, we rewrite the minimum height h^* as

$$h^* = 2r^* \left(1 + \left(\frac{\gamma_{oc} - \gamma_{oa}}{\gamma_{ac}} - 1 \right) \cdot e^{-d/4l_0} \right). \tag{8.12}$$

Using the numerical values available in the work of Persan *et al.*, we expect no crystallization for layer thickness $d < d_{min} \approx 1.6$ nm for the a-Si/SiO$_2$ system irrespective of the temperature. This is in good agreement with our measurements of the Si/SiO$_2$ system, where crystallization has been observed up to $d \approx 1.9$ nm. The value of $d_{min} = 1.6$ nm is larger than the one expected from purely geometrical considerations, which yields $d_{min} = 1.1$ nm. The higher value of $d_{min} \approx 1.6$ nm is clearly a consequence of a strong increase in the nucleation barrier caused by continuous increase in the effective specific interface energy when the a/c and the o/c interfaces approach each other. In our experiments with SiO/SiO₂ superlattices, we tried the crystallization of SiO layers with only 1 nm thickness. The observed diameter of the nanocrystals, however, was still in the range of 2 nm. The number of grown crystals was reduced as was evident by the reduced intensity of the photoluminescence (PL) that scales directly to the number of the crystals (see below). The above model was developed for the crystallization of amorphous silicon embedded into amorphous SiO₂. It can also be used in similar cases for embedded amorphous layers in an oxide or nitride. However, the disadvantage of the crystallization of such amorphous layers is that at the end of the crystallization finally the nanocrystals touch each others establishing grain boundaries in between. Such grain boundaries strongly affect the luminescence (and electronic) properties. The potential barrier needed for quantum confinement is missed between the crystals of the layers where grain boundaries dominate the recombination behavior by defects. The solution for this problem is the above-described SiO/SiO₂ superlattice that transforms into well-separated nanocrystals within the layers. The phase separation, which also needs energy, into amorphous Si and SiO₂ following Eq. (8.1) is not taken into account in the above equations describing the crystallization behavior. Additional simulations in future might more accurately describe the process in these more complicated systems.

Nevertheless, we also observed higher temperatures for a complete crystallization in case of SiO/SiO₂ superlattices. As a rule, one can use the following: amorphous nanoclusters are formed up to 900 °C, 1000 °C will result in an incomplete crystallization, and temperatures of 1100 °C are needed for a complete crystallization. Please note that this is valid for equilibrium process (slow increase and decrease in the temperature). Other embedding interfaces such as the group of nitrides or oxynitrides require higher annealing temperatures (up to 1250 °C). A way out could be to use rapid thermal annealing that is, however, a nonequilibrium process and its temperature and time behavior drastically deviate from the above.

8.4
Defects and their Signatures in ESR

Si nanocrystals completely embedded in amorphous SiO₂ is desired to realize a high potential barrier and hence good confinement. Incomplete crystallization causes a remaining amorphous shell, and the reduced potential barrier leads to problems with the quantum confinement. Such a remaining shell will have imperfections in the atomic bonds resulting in defects and disordered or broken bonds. For today's Si technology, the good Si/SiO₂ interface and its high stability are crucial and one of the historic reasons for the amazing developments in microelectronics. But how good is the SiO₂ interface adjacent to the Si nanocrystals? Can we get any hints for that in experiments? The remaining incomplete SiO$_x$ shell was proven by using photon-in photon-out soft X-ray spectroscopy, the electronic structure of silicon nanoclusters embedded in an electrically insulating SiO₂ host matrix was investigated as a function of nanocluster size [23]. We found the nanoclusters to be of a core–shell structure with a crystalline Si core and an ultrathin transition layer of a suboxide. Effects of electronic quantum confinement are detected in the Si cores. For more information, see the related paper. Morphological methods such as TEM or HRTEM can investigate only a small fraction of the nanoparticles, present in a certain volume, and most of them cannot discriminate between a-Si and a-SiO$_2$, making it impossible to draw conclusions about the presence of a-Si NPs. Here, electron spin resonance (ESR) can provide an answer, a technique sensing the whole volume of the sample with the ability to distinguish between different materials by using paramagnetic defects as a probe. ESR can identify these paramagnetic defects and infer their location within the sample structure. We will now present a detailed investigation of defects at Si/SiO$_2$ interfaces using electron spin resonance measurements as a comprehensive summary of our recent studies [24–26].

Point defects are generally unwanted for technological applications. However, such defects, when paramagnetic in nature, may well serve as very local probes of utmost atomic sensitivity. The presence of any paramagnetic defect can be quantified independent of its involvement in photoluminescence. Various ESR studies have been performed on nano-Si entities in SiO$_2$ and generally five types of point defects, that is, D (Si dangling bonds (DB) in disordered Si), $P_{b(0)}$, P_{b1}, E'_γ, and EX are observed. All works report a generally featureless isotropic line commonly referred to as the Si DB signal with zero crossing g value (g_c) in the range of 2.005–2.006. A first series of ESR measurements investigated Si nanoparticles embedded in SiO$_2$ and phosphosilicate glass fabricated by Si$^+$ ion implantation [27]. A rather broad signal was observed at $g_c \sim 2.006$ with a peak to peak width of $\Delta B_{pp} = 7.0$–7.9 G, which was suggested to originate from P_b centers. The P_b centers are prototype interface defects that are characteristic of the Si/SiO$_2$ interface. In Figure 8.9a–c, the DB, $P_{b(0)}$, and P_{b1} are schematically presented. The presence of P_b-type defects would be important, as these are deep nonradiative recombination centers, known to (nonradiatively) quench the PL from Si nanocrystals in the 1.4–2.2 eV range, irrespective of whether the source of the PL is QC or interface states [28]. Thus, the analysis of whether such defects are observed in ESR for Si nanocrystals embedded in SiO$_2$ is quite an important question. To make sure that the signal really is related to the interface of the nanocrystals and not to the interface to the substrate, one should also investigate the angle dependence of the signal. The absence of any angle dependence in the P_b ESR signal indicates that the defects stem from the interface between Si nanocrystals and surrounding SiO$_2$ and not from the Si substrate.

A more detailed ESR investigation on the origin of the Si DB signal was reported by Bratus *et al.* [29] studying nano-Si particles formed in evaporated SiO films after

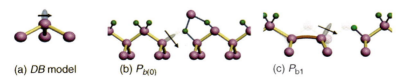

(a) *DB* model   (b) $P_{b(0)}$ (c) P_{b1}

Figure 8.9   (a) DB model, (b) $P_{b(0)}$, and (c) P_{b1}.

annealing. After annealing at 1100 °C (15 min), a Si DB signal was observed at $g_c \sim 2.0061$ that, via computer simulation, was described as a superposition of three components implying a strong DB line, modeled as a Lorentzian signal at $g_c = 2.0058$ with $\Delta B_{pp} = 6$ G, and powder pattern signals from chaotically oriented P_{b0} and P_{b1} defects in a ratio of $5.5 : 1 : 1$, respectively. Remarkably, the observed substantial density of DB centers would indicate that most of the Si NPs are, in fact, amorphous (or polycrystalline), contradicting the related morphological investigations. The applied X-band (9.4 GHz) spectroscopy clearly results in insufficient spectral resolution and disentanglement as in all previous reports. More in-depth and reliable signal discrimination requires experimental extension, for example, additional observation frequencies.

Let us state at this point the nature of the P_b centers: P_b was identified as trivalent interfacial Si ($Si_3 \equiv Si^{\bullet}$, where the dot represents an unpaired electron in a dangling Si sp³ hybrid) back bonded to three Si atoms in the bulk. A respective model is shown in Figure 8.9a. The technologically favored (100) Si/SiO₂ interface exhibits two types, termed P_{b0} and P_{b1}. The experimental evidence is that P_{b0} (Figure 8.9b) is chemically identical to P_b (Figure 8.9c) but now residing at microscopically (111)-oriented Si/SiO₂ facets. The P_{b1} center is assigned to a distorted interfacial Si–Si dimer ($\equiv Si–Si^{\bullet} = Si_2$ defect, where the em dash symbolizes a strained bond, with an approximately (211)-oriented unpaired Si sp³ hybrid) [30]. At the (110) Si/SiO₂ interface again, only one type, the P_b variant, is observed [31]. Thus, all three variants were shown to be interfacial trivalent Si centers [32], naturally occurring, for standard oxidation temperatures (800–960 °C), in areal densities of $[P_b] \sim 5 \times 10^{12}$ cm⁻² [33–36] and $[P_{b0}]$, $[P_{b1}] \sim 1 \times 10^{12}$ cm⁻² [36].

Hence, to take the ESR signal discrimination and assignment to the next decisive level, the experimental ESR range has to be extended to three observational frequencies, the X-band (9.5 GHz), K-band (20.5 GHz), and Q-band (34 GHz), usually driven in the adiabatic mode, and two detection modes (first and second harmonic) as we did in our papers. Routinely, conventional low-power first derivative absorption spectra were measured by applying sinusoidal modulation (100 kHz; amplitude $B_m \sim 0.8$ G) of the externally applied static field, B. These were complemented with high-power second harmonic dispersion mode measurements (see, Ref. [37], saturation spectroscopy) to unequivocally disentangle all the appearing signals. For microwave dissipation reasons, ESR experiments routinely had to be carried out at a temperature $T < 30$ K. While generally beneficial for signal resolution (Boltzmann factor, noise levels, etc.), low T measurements suffer from enhanced sensitivity to signal saturation (distortion) effects with decreasing entropy, which needs an appropriate reduction in incident microwave power P_μ, as well as B_m, to such levels that do not cause a visible first-derivative signal distortion. Signal-to-noise ratio can be improved by appropriate signal averaging (typically $\sim$100 scans). Paramagnetic defect densities were determined through double numerical integration of the detected $dP_{\mu x}/dB$ signals by comparison to the signal of a comounted point-like Si:P marker ($S = 1/2$ signal). The latter, with $g(4 \text{ K}) = 1.99869 \pm 0.00002$ and $g(30 \text{ K})$ 1.99876 ± 0.00002, also served as g marker. Proper corrections for the relative position of the sample and the marker in the cavity were taken into account when calculating the absolute intensity of detected signals. The attained relative and absolute accuracies on defect densities were estimated to be 5% and 15%, respectively.

With reliable attribution of the signals, DBs can be used as probes providing information about crystallinity and bordering of the Si NPs. As a major result, the occurring main DB signal could be disentangled as solely composed of a mix of the Si/SiO$_2$ interface-specific P$_{b0}$ and P$_{b1}$, resulting in particular conclusions on the Si nanoparticles state, morphology, and crucial defect nature. By using the inherent point defects as local probes, one can conclude on the nature and quality of the inner nc-Si/SiO$_2$ interfaces and embedding a-SiO$_2$ matrix.

However, as well known, Si DB-type defects, such as P$_b$-type centers at the Si/SiO$_2$ interface, D centers in amorphous or polycrystalline Si, and E' centers in SiO$_2$ may be readily ESR inactivated through bonding to hydrogen. So, while distinct systems of defect sites may occur, these may fully escape ESR detection. Here, vacuum ultraviolet (VUV) irradiation is very effective in cracking Si–hydrogen bonds, thus enabling to reveal defects through ESR observation. Appropriate photonic irradiation may reveal *weak* (strained) bonding sites at interfaces and amorphous networks through bond breaking, thus enabling through ESR identification to assess inherent interface and layer quality in terms of such less-than-ideal bonding sites. We do not use hydrogen in the preparation process of our superlattices; thus, instead of cracking Si–hydrogen bonds, irradiation process could be used to distinguish between P$_b$ centers at the Si/SiO$_2$ interface and the D centers in amorphous Si (if there is any).

So, in case of a particular interface defect at a particular c-Si/SiO$_2$ interface orientation, for example, P$_{b1}$ in (100) Si/SiO$_2$, with many (100) planes (facets) occurring spatially randomly oriented, a P$_{b1}$ powder pattern ESR shape will be observed. This, for example, would be expected for (100)-faceted Si NCs embedded with random orientations in the SiO$_2$ matrix, too. On the other hand, should the Si NCs occur in crystallographic registry, of course the anisotropic P$_{b1}$ ESR signal(s) as pertaining to the macroscopic (100) Si/SiO$_2$ interface would be observed. Hence, ESR may provide exclusive information on this matter. Please also note the above will apply for each type of interface defect at each type of interface orientation (facet). Thus, if we deal with an ensemble of faceted Si NCs randomly oriented within an otherwise a-SiO$_2$ matrix, we expect an overlap of several P$_b$-type powder patterns, that is, P$_b$ at (111) Si/SiO$_2$, P$_{b0}$ and P$_{b1}$ at (100) Si/SiO$_2$, and P$_b$ at (110) Si/SiO$_2$.

A typical low-power first derivative K-band spectrum measured at 4.2 K of our crystallized SiO/SiO$_2$ superlattice is shown in Figure 8.10. Two optimized fittings to the experimental K-band spectrum are shown in the figure, each based on two overlapping components, that is, P$_{b(0)}$ and D centers (red curve) and P$_{b(0)}$ and P$_{b1}$ signals (green curve), using the principal g values listed in the figure caption. The D signal is fitted with a Lorentzian shape of $\Delta B_{pp} = 7.6$ G and $g_c = 2.0055$, while for the simulation of the P$_b$-type defect powder patterns, Gaussian broadening functions of $\Delta B_{pp} = 6$ G, 11 G, and 11 G for P$_{b(0)}$, and 12 G, 6 G, and 7 G for P$_{b1}$ at g_1, g_2, and g_3, respectively, have been used. These appeared the best simulations attainable. Both simulations may seem reasonable, with successful fitting parts for each, the latter one appearing perhaps somewhat better, yet not decisively. However, fittings based

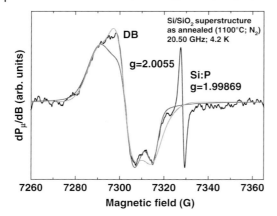

Figure 8.10 K-band ESR spectrum of nc-Si/SiO$_2$ superlattice (41 scans) measured at 4.2 K using $B_m = 1.0$ G and a reduced $P_\mu = 7.9 \times 10^{-10}$ W to avoid signal distortion. The colored curves represent optimized fittings of the DB signals through combining overlapping powder patterns corresponding to randomly oriented P$_{b(0)}$ and D centers (red curve) and randomly oriented P$_{b(0)}$ with P$_{b1}$ centers (green curve), respectively. The signal at g approximately 1.99869 results from a comounted Si:P marker sample. Principal g values used for the fit are $g_1 = 2.0015$, $g_2 = 2.0080$, and $g_3 = 2.0087$ for P$_{b(0)}$, $g = 2.0055$ for D centers, and $g_1 = 2.0020$, $g_2 = 2.0055$, and $g_3 = 2.0070$ for P$_{b1}$.

on the inclusion of all three types of signals, with even more fitting parameters, were not of much help in achieving a more conclusive result. So, it turned out that this "kind of best fitting approach" to ESR spectra *did not help to distinguish between the occurring signals.* The presence or absence of one or the other type could not be confirmed or disproved using this routine. Thus, any progress has to come from additional experiment. However, although not perfect, as the first main result, the occurrence of at least one P$_b$-type center was proven.

Hence, we extended our ESR experiments in two ways: (1) using low-temperature X and Q-band observations and (2) using out-of-phase second harmonic measurements under signal saturation conditions (saturation spectroscopy) in addition to the first harmonic detection. The potential benefit for disentanglement of overlapping signals from ESR observations at different observational frequency is obvious and further adds to the reliability of signal defect identification. The high-power second harmonic spectra are known, under appropriate experimental conditions, to resemble direct absorption shapes (χ''), as empirically evident [38]. So, peak maxima here would correspond to zero crossing g values of the regular dP_μ/dB spectra. In addition, at this point we subjected the sample to VUV irradiation for reasons given above. The respective ESR measurements are shown in Figure 8.11. This shows a representative conventional X-band dP_μ/dB spectrum, which may be compared with the similarly observed K-band result (see Figure 8.10). Apart from the main DB feature, we see the appearance of two more signals, that is, E'_γ and EX, which will be addressed later. Clearly, with respect to resolution enhancement of the DB signal, X-band observations provide no progress.

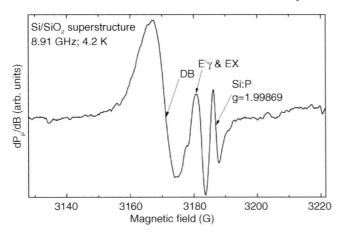

Figure 8.11 First derivative *X*-band (8.91 GHz) spectrum measured at 4.2 K using $P_\mu = 3.95 \times 10^{-7}$ W on a phase-separated SiO/SiO_2 superstructure subjected to VUV irradiation. The DB response appears as a single, rather featureless, signal, indicating the limited appropriateness of lower frequency observations regarding signal discrimination (cf. Figure 8.11).

Figure 8.12 shows a typical out-of-phase high-power second harmonic signal observed on the sample, finally enabling us to discriminate the signals contributing to the DB feature. Apart from the Si:P marker signal, four types of absorption signals are observed. These include the readily assigned E'' and EX signals (at $g_c = 2.00055$ and 2.0025, respectively), and two more prominent, broader signals characterized by $g_c = 2.0072$ and 2.0045, respectively, which correspond to the zero crossing g_c values of the P_{b0} and P_{b1} powder patterns. There is no evidence of the presence of a D signal at $g_c \sim 2.0055$, allowing us now to conclude that on ESR basis the Si nanoparticles are predominantly crystalline, delimited by (100), (110), or (111) Si/SiO_2 interfaces (and possibly others since P_b-type centers might as well occur at other c-Si/SiO_2 interface types, but with respect to ESR this remains so far unexplored). A majority (if not all) of the DB signals appear to be composed of $P_{b(0)}$ and P_{b1} powder pattern signals – a result already suggested by the successful fitting (blue curve) illustrated in Figure 8.10. In addition, if some amorphous (disordered) NPs would occur too, their number remains below, albeit rather impressive, the detection limit – so such amorphous NPs can be disregarded. Hence, firm ESR evidence for the crystalline state of the majority of the Si-nanoparticles (now called Si-nanocrystals) was proven in agreement with our TEM and XRD results.

At this point, the nature of the E'_γ and EX signals (at $g_c = 2.00055$ and 2.0025, respectively) seen after VUV irradiation will be discussed. Both the EX and E'_γ type of defects are related to the SiO_2 matrix in which the Si NCs are embedded. Their appearance has been anticipated and *is forced by the VUV treatment*. An estimation of their density would give some values for the quality of the matrix the nanocrystals are embedded in.

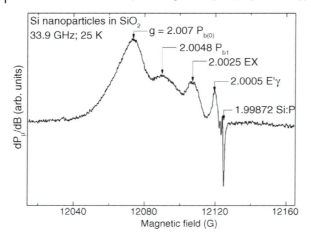

Figure 8.12 High-power (saturation) second harmonic out-of-phase Q-band spectrum taken at approximately 25 K on a SiO/SiO$_2$ sample subjected to VUV irradiation. The absorption-type signals provided by this detection mode, in combination with dissimilar responses to the spectroscopical settings, reveals the various constituent signals in the spectrum as marked by the maximum absorption g values (corresponding to the respective g$_c$ values in the first harmonic spectra). The spectrometer settings ($B_m = 0.37$ G and $P_\mu = 1.7 \times 10^{-6}$ W) were chosen to optimize resolution of the various spectral components. In particular, the nature of the DB signal (first harmonic detection, see Figure 8.10) as predominantly composed of P$_{b(0)}$ and P$_{b1}$ components is exposed.

The EX center appears first after 362 nm (filtered Xe lamp) irradiation with a density of $(2.5 \pm 0.4) \times 10^{11}$ cm^{-2} and increases after further full spectrum Xe lamp irradiation to $(6.3 \pm 1.0) \times 10^{11}$ cm^{-2} (corresponding to a volumetric value of [EX] = $\sim 3 \times 10^{16}$ cm^{-3}) remaining then constant after subsequent VUV illumination. An atomic model of EX is presented in Figure 8.13a. The presence of EX centers, albeit in rather low intensity, may indicate a somewhat lower quality of the ultimate SiO$_2$ matrix, which may be linked to the kind of deposition or phase separation annealing procedures applied. So far, nothing is known about the impact on PL or optical activity of these centers. Yet, electrically, there are some hints that these are in

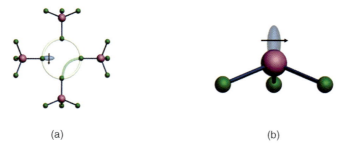

(a)   (b)

Figure 8.13   (a) Atomistic model of the EX; (b) model of the E'_γ center.

the positive charge state when ESR active, so because of the charge trapping nature, their presence may play a role.

The E'_γ center was identified as an unpaired electron at a threefold coordinated Si (generic entity $O_3 \equiv Si^\bullet$). It may be taken as a measure for the presence of oxygen vacancies, or at least, oxygen deficiencies of the SiO_2 matrix, establishing a quality label in that sense. A model is given in Figure 8.13b. The upper limit of the activation for the E'_γ centers is reached after VUV irradiation, that is, related to substantial increase in $h\nu$, not the exposure time. Volumetrically, based on the estimated amount of SiO_2, the observed E'_γ density corresponds to $[E'_\gamma] \sim 3 \times 10^{17}\,cm^{-3}$ of SiO_2, which may be compared with the value of $\sim 3 \times 10^{17}\,cm^{-3}$ found for the standard thin thermal SiO_2 on Si after exhaustive VUV irradiation [39]. Assuming 10 min VUV irradiation to exhaustively activate the E'_γ centers in this structure, this would indicate the quality of the current SiO_2 matrix reaching that of the standard thermal SiO_2, a rather amazing result.

As demonstrated in the chapter above, the detailed investigation of ESR active centers is discussed based on as-crystallized and VUV-treated samples. Only part of the rather extended analysis was selected here; more details are in the related papers. The selected details should help understand both the principal way of the analysis and the nature of the related defects (and their relation to values observed in other process including standard Si technology interfaces). Let us summarize at the end the main conclusions because some of them are amazing.

For the as-crystallized sample state, the main result is that the principal Si DB signal is exclusively composed of overlapping powder patterns of $P_{b(0)}$ and P_{b1} Si/SiO_2 interface-specific defects, hinting the NC-Si/SiO_2 interfaces. The absence of a D signal provides strong independent ESR evidence that for most part, if not all, the Si particles are crystalline in nature but randomly oriented. Assuming that the known properties of the standard thermal microscopic Si/SiO_2 interfaces may be transferred down to the nanoscale, the specific signatures of the revealed $P_{b(0)}/P_{b1}$ systems would morphologically agree with Si NCs, in average, being principally bordered by (111) and (100) facets, for example, in a (100) truncated (111) octahedron, in agreement with theoretical modeling. Such an idealized model nanocrystals is shown in Figure 8.14.

This P_b-type system density remains unchanged under subsequent various UV/VUV irradiations for substantial times, indicating an inherent defect system with no involvement of H passivation, as might have been expected from samples prepared without (deliberate) involvement of H, subjected to the final high-T phase separation annealing in N_2 at $1100\,°C$. Disregarding the occurrence of more than one defect at a Si NC, we find about one P_b-type defect every 1.4 Si particle, indicating the Si NC system to be comprised of two subsystems, one with and one without having incorporated a strain relaxing P_b-type defect. Adopting the general view of the macroscopic Si/SiO_2 system to the nanoscale, we expect that P_b-type centers are introduced due to interface lattice (c-Si)/network (a-SiO_2) mismatch.

Given the known PL quenching properties of Si DB type defects in general, we are left with two subsystems with drastically different PL behavior, as has indeed been demonstrated recently [24]. In addition to qualitative aspects, the absolute density

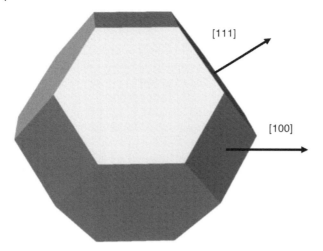

[111]

[100]

Figure 8.14 Idealized average morphology ((100) truncated (111) octahedron) of embedded Si NCs in phase-separated SiO/SiO₂ entities as inferred from the occurring $P_{b(0)}$ and P_{b1} (Si/SiO₂) interface defects, based on microscopic defect properties.

of P_b-type defects per effective Si NC (interface) area agrees well with the range of standard macroscopic Si/SiO₂ entities. One can conclude that the Si NC interfaces are of standard thermal (high) quality, meaning that even with growing Si nanocrystallites from within the SiO₂ environment one ends up with an interface quality competing with the best of the classical macroscopic growing techniques. That was for sure not a straightforward matter given the totally "opposite" ways of the Si/SiO₂ interfacial growth. Intriguingly, the P_b-type defects evolve in the same fashion. It seems to underscore the *universality* of the P_b-type centers as archetypical Si/SiO₂ interface defects unavoidably and inherently connected to the natural Si/SiO₂ mismatch, which needs to be incorporated in theoretical modeling. Thus, a "magic" recipe to realize a P_b-type defect lean Si/SiO₂ interface does not exist – not in case of the classical thermal Si/SiO₂ one and not in the current nc-Si/SiO₂ case. And if no postformation action is taken (i.e., inactivation), these defects will show up as detrimental unpaired-spin defects. Two SiO₂-associated defects (E'_γ and EX) were observed after UV/VUV irradiation. On the basis of the measured exhaustive volume densities of E'_γ, the attained quality of the Si oxide in the superstructure would match that of standard thermal SiO₂ on Si – quite remarkable, indeed, for the process of phase separation of SiO and crystallization used here. So, all together, the confirmed properties of the inherent paramagnetic point defects, serving as atomic probes, point to (top) standard quality of the inner Si NC/SiO₂ interfaces and the SiO₂ matrix. Thus, as to the control of inherent point defects (detrimental charge traps), it may be hard to further improve this NC-Si-in-SiO₂ structure. The EX centers, typically observed in densities one order of magnitude less than E'_γ, are ESR activated at lower photon energy than E'_γ. Since its atomic attributes are less well known, the relevance of the EX center, concluded to be in the positive charge state when ESR active, to the oxide quality is less clear.

8.5
Optical Properties

In this section, the optical properties of phase-separated and crystallized SiO/SiO_2 superlattices will be discussed in detail. In the first step, the change in PL as a function of annealing temperature is presented. As discussed above, the as-prepared amorphous SiO/SiO_2 superlattice transfers first into a structure containing amorphous Si nanocluster within the former SiO layer controlled in size by the layer thickness. The process is correlated with reordering of atoms and the related bonding structure. For this process, a signature in photoluminescence is expected. In Figure 8.15, the development of the PL as a function of annealing temperature is presented measured at room temperature using a HeCd laser as excitation source [10]. As can be seen here, the atomic reordering in the first step is correlated with a PL band at a peak position around 560 nm. The peak position does not change, but the intensity first increases up to 300 °C, decreases again, and vanishes at 700 °C. This PL band is related to the rearrangement of Si and O atoms with the ongoing phase separation into amorphous Si cluster and the stable SiO_2 matrix. In IR spectra, we see the 880 cm^{-1} band having a similar temperature behavior that we related to the establishment of the first Si rings and their growth into nanocluster. A model is given in Figure 8.4 that also shows that such intermediate states will be correlated with broken O bond (nonbridging oxygen) that from our simulations is expected to be the origin of the defect luminescence at 560 nm. Higher annealing

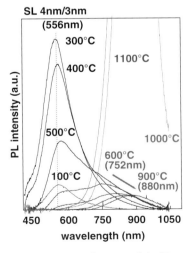

Figure 8.15 Development of the PL as a function of annealing temperature.

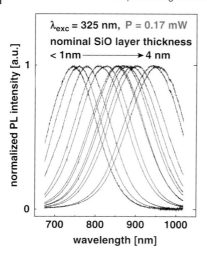

Figure 8.16 PL spectra after 1100 °C annealing normalized to peak intensity.

will heal out such defects and re-establish the stable Si−O−Si bridging bond configuration, as seen in the IR spectra (Figure 8.3). At the same time, the size of the amorphous Si clusters grows and the related signal of the amorphous Si shows up in PL, too, as a band shifting with increasing temperature to higher wavelength from around 700 nm (at 500 °C) to 880 nm (at 900 °C) as expected for larger Si nanoclusters. The phase separation is completed at 900 °C as we learned from IR spectroscopy. Above 900 °C, the amorphous clusters crystallize and the related PL is now a signature of the Si nanocrystals with an increase of intensity by one to two orders of magnitude. The PL peak position is strongly size dependent as demonstrated in Figure 8.16 with peak positions between 730 and 960 nm. The PL spectra after 1100 °C annealing in Figure 8.16 are normalized to peak intensity. Please note that in all our experiments, samples with a nanocrystal diameter of around 3–3.5 nm showed the strongest PL intensity, a fact that was observed from other experiments, too, but no conclusive explanation has been given so far. The most commonly used explanation for the energy shift of the peak position is quantum confinement of electron–hole pairs.

These first results indicated quantum confinement of electron–hole pairs in the Si nanocrystals as PL origin. We investigated the PL in detail at various temperatures down to 4 K as a function of NC size (Figure 8.17). The results are summarized by J. Heitmann *et al.* [40], describing the interplay of confinement and migration of carriers. Special care has to be taken during the measurements. The excitation power has to be low to avoid saturation effects due to the strong increase in exciton lifetime at low temperatures. We measured the temperature-dependent photoluminescence properties for samples of different nanocrystal sizes some years ago. Only one sample was presented in our paper, so I include three different samples and their temperature behavior for comparison. Clearly seen is the increase in the bandgap for

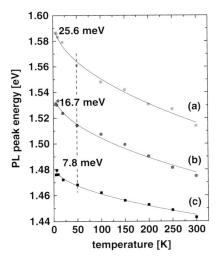

Figure 8.17 PL as a function of temperature down to 4 K and for different NC sizes.

lower temperature. For bulk Si this behavior goes into saturation at low temperature $(E_g(0))$ and can be fitted by the Varshni equation:

$$E_g(T) = E_g(0) + A \cdot \left(\frac{2}{\exp[E_{\mathrm{photon}}/kT]-1} + 1 \right). \tag{8.13}$$

The energy of bandgap of bulk semiconductors tends to decrease with increasing temperature as is well known from textbooks [41]. In the past, the temperature dependence of the energy bandgap was also fitted using the following expression for E_g as a function of temperature T:

$$E_g(T) = E_g(0) - \frac{\alpha \cdot T^2}{T + \beta}, \tag{8.14}$$

where $E_g(0)$, α and β are the fitting parameters. From atomistic point of view, one can understand this behavior better if we consider that the interatomic spacing increases when the amplitude of the atomic vibrations increases due to the increased thermal energy (i.e., temperature). This effect is quantified by the linear expansion coefficient of a material. If the interatomic spacing increases, then the potential seen by the electrons in the material decreases, and hence the bandgap energy is reduced. In addition, a direct modulation of the interatomic distance by applying high compressive (tensile) stress also causes an increase (decrease) in the bandgap.

We used Eq. (8.13) in the past to fit the behavior in the temperature range 50–300 K [40]. However, in case of the Si NC containing samples, there are significant deviations from the bulk silicon behavior for temperatures below 50 K. Instead of saturation to the 0 K value (as in case of bulk Si), an additional blueshift is found. With bigger crystal diameter, the effect gets smaller. The respective shift related to the 50 K value decreases from 25.6 to 16.7 meV and finally to only 7.8 meV. Please note that the overall shift in the range from 300 to 5 K also seems to be size dependent. Again, with increasing size the total shift observed is 72.2 (smallest size), 55.3, and 36.4 meV (biggest size). From the point of size control, sample (a) has the smallest nanocrystals that are in the range of 2 nm and a very well-controlled size distribution. With increasing SiO layer thickness, the size control is reduced and the deviations get bigger (sample (c)) but still are much better than in the case of randomly crystallized bulk SiO films. Instead of the rather complicated fitting, one could easily use other fitting functions such as power law, which then even fits the complete temperature range quite well. We did this in Figure 8.17 using the equation

$$E_g(T) = E_g(0\ K) - A \cdot T^{1/2} \tag{8.15}$$

taking into account only two parameters E_g (0 K) and A for fitting the respective samples: (a) 1.597 and 4.7 meV, (b) 1.541 and 3.7 meV, and (c) 1.483 and 2.2 meV. There might be different processes contributing to the reported behavior. One main part could be the different expansion coefficient of the two contributing materials: the Si nanocrystals and the amorphous SiO₂ matrix surrounding them. If this at low temperature results in high compressive stress, one could compensate the normally observed saturation of the bandgap at 0 K and further shift the bandgap to increased energies. For comparison also, the development of the integrated PL intensity is shown in Figure 8.18 that is obviously quite similar for the samples. All samples have their PL maximum at around 100 K. The photoluminescence can be observed at room temperature by naked eye in contrast to bulk Si.

In our 2004 published paper, we interpreted the deviation from the Varshni behavior as an interplay of migration and trapping of the excitons. Sample (a) is the sample with the best size control (with a deviation below ±0.3 nm). So, all particles having a nearly equal bandgap should contribute to the photoluminescence for that sample even at room temperature. There might be some migrations that, however, will not change the overall peak position in that case. Sample (a) thus represents

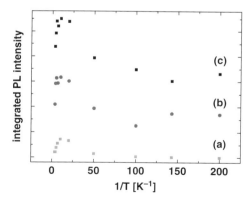

Figure 8.18 Integrated PL intensity for the same samples showing a maxima in PL intensity around 100 K.

the behavior of a *"monosized" Si NC* ensemble close to the critical crystallization size. In a way, one could also argue that the probability of a defect is less for smaller crystals. Hence, in view of our recent paper reporting the contributions of defects, sample (a) might have less contribution of such defects. In case of samples (b) and (c), the size distribution is larger (we do not have the values for these samples, but I would expect in the range of ±0.6 nm for the thickest sample) that will enable the relaxation of excitons at higher temperatures into larger crystals (smaller bandgap) having larger recombination times, but maybe also more defects. Recently, the interplay of defects and quantum confinement was shown by our group, please see the related paper [24]. However, in view of the above discussion, the different thermal expansion coefficients of Si and SiO_2 might also play a role and significantly contribute to the observed deviations from bulk silicon.

8.6
Applications of Si NCs and Concluding Remarks

The observation of efficient room-temperature luminescence in porous silicon in the 1990s initiated a strong interest in bandgap-engineered and luminescent silicon. Possible applications are silicon light sources that could replace III–V optoelectronic devices [42], flash memories allowing multilevel charge storage [9, 43], and solar cells [44]. The optical and electronic properties depend on the size of the silicon crystallites. Thus, a tight control of the nanocrystal size distribution is necessary in order to achieve well-defined optical and electronic properties. Si NCs obtained by phase separation of bulk nonstoichiometric SiO_x show a broad log-normal size distribution. In contrast, nanocrystals with a narrow Gaussian size distribution could be realized by the superlattice solid-phase crystallization approach discussed in this chapter. The role of defects and the origin of luminescence from undoped silicon nanoparticles surrounded by SiO_2 were clarified by combined PL and ESR experiments [24].

A major obstacle in the investigation of charge transport mechanisms through silicon nanocrystals has been the choice of insulator SiO_2 as the surrounding dielectric matrix. First experiments concerning the electrical behavior of undoped Si nanocrystal networks have been performed, but so far no clear picture of charge transport through superlattices has been developed [45]. Another open point is the role of dopants, that is, their location within the nanocrystals and/or the surrounding matrix [46] and their influence on the optical and electrical properties. Simulations on the optical properties of doped Si NCs have been performed [47]. First experiments on the electrical properties were done with doped plasma synthesized nanocrystals and showed that phosphorous and Si dangling bonds contribute to dark conductivity but act as recombination centers under illumination [48]. For photovoltaic applications, however, it is necessary to understand doping and electrical transport in Si NC solid-state devices, that is, Si NC in host matrixes. Si NCs devices that have been realized are mostly based on SiO_2 as a host matrix, thus suffering from the very low conductivity of SiO_2, and were developed without doping. The characterization of doped Si NC superlattices is still difficult due to uncertainties concerning the dopant location [49, 50]. In conclusion, research on Si NC is advancing rapidly, but many scientific issues are still unsolved.

References

1 Heitmann, J., Müller, F., Zacharias, M., and Gösele, U. (2005) *Adv. Mater.*, **17**, 795.

2 Lu, Z.-H., Lockwood, D.J., and Baribeau, J.-M. (1995) *Nature*, **378**, 258.

3 (a) Lockwood, D.J., Lu, Z.H., and Baribeau, J.-M. (1996) *Phys. Rev. Lett.*, **80**, 539; (b) Sullivan, B.T., Lockwood, D.J., Labbe, H.J., and Lu, Z.-H. (1996) *Appl. Phys. Lett.*, **69**, 3149.

4 Baribeau, J.-M., Lockwood, D.J., Lu, Z.H., Labbé, H.J., Rolfe, S.J., and Sproule, G.I. (1999) *J. Lumin.*, **80**, 417.

5 Modreanu, M., Gartner, M., and Cristea, D. (2002) *Mater. Sci. Eng. C*, **19**, 225.

6 Ioannou-Sougleridis, V., Nassiopoulou, A.G., Ouisse, T., Bassani, F., and d'Avitaya, F.A. (2001) *Appl. Phys. Lett.*, **79**, 2076.

7 Zacharias, M., Heitmann, J., Scholz, R., Kahler, U., Schmidt, M., and Bläsing, J. (2002) *Appl. Phys. Lett.*, **80**, 661.

8 Zacharias, M. (2002) DE 10104193A1. U.S. Patent 7,220,609.B2 (May 22, 2007).

9 Lu, .Z., Alexe, M., Scholz, R., Talalaev, V., Zhang, R.J., and Zacharias, M. (2006) *J. Appl. Phys.*, **100**, 014310.

10 (a) Yi, L.X., Heitmann, J., Scholz, R., and Zacharias, M. (2002) *Appl. Phys. Lett.*, **81**, 4248; (b) Yi, L.X., Heitmann, J., Scholz, R., and Zacharias, M. (2003) *J. Phys. Condens. Matter*, **15**, S2887.

11 Tsu, D.V., Lucovsky, G., and Davidson, B.N. (1989) *Phys. Rev. B*, **40**, 1795.

12 Williams, G.V.M., Bittar, A., and Trodahl, H.J. (1990) *J. Appl. Phys.*, **67**, 1874.

13 Persans, P.D., Ruppert, A., and Abeles, B. (1988) *J. Non-Cryst. Solids*, **102**, 130.

14 (a) Homma, I., Hotta, H., Kawai, K., Komiyama, H., and Tanaka, K. (1987) *J. Non-Cryst. Solids*, **97/98**, 947; (b) Homma, H., Schuller, I.K., Sevenhans, W., and Bruynseraede, Y. (1987) *Appl. Phys. Lett.*, **50**, 594.

15 Miyazaki, S., Ihara, Y., and Hirose, M. (1987) *J. Non-Cryst. Solids*, **97/98**, 887.

16 Zacharias, M., Bläsing, J., Veit, P., Tsybeskov, L., Hirschman, K., and Fauchet, P.M. (1999) *Appl. Phys. Lett.*, **74**, 2614.

17 Oki, F., Ogawa, Y., and Fujiki, Y. (1969) *Jpn. J. Appl. Phys.*, **8**, 1056.

18 Zacharias, M. and Streitenberg, P. (2000) *Phys. Rev. B*, **62**, 8391.

19 Xu, Q., Sharp, I.D., Yuan, C.W., Yi, D.O., Liao, C.Y., Glaeser, A.M., Minor, A.M., Beeman, J.W., Ridgway, M.C., Kluth, P., Ager, J.W., III, Chrzan, D.C., and Haller, E.E. (2006) *Phys. Rev. Lett.*, **97**, 155701.

20 Raoux, S., Jordan-Sweet, J.L., and Kellock, A.J. (2008) *J. Appl. Phys.*, **103**, 114310.

21 Persans, P.D., Ruppert, A., and Abeles, B. (1988) *J. Non-Cryst. Solids*, **102**, 130.

22 Spinella, C., Lombardo, S., and Priolo, F. (1998) *J. Appl. Phys.*, **84**, 5383.

23 Zimina, A., Eisebitt, St., Eberhardt, W., Heitmann, J., and Zacharias, M. (2006) *Appl. Phys. Lett.*, **88**, 163103.

24 Godefroo, S., Hayne, M., Jivanescu, M., Stesmans, A., Zacharias, M., Lebedev, O., Van Tendeloo, G., and Moshchalkov, V.V. (2008) *Nat. Nanotechnol.*, **3**, 174.

25 Jivanescu, M., Stesmans, A., and Zacharias, M. (2008) *J. Appl. Phys.*, **104**, 103518.

26 Jivanescu, M., Godefroo, S., Stesmans, A., and Zacharias, M. (2008) *Appl. Phys. Lett.*, **93**, 023123.

27 Fujii, M., Mimura, A., Hayashi, S., Yamamoto, K., Urakawa, C., and Ohta, H. (2000) *J. Appl. Phys.*, **87**, 1855. Sumida, K., Ninomiya, K., Fujii, M., Fujio, K., Hayashi, S., Kodama, M., and Ohta, H. (2007) *J. Appl. Phys.*, **101**, 033504.

28 Delerue, C., Allan, G., and Lannoo, M. (1993) *Phys. Rev. B*, **48**, 11024.

29 Bratus, V.Y., Yukhimchuk, V.A., Berezhinsky, L.I., Valakh, M.Y., Vorona, I.P., Indutnyi, I.Z., Petrenko, T.T., Shepeliavyi, P.E., and Yanchuk, I.B. (2001) *Semiconductors*, **35**, 821.

30 Stesmans, A., Nouwen, B., and Afanas'ev, V.V. (1998) *Phys. Rev. B*, **58**, 15801.

31 Helms, C.R. and Poindexter, E.H. (1994) *Rep. Prog. Phys.*, **57**, 791.

32 Poindexter, E.H. and Caplan, P.J. (1981) *Insulating Films on Semiconductors*, Springer, Berlin, p. 150.

33 Stesmans, A. and Afanas'ev, V.V. (1998) *J. Appl. Phys.*, **83**, 2449.

34 Stesmans, A. (1993) *Phys. Rev. B*, **48**, 2418.

35 Stesmans, A. and Afanas'ev, V.V. (1998) *Phys. Rev. B*, **57**, 10030.

36 Futako, W., Mizuochi, N., and Yamasaki, S. (2004) *Phys. Rev. Lett.*, **92**, 105505.

37 Griscom, D.L. (1979) *Phys. Rev. B*, **20**, 1823.

38 Griscom, D.L. (1979) *Phys. Rev. B*, **20**, 1823.

39 Stesmans, A. and Afanas'ev, V.V. (2005) *J. Appl. Phys.*, **97**, 033510.

40 Heitmann, J., Müller, F., Yi, L.X., Zacharias, M., Kovalev, D., and Eichhorn, F. (2004) *Phys. Rev. B*, **69**, 195309.

41 Sze, S.M. and Ng, K.K. (2007) *The Physics of Semiconductor Devices*, John Wiley & Sons, Inc., New York, p. 15.

42 Walters, R.J., Bourianoff, G.I., and Atwater, H.A. (2005) *Nat. Mater.*, **4**, 143.

43 Lu, T.Z., Alexe, M., Scholz, R., Talelaev, V., and Zacharias, M. (2005) *Appl. Phys. Lett.*, **87**, 202110.

44 Cho, E.-C., Park, S., Hao, X., Song, D., Conibeer, G., Park, S.C., and Green, M.A. (2008) *Nanotechnology*, **19**, 245201.

45 Rafiq, M.A., Tsuchiya, Y., Mizuta, H., Oda, S., Uno, S., Durrani, Z.A.K., and Milne, W.I. (2006) *J. Appl. Phys.*, **100**, 014303.

46 Pi, X.D., Gresback, R., Liptak, R.W., Campbell, S.A., and Kortshagen, U. (2008) *Appl. Phys. Lett.*, **92**, 123102.

47 Ramos, L.E., Degoli, E., Cantele, G., Ossicini, S., Ninno, D., Furthmuller, J., and Bechstedt, F. (2008) *Phys. Rev. B*, **78**, 235310.

48 Stegner, A.R., Pereira, R.N., Klein, K., Lechner, R., Dietmueller, R., Brandt, M.S., Stutzmann, M., and Wiggers, H. (2008) *Phys. Rev. Lett.*, **100**, 026803.

49 Hao, X.J., Cho, E.C., Flynn, C., Shen, Y.S., Park, S.C., Conibeer, G., and Green, M.A. (2009) *Sol. Energy Mater. Sol. Cells*, **93**, 273.

50 Song, D., Cho, E.-C., Conibeer, G., Huang, Y., and Green, M.A. (2007) *Appl. Phys. Lett.*, **91**, 123510.

9
The Synthesis of Silicon Nanocrystals by Ion Implantation
Robert Elliman

9.1
Introduction

Ion implantation can be defined as the controlled introduction of a chemical species into a material by bombardment with energetic ions [1]. In general, this involves bombardment with atomic and molecular ions of arbitrary energy distribution and includes applications such as plasma immersion ion implantation, but more commonly it involves bombardment with monoenergetic ions of a known species, as used in most semiconductor applications. In both cases, ions are directly injected into the material by virtue of their high velocity, independent of any thermodynamic constraints, as shown in Figure 9.1. Ion implantation is therefore a nonequilibrium process that can be used to create supersaturated solid solutions of implanted species from which nanocrystalline phases can be precipitated [2–8]. The ion irradiation process also induces compositional and structural changes in materials that provide additional routes to nanocrystal synthesis [4–8].

Metallic and semiconducting nanocrystals are most commonly formed in dielectric host materials by first forming a supersaturated solid solution of the component elements and then precipitating the secondary nanocrystalline-phase within the host material [2, 3]. Figure 9.1 outlines the basis of this process for ion-implanted samples. At low implant influences, the concentration of the implanted species is less than the solubility limit and they are incorporated within the host material as an atomic impurity or dopant. At higher implant influences, the local concentration exceeds the equilibrium solid solubility limit and the resulting solid solution is metastable. Phase separation can then occur during the ion implantation process, if there is sufficient atomic mobility, or during subsequent thermal annealing. At very high implant fluences, the concentration of the implanted species can become sufficient to form an interconnected network or a continuous buried layer of particular equilibrium phase. Under these conditions, subsequent annealing can lead to the formation of a uniform buried layer of a particular alloy or compound. (This effect forms the basis of the commercial SIMOX® process for fabricating silicon-on-insulator wafers.)

Silicon Nanocrystals: Fundamentals, Synthesis and Applications. Edited by Lorenzo Pavesi and Rasit Turan
Copyright © 2010 WILEY-VCH Verlag GmbH & Co. KGaA, Weinheim
ISBN: 978-3-527-32160-5

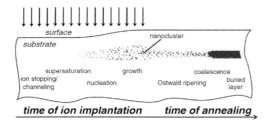

Figure 9.1 Schematic of the ion-beam synthesis process. High fluence implantation creates a supersaturated solid solution from which nanocrystals can nucleate and grow. Nanocrystals can form either during implantation, if the implanted atoms have sufficient atomic mobility, or during subsequent annealing. Very high implant fluences can lead to the formation of continuous buried layers. (From Ref. [18].)

The processes outlined above have been used to synthesize a broad range of nanostructures, including elemental, compound, and core-shell nanocrystals in a range of crystalline and amorphous host materials [2, 3]. This chapter provides an overview of the ion-beam synthesis of silicon nanocrystals. It concentrates on the conventional ion-beam synthesis route outlined above and also highlights other novel synthesis routes unique to ion implantation and ion irradiation. No attempt is made to provide a more general overview of ion-beam processing and its applications. Instead, the reader is referred to the comprehensive literature on this topic [2–8].

9.2
Ion Implantation

The most common commercial application of ion implantation is in the semiconductor industry where it is used for doping silicon wafers during integrated circuit manufacture [1]. Other semiconductor applications include doping and electrical isolation of III–V semiconductors and custom applications such as Smartcut® and SIMOX for producing silicon-on-insulator (SOI) material. These applications employ ion energies in the range of 200–3 MeV and irradiation influences in the range from 10^{11} ions cm^{-2}, for transistor threshold voltage adjustment, to around 10^{18} ions cm^{-2}, for the synthesis of SOI material. At present, there are around 4000 ion implanters employed worldwide for such processing, and modern production machines are very sophisticated tools designed for automated, batch-processing of hundreds of wafers per hour.

Despite their complexity, the basic functions of a modern ion implanter are well represented by a generic machine such as the one shown in Figure 9.2. This comprises several subsystems: (a) an ion source for generating positive or negative ions of the desired species, (b) an accelerating stage in which the extracted ions are accelerated through a fixed electrostatic potential, (c) a magnetic mass-spectrometer for filtering the desired species from other ions, (d) a beam or sample scanning system to allow uniform irradiation of the sample, (e) a target chamber to allow

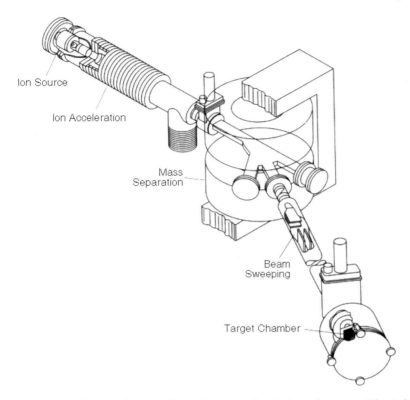

Ion Source

Ion Acceleration

Mass
Separation

Beam
Sweeping

Target Chamber

Figure 9.2 Schematic diagram of an ion implanter showing key subsystems. (After Ref. [6].)

sample positioning and temperature control, and (f) a vacuum system to maintain low operating pressures throughout the system, an essential requirement to minimize ion neutralization and maximize beam transmission. Each of these subsystems has undergone years of development and refinement to produce machines capable of delivering monoenergetic ions with currents in the range of 1–100 mA and capable of implanting 300 mm diameter wafers with a fluence uniformity of better than 1%. More complex accelerator configurations have also been developed for very low (<1 keV) and very high (>400 keV) energy implants, including linear-acelerator (LINAC) and tandem designs.

Ion-beam synthesis of silicon nanocrystals involves the implantation of silicon ions, and related dopants or impurities, into the surface region of bulk- or thin-film materials, such as silicon dioxide. Ion implantation has specific advantages for such processing, including the ability to implant any species into any host material independent of thermodynamic constraints, the ability to implant specific isotopes of an element, the ability to independently control the concentration and spatial distribution of the implanted species, the ability to implant preformed or prepatterned materials, and the ability to implant into substrates held at any temperature.

Its disadvantages include the fact that it is relatively expensive, particularly for applications that require irradiation of large areas to high ion fluences, that it creates radiation damage in the substrate material, and that the depth distribution of the implanted species is limited to less than a micrometer for most practical purposes. Within this framework ion implantation offers a valuable tool for nanocrystal synthesis and for exploring their structure, properties, and applications.

9.3
Spatial Distribution

The ability to control both the depth and the lateral distribution of nanocrystals is critical for many applications, including the fabrication of electronic, optoelectronic, and photonic devices and structures. For ion-implanted samples, the nanocrystal depth distribution is primarily determined by the Si-ion range distribution [6], with lateral variations defined either by suitable masking of the implanted ions or by subsequent etching.

9.3.1
Ion Range Distributions

When an energetic ion enters a solid, it loses energy through collisions with atoms and electrons and eventually comes to rest at some depth, as schematically shown in Figure 9.3a. Because this typically involves many uncorrelated collisions, these energy-loss processes can be considered to be independent and the total rate of energy loss, dE/dx, is then the sum of two contributions:

$$\frac{dE}{dx} = \left(\frac{dE}{dx}\right)_n + \left(\frac{dE}{dx}\right)_e, \tag{9.1}$$

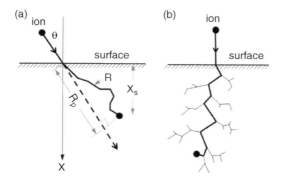

Figure 9.3 (a) Schematic of an implanted ion trajectory showing the relationship between the total range (path length) R, the projected range R_p, and the depth of penetration x_s for an ion implanted at an angle θ with respect to the surface normal. (b) Schematic of a collision cascade produced by an implanted ion and recoiling target atoms.

where $(dE/dx)_n$ is the rate of energy lost in nuclear collisions and $(dE/dx)_e$ is the rate of energy lost in electronic collisions. This is usually expressed in terms of a stopping cross section (or probability) S, where $S \equiv 1/N(dE/dx)$, and N is the atomic density of the target material. Then, the total stopping cross section S_t is given by

$$S_t = S_n + S_e, \tag{9.2}$$

where S_n is the nuclear stopping cross section and S_e is the electronic stopping cross section. Given the energy dependence of these stopping cross sections, it is then a simple matter to calculate the total ion path length R from the expression

$$R = \frac{1}{N} \int_{E_0}^{0} \frac{dE}{S_n + S_e}. \tag{9.3}$$

The calculation of the nuclear and electronic stopping cross sections and their energy dependence involves detailed consideration of the scattering processes [6]. The basic concepts underpinning these calculations were formulated in the first half of the twentieth century and work from this period is described in an extensive treatise published by Niels Bohr in 1948 (see Ref. [6] for a comprehensive review of range theory). For the energy range of relevance to ion implantation (typically heavy ions of an energy less than 1 MeV), the calculations are complicated by the fact that the ions do not have sufficient energy to fully penetrate the electron cloud of the scattering atoms and, therefore, the interaction cannot be described by a simple Coulomb potential. Also, the ions are not fully stripped of electrons during their passage through the material but have a mean charge state that is velocity dependent. These complexities were addressed by Lindhard and coworkers, in particular by Lindhard, Scharff, and Schiott, and their work, often referred to as LSS theory, continues to provide the basis for modern ion implantation range theory [6].

Suffice it to say, the nuclear and electronic stopping cross sections vary with ion energy (velocity), the basic form of which is shown in Figure 9.4. For low-energy ions, the energy loss is dominated by nuclear scattering, while at high energies it is dominated by electronic energy loss. The LSS formalism provides expressions for

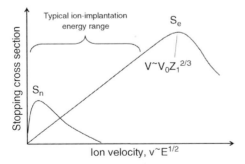

Figure 9.4 Schematic of the energy dependence of the nuclear (S_n) and electronic (S_e) stopping cross sections showing the dominance of nuclear stopping processes at low energies and the dominance of electronic stopping processes at high energies.

the nuclear and electronic stopping cross sections in terms of reduced range and energy parameters [6]. This yields a universal expression for the nuclear stopping cross section. It also predicts that the electronic energy loss is a linear function of ion velocity over the energy range typically employed for ion implantation but with a slope that depends on the ion–target combination. The electronic stopping cross section reaches a maximum for ion velocities v, around $v = v_0 z_1^{2/3}$, where z_1 is the atomic number of the ion and v_0 is the Bohr velocity of an electron, before decreasing again at higher velocities. (The maximum corresponds to the transition from partly to fully stripped ions.)

The LSS expressions for nuclear and electronic stopping cross sections can be used to calculate the total range of ions from Eq. (9.3). However, from an experimental viewpoint, it is not the total range or path length that is of interest but either the projected range along the incident ion trajectory or the average depth below the sample surface, as shown in Figure 9.3a. To address this need, Lindhard derived a set of integral equations from which the moments of the range distribution could be calculated. His treatment has subsequently been generalized within the framework of the Boltzmann transport equation to provide a comprehensive description of implanted ion and recoiled atom distributions [9].

From Figure 9.3a, it is clear that the depth distribution of implanted ions also depends on their angle of incidence with respect to the surface normal and that shallower depth distributions can be achieved by tilting to greater angles. For near-normal incidence, the penetration depth and projected range are similar and it is often sufficient to model the depth distribution by a Gaussian distribution of the form

$$N(x) = \frac{\phi}{(2\pi)^{1/2}\Delta R_p} \exp\left[-\frac{1}{2}\left(\frac{x-R_p}{\Delta R_p}\right)^2\right] \tag{9.4}$$

where $N(x)$ is the implanted impurity concentration, R_p is the mean projected range of the ions, ΔR_p is the projected range straggling (standard deviation of the distribution), ϕ is the total implanted ion fluence (typically expressed as ions cm^{-2}), and "x" is the depth. The peak impurity concentration can be found by setting $x = R_p$, which gives $N(R_p) \approx 0.4\phi/\Delta R_p$.

In practice, ion range distributions are most often calculated from Monte-Carlo simulations, using programs such as SRIM [10] or from semiempirical data sets, as used in semiconductor process simulators. Monte-Carlo simulations have the advantage that they can readily model complex multilayer, multielement materials and can follow both the implanted ion and the recoiling target atoms to build a comprehensive understanding of irradiation process. For example, Figure 9.5 shows the energy dependence of the rates of nuclear and electronic energy loss, and the calculated mean projected range and projected range straggling for Si ions implanted into SiO_2, as calculated by SRIM. These data show that the rates of nuclear and electronic energy loss are approximately equal for 100 keV Si ions, with nuclear energy loss dominating at lower energies and electronic energy loss dominating at higher energies. They also show that the mean projected range of Si increases approximately linearly with ion energy for energies up to 500 keV and that it is limited to less than 1 μm for commonly available implant energies (<400 keV).

Range distributions for Si in SiO_2 are shown in Figure 9.6a for selected ion energies together with Gaussian approximations to these distributions. Note that the

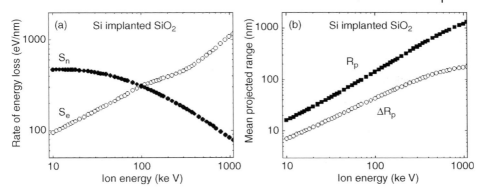

Figure 9.5 Energy loss and range parameters for Si ions implanted into SiO$_2$. (a) The rate of nuclear and electronic energy loss as a function of ion energy, and (b) the mean projected range and range straggling as a function of ion energy.

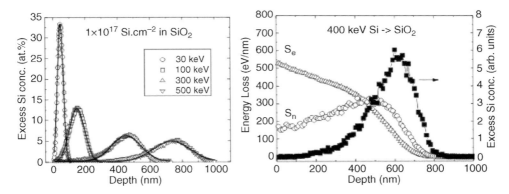

Figure 9.6 (a) Si range distribution in SiO$_2$ for ions implanted at energies of 30, 100, 300, and 500 keV, all for an implant fluence of 1×10^{17} Si cm^{-2}. The solid lines are Gaussian fits to the distributions. (b) The ion range and deposited energy distributions for 400 keV Si ions implanted into SiO$_2$.

low-energy distributions are well described by the Gaussian function, but as the energy increases the distributions become increasingly skewed due to the energy dependence of the energy cross sections. The relationship between the ion range and the energy deposition distributions is also shown in Figure 9.6b, highlighting the fact that the rate of nuclear and electronic energy loss changes with energy and therefore with depth.

9.3.2
Multiple Energy Implants

Although ion range distributions have specific forms, it is possible to construct more complex distribution by combining several implants of different energy and fluence.

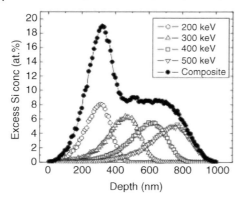

Figure 9.7 Si range distributions in SiO$_2$ for ions implanted at energies of 200, 300, 400, and 500 keV, all for an implant fluence of 1×10^{17} Si cm^{-2}, and a composite profile constructed from the separate implants by scaling the implant fluences by 2.0, 0.8, 0.8, and 1.2, respectively.

This can be used to construct uniform concentration profiles over selected regions or to construct more complex distributions for particular applications, such as the fabrication of optical waveguides. As an example, Figure 9.7 shows a composite profile constructed from four separate implants. In this case, the final silicon distribution includes a peak at a depth of 350 nm and a relatively uniform region ranging from 400 to 800 nm.

9.3.3
High-Fluence Effects

The calculation of ion ranges generally assumes that the target material is unperturbed by the implanted species and that its composition remains that of the starting materials. For semiconductor doping, where the implant fluences are relatively low (i.e., impurity concentrations $\ll 1$ at.%), this is a reasonable assumption, but for nanocrystal synthesis it is often not [6, 8].

High-fluence implants affect the range profile of ions in three main ways: (a) sputtering of the target material, (b) swelling of the implanted volume due to the incorporation of implanted ions, and (c) changes in the stopping power of ions caused by the modified composition of the target material. The relative significance of these effects varies with ion energy, with sputtering being a dominant effect for low-energy ions where much of the ion energy is deposited into nuclear collisions near the surface. In most cases, these effects lead to a broadening of the implanted ion distribution relative to a low-fluence distribution. In this case, it is necessary to use more complicated range calculations based either on semiempirical data or on Monte-Carlo simulations that take account of the sputtering, swelling, and stopping power changes [11], as shown in Figure 9.8.

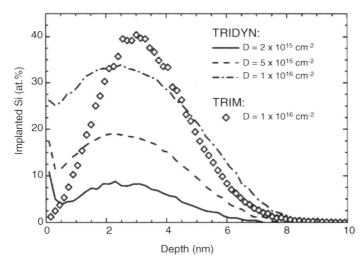

Figure 9.8 Comparison of Si depth distributions calculated by SRIM [10] and TRIDYN [11] for 1 keV Si implants into SiO$_2$. SRIM assumes that the substrate is unperturbed by the implant and TRIDYN accounts for compositional changes. The broadening of the implanted Si distribution is clearly evident in the latter case. (From Ref. [25].)

9.3.4
Lateral Patterning

The above discussion has been restricted to the depth distribution of Si nanocrystals, but it is also possible to control their lateral distribution by masking the region to be implanted with a suitable template [12, 13] or by subsequent patterning and etching of the implanted structure [14]. As an example, Figure 9.9 shows the photoluminescence (PL) emission from Si nanocrystals formed in SiO$_2$ by implantation through a Mo mask [13]. Using this technique, luminescent nanocrystal features as small as 0.5 µm were fabricated, thereby providing the basis for fabricating optical structures such as waveguides and gratings, as well as devices based on patterned luminescent material.

9.4
Size Distribution

The size distribution of nanocrystals is a function of ion implantation and annealing parameters and can be tailored to a particular mean size by suitable choice of these parameters. In general, however, the dependencies are more complicated than might be expected due to dynamic effects that occur during the implantation process. These arise from the competition between the formation and the irradiation-induced dissociation of clusters during the implantation process, and can lead to both the

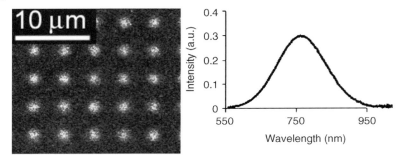

Figure 9.9 Photoluminescence image of a SiO$_2$ layer implanted with Si ions through a Mo contact mask to define a pattern of luminescent nanocrystals, together with a spectrum of the emission. This simple masking procedure was used to define luminescent features as small as 0.5 μm. (From Ref. [13].).

dependencies on the ion flux and the expected dependencies on ion energy, ion fluence, and substrate temperature. These dynamic effects are most evident for mobile species such as gold or silver in silica and can lead to the formation of nanocrystals during ion implantation at room temperature, without the need for further annealing. Such effects are much less significant for slow diffusing species, such as silicon in silica, where an annealing step is usually required for nanocrystal synthesis. However, even in this case it is not possible to rule out the possibility of some clustering or prenucleation during irradiation, particularly for elevated temperature or high-fluence implants. Clearly, such effects will influence any subsequent thermal evolution of the nanocrystals.

9.4.1
Annealing Temperature and Time

9.4.1.1 Nucleation and Growth of Nanocrystals
For low levels of supersaturation, the formation of nanocrystals during postimplantation annealing can be described by classical nucleation and growth theory. This predicts three stages of nanocrystal evolution: (a) nucleation of small but stable nanocrystals, (b) normal growth (relatively fast) of nanocrystals by the attachment of atoms supplied from the surrounding matrix, and (c) competitive growth (relatively slow) of nanocrystals by the exchange of atoms, during which larger nanocrystals grow at the expense of smaller nanocrystals (Ostwald ripening).

A supersaturated solid solution, such as the one produced by ion implantation, is metastable, and given sufficient atomic mobility (i.e., temperature and time), excess solute atoms diffuse and cluster, with the rate of cluster formation and the size distribution of clusters described by the classical theory of homogeneous nucleation [15]. In this model, the stability of clusters depends on their size, and only those greater than a critical size continue to grow. This critical size can be shown to be $r_c = -2\sigma/\Delta G_v$, where r_c is the critical radius, σ is the surface free energy, and ΔG_v is the change in volume free energy for an atom taken from the matrix and placed within the cluster. Clusters larger than r_c will grow and clusters smaller than r_c will dissociate and redissolve into the matrix.

Once stable nuclei have been formed, they grow by attachment of atoms from the surrounding matrix. The concentration of dissolved atoms therefore decreases as the diameter of the nanocrystals increases, with the nanocrystal radius r is given by $r^2 = r_0^2 + at$, where r_0 is the initial radius of the cluster, t is the annealing time, and "a" is a constant. This shows that the rate of growth slows with increasing time, and for long times and small initial cluster sizes is well described by a $t^{1/2}$ dependence.

As the surrounding matrix is denuded of dissolved atoms, the "normal" growth phase ceases and the mean size of the nanocrystals briefly saturates before a second stage of competitive growth begins to dominate. The theory of coarsening (Ostwald ripening) is based on the work of Liftshitz, Slyozov, and Wager (referred to as LSW) and its refinements [16]. This theory predicts that the nanocrystal size distribution should approach a time invariant form in which the mean particle size follows a $t^{1/3}$ dependence. This distribution has the form

$$P(u) = 3^4 2^{\frac{5}{3}} e u^2 (u+3)^{\frac{7}{3}} \left(\frac{3}{2} - u \right)^{-\frac{11}{3}} \exp \left[\left(\frac{2}{3} u - 1 \right)^{-1} \right], \tag{9.5}$$

where u is the nanocrystal radius normalized to the mean radius for which there is no net growth, r_m, and the time dependence of the mean particle radius is given by

$$r_m = \left(\frac{4 \alpha D t}{9} \right)^{\frac{1}{3}} \tag{9.6}$$

with $\alpha \sim T^{-1}$ and the diffusion coefficient D is defined as

$$D = D_0 \exp \left(\frac{\Delta E_D}{kT} \right), \tag{9.7}$$

where ΔE_D is the activation energy for atomic diffusion of the dissolved atoms in the matrix.

Classical nucleation and growth theory therefore provides a means of estimating the final nanocrystal density and size distribution, and their dependence on annealing temperature and time. Significantly, it also predicts that the size distribution is independent of the initial level of supersaturation. For a given annealing temperature and time, higher levels of supersaturation lead to higher nanocrystal densities but with the same size distribution.

This theory is, however, based on many assumptions, including that the system is dilute, homogeneous, and of infinite extent. In contrast, implanted samples have spatial distributions that are finite and depend on depth, with the added possibility of impurity clustering, or prenucleation, during implantation, particularly around the peak of the implant distribution where the concentration is highest. As a consequence of these effects, nanocrystals produced in ion-implanted samples are usually found to have spatially dependent size distributions, with larger precipitates forming around the peak of the implant distribution and smaller ones at the extremes of the distribution [17]. Some aspects of these distributions have been successfully modeled

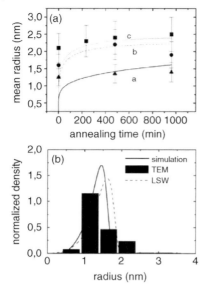

Figure 9.10 (a) The mean radius of Si nanocrystals formed in Si-implanted SiO_2 layers during annealing at $1100\,^{\circ}C$ as a function of annealing time for samples with peak excess silicon concentrations of 10 at.% (▲), 20 at.% (●), and 30 at.% (■). The solid curve "a" shows the predictions of LSW theory, and the curves labeled "b" and "c" correspond to an extended model. (b) The measured size distribution for the 10 at.% sample after annealing for 16 h at $1100\,^{\circ}C$, together with the distributions predicted by LSW and the extended model. Samples were implanted with 150 keV Si^+ ions. (After Ref. [20].)

by kinetic Monte-Carlo simulations [18], but unfortunately there are very few systematic experimental studies for comparison, largely due to the difficulty in imaging small Si nanocrystals in host materials such as SiO_2 [19].

Bonafos *et al.* [20, 21] have reported the temperature and time dependence of Si nanocrystal size distributions for samples implanted to different fluences with 150 keV Si^+ ions. Their results are summarized in Figure 9.10 and show that the mean size and size distribution of nanocrystals is in reasonable agreement with the LSW theory for low levels of supersaturation (≤10 at.%) and long anneal times. However, the theory is shown to underestimate the mean size of the nanocrystals at short annealing times and fails to account for an observed increase in mean nanocrystal size with increasing levels of supresaturation (>10 at.%). The latter is not unexpected as phase separation in this high supersaturation regime is expected to proceed via spinodal decomposition rather than conventional nucleation and growth [22]. The validity of the LSW theory for low levels of supersaturation and long anneal times is also supported by photoluminescence measurements [21]. Figure 9.11 shows a representative PL spectrum from samples with excess Si concentrations ranging from 1 to 10 at.% and the inset shows the dependence of the PL intensity on the excess Si concentration. The latter shows that the intensity increases linearly with concentration over this low concentration range, consistent with an increase in the nanocrystal density and with theoretical predictions [15]. For concentrations above

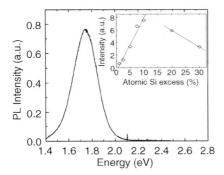

Figure 9.11 Measured size distribution of silicon nanocrystals formed in Si-implanted SiO_2 layers by annealing at 1100 °C. Samples were implanted with 100 keV Si^+ ions to fluences of (a) 6×10^{16} Si cm^{-2}, (b) 8×10^{16} Si cm^{-2}, and (c) 1×10^{17} Si cm^{-2}. The inset shows the mean nanocrystal diameter d_{av}. (After Ref. [64].)

10 at.%, the PL emission is redshifted (not shown), consistent with an increase in the average nanocrystal size, and the PL intensity decreases with increasing Si concentration. This is consistent with the formation of a more complex network of nanocrystals during spinodal decomposition [22], as noted above.

The failure of the theory to predict the nanocrystal size distributions for short annealing times remains unexplained but may have resulted from preclustering during the implantation process or from more complex nucleation and growth processes associated with the formation of silicon nanocrystals. Wang *et al.* [23] have studied the initial stages of phase separation in the Si:SiO_2 system and identified three stages of nanocrystal evolution in ion implanted samples annealed at 1100 °C. These included the formation of amorphous Si-rich SiO_x clusters, the formation of amorphous Si clusters, and the crystallization of these clusters to form nanocrystals. In this case, a complete transformation to the nanocrystalline phase was observed for annealing times of less than 30 min. By way of comparison, Iacona *et al.* undertook a comprehensive study of nanocrystal synthesis in SiO_x films prepared by plasma-enhanced chemical vapor deposition (PECVD) [24]. They observed that phase-separated Si initially consisted of both amorphous and crystalline clusters, consistent with Wang's results. However, they observed that the crystalline fraction increased with increasing temperature over the range 1000–1250 °C, eventually reaching 100% after a 1 h anneal at 1250 °C. This is a much higher temperature than that required to produce nanocrystals in ion implanted samples and is likely due to the lower Si diffusivity in PECVD films due to the presence of nitrogen. These examples serve to illustrate the complexity of the nanocrystal synthesis process and clearly demonstrate the need for more extensive work in this area.

9.4.2
Spinodal Decomposition

At high levels of supersaturation, phase separation can occur by spinodal decomposition in which there is little or no barrier to nanocrystal nucleation, and the system is driven by initial concentration fluctuations. In this case, the system forms an interconnected network of nonspherical precipitates and can exhibit complex self-assembly in the vicinity of surfaces and interfaces [25].

9.4.3
Effect of Surfaces and Interfaces

Surfaces and interfaces can act as sources and sinks for diffusing species and thereby affect nanocrystal evolution. For Si nanocrystals formed in SiO_2, the role of interfaces is generally limited to short-range effects due to the relatively low diffusivity of Si in SiO_2 [20, 26]. But even in this case important short-range effects have been observed, including the formation of two-dimensional nanocrystal arrays [22, 27] separated from a SiO_2 interface by a narrow ($\sim$5–10 nm) denuded zone. The effects of the interface are most dramatically illustrated by kinetic Monte-Carlo simulations, such as those shown in Figure 9.12. This shows the annealing behavior of nanocrystals in an 8 nm thick SiO_2 on Si after implantation with 1 keV Si ions to a fluence of 2×10^{15} Si cm^{-2}. As the nanocrystals evolve, a well-defined denuded zone develops in the vicinity of the Si/SiO_2 interface due to the influence of the Si substrate on Ostwald ripening [22]. A separate study has shown that the width of such denuded zones can be controlled by suitable choice of the implantation and annealing conditions [28]. This is an important consideration for nonvolatile memory (NVM) applications where the nanocrystals are charged and discharged by electron tunneling to the substrate [29].

9.4.4
Hot Implants

The nucleation and growth of nanocrystals during ion implantation relies on the competition between two processes: atomic diffusion and clustering, and ion

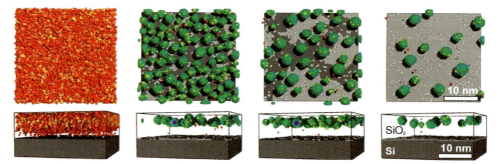

Figure 9.12 Kinetic Monte-Carlo simulation of silicon nanocrystal evolution in the vicinity of a Si/SiO$_2$ interface during annealing. The sample consists of an 8 nm thick SiO$_2$ layer on Si implanted with 1 keV ions to a fluence of 5×10^{15} Si cm^{-2}. The initial Si distribution is as shown in Figure 9.8. Upper panels show the interface in planview and the lower panels show cross sections. (From Ref. [22].)

irradiation-induced dissociation of the clusters [30]. Clearly, this competition can be influenced by the substrate temperature and this is particularly evident in systems where the implanted species has high a atomic mobility, such as Au- or Ag-implanted SiO_2. For example, irradiation of SiO_2 at room temperature with 2.75 MeV Au ions to a fluence of 1.5×10^{17} Au cm^{-2} produces Au clusters with diameters of order 1 nm [30]. The same irradiation at 400 °C produces nanocrystals with an average diameter of around 5 nm, and irradiation at 600 °C produces nanocrystals with an average diameter of around 10 nm. For very slow diffusers such as Si in SiO_2, the effect is far less significant and Si nanocrystals are not generally observed in as-irradiated samples, even when the substrate is heated to temperatures as high as 600 °C. Despite this, there is evidence that some clustering does occur during elevated temperature irradiation. For example, small changes in the final nanocrystal size distribution are observed in samples initially implanted with silicon at different temperatures when they are subsequently annealed at 1100 °C [31].

9.4.5
Annealing Ambient

The choice of annealing ambient, including the presence of trace impurities, can influence the diffusion and clustering of implanted species and thereby affect the size and spatial distribution of nanocrystals. Ge-implanted SiO_2 has been shown to be particularly sensitive to such effects, with Ge redistribution and clustering being very sensitive to trace amounts of oxidants (a few ppm of O_2 or H_2O) in the annealing ambient [32, 33]. This has been shown to lead to Ge redistribution to surfaces and interfaces and to produce Ge nanocrystal distributions that are narrower than the initial implant distribution. Such long-range diffusion is not observed for Si-implanted SiO_2 but the choice of annealing environment (Ar, N_2, or 5% H_2 in N_2) has been shown to affect the nanocrystal size distribution [34]. In this case, samples annealed in Ar contained larger nanocrystals than those annealed in nitrogen-containing ambients. Although the mechanisms responsible for this behavior remain unclear, it has been speculated that Si nanocrystals react with N_2 to form a thin oxynitride surface layer, a reaction possibly mediated by the presence of H_2 and O_2 [34]. This would act to reduce the effective diameter of the Si-nanocrystal core, consistent with the observations.

Direct oxidation of silicon nanocrystals can also be used to control their average size. In this case, the nanocrystals are annealed in an oxidizing ambient to produce SiO_2 surface layers, with the size of the Si nanocrystal controlled by the extent of oxidation [35].

9.5
Irradiation Effects

The above discussion has concentrated on the fate of implanted ions, but the collisions that act to slow these ions also impart energy to the atoms and electrons of the target material. This can act to change the topography, composition, and structure of the irradiated material through processes such as preferential sputtering, recoil implantation, radiation-enhanced diffusion, and radiation-induced impurity segregation [5–8]. These effects can be detrimental in some cases but offer additional flexibility for nanocrystal synthesis in other cases.

9.5.1
Radiation Damage

Irradiation of a solid with energetic ions creates radiation damage by displacing atoms in the solid, either directly through energetic nuclear collisions or indirectly through electronic excitation and ionization [6]. In metals and semiconductors, the damage results primarily from nuclear collision processes while in insulators it arises from both nuclear and electronic collision processes.

An energetic ion can transfer a large fraction of its energy to a target atom during a nuclear scattering event and as a result the target atom can recoil with energy sufficient to displace many additional atoms. These can in turn displace other atoms to create a collision cascade, as illustrated in Figure 9.3b. For low collision densities, the number of displaced atoms is proportional to energy deposited into atomic collision processes. However, at high densities the energy density can be sufficient to superheat or melt the cascade volume and in this case the number of displacements can far exceed the predictions of linear cascade theory. Electronic stopping excites and ionizes electrons in the host material, and in metals and semiconductors these generally relax and thermalize without causing atomic displacements. However, in highly ionic or insulating materials, ionization can lead to large Coulomb interactions between atoms, and this in turn can produce atomic displacements. For ion energies of relevance to ion implantation, the number of such defects is proportional to the rate of electronic energy loss. (For extreme rates of electronic energy loss, an ion may produce a track of modified or defective material due to a Coulomb explosion, in which a core of positively charged ions repel each other or through superheating of the volume due to efficient electron–phonon coupling.) As the ion-fluence increases, the radiation damage from individual ions accumulates, and depending on the nature and mobility of individual defects, may evolve into different forms.

For amorphous solids, such as SiO_2, radiation damage generally takes the form of bond distortions and compositional fluctuations, including the formation of defects such as nonbridging oxygen (NBOHCs) centers [36]. These microstructural changes can influence the properties of the host material and incorporated nanocrystals and can lead to macroscopic effects such as compaction or swelling of the material, which is directly associated with changes in refractive index. For example, as-implanted SiO_2 samples exhibit strong defect-related luminescence at around 660 nm (1.9 eV) that is generally attributed to NBOHCs [37]. The intensity of SiO_2 defect luminescence decreases during annealing at temperatures above 650 °C and is eventually dominated by emission from nanocrystals once they are formed at higher temperatures (>1000 °C), as shown in Figure 9.13 [36–38].

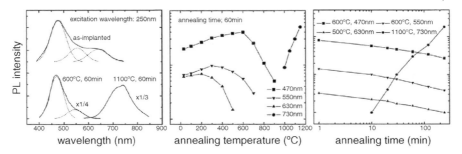

Figure 9.13 Photoluminescence spectra and the temperature and time dependence of the various emission peaks as a function of annealing temperature and time. (Note the log scales for the temperature and time dependence.) Samples were thermally grown SiO_2 layers implanted with 120 keV Si^+ ions to a fluence of 2×10^{16} Si cm^{-2}. (After Ref. [38].).

The photoluminescence emission from silicon nanocrystals is particularly sensitive to the presence of defects, both within the nanocrystal and the surrounding matrix. Indeed, even after high-temperature annealing the PL intensity can be increased by hydrogen passivation of remaining defects [39]. This makes PL emission a useful probe for studying radiation damage and its annealing behavior [40]. Ion irradiation of luminescent nanocrystals has been shown to quench the PL emission at relatively low ion fluences, and detailed studies have shown that nanocrystals accumulate defects more rapidly than their bulk counterparts [41]. The annealing kinetics of damaged or amorphized silicon nanocrystals are also very different from those of the bulk material [40]. For example, ion-irradiated nanocrystals only recover their full luminescence after annealing at temperatures as high as 1000 °C, whereas bulk amorphous silicon recrystallizes at temperatures around 650 °C. These observations highlight the importance of taking radiation damage into account when comparing the PL intensity from samples subjected to different irradiation and annealing conditions.

9.5.2
Ion-Beam Mixing

Ion irradiation can cause intermixing of the interface between thin films, and for Si/SiO_2 and $SiO_2/Si/SiO_2$ structures, this can lead to local compositional changes that are conducive to the formation of Si nanocrystals [42], as shown schematically in Figure 9.14. The evolution of such an intermixed structure during subsequent annealing is demonstrated by the kinetic Monte-Carlo simulations, as shown in Figure 9.15. These correspond to a structure consisting of a Si substrate/15 nm SiO_2/ 50 nm poly-Si, irradiated with 50 keV Si^+ ions to a fluence of 1×10^{16} cm^{-2} and then annealed to form nanocrystals. Silicon nanocrystals are observed to form in the vicinity of the interfaces but separated from them by 2–3 nm, similar to the effect discussed in Section 9.4.2. The difference in this case is that the source of excess silicon comes from intermixing rather than directly from the implantation of silicon.

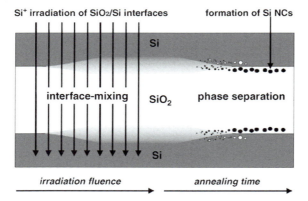

Figure 9.14 Schematic diagram showing ion-beam mixing of interfaces in a Si/SiO$_2$/Si structure and the evolution of the intermixed structure during subsequent annealing [65].

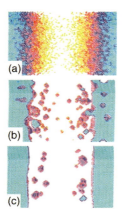

Figure 9.15 Kinetic Monte-Carlo simulation of nanocrystal evolution in an intermixed structure such as the one shown in Figure 9.13. The initial structure consists of a Si-substrate/15 nm SiO$_2$/ 50 nm poly-Si, irradiated with 50 keV Si$^+$ ions to a fluence of 1×10^{16} cm^{-2}, with the intermixing calculated from TRIDYN [11]. (After Ref. [66].)

In cases where Si ions are used to produce the intermixing, the final excess-Si distribution is a combination of the implanted and intermixed silicon and the ion implantation parameters can be used to control the final nanocrystal distribution [28].

9.5.3
Irradiation-Induced Precipitation

As previously discussed, supersaturated solid solutions are metastable and will phase separate given sufficient atomic mobility. Since the energy carried by energetic ions can be deposited into a small volume and can give rise to large local temperature rises, it seems reasonable that this might induce phase separation in such systems. However, as discussed in Section 9.4.3, the formation of silicon nanocrystals is usually compromised by the low diffusivity of silicon and by irradiation-induced dissociation of any clusters that do form. By irradiating a thin film with swift heavy ions, it is possible to reduce the collisional dissociation of clusters and still deposit sufficient energy to aid clustering. In this case, the ions must have sufficient energy to pass through the film with few nuclear collisions and have high enough electronic energy loss to cause local heating of the ion track. This typically requires very high energy heavy ions, such as 70 MeV Si ions, and as such is a rather extreme regime. Nonetheless, such irradiation has been shown to produce silicon nanocrystals in silicon-rich oxide layers [43]. Swift heavy ions have also proven of use for fabricating and modifying a range of other nanostructures [44].

9.6
Novel Structures and Applications

This section presents some novel structures and applications of silicon nanocrystals that exploit the flexibility of the ion-beam synthesis technique. A more detailed discussion of applications can be found in later chapters.

9.6.1
Alloying and Doping

The ability of ion implantation to implant any species or combination of species into any substrate material has been exploited by many researchers to study the effects of alloying [45–47] and doping on the physical properties of the nanocrystals. This includes the formation of SiGe [47] and SiC [46] nanocrystals, the doping of Si nanocrystals with species such as B, P, and Au [48], and the incorporation of optically active impurities such as Er into layers containing Si nanocrystals [49–51]. The latter has been a particularly active area of research due to the prospect of making an integrated optical amplifier based on Er, a prospect based on the fact that silicon nanocrystals have strong broadband optical absorption and act as very effective sensitizers for Er photoluminescence. In the present context, these examples simply serve to illustrate the flexibility of ion-implantation as a research tool for studying the synthesis, structure, properties, and application of silicon nanocrystals.

9.6.2
Choice of Substrate Material

Silicon nanocrystals are most commonly produced in SiO_2, and although ion implantation can readily produce nanocrystals in other host materials, the number of such studies is very limited [2, 52]. The preference for using SiO_2 stems from the fact that silicon dioxide has excellent physical and chemical properties and is extensively used in microelectronic and photonic applications. More significant, however, is the fact that silicon dioxide is very effective at passivating silicon surface states. This increases the luminescence efficiency of silicon nanocrystals and is therefore very important for photonic applications. In materials that do not provide the same degree of passivation, silicon nanocrystals offer little advantage over other nanocrystals. Indeed, for many applications, it may be preferable to use nanocrystals other than silicon in order to reduce processing temperatures or control chemical reactivity. As a consequence, most of the research on alternative host materials has involved the synthesis of nanocrystals other than silicon. For example, ion-beam synthesis of Ge nanocrystals in high-k dielectric films has been used for nonvolatile memory applications [53] largely due to the fact that they form at lower temperatures.

9.6.3
Light-Emitting Diodes

Much of the initial interest in silicon nanocrystals centers around the fact that they exhibited relatively strong photoluminescence and that the emission wavelength could be tuned by controlling the nanocrystal dimensions. This rapidly led to an interest in making electroluminescent devices such as light-emitting diodes. The first such structure was fabricated by ion-beam synthesis of silicon nanocrystals within the silicon dioxide insulating layer of a poly-$Si/SiO_2/Si$ device structure [54]. This device exhibited visible electroluminescence at room temperature under continuous operation and set the scene for subsequent studies, many of which exploited other methods of nanocrystal synthesis.

9.6.4
Waveguides and All-Optical Amplifiers

The presence of excess silicon increases the refractive index of silicon dioxide and can therefore be used to define a waveguide structure [55]. The near-Gaussian depth distribution produced by silicon implantation is well suited to such applications as it naturally forms an asymmetric slab waveguide that confines light at a known depth beneath the surface. In-plane confinement can also be achieved by appropriate masking or postimplant etching of the implanted region. Waveguides confined in one and two dimensions have been exploited to study linear and nonlinear optical effects, including self-guiding of light emitted by nanocrystals, photoinduced absorption, and optical gain [56].

The ability to codope the waveguides with other optically active impurities has also been exploited to extend the studies to other systems. Erbium-doped structures have been of particular interest in this regard due to their potential for all-optical amplification at wavelengths of interest for telecommunications and the fact that silicon nanocrystals act as an effective sensitizer for erbium excitation [57–60].

9.6.5
Nonvolatile Memory

Many nonvolatile flash memories exploit the fact that the switching response of a transistor can be modified by trapping or removing charge from a floating gate that is located between the substrate and the regular gate. However, the use of a continuous floating gate leads to its complete discharge in the event of dielectric breakdown, thereby causing device failure. This limitation can be overcome by storing the charge in discrete traps so that only a small fraction of the charge is lost in the event of breakdown. Silicon nanocrystals are well suited to this task and the first nanocrystal-based memory was demonstrated by Hori *et al.* [29] in 1992. Significantly, this device employed Si nanocrystals (NCs) produced by ion-beam synthesis. The device showed a significant improvement over conventional memories, including a relatively low operation voltage, high endurance, fast write/erase speeds, and better charge retention. Ion beam synthesis continues to play an important role in developing and understanding nanocrystal synthesis for such applications.

9.6.6
Microdisk Resonators and Lasers

Novel erbium-doped SiO_2 microdisk resonators have been fabricated on silicon using conventional planar processing technology and MeV Er-ion implantation. These resonators have optical quality factors of up to 6×10^7 and exhibit a gradual transition from spontaneous to stimulated emission at 1550 nm when resonantly pumped at 1450 nm [61]. Similar structures containing silicon nanocrystals and erbium have also been fabricated in both SiO_2 and Si_3N_4 using ion implantation. Significantly, these devices have the potential to generate coherent emission at 1550 nm under white light excitation due to the broadband absorption and sensitizing properties of the silicon nanocrystals [62].

9.6.7
Photonic Crystal Structures

The ability to ion-implant prepatterned materials raises the possibility of creating regular arrays of nanocrystals, including 3D structures. For example, Si nanocrystals have been formed within a regular 3D array of silica spheres (artificial opal) by ion implantation and annealing. For the implantation conditions employed, nanocrystals were formed only in the top-three sphere layers, but the resulting nanocrystal luminescence showed clear stop bands associated with the regular photonic crystal structure [63]. Time-resolved measurements show a rich distribution of decay rates, containing both shorter and longer decay components compared to the ordinary stretched exponential decay of Si nanocrystals, reflecting changes in the spontaneous emission rate of Si nanocrystals due to variations in the local density of states of real opal-containing defects.

9.7
Summary

Ion implantation has been shown to be an effective and flexible technique for the synthesis of silicon nanocrystals. In its simplest form, this involves implanting a suitable host material with excess silicon and heating it to form nanometer-scale precipitates. In this case, the spatial distribution of the nanocrystals can be controlled by a suitable choice of the implantation parameters, and their size distribution can be tailored by appropriate choice of the excess silicon concentration and the annealing time and temperature. In its more general form, ion irradiation can also be used to modify materials through processes such as recoil implantation and intermixing, some of which offer unique approaches to nanocrystal synthesis.

Ion implantation has several advantages over alternative techniques, including its ability to implant any species into any host material independent of thermodynamic constraints, its ability to implant preformed substrate materials and structures, its ability to independently control the concentration and spatial distribution of the implanted species, and its ability to implant into substrates held at any temperature. Its disadvantages include the fact that it is relatively expensive, that it creates radiation damage in the substrate material, and that the depth distribution of the nanocrystals is limited to less than a micrometer for most practical purposes. Within this framework, ion implantation offers a valuable tool for nanocrystal synthesis and for exploring their structure, properties, and applications, even in cases where it may not ultimately be the technique of choice for mass production.

References

1 Rimini, E. (1995) *Ion Implantation: Basics to Device Fabrication*, Springer.
2 Budai, J.D., White, C.W., Withrow, S.P. et al. (1997) *Nature*, **390**, 384–386.
3 Meldrum, A., Haglund, R.F., Boatner, L.A., and White, C.W. (2001) *Adv. Mater.*, **13**, 1431.
4 Dhara, S. (2007) *Crit. Rev. Solid State Mater. Sci.*, **32**, 1–50.
5 Nastasi, M. and Mayer, J.W. (1991) *Mater. Sci. Rep.*, **6**, 1–51.
6 Nastasi, M., Mayer, J.W., and Hirvonen, J.K. (1996) *Ion-Solid Interactions: Fundamentals and Applications*, Cambridge University Press, New York.
7 Williams, J.S. (1986) *Rep. Prog. Phys.*, **49**, 491–587.
8 Williams, J.S. and Poate, J.M. (eds) (1984) *Ion Implantation and Beam Processing*, Academic Press, Sydney.
9 Sigmund, P. (1983) *Phys. Scr.*, **28**, 257–267.

10 Ziegler, J.F., Biersack, J.P., and Ziegler, M.D. (2008) *SRIM: The Stopping and Range of Ions in Matter*, Lulu.
11 Moeller, W., Eckstein, W., and Biersack, J.P. (1988) *Comput. Phys. Commun.*, **51**, 355–368.
12 Nakama, Y., Nagamachi, S., Ohta, J., and Nunoshita, M. (2008) *Appl. Phys. Express*, **1**, 3.
13 Meldrum, A., Hryciw, A., Buchanan, K.S. et al. (2005) *Opt. Mater.*, **27**, 812–817.
14 Sychugov, I., Galeckas, A., Elfstrom, N. et al. (2006) *Appl. Phys. Lett.*, **89**, 3.
15 Porter, D.A. and Easterling, K.E. (1992) *Phase Transformations in Metals and Alloys*, CRC Press.
16 Baldan, A. (2002) *J. Mater. Sci.*, **37**, 2171–2202.
17 Espiau, R., de Lamaestre, R.E., and Bernas, H. (2006) *Phys. Rev. B*, **73**, 9.
18 Heinig, K.H., Muller, T., Schmidt, B. et al. (2003) *Appl. Phys. A*, **77**, 17–25.

19 Schamm, S., Bonafos, C., Coffin, H. *et al.* (2008) *Ultramicroscopy*, **108**, 346–357.

20 Bonafos, C., Colombeau, B., Altibelli, A. *et al.* (2001) *Nucl. Instr. Meth. B*, **178**, 17–24.

21 Lopez, M., Garrido, B., Bonafos, C. *et al.* (2001) *Nucl. Instr. Meth. B*, **178**, 89–92.

22 Muller, T., Heinig, K.H., and Moller, W. (2003) *Mater. Sci. Eng. B*, **101**, 49–54.

23 Wang, Y.Q., Smirani, R., and Ross, G.G. (2006) *J. Cryst. Growth*, **294**, 486–489.

24 Iacona, F., Bongiorno, C., Spinella, C. *et al.* (2004) *J. Appl. Phys.*, **95**, 3723–3732.

25 Muller, T., Heinig, K.H., and Moller, W. (2002) *Appl. Phys. Lett.*, **81**, 3049–3051.

26 Borodin, V.A., Heinig, K.H., and Reiss, S. (1997) *Phys. Rev. B*, **56**, 5332–5344.

27 Normand, P., Dimitrakis, P., Kapetanakis, E. *et al.* (2004) *Microelectron. Eng.*, **73–74**, 730–735.

28 Bonafos, C., Carrada, M., Cherkashin, N. *et al.* (2004) *J. Appl. Phys.*, **95**, 5696–5702.

29 Hori, T., Ohzone, T., Odark, Y., and Hirase, J. (1992) *IEEE IEDM Tech. Digest*, 469–472.

30 Strobel, M., Heinig, K.H., Moller, W. *et al.* (1999) *Nucl. Instr. Meth. B*, **147**, 343–349.

31 Sias, U.S., Behar, M., Boudinov, H., and Moreira, E.C. (2007) *J. Appl. Phys.*, **102**, 9.

32 Beyer, V. and von Borany, J. (2008) *Phys. Rev. B*, **77**, 13.

33 Heinig, K.H., Schmidt, B., Markwitz, A. *et al.* (1999) *Nucl. Instr. Meth. B*, **148**, 969–974.

34 Wilkinson, A.R. and Elliman, R.G. (2004) *J. Appl. Phys.*, **96**, 4018–4020.

35 Brongersma, M.L., Polman, A., Min, K.S. *et al.* (1998) *Appl. Phys. Lett.*, **72**, 2577–2579.

36 Devine, R.A.B. (1990) *Nucl. Instr. Meth. B*, **46**, 244–251.

37 Guha, S. (1998) *J. Appl. Phys.*, **84**, 5210–5217.

38 Song, H.Z. and Bao, X.M. (1997) *Phys. Rev. B*, **55**, 6988–6993.

39 Wilkinson, A.R. and Elliman, R.G. (2003) *Phys. Rev. B*, **68**, 8.

40 Pacifici, D., Moreira, E.C., Franzo, G. *et al.* (2002) *Phys. Rev. B*, **65**, 13.

41 Ridgway, M.C., Azevedo, G.D., Elliman, R.G. *et al.* (2005) *Phys. Rev. B*, **71**, 6.

42 Rontzsch, L., Heinig, K.H., Schmidt, B., and Mucklich, A. (2006) *Nucl. Instr. Meth. B*, **242**, 149–151.

43 Mohanty, T., Pradhan, A., Gupta, S., and Kanjilal, D. (2004) *Nanotechnology*, **15**, 1620–1624.

44 Toulemonde, M., Trautmann, C., Balanzat, E. *et al.* (2004) *Nucl. Instr. Meth. B*, **216**, 1–8.

45 Rebohle, L., Gebel, T., Frob, H. *et al.* (2001) *Appl. Surf. Sci.*, **184**, 156–160.

46 Wainstein, D., Kovalev, A., Tetelbaum, D. *et al.* (2008) *Surf. Interf. Anal.*, **40**, 571–574.

47 Zhu, J.G., White, C.W., Budai, J.D. *et al.* (1995) *J. Appl. Phys.*, **78**, 4386–4389.

48 Tchebotareva, A.L., de Dood, M.J.A., Biteen, J.S. *et al.* (2005) *J. Lumines.*, **114**, 137–144.

49 Wojdak, M., Klik, M., Forcales, M. *et al.* (2004) *Phys. Rev. B*, **69**, 4.

50 Polman, A. and van Veggel, F. (2004) *J. Opt. Soc. Am. B*, **21**, 871–892.

51 Priolo, F., Franzo, G., Iacona, F. *et al.* (2001) *Mater. Sci. Eng. B*, **81**, 9–15.

52 Choi, S., Yang, H., Chang, M. *et al.* (2005) *Appl. Phys. Lett.*, **86**, 3.

53 Lee, H.R., Choi, S., Cho, K., and Kim, S. (2007) *Thin Solid Films*, **516**, 412–416.

54 Lalic, N. and Linnros, J. (1999) *J. Lumin.*, **80**, 263–267.

55 Janda, P., Valenta, J., Ostatnicky, T. *et al.* (2006) *J. Lumin.*, **121**, 267–273.

56 Pelant, I., Tomasiunas, R., Sirutkaitis, V. *et al.* (2008) *J. Phys. D: Appl. Phys.*, **41**, 5.

57 Minissale, S., Gregorkiewicz, T., Forcales, M., and Elliman, R.G. (2006) *Appl. Phys. Lett.*, **89**, 3.

58 Iacona, F., Irrera, A., Franzo, G. *et al.* (2006) *IEEE J. Sel. Top. Quant. Electron.*, **12**, 1596–1606.

59 Daldosso, N., Navarro-Urrios, D., Melchiorri, M. *et al.* (2006) *IEEE J. Sel. Top. Quant. Electron.*, **12**, 1607–1617.

60 Kik, P.G. and Polman, A. (2002) *J. Appl. Phys.*, **91**, 534–536.

61 Kippenberg, T.J., Kalkman, J., Polman, A., and Vahala, K.J. (2006) *Phys. Rev. A*, **74**, 4.

62 Gardner, D.S. and Brongersma, M.L. (2005) *Opt. Mater.*, **27**, 804–811.

63 Janda, P., Valenta, J., Rehspringer, J.L. *et al.* (2007) *J. Phys. D*, **40**, 5847–5853.

64 Wang, Y.Q., Smirani, R., and Ross, G.G. (2004) *Nanotechnology*, **15**, 1554–1560.

65 Schmidt, B., Heinig, K.H., Rontzsch, L. *et al.* (2006) *Nucl. Instr. Meth. B*, **242**, 146–148.

66 Rontzsch, L., Heinig, K.H., and Schmidt, B. (2004) *Mater. Sci. Semicond. Process*, **7**, 357–362.

10
Structural and Optical Properties of Silicon Nanocrystals

Fabio Iacona, Giorgia Franzò, Alessia Irrera, Simona Boninelli, and Francesco Priolo

10.1
Introduction

During the past decade, most of the efforts made by the scientific community working in the field of silicon photonics were devoted to the development of a Si-based light source to be used for monolithic integration of optical and electrical functions on a single Si chip. Indeed, electrical interconnections based on metal lines impose today the most important limitation on the performance of Si-based microelectronic devices [1]. The parasitic capacities generated at the metal/insulating/metal capacitors present in complex multilevel metallization schemes constitute the main contribution to the delay in the signal propagation. The intrinsic resistivity of the metal lines, as well as the contact resistance at the various metal/metal interfaces, constitutes other relevant delay sources. A significant reduction in delay times has been achieved by replacing the well-established metallization schemes based on Al and SiO_2 with new materials, such as Cu-based metal films and low dielectric constant insulating layers, but, as soon as the size of the devices will further reduce, the delay due to metal interconnections will again represent an unacceptable bottleneck for device performance. An almost definitive solution to this problem could be the use of optical interconnections for the transfer of information inside a chip [1–4]. The most stringent requirement for the development of this strategy is the availability of Si-based or Si-compatible materials able to generate, guide, amplify, switch, modulate, and detect light. However, because Si is intrinsically unable to efficiently emit light due to its indirect bandgap, it is evident that the main limitation to the above-described strategy is the lack of an efficient Si-based light source. In this respect, the Si Raman laser recently discovered [5, 6], although an outstanding event, has a severe limitation since it cannot be electrically pumped.

Si nanocrystals (Si-ncs) are the most important candidates for the realization of efficient electrically pumped optical sources to be employed in silicon photonics. Light emission from Si nanostructures became a strategic field of research after strong visible photoluminescence (PL) at room temperature from porous Si was discovered by Canham in 1990 [7]. Most of the efforts of scientific

Silicon Nanocrystals: Fundamentals, Synthesis and Applications. Edited by Lorenzo Pavesi and Rasit Turan
Copyright © 2010 WILEY-VCH Verlag GmbH & Co. KGaA, Weinheim
ISBN: 978-3-527-32160-5

community progressively moved from porous Si, the bright emission of which was counterbalanced by its brittleness and optical instability, to Si-ncs embedded in SiO_2 films. Si-ncs are mechanically and optically very stable and, moreover, their synthesis and processing is fully compatible with the technology used for Si-based microelectronics. The deeper comprehension of the complex phenomena ruling the process of light emission from Si-ncs has led to a progressive optimization of the properties of this material [8–17]. In this respect, the most striking result was the discovery that Si-ncs exhibit optical gain [18–21]; in spite of this major breakthrough, Si-ncs have been considered for a long time not useful for developing efficient Si-based light-emitting devices because the embedding SiO_2 matrix is a very good insulating material and carrier injection through this layer could be very difficult. On the other hand, it is generally recognized that the peculiar properties of Si-ncs are closely linked to the very stable passivation ensured by the surrounding SiO_2, mainly by eliminating dangling bonds through the formation of Si=O bonds [22]; furthermore, SiO_2 also constitutes a low defect density medium in which the efficiency of non-radiative recombination processes is strongly reduced.

However, in spite of this drawback, several prototypes of light-emitting devices based on Si nanostructures, with more and more interesting performances, have been proposed. The electroluminescence (EL) of small Si clusters embedded in SiO_2 was first reported in the pioneering work of DiMaria *et al.* [23]. Thereafter, several devices based on porous Si were proposed, the most relevant one was probably the integrated bipolar transistor/porous Si-LED structure developed by Hirschman *et al.* [24]. Subsequent examples of Si-nc-based devices exhibiting a room-temperature EL in the visible/near-infrared region at reasonably low voltages overcome the difficulty of carrier injection in SiO_2 by confining Si-ncs to ultrathin layers [25–28] or by exploiting the tuning of Si-nc size and density that can be obtained in materials synthesized by thermal annealing of SiO_x layers [29]. The demonstration that the cross section for the electrical excitation of Si-ncs is about two orders of magnitude higher than the one observed under optical pumping has strongly supported the idea that these can also provide important perspectives for practical applications [30]. Recently, to improve the carrier injection efficiency in Si-ncs, a new device structure based on a field-effect EL mechanism was proposed [31]. This new approach is expected both to reduce nonradiative processes limiting EL efficiency, such as Auger recombination, and to improve the device reliability, owing to a reduced oxide wear-out with respect to the impact excitation by hot carriers.

A different approach used to efficiently produce photons from Si is based on the introduction of light-emitting impurities, such as Er. The 1.54 μm luminescence from Er-doped materials is due to an internal 4f shell transition of the rare earth ions. This wavelength is strategic in the telecommunication technology since it matches the window of maximum transmission for the silica optical fibers. Si nanoclusters act as efficient sensitizers for rare earth PL [32–36]; the energy transfer process from excited nanoclusters to the neighboring Er ions increases by about a factor of 100 the PL efficiency of the system [33, 35]. Efficient, stable, and room-temperature-operating

electroluminescent devices based on Er-doped Si nanoclusters have been proposed [37, 38].

The growing relevance of Si-ncs in the field of photonics has stimulated the search
for new approaches for the synthesis of materials with well-defined structural and
optical properties. At present, the most established process for the synthesis of Si-ncs
embedded in a SiO$_2$ matrix is based on the thermal annealing of a substoichiometric
silicon oxide film, which allows the thermodynamics-driven phase separation
between Si and SiO$_2$. Substoichiometric silicon oxide films are commonly indicated
with different, but essentially equivalent, acronyms, such as SRO (silicon-rich oxide),
SRSO (silicon-rich silicon oxide), SIPOS (semi-insulating polycrystalline silicon) or,
more simply and rigorously, with their chemical formula SiO$_x$ (with $0 < x < 2$).
Typically, a SiO$_x$ film can be obtained by chemical vapor deposition (CVD) or plasma-
enhanced chemical vapor deposition (PECVD) [12, 16, 21], Si ion implantation
in silica [9–11, 13–15, 18, 20], electron beam evaporation [19, 39], and magnetron
sputtering [40, 41]. Among the different techniques that can be used for SiO$_x$
synthesis, PECVD presents several advantages:

1) *Low-temperature processing.* PECVD usually works at a temperature of 400 °C or
   lower.
2) *Full compatibility with silicon standard technology.* Indeed, PECVD is widely used
   for thin-film deposition in the semiconductor industry. This makes the possibility of integrating layers containing Si-ncs inside a microelectronic device a
   reality in which electrical and optical functions coexist.
3) *High film quality.* Compared to sputter deposition or electron beam evaporation,
   where a certain degree of porosity is a relatively common problem, PECVD
   produces high-quality films.
4) The peculiar optical properties of Si-ncs strongly depend on structural properties
   (size, shape, volume density, crystallinity, defect density, surface passivation,
   etc.), which in turn are determined by excess concentration of Si in the SiO$_x$ film
   used as a precursor. PECVD allows a tight control over the film composition by
   simply changing the gas flow ratio of the precursors.

In this chapter, we will discuss the synthesis by PECVD and the structural and
optical properties of Si-ncs. Several analytical techniques will be used to investigate the
steps of the temperature-induced evolution of SiO$_x$ films leading to the formation of
Si-ncs. The PL properties of Si-ncs will be analyzed and the mechanisms of excitation
and de-excitation discussed. A correlation between structural and optical properties
will also be presented. Finally, Er-doping of Si nanoclusters will be proposed as an
interesting way to obtain a very efficient photon emission at 1.54 μm. The relevance
of the above results for the development of silicon photonics will be underlined.

10.2
Synthesis, Structure, and Thermal Evolution of SiO$_x$ Films

SiO$_x$ thin films discussed in this chapter were prepared by using a parallel plate PECVD
system, consisting of an ultrahigh vacuum chamber (base pressure 1×10^{-9} Torr)

and an RF generator (13.56 MHz), connected through a matching network to the top electrode of the reactor; the bottom electrode is grounded and also acts as sample holder. All deposition processes were performed by using a 50 W input power. The substrates, consisting of 5- or 6-in. (100) Czochralski Si wafers, were heated at 300 °C during the deposition. The source gases used were high-purity (99.99% or higher) SiH_4 and N_2O; the N_2O/SiH_4 flow ratio γ varied between 5 and 50, while the total gas flow rate was kept constant, at a value of about 140 sccm. The total pressure during deposition processes was kept constant at approximately 6×10^{-2} Torr. Typical deposition rates range from 15 to 20 nm/min.

The chemical composition of silicon oxide films deposited by using different values of γ has been determined by using the Rutherford backscattering spectrometry (RBS) technique. Figure 10.1 shows some representative RBS spectra, obtained in random configuration by using 2 MeV He$^+$ ions, and evidence marked differences in the Si content of the films. Figure 10.2a demonstrates Si concentration as a function of the γ value, as obtained from RBS data. The figure shows that a very large excess of N_2O (γ values ranging from 15 to 50) leads to the growth of a stoichiometric SiO_2 film, while for γ values lower than 15, films are substoichiometric. RBS data also demonstrate the presence of a clearly detectable (around 10 at.%) N concentration, due to the use of N_2O as gaseous precursor for deposition; N concentration is higher in substoichiometric films, even if a marked dependence on γ value cannot be assessed.

The refractive index (n) of the oxide films, measured by ellipsometry at 632.8 nm, is shown in Figure 10.2b as a function of γ. The trend closely follows that one of the Si concentration. Values of about 1.47 are found for SiO_2 films, while n increases up to about 1.8 in SiO_x films as a direct consequence of their high Si content.

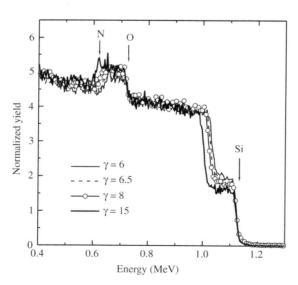

Figure 10.1 RBS spectra, obtained in random configuration, for silicon oxide films prepared by using γ values ranging from 6 to 15.

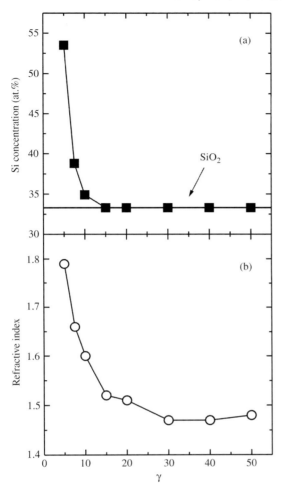

Figure 10.2 (a) Si concentration, derived from RBS measurements, and (b) refractive index (measured at $\lambda = 632.8$ nm) as a function of γ for silicon oxide films prepared by PECVD. The lines are drawn to guide the eye.

Detailed information about the structure of the SiO$_x$ films can be obtained by using the Fourier transform infrared (FTIR) spectroscopy. Figure 10.3 shows the 1400–600 cm^{-1} region of the FTIR spectra relative to as-deposited and annealed SiO$_x$ films containing 37 at.% of Si. The stretching vibration frequency of the as-deposited film can be found at about 1028 cm^{-1}, while a typical value for SiO$_2$ films is about 1075 cm^{-1} [42]. This behavior is consistent with literature data reporting an almost linear dependence of the Si–O–Si stretching frequency on the composition of SiO$_x$ films [42], and it is due to the induction effect resulting from replacing highly electronegative O atoms with Si atoms in the network of the substoichiometric film. The FTIR spectra confirm the presence of N in the oxide films; the region around

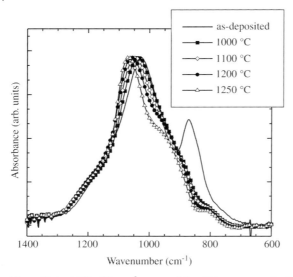

Figure 10.3 1400–600 cm^{-1} region of the FTIR spectrum of a SiO$_x$ film containing 37 at.% of Si, as-deposited, and after annealing processes performed at temperatures ranging from 1000 to 1250 °C.

850 cm^{-1} contains not only the absorption peak due to the stretching vibration of the Si—N bonds but also the peaks related to the presence of Si—H bonds [43]. Indeed, the FTIR spectrum is also characterized by another relevant absorption peak at about 2200 cm^{-1} due to the stretching vibration of Si—H bonds [43], which allows to estimate H concentrations as high as 1×10^{22} at./cm^3 [44].

Figure 10.3 also shows the FTIR spectra recorded after annealing the SiO$_x$ films at temperatures ranging between 1000 and 1250 °C for 1 h in N$_2$ atmosphere. The data demonstrate the increase in the frequency of the Si—O—Si stretching mode, from the initial value of 1028 cm^{-1} up to a value of 1068 cm^{-1} at 1250 °C. The interpretation of the low-wavenumber region of the spectra is complicated by the rearrangement N atoms undergo (RBS data demonstrate that the total N concentration does not appreciably change upon thermal annealing), mainly leading to the formation of a oxynitride phase, whose absorption is intermediate between that of the pure oxide and nitride phases since it is due to Si atoms simultaneously bonded to O and N atoms [45]. High-temperature annealing processes are also very effective in completely eliminating H from the films. The above-described behavior qualitatively holds for all the studied SiO$_x$ films.

The structure of the SiO$_x$ films, by neglecting their N and H contents, can be described as a simple Si/SiO$_2$ mixture (random mixture model, RMM [46]) or, alternatively, as a mixture of all Si oxidation states in the form of Si—Si$_x$O$_{4-x}$ tetrahedra ($0 \leq x \leq 4$), in which O atoms are replaced with Si atoms in order to respect the overall film stoichiometry (random bonding model, RBM [47]). The RMM is based on the assumption that the growth of a SiO$_x$ film is mainly ruled by chemistry: only the more stable phases Si and SiO$_2$ are formed, while the existence of the Si intermediate oxidation states is neglected; on the contrary, the RBM is a purely statistical model in which the Si—Si$_x$O$_{4-x}$ tetrahedra distribution depends only on the frequency of arrival of O and Si atoms on the substrate surface during deposition and therefore on their relative concentrations in the gaseous phase. Both "pure" models are never experimentally verified since usually the structure of SiO$_x$ films is intermediate between these two extreme situations. Deposition temperature plays a fundamental role in determining if the structure of a film is more similar to the RMM or to the RBM. Indeed, films prepared by thermally induced CVD processes have a structure more similar to the one predicted by the RMM with respect to films prepared at low temperature by PECVD [48] due to a phenomenon of temperature-induced ordering of the film, from a statistics-driven structure to a chemistry-driven one.

This effect works even if the energy needed for the system to reorganize itself according to chemical laws is supplied after the deposition. The Si—O—Si stretching peaks in Figure 10.3 demonstrate the occurrence of a temperature-induced transition from a system exhibiting a typical stretching frequency of a substoichiometric oxide [42] to a system with a stretching frequency more similar to that of stoichiometric SiO$_2$; this effect is due to the evolution of the initial structure toward a biphasic mixture in which Si coexists with a SiO$_x$ phase the composition of which is closer to the stoichiometry than that present in the as-deposited film. We can schematically illustrate the heat-induced disproportionation reaction the partially oxidized Si—Si$_x$O$_{4-x}$ tetrahedra undergo as

$$\text{Si}-\text{Si}_x\text{O}_{4-x} \rightarrow \frac{x}{4}\text{Si}-\text{Si}_4 + \left(1-\frac{x}{4}\right)\text{Si}-\text{O}_4 \quad (\text{with } x = 1, 2, 3), \tag{10.1}$$

where the Si—Si$_4$ tetrahedra constitute the building blocks of Si-ncs.

The temperature-induced decomposition of partially oxidized Si—Si$_x$O$_{4-x}$ tetrahedra into the more stable Si—Si$_4$ and Si—O$_4$ in SiO$_x$ films annealed at high temperatures has also been demonstrated on the basis of X-ray photoelectron spectroscopy data [49].

10.3
A Deeper Insight into the Thermal Evolution of SiO$_x$ Films

Figure 10.4 reports the planview energy-filtered transmission electron microscopy (EFTEM) images obtained from SiO$_x$ samples containing 46 at.% of Si as-deposited and annealed at 900, 1000, 1100, 1150, and 1250 °C for 1 h in N$_2$ environment. The principles of the EFTEM technique and the experimental procedure followed to obtain the images are both described in detail in Refs [50, 51]; we recall here only that the images were obtained by using an energy window centered on the Si plasmon loss at about 16 eV; under these conditions, the technique allows to quantitatively detect Si nanoclusters independent of the presence of a crystalline structure and its orientation, and therefore it represents a clear advancement with respect to the more traditional dark field transmission electron microscopy (DFTEM) technique, which is able to detect only a small fraction of the Si-nc population [50, 51].

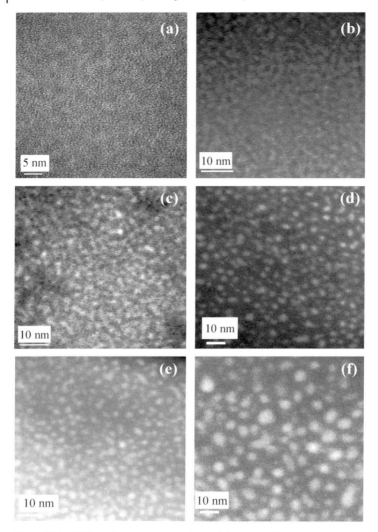

Figure 10.4 EFTEM planview images obtained from SiO_x films containing 46 at.% of Si (a) as-deposited, and annealed at (b) 900 °C, (c) 1000 °C, (d) 1100 °C, (e) 1150 °C, and (f) 1250 °C. The bright zones are associated with the presence of Si clusters.

Figure 10.4a shows a uniform gray background without any appreciable intensity contrast, suggesting the absence of phase separation effect between Si and SiO_2 in the as-deposited SiO_x samples. On the other hand, in the EFTEM image reported in Figure 10.4b, referring to a SiO_x sample annealed at 900 °C, some brighter zones can be clearly detected, indicating the occurrence of the first stages of the Si clustering, even if the image seems to suggest the formation of a Si network rather than of isolated clusters. The occurrence of phase separation between Si and SiO_2 becomes much more evident when the annealing temperature is increased, and well-defined Si nanoclusters embedded in the oxide matrix are distinctly visible in samples annealed at 1000 and 1100 °C (see Figure 10.4c and d). Further rises in the annealing temperature (1150 °C, see Figure 10.4e, and 1250 °C, see Figure 10.4f) lead to an increase in the nanocluster size. We underline that EFTEM images are elemental maps in which the detected Si nanoclusters can be either crystalline or amorphous. This is the reason why we use the more general term "nanocluster" for the discussion of EFTEM data.

Quantitative information on the SiO_x samples can be obtained by measuring the size of the Si nanoclusters detected in the planview EFTEM images of samples annealed at a temperature of 1000 °C or higher. The data are reported in Figure 10.5a

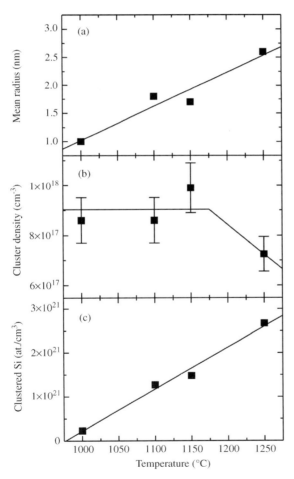

Figure 10.5 (a) Si nanocluster mean radius, (b) density of Si nanoclusters, (c) concentration of clustered Si atoms as a function of the annealing temperature. All data are obtained from the EFTEM images shown in Figure 10.4. The lines are drawn to guide the eye.

as a function of the annealing temperature; Si nanocluster mean radius increases by increasing the annealing temperature from 1.0 (at 1000 °C) to 2.6 nm (at 1250 °C). A similar trend has been observed in experiments in which, for a fixed annealing temperature, the Si concentration in the SiO_x films is increased [16].

Another important information that can be derived from planview EFTEM images is the volume density of Si nanoclusters at different temperatures. EFTEM is able to detect all the clusters present in a sample provided that the analysis is performed in regions thin enough to avoid shadowing effects between clusters placed at different depths and that the analyzed volume is known [50, 51]. The Si nanocluster density is shown in Figure 10.5b as a function of the annealing temperature; values are almost constant in the 1000–1150 °C range (around 9×10^{17} cm^{-3}), while a decrease (about 7×10^{17} cm^{-3}) is observed at 1250 °C. This result, coupled with the continuous increase in the cluster size with annealing temperature reported in Figure 10.5a, suggests that nanocluster growth is not only due to the inclusion of Si atoms diffusing from the oxide matrix but also due to Ostwald ripening effects, leading to the disappearance of small clusters.

From the above-reported data on the nanocluster density and mean radius, the number of clustered Si atoms per unit volume at different temperatures can be calculated. The data are reported in Figure 10.5c and demonstrate that the number of Si atoms forming detectable clusters remarkably increases with increasing annealing temperature. It is worth noting that, also at 1250 °C, a relevant fraction of excess Si atoms not detectable by EFTEM exists. Indeed, the value of 3×10^{21} cm^{-3} measured at this temperature has to be compared with an expected value for the excess Si content, deduced from RBS measurements, of about 1×10^{22} cm^{-3}. Very recent evidence supports the idea that the missing Si atoms remain dissolved in the SiO_x matrix; indeed, the quantitative analysis of the shape of the electron energy loss spectra of annealed SiO_x films, which is sensitive also to the presence of clusters with a size below the minimum size for TEM detection (about 0.7 nm), gives a very similar estimation for the clustered Si fraction [52]. The analysis of the energy loss spectra of SiO_x films has also demonstrated that the same concentration of clustered Si detected in annealed samples is already present in the as-deposited one [52]. This unexpected result is not in contrast with the EFTEM data shown above since the clusters present in as-deposited samples are smaller than the minimum size that can be detected by TEM. The thermal evolution of these clusters leads to their growth without substantially changing the fraction of clustered Si, which remains about 30%. In agreement with this picture, the process described by Eq. (10.1) and supported by the FTIR data in Figure 10.3 has to be considered still valid, but mainly occurring at the oxide/cluster interface. When clusters become larger, their surface/volume ratio decreases and, in turn, the concentration of partially oxidized tetrahedra, which constitute the interface between clusters and the surrounding oxide matrix, also decreases.

To gain a deeper knowledge of the structural properties of SiO_x films, we have also employed the DFTEM technique, which is sensitive to the presence of crystalline planes, and it is therefore able to map the system for the presence of Si-ncs. It was not possible to obtain any DFTEM image for samples as-deposited and annealed

Figure 10.6 DFTEM planview image obtained from a SiO$_x$ film containing 46 at.% of Si annealed at 1250 °C. The inset shows the electron diffraction pattern of the sample.

at 900 and 1000 °C. This evidence demonstrates that all the clusters shown in the EFTEM images in Figure 10.4b and c are amorphous. On the other hand, samples annealed at 1100 °C or at a higher temperature exhibit the presence of a diffraction pattern mainly consisting of three distinct rings corresponding to the (111), (220), and (311) planes of crystalline Si, as shown in the inset of Figure 10.6, demonstrating that a significant fraction of the clusters shown in Figure 10.4d–f are crystalline.

A typical planview DFTEM image obtained for a SiO$_x$ sample annealed at 1250 °C is shown in Figure 10.6; the image has been obtained by selecting a small portion (about 10%) of the diffraction ring due to the (111) Si planes and Si-ncs appear as bright spots on a dark background. The comparison between this image and the one obtained on the same sample by EFTEM (shown in Figure 10.4f) clearly demonstrates that EFTEM allows a much more complete characterization of SiO$_x$ samples.

For samples in which the amorphous and crystalline phases coexist, EFTEM and DFTEM data allow to estimate the crystalline fraction present at a given temperature. For this purpose, the ratio between the number of nanocrystals (as detected by DFTEM) and the total number of clusters (as detected by EFTEM) has been evaluated at different temperatures. The cluster count has been performed in the same sample region for both techniques; furthermore, to avoid shadowing effects, we selected for the analysis very thin regions. The data shown in Figure 10.7a demonstrate the continuous increase in the crystalline fraction by increasing the temperature. Under the reasonable hypothesis that all clusters are crystalline at 1250 °C [50, 51], the estimation of the nanocrystal fraction at lower temperatures leads to values of about 30% at 1100 °C (the temperature at which crystallization starts) and 60% at 1150 °C,

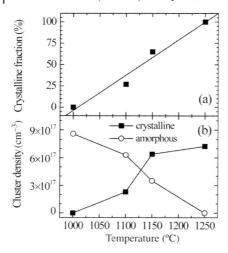

Figure 10.7 (a) Crystalline fraction, and (b) density of amorphous and crystalline nanoclusters, as a function of the annealing temperature. The data are obtained from EFTEM and DFTEM analysis. The lines are drawn to guide the eye.

clearly demonstrating that in this range the temperature plays a role not only by promoting the Si nanocluster growth but also by extensively inducing the amorphous to crystal transition [40].

Finally, we used the data on the Si nanocluster density reported in Figure 10.5b and those on the crystalline fraction in Figure 10.7a to calculate the concentration of Si-ncs and of amorphous clusters as a function of the annealing temperature. The data are reported in Figure 10.7b and demonstrate that the temperature progressively induces the complete transformation of the amorphous clusters ($9 \times 10^{17} \, \mathrm{cm}^{-3}$ at 1000 °C) in Si-ncs ($7 \times 10^{17} \, \mathrm{cm}^{-3}$ at 1250 °C), with the loss of about 20% of the clusters present at 1000 °C, due to the occurrence of Ostwald ripening phenomenon.

The behavior of the annealed SiO_x films described by Figures 10.4–10.7 has to be considered valid for any SiO_x composition, although the clustering and crystallization temperatures we have determined for a Si concentration of 46 at.% (1000 and 1100 °C, respectively) may vary in films with different compositions [16, 49]. On the other hand, very deep structural differences can be found in SiO_x films having the same composition but prepared by different techniques. The most striking example is the almost complete clustering of the Si excess found in SiO_x films prepared by magnetron sputtering [52].

10.4
Room-Temperature PL Properties of Si-ncs

The room-temperature PL properties of Si-ncs synthesized by thermal annealing of SiO_x films were examined by exciting the system with the 488 nm line of an Ar^+ laser. The pump power was fixed at 50 mW and the laser beam was chopped through an acousto-optic modulator at a frequency of 55 Hz. The PL signal was analyzed by a single-grating monochromator and detected by a photomultiplier tube or by a liquid N_2-cooled Ge detector. Spectra were recorded with a lock-in amplifier using the acousto-optic modulator frequency as a reference. All spectra were corrected for the spectral system response. Light emission from as-deposited samples was generally found in all cases to consist of weak signals, with a typical wavelength of 600–650 nm. On the other hand, SiO_x samples exhibit a strong room-temperature PL (several orders of magnitude higher than the signals coming from the as-deposited samples) in the 650–950 nm range after an annealing process at a temperature of 1000 °C or higher. The normalized PL spectra relative to a SiO_x film with a Si concentration of 37 at.% are shown in Figure 10.8a for samples annealed at 1100, 1200, and 1300 °C. The figure shows that the wavelength corresponding to the maximum intensity of the

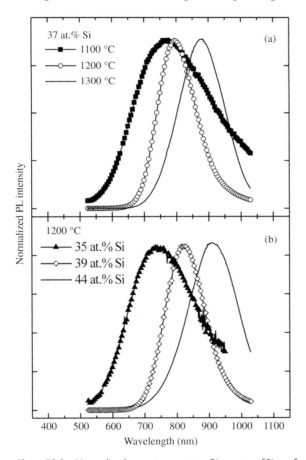

Figure 10.8 Normalized room-temperature PL spectra of Si-ncs formed by thermal annealing of (a) a SiO_x film with a Si concentration of 37 at.% after thermal annealing processes performed at 1100, 1200, and 1300 °C for 1 h; (b) SiO_x films with Si concentrations of 35, 39, and 44 at.% after a thermal annealing process per-formed at 1200 °C for 1 h.

PL peak (λ_{max}) increases with increasing annealing temperature from 770 nm at 1100 °C to 880 nm at 1300 °C. By keeping the annealing temperature constant, a dependence of λ_{max} on the sample composition can be established; indeed, Figure 10.8b shows the PL spectra of three different SiO_x samples after the same annealing process at 1200 °C. The PL peak is observed at 740 nm for sample with 35 at.% of Si, at 825 nm for sample with 39 at.% of Si, and finally at 910 nm for sample with 44 at.% of Si.

The PL measurements relative to SiO_x films with five different Si concentrations (from 35 to 44 at.%) annealed at five different temperatures (from 1000 to 1300 °C) are summarized in Figure 10.9a, where λ_{max} is plotted as a function of the annealing temperature. The data demonstrate that λ_{max} exhibits a marked redshift when annealing temperature is increased. The dependence is very strong for the sample with lower Si content, the PL peak of which was detected at about 650 nm at 1100 °C and at about 830 nm at 1300 °C (note that this sample does not show any appreciable PL signal after annealing at 1000 °C). The effect becomes less pronounced when the Si concentration is increased; for SiO_x films with 44 at.% of Si, all PL peaks are found between 900 and 950 nm.

The analysis of Figure 10.9a also demonstrates that the dependence of λ_{max} on the sample composition for a fixed annealing temperature shown in Figure 10.8b is a general effect. Indeed, at all temperatures the PL peak was found to redshift when the Si concentration was increased. The width of the wavelength range decreases with increasing temperature: while at 1100 °C all wavelengths are contained in about 250 nm-wide spectral range (650–900 nm), at higher temperatures this range remarkably narrows to about 170 nm (from 740 to 910 nm) at 1200 °C and 120 nm (from 830 to 950 nm) at 1300 °C.

The dependence of the integrated PL intensity on the annealing temperature has also been analyzed. The data in Figure 10.9b show that the PL intensity spans along two orders of magnitude; all signals increase with increasing annealing temperature from 1000 to 1250 °C, only for the highest Si concentration a slight decrease in the emission is noticed at 1250 °C. At 1300 °C, all PL signals remarkably decrease. It should be noted that the sample with 35 at.% of Si (not shown in Figure 10.9b) shows an almost constant intensity at 1250 and 1300 °C, exhibiting, differently from all the other samples, a very low PL intensity (or no light emission at all) in the 1000–1200 °C range. The figure also demonstrates that in the temperature range 1000–1100 °C, the maximum PL intensity is exhibited by films with the higher Si content; at 1200 °C, the intermediate concentrations are the most efficient, while at a temperature of 1250 °C or higher, the situation is totally reversed since the sample with 37 at.% of Si exhibits the most intense PL peaks.

10.5
Excitation and De-Excitation Properties of Si-ncs

The decay time of the room-temperature PL signal of Si-ncs is of particular interest since it depends on the confinement properties of nanocrystals. Figure 10.10 shows

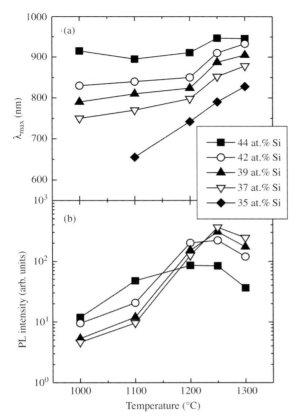

Figure 10.9 (a) Dependence of the wavelength of the PL peaks (λ_{max}) and (b) dependence of the integrated intensity of the PL peaks on the annealing temperature for Si-ncs synthesized from SiO$_x$ films having different Si concentrations. The lines are drawn to guide the eye.

the PL decay time at two detection wavelengths, 700 and 950 nm, for different Si-nc samples after shutting off through an acousto-optic modulator a 10 mW 488 nm laser beam at room temperature. The decay time at 700 nm (Figure 10.10a) increases with decreasing Si content and it is characterized by a stretched exponential shape that becomes more and more similar to a single exponential with decreasing Si content. On the other hand, the 950 nm signal decay time, shown in Figure 10.10b, is characterized by almost single exponentials with the same lifetime for all the Si contents.

Stretched exponential functions are commonly used to describe the decay time of Si-ncs [15]. In a stretched exponential, the decay line shape is given by

$$I(t) = I_0 \exp\left[-\left(\frac{t}{\tau}\right)^{\beta}\right], \tag{10.2}$$

where $I(t)$ and I_0 are the PL intensity as a function of time, and at $t = 0$, τ is the decay time and β is a dispersion factor ≤ 1. The smaller is β the more "stretched" is

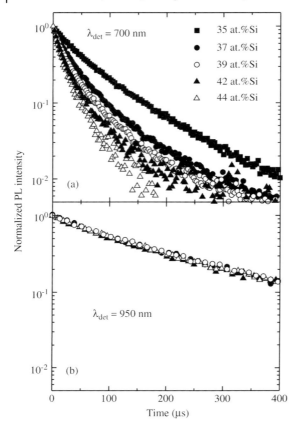

Figure 10.10 Measurements of the time decay of the PL signal at 700 nm (a) and at 950 nm (b) for Si-ncs formed by thermal annealing of SiO$_x$ films with different compositions annealed at 1250 °C for 1 h. Data were taken at room temperature.

the exponential. The results of the fits to the data in Figure 10.10a with Eq. (10.2) indicate that the factor β decreases from 0.85 to 0.63 and τ from 65 to 10 μs on increasing the Si content from 35 to 44 at.%. A decrease in β has been associated with a redistribution of the energy within the sample with a transfer from smaller nanocrystals (with larger gaps) to larger nanocrystals (with smaller gaps) [15]. This picture is consistent with the data in Figure 10.10a, showing smaller β and τ values in Si-rich samples in which the nanocrystal concentration is higher and hence the energy transfer is more probable. It should be stressed that since all measurements are performed at the same wavelength, they reveal the properties of ncs having the same size but embedded in a different environment. The markedly different behavior observed demonstrates that the environment of the ncs plays a quite important role in determining the decay time. The more the Si-ncs are isolated (larger β), the larger is the decay time τ since energy transfer becomes less probable. Moreover, larger Si-ncs (those emitting at 950 nm, see Figure 10.10b) cannot transfer their energy to the surrounding Si-ncs since their energy is not sufficient due to their smaller bandgaps. Therefore, they act as "isolated" Si-ncs in the sense that once excited they will re-emit the energy only radiatively. This explains the identical lifetime of 175 μs for all the studied samples and the almost single exponential behavior.

Another important issue to be investigated is the excitation cross section of Si-ncs and its dependence on the nc density. For this purpose, we studied the rise time of the PL intensity in Si-ncs. The PL intensity is in general given by

$$I \propto \frac{N^*}{\tau_R}, \tag{10.3}$$

where N^* is the concentration of excited ncs and τ_R denotes the radiative lifetime.

The rate equation for Si-nc excitation will be

$$\frac{dN^*}{dt} = \sigma\phi(N - N^*) - \frac{N^*}{\tau}, \tag{10.4}$$

where σ is the excitation cross section, ϕ is the photon flux, N is the total number of Si-ncs, and τ is the decay time, taking into account both radiative and nonradiative processes.

If a pumping laser pulse is turned on at $t = 0$, the PL intensity, according to Eqs. (10.3) and (10.4), will increase with the following law:

$$I(t) = I_0 \left\{ 1 - \exp\left[-\left(\sigma\phi + \frac{1}{\tau} \right) t \right] \right\}, \tag{10.5}$$

where I_0 is the steady-state PL intensity. The rise time τ_{on} will hence follow the relationship:

$$\frac{1}{\tau_{on}} = \sigma\phi + \frac{1}{\tau}. \tag{10.6}$$

A measure of the rise time as a function of photon flux ϕ will therefore give direct information on the excitation cross section.

For example, Figure 10.11a shows the room-temperature PL rise time for Si-ncs synthesized by annealing at 1250 °C for 1 h a SiO$_x$ film with a Si content of 35 at.%. The sample was pumped at 488 nm, with pump powers ranging from 1 to 30 mW, and the signal was detected at 850 nm. As predicted by Eqs. (10.5) and (10.6), the rise time becomes shorter and shorter by increasing pump power. By fitting these rise time curves with Eq. (10.5), the values of τ_{on} at the different pump powers are obtained. The reciprocal of τ_{on} is reported in Figure 10.11b as a function of the photon flux. The data follow a straight line according to Eq. (10.6) with a slope $\sigma \sim 1.8 \times 10^{-16}$ cm^2. The intercept of the fit straight line with the vertical axis gives the lifetime of the Si-ncs in the system at the measured wavelength. The obtained value (100 μs) is in agreement with decay time measurements at 850 nm on the same sample. In this way, we have been able to obtain a direct measurement of the excitation cross section of Si-ncs. We have performed such measurements for several

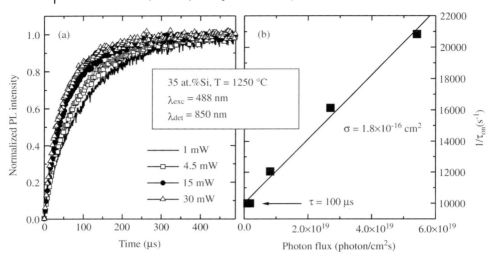

Figure 10.11 (a) Time-resolved PL intensity at 850 nm when switching on the pumping laser at $t = 0$ for Si-ncs formed by thermal annealing of a SiO_x film with a Si concentration of 35 at.% and annealed at 1250 °C. The excitation wavelength was 488 nm. Data were taken at room temperature and at different pump powers and are normalized to the maximum intensity. (b) Reciprocal of the τ_{on} time as obtained from the data in (a) as a function of the pump laser photon flux. The slope gives the cross section for Si-nc excitation.

different samples. Systems in which Si-ncs are almost isolated are characterized by cross section values very similar to that observed in the sample with 35 at.% of Si. A different behavior is observed in Si-nc systems in which a quite substantial energy transfer takes place. For example, for a SiO_x sample with 42 at.% of Si annealed at 1250 °C, the effective excitation cross section for Si-ncs is much higher than in the previous case being about 8×10^{-16} cm^2. This strong increase mirrors the fact that in the presence of energy transfer, the effective excitation cross section of each nanocrystal increases since excitation can occur not only through direct photon absorption but also through energy transfer from a nearby nanocrystal. A single-photon absorption can now excite several Si-ncs since the energy is transferred from one nc to the other. On the other hand, only one nc per incoming photon is excited at a time since energy transfer requires nonradiative de-excitation from the nc. In fact, an increase by a factor of 4 in the effective cross section corresponds to a decrease by a factor of 4 in the lifetime values shown in Figure 10.10a. This is a clear indication that the increase in cross section and decrease in time decay are different aspects of the same physical process.

10.6
Correlation between Structural and Optical Properties of Si-ncs

The mechanism of visible and near-infrared light emission from Si-ncs embedded in a SiO_2 matrix has been the object of an intense scientific debate for several years.

At present, it is commonly accepted that quantum confinement plays an important role in determining the intense room-temperature light emission of Si-ncs and its blueshift when size reduction of the nanostructures occurs [13, 14, 16]. On the other hand, the strong Stokes shift between absorption and emission in Si-ncs suggested that emission could be due to an interfacial radiative center [53, 54]. The quantum confinement picture and the interfacial state model have been reconciled by demonstrating that a Si=O double bond introduces size-dependent levels (both for electrons and for holes) within the gap [22]. According to this picture, the radiative recombination of an electron–hole pair trapped at the Si=O double bond or of a trapped electron and a free hole (depending on the Si-nc size) is the process responsible for the observed emission. This emission, however, is still size depen-dent due to quantum confinement effects, thus explaining the experimentally observed blueshift.

The EFTEM data shown in Figures 10.5 and 10.7 allow us to deepen our understanding of the complex phenomena ruling the process of light emission from Si-ncs; in particular, in addition to the above-discussed and well-established relationship between size and wavelength, it is possible to also explain the intensity of the PL signal on the basis of the structural properties of samples containing Si-ncs.

In the low pump-power regime, the PL intensity I is given by

$$I \propto \sigma\phi \frac{\tau}{\tau_R} N, \tag{10.7}$$

where τ is the lifetime, τ_R is the radiative lifetime, σ is the excitation cross section, ϕ is the photon flux, and N is the total number of emitting centers. According to Eq. (10.7), under the same excitation conditions, the PL intensity depends only on the lifetime τ and on the total number of emitting centers N. We measured the room-temperature PL intensity and the lifetime of the PL signal of SiO_x samples containing 46 at.% of Si annealed at different temperatures in the 1000–1250 °C range. Lifetime measure-ments were made by studying the decay time of the PL signal recorded at 850 nm. By fitting the experimental curves with Eq. (10.2), we found that for samples annealed in the temperature range of 1100–1250 °C, the lifetime is about 20–30 μs. On the other hand, the decay curve relative to the sample annealed at 1000 °C cannot be fitted by using Eq. (10.2) since it consists of two very different components, the first one corresponding to a very fast decay, with a lifetime shorter than 1 μs, and the second slower component, resembling that of the samples annealed at higher temperatures. The fast component of the lifetime could be an intrinsic property of the amorphous clusters or could be related to the presence of nonradiative defects in the matrix or within the clusters due to the low temperature of the annealing process. Since theoretical calculations have demonstrated that the radiative lifetime of amorphous Si clusters is very similar to that of Si-ncs [55], it is reasonable to conclude that nonradiative recombination centers (within the clusters or in the oxide matrix) influence the de-excitation properties of SiO_x systems annealed at relatively low temperature.

The normalized number N of emitting centers, as obtained from Eq. (10.7) by dividing the PL intensity I by the lifetime τ (and assuming an almost constant τ_R),

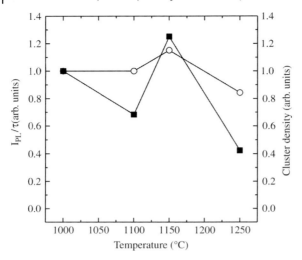

Figure 10.12 Concentration *N* of emitting centers, as obtained from Eq. (10.7) by dividing the PL intensity *I* by the lifetime τ (closed squares, left-hand side scale) and Si nanocluster density (open circles, right-hand side scale) as obtained from EFTEM data, as a function of the annealing temperature. Both sets of data are normalized to their respective values at 1000 °C. The lines are drawn to guide the eye.

is reported in arbitrary units in Figure 10.12 (closed squares, left-hand side scale) as a function of the annealing temperature. From the comparison between these data and the density of amorphous and crystalline clusters and their sum, shown in Figures 10.5b and 10.7b, it is possible to draw some important conclusions about the relationship between structural and optical properties of Si nanoclusters embedded in SiO_2 films. Figure 10.12 also reports (open circles, right-hand side scale) the normalized cluster density obtained from EFTEM data (corresponding to the sum of the amorphous and crystalline contributions), as a function of the annealing temperature. The figure demonstrates a good agreement between the two sets of data since the very fast PL decay observed in samples annealed at 1000 °C explains the presence of a low signal, and the decrease in the PL signal at 1250 °C is in agreement with the decrease in the cluster density due to the occurrence of Ostwald ripening phenomenon. It is possible to note that at 1250 °C the number of emitting centers is more than a factor of 2 lower than the nanocluster density, and this is due to the fact that at this temperature the nanocluster mean radius is very high (2.6 nm), and since our size distributions are typically characterized by a standard deviation of about 20%, a relevant fraction of the Si clusters is too large to be an efficient light emitter. Indeed, according to the effective mass approximation model [56], only clusters with a radius remarkably smaller than the Bohr radius of the exciton (4.3 nm for bulk Si) exhibit strong quantum confinement effects [57]. It is not possible to find a better agreement by correlating the PL data with only the amorphous fraction (it should be impossible to explain the PL signal at 1250 °C in a fully crystalline sample) or the crystalline fraction (the maximum PL intensity should be at 1250 °C and no PL signal should appear at 1000 °C). The above data constitute a clear evidence of the fact that the light emission process from annealed SiO_x films involves all Si clusters, regardless of their structure. Amorphous Si clusters constitute a relevant fraction of the overall population in samples annealed at temperatures lower than 1250 °C and therefore play a fundamental role in determining the PL properties of the system.

10.7
Er-Doped Si Nanoclusters

SiO_x films prepared by PECVD also represent an ideal precursor for the synthesis of Er-doped Si nanoclusters. For this purpose, since Er incorporation during the SiO_x growth requires the use of organometallic Er compounds, the handling of which can be very complicated, rare earth ions are commonly introduced by ion implantation [35–38, 58]. Er ions can be implanted in as-deposited SiO_x films or in films annealed either at temperatures allowing only the formation of amorphous clusters or at high temperatures needed for Si-nc formation. This thermal process will be referred to as "preannealing" hereinafter. In all cases, after the implantation step an annealing process at 900 °C, hereinafter referred to as "postannealing," is usually performed in order to recover the implantation damage and to optically activate Er ions.

Many points appear of great interest in this system. First of all, the nanoclusters seem to act as efficient sensitizers for the rare earth that is excited much more efficiently than in pure SiO_2. Indeed, room-temperature PL yields two orders of magnitude higher are observed for Er-doped SiO_2 in the presence of nanoclusters than in pure SiO_2 [33, 35]. Moreover, since Er is now embedded within a larger gap matrix, the nonradiative decay channels typically limiting Er luminescence in bulk Si (i.e. backtransfer, and Auger de-excitations) might be absent in this case. This further improves the luminescence yield.

The sensitizing action of Si nanoclusters is clearly demonstrated in Figure 10.13 where the room-temperature PL spectra of SiO_x samples in which Er has been implanted at different doses after Si-nc formation (as determined by a preannealing process at 1100 °C) are reported. Spectra were taken by exciting the samples with a 488 nm laser beam at a pump power of 50 mW. In the absence of Er, a PL signal at around 0.85 μm is observed coming from Si-ncs as a result of confined exciton recombination. As soon as Er is introduced, also at doses as small as $2 \times 10^{12}/cm^2$, the signal from Si-ncs at 0.85 μm is seen to decrease, while a new peak at around 1.54 μm, coming from the $^4I_{13/2} \rightarrow {}^4I_{15/2}$ intra-4f-shell Er transition, appears. With an increasing Er dose, this phenomenon becomes particularly evident with a quenching of the nc-related PL corresponding to a simultaneous enhancement in the Er-related PL.

The Si nanocluster-mediated excitation of the Er ions is further demonstrated in Figure 10.14 where the photoluminescence excitation (PLE) spectra of three different samples, normalized to the incoming photon fluxes, are reported as a function of the excitation wavelength. PLE spectra were measured by using a Xe lamp coupled to

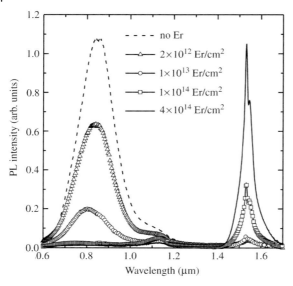

Figure 10.13 Room temperature PL spectra of Er-implanted Si nanoclusters at different Er doses. Samples have been preannealed at 1100 °C and postannealed at 900 °C.

a monochromator. The open squares and the solid line refer to the 1.54 μm line of Er in the presence of Si nanoclusters and in SiO$_2$, respectively. The Er concentration is the same for both samples, which were annealed at 900 °C for 1 h. It is worth noticing that for Er in SiO$_2$ (whose PLE spectrum has been multiplied by a factor of 4), photon

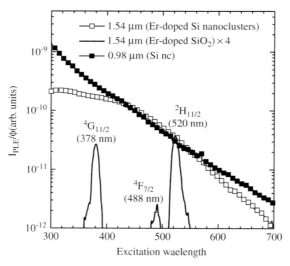

Figure 10.14 PLE spectra for Er-doped Si nanoclusters, Er-doped SiO$_2$ (multiplied by a factor of 4) and Si-ncs, normalized for the photon flux at each excitation wavelength. The PL signals are detected at 1.54 μm for both Er-doped samples and at 0.98 μm for Si-ncs.

emission can occur only at the resonant wavelengths of 378, 488, and 520 nm, corresponding to excited levels related to Er^{3+}. The presence of Si nanoclusters produces a quite interesting modification in the absorption spectrum of Er. Indeed now Er can be excited in a much broader wavelength range. If we now compare the PLE spectrum of Er in the presence of nanoclusters with that obtained for Si-ncs emitting at 0.98 μm (solid squares in Figure 10.14), we find a strong similarity in the trend over a wide range of excitation wavelengths. This, indeed, suggests that the Er excitation occurs via the Si nanoclusters that may absorb the incoming photons over a broad wavelength range and then transfer the energy to a nearby Er ion, by coupling with one of the Er excited levels.

The Er excitation mechanism in the presence of Si-nanoclusters is schematized in Figure 10.15. When we pump Er-doped Si-nanoclusters samples with a laser beam, photons are absorbed by the Si-nanoclusters and promote an electron from the conduction band (CB) to the valence band (VB). Theoretical calculations have demonstrated that the electron in CB is then trapped by a Si=O interfacial state (a) [22]. The recombination of the electron in the interfacial state with a hole in the valence band gives the typical light emission at about 0.8 μm (b). Alternatively, in the presence of Er, the energy can be transferred to the Er ion to excite it (c). Since the 0.8 μm wavelength (corresponding to ∼1.5 eV) couples well with the $^4I_{9/2}$ level of the Er manifold, it is reasonable to believe that this is indeed the Er level first excited by the nanocluster. From this level, a rapid relaxation occurs at the metastable $^4I_{13/2}$ level with the final emission of photons at 1.54 μm (c). Processes (b) and (c) are in competition with one another. When an Er ion is close to a

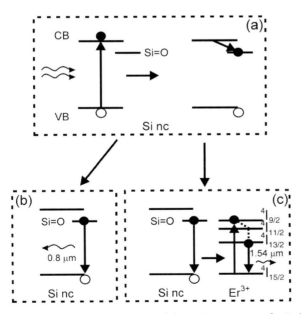

Figure 10.15 Schematic picture of the excitation process for Er-doped Si nanoclusters.

nanocluster, that nanocluster will become "dark" in the sense that the energy transfer to Er will be much more probable than the radiative emission.

Since the beginning of the research activity on Er-doped Si nanoclusters, it was commonly accepted that the formation of Si-ncs was a crucial step in the Er excitation [32–37]. However, in the case of Er-implanted samples, although Si-ncs are formed before Er introduction, the damage induced by the ion beam can strongly modify the structure of the nanograins even if a postannealing process is performed [58]. Indeed, Si-ncs are not the only efficient sensitizers for Er since small amorphous Si nanoclusters can also very efficiently excite Er, the PL intensities being even much higher. This is clearly demonstrated in Figure 10.16 where the room-temperature PL spectra of Er-doped SiO$_x$ layers preannealed at different temperatures (in the range 0–1250 °C) prior to Er implantation are shown. The temperature of the postannealing process was 900 °C for all samples. The PL intensity at 1.54 μm is identical for the SiO$_x$ samples as-deposited and preannealed at 500 °C, increases for preannealing at 800 °C and decreases again at higher temperatures. In particular, for a preannealing process at 1250 °C, the intensity is about a factor of 5 lower with respect to the maximum PL signal observed. This is a quite surprising result since at 1250 °C strong luminescent Si-ncs are formed, while at temperatures as low as 800 °C no crystalline phase and no relevant visible/near-infrared light emission are observed [50]. When a sample containing Si-ncs is implanted with Er and

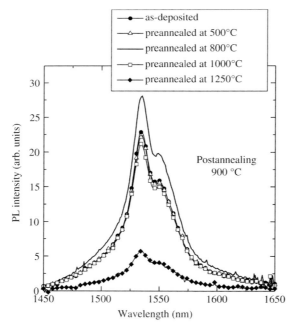

Figure 10.16 Room-temperature PL spectra for SiO$_x$ samples preannealed at different temperatures and then implanted with 5×10^{14} Er/cm^2. In all cases, the postimplantation thermal process was performed at 900 °C for 1 h.

postannealed at 900 °C, while the EFTEM technique still detects a Si nanocluster distribution very similar to the one observed before the implantation process, DFTEM does no longer detect Si-ncs. Therefore, after the Er implantation, the grains remain amorphous even if a thermal process at 900 °C is performed due to the fact that the crystallization temperature of amorphous nanometric structures is much higher than that of bulk material [40]. On the other hand, in the sample preannealed at 800 °C, implanted with Er and postannealed at 900 °C (that is the sample for which the PL signal at 1.54 μm is highest), in agreement with the data shown in Figure 10.4, the EFTEM analyses reveal a network of very small amorphous Si nanograins and there is no evidence for the presence of any crystalline phase. These data demonstrate that Er can be efficiently excited through an electron–hole-mediated process in the presence of amorphous Si nanoclusters. Moreover, the PL intensity is stronger in samples preannealed at lower temperatures (up to 800 °C) since in this case there are numerous and smaller Si nanograins that can absorb very efficiently the energy from the laser and transfer it to a nearby Er ion. This also increases the number of excitable Er centers.

Finally, it should be noted that the luminescence features of Er-implanted SiO_x films are similar to those observed in samples in which Er was introduced with different nondamaging techniques [33, 34], indicating that it is not important that the Si nanograins are crystalline or amorphous, since the only crucial step is the presence of a high density of small Si clusters that can act as both the absorbing medium and the sensitizers for Er.

10.8
Summary

In this chapter, we have presented an overview on the properties of Si-ncs embedded in SiO_2 synthesized by thermal annealing of SiO_x films prepared by PECVD. The deposition conditions needed to obtain SiO_x films with the suitable composition for Si-nc formation have been summarized. The thermally induced evolution of the SiO_x films has been studied in detail by means of different TEM techniques and it has been evidenced as the quantitative analysis of EFTEM and DFTEM data constitute an invaluable tool both for the comprehension of the structure of this material and for the optimization of its optical properties. SiO_x films prepared by PECVD constitute a unique precursor material for the controlled synthesis of Si-ncs exhibiting strong room-temperature PL; indeed, the wavelength of the PL signal of Si-ncs can be varied from 650 to 950 nm by increasing their mean radius through the increase in the annealing temperature or in the Si concentration. The excitation and de-excitation properties of Si-ncs have been studied and interpreted on the basis of structural considerations, and the role of amorphous clusters in the PL process elucidated. Finally, the basic properties and the potential of Er-doped Si nanoclusters have been reported and discussed.

The knowledge about Si-ncs is now mature enough to allow their applications in the fabrication of new generations of electroluminescent devices characterized by working conditions (operating voltage and current, stability, and reliability) and efficiency close to those required for practical applications and, more ambitiously, of silicon devices in which photonic and electronic functions are integrated by fully exploiting the unusual properties of Si-ncs. For this purpose, the integration of traditional LED structures with waveguides, nanocavities, and photonic crystals will probably represent a decisive step for the development of silicon photonics.

Acknowledgments

The contributions given by C. Spinella, C. Bongiorno, D. Pacifici, and M.G. Grimaldi to some of the experiments reported in this chapter are gratefully acknowledged.

References

1 Kimerling, L.C. (2000) *Appl. Surf. Sci.*, **159–160**, 8–13.

2 Polman, A. (2002) *Nat. Mater.*, **1**, 10–12.

3 Paniccia, M. and Koehl, S. (2005) *IEEE Spectr.*, **42**, 38–43.

4 Jalali, B., Paniccia, M., and Reed, G. (2006) *IEEE Microwave Mag.*, **7**, 58–68.

5 Rong, H., Liu, A., Jones, R., Cohen, O., Hak, D., Nicolaescu, R., Fang, A., and Paniccia, M. (2005) *Nature*, **433**, 292–294.

6 Rong, H., Jones, R., Liu, A., Cohen, O., Hak, D., Fang, A., and Paniccia, M. (2005) *Nature*, **433**, 725–728.

7 Canham, L.T. (1990) *Appl. Phys. Lett.*, **57**, 1046–1048.

8 Kanemitsu, Y., Ogawa, T., Shiraishi, K., and Takeda, K. (1993) *Phys. Rev. B*, **48**, 4883–4886.

9 Shimizu-Iwayama, T., Fujita, K., Nakao, S., Saitoh, K., Fujita, T., and Itoh, N. (1994) *J. Appl. Phys.*, **75**, 7779–7783.

10 Zhu, J.G., White, C.W., Budai, J.D., Withrow, S.P., and Chen, Y. (1995) *J. Appl. Phys.*, **78**, 4386–4389.

11 Min, K.S., Shcheglov, K.V., Yang, C.M., Atwater, H.A., Brongersma, M.L., and Polman, A. (1996) *Appl. Phys. Lett.*, **69**, 2033–2035.

12 Inokuma, T., Wakayama, Y., Muramoto, T., Aoki, R., Kurata, Y., and Hasegawa, S. (1998) *J. Appl. Phys.*, **83**, 2228–2234.

13 Brongersma, M.L., Polman, A., Min, K.S., Boer, E., Tambo, T., and Atwater, H.A. (1998) *Appl. Phys. Lett.*, **72**, 2577–2579.

14 Shimizu-Iwayama, T., Kurumado, N., Hole, D.E., and Townsend, P.D. (1998) *J. Appl. Phys.*, **83**, 6018–6022.

15 Linnros, J., Lalic, N., Galeckas, A., and Grivickas, V. (1999) *J. Appl. Phys.*, **86**, 6128–6134.

16 Iacona, F., Franzò, G., and Spinella, C. (2000) *J. Appl. Phys.*, **87**, 1295–1303.

17 Valenta, J., Juhasz, R., and Linnros, J. (2002) *Appl. Phys. Lett.*, **80**, 1070–1072.

18 Pavesi, L., Dal Negro, L., Mazzoleni, C., Franzò, G., and Priolo, F. (2000) *Nature*, **408**, 440–444.

19 Khriachtchev, L., Rasanen, M., Novikov, S., and Sinkkonen, J. (2001) *Appl. Phys. Lett.*, **79**, 1249–1251.

20 Luterova, K., Pelant, I., Mikulskas, I., Tomasiunas, R., Muller, D., Grob, J.-J., Rehspringer, J.-L., and Honerlage, B. (2002) *J. Appl. Phys.*, **91**, 2896–2900.

21 Dal Negro, L., Cazzanelli, M., Pavesi, L., Ossicini, S., Pacifici, D., Franzò, G., Priolo, F., and Iacona, F. (2003) *Appl. Phys. Lett.*, **82**, 4636–4638.

22 Wolkin, M.V., Jorne, J., Fauchet, P.M., Allan, G., and Delerue, C. (1999) *Phys. Rev. Lett.*, **82**, 197–200.

23 DiMaria, D.J., Kirtley, J.R., Pakulis, E.J., Dong, D.W., Kuan, T.S., Pesavento, F.L., Theis, T.N., Cutro, J.A., and Brorson, S.D. (1984) *J. Appl. Phys.*, **56**, 401–416.

24 Hirschman, K.D., Tsybeskov, L., Duttagupta, S.P., and Fauchet, P.M. (1996) *Nature*, **384**, 338–341.

25 Qin, G.G., Li, A.P., Zhang, B.R., and Li, B.-C. (1995) *J. Appl. Phys.*, **78**, 2006–2009.

26 Fujita, S. and Sugiyama, N. (1999) *Appl. Phys. Lett.*, **74**, 308–310.

27 Photopoulos, P. and Nassiopoulou, A.G. (2000) *Appl. Phys. Lett.*, **77**, 1816–1818.

28 Lalic, N. and Linnros, J. (1999) *J. Lumin.*, **80**, 263–267.

29 Franzò, G., Irrera, A., Moreira, E.C., Miritello, M., Iacona, F., Sanfilippo, D., Di Stefano, G., Fallica, P.G., and Priolo, F. (2002) *Appl. Phys. A: Mater. Sci. Process.*, **74**, 1–5.

30 Irrera, A., Pacifici, D., Miritello, M., Franzò, G., Priolo, F., Iacona, F., Sanfilippo, D., Di Stefano, G., and Fallica, P.G. (2002) *Appl. Phys. Lett.*, **81**, 1866–1868.

31 Walters, R.J., Bourianoff, G.I., and Atwater, H.A. (2005) *Nat. Mater.*, **4**, 143–146.

32 Kenyon, A.J., Trwoga, P.F., Federighi, M., and Pitt, C.W. (1994) *J. Phys. Condens. Matter*, **6**, L319–L324.

33 Fujii, M., Yoshida, M., Kanzawa, Y., Hayashi, S., and Yamamoto, K. (1997) *Appl. Phys. Lett.*, **71**, 1198–1200.

34 Shin, J.H., Kim, M.-J., Seo, S.-Y., and Lee, C. (1998) *Appl. Phys. Lett.*, **72**, 1092–1094.

35 Franzò, G., Vinciguerra, V., and Priolo, F. (1999) *Appl. Phys. A: Mater. Sci. Process.*, **69**, 3–12.

36 Kik, P.G., Brongersma, M.L., and Polman, A. (2000) *Appl. Phys. Lett.*, **76**, 2325–2327.

37 Iacona, F., Pacifici, D., Irrera, A., Miritello, M., Franzò, G., Priolo, F., Sanfilippo, D., Di Stefano, G., and Fallica, P.G. (2002) *Appl. Phys. Lett.*, **81**, 3242–3244.

38 Nazarov, A., Sun, J.M., Skorupa, W., Yankov, R.A., Osiyuk, I.N., Tjagulskii, I.P., Lysenko, V.S., and Gebel, T. (2005) *Appl. Phys. Lett.*, **86**, 151914.

39 Zacharias, M., Heitmann, J., Scholz, R., Kahler, U., Schmidt, M., and Blasing, J. (2002) *Appl. Phys. Lett.*, **80**, 661–663.

40 Zacharias, M., Blasing, J., Veit, P., Tsybeskov, L., Hirschman, K., and Fauchet, P.M. (1999) *Appl. Phys. Lett.*, **74**, 2614–2616.

41 Gourbilleau, F., Portier, X., Ternon, C., Voivenel, P., Madelon, R., and Rizk, R. (2001) *Appl. Phys. Lett.*, **78**, 3058–3060.

42 Pai, P.G., Chao, S.S., Takagi, Y., and Lukovsky, G. (1986) *J. Vac. Sci. Technol. A*, **3**, 689–694.

43 Mariotto, G., Das, G., Quaranta, A., Della Mea, G., Corni, F., and Tonini, R. (2005) *J. Appl. Phys.*, **97**, 113502.

44 Lanford, W.A. and Rand, M.J. (1978) *J. Appl. Phys.*, **49**, 2473–2477.

45 Olivares-Roza, J., Sanchez, O., and Albella, J.M. (1998) *J. Vac. Sci. Technol. A*, **16**, 2757–2761.

46 Temkin, R.J. (1975) *J. Non-Cryst. Solids*, **17**, 215–230.

47 Philipp, H.R. (1972) *J. Non-Cryst. Solids*, **8–10**, 627–632.

48 Knolle, W.R. and Maxwell, H.R., Jr. (1980) *J. Electrochem. Soc.*, **127**, 2254–2259.

49 Iacona, F., Lombardo, S., and Campisano, S.U. (1996) *J. Vac. Sci. Technol. B*, **14**, 2693–2700.

50 Iacona, F., Bongiorno, C., Spinella, C., Boninelli, S., and Priolo, F. (2004) *J. Appl. Phys.*, **95**, 3723–3732.

51 Boninelli, S., Iacona, F., Franzò, G., Bongiorno, C., Spinella, C., and Priolo, F. (2007) *J. Phys.: Condens. Matter*, **19**, 225003.

52 Franzò, G., Miritello, M., Boninelli, S., Lo Savio, R., Grimaldi, M.G., Priolo, F., Iacona, F., Nicotra, G., Spinella, C., and Coffa, S. (2008) *J. Appl. Phys.*, **104**, 094306.

53 Klimov, V.I., Schwarz, Ch.J., McBranch, D.W., and White, C.W. (1998) *Appl. Phys. Lett.*, **73**, 2603–2605.

54 Song, H. and Bao, X. (1997) *Phys. Rev. B*, **55**, 6988–6993.

55 Allan, G., Delerue, C., and Lannoo, M. (1997) *Phys. Rev. Lett.*, **78**, 3161–3164.

56 Efros, Al.L. and Efros, A.L. (1982) *Sov. Phys. Semicond.*, **16**, 772–775.

57 Trwoga, P.F., Kenyon, A.J., and Pitt, C.W. (1998) *J. Appl. Phys.*, **83**, 3789–3794.

58 Franzò, G., Boninelli, S., Pacifici, D., Priolo, F., Iacona, F., and Bongiorno, C. (2003) *Appl. Phys. Lett.*, **82**, 3871–3873.

11
Formation of Si-nc by Reactive Magnetron Sputtering

Fabrice Gourbilleau, Celine Ternon, Christian Dufour, Xavier Portier, and Richard Rizk

11.1
Introduction

Since the discovery of the efficient room-temperature photoluminescence (PL) of porous silicon [1], the possible development of Si-based optoelectronic device compatible with the mainstream silicon microelectronics has generated huge scientific activities focused on the nanoscale Si studies. Thus, a considerable effort was put in to produce and study Si nanostructures consisting of Si nanograins embedded in a silica matrix in order to understand the physical processes underlying the visible emission and to define future applications. Several methods have been used for fabricating such materials and consist in (i) implantation of Si^+ ions in thermally grown SiO_2 [2–10], (ii) laser-induced decomposition of gas precursors [11–14], (iii) plasma-enhanced chemical vapor deposition [15–20], (iv) magnetron cosputtering [21, 22], (v) porous silicon [1, 23–25], and (vi) evaporation [26, 27]. The visible emission observed has been attributed to a quantum confinement (QC) effect of photogenerated carriers in the nanoscale silicon particles [28] and has been thoroughly analyzed theoretically [29–32]. Besides this QC effect, some studies have also pointed out the concomitant role played by the $Si–SiO_2$ interface in the emission properties [2, 33–35]. Thus, the monitoring of both size and distribution of nanograins within the host matrix and the quality of the Si/SiO_2 interface appear to be the main parameters governing the emission properties of the Si-based nanomaterials. The most original way to control the Si grain size is the laser pyrolysis of silane in a gas flow reactor producing free Si nanoparticles as reported by Ledoux *et al.* [36] or the multilayer (ML) Si/SiO_2 approach in which the Si sublayer thickness should not exceed the QC-related critical value [37–41], evaluated to about 5 nm. Among the potential developments of Si-based nanostructured materials detailed in the literature, we report sensor applications [42], conception and design of efficient photonic structures [43, 44], nonvolatile memory devices [45], and tandem solar cells for the third-generation photovoltaics [46].

In this context, our group has developed an original approach by means of magnetron sputtering coupled with a reactive plasma that allows to grow composite

Silicon Nanocrystals: Fundamentals, Synthesis and Applications. Edited by Lorenzo Pavesi and Rasit Turan
Copyright © 2010 WILEY-VCH Verlag GmbH & Co. KGaA, Weinheim
ISBN: 978-3-527-32160-5

or multilayered materials in which the size and/or the density of the Si nanograins, and consequently the photoluminescent properties, can be monitored through either the deposition parameters or the postannealing treatment. This original approach will be described through the effect of deposition parameters on the microstructure and optical properties of the different layers fabricated.

11.2
Experimental

Single layers and multilayers were fabricated by RF reactive magnetron sputtering of a pure SiO$_2$ target under a mixture of Ar and H$_2$ plasma. The originality of our process lies in the use of hydrogen as a reactive gas instead of oxygen commonly employed for the fabrication of Si/SiO$_2$ MLs [37]. This way, we take advantage of the ability of hydrogen to reduce the oxygen species released from the sputtering of the pure SiO$_2$ target allowing the growth of silicon-rich silicon oxide (SRSO) layers. The Ar pressure was maintained constant at 1.5 Pa while the hydrogen partial pressure (P_{H_2}) was varied from 0.5 to 6.0 Pa. The power density applied for the sputtering of SiO$_2$ was 0.76 W/cm^2 whereas the substrate temperature (T_S) ranged from 60 to 600 °C. Depending on the characterization techniques, layers were deposited on (100) Si or fused quartz substrates.

Both the influence of the deposition parameters (P_{H_2} and T_S) and the effect of the annealing treatment on the multilayer structure were investigated through several characterization experiments. IR experiments were performed in the 500–4000 cm^{-1} range by means of a Nicolet 750-II spectrometer. The spectra were recorded at the Brewster angle of 65° that enables the simultaneous detection of the LO$_3$ and TO$_3$ vibrational modes of silica at about 1250 and 1080 cm^{-1}, respectively. Optical transmission measurements were performed in the 300–3000 nm range using a Perking Elmer UV-visible spectrophotometer. Transmission and high-resolution electron microscopies (TEM and HREM) were conducted on samples prepared in the cross-sectional configuration using a JEOL 2010 and an ABT EM002B microscope. X-ray diffraction (XRD) and reflectometry measurements were carried out on a Philips XPERT MPD PRO diffractometer. For reflectometry experiments, the X-ray beam of the Cu Kα radiation struck the film surface under a grazing incidence angle ω in the 0–1° range. The light emission properties were analyzed by photoluminescence measurements using different excitation lines: (i) the 488 nm line of an Ar$^+$ laser, (ii) the 300–400 band of a high-pressure Hg lamp, and (iii) the 325 nm line of a HeCd laser.

11.3
Results

In the first part of this section, the nature of the deposited materials with different hydrogen partial pressures P_{H_2} and substrate temperatures T_S will be analyzed to determine the optimized fabrication conditions for layers containing Si nanograins. Then, in the second part, composite and multilayer deposition and properties will be described.

11.3.1
Single Layer

Typical IR spectra have been recorded in the 900–1400 cm^{-1} range for layers deposited at $T_S = 60\,°C$ (Figure 11.1a) or $T_S = 500\,°C$ (Figure 11.1b) with different hydrogen contents after an annealing treatment at 1100 °C. The choice of such a high annealing temperature is linked to previous studies on Si-SiO$_2$ layers deposited by cosputtering that have shown a bright PL emission upon annealing at 1100 °C for 1 h [21]. The spectra have been normalized to the TO$_3$ vibration band to observe the effect of the deposition condition on the LO$_3$ peak. It has to be noticed that the hydrogen-related bonds peaking at about 640, 820, and 2100 cm^{-1} have not been

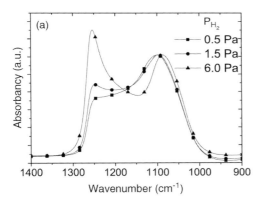

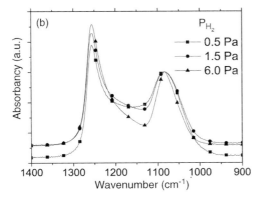

Figure 11.1 Infrared spectra recorded at a Brewster angle of 65° on the annealed single layers deposited at (a) $T_S = 60\,°C$ and (b) $T_S = 500\,°C$ for three different values of P_{H_2}.

observed in the IR spectra. Such a result is consistent with the expected total desorption of hydrogen after high-temperature annealing treatment. For the layers deposited at a low T_S, the increase in the hydrogen partial pressure leads to an increase in the LO_3 vibration band. The same but less pronounced effect can be observed for the highest T_S (Figure 11.1b). Based on the work of Olsen and Shimura [47] who ascribed this LO_3 mode to the $Si-O-Si$ bond at 180 ° at the interface between Si and SiO_2, the evolution of the intensity of this vibration band is therefore the result of the increase in the $Si-O-Si$ bond density with P_{H_2} in our layers. Moreover, the increase in P_{H_2} and the one in T_S favors the phase separation between Si and SiO_2 as evidenced by the decrease in the LO_4-TO_4 vibration band intensities peaking at 1125 and 1150 cm^{-1} and related to the disorder in the SiO_2 matrix. The concomitant increase in the LO_3 mode and the decrease in the LO_4-TO_4 vibration bands with the increase in P_{H_2} and/or T_S could be the result of the raising Si excess in the fabricated layer. To confirm this hypothesis, optical transmission spectroscopy measurements were performed. Figure 11.2 displays the evolution of the static refractive index measured at 2.0 μm as a function of T_S for the layer deposited under different values of P_{H_2} after annealing at 1100 °C. This figure demonstrates the systematic increase in the refractive index with both deposition parameters P_{H_2} and T_S to reach $T_S = 500$ °C and $P_{H_2} = 6$ Pa, a value close to that of the silicon (single crystal: 3.2). These results attest to the reactive character of the process allowing to control the Si incorporation with deposition parameters. To get some insight into the microstructure of the deposited layer, TEM observations were performed on some specific samples and reported in Figures 11.3 and 11.4. For low deposition temperature, the presence of Si nanocrystals within the films can be deduced from the corresponding electron diffraction pattern (EDP) that reveals a large (111) ring characteristic of the randomly oriented Si nanograins. When the substrate temperature is increased up to 500 °C, one can notice the presence of some small black contrast attesting the presence of Si nanocrystals in the layers deposited with the lowest P_{H_2} and confirmed by a well-defined (111) ring in the EDP

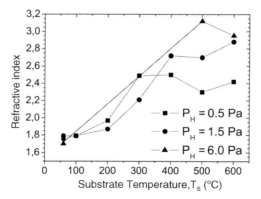

Figure 11.2 Evolution of the static refractive index measured at 2.0 μm as a function of the substrate temperature T_S for single layers deposited under different hydrogen partial pressures P_{H_2} and after an annealing treatment at 1100 °C.

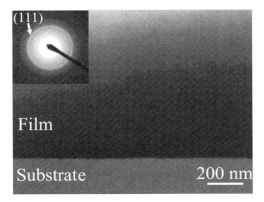

Figure 11.3 Bright field TEM image of an annealed single layer deposited at $T_S = 60\,^\circ$C and $P_{H_2} = 6.0$ Pa.

(Figure 11.4a). The increasing hydrogen content in the plasma favors the silicon incorporation as revealed by the evolution of the microstructure displayed in Figure 11.4b and c in which a columnar structure appears to be more and more pronounced. Thus, a high P_{H_2} combined with a high T_S value leads to the growth of large crystalline Si grains whereas under low P_{H_2} and T_S, the growing film presents a nanocomposite structure with Si nanograins embedded in the silica host matrix.

The first part of this section clearly shows the potential of this original approach dealing with the reactive magnetron sputtering of a pure silica cathode to grow different microstructures from nanocomposite layers to almost pure silicon film.

11.3.2
Composite Layer

Si–SiO$_2$ composite layers have been grown at $T_S = 60\,^\circ$C for three different P_{H_2} values ranging from 0.6 to 6.0 Pa. An annealing treatment at 1100 $^\circ$C is applied after deposition to allow the phase separation between Si and SiO$_2$ and to favor the Si nanograin formation. The evolution of the microstructure with P_{H_2} is displayed on Figure 11.5 in which the RX spectra are reported with their corresponding fitting curves. When the hydrogen partial pressure increases from 0.6 to 1.5 Pa, the width of the peaks increases and then decreases for the highest value of P_{H_2} (Table 11.1). The corresponding sizes deduced from the RX spectra are listed in Table 11.1. HREM observations performed on these layers (not shown here) evidence an increase in the average Si grain size from about 3 nm to about 4 nm when P_{H_2} is increased (Figure 11.6). The corresponding characteristics of the size distribution is reported in Table 11.2. It shows that the increase in P_{H_2} favors both the increase in Si grain density and, to a less extent, the increase in Si grain size. The discrepancies with the XRD estimation lies in the fact that XRD experiments give information on the coherent size corresponding to either a grain size if well crystallized or a distance between defects such as grain boundaries or twins. To confirm the measured Si size, photoluminescence experiments have been performed since the conditions of the

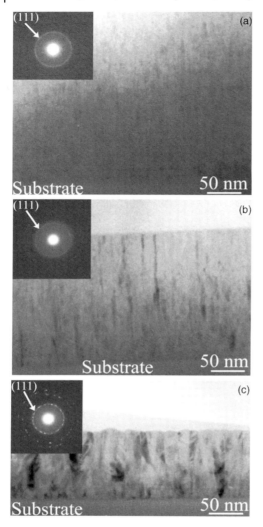

Figure 11.4 Bright field TEM images of the annealed films deposited at $T_S = 500\,^\circ$C and for P_{H_2} values of (a) 0.5 Pa; (b) 1.5 Pa; (c) 6.0 Pa.

quantum confinement of carriers are fulfilled. Figure 11.7 reports the PL spectra recorded for the annealed layers in the visible range. Each spectrum has been normalized to the thickness to allow the comparison of the PL intensities. One can observe that the PL intensity increases with P_{H_2} that is coherent with increasing density of Si nanocrystals estimated from the HREM images and reported in Table 11.2. The spectra are well fitted by a combination of two Gaussian curves allowing the determination of two PL peak positions depending on the fabrication conditions (Figure 11.8). One of these PL maxima peaks at the same wavelength

Table 11.1 Characteristics of the fitting and Si nanograin size.

P_{H_2} (Pa)	(111)			(220)			(311)		
	θ_c (°)	FWMH (°)	Size (nm)	θ_c (°)	FWMH (°)	Size (nm)	θ_c (°)	FWMH (°)	Size (nm)
0.6	28.4	1.6	4	47.1	2.2	3	55.6	3.9	1.5
1.5	28.4	0.8	10	47.3	1.3	6	55.7	1.9	3.5
6.0	28.3	2.5	2.5	47.5	2.8	2.5	55.1	3.5	2

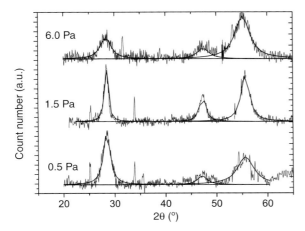

Figure 11.5 Diffraction spectra of composite layers deposited with different values of P_{H_2}. Each spectrum is fitted with three Lorentzian curves the characteristics of which are reported in Table 11.1.

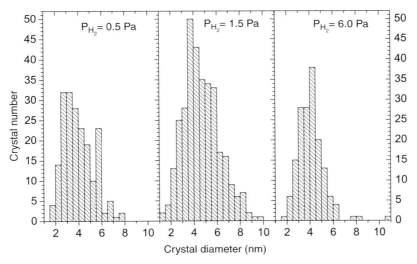

Figure 11.6 Size distribution of the Si grains deduced from the HREM observations of composite single layers deposited with different hydrogen partial pressures.

Table 11.2 Characteristics of the size distribution and density of Si nanograins.

P_{H_2} (Pa)	Grain number	Mean size (nm)	Density of nanograins (grains/μm^2)
0.6	195	3.1 ± 1	3800
1.5	326	4.7 ± 1	5800
6.0	162	4.1 ± 1	7400

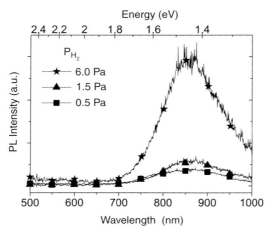

Figure 11.7 Photoluminescence spectra of the annealed single layers deposited with different hydrogen partial pressures. For clarity, the baseline of the spectra has been shifted.

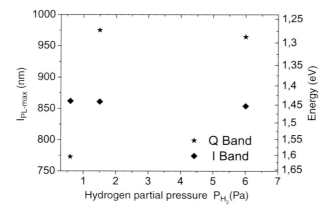

Figure 11.8 Photoluminescence peak position as a function of the hydrogen partial pressure.

(~ 875 nm, 1.45 eV) for all the P_{H_2} and is denoted I. The second one, labeled Q, depends on the hydrogen content in the plasma with a position at 775 nm (1.6 eV) for the lowest P_{H_2} value and an emission at about 970 nm (1.3 eV) for the highest P_{H_2} values. The presence of the I and Q bands has already been reported [48, 49] and has been ascribed to interface states for the emission band at around 850 nm (I band) and to the quantum confinement of the carriers for the second one (Q band). The evolution of the Q band position with the hydrogen content in the plasma thus confirms the increase in the Si grain size observed by HREM up to the critical size (~ 5 nm) above which the quantum confinement is not as much effective.

This first part has shown the possibilities offered by the reactive approach of the sputtering of a pure SiO_2 target. We have demonstrated that this process allows to grow layers in which the size and the density of Si nanocrystals can be monitored through the hydrogen partial pressure used during the deposition. Nevertheless, for an accurate control of the Si grain size, a more appropriate approach consists in the multilayer approach. We will now describe the fabrication of $SRSO/SiO_2$ multilayers by means of reactive magnetron sputtering that enables the growth of 3D silicon nanograins within the silicon-rich silicon oxide sublayer.

11.3.3
Multilayer

Figure 11.9 shows a typical TEM image of a multilayer produced by the reactive magnetron sputtering. The multilayer structure is produced through alternating depositions under pure Ar and Ar/H_2 plasma. Thus, MLs consist of an alternate stacking of silicon-rich silicon oxide sublayer and silicon oxide sublayer with controlled thicknesses. The samples were deposited at 500 °C and, for the SRSO

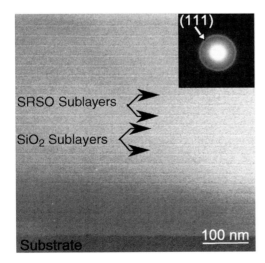

Figure 11.9 Typical TEM image of a multilayer structure with the corresponding electron diffraction pattern. The SRSO and SiO_2 sublayer thicknesses are about 3 and 19 nm, respectively.

sublayer, under a plasma containing 80% of hydrogen that corresponds to a partial pressure of 6.0 Pa. After deposition, the layers were annealed for 1 h under nitrogen flow at various temperatures.

11.3.3.1 Effect of the Sublayer Thickness

Effects of the SiO_2 sublayer thickness (t_{SiO_2}) on the structural and optical properties of 20-period annealed MLs are first studied. The thickness of the SRSO sublayer is fixed at 3 nm. The evolution of the PL spectra for different SiO_2 sublayer thicknesses is presented in Figure 11.10 in the 500–1800 nm range. The spectra of thicker samples present some oscillations that can be attributed to an interference phenomenon. This phenomenon originates from the formation of an optical microcavity between the top surface of the film and the film/substrate interface [50]. This is confirmed by the increasing number of oscillations for increasingly thicker layers. These interferences make the comparison of the spectra difficult. Nevertheless, one can notice that if an effect of the SiO_2 sublayer thickness exists, it has no detrimental effect on the recorded PL emission intensity. Zacharias *et al.* [39] pointed out the possible formation of silicon nanograins larger than the initial SRSO sublayer thickness in the case of thin SiO_2 sublayer. It is therefore necessary to consider that a minimum t_{SiO_2} is required. Consequently, we have chosen a $t_{SiO_2} = 19$ nm to avoid any overgrowth of the Si grains.

In the following, the effect of the SRSO sublayer thickness (t_{SRSO}) will be examined for the same 20-period ML for which the SiO_2 sublayer thickness has been fixed at 19 nm. After an annealing at 1100 °C for 1 h, PL experiments have been performed in the same wavelength range and the corresponding spectra are reported in Figure 11.11. The noise present in the infrared region is due to the change from the

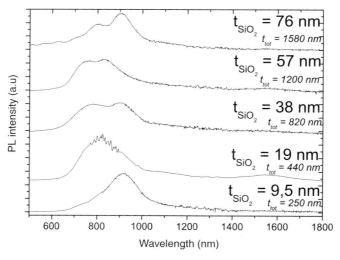

Figure 11.10 PL spectra of multilayers with different SiO_2 sublayer thicknesses, t_{SiO_2}. The thickness of the SRSO sublayer is fixed at 3 nm.

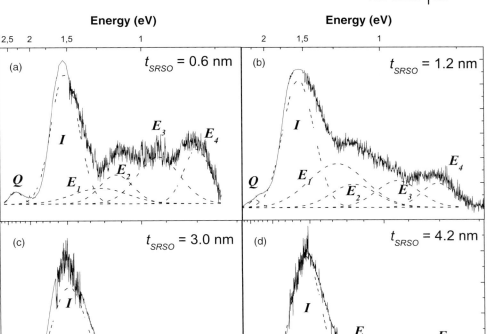

Figure 11.11 PL spectra of multilayers in the 500–1800 nm range for t_{SRSO}: (a) 0.6 nm, (b) 1.2 nm, (c) 3 nm, and (d) 4.2 nm. The thickness of the SiO_2 sublayer is fixed at 19 nm. The noise appearing in the spectra is due to the change of detector. The spectra are fitted with six Gaussian bands called Q, I, E_1, E_2, E_3, E_4.

photomultiplier to a Ge detector that is less sensitive in the visible range. The low-energy region of the spectra can be decomposed into four bands called hereafter E_1, E_2, E_3, and E_4, while in the visible region, the decomposition of each PL spectrum is optimized by two bands labeled Q and I. The evolution of the peak position with the SRSO sublayer thickness reveals that the maxima of the four E_i and I bands are rather independent of t_{SRSO}, whereas that of the Q band appears affected. The locations of these four E_i bands are very close to those reported by Street in the case of the amorphous silicon [51]. This could be a signature of the presence of amorphous tissue in our MLs. Figure 11.12 presents the evolution of the peak position obtained for the I and Q bands with respect to the thickness of the SRSO sublayer. These variations are compared with three models or simulations reported in the literature: the quantum confinement in amorphous silicon [52], the QC in nanocrystalline silicon [29], and the QC in nanocrystalline Si assisted by a Si-O phonon [33]. One can

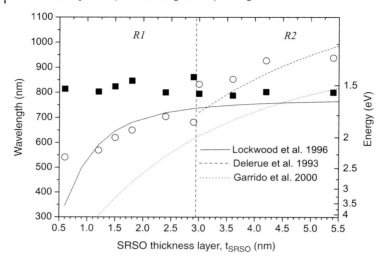

Figure 11.12 Peak positions of the Q (open circles) and I (black square) bands as a function of the SRSO layer thickness, t_{SRSO}. The lines correspond to the quantum confinement models or simulations in amorphous silicon (—) [52], in nanocrystalline silicon ($\cdots$) [29], and in Si nanocrystals assisted by a Si-O phonons (- - -) [33].

note that when t_{SRSO} is lower than 3 nm, the Q band position is highly energetic with respect to the I band while for $t_{SRSO} \geq 3$ nm, this is the opposite. This value close to 3 nm seems to be a singularity allowing to delimit the two regions labeled R_1 and R_2 for low and high thickness range, respectively. The comparison of our data with the simulations shows that none of them satisfactorily fits the evolution of the peak positions over the whole range of SRSO thicknesses. The simulation proposed by Lockwood *et al.* [52] fits well with the Q band in the R_1 region while the one described by Garrido *et al.* [33] reproduces closely the Q band evolution in the R_2 region. The former case suggests that the Si present within the sublayer is amorphous for $t_{SRSO} \leq 3$ nm. Above this value, silicon nanoparticles crystallize. To give some insight into these hypotheses, HREM observations were performed for one sample of each region (Figure 11.13). No Si nanocrystallites were detected in the ML with an SRSO sublayer thickness of 1.2 nm (Figure 11.13a) suggesting that silicon is amorphous as attested by the halo ring on the corresponding EDP. The diffraction dots originate from the substrate observed along the [110] direction. Figure 11.13b shows a silicon nanocrystal that is enlarged in the inset and whose size is comparable with t_{SRSO} fixed here at 3 nm. The (111) large diffraction ring also confirms the presence of Si nanocrystals in the ML.

As in the case of the composite layer, these two bands are ascribed to the electron–hole recombination (i) at silicon/silica interface for the I band and (ii) the silicon nanograins within the frame of quantum confinement model for the case of the Q band. A theoretical study [53] based on these experimental results has shown that both I and Q bands are superimposed on one another when some oxygen atoms are missing at the Si/SiO_2 interface saturated with Si=O double bonds.

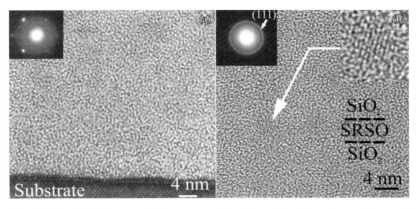

Figure 11.13 HREM images with the corresponding electron diffraction patterns of two multilayers from (a) region R_1 and (b) region R_2.

Both photoluminescence and microstructure results are consistent with the occurrence of transition from an amorphous to a crystalline phase for Si nanoparticles when the SRSO sublayer thickness reaches 3 nm. In Figure 11.12, the region R_1 is characterized by the formation of only amorphous silicon phase in the SRSO sublayer, whereas both amorphous and crystalline silicon phases should coexist in the R_2 region. The absence of any crystallization of the Si nanoparticles within the thinnest SRSO layers was thermodynamically predicted by Veprek *et al.* [54] who showed that the silicon amorphous phase is stable for grain sizes smaller than 3 nm. The observation of the E_i band whatever t_{SRSO} can be explained by the presence of an amorphous Si phase in the SRSO sublayer even for $t_{SRSO} \geq 3$ nm. Since these bands originate from the transition between localized states, they are independent of t_{SRSO}. To interpret our experiments, a model based on a square well potential has been developed and recently described in Ref. [35]. This model has allowed us to reproduce the evolution of both bands with t_{SRSO} and to demonstrate the presence of a 0.8 nm thick interfacial SiO_x layer between the Si nanograin and the silica host matrix.

11.3.3.2 Effect of the Annealing Temperature

In the light of the results described above, the annealing temperature T_A can be considered as one of the main parameters for monitoring the growth and the phase nature (crystalline or amorphous) of Si nanoparticles within the SRSO sublayer. Thus, the effect of T_A on a multilayer constituted of 20 periods of 3 nm thick SRSO and 19 nm thick SiO_2 sublayers has been studied in the 600–1100 °C temperature range.

Figure 11.14 displays the evolution of the two main vibrational modes, LO_3 and TO_3 against the annealing temperature T_A, while the inset shows a typical FTIR spectrum recorded in the 900–1400 cm^{-1} range. It can be seen that the two LO_3 and TO_3 modes shift with increasing T_A to higher frequencies, from 1225 to 1255 cm^{-1} and from 1075 to 1095 cm^{-1}, respectively. Such a behavior can be explained by the structural rearrangement of either or both silica and SRSO sublayers. To separate the

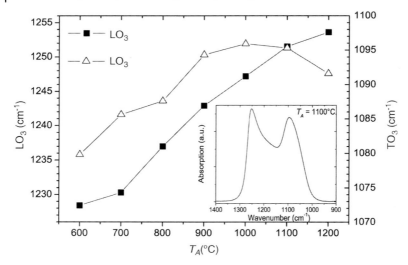

Figure 11.14 Variation in the LO$_3$ (left scale) and TO$_3$ (right scale) peak positions as a function of the annealing temperature T_A. The inset shows a typical infrared spectrum for $T_A = 1100\,°$C.

possible contribution of each sublayer, two reference samples consisting of a single thick silica or SRSO layer were fabricated. The thickness of each film was comparable to the total thickness of all corresponding sublayers (silica or SRSO) in the multilayer structure. These two thick films were fabricated under the same conditions and annealed at the same temperatures than for the ML. The evolution of their LO$_3$ and TO$_3$ modes are plotted in Figure 11.15 for the silica single layer and in the inset (Figure 11.15) for the SRSO film. The comparison of these two plots for either silica or SRSO thick layers with those of Figure 11.14 for the MLs leads to the following conclusions. Some similarities exist between the steep increase in the frequencies of both the LO$_3$ and TO$_3$ modes of MLs (Figure 11.14) and the shifts of these modes in case of a single silica layer (Figure 11.15), before becoming temperature independent for $T_A \geq 1100\,°$C. The corresponding modes for the single SRSO layer (inset of Figure 11.15) are nearly unaffected by T_A. Such observations suggest that the evolution of the vibrational modes in the MLs mainly originates from the silica sublayers as a result of the progressive recovering of the silica defects that are completely annealed at 1100 °C.

Figure 11.16 shows the HREM images of the MLs annealed at the indicated temperatures. For $T_A < 900\,°$C, no crystallization was detected (not shown), whereas Si nanocrystals (Si-ncs) start to form within the SRSO sublayers when T_A reaches and exceeds 900 °C, in perfect agreement with the observations of Zacharias *et al.* [39]. The average particle size is about 3 nm and appears unaffected by the increase in T_A until 1100 °C. This strongly indicates the role played by the thickness of the SRSO sublayer, sandwiched between adjacent silica sublayers, allowing the grain size to be controlled. However, the grains grow beyond the SRSO/silica interfaces and reach a size of about 5 nm when the films are annealed at 1200 °C, a temperature at which twins appear within the Si grains (as indicated by the small arrow in Figure 11.16).

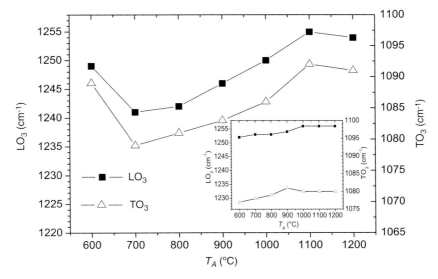

Figure 11.15 Evolution of the LO$_3$ (left scale) and TO$_3$ (right scale) peak positions with the annealing temperature T_A for a thick silica layer. The inset shows the corresponding positions recorded for a thick SRSO layer.

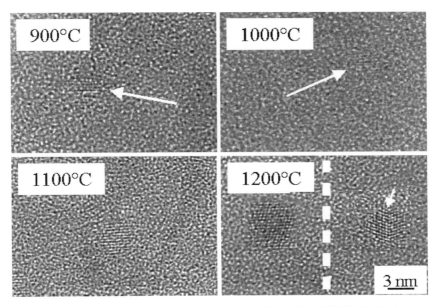

Figure 11.16 HREM images of the multilayers annealed from 900 to 1200 °C. For the highest temperature, the arrow indicates the presence of twin boundaries appearing within the nanocrystals.

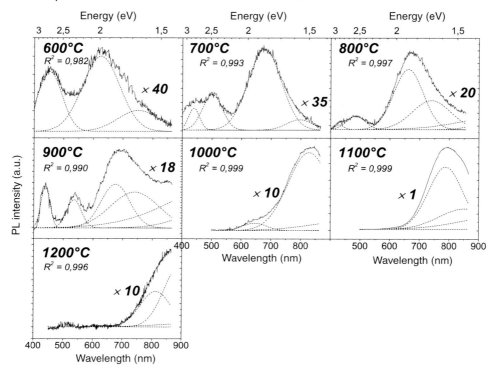

Figure 11.17 PL spectra recorded in the 400–850 nm range of the 3 nm-SRSO/19 nm-SiO$_2$ multilayer samples for the indicated T_A values. The multiplying factors reported on the spectra are relative to the most intense spectrum recorded after an annealing treatment at 1100 °C.

The PL spectra reported in Figure 11.17 for the 400–850 nm range were recorded on MLs annealed at various T_A values. These spectra show that increasing T_A up to 1100 °C leads to a systematic increase in the PL emission intensity I_{PL} (note the multiplication factor on each spectrum). I_{PL} decrease is thus observed for $T_A = 1200$ °C. After deconvolution, the 600 °C related PL spectrum reveals three emission band components: at 450 nm (2.76 eV), at 625 nm (1.99 eV), and at 750 nm (1.65 eV). When T_A increases, the relative intensities of these bands vary and a fourth band appears at 500 nm (2.48 eV) after an annealing at 700 °C. Only two of these bands remain after a treatment at 1000 °C, the weaker one shifting toward lower energies when T_A increases. The PL emission from the single thick SRSO and SiO$_2$ layers was also studied after identical annealing treatments. As expected, no luminescence properties were obtained for the SRSO structure since the microstructure of the deposited layers is similar to the one presented in Figure 11.4c for which the grain size is too big to exhibit visible PL. Concerning the silica thick layer, PL spectrum exhibits a weak signal located at about 630 nm (2.0 eV) that disappears when T_A is higher than 1000 °C (spectra not shown).

The comparison of the PL spectra recorded for the sputtered single silica layers and the MLs suggest that some bands observed for these originate from the silica matrix.

O'Reilly and Robertson [55] have determined the electronic structure of each possible defect in silica and have deduced the corresponding optical transitions. On the basis of this work, several authors have experimentally explored these optical transitions, particularly those located in the visible–near-infrared regions [56–58]. Three spectral regions can be associated with defects in silica: (i) from 1.85 eV (670 nm) to 2.0 eV (620 nm), (ii) at about 2.2 eV (564 nm), and (iii) from 2.4 eV (517 nm) to 2.8 eV (443 nm). The peak positions of the emission bands obtained from the MLs (full symbols) and the silica layers (open symbols) have been plotted in Figure 11.18 as a function of the annealing temperature T_A. The above-mentioned three spectral regions are also indicated through the hatched areas. This figure shows that some of the emission bands obtained for our MLs annealed at T_A below 1100 °C originate from the silica defects. When T_A reaches 1100 °C, these bands disappear confirming the recovery of the silica defects suggested by the evolution of the FTIR spectra (Figure 11.15). Regarding the I and Q bands described in the preceding section and attributed to the formation of Si nanograins, only the Q band is affected by the annealing temperature through a shift toward lower energies when $T_A = 1200$ °C. This result can be explained by an increase in the Si grain size in the SRSO sublayers, as evidenced by HRTEM observations (Figure 11.16).

The last point deals with the effect of T_A on the emission bands (Figure 11.17). On the one hand, the intensity of the band associated with silica defects increases from 600 to 800 °C and then drops until disappearing for $T_A = 1100$ °C. At low T_A, the numerous nonradiative defects explain the low emission efficiency, and their recovery with the annealing treatment favors the emission from the radiative defects of silica which, in turn, disappears above 1000 °C. On the other hand, the I and Q bands reach their maximum intensities for the samples annealed at 1100 °C. Below

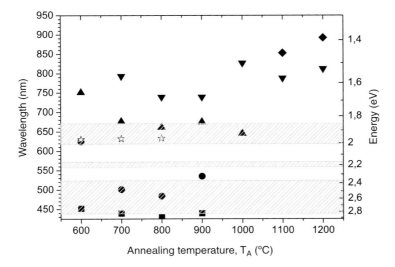

Figure 11.18 PL peak positions as a function of the annealing temperature T_A for the multilayers (full symbols) and silica thick films (open stars). The hachured areas represent the emission range of silica defects.

this temperature, the decrease in the defects density and the improvement in the interface quality between silicon and silica are responsible for a more pronounced increase in the luminescence. Above 1100 °C, both the appearance of new non-radiative recombination sites such as the twin boundaries evidenced by HREM experiments and the increase in the grain size might be responsible for the drop of the emission intensity.

11.4
Conclusion

In this paper, we have investigated the role of different fabrication parameters in the growth of Si nanograins within a SiO_2 host matrix using either a composite or a multilayer approach. Our goal was both to achieve the best structures presenting optimum photoluminescence properties and to understand the underlying physical mechanisms governing the emission process. The reactive process used for fabricating the layers allows to monitor the size and the density of the Si grains as it has been shown in the case of the composite $Si-SiO_2$ layers. In the case of the multilayer structure, the silica sublayer plays the double role of potential barrier and interdiffusion barrier and is therefore responsible for the confinement of the silicon grain inside the SRSO sublayer. This appears possible under our processing conditions with a minimum SiO_2 thickness of 19 nm. The SRSO sublayer thickness plays a key role in the luminescence and microstructure properties of the fabricated systems. Indeed, when the SRSO sublayer is thinner than 3 nm, silicon remains amorphous whereas for thicker sublayers, some silicon nanocrystals are formed. The particular process used to fabricate these multilayers allows the capping of the silicon nanograin surface by silicon oxide, leading to the observation of several emission bands in the near-infrared and visible regions. These bands can be ascribed to the carrier quantum confinement in the silicon nanostructures, to Si/SiO_2 interface states, and to levels localized in the bandgap of an amorphous silicon. The annealing temperature plays a major role: at low annealing temperature, silica defects act as luminescent centers in the high-energy range, before being recovered by high-temperature treatment. Such a thermal treatment favors the emission from Si nanocrystals that reaches a maximum after annealing at 1100 °C, hence fulfilling the conditions for optimum emission.

References

1 Canham, L.T. (1990) Silicon quantum wire array fabrication by electrochemical and chemical dissolution of wafers. *Appl. Phys. Lett.*, **57**, 1046–1048.

2 Kanemitsu, Y., Shimizu, N., Komoda, T., Hemment, P.L.F., and Sealey, B.J. (1996) Photoluminescent spectrum and dynamics of Si^+-ion-implanted and thermally annealed SiO_2 glasses. *Phys. Rev. B*, **54**, R14329–R14332.

3 Shimizu-Iwayama, T., Fujita, K., Nakao, S., Saitoh, S., Fujita, T., and Itoh, N. (1994) Visible photoluminescence in

Si$^+$-implanted silica glass. *J. Appl. Phys.*, **75**, 7779–7783.

4 Min, K.S., Shchegtov, K.V., Yang, C.M., Atwater, H., Brongersma, M.L., and Polman, A. (1996) Defect-related versus excitonic visible light emission from ion beam synthesized Si nanocrystals in SiO$_2$. *Appl. Phys. Lett.*, **69**, 2033–2035.

5 Liao, L.S., Bao, X.M., Zheng, X.Q., Li, N.S., and Min, B. (1996) Blue luminescence from Si$^+$-implanted SiO$_2$ films thermally grown on crystalline silicon. *Appl. Phys. Lett.*, **68**, 850–852.

6 Rebohle, L., Von Boramy, J., Skorupa, W., Tyschenko, I.E., and Frb, H. (1999) Photoluminescence and electroluminescence investigations at Ge-rich SiO$_2$ layers. *J. Lumin.*, **80**, 275–279.

7 Linros, J., Lalic, N., Galeckas, A., and Grivickas, V. (1999) Analysis of the stretched exponential photoluminescence decay from nanometer-sized silicon crystals in SiO$_2$. *J. Appl. Phys.*, **86**, 6128–6134.

8 Brongersma, M.L., Kik, P.G., Polman, A., Min, K.S., and Atwater, H.A. (2000) Size-dependent electron–hole exchange interaction in Si nanocrystals. *Appl. Phys. Lett.*, **76**, 351–353.

9 Garrido, B., Lopez, M., Garcia, C., Perez-Rodriguez, A., Morante, J.R., Bonafos, C., Carrada, M., and Claverie, A. (2002) Influence of average size and interface passivation on the spectral emission of Si nanocrystals embedded in SiO$_2$. *J. Appl. Phys.*, **91**, 798–807.

10 Salh, R., Fitting, L., Kolesnikova, E.V., Sitnikova, A.A., Zamoryanskaya, M.V., Schmidt, B., and Fitting, H.J. (2007) Si and Ge nanocluster formation in silica matrix. *Semiconductors*, **41**, 381–386.

11 Botti, S., Coppola, R., Gourbilleau, F., and Rizk, R. (2000) Photoluminescence from silicon nano-particles synthesized by laser-induced decomposition of silane. *J. Appl. Phys.*, **88**, 3396–3401.

12 Huisken, F., Hofmeister, H., Kohn, B., Laguna, M.A., and Paillard, V. (2000) Laser production and deposition of light-emitting silicon nanoparticles. *Appl. Surf. Sci.*, **154–155**, 305–313.

13 Li, X., He, Y., Talukdar, S.S., and Swihart, M.T. (2003) Process for preparing macroscopic quantities of brightly photoluminescent silicon nanoparticles with emission spanning the visible spectrum. *Langmuir*, **19**, 8490–8496.

14 Trave, E., Bello, V., Mattei, G., Mattiazzi, M., Borsella, E., Carpanese, M., Fabbri, F., Falconieri, M., D'Amato, R., and Herlin-Boime, N. (2006) Surface control of optical properties in silicon nanocrystals produced by laser pyrolysis. *Appl. Surf. Sci.*, **252**, 4467–4471.

15 Kenyon, A.J., Trwoga, P.F., and Pitt, C.W. (1996) The origin of photoluminescence from thin films of silicon-rich silica. *J. Appl. Phys.*, **79**, 9291–9300.

16 Wu, X.L., Siu, G.G., Tong, S., Liu, X.N., Yan, F., Jiang, S.S., Zhang, X.K., and Feng, D. (1996) Raman scattering of alternating nanocrystalline silicon/amorphous silicon multilayers. *Appl. Phys. Lett.*, **69**, 523–525.

17 Inokuma, T., Wakayama, Y., Muramoto, T., Aoki, R., Kurata, Y., and Hasegawa, S. (1998) Optical properties of Si clusters and Si nanocrystallites in high-temperature annealed SiO$_x$ films. *J. Appl. Phys.*, **83**, 2228–2234.

18 Wu, X., Bittner, A.M., Kern, K., Eggs, Ch., and Veprek, S. (2000) Kinetic oscillations of red photoluminescence from nanocrystalline Si/SiO$_2$ films. *Appl. Phys. Lett.*, **77** (5), 645–647.

19 Iacona, F., Franz, G., and Spinella, C. (2000) Correlation between luminescence and structural properties of Si nanocrystals. *J. Appl. Phys.*, **87**, 1295–1303.

20 Daldosso, N., Das, G., Larcheri, S., Dalba, G., Pavesi, L., Irrera, A., Priolo, F., Iacona, F., and Rocca, F. (2007) Silicon nanocrystal formation in annealed silicon-rich silicon oxide films prepared by plasma enhanced chemical vapor deposition. *J. Appl. Phys.*, **101**, 113510-1–113510-7.

21 Charvet, S., Madelon, R., Gourbilleau, F., and Rizk, R. (1999) Spectroscopic ellipsometry analyses of sputtered Si/SiO$_2$ nanostructures. *J. Appl. Phys*, **85**, 4032–4039.

22 Takeoka, S., Fujii, M., and Hayashi, S. (2000) Size-dependent

photoluminescence from surface-oxidized Si nanocrystals in a weak confinement regime. *Phys. Rev. B*, **62** (24), 16820–16825.

23 Kanemitsu, Y., Uto, H., Masumoto, Y., Matsumoto, T., Futagi, T., and Mimura, H. (1993) Microstructure and optical properties of free-standing porous silicon films: size dependence of absorption spectra in Si nanometer-sized crystallites. *Phys. Rev. B*, **48**, 2827–2830.

24 Schuppler, S., Friedman, S.L., Marcus, M.A., Adler, D.L., Xie, Y.H., Ross, F.M., Chabal, Y.J., Harris, T.D., Brus, L.E., Brown, W.L., Chaban, E.E., Szajowski, P.F., Christman, S.B., and Citrin, P.H. (1995) Size, shape and composition of luminescent species in oxidized Si nanocrystals and H-passivated porous Si. *Phys. Rev. B*, **52**, 4910–4925.

25 Coulthard, I., Jr., Antel, W.J., Freeland, J.W., Sham, T.K., Naftel, S.J., and Zhang, P. (2000) Influence of sample oxidation on the nature of optical luminescence from porous silicon. *Appl. Phys. Lett.*, **77** (4), 498–500.

26 Rinnert, H., Vergnat, M., and Burneau, A. (2001) Evidence of light-emitting amorphous silicon clusters confined in a silicon oxide matrix. *J. Appl. Phys.*, **89** (1), 237–243.

27 Cheylan, S., Elliman, R.G., Gaff, K., and Durandet, A. (2001) Luminescence from Si nanocrystals in silica deposited by helicon activated reactive evaporation. *Appl. Phys. Lett.*, **78**, 1670–1672.

28 Cullis, A.G., Canham, L.T., and Calcott, P.D.J. (1997) The structural and luminescence properties of porous silicon. *J. Appl. Phys.*, **82**, 909–965.

29 Delerue, C., Allan, G., and Lannoo, M. (1993) Theoretical aspects of the luminescence of porous silicon. *Phys. Rev. B*, **48**, 11024–11036.

30 Hill, N.A. and Whaley, K.B. (1995) Size dependence of excitons in silicon nanocrystals. *Phys. Rev. Lett.*, **75**, 1130–1133.

31 Kanemitsu, Y. (1996) Photoluminescence spectrum and dynamics in oxidized silicon nanocrystals: a nanoscopic disorder system. *Phys. Rev. B*, **53**, 13515–13520.

32 Allan, G., Delerue, C., and Lannoo, M. (1996) Nature of luminescent surface states of semiconductor nanocrystallites. *Phys. Rev. Lett.*, **76**, 2961–2964.

33 Garrido, B., Lopez, M., Gonzales, O., Perez-Rodriguez, A., Morante, J.R., and Bonafos, C. (2000) Correlation between structural and optical properties of Si nanocrystals embedded in SiO_2: the mechanism of visible light emission. *Appl. Phys. Lett.*, **77**, 3143–3145.

34 Daldosso, N., Luppi, M., Ossicini, S., degoli, E., Magri, R., Dalba, G., Fornasini, P., Grisenti, R., Rocca, F., Pavesi, L., Bonelli, S., Priolo, F., Spinella, C., and Iacona, F. (2003) Role of the interface region on the optoelectronic properties of silicon nanocrystals embedded in SiO_2. *Phys. Rev. B*, **68**, 085327-1–085327-8.

35 Ternon, C., Dufour, C., Gourbilleau, F., and Rizk, R. (2004) Roles of interfaces in nanostructured silicon luminescence. *Eur. Phys. J. B*, **41**, 325–332.

36 Ledoux, G., Guillois, O., Porterat, D., Reynaud, C., Huisken, F., Kohn, B., and Paillard, V. (2000) Photoluminescence properties of silicon nanocrystals as a function of their size. *Phys. Rev. B*, **62**, 15942–15951.

37 Sullivan, B.T., Lockwood, D.J., Labbe, H.J., and Lu, Z.-H. (1996) Photoluminescence in amorphous Si/SiO_2 superlattices fabricated by magnetron sputtering. *Appl. Phys. Lett.*, **69**, 3149–3151.

38 Tsybeskov, L., Hirshman, K.D., Dutlagupta, S.P., Zacharias, M., Fauchet, P.M., McCaffrey, J.P., and Lockwood, D.J. (1998) Nanocrystalline-silicon superlattice produced by controlled recrystallization. *Appl. Phys. Lett.*, **72**, 43–45.

39 Zacharias, M., Blsing, J., Veit, P., Tsybeskov, L., Hirschman, K., and Fauchet, P. (1999) Thermal crystallization of amorphous Si/SiO_2 superlattices. *Appl. Phys. Lett.*, **74**, 2614–2616.

40 Vinciguerra, V., Franz, G., Priolo, F., Iacona, F., and Spinella, C. (2000) Quantum confinement and recombination dynamics in silicon nanocrystals embedded in Si/SiO_2 superlattices. *J. Appl. Phys.*, **87**, 8165–8173.

41 Gourbilleau, F., Portier, X., Ternon, C., Voivenel, P., Madelon, R., and Rizk, R.

(2001) Si-rich/SiO$_2$ nanostructured multilayers by reactive magnetron sputtering. *Appl. Phys. Lett.*, **78**, 3058–3060.

42 Gaburro, Z., Oton, C.J., and Pavesi, L. (2004) Opposite effects of NO$_2$ on electrical injection in porous silicon. *Appl. Phys. Lett.*, **84**, 4388–4390.

43 Ghulinyan, M., Oton, C.J., Gaburro, Z., Bettoti, P., and Pavesi, L. (2003) Porous silicon free-standing coupled microcavities. *Appl. Phys. Lett.*, **82**, 1550–1552.

44 Belarouci, A. and Gourbilleau, F. (2007) Microcavity enhanced spontaneous emission from silicon nanocrystals. *J. Appl. Phys.*, **101**, 073108-1–073108-4.

45 Bonafos, C., Coffin, H., Schamm, S., Cherkashin, N., Ben Assayag, G., Dimitrakis, P., Normand, P., Carrada, M., Paillard, V., and Claverie, A. (2005) Si nanocrystals by ultra-low-energy ion beam-synthesis for non-volatile memory applications. *Solid-State Electr.*, **49**, 1734–1744.

46 Cho, E.C., Park, S., Hao, X., Song, D., Conibeer, G., Park, S.C., and Green, M.A. (2008) Silicon quantum dot/crystalline silicon solar cells. *Nanotechnology*, **19**, 245201–245206.

47 Olsen, J.E. and Shimura, F. (1989) Infrared reflection spectroscopy of the SiO$_2$-silicon interface. *J. Appl. Phys.*, **66**, 1353–1358.

48 Okamoto, S. and Kanemitsu, Y. (1997) Quantum confinement and interface effects on photoluminescence from silicon single quantum wells. *Solid State Commun.*, **103**, 573–576.

49 Kanemitsu, Y. and Okamoto, S. (1997) Photoluminescence from Si/SiO$_2$ single quantum wells by selective excitation. *Phys. Rev. B*, **56**, R15561–R15564.

50 Marra, D.C., Aydil, E.S., Joo, S.-J., Yoon, E., and Srdanov, V.I. (2000) Angle-dependent photoluminescence spectra of hydrogenated amorphous silicon thin films. *Appl. Phys. Lett.*, **77**, 3346–3348.

51 Street, R.A. (1976) Luminescence in amorphous semiconductors. *Adv. Phys.*, **25**, 397–453.

52 Lockwood, D.J., Lu, Z.H., and Baribeau, J.-M. (1996) Quantum confined luminescence in Si/SiO$_2$ superlattices. *Phys. Rev. Lett.*, **76**, 539–541.

53 Degoli, E. and Ossicini, S. (2000) The electronic and optical properties of Si/SiO$_2$ superlattices: role of confined and defect states. *Surf. Sci.*, **470**, 32–42.

54 Veprek, S., Iqbal, Z., and Sarott, F.A. (1982) A thermodynamic criterion of the crystalline-to-amorphous transition in silicon. *Phil. Mag. B*, **45**, 137–145.

55 O'Reilly, E.P. and Robertson, J. (1983) Theory of defects in vitreous silicon dioxide. *Phys. Rev. B*, **27**, 3780–3795.

56 Tohmon, R., Shimogaichi, Y., Mizuno, H., Ohki, Y., Nagasawa, K., and Hama, Y. (1989) 2.7-eV luminescence in as-manufactured high-purity silica glass. *Phys. Rev. Lett.*, **62**, 1388–1391.

57 Munekuni, S., Yamanaka, T., Shimogaichi, Y., Tohmon, R., Ohki, Y., Nagasawa, K., and Hama, Y. (1990) Various types of nonbridging oxygen hole center in high-purity silica glass. *J. Appl. Phys.*, **68**, 1212–1217.

58 Nishikawa, H., Watanabe, E., Ito, D., and Ohki, Y. (1994) Decay kinetics of the 4.4-eV photoluminescence associated with the two states of oxygen-deficient-type defect in amorphous SiO$_2$. *Phys. Rev. Lett.*, **72**, 2101–2104.

12
Si and SiC Nanocrystals by Pyrolysis of Sol–Gel-Derived Precursors

Aylin Karakuscu and Gian Domenico Soraru

Optical gain from silicon has become one of the hottest research topics in recent years due to Si abundance and its technological production easiness. One of the most interesting approaches is to produce a suitable dispersion of Si nanocrystals inside SiO_2 matrix. Accordingly, ion implantation, sputtering, and different types of deposition methods were reported to produce Si-nc/SiO_2 systems. In all these methods, ion implantation is used to form Si-rich SiO_2 system (usually called SiO_x-substoichiometric silica) that subsequently leads to the *in situ* formation of Si clusters inside the SiO_2 matrix. However, most of these processes require complicated equipment and the efficiency of these systems is still quite low due to both low density of Si-ncs and the large size distribution. Overcoming these problems may not be easy since increasing Si-nc concentration by increasing the implantation dose both produces defects due to high-energy usage and increases the Si-ncs size, thus finally reducing the optical gain.

One approach that could solve these problems is to focus on simpler production methods such as the polymer pyrolysis process. According to this route, Si—H containing polymer-based precursors are converted into a ceramic material via pyrolysis in controlled atmosphere at a temperature above 800–1000 °C. During the pyrolysis, Si—H bonds react to form Si—Si bonds with the evolution of H_2 gas. The synthesis of the Si—H polymer precursor can be done either by a sol–gel route or by an inorganic polymerization process. The use of the sol–gel process, due to the large availability of many kinds of commercial hybrid silicon alkoxides, is more flexible and more suitable to precisely tailor the composition of the resulting hybrid silicon network. Accordingly, the sol–gel method has been reported for the production of bulk and thin films of the Si-nc/SiO_2 systems and for doped systems such as Eu^{2+}-, Er^{3+}-doped Si-nc/SiO_2.

In this chapter, first a brief description of the sol–gel process will be provided followed by an introduction to the polymer pyrolysis to polymer-derived ceramics (PDCs). Then, the synthesis of Si-nc/SiO_2 materials, either as bulk or thin films, will be described, followed by the synthesis of Si-nc and SiC-nc containing SiO_2. At last,

Silicon Nanocrystals: Fundamentals, Synthesis and Applications. Edited by Lorenzo Pavesi and Rasit Turan
Copyright © 2010 WILEY-VCH Verlag GmbH & Co. KGaA, Weinheim
ISBN: 978-3-527-32160-5

some recent papers on the optical characteristics of multicomponent Si−C−O−N ceramics will be presented and discussed. In the conclusion, the possible evolution of Si nanocrystals from sol–gel-derived precursors will be discussed.

12.1
The Sol–Gel and PDC Processes

A detailed description of the sol–gel process to produce novel nanostructured materials is beyond the scope of this chapter. Here, only concepts important for understanding the application of this method for the synthesis of tailored preceramic networks for producing Si and SiC nanocrystals will be presented. In its essential steps, the sol–gel process is based on the hydrolysis and condensation reactions of silicon alkoxides. Alkoxides are metal organic compounds of the general formula, $Me(OR)_n$, where Me is a metal such as Si, Al, Ti, and so on and OR is the alkoxy group with $R = CH_3, CH_2CH_3, CH_2CH_2CH_3$, and so on. One of the most popular alkoxides in the sol–gel technology is TEOS or tetraethylorthosilicate with the formula $Si(OCH_2CH_3)_4$. Since CH_2CH_3 is the "ethyl" group, the OR group is reported as OEt, or ethoxy, group.

The first reaction of the sol–gel process is the *hydrolysis* that induces the substitution of the OR groups linked to silicon with silanol Si−OH groups. Then, these chemical species react together to form Si−O−Si (siloxane) bonds that lead to the formation of the silica network. This second reaction is called *condensation*. The advancement of the hydrolysis and condensation reactions leads to the formation of the tridimensional silica network. Usually, these two reactions are performed at room temperature by dissolving the alkoxides and water into a common solvent, such as an alcohol, which remains in the pores of the solid network at the end of the reactions. The "wet gel" must therefore be dried to obtain what is called "xerogel." The characteristic features of the resulting xerogel, among others, are the porosity (where pore size and pore size distribution are important), the amount of residual nonhydrolyzed alkoxy groups and residual noncondensed silanol. Moreover, the chemical homogeneity of the network can be obtained when a multicomponent gel is prepared. These features can be controlled to a certain extent by the proper choice of the following chemical parameters: the amount of water used for the hydrolysis, usually reported as the hydrolysis molar ratio, $H_2O/(metal\ alkoxide)$; the amount of solvent, again reported as the molar ratio of solvent and alkoxide and the type and amount of catalyst (acid or basis). A *hybrid silicon alkoxide* is an alkoxide in which one (or more) alkoxy group is substituted by an organic group directly bonded to the Si atom via a Si−C bond. The general formula of the hybrid silicon alkoxides is R_x−Si$(OR')_{4-x}$, $x = 1, 2, 3$; R and R' are organic groups. If R = H, the hybrid silicon alkoxides bear Si−H bonds. Since the Si−C bonds are stable under the hydrolysis conditions, the use of hybrid silicon alkoxides affords the opportunity to produce hybrid silica networks that are modified by organic moieties directly bonded to silicon atoms. A comprehensive report of the fundamental principles of sol–gel process can be found in the Brinker and Scherer book, *Sol–Gel Science* [1], while both the basic sciences and the technological applications are well presented in the *Handbook of Sol–Gel Science and Technology* edited by Professor Sumio Sakka [2].

Polymer-derived ceramics are a new family of nanostructured ceramic materials derived from inorganic polymeric precursors [3]. According to this route, the preceramic polymer is shaped, cross-linked, and pyrolized in controlled atmosphere (inert or reactive) at temperature exceeding 800 °C and converted into self-similar ceramic devices and components [4]. PDCs have shown extraordinary properties, especially at ultrahigh temperature such as oxidation and creep resistance [5, 6], chemical durability [7], electrical conductivity (from insulating up to semiconducting behavior) [8], and photo- and electroluminescence [9]. Such unusual properties originate from the atypical nanostructure of these ceramics in which various nanocrystalline phases grow *in situ* into an amorphous C-containing matrix [10, 11]. Two PDC systems have been widely studied: the silicon oxycarbides and the silicon carbonitrides. The composition regimes where they can be synthesized are shown in the composition diagrams in Figure 12.1. In both cases, the amorphous phases are generally formed in carbon-rich regimes, relative to the stoichiometric mixtures of the crystalline forms. The possibility of synthesizing PDCs of different composition in the ternary $Si-C-O$ and $Si-C-N$ phase diagrams depends on the possibility of synthesizing preceramic precursors with tailored architecture and chemical composition [12, 13]. Accordingly, stoichiometric silicon oxycarbide glasses, the composition of which depends on the tie line between SiC and SiO_2 (indicated with arrow in Figure 12.1), have been obtained from preceramic polymers with tailored composition [14]. This result has been achieved using the sol–gel method for the synthesis of the preceramic network. Indeed, the sol–gel process affords the possibility, through

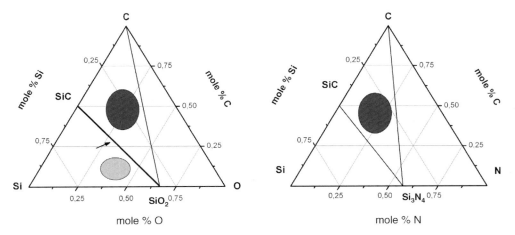

Figure 12.1 Composition regimes for two families of polymer-derived ceramics: silicon oxycarbides and silicon carbonitrides. Circles with dark gray color represent the well-studied areas of each ceramic. In SiOC phase diagram, additional circle with light gray color indicates the Si-rich SiOC region, discussed in the chapter. The arrow shows the stoichiometric SiOC line.

the use of different molecular precursors, to precisely tailor the composition of the resulting polymeric network [15]. Moreover, the sol–gel method allowed the formation of PDC compositions inside the $Si-SiO_2-SiC$ compatibility triangle [16] and even in the binary $Si-SiO_2$ system [17]. These last compositions are actually used for the synthesis of Si/SiC nanocrystals and of Si nanocrystals, respectively, embedded in an amorphous SiO_2 matrix.

12.2
Si nc/SiO₂ Glasses and Films

The attempt to produce sol–gel-derived Si-nc/SiO₂ bulk glasses was first made by Soraru *et al.* [18], using a pure triethoxysilane ($HSi(OEt)_3$) precursor. Triethoxysilane (T^H – also known as TREOS), as a preceramic precursor, was chosen due to the presence in its structure of Si–H bonds that form Si–Si bonds during pyrolysis in inert atmosphere and leads to the *in situ* formation of Si nanocrystals in the silica matrix. The sol–gel steps for the synthesis of the Si–H containing precursor are shown in Figure 12.2.

During pyrolysis, the Si–H moieties present in the precursor reacted to produce H_2 and Si–Si bonds that above 1000 °C reorganize and form Si nanocrystals embedded in the residual silica matrix. The formation of Si-ncs is verified by means of XRD and µ-Raman measurements between 1000 and 1200 °C. Meanwhile, a broad photoluminescence peak was observed, showing a redshift from 600 to 800 nm by increasing the heating temperature. The increase in the peak emission wavelength, together with the corresponding decrease in the FWHM, was assigned to the increase in the nanocrystal's size, from 3.0 to 4.4 nm, as measured by XRD (see Figure 12.3) [18].

Sol–gel process is well suited for preceramic film production by dip- or spin-coating methods. The resultant preceramic films can be converted into the corresponding ceramic films provided that the substrate is stable up to the maximum pyrolysis temperature. Accordingly, Das *et al.* [19] obtained Si-nc/SiO₂ thin films on Si substrate by spin coating a sol–gel solution of T^H. Films were annealed at different temperatures between 600 and 1300 °C, under N_2 gas atmosphere. The PL spectrum for 1200 °C shows an extremely intense band at around 790 nm, attributed to the

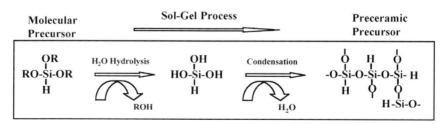

Figure 12.2 Sol–gel synthesis of the Si–H containing gel, precursor for Si-nc/SiO₂ materials.

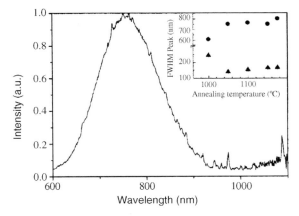

Figure 12.3 Photoluminescence spectrum of a typical bulk T^H-derived Si-nc/SiO₂ sample annealed at 1100 °C for 1 h. The inset shows the evolution with the pyrolysis temperature of the peak position and of the full width at half maximum. Reprinted with permission from the American Institute of Physics. Copyright (2003) by Soraru.

formation of Si-ncs. By further increasing the annealing temperature, a redshift is observed due to the increase in particle size, as shown in Figure 12.4.

Pivin *et al.* extensively studied the formation and luminescence behavior of sol–gel and polymer-derived films treated with ion irradiation [20–23]. This process does not involve heating of the sample and it is a valid alternative to the thermal conversion in case a limitation in the maximum temperature exists. Accordingly, Si-nc/SiO₂ thin films have been obtained by irradiating with Au ions a T^H-derived sol–gel film. Photoluminescence study indicated that the intensity of the luminescence did not decrease at high ion irradiation under the effect of the growth of Si clusters. In addition, it has been shown that films converted by thermal treatment at 1000 °C gave higher luminescence than ion-irradiated ones [20].

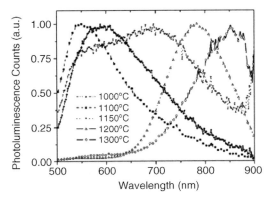

Figure 12.4 Normalized PL spectrum for the Si-nc/SiO₂ thin films (annealed at different temperatures) in the frequency range between 500 and 900 nm. Reprinted with permission of Elsevier. Copyright (2008) by Das.

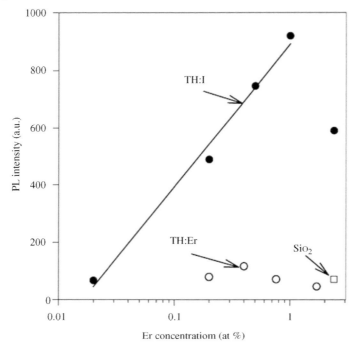

Figure 12.5 Comparison of intensities of infrared PL in Er-implanted TH (TH-I), Er-implanted SiO$_2$ and TH:Er films, all annealed at 1050 °C. Reprinted with permission of Springer. Copyright (2003) by Piven.

Piven *et al.* also studied the formation of Er-doped Si-nc/SiO$_2$ films produced by sol–gel-derived precursors with the aim of enhancing the excitation of Er atoms. Accordingly, Pivin showed that Si-nc/SiO$_2$ composite ceramics, pyrolyzed from pure triethoxysilane gel precursor, is an excellent host for Er ions. With the presence of Si-ncs, luminescence of Er ions was improved by a factor of 10 compared to pure silica matrix, shown in Figure 12.5. The best luminescence results were found when Er/T^H ratio was limited to less than 2% or annealing temperature was lower than 1050 °C [24].

12.3
(Si-nc + SiC-nc)/SiO$_2$ Glasses and Films

Modified silica gels obtained from cohydrolysis of triethoxysilane (T^H) and methyl-diethoxysilane, HMeSi(OEt)$_2$ (D^H, also known as MDES), afford the possibility of precisely controlling the oxygen and carbon contents in the gel by varying the relative amount of the two alkoxides. For these gels, the amount of carbon is related to the amount of oxygen according to the solid line in Figure 12.6 and it ranges from

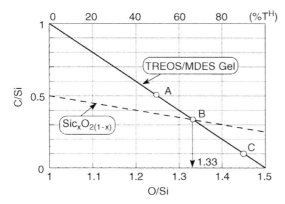

Figure 12.6 C/Si versus O/Si for gel samples obtained from cohydrolysis of TREOS and MDES (solid line) and for the stoichioimetric oxycarbide phase (dotted line). The compositions of the studied gels (A, B, and C) are also reported. Reprinted with permission from Blackwell-Wiley. Copyright (1995) by Soraru.

0 (pure T^H) to 1 ($T^H/D^H = 1$) [14]. In the same figure, the dotted line shows the relationship between the C and O contents in the stoichiometric oxycarbide phase. The two lines cross each other for a value of O/Si = 1.33, corresponding to a T^H/D^H molar ratio of 2. In this case, the carbon content of the gel matches the carbon amount of the corresponding oxycarbide phase and therefore this composition (B in Figure 12.6) has been selected for the synthesis of stoichiometric, transparent SiCO glass. If the T^H/D^H molar ratio is higher than two (sample C), the corresponding SiCO glass contains Si−Si bonds, while if the T^H/D^H ratio is lower than 2, such as in sample A, the final SiCO network contains C−C bonds and shows the tendency to form a free carbon phase. Gel precursors obtained from pure T^H lead to the formation of silica-based network with Si−Si bonds and no Si−C bonds, essentially a sub-stoichiometric silica glass (explained in Section 12.2).

The high-temperature annealing at 1200–1400 °C of amorphous silicon oxycarbide glasses activates a phase separation process that leads to the formation of nanocrystals of SiC and/or Si, depending on the composition of the starting glass [16, 25].

In order to investigate the luminescence properties, SiCO thin films were synthesized by the polymer pyrolysis method from sol–gel-derived precursors [26]. The precursor ratio is chosen ($T^H/D^H = 2$) in order to get stoichiometric SiCO composition without free carbon. According to photoluminescence analysis, films pyrolyzed at low temperatures (800–1000 °C) and high temperatures (1100–1250 °C) showed two different behaviors. Low-temperature pyrolysis caused intense blue luminescence that is believed to originate from defect-related centers appearing in amorphous SiCO structure. By increasing the pyrolysis temperature, defects were quenched; therefore, blue emission peak disappeared above 1000 °C. Starting from this temperature, a green-yellow peak centered at 560 nm (2.25 eV) appeared and reached a maximum at 1200 °C. This broadband was assigned to *in situ* formation of SiC-nc from the SiCO phase (Figure 12.7).

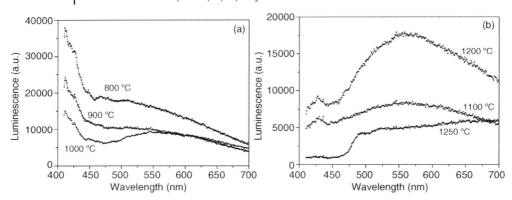

Figure 12.7 PL spectra of the stoichiometric SiOC films pyrolyzed at (a) 800–1000 °C and (b) 1100–1250 °C. Reprinted with permission from American Ceramic Society. Copyright (2008) by Karakuscu.

Both Si- and C-rich SiCO system and SiO_x system have also been studied by Pivin *et al.* Polysiloxanes, polycarbosilanes, and sol–gel-derived films were converted into ceramic films by irradiating them with C or Au ions: they showed a strong yellow PL emission [27–30]. The energy transferred by the ions to electron shells of target atoms is given within the range of 20–60 eV/atom. This PL was assigned to the segregation of C clusters into an amorphous $SiO_xC_yH_z$ matrix. The samples prepared by stoichiometric SiCO composition $(T^H/D^H = 2)$ showed luminescence only with traditional annealing since cluster formation could not be achieved with ion irradiation. They showed good luminescence properties compared to polysiloxanes, in which the C cluster formation was higher leading to a percolation effect and a consequent quenching of the photoluminescence. Correspondingly, for larger energy transfers the photoluminescence shifts toward red and then vanishes because of the growth and percolation of the C quantum dots [29].

SiCO system was found to be a suitable host for erbium and europium since it provides an inert/reducing matrix, stable in extreme environments (high temperature, aggressive media, etc.), and may even increase the light emission [31–33]. Moreover, the synthesis of the preceramic networks through the sol–gel process facilitated the introduction of these extra elements (Er and Eu) by dissolving them in the starting solution. Accordingly, transparent nanostructured erbium-activated SiC/SiO_2 nanocomposites were synthesized by annealing an Er-containing stoichiometric SiCO glass $(T^H/D^H = 2)$ at 1200–1300 °C. The Er^{3+} was introduced in the sol–gel solution using $ErCl_3 \cdot 6H_2O$. The characteristic emission peak of the $^4I_{13/2} \rightarrow {}^4I_{15/2}$ Er^{3+} transition at 1.55 μm was observed, and its intensity increased both with the pyrolysis temperature and with the Er concentration (at least up to 4% mol). The latter result indicated that the SiCO network can accommodate Er^{3+} without appreciable matrix strains. Moreover, the luminescence peak shape is not a characteristic of a crystalline environment, suggesting that the majority of the Er ions are embedded into the glass matrix [31]. In addition, $Eu(NO_3)_3$ is introduced in a stoichiometric SiCO sol–gel solution with the aim of synthesizing a Eu-containing transparent SiCO glass [33]. In this study, the pyrolysis was carried in Ar flow at different temperatures, and Eu^{3+}-based luminescence was observed up to 400 °C. After this temperature, a broad blue emission band, centered around 450 nm, was formed due to the reduction of Eu^{3+} to Eu^{2+}. These results suggest that the SiCO network led *in situ* formation of Eu^{2+} ions and provided reducing species during the pyrolysis.

12.4
Optical properties of Multicomponent Si-C-O-N Ceramics

The synthesis and the luminescence behavior of SiCN and SiOCN polymer-derived ceramic powders from commercially available Si-based polymers have been recently investigated. Liquid poly(ureamethylvinyl)silazane (known as Ceraset) and solid polymethylsilsesquioxane (known as MK) have been used as preceramic precursors, to propose a new transparent, thermally stable, and formable photoluminescent material for possible LED applications. Both Si-based polymers showed fluorescence in the UV range when heat treated at low temperatures (up to 500 °C). Annealing at 500 and 600 °C caused a white luminescence for the polysiloxane MK-derived sample unlike the polysilazane-derived sample that gives an emission in the blue-green range. The emission at low temperatures and the redshift with increasing heat treatment temperature were supposed to arise from dangling bonds and were caused by sp^2 carbon. The presence of free carbon diminished the luminescence properties at higher temperatures [34].

Similarly, Ferraioli *et al.* showed that SilconOxyCarboNitrides (SiCNO), produced from thermal treatments of highly cross-linked Si-, N-, and O-containing polymers, exhibited intense luminescence. Effects of the composition and heat treatment temperature on the luminescence were examined. High and intense emission was

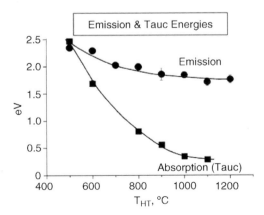

Figure 12.8 Comparison of emission energy and absorption bandgap with respect to annealing temperature. Reprinted with permission from Blackwell-Wiley. Copyright (2008) by Ferraioli.

observed in the broad range of 500–800 nm. Changing the annealing temperature led a redshift in the emission peak from 2.5 to 2 eV, whereas the absorption coefficient drastically decreased suggesting that photoluminescence and absorption may be caused by two different, spatially separated, molecular complexes (Figure 12.8). Since the network contains mixed bonds of Si−C−N−O tetrahedral, photoluminescence was assumed to arise from the mixed bond tetrahedra, whereas the absorption was assumed to occur in the graphene network [35].

Acknowledgments

The support of the EU through the Marie Curie Research Training network, PolyCerNet, CT 019601, is gratefully acknowledged.

References

1 Brinker, C.J. and Scherer, G.W. (1990) *Sol–Gel Science*, Academic Press, San Diego, CA.

2 Sakka, S. (2005) *Handbook of Sol–Gel Science and Technology: Processing, Characterization and Applications*, vols 2–3, Kluwer Academic Publishers, Boston.

3 Raj, R., Riedel, R., and Soraru, G.D. (2001) Introduction to the special topical issue on ultrahigh-temperature polymer-derived ceramics. *J. Am. Ceram. Soc.*, **84** (10), 2158–2159.

4 Liu, X., Li, Y.-L., and Hou, F. (2009) Fabrication of SiOC ceramic microparts and patterned structures from polysiloxanes via liquid cast and pyrolysis. *J. Am. Ceram. Soc.*, **92** (1), 49–53.

5 Modena, S., Soraru, G.D., Blum, Y., and Raj, R. (2005) Passive oxidation of an effluent system: the case of a polymer derived SiOC. *J. Am. Ceram. Soc.*, **88** (2), 339–345.

6 An, L., Riedel, R., Konetschny, C., Kleebe, H.-J., and Raj, Rishi (1998) Newtonian viscosity of amorphous silicon carbonitride at high temperature. *J. Am. Ceram. Soc.*, **81** (5), 1349–1352.

7 Soraru, G.D., Modena, S., Guadagnino, E., Colombo, P., Egan, J., and Pantano, C. (2002) Chemical durability of silicon-oxycarbide glasses. *J. Am. Ceram. Soc.*, **85** (6), 1529–1536.

8 Trassl, S., Puchingera, M., Rossler, E., and Ziegler, G. (2003) Electrical properties of amorphous $SiC_xN_yH_z$-ceramics derived from polyvinylsilazane. *J. Eur. Ceram. Soc.*, **23**, 781–789.

9 Loner, S. (2001) Vetri Ossicarburi Trasparenti, Tesi di Laurea in Ingegneria dei Materiali, Università di Trento, Italy.

10 Saha, A., Raj, R., and Williamson, D.L. (2006) A model for the nanodomains in polymer-derived SiCO. *J. Am. Ceram. Soc.*, **89** (7), 2188–2195.

11 Gregori, G., Kleebe, H.-J., Blum, Y., and Babonneau, F. (2006) Evolution of C-rich SiOC ceramics: Part II, characterization by high lateral resolution techniques: electron energy loss spectroscopy, high-resolution TEM and energy-filtered TEM. *Int. J. Mater. Res.*, **97**, 710–720.

12 Dirè, S., Oliver, M., and Soraru, G.D. (1999) Effect of polymer architecture on the formation of Si-O-C glasses, in *Advanced Synthesis and Processing of Composites and Advanced Ceramics II (Ceramic Transactions)* vol. 79 (eds K.V. Logan, Z. Munir, and R.M. Spriggs), American Ceramic Society, Westerville, OH, pp. 251–261.

13 Soraru, G.D. (1994) Silicon oxycarbide glasses from gels. *J. Sol–Gel Sci. Technol.*, **2**, 843–848.

14 Soraru, G.D., D'Andrea, G., Campostrini, R., Babonneau, F., and Mariotto, G. (1995) Structural characterization and high

temperature behaviour of silicon oxycarbide glasses prepared from sol–gel precursors containing Si–H bonds. *J. Am. Ceram. Soc.*, **78**, 379–387.

15 Sorarù, G.D., D'Andrea, G., Campostrini, R., and Babonneau, F. (1995) Si–O–C glasses from gels, in *Proceedings of the International Symposium on Sol–Gel Science and Technology (Sol–Gel Science and Technology)*, American Ceramic Society, pp. 135–146.

16 Bréquel, H., Parmentier, J., Walter, S., Badheka, R., Trimmel, G., Masse, S., Latournerie, J., Dempsey, P., Turquat, C., Desmartin-Chomel, A., Le Neindre-Prum, L., Jayasooriya, U.A., Hourlier, D., Kleebe, H.-J., Sorarù, G.D., Enzo, S., and Babonneau, F. (2004) Systematic structural characterisation of the high temperature behaviour of nearly-stoichiometric silicon oxycarbide glasses. *Chem. Mater.*, **16**, 2585–2598.

17 D'Andrea, G. (1995) Vetri SiOC via Sol–Gel, Tesi di Dottorato di Ricerca in Ingegneria dei Materiali, Università di Trento, Italy.

18 Soraru, G.D., Modena, S., Bettotti, P., Das, G., Mariotto, G., and Pavesi, L. (2003) Si nanocrystals obtained through polymer pyrolysis. *Appl. Phys. Lett.*, **83**, 749–751.

19 Das, G., Ferraioli, L., Bettotti, P., De Angelis, F., Mariotto, G., Pavesi, L., Di Fabrizio, E., and Soraru, G.D. (2008) Si-nanocrystals/SiO$_2$ thin films obtained by pyrolysis of sol–gel precursors. *Thin Solid Films*, **516**, 6804–6807.

20 Pivin, J.C., Colombo, P., Martucci, A., Soraru, G.D., Pippel, E., and Sendova-Vassileva, M. (2003) Ion beam induced conversion of Si-based polymers and gels layers into ceramics coatings. *J. Sol.-Gel. Sci. Technol.*, **26**, 251–255.

21 Pivin, J.C. and Jimenez de Castro, M. (2003) Erbium luminescence in suboxide films derived from triethoxysilane. *J. Sol.-Gel. Sci. Technol.*, **28**, 37–43.

22 Pivin, J.C., Sendova-Vassileva, M., Colombo, P., and Martucci, A. (2000) Photoluminescence of composite ceramics derived from polysiloxanes and polycarbosilanes by ion irradiation. *Mater. Sci. Eng. B*, **69–70**, 574–577.

23 Pivin, J.C., Jimenez de Castro, M., Hofmeister, H., and Sendova-Vassileva, M. (2003) Exciton-erbium coupling in SiO$_x$ suboxide films prepared by combining sol–gel chemistry and ion implantation. *Mater. Sci. Eng. B*, **97**, 13–19.

24 Pivin, J.C., Jimenez de Castro, M., and Sendova-Vassileva, M. (2003) Optical activation of Er ions by Si nanocrystals in films synthesized by sol–gel chemistry and ion implantation. *J. Mater. Sci. Mater. Electron.*, **14**, 661–664.

25 Burns, G.T., Taylor, R.B., Xu, Y., Zangvil, A., and Zank, G.A. (1992) High-temperature chemistry of the conversion of siloxanes to silicon carbide. *Chem. Mater.*, **4**, 1313–1323.

26 Karakuscu, A., Guider, R., Pavesi, L., and Soraru, G.D. (2008) Synthesis and optical properties of SiCn/SiO2 nanocomposite thin films, in *Proceedings of the Nanostructured Materials and Nanotechnology II, Ceramic Engineering and Science*, vol. 29 (eds S. Mathur and M. Singh), American Ceramic Society, pp. 85–91.

27 Pivin, J.C. and Colombo, P. (1997) Ceramic coatings by ion irradiation of polycarbosilanes and polysiloxanes: 1. Conversion mechanism. *J. Mater. Sci.*, **32**, 6163–6173.

28 Pivin, J.C., Colombo, P., and Sendova-Vassileva, M. (1998) Ion-induced conversion of polysiloxanes and polycarbosilanes into ceramics: mechanisms and properties. *Nucl. Instrum. Methods Phys. Res.*, **141**, 652–662.

29 Pivin, J.C., Colombo, P., and Soraru, G.D. (2000) Comparison of ion irradiation effects in silicon-based preceramic thin films. *J. Am. Ceram. Soc.*, **83**, 713–720.

30 Pivin, J.C., Sendova-Vassileva, M., and Colombo, P. (2000) Photoluminescence of composite ceramics derived from polysiloxanes and polycarbosilanes by ion irradiation. *Mater. Sci. Eng. B*, **69**, 574–577.

31 Soraru, G.D., Zhang, Y., Ferrari, M., Zampedri, L., and Goncalves, R.R. (2005) Novel Er-doped SiC/SiO$_2$ nanocomposites: synthesis via polymer pyrolysis and their optical

characterization. *J. Eur. Ceram. Soc.*, **25**, 277–281.

32 Pivin, J.C., Gaponenko, N.V., and Mudryi, A.V. (2000) Photoluminescence of Er-implanted silica, polysiloxane and porous silicon films. *Mater. Sci. Eng. B*, **69**, 215–218.

33 Zhang, Y., Quaranta, A., and Soraru, G.D. (2004) Synthesis and luminescent properties of novel Eu^{2+}-doped silicon oxycarbide glasses. *Opt. Mater.*, **24**, 601–605.

34 Menapace, I., Mera, G., Riedel, R., Erdem, E., Eichel, R.A., Pauletti, A., and Appleby, G.A. (2008) Luminescence of heat-treated silicon-based polymers: promising materials for LED applications. *J. Mater. Sci.*, **43**, 5790–5796.

35 Ferraioli, L., Ahn, D., Saha, A., Pavesi, L., and Raj, R. (2008) Intensely photoluminescent pseudo-amorphous SiliconOxyCarboNitride polymer–ceramic hybrids. *J. Am. Ceram. Soc.*, **91**, 2422–2424.

13
Nonthermal Plasma Synthesis of Silicon Nanocrystals

Lorenzo Mangolini and Uwe Kortshagen

13.1
Introduction

Nanoparticle formation in plasmas has been intensively studied for almost 20 years [1, 2]. In its early years, this work was motivated by particle formation occurring as a contamination problem both in the semiconductor processing and in the manufacture of solar cells. Hence, initially the main thrust of studying particle formation in plasmas was to avoid the nucleation of particles rather than harnessing its potential for the controlled synthesis of nanoparticles with desirable properties.

Much of the initial work on particle formation in plasmas addressed the nucleation and growth of silicon (Si) particles in silane plasmas since these plasmas are widely used for the deposition of amorphous silicon films for photovoltaic applications. Significant contributions in identifying the important processes in the kinetics of particle growth in silane plasmas – including the plasma chemistry leading to nucleation, particle growth by agglomeration, and particle charging – were made by several groups around the world [3–20].

The ability of nonthermal plasmas to form nanocrystals of materials such as silicon has been known for more than a decade. An early demonstration was the synthesis of crystalline silicon and germanium particles in a high-density helical resonator plasma by Gorla *et al.* [21]. Freestanding crystalline silicon particles of about 10 nm were also reported by Viera *et al.* [22, 23]. Most intriguingly, nanocrystals were generated in this study at gas temperatures as low as room temperature. Our own group also succeeded in demonstrating several plasma processes, coupled both inductively and capacitively that yielded silicon nanocrystals several tens of nanometers in size [24–27]. Due to their size, these nanocrystals were investigated for electronic device applications in nanoparticle-based transistors [24, 25].

Among the first to study useful applications of plasma-synthesized silicon nanocrystals was the group of Roca i Cabarrocas, which demonstrated that the deliberate deposition of silicon nanocrystals of about 2–5 nm in size in a matrix of amorphous silicon leads to a mixed-phase material, named polymorphous silicon (pm-Si:H), with improved properties for photovoltaic applications [26–29]. Compared to amorphous silicon films without nanocrystal inclusions [30], which are the material of choice for producing low-cost solar cells, cells based on pm-Si:H showed an improved stability with regard to the light-induced formation of metastable defects, known as the Staebler–Wronski effect [31], which is the main cause of performance deterioration in solar cells.

A few years later, Oda's group developed a plasma process [32–34] specifically with the goal of synthesizing silicon nanocrystals. The group used an ultrahigh frequency, capacitively coupled discharge to synthesize silicon nanocrystals and extract them from the plasma through injecting periodic pulses of hydrogen gas [32, 35]. Oda's work mainly focused on applications of silicon nanocrystals in electronic devices. The group demonstrated an impressive set of innovative nanoparticle electronic devices including single electron transistors [36, 37], nanocrystal memory devices [38, 39], and electron emitters [40].

Over the past few years, other groups also joined this field using plasmas for synthesizing silicon nanocrystals. New plasma approaches were developed, including microwave discharges operated at low pressure [41, 42] and microdischarges at atmospheric pressure [43]. Nonthermal plasmas by now are well established for the synthesis of nanocrystals of a wide range of materials including, but not limited to, silicon. In Section 13.2 of this chapter, we will describe the basics of nanocrystal formation in nonthermal plasmas. We will also discuss the processes that make plasmas unique among gas-phase media for nanocrystal synthesis: the charging of particles that suppresses particle coagulation and the selective heating of particles through energetic surface reactions. In Section 13.3, we will describe a plasma approach for the efficient synthesis of luminescent silicon nanocrystals with monodisperse sizes. In Section 13.4, we discuss the surface functionalization of silicon nanocrystals both in the liquid and in the plasma phase and the influence of the surface functionalization on optical properties of nanocrystals. The optical properties of surface-functionalized silicon nanocrystals are discussed in Section 13.5. Section 13.6 presents a brief summary.

13.2
Basics of Nanocrystal Formation in Plasmas

13.2.1
Nanoparticle Nucleation in Nonthermal Plasmas

Nonthermal plasmas are characterized by a strong thermal nonequilibrium among the species (for an introduction see, for instance, Refs [44, 45]). When an electric field is applied to a gas under low pressure, electrons easily get accelerated to energies sufficient to electronically excite and ionize the surrounding gas atoms. In steady state, the ionization produced by energetic electrons will precisely balance the charge carrier losses through diffusion to the wall, recombination, and other processes. Nonthermal plasmas are typically characterized by very hot electrons with temperatures between 20 000–50 000 K ($\sim$2–5 eV) and significantly colder positive ions and neutral molecules, whose temperatures remain close to room temperature. The energetic plasma electrons are very effective in dissociating precursor gas molecules. They also lead to a buildup of electric fields within the plasma wherever the plasma is in contact with boundaries such as reactor walls or the surfaces of nanoparticles. As the thermal velocity of electrons is much higher than that of the ions, all surfaces bounding the plasma are typically negatively charged. This negative charge balances the flux of negative electrons and positive ions to the surface by repelling a large fraction of the electrons while accelerating the positive ions. Without the presence of a negative potential with respect to the potential of the plasma on all bounding surfaces, the flux of electrons would exceed the ion flux by orders of magnitude.

During the course of plasma synthesis of silicon nanocrystals, the presence of hot electrons in the plasma leads to an effective dissociation of a silicon precursor such as silane (SiH_4). Particle formation occurs as a result of chemical clustering of silane radicals. An important aspect of clustering in silane plasmas is that unsaturated Si_nH_m clusters have positive electron affinities [46]. Thus, these clusters can readily attach electrons to become negatively charged and can thereby be electrostatically trapped in the plasma. Hollenstein and coworkers performed mass spectrometric studies in which they found that negatively charged clusters (anions) grew to larger sizes than positive (cation) or neutral clusters [47–51], as shown in Figure 13.1 [49]. They reasoned that only anionic clusters, being electrostatically trapped, would have time to grow, whereas cations and neutrals would tend to diffuse to the walls before growing to large sizes.

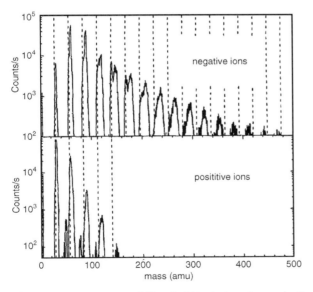

Figure 13.1 Mass spectra of silicon hydride cluster anions and cations in an RF silane plasma, for pure silane at 76 mTorr. (From Ref. [51].)

The picture thus emerged suggested that the likely path to particle nucleation was through anion–molecule reactions, for example, through sequential hydrogen elimination reactions of the form $Si_nH^-_{2n+1} + SiH_4 \rightarrow Si_{n+1}H^-_{2n+3} + H_2$ [20]. These ion–molecule reactions would be expected to be faster than neutral chemistry, especially as heavy-species temperatures in these plasmas are modest.

The model of negative ion-induced clustering is now supported by a large number of studies, including modeling studies based on very comprehensive, chemical-kinetics computations [19, 20, 52]. However, alternative models have been proposed as well. Watanabe *et al.* concluded that clustering proceeds mainly through neutral reaction chemistry involving the insertion of SiH_2 radicals into neutral clusters [53–55]. Gallagher [56] developed a model that considered cluster growth and charging by several generic processes including attachment of radicals, silane, electrons, and cations; detachment of electrons; and diffusive loss. His main conclusion was that SiH_3 is the key particle growth species [56, 57]. He also concluded that electron attachment to SiH_3 was an important process in initiating clustering. The diversity of views demonstrates that it is dangerous to generalize the mechanism of particle formation as it likely depends sensitively on the particular conditions prevailing in any particular plasma process.

13.2.2
Nanoparticle Charging

To understand nanoparticle formation in plasmas, it is important to understand the processes that determine nanoparticle charging. As particles in plasmas are negatively charged under many circumstances, electrons experience a repulsive potential when approaching the particle while ions are attracted by the particle. Assuming a Maxwell–Boltzmann velocity distribution for electrons, the electron current is given by the thermal flux reduced by the Boltzmann factor:

$$I_e = en_e S \sqrt{\frac{k_B T_e}{2\pi m_e}} \exp\left(-\frac{e|\Phi|}{k_B T_e}\right), \quad \Phi < 0. \tag{13.1}$$

Here I_e is the electron current, e is the elementary charge, S is the particle surface area, k_B is the Boltzmann constant, T_e is the electron temperature, m_e is the electron mass, and Φ denotes the particle potential. The ion current is usually described within the framework of the orbital motion limited (OML) theory [58, 59]. This theory applies to ions that experience an attractive potential in a strictly collisionless environment and accounts for the fact that only the fraction of ions will be collected by the particles whose angular momentum while approaching the particle is below a certain threshold. The ion current is given by

$$I_i = en_i S \sqrt{\frac{k_B T_i}{2\pi m_i}} \left(1 + \frac{e|\Phi|}{k_B T_i}\right), \quad \Phi < 0. \tag{13.2}$$

Here I_i is the ion current, T_i denotes the ion temperature, and m_i is the ion mass. By balancing the ion current and the electron current, one can find the average particle potential and particle charge. However, as nanometer-sized particles in plasmas carry only a few elementary charges, the charging processes need to be considered stochastically. Equations (13.1) and (13.2) can be used to define frequencies of electron and ion capture by particles at a certain potential. By using the principle of detailed balancing for particles of a given charge state, the particle charge distribution can be derived.

In Ref. [60], we considered the situation of nanoparticles being immersed in a plasma. Particles were allowed to get charged and to agglomerate while being considered confined in the reactor. While this situation is not necessarily characteristic to many plasma synthesis reactors, it is still instructive as it reveals some important physical aspects of nanoparticle charging. Figure 13.2 shows results of the simulations [60] for the case of nanoparticles with an initial density $(t=0)$ of 10^{12} cm^{-3} in a plasma with a positive ion density of 3×10^{9} cm^{-3}. A universal property of all plasmas is their "quasineutrality," which means that the overall density of positive and negative charge carriers within any volume is (approximately) equal. Hence, particles initially are bipolarly charged since the particle density is higher than the positive ion density in the plasmas, which implies that there are not enough electrons present to charge all particles negatively. In the simulation in Ref. [60], we allowed these particles to coagulate, which reduces the particle density, while we also assumed that agglomerates immediately coalesce into larger, spherical particles. Once the particle density reduces to a density less than the ion density, all particles can become negatively charged and particle coagulation ceases. The positive ion density in the plasma thus plays an important role as a threshold for the particle density, below which particles are negatively charged and particle agglomeration is prevented.

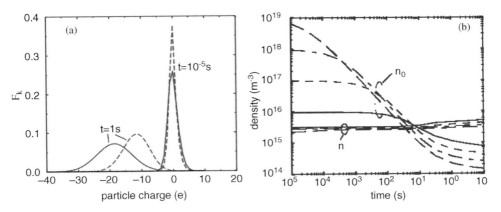

Figure 13.2 (a) Particle charge distribution for particles with an initial density of 10^{12} cm^{-3} and a final density of 3×10^{8} cm^{-3} in a plasma with a positive ion density of 3×10^{9} cm^{-3}. (b) Evolution of the particle density for various initial particle densities of 1 nm radius particles: n_p is the particle density and n_i the self-consistent ion density for the same initial density of particles. (See Ref. [61] for more details.)

13.2.3
Nanoparticle Heating in Plasmas

An interesting and still quite underappreciated feature of nonthermal plasmas is their ability to selectively heat immersed nanoparticles to temperatures several hundreds of Kelvin beyond the gas temperature. This makes nonthermal plasmas particularly attractive for the synthesis of nanocrystals of covalently bonded semiconductors such as the group IV and III–V semiconductors. Until recently, this feature was little known and barely understood. Hence, the observation of the formation of crystalline nanoparticles in many chemically active plasma presented a rather persistent puzzle [61–63]. For instance, studies on silicon nanoparticles suggested that the crystallization temperature of particles with diameters of 4, 6, 8, and 10 nm are 773, 1073, 1173, and 1273 K, respectively [64]. It is thus quite surprising that nanocrystals requiring such high crystallization temperatures are formed in plasmas at low gas temperature. For instance, our group observed crystalline silicon nanoparticles formed in a flow-through-type plasma setup, in which the nanoparticles reside in the plasma only for a few milliseconds (typically 2–6 ms), as discussed in Section 13.3 [63]. In the same study, we also determined that the gas temperatures were between 420 and 520 K, that is, significantly lower than the crystallization temperatures reported for small silicon nanoparticles; yet nanocrystals were consistently found. The only logical conclusion of this observation has to be that the nanoparticles in the plasma are at significantly higher temperatures than the surrounding gas.

In a recent study, we investigated the particle heating using a stochastic Monte-Carlo model to study the particle temperature. In our work, a nonsteady formulation of the particle energy balance was used:

$$\frac{4}{3}\pi r_p^3 \varrho C \frac{dT_p}{dt} = G - L, \tag{13.3}$$

where ϱ is the silicon density and C is its specific heat, r_p and T_p are the particle radius and temperature, respectively, G is the generic heat release term, and L is the heat sink term. The silicon density and specific heat are assumed to be equal to those of the bulk material. The equation is discretized using a typical time step of 10^{-10} s. The cooling term L is calculated in the model as

$$\frac{1}{4} n_{gas} S \sqrt{\frac{8 k_B T_{gas}}{\pi m_{gas}}} \cdot \frac{3}{2} k_B (T_p - T_{gas}). \tag{13.4}$$

This term represents the conduction losses to the background gas. Here, n_{gas} is the background gas density, T_{gas} is the background gas temperature, S is the particle surface area $4\pi r_p^2$, m_{gas} is the atomic mass of the background gas. This term is implemented in the model using a continuous nonstochastic approach since the collision frequency between the nanoparticles and the background gas atoms is much larger than the collision frequency with ions and radicals. The background gas temperature is assumed to be equal to 300 K, which is a good assumption for many nonthermal plasmas. The generation term G in Eq. (13.3) describes the heat released by the ion recombination at the particle surface. Moreover, in the case of the synthesis of silicon nanocrystals from silane (SiH_4), it is necessary to consider exothermic and endothermic reactions of hydrogen with the particle surface.

Collisions of electrons and ions with the nanocrystal can be described by collision frequencies, which can be easily derived from the electron and ion currents to the particles given by Eqs. (13.1) and (13.2): $v_{e,i} = I_{e,i}/e$. Each electron–ion recombination event is assumed to release 15.76 eV of energy to the particles since this is the ionization energy of argon, which is used as a carrier gas in our experiments.

A collision between atomic hydrogen and the particle surface can lead to different energy transfer mechanisms depending on the fractional hydrogen surface coverage. The number of silicon surface atoms as a function of size is calculated using the relation given in Ref. [65]. Since hydrogen is bonded to the silicon nanocrystal surface with mono-, di-, and trihydride bonds, the assumption that on average two hydrogen atoms are bonded to every surface silicon atom is used. In case the surface coverage is <100%, there is a finite probability of a hydrogen atom landing on an unterminated surface site, that is, a dangling bond. The reaction of a hydrogen atom with a dangling bond is assumed to have a probability of 100% (sticking coefficient equal to 1), and the energy released in this event is equal to the binding energy of the Si−H bond, which is 3.1 eV. A hydrogen radical landing on the hydrogen-terminated fraction of the surface has a 11% probability of abstracting a hydrogen atom from the surface through the Eley–Rideal mechanism [66], and the heat released in this case is equal to the bond energy of the hydrogen molecule (4.51 eV) minus the energy of the broken Si−H bond (3.1 eV), resulting in an energy release to the particle of 1.41 eV. In the remaining 89% of the cases, hydrogen can diffuse on the particle surface. The diffusion coefficient of atomic hydrogen on a hydrogen-terminated silicon surface follows an Arrhenius expression with a small activation energy of 0.1 eV and a pre-exponential factor equal to 2.27×10^{-4} cm^2/s [67]. The physisorbed hydrogen can either diffuse to a dangling bond, releasing an energy of 3.1 eV, or it can react with another incoming hydrogen radical. The hydrogen recombination event releases 4.51 eV. Finally, in case the particle temperature is particularly high, thermal desorption of hydrogen through the Langmuir–Hinshelwood mechanism occurs. This involves the formation of the H−H bond and the breaking of two Si−H bonds. The energy lost in this process is equal to 1.69 eV. Thermal desorption of hydrogen solely depends on the particle temperature and hydrogen surface coverage [68], it has an activation energy of 1.86 eV and starts to play a role around a temperature of 900 K.

A typical result of these simulations is shown in Figure 13.3 for the case of low particle concentration ($n_p \ll n_{ion}$). The main result of this model is that very small particles in a plasma experience a very unsteady temperature history, with the instantaneous temperature largely exceeding the background gas temperature. Each of the spikes in particle temperature observed in Figure 13.3 is the result of an exothermic surface reaction event. While the extent of the temperature spikes strongly depends on the particle size, the average particle temperature does not since heating terms and cooling terms both scale linearly with the particle surface area. An important result from Figure 13.3 is that the instantaneous particle temperature observed in our simulations can easily exceed the crystallization

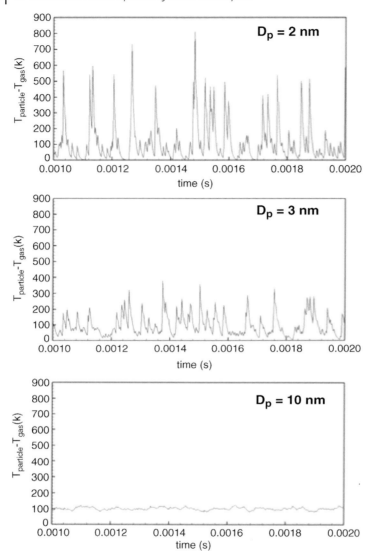

Figure 13.3 Simulated temperature history of small nanoparticles in a nonthermal plasma. The plasma density is $5 \times 10^{10}\,\text{cm}^{-3}$ and the gas temperature is 300 K.

temperature for small silicon particles that were reported in Ref. [64]. We thus suggest that the temporary strong heating of nanoparticles through energetic surface reactions, coupled with the slow cooling through conduction and convection at low pressure, is likely responsible for the ability of nonthermal plasmas to generate nanocrystals even of high crystallization point materials.

Given the unsteady temperature of small particles, and since at any given instant one particle has a temperature that is independent of the temperature of other

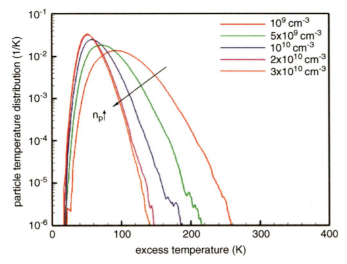

Figure 13.4 Particle temperature distribution function for a nanoparticle of 5 nm in diameter for various particle densities in the plasma.

particles, it is reasonable to introduce a particle temperature distribution function. This is defined as the fraction of particles that at any time can have a certain excess temperature with respect to the background gas. The particle temperature distribution function is shown in Figure 13.4 for a nanoparticle of 5 nm in diameter.

Figure 13.4 also investigates the influence of the particle loading on the particle temperature distribution function. It demonstrates that the particle temperature overall will decrease with increasing particle density. This is a result of changes in plasma properties due to the presence of particles. For a larger particle density, the free electron density in the plasma decreases since an increasingly larger fraction of the plasma electrons get attached to nanoparticles. As a consequence, the electron–ion recombination frequency decreases leading to reduced nanoparticle heating. For a given particle density, the particle temperature distribution can also be changed by the power coupled into the plasma since larger power will lead to larger electron and ion densities and hence enhanced nanoparticle heating through electron–ion recombination.

13.3
Silicon Nanocrystal Synthesis in Nonthermal Plasmas

As discussed in Section 13.1, there are a number of plasma approaches that have been used for synthesizing silicon nanocrystals. We will focus on the discussion of one approach proposed by our group, as the principles of silicon nanocrystal synthesis are similar in all approaches. To date, the optical properties of silicon nanocrystals produced following the approach introduced here have likely been characterized most completely.

13.3.1
Experimental Apparatus

The plasma reactor design described here had originally been reported in Ref. [63]. The reactor consists of a quartz tube with 3/8 in. (~9.5 mm) outer diameter and 1/4 in. (~6.8 mm) inner diameter, mounted on a gas delivery line and exhaust system with ultratorr connectors. Radiofrequency (RF) power is fed into the system using two ring electrodes and a modified T-type matching network. The plasma is typically generated at a pressure of 1.4 Torr (187 Pa), but luminescent nanocrystals have also been obtained at pressures as high as 15 Torr (~2000 Pa). Most of the experimental results presented here were obtained at a pressure of 1.4 Torr since it was found that under these conditions the largest yield of nanocrystals can be achieved. Typical gas flow rates at 1.4 Torr are up to 100 sccm of Ar, around 15 sccm of SiH_4 (5% dilution in either He or Ar), and few sccm of additional hydrogen. The residence times of gas in the plasma, calculated on the basis of the gas flow velocity, are between a few tens of milliseconds, to less than 5 ms. In the photo of the discharge in Figure 13.5, it can be seen that the plasma consists of two regions. In the part upstream of the electrodes, the plasma emission appears weaker than in the region downstream of the two

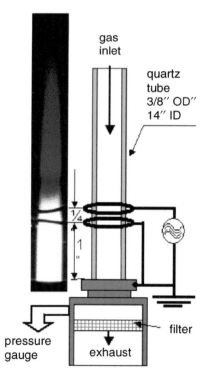

Figure 13.5 Schematic of a radiofrequency flow-through reactor for the synthesis of silicon nanocrystals.

electrodes. Also, significant growth of a silicon film in the reactor tube can be observed that always appears upstream of the electrodes, while the growth is much less significant downstream of the ring electrodes. This suggests that very fast nucleation and precursor consumption occurs upstream of the electrodes, while the exact evolution of particle size and structure in the plasma downstream has yet to be understood.

Plasma power plays a crucial role in this process. RF current and voltage have been measured to estimate the actual power consumption in the discharge, giving a value of $\sim$30 W. Given the very small discharge volume ($\sim$2 cm^3), a considerable power density is achieved in the plasma region. It has been found that operating at too little a power generally leads to the production of amorphous particles, while operating at too high a power leads to a dramatic decrease in particle mass yield. As discussed above, intense particle heating of the particles occurs via surface reaction, the most important of which is ion–electron recombination at the particle surface. Hence, we hypothesize that too intense a discharge might actually even induce evaporation of small clusters, and the precursor might then be efficiently lost as a film deposited on the reactor walls.

A filter placed downstream of the discharge collects the particles produced in the plasma. The filter is made of a fine stainless steel mesh ($\sim$400 wires/in.) and few minutes of deposition are sufficient to completely coat the filter. For the transition electron microscope (TEM) characterization, a TEM grid is attached to the stainless steel filter with a double-sided carbon tape and exposed to the nanoparticle stream for a few seconds. After the mesh is extracted from the system, the particles can be left on the mesh to be oxidized in air, or rinsed with various solvents. The dispersion in a solvent does not prevent oxidation, unless oxygen and water dissolved in the solvent are carefully removed.

One interesting feature of this process is its high precursor conversion efficiency, as seen in Figure 13.6. A residual gas analyzer was used to monitor the signal at 30

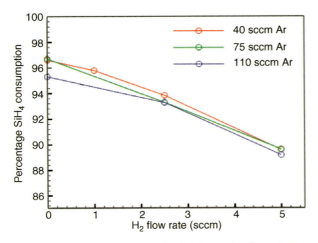

Figure 13.6 Precursor consumption for the reactor shown in Figure 13.5.

and 31 amu at the system exhaust, with and without striking the discharge. The signal at 30 and 31 amu is produced by silane in the gas flow, and this measurement is carried out to quantify the percentage of consumption of the precursor by the nonthermal plasma process. The fractional consumption of silane does not depend on the gas carrier flow rate, which implies that the nucleation and growth of particles occurs very rapidly. For an Argon flow rate of 40 sccm, the residence time is approximately 5 ms, while it is 2 ms for a flow rate of 110 sccm. A smaller fraction of the silane is consumed when hydrogen is added to the plasma. This implies that hydrogen is quenching the particle nucleation and reducing the density of silane radicals (SiH_3, SiH_2, etc.) that contribute to film growth.

13.3.2
Nanocrystal Characterization

13.3.2.1 TEM Characterization
The particle size, shape, and structure were assessed via an FEI Tecnai T12 microscope, which operates at an accelerating voltage of 120 kV and uses a thermionic emission gun with a LaB_6 filament. This microscope's resolution is high enough to easily resolve lattice fringes of the {111} plane of silicon, with a spacing of 0.314 nm. For high-resolution TEM, an FEI Tecnai F30 was used.

TEM images of silicon nanoparticles deposited on the grid directly from the gas phase a few inches downstream of the plasma are shown in Figure 13.7. The sample was produced at a pressure of 14 Torr, with 200 sccm of Ar and 1.23 sccm of SiH_4 (5% in helium). The grid was exposed to the aerosol for 5 min, which resulted in the formation of a thick deposit with multiple layers of particles on top of each other. The TEM image clearly shows that the particles are very monodisperse. Particle sizes around 5–8 nm are observed. The dark field image in Figure 13.7b clearly shows that the vast fraction of particles is crystalline. A higher magnification TEM image for this sample is shown in Figure 13.7c, in which a hole in the thin film amorphous carbon grid is imaged with an agglomerate of particles hanging from the film. The particles appear to be spherical, even though facets seem to be present for some of the particles in Figure 13.7c. This, in general, has been observed for "big" particles as the ones in this sample. In this image, {111}-lattice fringes of the cubic diamond structure of silicon are clearly seen; the lattice constant was measured at 0.315 nm, which corresponds well to the expected lattice spacing of 0.314 nm. Around the particles, an amorphous shell is seen that is very likely the native oxide shell. In some experiments, nanocrystals were introduced into the TEM while avoiding exposure to ambient air and the amorphous layer was absent.

13.3.2.2 Particle Size Distribution
Particle size distributions were determined from the TEM images by analyzing several hundreds of particles. While the standard way of measuring particle size distributions in the aerosol community is the use of differential mobility analyzers (DMAs) [69], this has not been possible for the plasma system described here. The

(a)

(b)

(c)

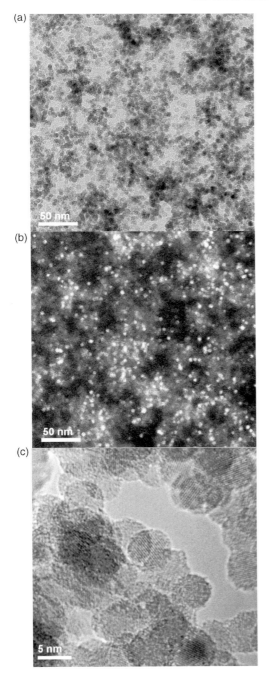

Figure 13.7 Bright field TEM image (a) and dark field image (b) of silicon nanocrystals produced under the following conditions: 200 sccm Ar, 1.23 sccm SiH_4 (5% in Helium), $p = 14$ Torr. (c) High magnification image for the same sample. A hole in the thin film carbon grid is used to image the particles without any amorphous background.

main reason is that DMAs typically operate at atmospheric pressure, thus the aerosol needs to be extracted from the system and its pressure needs to be raised to perform the measurement. Although this is possible, it is difficult to achieve while avoiding agglomeration of particles in the process.

To determine particle size distributions, many different TEM images were analyzed so that the particle count was sufficiently high to ensure an acceptable statistical accuracy. The particle size distribution for two different samples is shown in Figure 13.8. These two samples were produced at a pressure of 1.4 Torr, with an Argon flow rate of 40 sccm and a SiH_4 (5% in Ar) flow rate of 8 sccm, for the first sample, and with the same flow rate reduced by a factor of 4 for the second sample. For these two samples, the ratio of the Ar to SiH_4 flow rate is kept constant, showing how the decreased residence time leads to the production of smaller particles, going from an average size of 5.72 nm down to 3 nm. In general, we find the size distributions are fairly narrow with standard deviations typically about 15% of the average particle size.

The main parameter affecting the size distribution is the residence time of particles in the reactor. As seen from Figure 13.9, the silane partial pressure has only little influence on the particle size. This implies that a higher precursor flow leads to a stronger nucleation rate while not significantly affecting the final particle size. As already mentioned, adding hydrogen also leads to the formation of smaller particles.

13.3.2.3 Nanocrystal Surface Conditions

An FTIR analysis was used to investigate the type of chemical bonding in the nanocrystal samples and, in particular, to monitor the bonding present at the particle surface as a function of the various post-treatments that have been tested by our

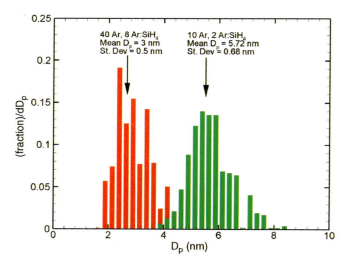

Figure 13.8 Particle size distributions.

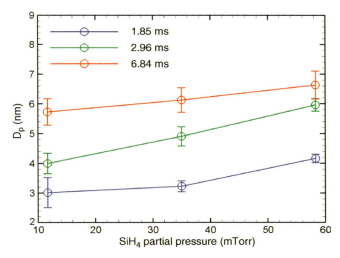

Figure 13.9 Average particle size as function of the gas flow rate and silane partial pressure.

group. The coverage of the silicon nanocrystal surfaces both determines their optical properties and the suitability of the surface for certain surface functionalization schemes.

Figure 13.10 shows the FTIR spectrum of as-produced silicon nanoparticles collected on the regular stainless steel mesh. The sample was prepared by extracting it from the system within a nitrogen-purged glove bag. The filter was then sealed between two quick flange blanks, transferred to the FTIR spectrometer, and opened

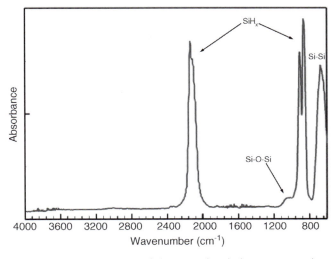

Figure 13.10 FTIR spectrum of plasma-produced silicon nanocrystals.

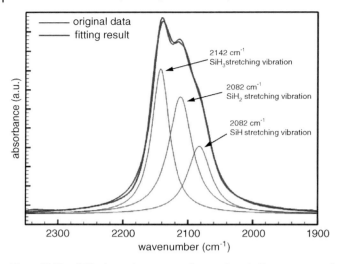

Figure 13.11 FTIR absorption spectra of as-produced silicon nanocrystals. The sample was deposited with the reactor of Figure 2.2, at a pressure of 1.4 Torr, with 70 sccm of Argon and 15 sccm SiH₄ (5% in helium).

to the air. The sample compartment of the FTIR spectrometer is also nitrogen purged, but very brief exposure to air is inevitable, which very likely leads to the appearance of a very small oxide peak around $1100 \, \text{cm}^{-1}$.

The main peaks in the absorption spectra are those related to Si–Si bond around $660 \, \text{cm}^{-1}$, and the silicon hydride peaks around 2100 and $900 \, \text{cm}^{-1}$. The particles are mainly hydrogen terminated, as expected from particles produced using SiH₄ as a precursor. More information can be collected from a detailed study of the peak shapes at 2100 and $900 \, \text{cm}^{-1}$. Contribution from Si–H, Si–H₂, and Si–H₃ bonds contribute to the signals in these two different regions. This is shown in Figure 13.11.

The assignment of the various surface hydrides was established on the basis of the work of Marra and Aydil [70, 71]. The peak at $2100 \, \text{cm}^{-1}$ can be deconvoluted into three components due to Si–H, Si–H₂, and Si–H₃ stretching vibration. In Figure 13.11, the experimental profile is fitted to three Lorentzian peaks centered at $2142 \, \text{cm}^{-1}$ for the Si–H₃ component, at $2111 \, \text{cm}^{-1}$ for the Si–H₂ component, and at $2082 \, \text{cm}^{-1}$ for the Si–H component. The peak position is in excellent agreement with what is found in the literature [70–73]. The line widths are between 39 and $44 \, \text{cm}^{-1}$ for these three peaks and are lower than what is typically found in the case of hydrogenated amorphous silicon ($>80 \, \text{cm}^{-1}$) [70]. This is a strong indication that the particle core is not hydrogenated, and the hydrogen is bonded to the particle surface. The same fitting procedure can be used to identify silicon dihydride and trihydride contribution to the ~$900 \, \text{cm}^{-1}$ peak. Again, on the basis of the work in Ref. [70, 71], the peaks at 912 and $864 \, \text{cm}^{-1}$ are assigned to the symmetric and degenerate deformation of Si–H₃ and the 898 and $848 \, \text{cm}^{-1}$ to the scissor and wag modes of Si–H₂, respectively.

It is obvious that the surface of plasma-produced silicon nanocrystals is covered with hydrogen and that higher and lower hydrides contribute to the absorption signal. This is to be expected for the case of a silicon nanocrystal with unreconstructed surface. However, the high signal of SiH$_3$ stretches is surprising and requires further investigation.

There has been speculation whether hydrogen termination of the silicon nano-crystal surfaces can provide stable surface passivation. In the case of flat silicon wafers, a perfectly hydrogen-terminated surface can be stable for many hours to days. Time-resolved studies give information on the oxidation kinetics of small silicon crystals. A series of FTIR spectra were acquired over a range of 24 h using an ATR-FTIR spectrometer. The sample was then extracted from the plasma reactor under nitrogen atmosphere using a glove bag and inserted into a vial closed by a septum. Particles were sonicated into degassed and dried toluene and few droplets were dropped onto the Ge crystal of the spectrometer. Different from the spectrometer used to acquire the spectra in Figures 13.10 and 13.11, this particular ATR-FTIR spectrometer does not have a nitrogen-purged sample compartment. The compartment is purged with air, which is flown through a water-removal filter. In this way, the continuous increase in the intensity of the Si—O—Si peak can be monitored, and the effect of moisture on the oxidation kinetic can be investigated by simply opening the sample compartment to the room air. In Figure 13.12, the time evolution of the absorption scans is reported. The first spectrum is the one taken 11 min after dropping the slurry onto the germanium crystal; this is done to let the toluene evaporate. Residual peaks from the absorbed solvent molecules can be observed in all the spectra around 2900 and 1450 cm^{-1}, which are respectively assigned to CH$_x$ stretching modes and ring stretch modes in aromatic compounds [74].

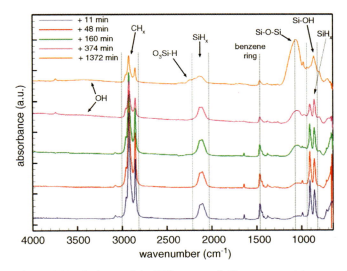

Figure 13.12 Evolution of the FTIR spectra of silicon nanocrystals exposed to air.

The most noticeable feature is the growth of the oxide peak around $1050\,\mathrm{cm}^{-1}$. The height of the silicon hydride peaks at 2100 and around $900\,\mathrm{cm}^{-1}$ remains constant for several hours, which implies that oxygen and moisture are attacking the Si–Si backbonds that are not protected by the presence of hydrogen at the surface. A slight change in shape of the SiH_x peaks can be seen in the spectra at $+374$ min, and it is much more noticeable after a full day of oxidation ($+1372$ min). In the last spectrum, the shape of the SiH_x peak at $2100\,\mathrm{cm}^{-1}$ has significantly changed; the peak at $2250\,\mathrm{cm}^{-1}$ is assigned to the vibrational stretching of the Si–H bonds in the presence of oxygen backbonding. In the last spectrum, a broad feature at $3400\,\mathrm{cm}^{-1}$ is present, which is due to the O–H stretching mode. Comparisons of the oxidation rates in dry air and under the presence of moisture show a significantly, four to five times, faster oxidation in the presence of moisture.

From this series of FTIR studies, two conclusions can be drawn: hydrogen is not a good protection against the growth of a native oxide and the combination of oxygen with moisture present in the air leads to a quick oxidation of the particles. This is very different from the case of hydrogen-terminated flat silicon surfaces, which are known to be chemically stable in air for an extended period of time. The difference is very likely due to the curvature of the nanocrystal surface.

13.3.2.4 Optical Properties of Surface-Oxidized Silicon Nanocrystals

The observation of visible photoluminescence (PL) from plasma-produced silicon nanoparticles confirms that the process is capable of producing very small crystallites. The phenomenon of photoluminescence from semiconductor quantum dots is influenced by many properties of the sample, such as crystal size, surface passivation, interparticle distance, and chemical environment in which the sample is studied.

The particle luminescence spectra for silicon are characterized by a broad emission, as shown in Figure 13.13, the FHWM of the peak being as large as

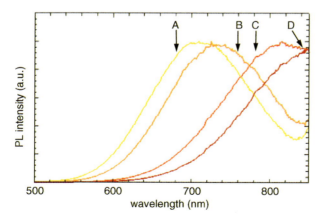

Figure 13.13 Photoluminescence emission spectra from silicon nanocrystals produced for different plasma conditions. Excitation wavelength is 360 nm. Spectra are normalized to their emission maximum.

100 nm. This is a consequence of the fairly broad particle size distribution, but it is also related to the mechanism that leads to visible emission in the case of silicon. In fact, single-particle spectroscopy experiments have shown that the spectra of a single crystal do not differ significantly from the broad spectra of an ensemble of particles [75]. Such a broad emission suggests that the band structure is still indirect. Radiative processes are coupled to phonon vibrations or to radiative centers likely present at the particle surface.

13.4
Surface Functionalization of Silicon Nanocrystals

13.4.1
Liquid-Phase Functionalization

It is well known that the growth of an approximately 1 nm thick oxide shell improves the optical properties of silicon quantum dot with respect to the as-produced material [76, 77]. Still, the quantum yield of surface-oxidized material is generally low. The need for a reproducible approach to synthesize silicon quantum dots with bright photoluminescence has led to investigations of more sophisticated surface passivation techniques. Liquid-phase passivation techniques are well known and have already been applied to flat silicon surfaces [78–81], porous silicon [81–85], freestanding silicon nanocrystals obtained by liquid-phase synthesis [86–90], and freestanding silicon nanocrystals etched with a HF/HNO_3 mixture [91–93]. Liquid-phase-treated organically capped porous silicon surfaces or silicon nanoparticles show intense visible photoluminescence and are fairly resistant to oxidation. In the case of freestanding particles, stable colloidal solutions of silicon nanocrystals can be obtained. This greatly improves the liquid-phase processability of the material, which is of fundamental importance to several applications, such as organic light-emitting devices (OLEDs) and printable electronics.

In general, organic chains can be grafted onto silicon surfaces through reaction with different functional groups, such as unsaturated carbon groups in the case of alkenes ($-CH=CH_2$) and alkynes ($-C\equiv CH$), amino groups in the case of amines ($-NH_2$) [78, 79], and hydroxyl groups for alcohols ($-OH$) [80]. After reaction, the organic chain is bonded to the silicon surface through $Si-C$, $Si-N-C$, and $Si-O-C$ links, respectively. Hydrosilylation with alkenes and alkynes requires a hydrogen-terminated silicon surface as is naturally the case with plasma-synthesized silicon nanocrystals. Hydrosilylation with alkenes is thus the easiest approach to organically passivate the nanoparticles. Our group succeeded both in adapting known liquid-phase hydrosilylation procedures to plasma-produced nanoparticles and in developing new plasma-based hydrosilylation approaches. The former method is characterized by the excellent optical quality of the material produced while the latter method is characterized by its simplicity and efficiency. We will thus introduce here both approaches.

13.4.2
Liquid-Phase Hydrosilylation Procedure

The hydrosilylation reaction with alkenes is a well-established technique that involves the linking of the organic molecule to the particle surface via the formation of a Si—C covalent bond. An excellent review on this reaction scheme is given in Ref. [81]. The most important step in the hydrosilylation reaction is the cleavage of a surface Si—H bond and the reaction of the surface-dangling bond with the unsaturated carbon group of the reactant.

For plasma-produced silicon nanocrystals, our group has established the following liquid-phase passivation procedure: First, nanocrystals are collected on filter meshes in the plasma system. They are then extracted in an air-free environment from the plasma system. For this purpose, a nitrogen-purged glove bag intersects the system exhaust line so that the sample can be removed from the system in a nitrogen atmosphere, placed into a vial, and sealed with a rubber septum. Added then to the vial in a canular transfer are the organic monolayer precursor (e.g., 1-dodecene) and a suitable solvent (e.g., mesitylene), which have previously been carefully degassed and dried (i.e., removal of water). Now sonication is used to remove nanocrystals from the filter and transfer them into the mixture of precursor molecules and solvent. The nanocrystal slurry is now transferred into a reaction flask such as a 200 ml Pyrex flask with a water-cooled refluxer mounted on it. Then the hydrosilylation reaction is initiated by heating the solution to the solvent boiling point and keeping it at that temperature for several hours. The reactor is constantly purged with nitrogen during the reaction. After the reaction, a clear solution is obtained, indicating that initial large agglomerates of nanocrystals have dissolved and a colloidal of individual, functionalized nanocrystals has been formed. In the next step called drying, the sample may be heated at approximately 80 °C and kept at that temperature under vacuum over night in order to remove the solvent and nonreacted precursor molecules. The purpose of drying the nanocrystals is to be able to redisperse them in any solvent to meet specific needs.

Different alkenes have been grafted onto the plasma-produced silicon nanoparticles. 1-Octadecene was the first molecule to be studied because of the many successful reports of functionalization of silicon nanocrystals with this molecule [92, 93]. However, 1-octadecene has a high boiling point of 315 °C, which complicates drying. For this reason, shorter ligands such as 1-dodecene and 1-octene were used as well. For shorter ligands, the ligand–solvent interaction is insufficient to overcome the van der Waals forces among the initially agglomerated silicon nanocrystals, and clear colloids cannot be obtained.

Success of the hydrosilylation procedure can be verified with FTIR. The spectra in Figure 13.14 were acquired on the attenuated total reflectance ATR-FTIR. For all these spectra, the samples were carefully kept air-free and redispersed in either chloroform or hexane before being drop-cast onto the germanium ATR crystal. The spectrum of the as-produced silicon nanocrystals shows strong contribution from silicon hydride bonding. The peak centered at $2100 \, \text{cm}^{-1}$ corresponds to the stretching vibrational mode of the silicon–hydrogen bond in SiH, SiH_2, and SiH_3

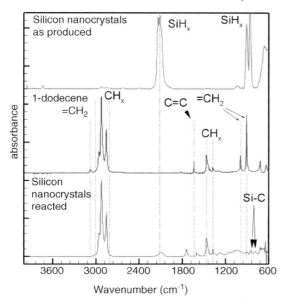

Figure 13.14 FTIR spectra of as-produced silicon nanocrystals, neat 1-dodecene, and silicon nanocrystals reacted with 1-dodecene.

groups. The peaks around 900 and 850 cm^{-1} correspond to SiH$_3$ degenerate and symmetric deformation modes and to SiH$_2$ scissor- and wag-vibration modes. The small bump around 1050 cm^{-1} is very likely due to some oxidation that takes place during the sample transfer and during the FTIR spectrum acquisition since the sample compartment in the IR spectrometer that has been used for this measurement is not nitrogen purged. The peak assignment for the organic molecules is taken from [74]. The main peaks in the neat 1-dodecene spectrum can be assigned to CH stretching vibration in CH$_x$ groups for the features around 2800 cm^{-1}. Further peaks are related to the CH$_2$ scissor deformation at 1475 cm^{-1} and the symmetric deformation of CH$_3$ around 1380 cm^{-1}. The presence of the unsaturated carbon group is confirmed by the presence a peak corresponding to the C=C stretching mode at 1645 cm^{-1}. The peaks at 914 and 991 cm^{-1} are due to the =CH out-of-plane deformation and the peak at 3085 cm^{-1} is due to the =CH stretch. It is important to note that all peaks related to the unsaturated carbon bonds are absent in the spectrum of the hydrosilylated silicon nanoparticles, while the strong CH$_x$ peak at 2800 cm^{-1} is dominant. This is a strong indication that the liquid-phase reaction has successfully capped the hydrogen-terminated silicon nanocrystals with 1-dodecene and that the hydrogenated silicon surface has reacted with the unsaturated group of the capping molecule. Similar spectra have been reported in the literature for other hydrosilylated silicon nanoparticles [92]. Small peaks around 720 and 800 cm^{-1} can be assigned to the formation of Si−C bonds, and they provide a direct proof that the reaction was successful.

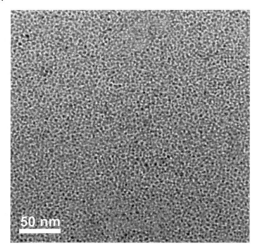

Figure 13.15 TEM image of alkylated silicon nanoparticles. The particles were redispersed in toluene after liquid-phase passivation and drying. The sample was drop-cast on a regular thin film carbon grid.

Another direct proof of a successful hydrosilylation reaction is the fact that the functionalized nanocrystals are perfectly soluble in nonpolar solvents, such as toluene, chloroform, or hexane. The solution passes easily through a 0.2 μm filter without any loss of coloration and is stable for an indefinite amount of time. In fact, TEM studies confirm that a colloid of individual, nonagglomerated silicon nanocrystals is obtained. A bright field TEM image of film of hydrosilylated particles is shown in Figure 13.15. The sample was hydrosilylated with 1-dodecene, dried to remove the unreacted alkene and redispersed in toluene. Few droplets were then dropped onto the thin-film carbon grid. The particles form a very uniform layer on the grid surface. Each of the dark specks in the image in Figure 13.15 is an individual silicon nanocrystal, as can be confirmed by dark field TEM and high-resolution studies. The lighter halo around the nanocrystals is likely due to the organic ligand shell. The nanocrystal film is close to a monolayer, since occasional uncovered regions can be found on the grid. The particles do not self-organize in an ordered pattern, as has been observed with some other nanocrystal films. The formation of crystalline nanoparticle superlattices is of great interest for particle-based devices [94], but it is known that particles with a very narrow size distribution are needed to organize in an ordered fashion. Unfortunately, the particle size distribution in the case of plasma-produced silicon quantum dots is fairly broad, with a typical standard deviation of approximately 1 nm.

13.4.3
Plasma Grafting of Silicon Nanocrystals

Although the liquid-phase hydrosilylation of silicon nanocrystals provides material with excellent optical properties, as discussed in this section, it also suffers from a number of drawbacks. The liquid-phase reaction is slow, typically requiring several hours of reaction time. Attaching short organic chains is difficult since the lower boiling point of smaller molecules reduces the reaction temperature and increases the reaction time. Moreover, the process of dissolution of the particle agglomerates is controlled by the competition between the van der Walls forces holding the particles together and the interaction between the organic chain and the solvent, which tends to stabilize the particle in solution. Hence, a clear solution cannot be obtained, for instance, when reacting approximately 4 nm silicon particles with 1-hexene even for many hours of thermal reaction because the 6-carbon-long chain is too short to pull the particles apart from each other. Liquid-phase reactions also involve all the complications of air-sensitive chemistry, usually requiring access to a Schlenk line.

The field of the in-flight, aerosol processing of nanoparticles is quickly growing. In Ref. [95], aluminum particles were in-flight coated with a carbon shell using acetylene as a precursor. In Ref. 96, silicon nanoparticles were hydrosilylated with 6- and 4-carbon-long chains in a thermally induced aerosol-phase reaction. However, aerosol processes using neutral particles face problems due to rapid particle agglomeration. By functionalizing particles in the plasma, particles can be maintained at negative charge during the surface reaction process, thus suppressing coagulation as during the synthesis step. The plasma may also provide for particle heating that may aid thermal activation of surface reactions and at the same time prevent significant diffusion losses to the reactor walls due to negative charging of both particles and reactor walls. The group of Matsoukas has already shown that micron-sized silica particles can be dispersed in a parallel plate PECVD reactor and then coated with an amorphous carbon shell [97, 98].

The reactor used for plasma functionalization of silicon nanocrystals is composed of two stages as shown in the schematic in Figure 13.16. The first stage is identical to the silicon nanocrystal synthesis reactor described in Section 13.3 [63].

The second stage is a 5 in. diameter parallel plate reactor with a 2 in. gap. Particles are nucleated and grown in the first stage; the gas flow drags them into the second stage where an additional argon inlet is present. The secondary argon flow is passed through a bubbler filled with organic molecules to be grafted onto the nanocrystal surface, which is evaporated and fed into the plasma system. The bubbler pressure and the secondary argon flow rate can be adjusted to control the organic molecule vapor pressure in the second stage. The parallel plate reactor is used as a continuous flow reactor, with particles being carried below the bottom plate and to the exhaust line where they are collected with a filter. While the nanocrystals are collected as previously in the form of an agglomerated powder, their surfaces are now already functionalized with an organic monolayer if the plasma grafting reaction was successful.

Initial tests of this approach were performed with 1-dodecene, as hydrosilylation with this molecule in the liquid phase has been one of our benchmarks. For the gas-phase reaction, 30 sccm of argon and 8 sccm of SiH_4 (5% in He) are flown in the first stage. Fifteen sccm of argon is flown though the bubbler, which is kept at the same pressure of the reactor (1.4 Torr). A power of 60 W is used in the first reactor, while very low power (10–15 W) is needed in the second stage to successfully functionalize

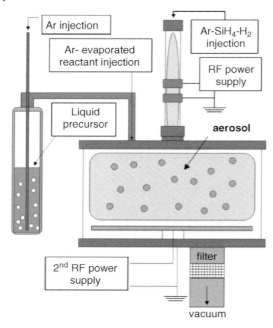

Figure 13.16 Schematic of the apparatus used for the in-flight grafting of silicon quantum dots with organic molecules.

the silicon particle surface. The resulting deposit on the filter is very different from the typical orange and fluffy powder film and it appears like a much more compact brown layer covering the stainless steel mesh. When the mesh is dipped into toluene, the deposit starts dissolving without any sonication, which is very different from the behavior of nonfunctionalized nanoparticles, and a clear colloidal dispersion of silicon nanocrystals is obtained, as shown in Figure 13.17. A control sample was deposited under exactly the same conditions but without igniting the plasma in the second stage. This sample showed no tendency to come of the mesh filter without sonication and the resulting colloid is turbid, indicating the presence of large, micrometer-sized agglomerates. This is a clear indication of the lack of success of the surface functionalization without the presence of the second plasma. The clear solution of sample A does not show any sign of flocculation even many weeks after production, and can be passed through a 0.22 μm filter without any loss of coloration. This experiment strongly suggests that the second nonthermal plasma provides the necessary activation energy to trigger the gas-phase hydrosilylation reaction with 1-dodecene.

It is interesting to compare the effectiveness of plasma functionalization and liquid-phase functionalization of silicon nanocrystals. FTIR is one tool to obtain some insight into the surface coverages obtained with both methods. Figure 13.18 compares the FTIP spectra of liquid-phase reacted and plasma-phase functionalized silicon nanocrystals. The main features present in the spectra of the as-produced silicon nanocrystals and neat dodecene have already been discussed. As already

Figure 13.17 Effect of the second discharge. The clear solution in the left vial is obtained with the second plasma on, while the cloudy solution in the right vial is produced under exactly the same conditions but with the second plasma off.

shown in Section 13.3, the FTIR spectrum of the hydrosilylated nanoparticles shows strong contributions from CH_x bonds between 2800 and 3000 cm^{-1}, but the peaks related to C=C bonds that are present in the spectrum of 1-dodecene do not appear in the spectrum of the treated particles. This is a strong indication that the carbon double bond reacted with the hydrogenated silicon surface, in agreement with the general understanding of the mechanism with which hydrosilylation proceeds [81]. The spectrum from the gas-phase hydrosilylated nanocrystals also does not show any

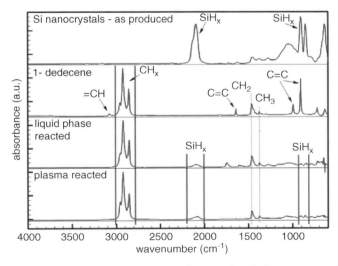

Figure 13.18 FTIR absorption spectra for as-produced silicon nanocrystals, 1-dodecene, liquid-phase hydrosilylated, and gas-phase hydrosilylated silicon nanoparticles.

peak from the C=C bond, and it is almost identical to the spectrum of the liquid-phase passivated nanoparticles. The ratio of the integrated signal between 2800 and 3000 cm^{-1}, which is due to CH$_x$ bonds, and the signal around 2100 cm^{-1}, due to SiH$_x$ bonds, is equal to 12.88 for the liquid-phase passivation and it is equal to 12.34 for the gas-phase passivation, indicating that the hydrocarbon surface coverage obtained from the two techniques is very similar. This is a quite intriguing result as the liquid-phase functionalization of the sample typically takes a few hours of reaction time, while the plasma functionalization proceeds during less than 1 s of residence time of the silicon nanocrystals in the second plasma process.

Plasma functionalization of silicon nanocrystals was successfully achieved with a wide range of molecules. These molecules are listed in Table 13.1. It is particularly interesting both to study the reactions with short molecules, which cannot be successfully attached via liquid-phase reactions, and to investigate the importance of different functional groups. The molecules used for this study are all 6-carbon long and are hexane, 1-hexyne, 1-hexene, hexyl-amine, and hexyl-alcohol. The structure of these molecules is shown in Table 13.1. Molecules with short 6-carbon chains cannot be successfully reacted with silicon crystals in liquid-phase reactions. Even after extended liquid-phase reaction, a clear solution could not be obtained and significant flocculation and sedimentation was observed. The particle size for this experiment was approximately 4 nm. On the contrary, particles of the same size can be efficiently grafted with 1-hexene in the plasma, and the resulting material is soluble in chloroform for an indefinite amount of time.

The FTIR spectra of the silicon nanocrystals treated with different molecules are shown in Figure 13.19. The assignment of the peaks is based upon Ref. [74]. The spectra have been normalized with respect to the area of the 2800–3000 cm^{-1} peak, which is due to CH$_x$ stretching vibrations. This peak is prominent in all the spectra of reacted particles, proving that all the organic molecules have reacted, to various

Table 13.1 List molecules that were successfully grafted to silicon nanocrystals using plasma functionalization.

Name	Structure
1-Dodecene	$H_3C-(CH_2)_9-CH=CH_2$
Dodecane	$H_3C-(CH_2)_{10}-CH_3$
Octyl-alcohol	$H_3C-(CH_2)_7-OH$
1-Hexene	$H_3C-(CH_2)_3-CH=CH_2$
1-Hexyne	$H_3C-(CH_2)_3-C≡CH_2$
Hexane	$H_3C-(CH_2)_4-CH_3$
Hexyl-alcohol	$H_3C-(CH_2)_5-OH$
Hexyl-amine	$H_3C-(CH_2)_5-NH_2$
1-Pentene	$H_3C-(CH_2)_2-CH=CH_2$
Acrylic acid	$H_2C=CH-COOH$
Allylamine	$H_2C=CH-CH_2-NH_2$
Ethylene-diamine	$H_2N-CH_2-CH_2-NH_2$
Ethylene-glycol	$HO-CH_2-CH_2-OH$

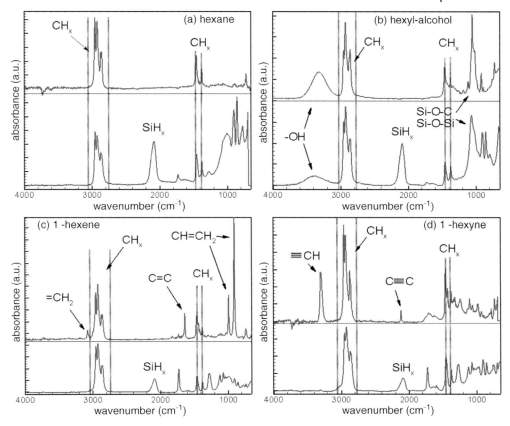

Figure 13.19 FTIR spectra of silicon quantum dots reacted with various 6-carbon-long organic molecules.

extents, with the silicon nanoparticles produced in the first stage. The as-produced silicon nanocrystal spectrum is not shown since it is equal to the one shown in Figure 13.18. In the case of 1-hexene and 1-hexyne, the different stretching modes due to carbon double bonds and triple bonds are not present in the spectra of the grafted nanoparticles. For 1-hexene, the peak at $3100 \, \text{cm}^{-1}$ corresponds to the $=C-H$ stretch mode, the peak at $1650 \, \text{cm}^{-1}$ corresponds to the $C=C$ stretch, and the two sharp features at 900 and $1000 \, \text{cm}^{-1}$ are due to $CH=CH_2$ out of plane deformations. These features are absent in the spectrum of the grafted silicon nanoparticles. The same behavior can be observed in the case of 1-hexyne since the $C \equiv C$ stretch at $2200 \, \text{cm}^{-1}$ and the $\equiv CH$ stretch at $3300 \, \text{cm}^{-1}$ are absent in the treated silicon quantum dot spectrum.

The spectra in Figure 13.19d are slightly more difficult to interpret. In the case of hexyl-alcohol, the linkage with the particle surface is expected to occur via $Si-O-C$ bonds. The broad feature at $3400 \, \text{cm}^{-1}$ is the signature of the alcohol functional group. This does not disappear completely after in-flight reaction, but its intensity decreases significantly with respect to the CH_x stretching signal at 2900 cm^{-1}. The feature at 1050 cm^{-1} in the hexyl-alcohol spectrum is due to the C$-$O stretch, but the shape of the signal changes after reaction with the particles. The broader signal between 1000 and 1100 cm^{-1} in the produced material spectrum can, in fact, be attributed not only to Si$-$O$-$C but also to Si$-$O$-$Si and to Si$-$C covalent bond. Overall, we can conclude that hexyl-alcohol has been successfully grafted onto the particle surface, but that due to problems with the selectivity of the process, the reacting molecule has also bonded to the particle via Si$-$C linkage. The fact that $-$OH signal is still present in the spectrum strongly supports this hypothesis, together with the fact that hexane also reacts with the particles, as shown in Figure 13.19a. Particles reacted with hexane, 1-hexene, 1-hexyne, and hexyl-alcohol are soluble in chloroform. No sedimentation is noticeable, even many days after production. Particle reacted with 1-hexene and 1-hexyne give a clear solution in toluene after sonication, but significant sedimentation takes place in few hours. As a comparison, particles reacted with 1-dodecene are indefinitely stable in toluene, very likely due to the presence of a longer ligand at the surface. The reaction with hexyl-amine was found to be not successful.

The ratio of the integrated CH_x signal at 2900 cm^{-1} to the integrated SiH_x signal at 2100 cm^{-1} is shown in Figure 13.20 for the case of 1-hexyne, 1-hexene, hexane, and hexyl-alcohol. This ratio is used to monitor the relative surface coverage of organic molecules onto the particle surface, and to study the relative reactivity of different functional groups for the case of the plasma-initiated in-flight hydrosilylation.

1-Hexyne is the molecule that most readily reacts with the particle surface, followed by 1-hexene and hexyl-alcohol. Even some reaction with the saturated alkane (hexane) occurs, which would not be the case in the liquid phase. In this sense, the plasma process is less selective than liquid-phase surface functionalization since some radicals are produced indiscriminately from usually inert precursors. Nevertheless, a significant amount of selectivity defined by the functional group is retained, as the reactivity with alkynes and alkenes is much larger than the reactivity with alkanes.

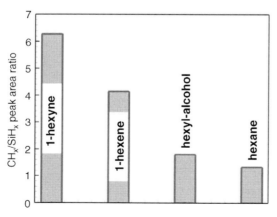

Figure 13.20 Ratio of the integrated area for the CH_x peak around 2900 cm^{-1} to the SiH_x peak at 2100 cm^{-1} for different organic molecules.

Alcohols also react more readily than alkanes but not as efficiently as for molecules with unsaturated carbon bonds.

In summary, we have shown in this section that plasma functionalization is an efficient route to assembling organic monolayers on the surfaces of silicon nanocrystals. In particular with organic molecules with functional carbon double or triple bonds, almost the same surface coverage with organic molecules can be achieved within one second of plasma treatment as compared to several hours of liquid-phase treatment.

13.5
Optical Properties of Plasma-Synthesized and Surface-Functionalized Silicon Nanocrystals

As an indirect bandgap semiconductor, silicon in its bulk form is an inefficient optical emitter and absorber. Hence, the observation of strong room-temperature photoluminescence from porous silicon, which contained small nanocrystals, attracted significant attention [99, 100]. It is now widely believed that the improved optical properties of nanocrystalline silicon result from mainly two physical phenomena: the enhanced recombination rate of electrons and holes due to the increased overlap of the electron and hole wave function confined to the nanocrystal [101], and the reduction of the rate of nonradiative events [102].

There have been various reports of high optical activity of silicon nanocrystals with various surface functionalizations. It is generally believed that silicon nanocrystals passivated by native oxide layers can have PL quantum yields, defined as the number of photons emitted divided by the number of exciting photons absorbed, of up to about 10%. There are limited reports of even higher quantum yields [76]. High quantum efficiencies have also been reported for the silicon nanocrystals embedded in high-quality thermal oxides [103]. However, these studies reported the quantum efficiency of the fraction of "bright," that is, emitting, nanocrystals, while the fraction of "dark," that is, nonemitting, nanocrystals remained unknown. High quantum yields of ensembles of nanoparticles have also been achieved with organically terminated nanocrystals, achieving ensemble quantum yields as high as 23% [104] and 30% [43]. Moreover, in single-quantum dot experiments PL quantum yields as high as 88% were observed for individual quantum dots [105]. However, these studies also pointed to the often large difference in the quantum yields of individual particles and nanocrystal ensembles, which often were found to be on the order of just a small percentage due to the large fraction of dark particles.

In this section, we discuss quantum yields of liquid and plasma-functionalized silicon nanocrystals and focus on their differences. We also discuss the radiative decay rates observed in very small silicon nanocrystals, which hint toward a transition from, indirect to direct bandgap behavior.

Accurate measurements of the ensemble quantum yield can be achieved with the use of an integrating sphere [106]. In our procedure, silicon nanocrystal PL is excited by illumination with a blue LED with the emission peak centered at 390 nm. Both the

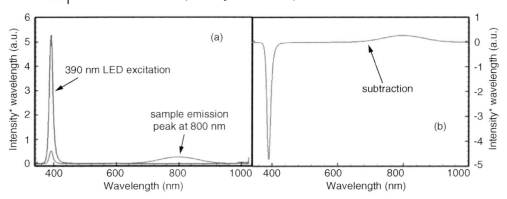

Figure 13.21 Scheme for performing photoluminescence quantum yield measurements. See text for details.

excitation source and the PL emission of the sample are simultaneously recorded with a low-resolution spectrometer with a broadband detector. The procedure used for measuring the quantum yield is as follows (as illustrated in Figure 13.21): The procedure used for measuring the quantum yield is as follows: first, as illustrated in figure 13.21(a), a baseline measurement is taken introducing a vial with the solvent used in the nanocrystals colloid with the excitation turned on. This spectrum presents one peak at 390 nm. Second, a vial containing colloidal silicon quantum dots is introduced into the integrating sphere. This second sample spectrum exhibits a strongly reduced excitation peak at 390 nm due to the absorption by the silicon quantum dots and the broad PL emission peak originating from the silicon quantum dot luminescence. By subtracting the baseline from the sample spectrum, one obtains a spectrum showing two peaks for the absorption and emission of the silicon quantum dots, as illustrated in figure 13.21(b). Depending on spectroscopic system, it may be necessary to convert the signals into signals that are proportional to the number of photons. For instance, if the detector is calibrated to provide a signal proportional to the radiant intensity, multiplying this signal with the emission wavelength will give a quantity proportional to the number of detected photons. The quantum yield can now be found as the ratio of the integrated PL emission over the integrated absorption. As we discussed in great detail in Ref. [107], by using an integrating sphere this procedure will automatically yield the ensemble quantum yield as it accounts for the absorption by both "bright" and "dark" nanocrystals. In addition, the method is very robust as it is independent of the exact geometrical arrangement between the excitation source and the sample, for example, the fraction of excitation that may hit or miss the sample, as well as potential effects of scattering from particle agglomerates.

Quantum yield measurements were performed for particles that were passivated in the liquid phase with either 1-dodecene or 1-octadecene over a range of emission wavelengths, that is, particle sizes. Results are shown in Figure 13.22. This figure shows that very high ensemble quantum yields, up to approximately 70%, can be achieved for silicon nanocrystals emitting at wavelengths >800 nm. This is an

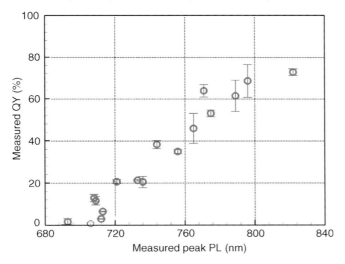

Figure 13.22 Measured quantum yield versus peak emission wavelength for alkylated silicon quantum dots.

extraordinarily high value for an ensemble of silicon crystals, which has only been exceeded by single-quantum dot experiments of particularly efficient, "bright" silicon crystals. It demonstrates that the liquid-phase surface functionalization with organic molecules leads to a very good termination of surface states, with which silicon nanocrystals can achieve quantum yields rivaling those of direct bandgap semiconductor nanocrystals. For comparison, the first reports of high efficiency photoluminescence from II to VI semiconductor quantum dots [108, 109] were for CdSe particles overcoated with an inorganic ZnS shell. The quantum yield of these core-shell dots was around 50%. Successive improvements in the synthesis techniques allowed production of CdSe/ZnS quantum dots with yields as high as 80% [110]. Unfortunately, the high quantum yield is limited to samples with a peak PL wavelength close to 800 nm. The quantum yield drops quickly at shorter emission wavelength, and emission below 700 nm becomes extremely inefficient for reasons yet to be understood.

In Ref. [111], we studied the photoluminescence of these liquid-phase organically functionalized silicon nanocrystals with picosecond and femtosecond resolution. We found a remarkable temporal evolution of the time-resolved PL signal. The early-time PL spectra (<1 ns after a femtosecond pump-pulse) show strong dependence on nanocrystal size, which is attributed to emission involving nanocrystal-quantized states. The PL spectra recorded for long delays after the pump-pulse (>10 ns) are almost independent of nanocrystal size and are likely due to surface-related recombination. Intrinsic radiative rate constants for nanocrystals of different sizes were determined from the instantaneous PL intensities measured 2 ps after excitation. We found that the radiative decay rates sharply increase for confinement energies greater than 1 eV, indicating a fast, exponential growth of the oscillator strength of zero-phonon, pseudodirect transitions.

It is interesting to compare the optical properties of plasma-phase and liquid-phase functionalized silicon nanocrystals. While we discussed above that the surface coverages of plasma-functionalized and liquid-phase functionalized silicon nano-crystals are almost identical, there is a significant difference in photoluminescence quantum yield. The plasma-grafted silicon nanocrystals show visible photolumines-cence with typical quantum yields below 10% for particles that have a peak emission wavelength between 800 and 850 nm. This is significantly lower than the 60% quantum yield that can be reproducibly obtained from liquid-phase treated samples that emit at the same wavelength.

To check the influence of the liquid-phase passivation on the optical properties of the particles, a gas-phase treated sample was refluxed for few hours in the usual dodecene–mesitylene mixture that is used for the liquid-phase passivation. The quantum yield of the as-produced plasma-functionalized sample was 1%. After liquid-phase post-treatment, the quantum yield of the sample increased to 45%. The photoluminescence spectrum did not change significantly after liquid-phase post-treatment. FTIR analysis of this sample also did not show a significant qualitative difference. However, it is important to note that the ratio of the integrated CH_x signal to the integrated SiH_x signal increased to 17.1 from an initial value of 12.34 found for the in-flight grafted material. This suggests that the hydrocarbon coverage has increased during the liquid-phase post-treatment, which may contribute to the increase in PL quantum yield from 1 to 45%. However, a control experiment that was performed suggests that the added surface coverage with surfactant molecules is not the only reason for the improved quantum yield. In this control experiment, plasma-grafted particles were refluxed in the gas phase with mesitylene alone for 3 h. No 1-dodecene was added to the liquid mixture this time. After initial plasma treatment, this sample had a QY of 7.5%. After refluxing in mesitylene, the QY increased to 53%. Obviously, the hydrocarbon surface coverage did not increase since no additional 1-dodecene is present to react with the particle surface.

A more careful FTIR study of the various components of the SiH_x peaks at approximately $2100\,cm^{-1}$ provides some insight into the role of the liquid-phase post-treatment. The SiH_x peak at $2100\,cm^{-1}$ was fitted with three Lorentzian profiles centered around 2135, 2110, and $2080\,cm^{-1}$, corresponding to silicon tri-, di-, and monohydride groups at the nanocrystal surfaces [70]. In Figure 13.23, we summarize the results and show the variation in relative intensities of different silicon hydride modes as a function of the processing conditions.

Figure 13.23 demonstrates that the liquid-phase treatment of silicon nanocrystals leads to a remarkable redistribution of silicon-hydrides from SiH_2 to SiH, regardless of the presence of the organic capping agent 1-dodecene. The fact that the quantum yield of liquid-phase processed nanocrystals also increases regardless of the presence of 1-dodecene during the liquid-phase post-treatment suggests that the reorganiza-tion of hydrogen on the silicon nanocrystal surface plays an important role in the quantum yield increase. The temperature applied during liquid-phase processing, $169\,°C$ in the case of mesitylene, may be sufficient to lead to a redistribution of hydrogen from dihydrides to saturate dangling bonds present at the surface. These dangling bonds may be sites of nonradiative recombination and the passivation with

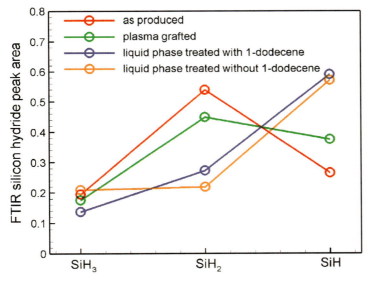

Figure 13.23 Relative variation of the contribution of different hydrides to the profile of the $2100\,cm^{-1}$ SiH_x peak. The contribution is calculated on the basis of the area of the fitted Lorentzian peaks and normalized over the total area of the three peaks.

hydrogen may explain the increase in quantum yield. This hypothesis is supported by electron paramagnetic resonance (EPR) measurements, which point to a significant reduction in the free spin density. The exact role played by the organic surfactants compared to the hydrogen reorganization at the silicon nanocrystals surfaces deserves further investigation.

13.6
Summary and Conclusions

In this chapter, we discussed the synthesis and surface functionalization of silicon nanocrystals with nonthermal plasmas. In nonthermal plasmas, energetic electrons provide the source of dissociation of gaseous precursors leading to the formation of nanocrystals through chemical clustering of precursor radicals. Initial clusters grow into particles of several nanometers in size. Clusters and particles are primarily negatively charged due to the presence of hot and mobile free electrons in the plasma. Strongly exothermic surface reactions at the nanoparticle surface, including electron–ion recombination and hydrogen recombination, combined with slow convective–conductive cooling of the particles at low pressure results in particle temperatures that can exceed the temperature of the surrounding gas by several hundreds of Kelvin. We suggest that these elevated particle temperatures are essential for the formation of high-quality nanocrystals in nonthermal plasmas.

A nonthermal plasma approach was introduced that synthesizes silicon nano-crystals at a high rate in a flow-through reactor. Silicon crystals reside in the plasma for a few milliseconds and grow during this time to several nanometers in size. The nanocrystal size can be tuned by adjusting the residence time of particles in the plasma. Particles are fairly monodisperse with standard deviations in the size distribution of about 15–20% of the average nanocrystal size. Surfaces of the silicon crystals are hydrogen-terminated that is ideal for the subsequent surface functionalization.

Surface properties of plasma-produced silicon nanocrystals are essential for their optical properties. Surface functionalization with organic molecules can be achieved both in the liquid- and in the plasma phase. Organic molecules are reacted through a terminal carbon double bond to form a covalent silicon–carbon bond with the silicon nanocrystal surface. While liquid-phase reaction requires transfer of the silicon crystals from a synthesis reactor into a liquid-phase reaction vessel and time-consuming reaction of several hours, the plasma-phase reaction can be performed in a simple two-stage reactor and it proceeds with reaction times of less than a second. FTIR demonstrated that surface coverages with organic molecules achieved with both methods are almost identical.

Optical properties of silicon nanocrystals are strongly affected by the surface functionalization. While nonfunctionalized silicon nanocrystals surface-passivated with a native silicon oxide often have ensemble quantum yields of photoluminescence of approximately 10%, significantly higher quantum yields can be achieved with organically functionalized silicon nanocrystals processed in the liquid-phase. In the deep red, at wavelengths larger than 800 nm, quantum yields as high as 70% were achieved, rivaling those of direct bandgap semiconductor quantum dots. Surprisingly, plasma-functionalized silicon nanocrystals with almost identical organic surface coverage showed quantum yields of typically less than 10%. Through liquid-phase post-treatment of such nanocrystals, it was determined that the reorganization of hydrogen atoms on the nanocrystal surfaces is at least as important as the organic molecule layer.

In conclusion, plasma synthesis and the appropriate surface treatment have shown the potential of silicon nanocrystals to lead to optical properties that are rival those of other direct bandgap semiconductor nanocrystals.

References

1 Roth, R.M., Spears, K.G., Stein, G.D., and Wong, G. (1985) Spatial dependence of particle light scattering in an RF silane discharge. *Appl. Phys. Lett.*, **46** (3), 253–255.

2 Bouchoule, A. (ed.) (1999) *Dusty Plasmas: Physics, Chemistry and Technological Impacts in Plasma Processing*, John Wiley & Sons, Ltd, Chichester, UK.

3 Boufendi, L. and Bouchoule, A. (1994) Particle nucleation and growth in a low-pressure argon-silane discharge. *Plasma Sources Sci. Technol.*, **3**, 263.

4 Boufendi, L., Bouchoule, A., Porteous, R.K., Blondeau, J.P., Plain, A., and Laure, C. (1993) Particle–particle interaction in dusty plasmas. *J. Appl. Phys.*, **73**, 2160.

5 Boufendi, L., Plain, A., Bloundeau, J.P., Bouchoule, A., Laure, C., and Toogood, M. (1992) Measurements of particle size kinetics from nanometer to micrometer scale in a low-pressure argon-silane radio-frequency discharge. *Appl. Phys. Lett.*, **60**, 169.

6 Bouchoule, A., Plain, A., Boufendi, L., Bloundeau, J.P., and Laure, C. (1991) Particle generation and behavior in a silane-argon low-pressure discharge under continuous or pulsed radio-frequency excitation. *J. Appl. Phys.*, **70**, 1991.

7 Courteille, C., Dorier, J.-L., Dutta, J., Hollenstein, C., Howling, A.A., and Stoto, T. (1995) Visible photolumine-scence from hydrogenated silicon particles suspended in a silane plasma. *J. Appl. Phys.*, **78** (1), 61–66.

8 Howling, A.A., Sansonnens, L., Dorier, J.-L., and Hollenstein, C. (1994) Time-resolved measurements of highly polymerized negative ions in radio frequency silane plasma deposition experiments. *J. Appl. Phys.*, **75**, 1340.

9 Howling, A.A., Sansonnens, L., Dorier, J.-L., and Hollenstein, C. (1993) Negative hydrogenated silicon ion clusters as particle precursors in RF silane plasma deposition experiments. *J. Phys. D: Appl. Phys.*, **26**, 1003.

10 Howling, A.A., Dorier, J.-L., and Hollenstein, C. (1993) Negative ion mass spectra and particulate formation in radio frequency silane plasma deposition experiments. *Appl. Phys. Lett.*, **62**, 1341.

11 Stoffels, W.W., Stoffels, E., Kroesen, G.M.W., and Hoog, F.J.d. (1996) Detection of dust particles in the plasma by laser-induced heating. *J. Vac. Sci. Technol. A*, **14**, 588.

12 Stoffels, W.W., Stoffels, E., Kroesen, G.M.W., and Hoog, F.J.d. (1993) Detection of particulates in a RF plasma by laser evaporation and subsequent discharge formation. *J. Appl. Phys.*, **74**, 2959.

13 Shiratani, M., Kishigaki, N., Kawasaki, H., Fukuzawa, T., and Watanabe, Y. (1996) Two-dimensional spatial profiles of size and density of particulates grown in RF silane plasmas. *IEEE Trans. Plasma Sci.*, **24**, 99.

14 Watanabe, Y. and Shiratani, M. (1993) Growth kinetics and behavior of dust particles in silane plasmas. *Jpn. J. Appl. Phys.*, **32**, 3074.

15 Shiratani, M., Yamashita, M., Tsuruoka, H., and Watanabe, Y. (1992) Development

of two laser light scattering methods for observation of particle growing process in a helium-silane RF plasma. *Proc. Jpn. Symp. Plasma Chem.*, **5**, 93.

16 Gallagher, A., Howling, A.A., and Hollenstein, C. (2002) Anion reactions in silane plasma. *J. Appl. Phys.*, **91** (9), 5571–5580.

17 Gallagher, A. (2000) Model of particle growth in silane discharges. *Phys. Rev. E*, **62**, 2690.

18 Childs, M.A. and Gallagher, A. (2000) Plasma charge–density ratios in a dusty plasma. *J. Appl. Phys.*, **87**, 1086.

19 Bhandarkar, U., Kortshagen, U., and Girshick, S.L. (2003) Gas temperature effects on particle generation in an argon-silane low-pressure plasma. *J. Phys. D: Appl. Phys.*, **36**, 1399–1408.

20 Bhandarkar, U.V., Swihart, M.T., Girshick, S.L., and Kortshagen, U.R. (2000) Modelling of particle nucleation in low-pressure silane plasma. *J. Phys. D: Appl. Phys.*, **33**, 2731–2746.

21 Gorla, C.R., Liang, S., Tompa, G.S., Mayo, W.E., and Lu, Y. (1997) Silicon and germanium nanoparticle formation in an inductively coupled plasma reactor. *J. Vac. Sci. Technol. A*, **15**, 860.

22 Viera, G., Huet, S., Mikikian, M., and Boufendi, L. (2002) Electron diffraction and high-resolution transmission microscopy studies of nanostructured Si thin films deposited by radiofrequency dusty plasmas. *Thin Solid Films*, **403**, 467–470.

23 Viera, G., Mikikian, M., Bertran, E., Cabarrocas, P.R.i., and Boufendi, L. (2002) Atomic structure of the nanocrystalline Si particles appearing in nanostructured Si thin films produced in low-temperature radiofrequency plasmas. *J. Appl. Phys.*, **92** (8), 4684–4694.

24 Bapat, A., Gatti, M., Ding, Y., Campell, S.A., and Kortshagen, U. (2007) A plasma process for the synthesis of cubic-shaped silicon nanocrystals for nanoelectronic devices. *J. Phys. D: Appl. Phys.*, **40** (8), 2247–2257.

25 Ding, Y., Dong, Y., Bapat, A., Perrey, C.R., Carter, C.B., Kortshagen, U.R., and Campbell, S.A. (2006) Single nanoparticle semiconductor devices.

IEEE Trans. Electron. Devices, **53** (10), 2525.

26 Longeaud, C., Kleider, J.P., Cabarrocas, P.R.i, Hamma, S., Meaudre, R., and Meaudre, M. (1998) Properties of a new a-Si:H-like material: hydrogenated polymorphous silicon. *J. Non-Cryst. Solids*, **227–230**, 96–99.

27 Butte, R., Meaudre, R., Meaudre, M., Vignoli, S., Longeaud, C., Kleider, J.P., and Cabarrocas, P.R.i. (1999) Some electronic and metastability properties of a new nanostructured material: hydrogenated polymorphous silicon. *Philos. Mag. B*, **79** (7), 1079.

28 Butte, R., Vignoli, S., Meaudre, M., Meaudre, R., Marty, O., Saviot, L., and Cabarrocas, P.R.i. (2000) Structural, optical and electronic properties of hydrogenated polymorphous silicon films deposited at 150 degree C. *J. Non-Cryst. Solids*, **266** (A), 263–268.

29 Meaudre, M., Meaudre, R., Butte, R., Vignoli, S., Longeaud, C., Kleider, J.P., and Cabarrocas, P.R.i. (1999) Midgap density of states in hydrogenated polymorphous silicon. *J. Appl. Phys.*, **86** (2), 946–950.

30 Street, R.A. (1991) *Hydrogenated Amorphous Silicon*, Cambridge University Press, Cambridge.

31 Staebler, D.L., and Wronski, C.R. (1976) Reversible conductivity changes in discharge-produced amorphous Si. *Appl. Phys. Lett.*, **31**, 292.

32 Oda, S. and Otobe, M. (1995) Preparation of nanocrystalline silicon by pulsed plasma processing. *Mater. Res. Soc. Symp. Proc.*, **358**, 721–731.

33 Otobe, M., Kanai, T., Ifuku, T., Yajima, H., and Oda, S. (1996) Nanocrystalline silicon formation in a SiH_4 plasma cell. *J. Non-Cryst. Solids*, **198–200**, 875–878.

34 Oda, S. (1997) Preparation of nanocrystalline silicon quantum dot structure by a digital plasma process. *Adv. Colloid Interf. Sci.*, **71–72**, 31–37.

35 Oda, S. (2003) Neosilicon materials and silicon nanodevices. *Mater. Sci. Eng.*, **B101**, 19–23.

36 Nishiguchi, K. and Oda, S. (2000) Electron transport in a single silicon

quantum structure using a vertical silicon probe. *J. Appl. Phys.*, **88** (7), 4186–4190.

37 Fu, Y., Willander, M., Dutta, A., and Oda, S. (2000) Carrier conduction in a Si-nanocrystal-based single-electron transistor-I. Effect of gate bias. *Superlatt. Microstruct.*, **28** (3), 177–187.

38 Banerjee, S., Huang, S., Yamanaka, T., and Oda, S. (2002) Evidence of storing and erasing of electrons in a nanocrystalline-Si based memory device at 77 K. *J. Vac. Sci. Technol. B*, **20** (3), 1135–1138.

39 Hinds, B.J., Yamanaka, T., and Oda, S. (2001) Emission lifetime of polarizable charge stored in nano-crystalline Si based single-electron memory. *J. Appl. Phys.*, **90** (12), 6402–6408.

40 Nishiguchi, K., Zhao, X., and Oda, S. (2002) Nanocrystalline silicon electron emitter with a high efficiency enhanced by a planarization technique. *J. Appl. Phys.*, **92** (5), 2748–2757.

41 Knipping, J., Wiggers, H., Rellinghaus, B., Roth, P., Konjhodzic, D., and Meier, C. (2004) Synthesis of high purity silicon nanoparticles in a low pressure microwave reactor. *J. Nanosci. Nanotechnol.*, **4** (8), 1039–1044.

42 Lechner, R., Wiggers, H., Ebbers, A., Steiger, J., Brandt, M.S., and Stutzmann, M. (2007) Thermoelectric effect in laser annealed printed nanocrystalline silicon layers. *Phys. Stat. Sol. (RRL)*, **1** (5), 262–264.

43 Sankaran, R.M., Holunga, D., Flagan, R.C., and Giapis, K.P. (2005) Synthesis of blue luminescent Si nanoparticles using atmospheric-pressure microdischarges. *Nano Lett.*, **5** (3), 531–535.

44 Lieberman, M.A. and Lichtenberg, A.J. (1994) *Principles of Plasma Discharges and Materials Processing*, John Wiley & Sons, Inc., New York, NY.

45 Raizer, Y.P. (1991) *Gas Discharge Physics*, Springer, Berlin.

46 Swihart, M.T. (2000) Electron affinities of selected hydrogenated silicon clusters (Si_xH_y, $x = 1$–7, $y = 0$–15) from density functional theory calculations. *J. Phys. Chem. A.*, **104** (25), 6083–6087.

47 Howling, A.A., Dorier, J.-L., and Hollenstein, C. (1993) Negative ion mass

spectra and particulate formation in radio frequency silane plasma deposition experiments. *Appl. Phys. Lett.*, **62** (12), 1341–1343.

48 Howling, A.A., Sansonnens, L., Dorier, J.-L., and Hollenstein, C. (1993) Negative hydrogenated silicon ion clusters as particle precursors in RF silane plasma deposition experiments. *J. Phys. D: Appl. Phys.*, **26**, 1003–1006.

49 Howling, A.A., Sansonnens, L., Dorier, J.-L., and Hollenstein, C. (1994) Time-resolved measurements of highly polymerized negative ions in radio frequency silane plasma deposition experiments. *J. Appl. Phys.*, **75**, 1340–1353.

50 Hollenstein, C., Dorier, J.-L., Dutta, J., Sansonnens, L., and Howling, A.A. (1994) Diagnostics of particle genesis and growth in RF silane plasmas by ion mass spectrometry and light scattering. *Plasma Sources Sci. Technol.*, **3**, 278–285.

51 Howling, A.A., Courteille, C., Dorier, J.-L., Sansonnens, L., and Hollenstein, C. (1996) From molecules to particles in silane plasmas. *Pure Appl. Chem.*, **68** (5), 1017–1022.

52 Kortshagen, U.R., Bhandarkar, U.V., Girshick, S.L., and Swihart, M.T. (1999) Generation and growth of nanoparticles in low-pressure plasmas. *Pure Appl. Chem.*, **71**, 1871.

53 Watanabe, Y., Shiratani, M., Kawasaki, H., Singh, S., Fukuzawa, T., Ueda, Y., and Ohkura, H. (1996) Growth processes of particles in high frequency silane plasmas. *J. Vac. Sci. Technol. A*, **14** (2), 540–545.

54 Watanabe, Y., Shiratani, M., Fukuzawa, T., Kawasaki, H., Ueda, Y., Singh, S., and Ohkura, H. (1996) Contribution of short lifetime radicals to growth of particles in SiH$_4$ HF discharges and effects of particles on deposited films. *J. Vac. Sci. Technol. A*, **14** (3), 995–1001.

55 Fukuzawa, T., Obata, K., Kawasaki, H., Shiratani, M., and Watanabe, Y. (1996) Detection of particles in RF silane plasmas using photoemission method. *J. Appl. Phys.*, **80** (6), 3202–3207.

56 Gallagher, A. (2000) A model of particle growth in silane discharges. *Phys. Rev. A*, **62** (2 part B), 2690–2706.

57 Gallagher, A., Howling, A.A., and Hollenstein, C. (2002) Anion reactions in silane plasma. *J. Appl. Phys.*, **91** (9), 5571–5580.

58 Bernstein, I.B. and Rabinovitz, I.N. (1959) Theory of electrostatic probes in a low-density plasma. *Phys. Fluids*, **2**, 112.

59 Allen, J.E., Annaratone, B.M., and de Angelis, U. (2000) On the orbital motion limited theory for a small body at floating potential in a Maxwellian plasma. *J. Plasma Physics*, **63** (4), 299–309.

60 Kortshagen, U. and Bhandarkar, U. (1999) Modeling of particulate coagulation in low pressure plasmas. *Phys. Rev. E*, **60** (1), 887.

61 Cabarrocas, P.R.i., Gay, P., and Hadjadj, A. (1996) Experimental evidence for nanoparticle deposition in continuous argon-silane plasmas: effects of silicon nanoparticles on film properties. *J. Vac. Sci. Technol. A*, **14**, 655.

62 Bapat, A., Perrey, C.R., Campbell, S.A., Carter, C.B., and Kortshagen, U. (2003) Synthesis of highly oriented, single-crystal silicon nanoparticles in a low-pressure, inductively coupled plasma. *J. Appl. Phys.*, **94** (3), 1969–1974.

63 Mangolini, L., Thimsen, E., and Kortshagen, U. (2005) High-yield plasma synthesis of luminescent silicon nanocrystals. *Nano Lett.*, **5** (4), 655–659.

64 Hirasawa, M., Orii, T., and Seto, T. (2006) Size-dependent crystallization of Si nanoparticles. *Appl. Phys. Lett.*, **88** (9), 093119.

65 Bouchoule, A. (1999) *Dusty Plasmas*, John Wiley & Sons, Ltd, West Sussex, England.

66 Koleske, D.D., Gates, S.M., and Jackson, B. (1994) Atomic H abstraction of surface H on Si: an Eley-Rideal mechanism? *J. Chem. Phys.*, **101** (4), 3301–3309.

67 Valipa, M.S. and Maroudas, D. (2005) Atomistic analysis of the mechanism of hydrogen diffusion in plasma-deposited amorphous silicon thin films. *Appl. Phys. Lett.*, **87**, 261911.

68 Sinniah, K., Sherman, M.G., Lewis, L.B., Weinberg, W.H., Yates, J.T., Jr., and

Janda, K.C. (1990) Hydrogen desoprtion from the monohydride phase on Si(100). *J. Chem. Phys.*, **92** (9), 5700–5711.

69 Whitby, K.T. and Clark, W.E. (1966) Electric aerosol particle counting and size distribution measuring system for the 0.015 to 1 µm size range. *Tellus*, **18** (2–3), 573–586.

70 Marra, D.C., Edelberg, E.A., Naone, R.L., and Aydil, E.S. (1998) Silicon hydride composition of plasma-deposited hydrogenated amorphous and nanocrystalline silicon films and surfaces. *J. Vac. Sci. Technol. A*, **16** (6), 3199–3210.

71 Marra, D.C. (2000) Plasma deposition of hydrogenated amorphous silicon studied using *in situ* multiple total internal reflection infrared spectroscopy, in *Chemical Engineering*, University of California, Santa Barbara.

72 Kessels, W.M.M., Marra, D.C., Van de Sanden, M.C.M., and Aydil, E.S. (2002) *In situ* probing of surface hydrides on hydrogenated amorphous silicon using attenuated total reflection infrared spectroscopy. *J. Vac. Sci. Technol. A*, **20** (3), 781–789.

73 Kessels, W.M.M., Smets, A.H.M., Marra, D.C., Aydil, E.S., Schram, D.C., and Van de Sanden, M.C.M. (2001) On the growth mechanism of a-Si:H. *Thin Solid Films*, **383**, 154–160.

74 Lambert, J.B., Shurvell, H.F., Lightner, D., and Cooks, R.G. (1987) *Introduction to Organic Spectroscopy*, MacMillan Publishing Company, New York.

75 Valenta, J., Juhasz, R., and Linnros, J. (2002) Photoluminescence spectroscopy of single silicon quantum dots. *Appl. Phys. Lett.*, **80** (6), 1070–1072.

76 Ledoux, G., Gong, J., Huisken, F., Guillois, O., and Reynaud, C. (2002) Photoluminescence of size-separated silicon nanocrystals: confirmation of quantum confinement. *Appl. Phys. Lett.*, **80** (25), 4834–4836.

77 Ledoux, G., Guillois, O., Porterat, D., Reynaud, C., Huisken, F., Kohn, B., and Paillard, V. (2000) Photoluminescence properties of silicon nanocrystals as a function of their size. *Phys. Rev. B*, **62** (23), 15942–15951.

78 Bergerson, W.F., Mulder, J.A., Hsung, R.P., and Zhu, X.-Y. (1999) Assembly of organic molecules on silicon surfaces via the Si–N linkage. *J. Am. Chem. Soc.*, **121**, 454–455.

79 Zhu, X.-Y., Mulder, J.A., and Bergerson, W.F. (1999) Chemical vapor deposition of organic monolayers on Si(100) via Si–N linkages. *Langmuir*, **15**, 8147–8154.

80 Zhu, X.-Y., Boiadjiev, V., Mulder, J.A., Hsung, R.P., and Major, R.C. (2000) Molecular assemblies on silicon surface via Si–O linkages. *Langmuir*, **16**, 6766–6772.

81 Buriak, J.M. (2002) Organometallic chemistry on silicon and germanium surfaces. *Chem. Rev.*, **102** (5), 1271–1308.

82 Buriak, J.M., Stewart, M.P., Geders, T.W., Allen, M.J., Choi, H.C., Smith, J., Raftery, D., and Canham, L.T. (1999) Lewis acid mediated hydrosilylation on porous silicon surface. *J. Am. Chem. Soc.*, **121**, 11491–11502.

83 Schmeltzer, J.M., Porter, L.A., Jr., Stewart, M.P., and Buriak, J.M. (2002) Hydride abstraction initiated hydrosilylation of terminal alkenes and alkynes on porous silicon. *Langmuir*, **18**, 2971–2974.

84 Stewart, M.P. and Buriak, J.M. (2001) Exciton-mediated hydrosilylation on photoluminescent nanocrystalline silicon. *J. Am. Chem. Soc.*, **123**, 7821–7830.

85 de Smet, L.C.P.M., Zuilhof, H., Sudholter, E.J.R., Lie, L.H., Houlton, A., and Horrocks, B.R. (2005) Mechanism of the hydrosilylation reaction of alkenes at porous silicon: experimental and computational deuterium labeling studies. *J. Phys. Chem. B*, **109**, 12020–12031.

86 Baldwin, R.K., Pettigrew, K.A., Ratai, E., Augustine, M.P., and Kauzlarich, S.M. (2002) Solution reduction synthesis of surface stabilized silicon nanoparticles. *Chem. Commun.*, 1822–1823.

87 Tilley, R.D., Warner, J.H., Yamamoto, K., Matsui, I., and Fujimori, H. (2005) Microemulsion synthesis of monodisperse surface stabilized silicon nanocrystals. *Chem. Commun.*, **14**, 1833–1835.

88 Tilley, R.D. and Yamamoto, K. (2006) The microemulsion synthesis of hydrophobic and hydrophilic silicon nanocrystals. *Adv. Mater.*, **18**, 2053–2056.

89 Warner, J.H., Hoshino, A., Yamamoto, K., and Tilley, R.D. (2005) Water-soluble photoluminescent silicon quantum dots. *Angew. Chem. Int. Ed.*, **44**, 4550–4554.

90 Rogozhina, E.V., Eckhoff, D.A., Gratton, E., and Braun, P.V. (2006) Carboxyl functionalization of ultrasmall luminescent silicon nanoparticles through thermal hydrosilylation. *J. Mater. Chem.*, **16**, 1421–1430.

91 Li, X., He, Y., Talukdar, S.S., and Swihart, M.T. (2003) Process for preparing macroscopic quantities of brightly photoluminescent silicon nanoparticles with emission spanning the visible spectrum. *Langmuir*, **19**, 8490–8496.

92 Walters, R.J., Kik, P.G., Casperson, J.D., Atwater, H.A., Lindstedt, R., Giorgi, M., and Bourianoff, G. (2004) Silicon optical nanocrystal memory. *Appl. Phys. Lett.*, **85** (13), 2622–2624.

93 Hua, F., Swihart, M.T., and Ruckenstein, E. (2005) Efficient surface grafting of luminescent silicon quantum dots by photoinitiated hydrosilylation. *Langmuir*, **21** (13), 6054–6062.

94 Murray, C.B., Kagan, C.R., and Bawendi, M.G. (1995) Self-organization of CdSe nanocrystallites into three-dimensional quantum dot superlattices. *Science*, **270** (5240), 1335–1338.

95 Park, K., Rai, A., and Zachariah, M.R. (2006) Characterizing the coating and size-resolved oxidative stability of carbon-coated aluminum nanoparticles by single-particle mass-spectrometry. *J. Nanopart. Res.*, **8**, 455–464.

96 Liao, Y.-C. and Roberts, J.T. (2006) Self-assembly of organic monolayers on aerosolized silicon nanoparticles. *J. Am. Chem. Soc.*, **128**, 9061–9065.

97 Cao, J. and Matsoukas, T. (2002) Deposition kinetics on particles in a dusty plasma reactor. *J. Appl. Phys.*, **92** (5), 2916–2922.

98 Cao, J. and Matsoukas, T. (2003) Particle coating in seeded dusty plasma reactor: distribution of deposition rates. *J. Vac. Sci. Technol. B*, **21** (5), 2011–2017.

99 Cullis, A.G. and Canham, L.T. (1991) Visible light emission due to quantum size effects in highly porous crystalline silicon. *Nature*, **335**, 335–338.

100 Canham, L.T. (1990) Silicon quantum wire array fabrication by electrochemical and chemical dissolution of wafers. *Appl. Phys. Lett.*, **57**, 1046.

101 Delerue, C., Allan, G., and Lannoo, M. (2001) Electron–phonon coupling and optical transitions for indirect-gap semiconductor nanocrystals. *Phys. Rev. B*, **64** (19), 193402.

102 Brus, L.E., Szajowski, P.J., Wilson, W.L., Harris, T.D., Schuppler, S., and Citrin, P.H. (1995) Electronic spectroscopy and photophysics of Si nanocrystals: relationship to bulk c-Si and porous Si. *J. Am. Chem. Soc.*, **117**, 2915–2922.

103 Walters, R.J., Kalkman, J., Polman, A., Atwater, H.A., and de Dood, M.J.A. (2006) Photoluminescence quantum efficiency of dense silicon nanocrystal ensembles in SiO_2. *Phys. Rev. B*, **73** (13), 132302.

104 Holmes, J.D., Ziegler, K.J., Doty, C., Pell, L.E., Johnston, K.P., and Korgel, B.A. (2001) Highly luminescent silicon nanocrystals with discrete optical transitions. *J. Am. Chem. Soc.*, **123**, 3743–3748.

105 Credo, G.M., Mason, M.D., and Buratto, S.K. (1999) External quantum efficiency of single porous silicon nanoparticles. *Appl. Phys. Lett.*, **74** (14), 1978–1980.

106 de Mello, J.C., Wittmann, H.F., and Friend, R.H. (1997) An improved experimental determination of the external photoluminescence quantum efficiency. *Adv. Mater.*, **9** (3), 230–232.

107 Mangolini, L., Jurbergs, D., Rogojina, E., and Kortshagen, U. (2006) Plasma synthesis and liquid-phase surface passivation of brightly luminescent Si nanocrystals. *J. Luminescence*, **121**, 327–334.

108 Hines, M.A. and Guyot-Sionnest, P. (1996) Synthesis and characterization of strongly luminescing ZnS-capped CdSe nanocrystals. *J. Phys. Chem.*, **100**, 468–471.

109 Dabbousi, B.O., Rodriguez-Viejo, J., Mikulec, F.V., Heine, J.R., Mattoussi, H., Ober, R., Jensen, K.F., and Bawendi, M.G.

(1997) (CdSe)ZnS core-shell quantum dots: synthesis and characterization of a size series of highly luminescent nanocrystallites. *J. Phys. Chem. B*, **101**, 9463–9475.

110 Reiss, P., Bleuse, J., and Pron, A. (2002) Highly luminescent CdSe/ZnSe core/shell nanocrystals of low size dispersion. *Nano Lett.*, **2** (7), 781–784.

111 Sykora, M., Mangolini, L., Schaller, R.D., Kortshagen, U., Jurbergs, D., and Klimov, V.I. (2008) Size-dependent intrinsic radiative decay rates of silicon nanocrystals at large confinement energies. *Phys. Rev. Lett.*, **100**, 067401.

14
Silicon Nanocrystals in Porous Silicon and Applications

Bernard Gelloz

14.1
Introduction

The applications of porous silicon (PSi) were mostly limited to thick oxide manufacturing until the suggestion in 1990 that quantum confinement could take place in its nanostructure [1, 2]. The visible photoluminescence (PL) of PSi, reported by L.T. Canham [1], generated an immense interest in the material, particularly for optoelectronic purposes. As a result, PSi has been extensively studied. In spite of the fact that commercially viable light-emitting devices have not yet been achieved, the interest in PSi has kept on growing in the past 18 years because of the many useful properties, originating from quantum confinement and the porous nature of the material, that were found in PSi. These properties expand the range of applications available for silicon, in particular in photonics, sensing, biotechnology, and acoustics.

The PSi formation has been described in detail in a book by V. Lehmann [3]. The structural and optical properties of PSi were reviewed in 1997 [4, 5]. The same year, a data review book about PSi was published [6]. PSi was again reviewed in 2000 [7]. The electroluminescence (EL) of PSi was reviewed in detail more recently [8–10]. Finally, the potential of PSi in drug delivery devices and materials was reviewed in 2008 [11]. This chapter summarizes the most important knowledge and expands on new developments.

PSi can exhibit different morphologies and typical pore sizes. It is characterized by its thickness, its porosity (volumetric fraction of void within PSi), its internal surface (taking into account the surface of pores), and its morphology (pore size, shape, and density; anisotropy). It is categorized according to the dominant pore diameter as microporous (≤ 2 nm), mesoporous (2–50 nm), and macroporous (>50 nm) [12]. Si nanocrystals may be mostly found not only in microporous Si but also in mesoporous Si, depending on the porosity and sample history. This chapter deals only with PSi that includes Si nanocrystals in its structure. Besides, as applications of PSi as a sensitizer for oxygen molecules and explosives are addressed in Chapter 19 of this book, they are not described here.

Silicon Nanocrystals: Fundamentals, Synthesis and Applications. Edited by Lorenzo Pavesi and Rasit Turan
Copyright © 2010 WILEY-VCH Verlag GmbH & Co. KGaA, Weinheim
ISBN: 978-3-527-32160-5

This chapter first describes the different PSi formation methods (Section 14.2). The influence of the formation parameters on PSi properties is also discussed. In Section 14.3, the structural properties of PSi are described and diverse postformation treatments are introduced. Then, in Section 14.4, various physical properties of PSi and corresponding applications are discussed.

14.2
Preparation of Porous Si Layers

PSi is generally formed by chemical or electrochemical etching of Si wafers. The most studied, well-controlled, and used technique is anodization of Si wafers in dilute hydrofluoric acid (HF) electrolytes (Section 14.2.1) [3, 4, 7, 13]. Most of this chapter deals with PSi prepared by this technique.

Galvanic etching (Section 14.2.2.1) [14–18], which does not require any external power source, can also be used to obtain well-controlled PSi layers. However, it is far less known and popular a technique than anodization.

Electroless etching (Section 14.2.2.2), in particular the so-called stain etching [19, 20], can lead to the formation of PSi. So far, it has not been a popular method because it offers little control over resulting PSi structures. However, recent progress in the control of PSi formation [21] may renew some interest in this technique.

In all cases, the formation of PSi is a wet process. Drying techniques are presented in Section 14.2.3.

14.2.1
Porous Si Prepared by Anodization

14.2.1.1 Conditions Leading to Porous Si Formation
Depending on experimental conditions, a Si wafer electrochemically polarized in an electrolyte containing hydrofluoric acid can either be progressively dissolved (electropolishing), or partially dissolved (PSi formation; the dissolution proceeds only at the PSi/substrate interface), or behave as an inert electrode. Critical experimental parameters are the current density, the electrolyte composition and characteristics (e.g., HF concentration, viscosity, and ability to wet the Si surface), the temperature, and the Si-doping density and type.

Figure 14.1 shows typical current–voltage characteristics for Si wafers in dilute aqueous HF. Hole supply by Si is the necessary initial step of the dissolution process. Therefore, the Si electrode is inert under cathodic polarization. In this case, the current observed for n-type Si and illuminated p-type Si results from hydrogen evolution. Under anodic polarization, holes are supplied by Si. For the fabrication of luminescent PSi from lightly doped n-type Si, the generation of holes is generally achieved by front-side or backside illumination. PSi formation is obtained for current densities lower than a critical current density J_{PS}, corresponding to potential V_{PS}. Above J_{PS}, complete electrochemical dissolution of Si (electropolishing) takes place.

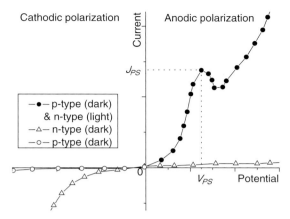

Figure 14.1 Typical shape of current–voltage characteristics of p- and n-type Si electrodes in aqueous hydrofluoric acid in the dark or under illumination (light). The origin of the potential axis is set to the flat band potential in both cases. PSi is formed under anodic polarization for a current density below J_{PS} (corresponding to potential V_{PS}). The characteristic of n-type Si is identical to that of p-type Si provided illumination powerful enough to generate the necessary holes is used.

The value of J_{PS} depends on several parameters. Because it is limited by mass transport and reaction kinetics, it exhibits a dependency on HF concentration and electrolyte temperature [3]. It also depends on the Si crystalline orientation [3, 22].

14.2.1.2 Porous Si Formation Mechanism

14.2.1.2.1 **Electrochemical Etching of Si**   The current peak characterized by J_{PS} indicates a transition between a charge supply-limited reaction (for $J < J_{PS}$) and a kinetic and mass transfer (in the electrolyte) limited reaction (for $J > J_{PS}$). For $J < J_{PS}$, a divalent reaction has been proposed [3]:

$$Si + 4HF_2^- + h^+ \rightarrow SiF_6^{2-} + 2HF + H_2 + e^- \tag{14.1}$$

For $J > J_{PS}$, the limitation by kinetic and mass transfer in the electrolyte results in Si dissolution taking place in a two-step process [3]. First, electrochemical oxidation of Si occurs via a tetravalent reaction:

$$Si + 2H_2O + 4h^+ \rightarrow SiO_2 + 4H^+$$

The second step is chemical dissolution of the oxide due to HF, $(HF)_2$ or HF_2^-:

$$SiO_2 + 2HF_2^- + 2HF \rightarrow SiF_6^{2-} + 2H_2O$$

The detailed actual reaction pathways on the atomic scale can be found in the book of Lehmann [3] or other reviews [7, 23].

Actually, the valence of dissolution in the PSi formation regime is not equal to 2, as would be expected from reaction (14.1), but rather generally found to be between 2 and 3, depending on anodization parameters, particularly current density. This fact can be understood if both the divalent and the tetravalent reactions take place simultaneously, but at different locations at the PSi/substrate interface, due to a significant current density distribution [3]. The valence increases when the global anodization current density increases and approaches J_{PS} because the contribution of local current densities at or above J_{PS} increases.

14.2.1.2.2 **Pore Initiation and Propagation in Microporous Si**   Reaction (14.1) and its pathways on the atomic scale do not explain why pores initiate at the surface of Si, even on atomically flat surfaces. The most supported mechanism for pore initiation is based on linear stability analyses [24]. Basically, if the propagating front of dissolution becomes rough (in the model, a sine wave is assumed) due to a perturbation of its flatness, then the surface may evolve in two possible ways. If the model shows a damping of the perturbation, the system will be stable. This case corresponds to electropolishing. Otherwise, pore initiation occurs, leading to PSi formation.

In order to form a stable porous layer, the dissolution has to take place at the PSi/substrate interface. Passivation of the pore walls and passivity breakdown at the pore tips are essential. This is achieved if holes supplied by the substrate are consumed at the pore tips and cannot penetrate into the already formed PSi layer. For $J < J_{PS}$, the dissolution reaction is limited by the supply of holes. This is consistent with consumption of all supplied holes at the PSi/substrate interface [25]. However, this simple consideration may not be sufficient. The passivation of the pore walls in microporous Si can be understood if an energy barrier prevents hole injection into PSi. This energy barrier has been attributed to a quantum confinement effect resulting from low-dimensional Si structures forming the PSi layer [2]. In mesoporous and macroporous Si, the space charge region at the PSi/substrate interface plays a significant passivating role [3].

14.2.1.3 **Effect of Anodization Conditions**
For PSi formation, anodization can be conducted either under galvanostatic (current-controlled) or potentiostatic (potential-controlled) conditions. In general, the former condition is preferred. It leads to better in-depth uniformity of PSi layers than the latter one since it keeps the rate of hole supply constant throughout the experiment. However, for long anodization time, typically over 20 min, significant porosity gradient may result because of extra chemical dissolution of PSi. Under galvanostatic condition, the etching rate stays mostly constant during anodization. Thus, the PSi thickness is proportional to the anodization time.

Well-controlled thick PSi layers can be readily obtained by anodization of p-type Si in the dark. For lightly doped n-type Si, the necessary holes are usually obtained by photogeneration. Front-side illumination through the electrolyte generally leads to PS layers exhibiting in-depth porosity gradients. This problem can be avoided by using backside illumination [3]. Furthermore, recent reports show that holes can be generated at the n-type Si/solution interface without illumination using either modified substrates (with a p-type backside region)[26], or Hall effect [27], or lateral potential fields [28]. Illumination is not required for heavily doped n-type Si because holes can be generated at the Si/electrolyte interface by a tunneling effect through the narrow space charge region of the Si substrate.

Reaction selectivity upon Si doping and type is observed [3, 13]. For a given potential, the anodic current increases with the doping level and in the following order of Si-doping type: n, p, p$^+$, and n$^+$. Thus, in Si substrates including regions of different doping levels or types, selective etching is achievable. This property is particularly useful for micromachining and photonic applications.

Anodization of Si generates gaseous H$_2$ molecules that must be evacuated from the pores. Silicon surface being hydrophobic, hydrogen gas can lead to local drying and hampering of the etching. A surfactant can be added to lower the surface tension of the solution. Ethanol is the most commonly used surfactant. Practically, at least 15 vol.% of ethanol is required for a good evacuation of H$_2$ bubbles. Less often used surfactants are Triton X-100$^®$ and Mirasol$^®$. A few drops are sufficient for an efficient removal of bubbles. Acetic acid (minimum of ∼5%) can also be used [13].

The effects of HF concentration, current density, temperature, and substrate doping density and type on the porosity, etching rate, and J_{PS} are summarized in Table 14.1. The closer the current density is to J_{PS}, the higher the porosity. Since the melting point of ethanol is −114.3 °C, the temperature of ethanolic HF solutions can be varied over a wide range. Lowering the temperature leads mainly to a decrease in J_{PS} (and therefore an increase in the porosity) and a decrease in the roughness of the PSi/substrate interface (viscosity effect), which is useful for photonic applications [29, 30].

Illumination during anodization can induce additional chemical etching that results in increased porosity. It generally leads to porosity gradients since (i) the top of the PSi layer under formation is exposed a longer time than the PSi part that is formed in the last moment of anodization and (ii) the absorption of the incident light by PSi depends on the layer depth.

The effect of a magnetic field with a strength of up to 1.7 T on the structure and PL of microporous Si has also been studied [31]. The magnetic field was applied perpendicularly to the substrate. The structural and optical isotropy was enhanced. The porosity and PL intensity were increased.

PSi layers can be removed from the Si substrate by applying a short high-current pulse (J higher than J_{PS}) at the end of the formation process. In this way, nanocrystalline PSi membranes can be produced.

Table 14.2 shows typical examples of formation conditions and resulting porosity.

Table 14.1 Effects of anodization conditions and substrate doping density and type on J_{PS} and PSi porosity and etching rate.

Effect of an increase in	Porosity	Formation rate	J_{PS}
HF concentration	Decreases	Decrease	Increase
Current density	Increases	Increase	—
Temperature (viscosity effect)	Decreases	Increase	Increase
Wafer doping (p-type)	Decreases	Increase	—
Wafer doping (n-type)	Increases	Increase	—

Table 14.2 Examples of porous Si formation conditions.

Substrate	Porosity (%)	Rate of formation (μm/min)	Current density (mA/cm^2)	[HF] (vol.%)
P	57	2.8	50	35
P	62	1.1	20	27.5
P	68	2.4	50	27.5
P	72	3.35	75	27.5
P	80	0.3	10	15
P$^+$	70	3.16	80	17.5
P$^+$	80	2.73	80	15
P$^+$	90	1.66	50	10

All solutions contain a maximum 50% (vol.) ethanol. Anodization is performed in the dark. Results are for p-type (4–8 Ω cm) and p$^+$-type ($\approx$0.01 Ω cm) silicon substrates.

14.2.1.4 Local Formation and Patterning of Porous Si

Local formation can be achieved either by (i) local anodization through a mask, (ii) anodization using etch stops, or (iii) anodization followed by local etching via lithography.

The first technique leads to undercutting (PSi formation proceeds also under the mask) because etching is almost isotropic (crystalline orientation induces a small anisotropy) in three dimensions. The mask must be HF resistant or stable enough to withstand HF during anodization. Standard photoresists can be used but are degraded during anodization leading to undercutting and anodization through the mask via pinholes [32–34]. Si dioxide can be used for short anodization times [32, 35]. Stoichiometric silicon nitride and silicon carbide induces stress problems leading to cracking [32, 36]. However, nonstoichiometric silicon nitride is a good masking material. The best conventional mask consists of a polycrystalline Si layer on top of Si dioxide [37, 38]. Besides, ultrathin (<1 nm) carbonaceous masking achieved by local electron beam-induced deposition led to a high selectivity of PSi formation [39].

The second technique relies on doping and type selectivity of anodization mentioned in Section 14.2.1.4. Selected areas with different doping levels or doping types can be formed by ion implantation through a mask [40, 41] or focused ion beam [42]. Selectivity is also achievable via illumination in lightly doped n-type Si. In this case, only illuminated areas are etched. Using near-field optical microscopy, nanometer-sized patterning is possible [43].

By playing on the local resistivity of the Si wafer, patterns of porosity and PSi thickness can be implemented. The technique of focused ion beam has led to several reports showing the patterning of PSi PL (Section 14.4.3.1.1) [44] and photonic structures (Section 14.4.2.2.2) [45–47].

In the third technique, pattern transfer to PSi can be done using general patterning of resist followed by wet or dry etching [48]. Writing nanostructures (20–50 nm in width) into PSi was achieved by scanning tunneling microscopy [49, 50].

14.2.1.5 Anodization Cells

Cells and counter electrodes should be inert in HF. Typically, Teflon is used as the cell material and Pt wires or sheets are used as counter electrodes [4, 7, 13]. An ideal cell should be designed to get uniform current density over all the Si substrate area in contact with the electrolyte in order to generate PSi with good lateral uniformity. In the absence of cells, HF-resistant tape may be used. The orientation of the Si wafer with respect to the vertical direction has an effect on the hydrogen bubbles path, which is always vertical. If controlled front-side illumination is needed, a cell where the Si substrate is vertical may be preferred for evacuation of the bubbles away from the optical path (horizontal). For backside illumination, the ideal solution is to use the double-tank cell (backside is in contact with another transparent solution). Alternatively, a conventional cell could be used if the backside of the Si wafer is not covered and the electrical contact is taken on a metal grid (or other pattern) letting a portion of the light through.

14.2.2
Formation of Porous Si Without External Power Supply

14.2.2.1 Galvanic Etching

PSi can be formed without using an external current or voltage source. A galvanic cell is formed by connecting an inert metal electrode to a Si wafer, both immersed in a solution containing aqueous HF and an oxidizing agent [14]. The principle of a galvanic element is shown in Figure 14.2. Curve (1) is the current–potential curve for p-type Si in dilute HF. It is similar to that shown in Figure 14.1 for the same type of Si wafer. Curve (2) corresponds to the inert metal in the same conditions. V_{Si} and V_M are the open-circuit potentials (OCP) for the Si wafer and the metal, respectively. Considering the metal under cathodic polarization with respect to a counter electrode

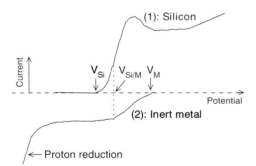

Figure 14.2 Schematic representation of galvanic cell formation. Curve (1) and (2) are the current–potential characteristics for p-type Si and an inert metal electrode, respectively, in an air-saturated HF solution. V_{Si} and V_M are the OCP when only Si or the metal is in the solution, respectively. When the Si and metal electrodes are simultaneously in the solution and short circuited, the OCP is at $V_{Si/M}$ and the anodic and cathodic currents are equal.

(curve 2), the oxidizing agent is reduced at the metal electrode, leading to an increase in cathodic current.

When the Si wafer is connected to the metal electrode, the rest potential of the system assumes a value $V_{Si/M}$ in between V_{Si} and V_M, where the anodic current (corresponding to Si dissolution) and the cathodic current (corresponding to the reduction of the oxidizing agent at the metal electrode) are equal. Electrons from the Si valence band flow to the metal where they participate in the reduction reaction. Holes left behind in the valence band are involved in the Si dissolution.

An important requirement for the formation of a galvanic element is that V_M must be positive with respect to V_{Si}. Then, depending on experimental conditions, $V_{Si/M}$ can be either lower than V_{PSi}, leading to PSi formation, or lower than V_{PSi}, leading to electropolishing. In the case of n-type Si, illumination is necessary to generate holes at the Si/electrolyte interface [15]. Using light intensities as low as 0.02 mW/cm^2 with very dilute HF solutions (0.1–0.3%) can lead to a luminescent PSi [18].

The etch rate is proportional to the reduction current. It can be increased either by raising the concentration of the oxidizing agent, or by increasing the surface area of the metal electrode, or by enhancing the diffusion of the oxidizing agent to the metal electrode.

The metal electrode can have an influence on the etching. For example, the current was found higher for Pt than for Au [14]. When the solution is in contact with air, oxygen is the natural oxidizing agent. Other oxidizing agents, in particular peroxodisulfate [14] and hydrogen peroxide [14, 17], may be added to enhance the etch rate.

For anodic etching, surfactants may be added to the solution to facilitate hydrogen bubble removal. Since the etching relies on an anodic current, most considerations detailed in Section 14.2.1 also apply to galvanic etching. Basically, the metal and oxidizing agent influence the current density, whereas the Si properties and HF concentration influence V_{PSi}.

This technique has several advantages. It does not require any external power source (chemical energy is converted into electrical energy). An electrochemical cell is not needed. The metal can be deposited on the backside of the Si substrate that is then merely immersed in the solution. It offers reasonably high etch rates. For example, for p$^+$-type Si, an etch rate of about 9 µm/min was shown [17]. PSi obtained by this method can show good lateral and in-depth uniformity.

14.2.2.2 Electroless Etching

In electroless etching, also called stain etching, Si is immersed into a solution containing HF and a strong oxidizing agent. The necessary initial step for Si dissolution, hole injection into the Si valence band, occurs via electron transfer from the valence band to the acceptor level of the oxidant species, which is reduced. The rate of hole injection must be equal to the rate at which holes are consumed in the Si dissolution reaction. Therefore, the kinetics of the reduction reaction is important and determines the etching rate.

Both p- and n-type Si can be etched in the dark when the acceptor level of the oxidizing agent overlaps the valence band of Si. When the acceptor level overlaps the conduction band of Si, illumination is necessary to bring about the Si dissolution.

The most used oxidizing agent is nitric acid [19, 20]. However, the reduction of nitric acid is a complicated multistep reaction, which implies an induction period prior to PSi formation and heavy bubble formation. As a result, stain etching lacks of control and suffers from irreproducibility problems.

Recently, other etchants have been proposed, leading to better results [20, 21, 51]. NH_4HF_2 has been used instead or in combination with HF. HCl was used to control the pH. HNO_3 was substituted by Fe^{3+}, Cr(VI), or Mn(VII). No bubbles are formed when $FeCl_3$ and $NaMnO_4$ are used. These studies showed that the PSi morphology, porosity, and PL characteristics depend on the etchant composition [21]. In accordance with the quantum confinement model of pore wall passivation, the energy of the PL peak was found correlated with the electrochemical potential of the oxidant. More positive potentials lead to bluer PL.

Finally, PSi powders prepared by stain etch are now commercially available (http://vestaceramics.net/).

A metal can be used to catalyze the dissolution process. This effect was used to create nanoscale 3D patterns of light-emitting PSi [52].

14.2.3
Drying of Porous Si

After formation in HF solution, PSi is rinsed, generally in ethanol or in water. Then, upon drying by only evaporation or under nitrogen flow, the PSi layers of high porosity (>75%) typically suffer from cracking, and the whole layer can be destroyed [4, 7]. The higher the porosity, the more fragile the nanostructure is. For a given porosity, the thicker the layer, the more prone it is to cracking. At comparable porosity, PSi formed from heavily doped Si is more resistant to cracking than that formed from lightly doped Si due to differences in the nanostructure morphology.

PSi cracking originates from high pressures in capillary liquid inside the pores, which are due to the surface tension at the liquid–gas interface. Recently, the evolution of real-time stress in PSi during drying was investigated by micro-Raman spectroscopy [53].

Using a liquid of low surface tension can help circumvent the cracking problem [4, 7]. Pentane works well for drying PSi formed from p^+-type Si, even for porosity up to 90%, but is less efficient for lightly doped Si. Because it is not miscible with water, ethanol is used as an intermediate liquid between water and pentane.

Freeze-drying can also help dry high-porosity PSi, though it has not been much studied [4, 7]. The temperature of wet PSi is lowered down to $50\,°C$, and then sublimation is carried out under vacuum. It seems more efficient with PSi formed from p^+-type Si.

The most efficient drying method is supercritical drying [4, 7]. The liquid (generally CO_2) from the pores is removed above the critical point in order to avoid interfacial tensions due to the presence of liquid–gas interfaces. PSi of very high porosity (>90%) can be dried without any microscopic damages. However, this technique is quite expensive and difficult to implement. As a result, it is used only when absolutely necessary.

14.3
Structural Properties of As-Formed and Modified Porous Si

14.3.1
As-Formed Porous Si

The crystalline nature of the Si substrate is mostly preserved upon anodization [4, 7]. Thus, microporous Si is mostly crystalline, but a lattice expansion Δa is observed compared to bulk Si. Typical values of $\Delta a/a$ range from 3×10^{-4} to 5×10^{-3} [4, 7]. The higher the porosity, the larger the expansion. PSi formed from lightly doped p-type Si undergoes larger expansion than that of the same porosity PSi formed from p^+-type Si, likely due to the difference in morphology (see below). The origin of the expansion is related to the hydrogen coverage of the Si surface. There is a correlation between hydrogen desorption and strain during thermal annealing [54–56]. Partial oxidation of PSi also leads to an increase in $\Delta a/a$ [56–58].

Diverse analyses [4, 6, 7] have shown that microporous Si consists of variously connected Si nanocrystals the size of which typically decreases as the porosity increases. Basically, lightly doped Si leads to mostly isotropic PSi layers having a sponge-like nanostructure. Typically, size ranges from 1 to 5 nm in radius, depending on porosity. Highly doped Si leads to a columnar structure that depends on crystalline orientation, with relatively large ($\sim$5–20 nm) straight but still more or less interconnected pores at the surface of which exist nanostructures of lower dimensions. In this case, the size distribution is large and can show a double-peak shape. The first peak corresponds to the fine structure (1–3 nm), similar to that of lightly doped Si, whereas the second peak relates to the coarser structure (7–15 nm) [59]. The dependence of the morphology on the doping level explains many differences in properties (lattice expansion, specific area, hardness, anisotropy, PL, absorption, etc.).

The total surface area exposed to air can be very high due to the porous nature of the material. It significantly depends on the porosity for lightly doped Si, and much less for high doping levels, where it was found to be about $300 \, \mathrm{m^2/cm^3}$ [60]. For PSi formed from lightly doped p-Si, it was measured from over $900 \, \mathrm{m^2/cm^3}$ for a porosity of 51% down to about $100 \, \mathrm{m^2/cm^3}$ for a porosity of 95% [61]. This property can be an advantage when high surfaces are needed in a small volume (e.g., for sensing applications or drug delivery) whereas it is generally a drawback for the material stability.

The surface of as-anodized PSi is terminated by $\mathrm{Si-H}_x$ ($x = 1, 2, 3$) bonds, as evidenced by infrared spectroscopy [4, 7]. Hydrogen desorbs from PSi for temperatures above $300 \,^\circ\mathrm{C}$ [62], resulting in the appearance of dangling bonds. The PL is then quenched as a result of defect-mediated nonradiative recombinations. Exposure of PSi to air leads to contamination, incorporation of oxygen atoms, and eventually to the growth of native oxide [4, 7]. Because the inner surface is so large, progressive growth of native oxide can considerably change the properties of PSi and results in poor stability of the as-prepared material.

14.3.2
Oxidation of Porous Si

PSi can be intentionally oxidized using thermal, chemical (in water vapor, in boiling water [63, 64], or photoinduced in oxidizing atmosphere [65–67]), or electrochemical oxidation techniques, in particular for stabilization purposes [4]. Using thermal or chemical oxidation, PSi can be either partially or fully oxidized depending on temperature, initial porosity, and treatment duration [4, 68], whereas electrochemical oxidation is limited by the electrical contact between the substrate and the PSi layer, which deteriorate upon oxidation [69–74]. However, the latter technique provides a high control over the extent of oxidation and resulting nanostructure and was used in various PSi devices [75–77].

Typically, thermal oxidation in oxygen at temperatures below 800 °C induces a rather low-quality oxide, with a relatively high density of nonradiative Pb centers, whereas higher temperatures lead to good-quality oxide, favorable to the luminescence [62, 78, 79]. For temperatures above 900 °C, it is often necessary to implement a preliminary partial oxidation step at 300 °C in order to avoid pore coalescence at high temperature. Rapid thermal oxidation can be used to prevent pore coalescence even with temperatures above 800 °C [79]. This technique leads to highly luminescent layers of Si nanocrystals melted into an oxide matrix.

Recently, high-pressure water vapor annealing was used to stabilize PSi as well as its PL and electroluminescence, and to obtain very highly efficient PL [80–86]. PSi is placed inside a closed chamber with liquid water. As the temperature is raised, the water is vaporized and a high pressure is induced in the chamber. Typical temperature, pressure, and duration are 150–300 °C, 1–4 MPa, and 2–6 h, respectively. The treatment produces a thin oxide layer at the surface of the PSi skeleton. The key advantage of the treatment compared to conventional oxidation technique is that the oxide is very stable and of good structural quality because it is relaxed and exhibits a very low density of defects (Pb centers). Moreover, this technique leads to completely stable layers whereas PSi oxidized by other techniques still suffers from drifting of properties upon aging.

14.3.3
Stabilization of the Porous Si Surface

Oxidation and particularly high-pressure water vapor annealing can produce very stable PSi layers [80–86]. However, oxidation changes the properties of the film. It may be useful to stabilize PSi without changing much its original properties. Basically, two approaches can be considered: (i) physical protection of PSi by encapsulation [87] or pore-filling techniques [88] and (ii) changing the chemistry of the PSi surface. Derivatization with long alkyl groups led to good stability of PL [89], EL [90], photonic properties [91], and biosensors [92]. A few reports deal with nitridation [93–96] and thermal carbonization [97] of PSi. The latter process was shown to stabilize passive photonic structures without inducing much property shift [98].

Table 14.3 Summary of main properties and applications of microporous and mesoporous silicon.

Property	Main applications
Luminescence	Optoelectronics (LEDs, markers, sensitization)
Porosity-dependent refractive index (1.1–3)	Photonics (waveguides, filters, mirrors, cavities, optical barcodes, optical processing)
Loading of compounds and binding/adsorption of molecules into the pores; large area and volume of pores	Optical switching and sensing; drug delivery; explosives
Nonlinear optical properties, anisotropy	Optical switching; birefringence; generation of harmonics
Electron cascade tunneling through Si nanodots/thin oxide chains	Cold cathodes; flat panel displays; lithography; ion sources; H_2 generation
Extremely low thermal conductivity	Sound/ultrasound emission; actuators; position sensing; digital transmission
Porosity-dependent acoustic properties	Acoustic band crystals
Biocompatibility	Drug delivery, implantable devices; tooth-paste
Sacrificial layers	MEMS; SOI fabrication

14.4
Physical Properties and Applications of Porous Si

PSi basically consists of a mixture of crystalline silicon (surface chemistry can vary) and air, characterized by the porosity. However, the size of the pores and the Si skeleton, as well as the large surface area, dramatically affects its physical properties. Thus, PSi exhibits a rather large number of new properties that considerably expand the practical potential of bulk Si. Table 14.3 summarizes the main properties and related applications of microporous and mesoporous silicon.

14.4.1
Electrical Properties

14.4.1.1 Electrical Conduction and Transport
The electrical resistivity of PSi is typically greater than $10^5 \, \Omega$ cm at room temperature [4, 7] because the PSi skeleton is depleted by free carriers. Two mechanisms are responsible for the depletion. The first is energy gap widening from quantum confinement that reduces the thermal generation of free carriers. However, high resistivity is also observed in mesoporous Si where the quantum confinement is not always strong. In fact, the main reason for the depletion is trapping of free carriers by surface states [7, 99, 100]. For p-type Si, doping atoms are still present in PSi at concentration similar to that of the Si substrate, but they are in neutral state [100].

Electrical transport in PSi occurs via mechanisms typically found in insulators, such as Poole–Frenkel, Fowler–Nordheim tunneling, space charge limitation, and ballistic emission. Depending on PSi type, porosity, and the applied voltage, one or several of these transport mechanisms may be involved, making a clear classification rather difficult.

Anisotropic conduction [101, 102] and Coulomb blockade [103, 104] have been reported in PSi.

14.4.1.2 Electro-Optic Memory

Negative resistance effect [105] and memory operation [106] have been shown in thermally oxidized PSi. These phenomena are based on trapping and subsequent release of carriers inside PSi under polarization. The nonvolatile electrical memory is based on the analysis of the hysteresis of the current–voltage curve that results from band distortion produced by the stored and released charge carriers.

With luminescent PSi, the device can operate as a nonvolatile light-emissive memory. It can be used both electrically and optically. The writing mode can be electrical or electro-optical, whereas the reading mode can be electrical and/or optical via EL. Furthermore, owing to the absence of lateral percolation of carriers, the PSi memory can be used as a nonpixel optical memory.

14.4.1.3 Sensing Based on Change of Conductivity

Since the specific surface of PSi (and the proportion of surface Si atoms in PSi) can be very large (see Section 14.3.1), its chemical state can strongly influence the PSi conduction. This property can be used for electrical sensing. Polar gases and liquids (e.g., water and ethanol) condensed into the pores lead to an increase in conductivity [107–109] due to an increase in effective dielectric constant [108]. Very strong and reversible interaction between the surface of PSi and NO_2 [110–112] has been found and sensors exhibiting sub-ppm sensitivity and more than an order of magnitude increase in conductivity (with p^+-type PSi) for a variation of 1 ppm of NO_2 were demonstrated. The mobility could be increased up to a value close to that of bulk Si. A chemisorption phenomenon has been suggested to explain the generation of free carriers in PS. The NO_2 adsorption effect on p^+–n Si junctions surrounded by a PSi layer has been investigated [113]. It was very selective relative to humidity and ethanol gas. An NO_2 concentration as low as 100 ppb produced a current variation of about an order of magnitude.

The intensity of the reverse current and of the open-circuit voltage of a Au/PSi/Si Schotky structure was used for sensing of humidity, CO, and H_2S gases [114]. The response time was about 60 s. Using the open-circuit voltage, sensitivities of 4 and 2 mV/ppm were obtained for CO and H_2S gases, respectively. The mechanism of the sensing was attributed to the presence of hydrogen and a proton exchange phenomenon similar to that of a hydrogen fuel cell.

A hydrogen sensor made of PSi covered by either TiO_2 or ZnO thin films has been proposed [115]. Pt was used on top of the structure as catalyst. It is based on resistivity changes relative to air atmosphere. It has been tested at room temperature and at 40 °C. The response time was about 20 s. Exposure of 5000 ppm of hydrogen produced a decrease in resistivity of 2.5–4 times depending on structure and temperature.

Electrical sensing can also make use of differences in the electrical or electrochemical behavior of what is inside the pores or attached to the pore walls. For example, a label-free nanoporous Si-based specific DNA biosensor for detecting *Salmonella enteritidis* has been demonstrated [116]. It uses redox indicators and cyclic voltammetry for sensing principle. It was more sensible than a planar sensor due to the large inner surface offered by the PSi structure. A sensitivity of 1 ng/ml was shown.

14.4.1.4 Photodetection

Zheng *et al.* [117] fabricated a highly sensitive photodetector made from a metal–PSi junction. Close to unity quantum efficiency could be obtained in the wavelength range of 630–900 nm without any antireflective coating. The detector response time was about 2 ns, with a 9 V reverse bias. Tsai *et al.* [118] used a metal–semiconductor–metal photoconductor and a p–n photodiode based on RTO-treated PSi. The photoconductor exhibited 2.8 higher responsivity at 350 nm than an UV-enhanced Si photodiode. The photodiode showed an external quantum efficiency of 75% at 740 nm.

Color-sensitive photodetectors have been fabricated [119–121]. The color sensitivity was achieved by using PSi Fabry–Perot filters on top of a p^+–n Si photodiode. The Fabry–Perot filter was made in the p^+ part of the initial p^+–n junction.

The geometry of the metal contact on PSi is important for the responsivity spectra of the PSi photodetectors [122]. Moreover, the length of the way traveled by separated electron–hole pairs should be as short as possible for better response. Faults and disorders in PSi also have important effects.

There is hope that PSi could be used to fabricate solar cells able to efficiently convert all the light emitted by the sun.

14.4.1.5 Ballistic Transport and Applications

14.4.1.5.1 Ballistic Electron Emission
A material consisting of Si nanocrystals interconnected by thin tunnel oxide can be prepared by partially oxidizing PSi. For this purpose, PSi layers are fabricated from heavily doped n-type either single-crystalline or polycrystalline substrates. PSi acts as a drift layer producing ballistic electrons (injected into the layer from the substrate) via multiple-tunneling cascade under an external electric field applied between the substrate and a top metal electrode. As a result, ballistic electron of several electron volts in energy (Figure 14.3a) can be emitted from the top metal electrode into a vacuum [123, 124], a gas (up to atmospheric pressures) [125, 126], or directly into a material deposited onto PSi [127–129]. When operated into vacuum, the emission efficiency (emission current divided by current flowing through the diode) has reached 12% [123].

The mechanism of ballistic electron emission (Figure 14.3b) has been confirmed by several experimental and theoretical studies. Major potential drops occur mainly through the thin oxide interconnecting the Si nanocrystals. The voltage dependence of the emission current is well described by the Fowler–Nordheim scheme. Time-of-flight measurements indicate an extremely large electron drift length (~2 μm or more) in such PSi layers under high electric field [130, 131]. The features of the

(a)

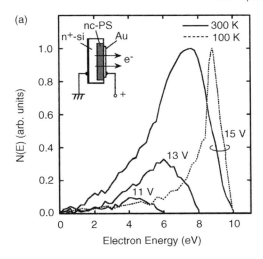

(b)

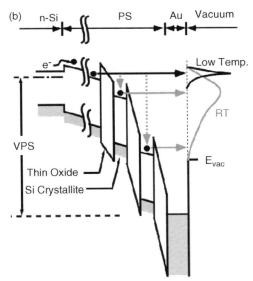

Figure 14.3 (a) Energy distribution of electrons emitted from a PSi diode for different applied voltages at room and low temperatures. Inset shows a schematic representation of the device. (b) Mechanism of ballistic electron emission [127]. Reproduced with permission from Ref. [127]. Copyright (2002) by the American Institute of Physics.

energy distribution of emitted electrons (at low and room temperatures) also support the model of multiple tunneling through the thin oxide surrounding the Si nano-crystals [132–134].

This type of surface-emitting cold cathode has several advantages compared to conventional ones. It is less sensible to ambient pressure [125, 126], exhibits a smaller emission angle dispersion [134], requires lower power consumption, and is compatible with monolithic processing.

14.4.1.5.2 Application as Flat Panel Displays
PSi electron emitters are available for the excitation of fluorescent screens. A prototype of 4×4 pixels display panel was demonstrated [135, 136]. Another prototype was fabricated on a glass substrate using a low-temperature process (oxidation was performed electrochemically) [77]. This flat panel display was 168×126 pixels, 2.6 in. diagonal full color. Subpixel size was $320 \times 107\,\mu m$. The device could operate at relatively low vacuum level (10 Pa) and without any focusing electrodes even the distance between pixels was only $40\,\mu m$. The same group then reported a prototype of a larger display (7.6 in. in diagonal). [137]

A principle of planar-type visible light emission has been reported using ballistic electrons as excitation source [127, 128, 138, 139]. The device is composed of a surface-emitting cold cathode made of PSi and a luminescent material directly deposited onto the electron emitter [127]. The diode structure is n^+-Si/PSi/luminescent film/Au. Any type of luminescent material could be used in order to produce any color. Red, green, and blue emission from organic compounds has been demonstrated [127, 128]. Electrons are emitted into the luminescent film where they generate light emission by impact. No vacuum spacing is required. Both polycrystalline Si [138] and crystalline Si [127, 139] have been used.

A cavity effect was obtained by sandwiching the luminescent film between two Ag mirrors that were also used as electrodes [129]. A narrowing of the emission spectrum was obtained (luminescent film was ZnS:Mn). In addition, this device allows the independent control of the voltage across the PSi electron emitter and that across the luminescent film. The mechanism of EL generation was confirmed using this injection configuration.

A device in which the luminescent layer is also a PSi layer has been investigated [140]. The intrinsic EL of the nc-Si diode was significantly enhanced by the introduction of the ballistic excitation mode.

14.4.1.5.3 Ballistic Electron Emission in Gas and Liquids
Electron emission in air and different gases at various pressures (up to atmospheric pressures) has been reported [125, 126]. In air, about $1\,\mu A/cm^2$ was detected at the collector electrode located at a distance of 1 mm from the emitter surface. This current was attributed to electron attachment onto oxygen molecules in the vicinity of the front surface. The operation stability in air has been improved combining several surface treatments: high-pressure water vapor annealing, forming gas ($N_2 + H_2$) annealing, and chemical modification [126]. An application is surface charging of an insulating polymer [125]. The polymer can be quickly charged up to a certain negative potential determined by the collector voltage. Parallel electron beam lithography can also be considered as a potential application.

The ballistic emitter can also operate in pure water by simply immersing the diode, letting only the top metal electrode in contact with the liquid. It may be used to produce hydrogen by reducing H^+ ions without the need of a counter electrode (necessary in electrolysis of water) and without generating by-products such as oxygen [141]. A hydrogen generation rate of $2\,\mu mol/h$ was obtained, in accordance with the diode emission current. This value is rather low compared to other techniques. Moreover, the efficiency of this device, limited by that of the ballistic emission (typically 1–5%), is much lower than that of the commercial electrolysis of electrolyte-dissolved water (50–70%), but could be enhanced by optimization. The device was also operated in dilute aqueous H_2SO_4 [142]. The generation of hydrogen was enhanced and the pH was increased as a result of H^+ consumption. The ballistic emitter could have applications in general electrochemistry as well.

14.4.2
Photonics

14.4.2.1 Refractive Index and Absorption Coefficient

The refractive index of PSi is generally not merely obtained by taking that of crystalline Si and applying a correction based on the porosity alone, though it monotonically varies in accordance with porosity. The interaction of light with PSi can be affected both by the characteristic size of pores and Si skeleton in PSi and by the quantum confinement effects. In PSi, the typical size ranges from 1 to 20 nm. Therefore, for wavelength much longer than this typical size, and in particular for visible and infrared light, PSi can be approximated as a continuous effective medium. Effective medium theories [4, 5, 7] that are often used for PSi are listed in Table 14.4. However, satisfactory fits of reflection or transmission spectra over the whole visible range are not always achievable with these theories, and using an arbitrary dispersion relation is sometimes necessary. The roughness and waviness of the different interfaces may be taken into account for better accuracy [30, 143]. In the near-infrared visible spectral region, the real part of the refractive index of PSi can be tuned from about 1.3 to 2.6 and 1.3 to 1.8, for heavily and lightly doped p-type Si, respectively, simply by changing the anodization current density [5, 144].

The absorption coefficient of PSi can also be conveniently measured by reflection of PSi [5] or by transmission of freestanding layers [4, 5, 7, 145]. The latter showed the influence of the quantum confinement at high porosities (>70%) and the indirect

Table 14.4 Effective medium approximation for the dielectric function of PSi.

Theory	Formula
Bruggeman	$P\dfrac{\varepsilon_M - \varepsilon_{eff}}{\varepsilon_M + 2\varepsilon_{eff}} + (1-P)\dfrac{\varepsilon - \varepsilon_{eff}}{\varepsilon + 2\varepsilon_{eff}} = 0$
Maxwell Garnett	$\dfrac{\varepsilon_{eff} - \varepsilon_M}{\varepsilon_{eff} + 2\varepsilon_M} = (1-P)\dfrac{\varepsilon - \varepsilon_M}{\varepsilon + 2\varepsilon_M}$
Looyenga	$\varepsilon_{eff}^{1/3} = (1-P)\varepsilon^{1/3} + P\varepsilon_M^{1/3}$

P is the porosity. ε and ε_M are the dielectric functions of Si and the material inside the pores, respectively. ε_{eff} is the calculated dielectric function.

nature of PSi bandgap. It also showed that quantum confinement is more effective in PSi formed from Si substrates of low doping level. Typically, the absorption of PSi is very much decreased in the visible region as a result of lowered density and quantum confinement. Thus, photonic structures can be considered from the near infrared to the visible and even near UV.

The variation in refractive index, coupled to the possibility of forming very thick PSi multilayers by simply varying the anodization current density over time, allows the easy fabrication of various photonic structures in a single anodization step. Practically, high doping is generally preferred to light doping because it allows higher index contrasts and leads to mechanically more stable PSi layers. Only p-type PSi is used in photonics because PSi formed from n-type Si is not homogeneous enough in its depth.

14.4.2.2 Passive Photonic Structures

Table 14.5 summarizes the most usual photonic structures together with their properties and potential applications.

Table 14.5 Examples of photonic structures, properties, and functions.

Name	Refractive index modulation	Optical properties	Applications
Black silicon	Gradient-index multilayer	Low reflection	Antireflection coatings
Distributed Bragg reflector	$(LH)_n$	Photonic bandgap surrounded by sidelobes	Mirrors; dichroic mirrors
Omnidirectional reflector	Chirped DBR; Gaussian profiles; superposition of bandgaps	Omnidirectional bandgap	Same as DBR; solar cells
Microcavity	$(LH)_n H(LH)_n$	One or more modes allowed inside photonic bandgap	Interference filters, sensing, switching; second- and third-harmonic generation using birefringence
Rugate	Sine wave	Photonic bandgap without sidelobes	Mirror, pass-band filters, optical barcodes, sensing
Coupled microcavities	$[(LH)_n/H]_m/(LH)_n/L$	Photonic minibands; nanometer linewidth transmission/reflection; manipulation of light	Mutliplexing, sensing, switching

Examples of structures are given in some cases, with H and L representing high- and low-porosity PSi layers, respectively, while n and m are integers. All these structures could be combined together. Apodization and index matching could be implemented to further tailor the photonic response.

14.4.2.2.1 **Waveguides** Optical waveguides using PSi multilayers were demonstrated for the IR and the visible [146–149]. However, the losses are quite high compared to conventional waveguides due to roughness of various interfaces. The difference in structure, and thus refractive index, between PSi formed from p^+-type and p-type Si was used to fabricate the core and clad, respectively, of a buried waveguide that included bents exhibiting losses lower than in a conventional optical fiber scheme [150].

PSi waveguides can be a useful host for the material exhibiting optical gain. Optical gain (max: 40 dB/cm) was reported in dye-impregnated (Nile blue) oxidized PSi waveguides [151]. PSi waveguides codoped by erbium and ytterbium ions exhibited internal optical gain of 6.4 dB/cm at 1.53 μm [152].

14.4.2.2.2 **Manipulation of Light Using Index Modulation** Distributed Bragg reflectors (DBRs) can be easily fabricated in PSi [4, 5, 7, 153, 154]. The period is produced by forming two consecutive quarter wavelength PSi layers of two different porosities. The photonic bandgap of these 1D photonic crystals can be tailored by using superposition of several DBRs [155]. This superposition principle also applies to other photonic structures (e.g., rugates, see below).

The reflectivity of DBRs exhibits a large angular dependence. However, omnidirectional reflectors with typical bandwidths of 200–340 nm in the near infrared can be achieved. One possibility is using a chirping process (gradual increase in the spatial period of the structure) [156]. Such chirped PSi reflectors have been integrated into a thin-film epitaxial solar cell, significantly improving the cell performance [157]. A unit cell formed by varying the refractive index of the multilayers according to the envelope of a Gaussian function has also been used [158]. Omnidirectional reflectors have also been obtained using low refractive index contrast PSi layers [159]. The principle is based on the fact that the bandgaps of the substructures are centered at different wavelengths and overlap to span a common wavelength range for any incidence angle.

Fabry–Perot microcavities, consisting of a PSi layer (a defect layer in the otherwise perfect photonic crystal) sandwiched in between two DBRs can also be fabricated. A resonant optical mode can be created in the photonic bandgap. The performance, including the Q factor of the resonance, of such filters can be improved by decreasing the roughness of the interfaces by using low-temperature anodization [160] and etch stops [161], and correcting for the in-depth porosity drift observed for thick layers [162, 163]. Subnanometer linewidths of the resonance have been reported [160, 162, 163].

Rugate filters (using sine wave index profiles with or without additional apodization) and structures including index matching characteristics in order to minimize the sidelobes found in the reflection and transmission spectra of DBRs and Fabry–Perot microcavities were demonstrated (Figure 14.4) [164]. These rugate structures can have reflectivity bands as narrow as 11 nm [165]. By superposing several rugate filters in a 1D structure, one can fabricate multiple stop-bands [164] and also optical barcodes (Figure 14.5) [166]. Using such micrometer-sized pieces of

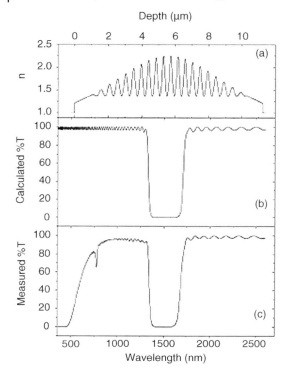

Figure 14.4 Twenty-period apodized rugate filter with index-matching layers. (a) Nominal index profile versus physical depth, (b) calculated transmission spectrum, and (c) measured transmission spectrum. Sidelobes and ripples are mostly absent. Reproduced with permission from Ref. [164]. Copyright (2005) by the Optical Society of America.

barcodes, biomolecular screening has been demonstrated [165]. This technique has a number of advantages compared to present screening methods in terms of bandwidth, toxicity, and cost. Appropriate combination of rugate filters can lead to broad (>1000 nm in the near-infrared) stop-bands and pass-bands [167]. Furthermore, by using birefringent PSi obtained from (110)-oriented Si wafers, dichroic rugate filters have been demonstrated [168].

By stacking several identical microcavities, coupling of the single resonance states occurs and a miniband is formed. By using a few stacked microcavities, a clear splitting of the single states is observed [162, 163, 169]. Since individual peaks of the resonance band have their linewidth in the nanometer range, the miniband is extremely sensitive to structure modification upon contamination and aging. Stabilization was achieved by surface chemical modification [91].

Coupled microcavities exhibit similarities with coupled quantum wells in electronics. Thus, phenomena observed in quantum well superlattices have been reproduced with optical waves in PSi photonic structures. Optical analogue of electronic Bloch oscillations [170] and Zener tunneling of light waves in an optical superlattice [171, 172] have been observed. For these purposes, the formation of a

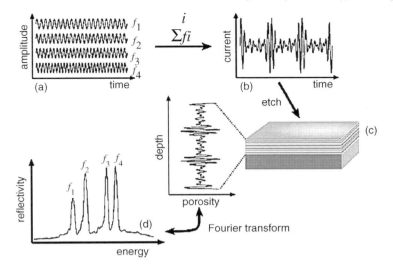

Figure 14.5 Waveform superposition method for barcode design. Four sine waves with different frequencies (a) are added together to generate a composite waveform that is then converted into an anodization current–time waveform (b), resulting in a porosity-depth profile into the Si substrate (c). The Fourier transform of the composite waveform represents the resulting optical reflectivity spectrum (d). The four frequency components appear as separate spectral peaks. The position and intensity of each peak in the spectrum are determined by the frequency and amplitude, respectively, of its corresponding sine component. Reproduced with permission from Ref. [166]. Copyright the Royal Society of Chemistry.

Wannier–Stark ladder was obtained by using a stack of microcavities that individual resonant states were appropriately chosen. The Zener tunneling phenomenon is illustrated in Figure 14.6. Functional devices can be achieved by tuning the individual photonic states. This can be done by changing the refractive index either locally or throughout the multilayer structure, by using gas infiltration or vapor condensation. This phenomenon can be used for sensing and optical switching [171, 173–175].

This electronic similarity can be extended to random systems. Random photonic structures can be easily formed with PSi by introducing disorder in the anodization process, for example, using a Fibonacci series [176]. Characteristic high transmission peaks due to resonances inside the sample were observed and the localization length was determined. Such random system was used to observe optical necklace states in Anderson localized 1D systems [177].

Using a gradient-index multilayer structure, Ma *et al.* [178] fabricated "black silicon," that is, a coating that exhibits reflection below 5% in the range 3000–28 000 cm^{-1}. The field of solar cells may benefit from such structures.

By making use of the in-plane anisotropy of the refractive index (birefringence) resulting from structure anisotropy, dichroic behaviors were demonstrated in DBRs, microcavities [179, 180], and rugate filters [168]. The reflection and transmission spectra depend on the polarization of incident light. Second- and third-harmonic generation in birefringent PSi single layer and photonic crystals has been proposed [181–183].

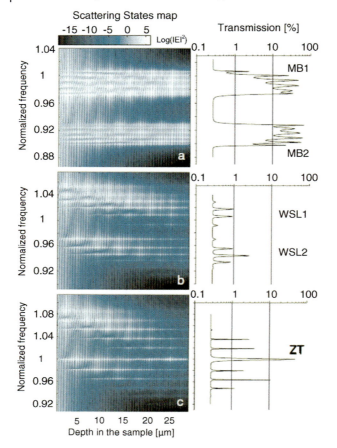

Figure 14.6 Calculation of the intensity distribution of the light inside three different samples. The intensity is plotted as a color scale versus the normalized frequency ω/ω_0 and depth inside the sample, with $\omega_0 = 192.3$ THz. (a) Flat miniband situation. (b) Optical Wannier–Stark ladders. (c) Resonant Zener tunneling. Right panels show the calculated transmission spectra. Reprinted with permission from Ref. [172]. Copyright (2005) by the American Physical Society.

Photonic structures can also be built laterally. Lateral patterns can be produced by using the selectivity offered by different doping levels or types [184, 185], or simply by masking techniques. A technique useful for large-scale processing is based on interference patterns produced by two or more incident coherent light beams. A simple two-step lithographic technique consisting of local photooxidation (which only slightly changes the photonic properties of PSi) followed by exposure to methanol (which suppress the resonant modes of nonoxidized PSi) was proposed to spatially localize PSi photonic crystals [186]. DBRs with submicrometer lateral dimensions have been demonstrated using focused ion beam technology to locally alter the Si substrate resistivity prior to anodization [45]. The technique has also been applied to microcavities [46]. The same group also demonstrated three-dimensional DBRs with micrometer resolution [47]. Such DBR pixels can reflect all visible light through a wide angular range, resulting in white reflective surfaces. Patterning of sizes and shapes allows various contrast effects.

14.4.2.3 Active Photonic Structures: Switching, Sensing

Active photonic devices (e.g., optical switches, sensors, etc.) can be achieved by enabling the change of the refractive index of either PSi or the material filling the pores, by an external excitation.

14.4.2.3.1 Optical Switching Weiss *et al.* [187, 188] have demonstrated electrically tunable PSi active mirrors using liquid crystals impregnated into a PSi microcavity. The voltage applied to the diode rotates the molecules of the liquid crystal. Experimental change of 40% in reflectance and a reversible 12 nm shift of the reflectance spectrum have been demonstrated.

14.4.2.3.2 Optical Gas/Liquid/Biomolecule Sensing and Drug Delivery Optical sensing is based on a shift of reflectivity due to a shift of the refractive index of PSi, which is related to its composition (more particularly the composition of its pores assuming the PSi structure itself is not affected).

Such sensors can be used for molecular binding assays and monitoring *in vivo* and in real time the drug load of a PSi layer. Barcodes [165] discussed in Section 14.4.2.2.2 were also used for these purposes. Besides, PSi can be used as stratified nanoreactors for enzyme compartmentalization [189], reagent delivery [190], protein separation [191], and reactant heating [192]. The feasibility of stable *in vitro* biosensors was addressed in a study of PSi stabilized using different types of surface chemistries [92]. Drug delivery and cancer treatments using PSi has been reviewed in 2008 [11]. The performance and sensitivity of PSi biosensors based on microcavities was recently investigated [193].

Gas sensing can also be done using various PSi photonic structures. The optical Zener tunneling described in Section (Figure 14.8) can be triggered externally by tuning the resonance states of individual photonic microcavities. This can be done by changing the refractive index of the multilayer structure, by using gas infiltration or vapor condensation. This phenomenon can be used for sensing and optical switching [171, 173–175]. Due to porosity gradient across the structure, such sensor (freestanding) has a different response depending on the direction of the gas flow.

Recently, the sensitivity of rugate filters for gas sensing has been studied. Time response and penetration of the gas in the depth of PSi was estimated [194]. A double-stack structure, in which one photonic crystal (rendered hydrophilic) is used to monitor humidity and another one (rendered hydrophobic) to ensure an organic compound has been employed in order to null the effect of changing humidity [195]. A polymer–PSi microcavity was proposed to detect volatile explosives such as trinitrotoluene [196]. A microsystem integrating a PSi microcavity and a microfluidic system was developed and applied to sensing of organic molecules [197]. The anisotropy of the transmission of PSi can also be used for sensing [198]. A DNA sensor using a PSi waveguide and operating in the Kretschmann configuration was proposed with a detection limit of 42 nM [199]. A principle of sensing organic molecules using surface electromagnetic waves interacting with a 1D PSi photonic crystal was also shown [200]. Similar concept was reported by Guillermain *et al.* [201]. Glucose oxidase detection with a resolution of 25 nM has been demonstrated using stabilized and functionalized PSi microcavities [202].

14.4.2.3.3 **Nonlinear Properties** PSi also behaves as a nonlinear medium. Carrier population dynamics in Si nanocrystals under photoexcitation [203, 204] or carrier injection [205] can be followed by significant changes of absorption and refractive index. Using the high sensitivity of the resonance of a PSi microcavity, a switching function becomes available by monitoring the transmission or reflection of light at the resonance wavelength of the cavity. Second- and third-harmonic generation in birefringent PSi single layer and photonic crystals has been proposed [181–183]. The study of nonlinearities in PSi (porosity up to 70%) optical waveguides at 1550 nm showed that two-photon absorption and self-phase modulation coefficients are comparable to those of bulk Si, whereas free-carrier absorption and dispersion are significantly faster and stronger in PSi [206].

14.4.3
Luminescence

14.4.3.1 **Photoluminescence**
PSi has been reported to luminesce efficiently from the near infrared (~1.5 μm) to the near UV as a result of distinct emission bands with different origins (Table 14.6) [4, 7]. Very recently, Gelloz and Koshida found efficient blue phosphorescence emission with a lifetime of several seconds in oxidized PSi up to about 200 K. The mechanism and origin of this luminescence are under investigation.

14.4.3.1.1 **Characteristics of the S-Band** The S-band (S for slow) is the most studied band for its potential in optoelectronics, photonics, and sensing. Other bands are related to surface states or oxide-related luminescence centers. The S-band mostly originates from exciton recombinations in Si nanocrystals as indicated by polarization memory of PL, PL saturation under high excitation due to Auger recombinations,

Table 14.6 Porous silicon luminescence bands.

Spectral range	Peak wavelength (nm)	Label	Directly electrically excitable
Near IR	1100–1500	IR-band	No
From red to blue	400–800	S-band	Yes
Blue-green	~470	F-band	No
UV	~350	UV-band	No

and resonant excitation and hole-burning experiments (evidencing phonon-mediated recombinations and singlet–triplet exciton-state splitting) [4, 7]. Very high confinement energy ($>0.7\,eV$) results in the break of k-conservation rules and direct recombinations become possible [207].

Figure 14.7a shows the progressive shift of the PL band of PSi having its surface terminated by silicon–hydrogen bonds (oxygen-free) across the visible region obtained by changing the anodization conditions [208]. In general, its intensity increases and its peak wavelength decreases when the porosity increases since the higher the porosity, the smaller the average size and the higher the density of luminescent Si nanocrsytals. Typically, a minimum porosity of 70% is necessary to get emission from the S-band. Moreover, the efficiency usually decreases in the order n-type, p-type, n^+-type, and p^+-type PSi due to differences in nanostructures. Lateral changes in PL intensity and peak wavelength can be achieved by changing laterally the PS structure or porosity. For example, one could first set up a resistivity pattern (e. g., using focused ion beam [44]) and then perform the anodization, or first make PSi and then induce a two-dimensional pattern by local dissolution of PSi blocks (e.g., using photodissolution with light interference or masking techniques).

However, surface states, in particular those introduced by oxygen via Si=O bonds, can greatly influence the peak wavelength, as shown in Figure 14.7b. In zone II and III, localized states lie inside the bandgap, leading to red-orange PL instead of the green and blue emission expected from band to band recombinations [208].

The PL lifetime typically ranges from a few nanoseconds for blue emission to $100\,\mu s$ for red emission at room temperature. It increases with emission wavelength

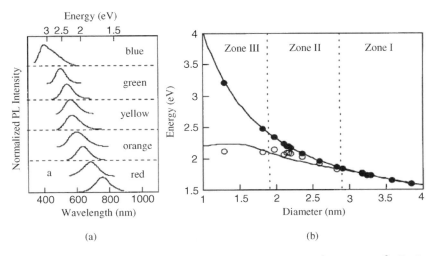

(a) (b)

Figure 14.7 (a) Room-temperature PL spectra from PSi samples with different porosities kept under Ar atmosphere. (b) Experimental and theoretical PL energies as a function of crystallite size. The upper line is the free exciton bandgap and the lower line is the lowest transition energy in the presence of a Si=O bond. The solid and open dots are the peak PL energies of PSi samples kept in Ar and air, respectively. Reprinted with permission from Ref. [208]. Copyright (1999) by the American Physical Society.

Table 14.7 Some spectral characteristics of the S-band.

Property	Typical values	Comments
Peak wavelength	1300–400 nm	Porosity dependent
External quantum efficiency	PL: 5–10%; up to 23%	At 300 K; depends on porosity and surface passivation
	EL: up to 1.1%	
FWHM	150–180 nm	At 300 K; due to size distribution; nanometer range in PSi microcavities
Decay time	PL: ns to 100 μs (from blue to red)	Depends on wavelength, temperature, and sample history; EL: can be limited by capacitive effects
	EL: ≥ PL decay time	
Degree of polarization	≤ 0.2	Can be anisotropic (effect of crystalline orientations)
Fine structure under resonant excitation	Phonon replica at 56 and 19 meV	Consistent with Si phonons

in agreement with the size dependence of quantum confinement strength. It increases at low temperature due to singlet–triplet exciton-state splitting [4, 7]. The PL decay follows a stretched exponential curve due to the inhomogeneous nature of the emission. This behavior has been attributed to exciton migration, carrier escape from Si dots, or distribution of dot shape emitting at the same energy [4].

The S-band exhibits a FWHM of about 150 nm, resulting from the size distribution of luminescent nanocrystals. The linewidth can be drastically decreased by placing the luminescent PSi layer in between two DBRs. Thus, a luminescent microcavity is formed. The PL FWHM can be as low as a few nanometers [209]. Moreover, highly directional tunable PL and EL emission can be obtained [209–211].

The main characteristics of the S-band are summarized in Table 14.7. Some EL data are also given. Table 14.8 shows the effects of various useful treatments on the PL and EL of this band.

14.4.3.1.2 **Photoluminescence Efficiency**

A single Si dot can have high quantum efficiency (>35%) [212]. However, the PL quantum efficiency of PSi layers is affected by the dot density, the proportion of luminescent dots, and the absorption by nonluminescent large structures in PSi.

The PL efficiency can be improved by (i) increasing the number of luminescent nanocrystals, (ii) decreasing the light absorption by nonluminescent Si structures in PSi, (iii) providing good surface passivation to the Si nanocrystals, and (iv) increasing the absorption of luminescent nanocrystals, for example, by using plasmonic effects.

When the anodization is carried out in the presence of a magnetic field (up to 1.9 T), the PL is found more efficient than without magnetic field. The magnetic

Table 14.8 General effects of various useful treatments on the PL and EL of PSi of medium porosity (<75%).

Treatment	Effect on PL		Effect on EL		Mechanism of the changes
	Efficiency	Stability	Efficiency	Stability	
Thermal oxidation ($T < 600$ C) [62, 78, 79]	Decreases	Decreases	Decreases	Decreases	Nonradiative defects at Si/SiO$_2$ interface
Rapid thermal oxidation ($T > 800\,°C$) [79]	Increases	Increases	Increases	Increases	Good quality of Si/SiO$_2$ interface; enhanced confinement
ECO [70, 75, 76, 240]	Increases (~10%)	Increases	Increases (1.1%)	Increases	Enhanced confinement and surface passivation
SCM [89, 90]	Almost no change	Good (months)	Decreases or no change	Good (months)	Enhanced surface passivation; physical protection of surface
HWA [83, 85, 86]	Increases (red: 23%; possible blue emission)	Very good (years)	Increases	Very good (years)	Relaxed SiO$_2$; good quality of Si/SiO$_2$ interface
ECO followed by HWA [82, 86]	Increases (red only)	Very good (years)	—	—	Same as ECO and HWA; better preservation of Si dots
Combination of HWA and SCM [81]	Increases	Very good	—	—	Same as ECO and SCM; SCM before HWA: better preservation of Si dots; good also for high porosity

Oxidation treatments may deteriorate the PL and EL efficiency of high-porosity PSi because it oxidizes fully or too much the PSi skeleton. Values given in percentages are record external quantum efficiencies. ECO, SCM, and HWA stand for electrochemical oxidation, surface chemical modification, and high-pressure water vapor annealing, respectively.

field leads to the formation of more spherical nanocrystals (more isotropic nanostructures) [31].

Oxidation of PSi decreases the nanocrystals size, resulting in enhanced quantum confinement. Size decrease can increase the density of luminescent nanocrystals (increasing the PL intensity) while it induces a blueshift of the emission of already luminescent nanocrystals. Oxidation also affects the nanocrystal passivation via the quality of the surface oxide. The effect of oxidation also depends very much on the initial PSi porosity [68]. Indeed, the same oxidation treatment could lead to a PL intensity enhancement in a low porosity PSi layer and to a degradation of PL intensity in a highly porous layer due to full oxidation of most nanocrystals. Oxidation methods are discussed in Section 14.3.2. Anodic oxidation [70, 213], rapid thermal oxidation above 800 °C [79], and high-pressure water vapor annealing [85, 86] are particularly useful for getting highly efficient PL. The last approach can be used to produce efficient blue PL [214]. By using a combination of thermal annealing and high-pressure water vapor annealing, samples emitting only blue (without any contribution of the red-emitting S-band) bright PL can be produced [215]. This blue emission can be divided in two main contributions. One is very fast (nanoseconds) and the other one is unusually slow (lifetimes of seconds). This phosphorescence, not due to quantum confinement in Si nanocrystals, is under investigation.

The highest reported external quantum efficiencies at room temperature range from 10 [216, 217] to 23% [85, 86]. The highest value was obtained by passivating the PSi surface with very good quality thin oxide obtained by high-pressure water vapor annealing [85, 86]. This technique also stabilizes the PSi structure, PL [85, 86], EL (Section 14.4.3.3) [83], and photonic structures [80].

A recent report shows an increase in PL intensity by 67% by depositing a thin Al layer (5.2 nm) by sputtering onto PSi [218]. The results were explained by the formation of Si–Al bonds. However, the actual mechanism of the enhancement and whether or not the Al penetrates inside the PSi pores remain unclear.

14.4.3.1.3 **Photoluminescence Stabilization** Passivation of Si nanocrystals is absolutely necessary to prevent competitive nonradiative recombinations (e.g., trapping by defects, charge exchanges with the surface and the environment) and to prevent the PL degradation due to chemical evolution of PSi surface (e.g., oxidation of Si–H bonds in air, contamination by carbon leading to blue emission [219]).

The PL can be very efficiently stabilized by high-pressure water vapor annealing [85, 86]. A less efficient but still very useful technique is chemical or electrochemical derivatization with organic molecules [89]. The latter method does not oxidize PSi that thus retains its original properties after the stabilization treatment. This method was also used to stabilize the EL (Section 14.4.3.3) [90] and photonic coupled optical microcavities (Section 14.4.2.2.1) [91]. Silver atoms deposited inside PSi may also help in passivating the Si nanocrystals [220, 221].

14.4.3.2 **Energy Transfer, Sensing, and Imaging Using Porous Si Photoluminescence**
Some gases can change the PSi PL either reversibly or irreversibly [7]. Complete irreversible quenching of the PL is observed when the PSi surface

is exposed to reactive molecules. Reversible quenching of the PL is obtained when no chemical reactions (or reversible reactions) occurs at the PSi surface. It was observed with adsorption of diverse organic molecules and solvents [222–224]. The PL quenching occurs either via dielectric constant change or energy transfer.

Energy transfer from Si nanocrystals to luminescent dyes incorporated into PSi pores has been suggested [225]. Recently, it was demonstrated by using polarization memory of PL [226]. Generation of singlet oxygen at room temperature mediated by energy transfer from photoexcited PSi is dealt with in Chapter 19.

PSi is nontoxic and biodegradable. Furthermore, it was recently demonstrated (via a tumor imaging test) that biodegradable luminescent PSi could be used for *in vivo* applications [227].

14.4.3.3 Electroluminescence

The EL of PSi has been reviewed in detail recently [8–10]. Efficient, well-controlled, and reproducible EL can be observed in PSi contacted by a liquid electrolyte during anodic oxidation [70, 71] or using redox species able to inject carriers inside Si nanocrystals (e.g., peroxodisulfate can inject holes) [228–230]. Voltage-tunable EL [231] and PL [232, 233] were observed in such systems. The rather high efficiency of this type of EL is explained by the carrier injection that is possible at low voltages ($\sim$1–2 V) in the whole or part of the PSi skeleton [25] without significant voltage drop across the PSi layer [234, 235].

On the contrary, in solid-state devices, the voltage is applied between the substrate backside and a top semitransparent electrode (usually gold or indium tin oxide). Due to the high resistivity of PSi (see Section 14.4.1.1), carrier transport and injection in Si nanocrystals is very difficult. As a result, the EL mechanism in PSi mainly involves impact processes, even in porosified p–n junctions. PSi EL suffers from poor efficiency and stability. Its speed is limited to the MHz regime by the luminescence lifetime.

Many strategies have been implemented in order to improve the efficiency and stability of PSi EL [8–10]. Porosified p–n junctions, partially oxidized PSi, PSi impregnated by another conductive material (mostly polymers and metals), different configuration of the top contact, layer capping, and surface modification have been investigated. The most noticeable results (Table 14.9) were obtained using thin p–n junctions [236, 237], thermal [238, 239] or electrochemical [75, 76, 240] oxidation, and high-pressure water vapor annealing [83].

Electrochemical oxidation is very powerful for decreasing the leakage current by selectively oxidizing the coarser regions of the PSi skeleton without affecting much the Si nanocrystals. It also optimizes the carrier injection into the nanocrystals [75, 76].

The stability is a major challenge of PSi EL. It can be very much improved by surface modification with long organic molecules [90]. More recently, high-pressure water vapor annealing was shown to completely stabilize not only the PL [85, 86] but also the EL [83]. This technique is very promising for improving all the optoelectronic devices based on PSi.

Table 14.9 Some characteristics of noticeable EL devices.

Contact	Structure	Post-treatment	EL threshold voltage (mA/cm²)	Stability of EL intensity	Emission peak (nm)	Highest efficiency EQE/EPE (%)	Year	Reference
ITO	p^+n(L)	1 min under illumination	$2.3/10^{-3}$	Hours	600	0.18/—	1995	[236]
Al-poly Si	p^+p(D)	Anneal in N_2 or in 10% O_2 in N_2	1.5/2	1 month	620–770	0.1/—	1995	[238]
Au	p^+n(L)		3/1	Seconds	670–780	0.2/—	1996	[237]
ITO	n^+(L)	ECO	$3.5/4 \times 10^{-4}$	Hours	640	0.51/0.05	1998	[76, 240]
ITO	p^+n^+(L)	ECO	$5/1.5 \times 10^{-4}$	Hours	650	1.1/0.08	1998	[240]
ITO	n^+(D)n^+(L)	ECO	$2.2/7 \times 10^{-4}$	Days, stable EQE	680	1.07/0.37	2000	[75]
ITO/a-C	n^+(D)n^+(L)	ECO	$0.5/3 \times 10^{-2}$	Minutes	Voltage tunable	—	2004	[241]
ITO	n^+(D)n^+(L)	ECO + HWA	$2/10^{-4}$	Stable	820	—	2006	[83]

D and L mean that anodization has been conducted in the dark and under illumination, respectively. ECO and HWA stand for electrochemical oxidation and high-pressure water vapor annealing, respectively. EQE and EPE mean external quantum efficiency and external power efficiency, respectively.

The EL spectrum is generally identical to the PL one except for devices where the EL mechanism involves other luminescence bands rather than the S-band. One-peak voltage-tunable EL between 2 and 5 V was observed using a thin and efficient PSi device [241]. As for PL, the EL spectrum can be narrowed down to a few nanometers in FWHM using microcavities [209–211].

Integration issues have also been investigated [242]. An EL device was integrated and driven by a bipolar transistor [239]. The use of polycrystalline PSi [243] and SOI technology [244] was also demonstrated.

14.4.4
Acoustic and Thermal Properties

14.4.4.1 Acoustic Properties and Acoustic Band Crystals
14.4.4.1.1 Acoustic Properties of Porous Si
Acoustic waves can exhibit both transverse and longitudinal vibrations. The latter corresponds to sound waves. These waves are affected by the elasticity (relating stresses generated by any small deformation of the medium to displacements) and the density ϱ (corresponding to the mechanical inertia of the medium). The velocity of acoustic waves is given by $V = (c/\varrho)^{0.5}$, where c is a component of the elastic tensor. An important parameter is the acoustic impedance given by $Z = \varrho V = (\varrho c)^{0.5}$. It is very sensitive to the density of the material. Therefore, a significant dependence on PSi porosity is expected.

Only a few data of acoustic velocity of PSi can be found in the literature [245, 246]. For mesoporous p^+-type PSi, an empirical formula was derived: $V = V_0(1-\text{porosity})^k$ [247], where k and V_0 are equal to 1.095 and 8570 m/s for longitudinal waves and 1.19 and 5840 m/s for shear waves, respectively. Fan *et al.* [246] obtained $k = 1.083$ for longitudinal waves while studying porosities in the range 57–83%. This formula (valid for mesoporous p^+-type PSi) cannot be generalized for all PSi morphologies.

The density of PSi is simply obtained from the porosity using $\varrho = (1-\text{porosity})\varrho_0$, where ϱ_0 is the density of silicon, neglecting air.

14.4.4.1.2 Acoustic Band Crystals
Acoustic crystals can take many forms, similar to photonic crystals (DBR, cavities, rugates, etc.). The role of the refractive index variation for optics is played by the inverse of the acoustic impedance Z. Only two reports exist in the literature. In the first one, the band structure of a simple acoustic multilayer reflector was calculated [248]. The influence of porosity and velocity contrast was studied. This first report shows the possibility of acoustic bandgaps in PSi multilayers in the 100–500 MHz range.

In the second report, a matrix transfer formalism was used [249]. The reflection of DBRs and rugates with optional apodization and impedance matching were calculated. The report also showed the corresponding optical response since the acoustic crystal is also a photonic crystal, but for different frequencies. Typically, acoustic crystals can be used in the GHz range whereas the photonic bands were in the 100–500 THz range.

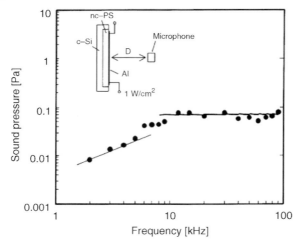

Figure 14.8 Frequency dependence of the sound pressure amplitude at an electrical input power of 1 W/cm². The device structure measurement setup is shown in the inset. The PSi layer was p-type, 70% porosity, and 30 µm thick.

14.4.4.2 Thermoacoustic Emission

Because of depletion of free carriers and phonon confinement in Si nanocrystals, the thermal conductivity of PSi is two orders of magnitude lower than that of crystalline silicon [250] and the heat capacity per unit volume is also much lower. These properties have been taken advantage of to electrically generate ultrasound emission [251].

The device (inset of Figure 14.8) consists of a metal electrode on a p-type PSi layer (10–50 µm thick) that is left on the Si substrate. An AC electrical current (typically a sine wave signal) is applied to the metal electrode. The electrode is heated by Joule effect. The temperature is then efficiently transferred to the air because of the low heat transfer efficiency in the PSi layer. Figure 14.8 shows the frequency dependence of the acoustic pressure generated by the device. Sound and ultrasound can be emitted from the device. Its frequency response is flat from 10 kHz up to a limit of about 1 GHz. This means that accurate signals can be generated, even by pulse operation rather than using typical sine waves. Furthermore, phased array integration of several individual emitters was demonstrated, and the emission directionality could be controlled [252]. This device is very stable, even without stabilization treatments, because the thermal characteristics of PSi do not change much upon aging.

Various applications have been suggested. These include wideband digital speakers, distance sensing [253], noncontact actuators, imitation of natural and animal ultrasound emission (e.g., mouse-pup ultrasonic vocalization [254]), and acoustic digital transmission.

Table 14.10 compares the characteristics of the PSi emitter with that of conventional piezoelectric or mechanical ones [124].

Table 14.10 Main characteristics of PSi and conventional ultrasonic emitters [124].

Specific items	PSi emitter	Conventional
Principle	Thermoacoustic	Electroacoustic
Operation mode	Stable surface	Mechanical vibration
Acoustic output	αP_{in}	αV_{in}
Frequency characteristics	Flat	Resonant
Impulse signal	Ideal	—
Scaling merit (size and time)	Available	Nothing
Total harmonic distortion	<0.1%	—
Dynamic range 10–100 kHz	>60 dB	—
Arrayed integration	Easy	Difficult

14.5
Conclusions

The PSi formation by anodization, galvanic etching, and stain etching has been presented. The most used and controlled method is anodization. The two other methods have the advantage of not requiring any external power source. However, they do not offer enough control to enable the formation of stratified PSi. Stain etching also suffers from rather poor control.

The PSi structure (typical size, porosity, and morphology) can be tailored over a wide range depending on parameters such as doping type and level and anodization conditions.

PSi exhibits several properties that are not available in bulk silicon, expanding the range of application of the silicon technology. The most important properties and corresponding applications are summarized in Table 14.3. PSi is particularly useful in photonics for various types of optical processing and biosensing (Table 14.5). Other remarkable applications include electron and ultrasound emission, acoustic processing, and drug delivery.

The field of light emission very much progressed in the past few years, with the achievement of highly efficient and stable PL (Table 14.8). Efforts are still needed in the field of EL.

PSi suffers from its poor stability, rendering any application difficult. However, recent developments in PSi stabilization, particularly chemical derivatization and high-pressure water vapor annealing, are very promising.

References

1 Canham, L.T. (1990) Silicon quantum wire array fabrication by electrochemical and chemical dissolution of wafers. *Appl. Phys. Lett.*, **57**, 1046–1048.

2 Lehmann, V. and Gosele, U. (1991) Porous silicon formation: a quantum wire effect. *Appl. Phys. Lett.*, **58**, 856–858.

3 Lehmann, V. (2002) *Electrochemistry of Silicon: Instrumentation, Science, Materials and Applications*, Wiley-VCH Verlag GmbH, Weinheim

4 Cullis, A.G., Canham, L.T., and Calcott, P.D.J. (1997) The structural and luminescence properties of porous silicon. *J. Appl. Phys.*, **82**, 909–965.

5 Theiss, W. (1997) Optical properties of porous silicon. *Surf. Sci. Rep.*, **29**, 95–192.

6 Canham, L.T. (ed.) (1997) *Properties of Porous Silicon*, vol. **18**, INSPEC, The Institution of Electrical Engineers, London.

7 Bisi, O., Ossicini, S., and Pavesi, L. (2000) Porous silicon: a quantum sponge structure for silicon based optoelectronics. *Surf. Sci. Rep.*, **38**, 5–126.

8 Gelloz, B. and Koshida, N. (2004) in *The Handbook of Electroluminescent Materials* (ed. D.R. Vij), Institute of Physics Publishing, Bristol and Philadelphia, pp. 393–475.

9 Gelloz, B. and Koshida, N. (2003) in *Handbook of Luminescence, Display Materials, and Devices* (eds H.S. Nalwa and L.S. Rohwer), American Scientific Publishers, Stevenson Ranch, CA, pp. 128–156.

10 Gelloz, B. and Koshida, N. (2008) in *Device Applications of Silicon Nanocrystals and Nanostructures* (ed. N., Koshida), Springer, pp. 25–70.

11 Anglin, E.J., Cheng, L.Y., Freeman, W.R., and Sailor, M.J. (2008) Porous silicon in drug delivery devices and materials. *Adv. Drug. Delivery Rev.*, **60**, 1266–1277.

12 Rouquerol, J., Avnir, D., Fairbridge, C.W., Everett, D.H., Haynes, J.H., Pernicone, N., Ramsay, J.D.F., Sing, K.S.W., and Unger, K.K. (1994) Recommendations for the characterization of porous solids. *Pure Appl. Chem.*, **66**, 1739–1758.

13 Halimaoui, A. (1997) in *Properties of Porous Silicon* (ed. L.T. Canham), INSPEC, The Institution of Electrical Engineers, London, pp. 12–22.

14 Ashruf, C.M.A., French, P.J., Bressers, P.M.M.C., and Kelly, J.J. (1999) Galvanic porous silicon formation without external contacts. *Sens. Actuators A*, **74**, 118–122.

15 Zhang, Z.W., Lerner, M.M., Alekel, T., and Keszler, D.A. (1993) Formation of a photoluminescent surface on N-Si by irradiation without an externally applied potential. *J. Electrochem. Soc.*, **140**, L97–L98.

16 Splinter, A., Sturmann, J., and Benecke, W. (2001) Novel porous silicon formation technology using internal current generation. *Mater. Sci. Eng. C*, **15**, 109–112.

17 Splinter, A., Sturmann, J., and Benecke, W. (2001) New porous silicon formation technology using internal current generation with galvanic elements. *Sens. Actuators A*, **92**, 394–399.

18 Koyama, H. and Takemura, K. (2004) Fabrication of luminescent porous silicon layers using extremely dilute HF solutions. *Electrochem. Soc. Proc.*, **13**, 289.

19 Fathauer, R.W., George, T., Ksendzov, A., and Vasquez, R.P. (1992) Visible luminescence from silicon-wafers subjected to stain etches. *Appl. Phys. Lett.*, **60**, 995–997.

20 Kolasinski, K.W. (2005) Silicon nano-structures from electroless electro-chemical etching. *Curr. Opin. Solid State Mater. Sci.*, **9**, 73–83.

21 Nahidi, M. and Kolasinski, K.W. (2006) Effects of stain etchant composition on the photoluminescence and morphology of porous silicon. *J. Electrochem. Soc.*, **153**, C19–C26.

22 Lehmann, V. (1993) The physics of macropore formation in low doped N-type silicon. *J. Electrochem. Soc.*, **140**, 2836–2843.

23 Smith, R.L. and Collins, S.D. (1992) Porous silicon formation mechanisms. *J. Appl. Phys.*, **71**, R1–R22.

24 Wehrspohn, R.B., Ozanam, F., and Chazalviel, J.N. (1999) Nano- and macropore formation in p-type silicon. *J. Electrochem. Soc.*, **146**, 3309–3314.

25 Gelloz, B., Bsiesy, A., Gaspard, F., and Muller, F. (1996) Conduction in porous silicon contacted by a liquid phase. *Thin Solid Films*, **276**, 175–178.

26 Lin, J.C., Tsai, W.C., and Chen, W.L. (2007) Light emission and negative differential conductance of n-type

nanoporous silicon with buried p-layer assistance. *Appl. Phys. Lett.*, **90**, 091117.

27 Lin, J.C., Lee, P.W., and Tsai, W.C. (2006) Manufacturing method for n-type porous silicon based on Hall effect without illumination. *Appl. Phys. Lett.*, **89**, 121119.

28 Li, S.Q., Wijesinghe, T.L.S.L., and Blackwood, D.J. (2008) Photoluminescent n-type porous silicon fabricated in the dark. *Adv. Mater.*, **20**, 3165–3168.

29 Servidori, M., Ferrero, C., Lequien, S., Milita, S., Parisini, A., Romestain, R., Sama, S., Setzu, S., and Thiaudiere, D. (2001) Influence of the electrolyte viscosity on the structural features of porous silicon. *Solid State Commun.*, **118**, 85–90.

30 Setzu, S., Lerondel, G., and Romestain, R. (1998) Temperature effect on the roughness of the formation interface of p-type porous silicon. *J. Appl. Phys.*, **84**, 3129–3133.

31 Nakagawa, T., Koyama, H., and Koshida, N. (1996) Control of structure and optical anisotropy in porous Si by magnetic-field assisted anodization. *Appl. Phys. Lett.*, **69**, 3206–3208.

32 Nassiopoulos, A.G., Grigoropoulos, S., Canham, L., Halimaoui, A., Berbezier, I., Gogolides, E., and Papadimitriou, D. (1995) Submicrometer luminescent porous silicon structures using lithographically patterned substrates. *Thin Solid Films*, **255**, 329–333.

33 Steiner, P. and Lang, W. (1995) Micromachining applications of porous silicon. *Thin Solid Films*, **255**, 52–58.

34 Steckl, A.J., Su, J.N., Xu, J., Li, J.P., Yuan, C., Yih, P.H., and Mogul, H.C. (1994) Selective-area room-temperature visible photoluminescence from Sic/Si heterostructures. *Appl. Phys. Lett.*, **64**, 1419–1420.

35 Noguchi, N. and Suemune, I. (1994) High-resolution patterning of luminescent porous silicon with photoirradiation. *Jpn. J. Appl. Phys., Part 1*, **33**, 590–593.

36 Kruger, M., Arens Fischer, R., Thonissen, M., Munder, H., Berger, M.G., Luth, H., Hilbrich, S., and Theiss, W. (1996) Formation of porous silicon on patterned substrates. *Thin Solid Films*, **276**, 257–260.

37 Kaltsas, G. and Nassiopoulos, A.G. (1997) Bulk silicon micromachining using porous silicon sacrificial layers. *Microelectron. Eng.*, **35**, 397–400.

38 Lang, W., Steiner, P., Richter, A., Maruszyk, K., Weimann, G., and Sandmaier, H. (1994) Application of porous silicon as a sacrificial layer. *Sens. Actuators A*, **43**, 239–242.

39 Djenizian, T., Santinacci, L., Hildebrand, H., and Schmuki, P. (2003) Electron beam induced carbon deposition used as a negative resist for selective porous silicon formation. *Surf. Sci.*, **524**, 40–48.

40 Lang, W., Steiner, P., and Sandmaier, H. (1995) Porous silicon: a novel material for microsystems. *Sens. Actuators A*, **51**, 31–36.

41 Higa, K. and Asano, T. (1996) Fabrication of single-crystal Si microstructures by anodization. *Jpn. J. Appl. Phys., Part 1*, **35**, 6648–6651.

42 Xu, J. and Steckl, A.J. (1994) Fabrication of visibly photoluminescent Si microstructures by focused ion-beam implantation and wet etching. *Appl. Phys. Lett.*, **65**, 2081–2083.

43 Diesinger, H., Bsiesy, A., and Herino, R. (2003) Nano-structuring of silicon and porous silicon by photo-etching using near field optics. *Phys. Status Solidi A*, **197**, 561–565.

44 Teo, E.J., Breese, M.B.H., Bettiol, A.A., Mangaiyarkasi, D., Champeaux, F., Watt, F., and Blackwood, D.J. (2006) Multicolor photoluminescence from porous silicon using focused, high-energy helium ions. *Adv. Mater.*, **18**, 51–55.

45 Breese, M.B.H. and Mangaiyarkasi, D. (2007) Porous silicon Bragg reflectors with sub-micrometer lateral dimensions. *Opt. Express*, **15**, 5537–5542.

46 Mangaiyarkarasi, D., Breese, M.B.H., Ow, Y.S., and Vijila, C. (2006) Controlled blueshift of the resonant wavelength in porous silicon microcavities using ion irradiation. *Appl. Phys. Lett.*, **89**, 021910.

47 Mangaiyarkarasi, D., Breese, M.B.H., and Ow, Y.S. (2008) Fabrication of three dimensional porous silicon distributed Bragg reflectors. *Appl. Phys. Lett.*, **93**, 221905.

48 Couillard, J.G. and Craighead, H.G. (1994) Photolithographic patterning of porous silicon. *J. Vac. Sci. Technol. B*, **12**, 161–162.

49 Enachescu, M., Hartmann, E., and Koch, F. (1994) Writing electronic nanometer structures into porous Si films by scanning-tunneling-microscopy. *Appl. Phys. Lett.*, **64**, 2253–2255.

50 Enachescu, M., Hartmann, E., and Koch, F. (1996) Stable nanostructuring of ultrathin porous silicon films by scanning tunneling microscopy. *J. Appl. Phys.*, **79**, 2948–2953.

51 de Vasconcelos, E.A., da Silva, E.F., dos Santos, B.E.C.A., de Azevedo, W.M., and Freire, J.A.K. (2005) A new method for luminescent porous silicon formation: reaction-induced vapor-phase stain etch. *Phys. Status Solidi A*, **202**, 1539–1542.

52 Chun, I.S., Chow, E.K., and Li, X.L. (2008) Nanoscale three dimensional pattern formation in light emitting porous silicon. *Appl. Phys. Lett.*, **92**, 191113.

53 Qiu, W., Kang, Y.L., Li, Q., Lei, Z.K., and Qin, Q.H. (2008) Experimental analysis for the effect of dynamic capillarity on stress transformation in porous silicon. *Appl. Phys. Lett.*, **92**, 041906.

54 Sugiyama, H. and Nittono, O. (1989) Annealing effect on lattice distortion in anodized porous silicon layers. *Jpn. J. Appl. Phys., Part 2*, **28**, L2013–L2016.

55 Buttard, D., Bellet, D., and Dolino, G. (1996) X-ray-diffraction investigation of the anodic oxidation of porous silicon. *J. Appl. Phys.*, **79**, 8060–8070.

56 Sugiyama, H. and Nittono, O. (1990) Microstructure and lattice distortion of anodized porous silicon layers. *J. Cryst. Growth*, **103**, 156–163.

57 Gardelis, S., Bangert, U., Harvey, A.J., and Hamilton, B. (1995) Double-crystal X-ray-diffraction, electron-diffraction, and high-resolution electron-microscopy of luminescent porous silicon. *J. Electrochem. Soc.*, **142**, 2094–2101.

58 Kim, K.H., Bai, G., Nicolet, M.A., and Venezia, A. (1991) Strain in porous Si with and without capping layers. *J. Appl. Phys.*, **69**, 2201–2205.

59 Munder, H., Berger, M.G., Frohnhoff, S., Thonissen, M., and Luth, H. (1993) A nondestructive study of the microscopic structure of porous Si. *J. Lumin.*, **57**, 5–8.

60 Herino, R., Bomchil, G., Barla, K., Bertrand, C., and Ginoux, J.L. (1987) Porosity and pore-size distributions of porous silicon layers. *J. Electrochem. Soc.*, **134**, 1994–2000.

61 Halimaoui, A. (1994) Determination of the specific surface-area of porous silicon from its etch rate in HF solutions. *Surf. Sci.*, **306**, L550–L554.

62 Yamada, M. and Kondo, K. (1992) Comparing effects of vacuum annealing and dry oxidation on the photoluminescence of porous Si. *Jpn. J. Appl. Phys., Part 2*, **31**, L993–L996.

63 Hou, X.Y., Shi, G., Wang, W., Zhang, F.L., Hao, P.H., Huang, D.M., and Wang, X. (1993) Large blue shift of light-emitting porous silicon by boiling water-treatment. *Appl. Phys. Lett.*, **62**, 1097–1098.

64 Li, K.H., Tsai, C., Shih, S., Hsu, T., Kwong, D.L., and Campbell, J.C. (1992) The photoluminescence spectra of porous silicon boiled in water. *J. Appl. Phys.*, **72**, 3816–3817.

65 Baba, M., Kuwano, G., Miwa, T., and Taniguchi, H. (1994) *In-situ* measurements of water immersion and UV irradiation effects on intensity and blue-shift of visible photoluminescence in porous Si. *Jpn. J. Appl. Phys., Part 2*, **33**, L483–L486.

66 Chang, I.M., Chuo, G.S., Chang, D.C., and Chen, Y.F. (1995) Evolution of photoluminescence of porous silicon under light exposure. *J. Appl. Phys.*, **77**, 5365–5368.

67 Tischler, M.A., Collins, R.T., Stathis, J.H., and Tsang, J.C. (1992) Luminescence degradation in porous silicon. *Appl. Phys. Lett.*, **60**, 639–641.

68 Gelloz, B. (1997) Possible explanation of the contradictory results on the porous silicon photoluminescence evolution after low temperature treatments. *Appl. Surf. Sci.*, **108**, 449–454.

69 Bsiesy, A., Gaspard, F., Herino, R., Ligeon, M., Muller, F., and Oberlin, J.C. (1991) Anodic-oxidation of porous silicon layers formed on lightly P-doped substrates. *J. Electrochem. Soc.*, **138**, 3450–3456.

70 Billat, S. (1996) Electroluminescence of heavily doped p-type porous silicon under electrochemical oxidation in galvanostatic regime. *J. Electrochem. Soc.*, **143**, 1055–1061.

71 Halimaoui, A., Oules, C., Bomchil, G., Bsiesy, A., Gaspard, F., Herino, R., Ligeon, M., and Muller, F. (1991) Electroluminescence in the visible range during anodic-oxidation of porous silicon films. *Appl. Phys. Lett.*, **59**, 304–306.

72 Grosman, A., Chamarro, M., Morazzani, V., Ortega, C., Rigo, S., Siejka, J., and Vonbardeleben, H.J. (1993) Study of anodic-oxidation of porous silicon: relation between growth and physical-properties. *J. Lumin.*, **57**, 13–18.

73 Vial, J.C. *et al.* (1993) Bright visible-light emission from electrooxidized porous silicon: a quantum confinement effect. *Physica B*, **185**, 593–602.

74 Vazsonyi, E.B., Koos, M., Jalsovszky, G., and Pocsik, I. (1993) The role of hydrogen in luminescence of electrochemically oxidized porous Si layer. *J. Lumin.*, **57**, 121–124.

75 Gelloz, B. and Koshida, N. (2000) Electroluminescence with high and stable quantum efficiency and low threshold voltage from anodically oxidized thin porous silicon diode. *J. Appl. Phys.*, **88**, 4319–4324.

76 Gelloz, B., Nakagawa, T., and Koshida, N. (1998) Enhancement of the quantum efficiency and stability of electroluminescence from porous silicon by anodic passivation. *Appl. Phys. Lett.*, **73**, 2021–2023.

77 Komoda, T., Honda, Y., Ichihara, T., Hatai, T., Takegawa, Y., Watanabe, Y., Aizawa, K., Vezin, V., and Koshida, N. (2002) *SID'02 Digest*, **33**, 1128.

78 Takazawa, A., Tamura, T., and Yamada, M. (1994) Photoluminescence mechanisms of porous Si oxidized by dry oxygen. *J. Appl. Phys.*, **75**, 2489–2495.

79 Petrovakoch, V., Muschik, T., Kux, A., Meyer, B.K., Koch, F., and Lehmann, V. (1992) Rapid-thermal-oxidized porous Si: the superior photoluminescent Si. *Appl. Phys. Lett.*, **61**, 943–945.

80 Gelloz, B., Shibata, T., Mentek, R., and Koshida, N. (2007) Pronounced photonic effects of high-pressure water vapor annealing on nanocrystalline porous silicon. *Mater. Res. Soc. Symp. Proc.*, **958**, 227–232.

81 Gelloz, B. and Koshida, N. (2007) Highly efficient and stable photoluminescence of nanocrystalline porous silicon by combination of chemical modification and oxidation under high pressure. *Jpn. J. Appl. Phys., Part 1*, **46**, 2429–2433.

82 Gelloz, B. and Koshida, N. (2006) Highly enhanced photoluminescence of as-anodized and electrochemically oxidized nanocrystalline p-type porous silicon treated by high-pressure water vapor annealing. *Thin Solid Films*, **508**, 406–409.

83 Gelloz, B., Shibata, T., and Koshida, N. (2006) Stable electroluminescence of nanocrystalline silicon device activated by high pressure water vapor annealing. *Appl. Phys. Lett.*, **89**, 191103.

84 Gelloz, B. and Koshida, N. (2006) Highly enhanced efficiency and stability of photo- and electro-luminescence of nano-crystalline porous silicon by high-pressure water vapor annealing. *Jpn. J. Appl. Phys., Part 1*, **45**, 3462–3465.

85 Gelloz, B., Kojima, A., and Koshida, N. (2005) Highly efficient and stable luminescence of nanocrystalline porous silicon treated by high-pressure water vapor annealing. *Appl. Phys. Lett.*, **87**, 031107.

86 Gelloz, B. and Koshida, N. (2005) Mechanism of a remarkable enhancement in the light emission from nanocrystalline porous silicon annealed in high-pressure water vapor. *J. Appl. Phys.*, **98**, 123509.

87 Giaddui, T., Forcey, K.S., Earwaker, L.G., Loni, A., Canham, L.T., and Halimaoui, A. (1996) Reduction of ion beam induced and atmospheric ageing of porous silicon using Al and SiO$_2$ caps. *J. Phys. D: Appl. Phys.*, **29**, 1580–1586.

88 Halimaoui, A., Campidelli, Y., Badoz, P.A., and Bensahel, D. (1995) Covering and filling of porous silicon pores with Ge and Si using chemical-vapor-deposition. *J. Appl. Phys.*, **78**, 3428–3430.

89 Buriak, J.M. (2002) Organometallic chemistry on silicon and germanium surfaces. *Chem. Rev.*, **102**, 1271–1308.

90 Gelloz, B., Sano, H., Boukherroub, R., Wayner, D.D.M., Lockwood, D.J., and Koshida, N. (2003) Stabilization of porous silicon electroluminescence by surface passivation with controlled covalent bonds. *Appl. Phys. Lett.*, **83**, 2342–2344.

91 Ghulinyan, M., Gelloz, B., Ohta, T., Pavesi, L., Lockwood, D.J., and Koshida, N. (2008) Stabilized porous silicon optical superlattices with controlled surface passivation. *Appl. Phys. Lett.*, **93**, 061113.

92 Alvarez, S.D., Derfus, A.M., Schwartz, M.P., Bhatia, S.N., and Sailor, M.J. (2009) The compatibility of hepatocytes with chemically modified porous silicon with reference to *in vitro* biosensors. *Biomaterials*, **30**, 26–34.

93 Li, G.B., Hou, X.Y., Yuan, S., Chen, H.J., Zhang, F.L., Fan, H.L., and Wang, X. (1996) Passivation of light-emitting porous silicon by rapid thermal treatment in NH$_3$. *J. Appl. Phys.*, **80**, 5967–5970.

94 Anderson, R.C., Muller, R.S., and Tobias, C.W. (1993) Chemical surface modification of porous silicon. *J. Electrochem. Soc.*, **140**, 1393–1396.

95 Dillon, A.C., Gupta, P., Robinson, M.B., Bracker, A.S., and George, S.M. (1991) Ammonia decomposition on silicon surfaces studied using transmission Fourier-transform infrared-spectroscopy. *J. Vac. Sci. Technol. A.*, **9**, 2222–2230.

96 Morazzani, V., Cantin, J.L., Ortega, C., Pajot, B., Rahbi, R., Rosenbauer, M., Von Bardeleben, H.J., and Vazsonyi, E. (1996) Thermal nitridation of p-type porous silicon in ammonia. *Thin Solid Films*, **276**, 32–35.

97 Salonen, J., Laine, E., and Niinisto, L. (2002) Thermal carbonization of porous silicon surface by acetylene. *J. Appl. Phys.*, **91**, 456–461.

98 Torres-Costa, V., Martin-Palma, R.J., Martinez-Duart, J.M., Salonen, J., and Lehto, V.P. (2008) Effective passivation of porous silicon optical devices by thermal carbonization. *J. Appl. Phys.*, **103**, 083124.

99 Lehmann, V., Hofmann, F., Moller, F., and Gruning, U. (1995) Resistivity of porous silicon: a surface effect. *Thin Solid Films*, **255**, 20–22.

100 Grosman, A. and Ortega, C. (1997) in *Properties of Porous Silicon* (ed. L.T. Canham), INSPEC, The Institution of Electrical Engineers, London, pp. 328–335.

101 Borini, S., Boarino, L., and Amato, G. (2006) Anisotropic resistivity of (100)-oriented mesoporous silicon. *Appl. Phys. Lett.*, **89**, 132111.

102 Forsh, P.A., Martyshov, M.N., Timoshenko, V.Y., and Kashkarov, P.K. (2006) Alternating current conductivity of anisotropically nanostructured silicon. *Semiconductors*, **40**, 471–475.

103 Borini, S., Boarino, L., and Amato, G. (2006) Coulomb blockade tuned by NO$_2$ molecules in nanostructured silicon. *Adv. Mater.*, **18**, 2422.

104 Hamilton, B., Jacobs, J., Hill, D.A., Pettifer, R.F., Teehan, D., and Canham, L.T. (1998) Size-controlled percolation pathways for electrical conduction in porous silicon. *Nature*, **393**, 443–445.

105 Ueno, K. and Koshida, N. (1998) Negative-resistance effects in light-emitting porous silicon diodes. *Jpn. J. Appl. Phys., Part 1*, **37**, 1096–1099.

106 Ueno, K. and Koshida, N. (1999) Light-emissive nonvolatile memory effects in porous silicon diodes. *Appl. Phys. Lett.*, **74**, 93–95.

107 Benchorin, M., Kux, A., and Schechter, I. (1994) Adsorbate effects on photoluminescence and electrical-conductivity of porous silicon. *Appl. Phys. Lett.*, **64**, 481–483.

108 Tsu, R. and Babic, D. (1994) Doping of a quantum-dot. *Appl. Phys. Lett.*, **64**, 1806–1808.

109 Timoshenko, V.Y., Dittrich, T., Lysenko, V., Lisachenko, M.G., and Koch, F. (2001) Free charge carriers in mesoporous silicon. *Phys. Rev. B*, **6408**, art. no. 085314.

110 Harper, J. and Sailor, M.J. (1996) Detection of nitric oxide and nitrogen dioxide with photoluminescent porous silicon. *Anal. Chem.*, **68**, 3713–3717.

111 Boarino, L., Geobaldo, F., Borini, S., Rossi, A.M., Rivolo, P., Rocchia, M., Garrone, E., and Amato, G. (2001) Local environment of boron impurities in porous silicon and their interaction with

NO₂ molecules. *Phys. Rev. B*, **6420**, art. no. 205308.

112 Geobaldo, F., Onida, B., Rivolo, P., Borini, S., Boarino, L., Rossi, A., Amato, G., and Garrone, E. (2001) IR detection of NO₂ using p(+) porous silicon as a high sensitivity sensor. *Chem. Commun.*, 2196–2197.

113 Barillaro, G., Diligenti, A., Strambini, L.M., Comini, E., and Faglia, G. (2008) NO₂ adsorption effects on p(+)–n silicon junctions surrounded by a porous layer. *Sens. Actuators B*, **134**, 922–927.

114 Dzhafarov, T., Yuksel, S.A., and Lus, C.O. (2008) Porous silicon-based gas sensors and miniature hydrogen cells. *Jpn. J. Appl. Phys.*, **47**, 8204–8207.

115 Aroutiounian, V., Arakelyan, V., Galstyan, V., Martirosyan, K., and Soukiassian, P. (2009) Hydrogen sensor made of porous silicon and covered by TiO2-*x* or ZnO< Al > thin film. *IEEE Sens. J.*, **9**, 9–12.

116 Zhang, D. and Alocilja, E.C. (2008) Characterization of nanoporous silicon-based DNA biosensor for the detection of *Salmonella enteritidis*. *IEEE Sens. J.*, **8**, 775–780.

117 Zheng, J.P., Jiao, K.L., Shen, W.P., Anderson, W.A., and Kwok, H.S. (1992) Highly sensitive photodetector using porous silicon. *Appl. Phys. Lett.*, **61**, 459–461.

118 Tsai, C.C., Li, K.H., Campbell, J.C., and Tasch, A. (1993) Photodetectors fabricated from rapid-thermal-oxidized porous Si. *Appl. Phys. Lett.*, **62**, 2818–2820.

119 Kruger, M. *et al.* (1997) Color-sensitive Si-photodiode using porous silicon interference filters. *Jpn. J. Appl. Phys.*, *Part 2*, **36**, L24–L26.

120 Kruger, M. *et al.* (1997) Color-sensitive photodetector based on porous silicon superlattices. *Thin Solid Films*, **297**, 241–244.

121 Torres-Costa, V., Martin-Palma, R.J., and Martinez-Duart, J.M. (2007) All-silicon color-sensitive photodetectors in the visible. *Mater. Sci. Eng. C*, **27**, 954–956.

122 Salgado, G.G., Hernandez, R., Martinez, J., Diaz, T., Juarez, H., Rosendo, E., Galeazzi, R., Garcia, A., and Juarez, G. (2008) Fabrication, characterization, and

analysis of photodetectors metal-porous silicon with different geometry and thickness of the porous silicon layer. *Microelectron. J.*, **39**, 489–493.

123 Koshida, N., Ozaki, T., Sheng, X., and Koyama, H. (1995) Cold electron-emission from electroluminescent porous silicon diodes. *Jpn. J. Appl. Phys., Part 2*, **34**, L705–L707.

124 Koshida, N. and Matsumoto, N. (2003) Fabrication and quantum properties of nanostructured silicon. *Mater. Sci. Eng. R.*, **40**, 169–205.

125 Ohta, T., Kojima, A., Hirakawa, H., Iwamatsu, T., and Koshida, N. (2005) Operation of nanocrystalline silicon ballistic emitter in low vacuum and atmospheric pressures. *J. Vac. Sci. Technol. B*, **23**, 2336–2339.

126 Ohta, T., Kojima, A., and Koshida, N. (2007) Emission characteristics of nanocrystalline porous silicon ballistic cold cathode in atmospheric ambience. *J. Vac. Sci. Technol. B*, **25**, 524–527.

127 Nakajima, Y., Kojima, A., and Koshida, N. (2002) Generation of ballistic electrons in nanocrystalline porous silicon layers and its application to a solid-state planar luminescent device. *Appl. Phys. Lett.*, **81**, 2472–2474.

128 Nakajima, Y., Uchida, T., Toyama, H., Kojima, A., Gelloz, B., and Koshida, N. (2004) A solid-state multicolor light-emitting device based on ballistic electron excitation. *Jpn. J. Appl. Phys., Part 1*, **43**, 2076–2079.

129 Gelloz, B., Sato, M., and Koshida, N. (2008) Cavity effect in nanocrystalline porous silicon ballistic lighting device. *Jpn. J. Appl. Phys.*, **47**, 2902–2905.

130 Kojima, A. and Koshida, N. (2001) Evidence of enlarged drift length in nanocrystalline porous silicon layers by time-of-flight measurements. *Jpn. J. Appl. Phys., Part 1*, **40**, 2779–2781.

131 Klima, O., Hlinomaz, P., Hospodkova, A., Oswald, J., and Kocka, J. (1993) Transport properties of self-supporting porous silicon. *J. Non-Cryst. Solids*, **166**, 961–964.

132 Sheng, X., Koyama, H., and Koshida, N. (1998) Efficient surface-emitting cold cathodes based on electroluminescent

porous silicon diodes. *J. Vac. Sci. Technol. B*, **16**, 793–795.

133 Koshida, N., Sheng, X., and Komoda, T. (1999) Quasiballistic electron emission from porous silicon diodes. *Appl. Surf. Sci.*, **146**, 371–376.

134 Sakai, D., Oshima, C., Ohta, T., and Koshida, N. (2008) *J. Vac. Sci. Technol. B*, **26**, 1782–1786.

135 Komoda, T., Honda, Y., Hatai, T., Watanabe, Y., Ichihara, T., Aizawa, K., Kondo, Y., Sheng, X., Kojima, A., and Koshida, N. (1999) IDW'99, p. 939.

136 Komoda, T., Ichihara, T., Honda, Y., Aizawa, K., and Koshida, N. (2001) *Mater. Res. Soc. Symp. Proc.*, **638**, F4.1.1.

137 Komoda, T., Ichihara, T., Honda, Y., Hatai, T., Baba, T., Takegawa, Y., Watabe, Y., Aizawa, K., and Koshida, N. (2004) Fabrication of a 7.6-in-diagonal prototype ballistic electron surface-emitting display on glass substrate. *J. Soc. Inf. Display*, **12**, 29–35.

138 Nakajima, Y., Kojima, A., Toyama, H., and Koshida, N. (2002) A solid-state light-emitting device based on excitations of ballistic electrons generated in nanocrystalline porous polysilicon films. *Jpn. J. Appl. Phys., Part 1*, **41**, 2707–2709.

139 Nakajima, Y., Kojima, A., and Koshida, N. (2001) *Mater. Res. Soc. Symp. Proc.*, **638**, F4.2.1.

140 Gelloz, B., Kanda, T., Uchida, T., Niibe, M., Kojima, A., and Koshida, N. (2005) Electroluminescence enhancement assisted with ballistic electron excitation in nanocrystalline silicon diodes. *Jpn. J. Appl. Phys., Part 1*, **44**, 2676–2679.

141 Koshida, N., Ohta, T., and Gelloz, B. (2007) Operation of nanosilicon ballistic electron emitter in liquid water and hydrogen generation effect. *Appl. Phys. Lett.*, **90**, 163505.

142 Ohta, T., Gelloz, B., and Koshida, N. (2008) Characteristics of nanosilicon ballistic cold cathode in aqueous solutions as an active electrode. *J. Vac. Sci. Technol. B*, **26**, 716–719.

143 Lerondel, G., Romestain, R., and Barret, S. (1997) Roughness of the porous silicon dissolution interface. *J. Appl. Phys.*, **81**, 6171–6178.

144 Theiss, W. and Hilbrich, S. (1997) in *Properties of Porous Silicon* (ed. L.T. Canham), INSPEC, The Institution of Electrical Engineers, London, pp. 223–228.

145 von Behren, J., van Buuren, T., Zacharias, M., Chimowitz, E.H., and Fauchet, P.M. (1998) Quantum confinement in nanoscale silicon: the correlation of size with bandgap and luminescence. *Solid State Commun.*, **105**, 317–322.

146 Loni, A., Canham, L.T., Berger, M.G., Arens-Fischer, R., Munder, H., Luth, H., Arrand, H.F., and Benson, T.M. (1996) Porous silicon multilayer optical waveguides. *Thin Solid Films*, **276**, 143–146.

147 Ferrand, P., Loi, D., and Romestain, R. (2001) Photonic band-gap guidance in high-porosity luminescent porous silicon. *Appl. Phys. Lett.*, **79**, 3017–3019.

148 Ferrand, P., Romestain, R., and Vial, J.C. (2001) Photonic band-gap properties of a porous silicon periodic planar waveguide. *Phys. Rev. B*, **63**, art. no. 115106.

149 Ferrand, P. and Romestain, R. (2000) Optical losses in porous silicon waveguides in the near-infrared: effects of scattering. *Appl. Phys. Lett.*, **77**, 3535–3537.

150 Takahashi, M. and Koshida, N. (1999) Fabrication and characteristics of three-dimensionally buried porous silicon optical waveguides. *J. Appl. Phys.*, **86**, 5274–5278.

151 Oton, C.J., Navarro-Urrios, D., Capuj, N.E., Ghulinyan, M., Pavesi, L., Gonzalez-Perez, S., Lahoz, F., and Martin, I.R. (2006) Optical gain in dye-impregnated oxidized porous silicon waveguides. *Appl. Phys. Lett.*, **89**, 011107.

152 Najar, A., Charrier, J., Lorrain, N., Haji, L., and Oueslati, M. (2007) Optical gain measurements in porous silicon planar waveguides codoped by erbium and ytterbium ions at 1.53 mu m. *Appl. Phys. Lett.*, **91**, 121120.

153 Thonissen, M. and Berger, M.G. (1997) in *Properties of Porous Silicon* (ed. L.T. Canham), INSPEC, The Institution of Electrical Engineers, London, pp. 30–37.

154 Thonissen, M., Kruger, M., Lerondel, G., and Romestain, R. (1997) in *Properties of Porous Silicon* (ed. L.T. Canham), INSPEC, The Institution of Electrical Engineers, London, pp. 349–355.

155 Agarwal, V. and del Rio, J.A. (2003) Tailoring the photonic band gap of a porous silicon dielectric mirror. *Appl. Phys. Lett.*, **82**, 1512–1514.

156 Bruyant, A., Lerondel, G., Reece, P.J., and Gal, M. (2003) All-silicon omnidirectional mirrors based on one-dimensional photonic crystals. *Appl. Phys. Lett.*, **82**, 3227–3229.

157 Kuzma-Filipek, I.J., Duerinckx, F., Van Kerschaver, E., Van Nieuwenhuysen, K., Beaucarne, G., and Poortmans, J. (2008) Chirped porous silicon reflectors for thin-film epitaxial silicon solar cells. *J. Appl. Phys.*, **104**, 073529.

158 Estevez, J.O., Arriaga, J., Blas, A.M., and Agarwal, V. (2009) Enlargement of omnidirectional photonic bandgap in porous silicon dielectric mirrors with a Gaussian profile refractive index. *Appl. Phys. Lett.*, **94**, 061914.

159 Xifre-Perez, E., Marsal, L.F., Ferre-Borrull, J., and Pallares, J. (2009) Low refractive index contrast porous silicon omnidirectional reflectors. *Appl. Phys. B*, **95**, 169–172.

160 Reece, P.J., Lerondel, G., Zheng, W.H., and Gal, M. (2002) Optical nicrocavities with subnanometer linewidths based on porous silicon. *Appl. Phys. Lett.*, **81**, 4895–4897.

161 Billat, S., Thonissen, M., ArensFischer, R., Berger, M.G., Kruger, M., and Luth, H. (1997) Influence of etch stops on the microstructure of porous silicon layers. *Thin Solid Films*, **297**, 22–25.

162 Ghulinyan, M., Oton, C.J., Bonetti, G., Gaburro, Z., and Pavesi, L. (2003) Free-standing porous silicon single and multiple optical cavities. *J. Appl. Phys.*, **93**, 9724–9729.

163 Ghulinyan, M., Oton, C.J., Gaburro, Z., Bettotti, P., and Pavesi, L. (2003) Porous silicon free-standing coupled micro-cavities. *Appl. Phys. Lett.*, **82**, 1550–1552.

164 Lorenzo, E., Oton, C.J., Capuj, N.E., Ghulinyan, M., Navarro-Urrios, D., Gaburro, Z., and Pavesi, L. (2005) Porous silicon-based rugate filters. *Appl. Opt.*, **44**, 5415–5421.

165 Cunin, F., Schmedake, T.A., Link, J.R., Li, Y.Y., Koh, J., Bhatia, S.N., and Sailor, M.J. (2002) Biomolecular screening with encoded porous-silicon photonic crystals. *Nat. Mater.*, **1**, 39–41.

166 Sailor, M.J. and Link, J.R. (2005) "Smart dust": nanostructured devices in a grain of sand. *Chem. Commun.*, 1375–1383.

167 Ishikura, N., Fujii, M., Nishida, K., Hayashi, S., Diener, J., Mizuhata, M., and Deki, S. (2008) Broadband rugate filters based on porous silicon. *Opt. Mater.*, **31**, 102–105.

168 Ishikura, N., Fujii, M., Nishida, K., Hayashi, S., and Diener, J. (2008) Dichroic rugate filters based on birefringent porous silicon. *Opt. Express*, **16**, 15531–15539.

169 Pavesi, L., Panzarini, G., and Andreani, L.C. (1998) All-porous silicon-coupled microcavities: experiment versus theory. *Phys. Rev. B*, **58**, 15794–15800.

170 Sapienza, R., Costantino, P., Wiersma, D., Ghulinyan, M., Oton, C.J., and Pavesi, L. (2003) Optical analogue of electronic Bloch oscillations. *Phys. Rev. Lett.*, **91**, 263902.

171 Ghulinyan, M., Gaburro, Z., Wiersma, D.S., and Pavesi, L. (2006) Tuning of resonant Zener tunneling by vapor diffusion and condensation in porous optical superlattices. *Phys. Rev. B*, **74**, 045118.

172 Ghulinyan, M., Oton, C.J., Gaburro, Z., Pavesi, L., Toninelli, C., and Wiersma, D.S. (2005) Zener tunneling of light waves in an optical superlattice. *Phys. Rev. Lett.*, **94**, 127401.

173 Ghulinyan, M., Gaburro, Z., Wiersma, D.S., and Pavesi, L. (2007) Vapor control of resonant Zener tunneling of light in a photonic crystal. *Phys. Status Solidi A*, **204**, 1351–1355.

174 Gaburro, Z., Ghulinyan, M., Pavesi, L., Barthelemy, P., Toninelli, C., and Wiersma, D. (2008) Dynamics of capillary condensation in bistable optical superlattices. *Phys. Rev. B*, **77**, 115354.

175 Barthelemy, P., Ghulinyan, M., Gaburro, Z., Toninelli, C., Pavesi, L., and Wiersma, D.S. (2007) Optical switching by capillary condensation. *Nat. Photonics,* **1**, 172–175.

176 Dal Negro, L., Oton, C.J., Gaburro, Z., Pavesi, L., Johnson, P., Lagendijk, A., Righini, R., Colocci, M., and Wiersma, D.S. (2003) Light transport through the band-edge states of Fibonacci quasicrystals. *Phys. Rev. Lett.,* **90**, 055501.

177 Bertolotti, J., Gottardo, S., Wiersma, D.S., Ghulinyan, M., and Pavesi, L. (2005) Optical necklace states in Anderson localized 1D systems. *Phys. Rev. Lett.,* **94**, 113903.

178 Ma, L.L., Zhou, Y.C., Jiang, N., Lu, X., Shao, J., Lu, W., Ge, J., Ding, X.M., and Hou, X.Y. (2006) Wide-band "black silicon" based on porous silicon. *Appl. Phys. Lett.,* **88**, 171907.

179 Diener, J., Kunzner, N., Kovalev, D., Gross, E., Timoshenko, V.Y., Polisski, G., and Koch, F. (2001) Dichroic Bragg reflectors based on birefringent porous silicon. *Appl. Phys. Lett.,* **78**, 3887–3889.

180 Diener, J., Kunzner, N., Kovalev, D., Gross, E., Koch, F., and Fujii, M. (2002) Dichroic behavior of multilayer structures based on anisotropically nanostructured silicon. *J. Appl. Phys.,* **91**, 6704–6709.

181 Golovan, L.A. *et al.* (2001) Phase matching of second-harmonic generation in birefringent porous silicon. *Appl. Phys. B,* **73**, 31–34.

182 Kunzner, N., Kovalev, D., Diener, J., Gross, E., Timoshenko, V.Y., Polisski, G., Koch, F., and Fujii, M. (2001) Giant birefringence in anisotropically nanostructured silicon. *Opt. Lett.,* **26**, 1265–1267.

183 Soboleva, I.V., Murchikova, E.M., Fedyanin, A.A., and Aktsipetrov, O.A. (2005) Second- and third-harmonic generation in birefringent photonic crystals and microcavities based on anisotropic porous silicon. *Appl. Phys. Lett.,* **87**, 241110.

184 Mangaiyarkarasi, D., Teo, E.J., Breese, M.B.H., Bettiol, A.A., and Blackwood, D.J. (2005) Controlled shift in emission wavelength from patterned porous silicon using focused ion beam irradiation. *J. Electrochem. Soc.,* **152**, D173–D176.

185 Teo, E.J., Mangaiyarkarasi, D., Breese, M.B.H., Bettiol, A.A., and Blackwood,

D.J. (2004) Controlled intensity emission from patterned porous silicon using focused proton beam irradiation. *Appl. Phys. Lett.,* **85**, 4370–4372.

186 Park, H., Dickerson, J.H., and Weiss, S.M. (2008) Spatially localized one-dimensional porous silicon photonic crystals. *Appl. Phys. Lett.,* **92**, 011113.

187 Weiss, S.M., Ouyang, H.M., Zhang, J.D., and Fauchet, P.M. (2005) Electrical and thermal modulation of silicon photonic bandgap microcavities containing liquid crystals. *Opt. Express,* **13**, 1090–1097.

188 Weiss, S.M. and Fauchet, P.M. (2003) Electrically tunable porous silicon active mirrors. *Phys. Status Solidi A,* **197**, 556–560.

189 Thomas, J.C., Pacholski, C., and Sailor, M.J. (2006) Delivery of nanogram payloads using magnetic porous silicon microcarriers. *Lab Chip,* **6**, 782–787.

190 Dorvee, J.R., Derfus, A.M., Bhatia, S.N., and Sailor, M.J. (2004) Manipulation of liquid droplets using amphiphilic, magnetic one-dimensional photonic crystal chaperones. *Nat. Mater.,* **3**, 896–899.

191 Pacholski, C., Sartor, M., Sailor, M.J., Cunin, F., and Miskelly, G.M. (2005) Biosensing using porous silicon double-layer interferometers: reflective interferometric Fourier transform spectroscopy. *J. Am. Chem. Soc.,* **127**, 11636–11645.

192 Park, J.H., Derfus, A.M., Segal, E., Vecchio, K.S., Bhatia, S.N., and Sailor, M.J. (2006) Local heating of discrete droplets using magnetic porous silicon-based photonic crystals. *J. Am. Chem. Soc.,* **128**, 7938–7946.

193 Ouyang, H., Striemer, C.C., and Fauchet, P.M. (2006) Quantitative analysis of the sensitivity of porous silicon optical biosensors. *Appl. Phys. Lett.,* **88**, 163108.

194 Salem, M.S., Sailor, M.J., Fukami, K., Sakka, T., and Ogata, Y.H. (2008) Sensitivity of porous silicon rugate filters for chemical vapor detection. *J. Appl. Phys.,* **103**, 083516.

195 Ruminski, A.M., Moore, M.M., and Sailor, M.J. (2008) Humidity-compensating sensor for volatile organic compounds using stacked porous silicon photonic crystals. *Adv. Funct. Mater.*, **18**, 3418–3426.

196 Levitsky, I.A., Euler, W.B., Tokranova, N., and Rose, A. (2007) Fluorescent polymer-porous silicon microcavity devices for explosive detection. *Appl. Phys. Lett.*, **90**, 041904.

197 De Stefano, L., Malecki, K., Della Corte, F.G., Moretti, L., Rea, I., Rotiroti, L., and Rendina, I. (2006) A microsystem based on porous silicon-glass anodic bonding for gas and liquid optical sensing. *Sensors*, **6**, 680–687.

198 Gross, E., Kovalev, D., Kunzner, N., Timoshenko, V.Y., Diener, J., and Koch, F. (2001) Highly sensitive recognition element based on birefringent porous silicon layers. *J. Appl. Phys.*, **90**, 3529–3532.

199 Rong, G., Najmaie, A., Sipe, J.E., and Weiss, S.M. (2008) Nanoscale porous silicon waveguide for label-free DNA sensing. *Biosens. Bioelectron.*, **23**, 1572–1576.

200 Descrovi, E., Frascella, F., Sciacca, B., Geobaldo, F., Dominici, L., and Michelotti, F. (2007) Coupling of surface waves in highly defined one-dimensional porous silicon photonic crystals for gas sensing applications. *Appl. Phys. Lett.*, **91**, 241109.

201 Guillermain, E., Lysenko, V., Orobtchouk, R., Benyattou, T., Roux, S., Pillonnet, A., and Perriat, P. (2007) Bragg surface wave device based on porous silicon and its application for sensing. *Appl. Phys. Lett.*, **90**, 241116.

202 Palestino, G., Legros, R., Agarwal, V., Perez, E., and Gergely, C. (2008) Functionalization of nanostructured porous silicon microcavities for glucose oxidase detection. *Sens. Actuators B*, **135**, 27–34.

203 Matsumoto, T., Daimon, M., Mimura, H., Kanemitsu, Y., and Koshida, N. (1995) Optically induced absorption in porous silicon and its application to logic gates. *J. Electrochem. Soc.*, **142**, 3528–3533.

204 Takahashi, M., Toriumi, Y., Matsumoto, T., Masumoto, Y., and Koshida, N. (2000) Significant photoinduced refractive index change observed in porous silicon Fabry–Perot resonators. *Appl. Phys. Lett.*, **76**, 1990–1992.

205 Takahashi, M., Toriumi, Y., and Koshida, N. (2000) Current-induced optical effect in porous silicon Fabry–Perot resonators. *Phys. Status Solidi A*, **182**, 567–571.

206 Apiratikul, P., Rossi, A.M., and Murphy, T.E. (2009) Nonlinearities in porous silicon optical waveguides at 1550 nm. *Opt. Express*, **17**, 3396–3406.

207 Kovalev, D., Heckler, H., Ben-Chorin, M., Polisski, G., Schwartzkopff, M., and Koch, F. (1998) Breakdown of the k-conservation rule in Si nanocrystals. *Phys. Rev. Lett.*, **81**, 2803–2806.

208 Wolkin, M.V., Jorne, J., Fauchet, P.M., Allan, G., and Delerue, C. (1999) Electronic states and luminescence in porous silicon quantum dots: the role of oxygen. *Phys. Rev. Lett.*, **82**, 197–200.

209 Pavesi, L., Guardini, R., and Mazzoleni, C. (1996) Porous silicon resonant cavity light emitting diodes. *Solid State Commun.*, **97**, 1051–1053.

210 Araki, M., Koyama, H., and Koshida, N. (1996) Controlled electroluminescence spectra of porous silicon diodes with a vertical optical cavity. *Appl. Phys. Lett.*, **69**, 2956–2958.

211 Chan, S. and Fauchet, P.M. (1999) Tunable, narrow, and directional luminescence from porous silicon light emitting devices. *Appl. Phys. Lett.*, **75**, 274–276.

212 Valenta, J., Juhasz, R., and Linnros, J. (2002) Photoluminescence spectroscopy of single silicon quantum dots. *Appl. Phys. Lett.*, **80**, 1070–1072.

213 Bsiesy, A., Vial, J.C., Gaspard, F., Herino, R., Ligeon, M., Muller, F., Romestain, R., Wasiela, A., Halimaoui, A., and Bomchil, G. (1991) Photoluminescence of high porosity and of electrochemically oxidized porous silicon layers. *Surf. Sci.*, **254**, 195–200.

214 Gelloz, B., Koyama, H., and Koshida, N. (2008) Polarization memory of blue and red luminescence from nanocrystalline porous silicon treated by high-pressure water vapor annealing. *Thin Solid Films*, **517**, 376–379.

215 Gelloz, B., Mentek, R., and Koshida, N. (2009) Specific blue light emission from nanocrystalline porous Si treated by high-pressure water vapor annealing. *Jpn. J. Appl. Phys., Part 1*, **48**, 04C119.

216 Vial, J.C., Bsiesy, A., Gaspard, F., Herino, R., Ligeon, M., Muller, F., Romestain, R., and Macfarlane, R.M. (1992) Mechanisms of visible-light emission from electrooxidized porous silicon. *Phys. Rev. B*, **45**, 14171–14176.

217 Skryshevsky, V.A., Laugier, A., Strikha, V.I., and Vikulov, V.A. (1996) Evaluation of quantum efficiency of porous silicon photoluminescence. *Mater. Sci. Eng. B*, **40**, 54–57.

218 Kim, H., Hong, C., and Lee, C. (2009) Enhanced photoluminescence from porous silicon passivated with an ultrathin aluminum film. *Mater. Lett.*, **63**, 434–436.

219 Loni, A., Simons, A.J., Calcott, P.D.J., and Canham, L.T. (1995) Blue photoluminescence from rapid thermally oxidized porous silicon following storage in ambient air. *J. Appl. Phys.*, **77**, 3557–3559.

220 Sun, J., Lu, Y.W., Du, X.W., and Kulinich, S.A. (2005) Improved visible photoluminescence from porous silicon with surface Si−Ag bonds. *Appl. Phys. Lett.*, **86**, 171905.

221 Lu, Y.W., Du, X.W., Sun, J., Han, X., and Kulinich, S.A. (2006) Influence of surface Si−Ag bonds on photoluminescence of porous silicon. *J. Appl. Phys.*, **100**, 063512.

222 Lauerhaas, J.M., Credo, G.M., Heinrich, J.L., and Sailor, M.J. (1992) Reversible luminescence quenching of porous Si by solvents. *J. Am. Chem. Soc.*, **114**, 1911–1912.

223 Shimura, M., Makino, N., Yoshida, K., Hattori, T., and Okumura, T. (1999) Reversible quenching of red photoluminescence of porous silicon with adsorption of alcohol OH-groups I. *Electrochemistry*, **67**, 63–67.

224 Fellah, S., Ozanam, F., Gabouze, N., and Chazalviel, J.N. (2000) Porous silicon in solvents: constant-lifetime PL quenching and confirmation of dielectric effects. *Phys. Status Solidi A*, **182**, 367–372.

225 Letant, S. and Vial, J.C. (1997) Energy transfer in dye impregnated porous silicon. *J. Appl. Phys.*, **82**, 397–401.

226 Chouket, A., Elhouichet, H., Oueslati, M., Koyama, H., Gelloz, B., and Koshida, N. (2007) Energy transfer in porous-silicon/ laser-dye composite evidenced by polarization memory of photoluminescence. *Appl. Phys. Lett.*, **91**, Artn 211902.

227 Park, J.H., Gu, L., von Maltzahn, G., Ruoslahti, E., Bhatia, S.N., and Sailor, M.J. (2009) Biodegradable luminescent porous silicon nanoparticles for *in vivo* applications. *Nat. Mater.*, **8**, 331–336.

228 Bressers, P.M.M.C., Knapen, J.W.J., Meulenkamp, E.A., and Kelly, J.J. (1992) Visible-light emission from a porous silicon solution diode. *Appl. Phys. Lett.*, **61**, 108–110.

229 Kooij, E.S., Despo, R.W., and Kelly, J.J. (1995) Electroluminescence from porous silicon due to electron injection from solution. *Appl. Phys. Lett.*, **66**, 2552–2554.

230 Green, W.H., Lee, E.J., Lauerhaas, J.M., Bitner, T.W., and Sailor, M.J. (1995) Electrochemiluminescence from porous silicon in formic-acid liquid-junction cells. *Appl. Phys. Lett.*, **67**, 1468–1470.

231 Bsiesy, A., Muller, F., Ligeon, M., Gaspard, F., Herino, R., Romestain, R., and Vial, J.C. (1993) Voltage-controlled spectral shift of porous silicon electroluminescence. *Phys. Rev. Lett.*, **71**, 637–640.

232 Romestain, R., Vial, J.C., Mihalcescu, I., and Bsiesy, A. (1995) Saturation and voltage quenching of the porous silicon luminescence and importance of the auger effect. *Phys. Status Solidi B*, **190**, 77–84.

233 Gelloz, B., Bsiesy, A., and Herino, R. (2003) Electrically induced luminescence quenching in p(+)-type and anodically oxidized n-type wet porous silicon. *J. Appl. Phys.*, **94**, 2381–2389.

234 Gelloz, B., Bsiesy, A., and Herino, R. (1999) Light-induced porous silicon photoluminescence quenching. *J. Lumin.*, **82**, 205–211.

235 Gelloz, B. and Bsiesy, A. (1998) Carrier transport mechanisms in porous silicon in contact with a liquid phase: a diffusion process. *Appl. Surf. Sci.*, **135**, 15–22.

236 Loni, A., Simons, A.J., Cox, T.I., Calcott, P.D.J., and Canham, L.T. (1995)

Electroluminescent porous silicon device with an external quantum efficiency greater-than 0.1-percent under CW operation. *Electron. Lett.*, **31**, 1288–1289.

237 Lalic, N. and Linnros, J. (1996) Characterization of a porous silicon diode with efficient and tunable electroluminescence. *J. Appl. Phys.*, **80**, 5971–5977.

238 Tsybeskov, L., Duttagupta, S.P., and Fauchet, P.M. (1995) Photoluminescence and electroluminescence in partially oxidized porous silicon. *Solid State Commun.*, **95**, 429–433.

239 Hirschman, K.D., Tsybeskov, L., Duttagupta, S.P., and Fauchet, P.M. (1996) Silicon-based visible light-emitting devices integrated into microelectronic circuits. *Nature*, **384**, 338–341.

240 Gelloz, B., Nakagawa, T., and Koshida, N. (1998) *Mater. Res. Soc. Symp. Proc.*, **536**, 15.

241 Gelloz, B. and Koshida, N. (2004) High performance electroluminescence from nanocrystalline silicon with carbon buffer. *Jpn. J. Appl. Phys., Part 1*, **43**, 1981–1985.

242 Barillaro, G., Diligenti, A., Pieri, F., Fuso, F., and Allegrini, M. (2001) Integrated porous-silicon light-emitting diodes: a fabrication process using graded doping profiles. *Appl. Phys. Lett.*, **78**, 4154–4156.

243 Koshida, N., Takizawa, E., Mizuno, H., Arai, S., Koyama, H., and Sameshima, T. (1998) *Mater. Res. Soc. Symp. Proc.*, **486**, 151.

244 El-Bahar, A. and Nemirovsky, Y. (2000) A technique to form a porous silicon layer with no backside contact by alternating current electrochemical process. *Appl. Phys. Lett*, **77**, 208–210.

245 Boumaiza, Y., Hadjoub, Z., Doghmane, A., and Deboub, L. (1999) Porosity effects on different measured acoustic parameters of porous silicon. *J. Mater. Sci. Lett.*, **18**, 295–297.

246 Fan, H.J., Kuok, M.H., Ng, S.C., Boukherroub, R., Baribeau, J.M., Fraser, J.W., and Lockwood, D.J. (2002) Brillouin spectroscopy of acoustic modes in porous silicon films. *Phys. Rev. B*, **65**, 165330.

247 Dafonseca, R.J.M., Saurel, J.M., Despaux, G., Foucaran, A., Massonne, E., Taliercio, T., and Lefebvre, P. (1994) Elastic characterization of porous silicon by acoustic microscopy. *Superlattices Microstruct.*, **16**, 21–23.

248 Kiuchi, A., Gelloz, B., Kojima, A., and Koshida, N. (2005) *Mater. Res. Soc. Symp. Proc.*, **832**, F3.7.1–F3.7.6.

249 Reinhardt, A. and Snow, P.A. (2007) Theoretical study of acoustic band-gap structures made of porous silicon. *Phys. Status Solidi A*, **204**, 1528–1535.

250 Lang, W. (1997) in *Properties of Porous Silicon* (ed. L.T. Canham), INSPEC, The Institution of Electrical Engineers, London, pp. 138–141.

251 Shinoda, H., Nakajima, T., Ueno, K., and Koshida, N. (1999) Thermally induced ultrasonic: emission from porous silicon. *Nature*, **400**, 853–855.

252 Gelloz, B., Sugawara, M., and Koshida, N. (2008) Acoustic wave manipulation by phased operation of two-dimensionally arrayed nanocrystalline silicon ultrasonic emitters. *Jpn. J. Appl. Phys.*, **47**, 3123–3126.

253 Tsubaki, K., Yamanaka, H., Kitada, K., Komoda, T., and Koshida, N. (2005) Three-dimensional image sensing in air by thermally induced ultrasonic emitter based on nanocrystalline porous silicon. *Jpn. J. Appl. Phys., Part 1*, **44**, 4436–4439.

254 Uematsu, A., Kikusui, T., Kihara, T., Harada, T., Kato, M., Nakano, K., Murakami, O., Koshida, N., Takeuchi, Y., and Mori, Y. (2007) Maternal approaches to pup ultrasonic vocalizations produced by a nanocrystalline silicon thermo-acoustic emitter. *Brain Res.*, **1163**, 91–99.

15
Silicon Nanocrystal Flash Memory

Shunri Oda and Shaoyun Huang

15.1
Introduction

Nonvolatile memory is an indispensable component of the electronic system in the twenty-first century. It is widely used in digital consumer durables and mobile industries, such as personal computers, cellular phones, portable music player, portable game player, digital cameras, networks, automotive systems, and portable global positioning systems. A small cell size combined with a fast in-system erasing capability has made flash architecture, which is fully compatible with the standard CMOS (complementary metal-oxide semiconductor) technology, the dominant technology for the nonvolatile memory [1]. Samsung [2] and Toshiba [3] have exemplified memory chips with 32 Gb densities fabricated by 40 nm technology. However, the improvements in evolutionary technology may no longer sustain the further scaling trends in the foreseeable future due to the scaling limits by the device structures on physical fundamental, material, circuit, and system [4, 5]. Break-throughs both in materials and in architectures are required to continue or get beyond the scaling-down paradigm [6–8].

15.1.1
Challenges in Silicon Flash Memory Technology

The scaling of integrated circuits continues to be a critical capability for lower fabrication cost per chip [9, 10]. The scaling in turn results in higher performance with faster cycling speed and lower power consumption. The conventional scaling of gate oxide thickness, source/drain extension (SDE), junction depths, and gate lengths has reduced the MOS gate dimension from 10 μm in the 1970s to the present size below 0.1 μm [11]. As early as in 1965, Dr. Gordon E. Moore observed that the number of transistors being manufactured on a silicon chip was doubling every 18 months, known as "Moore's law" [12]. Since then, the semiconductor industry has been following the Moore's law for the past 40 years. For the last few years, the pace of scaling has been accelerating, and we are approaching some of the fundamental

Silicon Nanocrystals: Fundamentals, Synthesis and Applications. Edited by Lorenzo Pavesi and Rasit Turan
Copyright © 2010 WILEY-VCH Verlag GmbH & Co. KGaA, Weinheim
ISBN: 978-3-527-32160-5

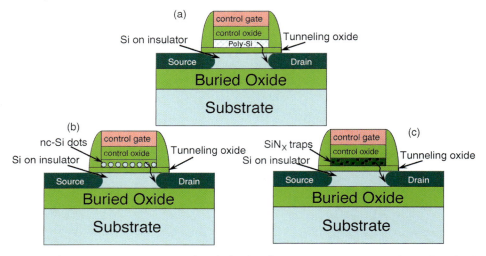

Figure 15.1 (a) A conventional stacked poly-Si floating gate memory, using electrically isolated charge storage nodes as a floating gate to solve the lateral leakage problem, based on (b) semiconductor QDs and (c) traps in a silicon nitride film.

limits. We will not go into details about them, which can be found in many recent reviews [4, 5, 13, 14]. The flash memory scaling is as rapid as CMOS logic device scaling [11].

A flash memory layout consists of two parts: the core memory cells and the peripheral microcontroller circuit for data storage and high-voltage generation, respectively. The core memory cell, shown in Figure 15.1a, is made by two gates. One is the control gate such as in an MOS transistor. The second is a floating gate that is insulated all around by an oxide film. Any carrier placed on the floating gate may get trapped there and thus store the information. The scaling of flash memory demands thinning of gate insulators to keep short channel effects under control and to maximize performance. But a tunneling oxide film of more than 8 nm is indispensable to warrant a 10-year data retention. Because of the thick tunneling oxide, a high voltage of more than 17 V is required for programming and erasing the memory. The endurance is limited within 10^5 cycles due to hot electron effects. If the tunnel oxide were to be scaled below 2 nm, the programming voltage could be reduced down to 4 V and the endurance could be increased infinitely due to the direct tunneling. However, the current leakage through the thin insulators may compromise the retention time, which would be reduced down to several seconds. On the other hand, the peripheral circuit consumes a lot of area and a lot of power as it produces the high voltage. The peripheral circuit scales more slowly than the core memory since the operation voltages have not been scaled down over the past several technology generations. Therefore, the challenge is to reduce the tunneling oxide film, while the retention can be maintained for 10 years.

Moreover, in the nanometer-scale region, the quantum mechanical effect – another crucial factor – may prevent feature sizes of tens of nanometers being achieved [4, 7, 15]. This fundamental physical limit is virtually an impenetrable barrier to future advances of Si technology because it is independent of the characteristics of any particular material, device structure, circuit configuration, or system architecture.

If we have to face these problems sooner or later, the challenge is to find alternative technologies, which may be based on a much thinner tunneling oxide film to allow lower voltage operations and on the quantum mechanical effect to allow new transport principle.

15.1.2
Emerging nc-Si Flash Memory Devices

In a conventional flash memory, as shown in Figure 15.1a, charge leakages through just one localized oxide defect may fail the memory because charge in the stacked poly-Si layer may escape through the leakage path completely. A novel memory structure using electrically isolated charge-storage nodes as a floating gate has been proposed, as shown in Figure 15.2b and c. The leakage path may fail the corresponding memory node only, hence the reliability and the retention time can be improved remarkably. On the other hand, a much thinner tunneling oxide film switches tunnel currents from the conventional Fowler–Nordheim (FN) tunneling to direct tunneling in programming and erasing processes, hence allowing a faster operation speed and a satisfied endurance [16]. The electrically isolated charge storage nodes, currently employed, rely on the localized defects in silicon nitride films [17] or the confinement in semiconductor quantum dots (QDs) [18]. The former has been the first realized nonvolatile memory and suffers from shallow traps. The latter received more attention since the potential well of Si/SiO_2 system is considerably deep and the quantum mechanical effect may change the fundamental performances [19].

Nanocrystalline silicon dots are very small semiconductor structures (on the order of nanometers or tens of nanometers in diameter) surrounded by a material with a wider bandgap (usually SiO_2). The three-dimensional confinement results in individual electronic energy levels when the dimension is smaller than the typical de-Broglie electronic wavelength [20–22]. Relying on the Coulomb blockade effect as a new principle for the controlled transfer of a small number of electrons, data are programmed with one single electron and even one single electron guarantees the memory operation, commonly referred to as single-electron memory (SEM) [23]. Single-electron charging with nc-Si dots has already been demonstrated in attaining discrete threshold shifts at low temperatures [24]. With the development of nanotechnology, the discrete threshold shift has also been realized at room temperature [25].

Addition to the successful demonstrations of memory operations with fast operation speed, the charge retention time of nc-Si memory is found insufficient to warrant 10-year storage [18, 26]. Multiple memory nodes have been exampled to improve charge retention time with consuming negligible programming/erasing time [27, 28].

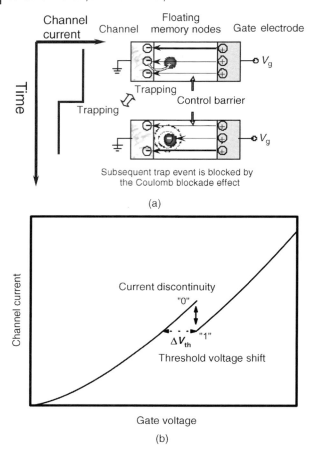

(a)

(b)

Figure 15.2 (a) Screening effects of the floating gate. The Coulomb blockade effect guarantees one single-electron operations in the floating dot. The channel current is suddenly decreased/increased due to electron trapping/ detrapping into/out of the floating gate. (b) Memory operation with a programming shows the current discontinuity to represent a bit by "0" or "1."

15.1.3
Outline

The chapter intends to review the features of nc-Si flash memory, encompassing memory structures, fabrication methodologies, operation mechanisms, and memory characteristics improvements. In Section 15.2, we study the general structure of nc-Si memory that is fully compatible with well-established CMOS technology. We try to emphasize the flash memory with floating gate structures. Fabrication of the nanometer-scale memory cells will be discussed in Section 15.3. Section 15.4, which discusses the memory characteristics, is the core of the chapter. General memory operations, electron charge–storage–discharge with interfacial states, retention characteristics, operation speed, and endurance will be discussed in detail. The nc-Si dot-based flash memory will be compared with other emerging competitors in Section 15.5. In the end, we conclude on nc-Si flash memory and discuss about the future prospects in Section 15.6.

15.2
Structure of nc-Si Flash Memory

Most of the nc-Si memory devices are in the floating gate metal-oxide semiconductor field-effect transistor (MOSFET) style, in which a layer of electrically isolated nc-Si dots replaces the stacked poly-Si layer, as shown in Figure 15.1b. The electrons stored in QDs screen the mobile charge in the channel, thus inducing a shift in the threshold voltage (V_{th}) [29, 30]. Though there is no packing density improvement compared to the conventional flash memory, the scalability due to charge trapped in the discrete nc-Si dots opens a path toward the Terabit era [23]. In the past decade, a variety of nc-Si memory devices have been fabricated and successfully operated, some of them at room temperature [23, 25, 26, 30, 31]. In the following section, we summarize the memory structures and the critical memory size, which allows the memory to work by the Coulomb blockade effect.

15.2.1
nc-Si Memory Structures

Most of the nc-Si memory devices have been classified by the structure with distributed nc-Si dots in a floating gate separated from the inversion surface by a thin tunneling oxide. The memory operations rely on storing charges in the floating gate and thereby screening the mobile charge in the channel, shifting its threshold voltage, hence two stable states can be detected electrically, as shown in Figure 15.2. At a low voltage applied to the control gate negatively, the nc-Si memory behaves like a conventional MOSFET. Upon application of a sufficiently high positive-charging voltage onto the control gate, electrons tunnel from the Si conduction band into the three-dimensional quantum-confined nc-Si dots, resulting in a negative charge accommodation and a positive shift of V_{th}. A large negative voltage pulse is applied to the control gate to drive the stored electrons back into the drain. The high V_{th} is thus switched to the low.

The nc-Si dots are surrounded via small tunnel junctions (SiO$_2$). Adding one electron to one dot consumes a charging energy, $E_c = e^2/2C$. Though the elementary charge $e \cong -1.6 \times 10^{-19}$ C is quite small on human scale, the dot capacitance C is proportional to its size and may become rather small in the nanometer-scale region. The charging energy becomes greater than the thermal kinetic energy at 300 K if the dot size is smaller than 10 nm [24]. Under such condition, the change by a single electron leads to a significant change by the potential $U_e = e/C$ on the dot. As a result, no carrier can flow to the charged nc-Si dot and the flow of current is blocked, the Coulomb blockade effect [32]. This self-limited charging can guarantee single-electron operations.

Lithography at nanometer precision is not required for device fabrication because the Coulomb blockade occurs in an nc-Si dot. However, the external environment, such as the gate and channel, is essential to warrant the device to work at room temperature because they control the electron charging/discharging with the nc-Si dots. In a memory structure, the single-electron charging and discharging consumes a charging energy of $e^2/2C_{tt}$, which must be much larger than a thermal perturbation. Here, C_{tt} is the total capacitance of the memory system. By an order estimate, one may find that the C_{tt} should be at 10^{-18} F order, which corresponds to nanometer-scale CMOS geometry. Therefore, both the gate and the channel engineering are required to realize a satisfactory performance.

Since the charging and discharging with the individual nc-Si dot take place by discrete charge, nc-Si memory operations are very sensitive to background charges, usually the charges in interfacial states. A high dot density, which is at least one order of magnitude larger than the density of interfacial states, is necessary to warrant better robustness [33]. The modern Si CMOS technology provides the interfacial state density as low as 10^{10} cm^{-2}. Therefore, the memory operations can be dominated by the embedded nc-Si dots if their density is greater than 10^{11} cm^{-2}. However, in order to reduce the possible lateral charge leakage and the interaction between the neighboring dots, one needs to separate each dot approximately. For instance, the Coulomb interaction between the two neighboring dots with the diameter of 10 nm should not be greater than the charging energy of a dot. That requires the mean separation of at least 24 nm corresponding to a density of 1.7×10^{11} cm^{-2} for a monolayer. Therefore, most of the nc-Si memory devices use the dot density of $1–2 \times 10^{11}$ cm^{-2} for monolayered nc-Si dots. An increase in nc-Si dot density may correspondingly enlarge the memory window [29, 34].

The position of nc-Si dots on a channel is random in many works. Moreover, the nc-Si dots vary in their size from several nanometers to several tens of nanometers and are separated from the channel surface at various distances. The variation in the screening efficiency of charged nc-Si dots leads to device-to-device fluctuations in threshold voltage shifts (ΔV_{th}). Based on the Hartree model, a self-consistent Poisson–Schrödinger calculation demonstrated that a QD capacitance could be no longer constant due to multielectron interactions within the nc-Si dot [35]. Theoretical simulations showed that the programming time is improved by several orders of magnitude when the shape of nc-Si dot changes from sphere to hemisphere [36]. More common considerations pointed out that the surface effects of nc-Si dots, such as strain and interfacial traps, may also influence or even dominate the electron charging/discharging with the nc-Si dots [37, 38]. Recent work also figured out that the change of crystallographic orientation from [010] to [111] resulted in changes in the programming time by a factor of 10 for a 5 nm hemispherical nc-Si dot [39]. To solve the problems, one needs to make the nc-Si dots' size uniform and position assembled. Molas *et al.* reported that the memory window may be impacted by the percolation paths, which induced an increase of about a factor of 4 with varying gate length from 50 to 200 nm [40]. Therefore, the underneath channel should also be narrow and short enough to restrict the percolation path. The channel length that can be controlled by one electron on the floating nc-Si dot is roughly equal to the Debye screen length, which is approximately 74 nm for the moderately doping channel at room temperature [41]. Since the screening efficiency strongly depends on the separation between each dot and the width of the channel, the dots out of the narrow channel make negligible contributions to device performance. Thus, the electrical response owing to trapping of charges arises from the nc-Si dots located in the active channel area.

The nc-Si memory presents fast programming/erasing with the advantage of the direct tunneling through an ultrathin tunneling oxide film. However, the thin barrier in turn leads to an insufficient storage of nc-Si memory. There is a tough trade-off between the programming/erasing time and the retention time if we rely only on regulating the tunnel barrier thickness. Some novel structures with modifications of both the floating memory nodes and the tunnel barriers were proposed in order to improve the retention time. Ohba *et al.* proposed doubly stacked nc-Si dots as a floating gate. Since the charge leakage between the upper dots and the channel can be suppressed by the energy barrier due to the Coulomb blockade in the lower dot, the charge retention time is improved exponentially by reducing the lower dot size. Meanwhile, the programming/erasing time could be maintained if the tunneling oxide thickness is simultaneously adjusted [27]. A multiple stacked floating gate was demonstrated by Zacharias *et al.* [34]. Oda *et al.* [28] demonstrated that surface-nitrided nc-Si dots could improve the charge retention time by three orders of magnitude, compared to the normal nc-Si. The nc-Si core provides the fast programming and the nitride film enables the long-term retention. Likharev proposed a "crested" energy barrier with the height peak in the middle. The "crested" barrier is much more sensitive to applied voltages than the uniform height [42]. Therefore, tunneling current can be greatly altered by slightly changing the gate voltage. The information processing by electron transportation can be greatly improved [43].

15.2.2
Representative nc-Si Memory Structures

Depending on the number of nc-Si dots involved, the nc-Si memory can be clarified by two main kinds of architectures: charge stored in multiple nc-Si dots and charge stored in one single nc-Si dot. Both of them were operated by the single-electron effect.

15.2.2.1 Storing Charges with Multiple nc-Si Dots
In 1996, Tiwari of IBM experimentally demonstrated a MOSFET structure with nc-Si dots physically separated from a channel, as shown in Figure 15.3 [29]. The thin tunneling oxide film (∼1.4 nm) and the monolayer of distributed nc-Si dots, covering the entire channel region, featured largely in the device performance by the uni-formed tunneling distance. The ΔV_{th} or the flat-band voltage shift (ΔV_{FB}) due to one-electron storage can be estimated by

$$\Delta V_{th(FB)} = \frac{e n_{dot}}{\varepsilon_{ox}}\left(t_{upper} + \frac{1}{2}\frac{\varepsilon_{ox}}{\varepsilon_{Si}}t_{dot}\right), \tag{15.1}$$

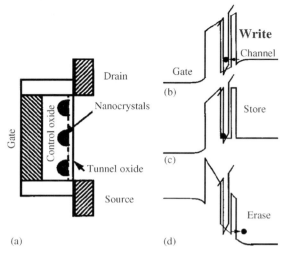

Figure 15.3 (a) A schematic cross section of MOSFET nc-Si memory, and band diagram during (b) injection, (c) storage, and (d) removal of an electron from a silicon nanocrystal. Reprinted with permission from Ref. 26. Copyright (1906) by the American Institute of Physics.

where n_{dot}, ε_{ox}, ε_{Si}, t_{upper}, and t_{dot} are the density of nc-Si dot, the permittivity of oxide and Si, the thickness of upper gate oxide, and the diameter of nc-Si dot, respectively. The equation has been developed for multiple-layered nc-Si dots by Zacharias as follows:

$$\Delta V_{th(FB)} = \frac{en_{dot}}{\varepsilon_{ox}}\left(t_{upper1} + \frac{1}{2}\frac{\varepsilon_{ox}}{\varepsilon_{Si}}t_{dot}N + t_{upper2} + \cdots + t_{upperN}\right), \qquad (15.2)$$

where all nc-Si layers are supposed to have the same areal density and thickness [34]. Though the charge is changed discretely, the $\Delta V_{th(FB)}$ gives a continued change due to the randomly charging/discharging with a huge number of nc-Si dots. Multiple charging with dots can increase the charge retention time. However, both the random distribution of nc-Si dots on a large channel area and the size fluctuations of nc-Si dots may also lead to dispersions in programming/erasing speeds and $\Delta V_{th(FB)}$. A narrow/short channel, which is smaller than Debye length of electron in silicon, was developed to compensate for the randomly deposited nc-Si dots by containing several size-uniformed nc-Si dots only [44].

Since nc-Si dots can be naturally formed without any need for sophisticated lithography technologies and can be compatibly assembled into flash memory, the nc-Si memory using multiple nc-Si dots is suitable for mass production.

15.2.2.2 Storing Charges with Single nc-Si Dot

In the early 1990s, Yano demonstrated that a MOSFET with a poly-Si wire channel could show memory effects at room temperature [45]. The novel feature of the device is the ultrathin poly-Si channel with 100 nm width. A large fluctuation from 1 to 5 nm in thickness takes place in the channel, giving rise to a lot of crystalline silicon grains.

The lateral grain size is approximately 10 nm. As a positive bias is applied to the gate, a narrow percolation channel and an isolated island appear. The observed memory effects are apparently related to the capture of single electron in some certain grains of the channel. It is likely that only one of the grains close to the conduction path is charged and acts as a floating gate. The position of that grain differs from device to device. There is also no control of the grain structure and channel width. The memory suffers from great V_{th} fluctuations.

In 1996, Chou [30] fabricated a more reproducible structure with a single poly-Si dot defined over a narrow silicon-on-insulator (SOI) MOSFET channel. Chou's memory enables a precise control of the actual transistor channel width, the floating gate dimension, and tunnel barrier thickness. Therefore, uniformed memory characteristics can be obtained. A similar work was reported by Nakajima, as shown in Figure 15.4 [25]. The ΔV_{th} due to n electron storage can be calculated by [41]

$$\Delta V_{th} = \frac{ne}{C_{fg} + \left(\frac{C_{\Sigma}}{C_{fc}}\right) C_{cg}}, \tag{15.3}$$

and

$$C_{\Sigma} = C_{fc} + C_{fg}, \tag{15.4}$$

(a)

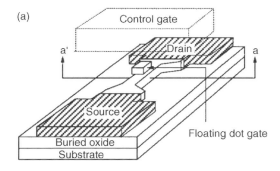

(b)

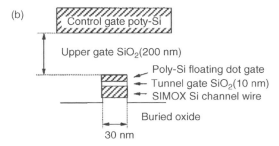

Figure 15.4 (a) Schematic diagram of Si single-electron FET with self-aligned floating gate. (b) Cross-sectional view along a–a' line. Reprinted with permission from Ref. [25]. Copyright (1997) by the American Institute of Physics.

where C_{fc}, C_{gc}, and C_{fg} are the mutual capacitance between the floating dot and the channel, the capacitance between the gate and the channel, and the capacitance between the floating dot and the gate, respectively. The integer number of electrons (n) leads to a discrete ΔV_{th}. However, these devices failed in mass production because sophisticated lithography technologies are required. In addition, the electron states in the poly-Si dot are unknown, making the analysis of memory characteristics complicated.

15.2.3
Summary

The mainstream nc-Si memory is a floating gate MOSFET structure. An overview on the representative structures shows that the nc-Si memory using multiple nc-Si dots to store charges is more suitable for mass production in the near future. One single-dot memory by the single-electron operation is an ultimate candidate. The random distribution of nc-Si dots may be acceptable if we can control the channel length and width precisely. A meaningful nc-Si memory may be based on a MOSFET with few size-uniformed nc-Si dots as a floating gate over a narrow and short channel.

15.3
Fabrication Methodologies

Nanocrystalline silicon dots used in the memory devices have typical sizes not larger than 10 nm. The memory device should limit the channel width and the gate length within 100 nm to work with the nc-Si dots. These will require an absolute nanometer revolution in all aspects of the Si industry, hence spurring the conventional micro-fabrication forward to the nanofabrication. Moreover, the prospective fabrication methods should eliminate complicated processes and build upon the well-established Si CMOS technology.

15.3.1
Fabrication of nc-Si Dots

Recent progress in lithography, colloidal chemistry, and epitaxial growth has made it possible to fabricate nc-Si dots, in which carriers or excitons are confined in all three dimensions [20]. Based on the lithography, however, mass production of nc-Si dots from the bulk remains a big challenge. On the other hand, the homogeneity of the dot size and position is also regarded to be crucial for many prospective applications. Therefore, self-assembling nc-Si dots from atomic levels are emerging to obtain control both in yield fabrication and position distribution due to their inherent formation mechanisms, which potentially assemble the floating gate into the memory without complicated fabrications [46]. In general, a nanoscale nc-Si memory is approached from both ends: top–down (bulk to particle) and bottom–up (atom to particle) methodologies.

15.3.1.1 **Top–Down Methodology**

Photolithography is a main tool for the microtechnology industry and allows large-scale, low-cost manufacturing of microstructures in silicon wafers. The method may potentially provide deep micrometer-scale fabrication [47]. Using an electron beam instead of light, electron beam lithography (EBL) even offers nanometer-scale revolution [48]. Following the lithography step, lift-off, etching, or drilling processes allow to create nanometer-sized dimensional structures. This approach is referred to as "top–down" methodology.

In this approach, increasing resolution is possible to use smaller wavelengths determined by the Rayleigh's criterion. At present, Si industry is working to make the 32 nm lithography technology to have been applied by 2009 using deep ultraviolet (UV) photons through a patterned mask. However, working at shorter wavelengths than the deep UV is very difficult because conventional glasses of lens absorb rather than transmit at the deep micro levels. The EBL is a direct analogy with the photolithography. The electron beam (EB) exposure alters the chemistry of the resist instead of the light exposure. It is capable of very high resolution, almost toward the atomic level. Recent advances demonstrated that scanning probe microscopy (SPM) could offer an ultimate nanostructure by atomic manipulations. Examples include EB exposure by digital scanning electron microscope [49] and nanopattern or manufacture by means of atomic force microscope (AFM) tips using field-induced oxidation techniques [50].

However, unlike the photolithography by which a whole wafer can be exposed simultaneously, the EBL and SPM require EBs to draw each feature one at a time. As a consequence, the processes involve time by many orders of magnitude slower than the photolithography. Meanwhile, they are expensive and complicated. Therefore, the fabrication of nc-Si dots based on the EB technology is not suitable for mass productions and not a realistic manufacturing technology. Nanoimprint method, alternatively, can realize sub-10 nm-scale large-area fabrication [51]. Yet, the true potential of nanoimprint has not been fully studied and it still faces problems of imprint machines and appropriate molds/resist. Another major problem with the top–down lithographic approach is that it requires very clean environment otherwise dust can mask part of the exposed area. The cost of lithography is also increasing at an exponential rate as the feature size decreases and may eventually hinder the progress of microelectronics [52].

15.3.1.2 **Bottom–Up Methodology**

An alternative approach is to use a technique by which the nc-Si dots can be self-assembled. Plasma-enhanced/low-pressure chemical vapor deposition (PECVD/LPCVD), metal-oxide CVD (MOCVD), molecular beam epitaxy (MBE), and hot wire methods produce a lot of individual atom clusters or QDs with nanometer dimensions. In brief, synthesizing atoms or molecules on a nanometer scale and then using more elegant techniques to fit and construct those materials into larger structures with a well-defined geometry is referred to as "bottom–up" methodology.

The PECVD/LPCVD from the source gas of silane or disilane is widely used to fabricate poly-Si films in current Si industries. By controlling the early growth stages, researchers have recently succeeded in the formation of nc-Si dots with the diameter of a few nanometers on SiO_2 or Si_3N_4 substrates [53, 54]. However, LFCVD deposition usually requires a substrate temperature as high as $600\,^\circ C$ to obtain crystallizations and eliminate amorphous issues. The growth of nc-Si dots, 5–6 nm in height and 20–30 nm in diameter, by MBE on a SiC (0001) substrate is possible to use the mismatch in lattice parameters to develop a strain at the interface [55]. Another preparation of nc-Si dots embedded within SiO_2 film is the Si ion implantation [56]. This method, unlike doping, uses the low-energy Si^+ implantation to reduce introduced defects. Size, density, and location of the nc-Si dots were demonstrated to correlate with the dose and acceleration of ion as well as the subsequent annealing conditions, however hard it may be to make precise control [57]. Other preparation methods that have been used for forming nc-Si dots rely on crystallization of amorphous silicon [58] or aerosol nc-Si dots from silane pyrolysis [31] and various sputtering techniques [59]. It is worthwhile to mention that most of sputtering approaches need postannealing a temperature up to $1000\,^\circ C$.

A unique method, as shown in Figure 15.5, was developed to control the dot size by manipulating the gas-phase nucleation and growth to provide high-dense nc-Si dots in very high-frequency (VHF) decomposition of silane plasma [60]. The flow rates of fed gases and the pulse sequences of fed gases can be varied according to the desired dot size and density. The prepared nc-Si dots can be deposited on a substrate at temperature as low as $100\,^\circ C$ by mono- or multilayers. The great advantage of synthesizing particles at a lower temperature benefits the following fabrications with more flexibility. Figure 15.6a shows the deposition system, composed of an ultrahigh vacuum (UHV) chamber equipped with two plasma cells. Nanocrystalline silicon dots are formed in one of the cells by VHF plasma decomposition of silane and coalescence of radicals [61]. The other cell is capable of the on-site nitridation by nitrogen plasma for surface modifications [62]. Si fine particles formed in the plasma cell are extracted through an orifice to the UHV chamber and deposited onto a substrate. Figure 15.6b shows a high-resolution transmission electron microscope (TEM) image of the nc-Si dot with a spherical shape. The lattice image clarifies that the dot is a single-domain crystal showing facets of the (111) plane.

Assembling nc-Si dots requires not only size-uniformed dots but also uniform interdot distances in order to control the interaction between them. The SPM technique allows to remove unwanted nc-Si dots and push the size-uniformed dots with the tip of AFM or scanning tunneling microscopy (STM) under the control of software. An accurate and reproducible positioning of nc-Si dots on a surface was demonstrated [63]. The disadvantage of these techniques, however, is that these are time consuming and have low productivities. Taking advantage of energetic and statistical forces, self-assembled growth offers a large area of the spontaneous formation of highly ordered nanostructures without prepatterned masks. The nanostructures can be selectively formed with inorganic materials on originally spontaneous self-assembled templates. Researchers at IBM have demonstrated that arrays of size-uniformed 20 nm diameter nc-Si dots were formed as a floating gate between a poly-Si gate and a Si substrate by self-assembled polymer molecules [64]. It must be stated that we are still far from understanding how to carry out a practical

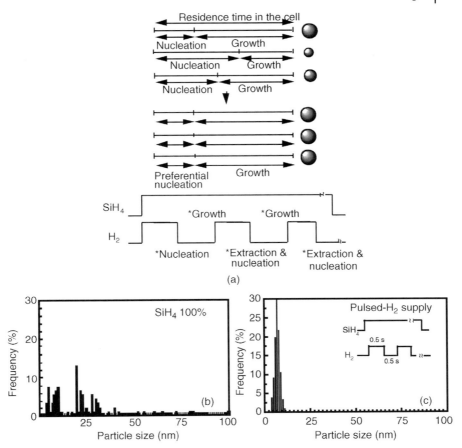

Figure 15.5 (a) Separation of nucleation and crystal growth by modulating H_2 gas pulses may control the formation of size-uniformed nc-Si dots. The size distribution in case of (b) no pulsed H_2 gas supply and (c) pulsed H_2 gas supply in SiH_4 plasma.

engineered self-assembling of device components to create a circuit, which is as complex and reliable as the modern photolithographic circuit.

15.3.2
Memory Cell Fabrications

The next challenge is to assemble the nc-Si dots into the memory structure. In Tiwari's proposal, fabrication of nc-Si memory is almost the same as the current flash memory, as shown in Figure 15.3. The only difference is to replace the deposition of stacked poly-Si with the deposition of nc-Si dots. A large number of nc-Si dots are involved in memory operation. A typical MOSFET nc-Si memory structure using one single nc-Si dot was demonstrated by Chou and Nakajima independently, as shown in Figure 15.4. In the Chou's memory, for instance, a small ($\sim 7 \times 7$ nm in area and 2 nm

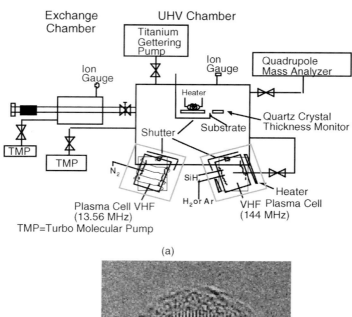

(a)

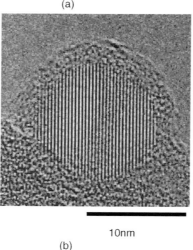

10nm

(b)

Figure 15.6 (a) Schematic of experimental setup used for nc-Si dot deposition and on-site nitridation; (b) a high-resolution TEM image of the prepared nc-Si dot.

thick) poly-Si dot, which was patterned by the EBL, lift-off, and reactive ion etching techniques, was well located over a narrow (~10 nm in width) crystal-Si MOSFET channel. The fabrication of Nakajima's memory started from silicon on insulator by the top–down approach, as shown in Figure 15.7. The advantage of the structure is that the floating poly-Si dot is naturally assembled into the gate by a self-align etching technique. The controllability and reproducibility of nc-Si dot dimensions are satisfied due to the exact location of floating gate.

A more flexible memory essentially demands a short/narrow channel in which only several size-uniformed nc-Si dots are located to take advantage of the two

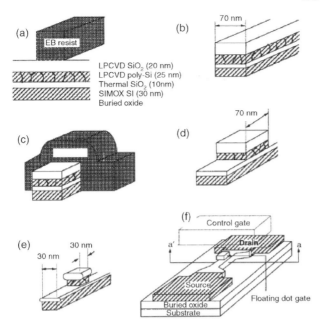

LPCVD SiO$_2$ (20 nm)
LPCVD poly-Si (25 nm)
Thermal SiO$_2$ (10nm)
SIMOX SI (30 nm)
Buried oxide

Figure 15.7 (a)–(e) Schematic drawing of fabrication process for Si single-electron FET memory with self-aligned floating gate. (f) Schematic diagram of Si single-electron FET memory with self-aligned floating gate. Reprinted with permission from Ref. [25]. Copyright (1997) by the American Institute of Physics.

kinds of aforementioned memories [18]. The fabrication began with the dry-oxidation thinning of separation-by implanted-oxygen (SIMOX) Si layer (p-type) to 30 nm and partial HF chemical etching of the top oxide layer. The pattern, as shown in Figure 15.8a, with the short and narrow Si channel (40 nm wide and 50 nm long) was written by the EBL. Next, electron cyclotron resonance reactive ion etching (ECR-RIE) was used to transfer the resist pattern to the SOI layer. A 2 nm SiO$_2$ film as a tunneling barrier was prepared by chemical oxidation with H$_2$SO$_4$/H$_2$O$_2$ (30:70). Then, the nc-Si dots with a diameter of 8 ± 1 nm were deposited by the VHF decomposition of silane with a density of $1–2 \times 10^{11}$ cm^{-2} [60]. This produced four to five nc-Si dots in the active area, as shown in Figure 15.8b. A 50 nm gate oxide film was deposited by TEOS PECVD at the substrate temperature of 300 °C, as shown in Figure 15.8c. The contact pad area was heavily doped by the P$^+$ ion-implantation technique and the self-aligned channel was simultaneously obtained. Subsequently, the sample was annealed in N$_2$ ambient at 1100 °C for 1 h to improve gate oxide quality and activate the dopants. Finally, the electrodes were formed by the aluminum lift-off technique, as shown in Figure 15.8d. The memory takes both top–down and bottom–up techniques to approach a reproducible nc-Si memory device.

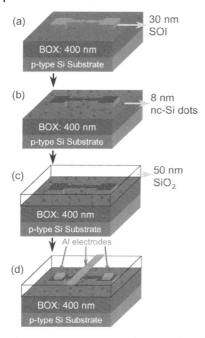

Figure 15.8 (a) A patterned narrow/short channel made on an SOI thin film, (b) deposition of nc-Si dots, (c) deposition of control gate SiO₂, and (d) fabrication of gate and source/drain electrodes.

15.3.3
Summary

As efforts in various areas of nanometer-scale science and technology continue to grow, it is certain that many new manufacturing techniques will be discovered. At present, a synergy of the top–down with the bottom–up technology may be a better way of approaching practicable fabrication. The approach combines the best aspects of top–down technology, such as deep-lithography, with the self-assembling of bottom–up technology and allows fabricating highly integrated two- and three-dimensional nc-Si memory devices.

15.4
Characteristics of nc-Si Flash Memory

Advantages of nc-Si flash memory, in fact, derive from the electron charging/discharging with the distributed nc-Si dots. For the sake of simplicity, we start to investigate nc-Si QDs using a SiO_2/nc-Si/SiO_2 capacitor memory structure, as shown in Figure 15.9. The structure is a primary unit of a MOSFET nc-Si memory as a floating gate. In this section, the characteristics of nc-Si memory are described in the

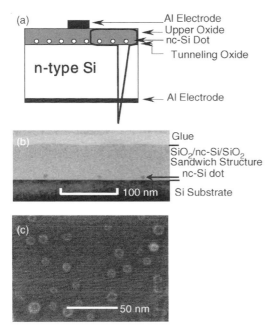

Figure 15.9 (a) Schematics of nc-Si dot capacitor memory; (b) the TEM image of sandwich structure; (c) the plane-view scanning electron microscopy image of nc-Si dot.

light of two kinds of structures: an nc-Si dot-embedded capacitor memory and a MOSFET memory using the former as a floating gate.

15.4.1
Memory Operations

The typical capacitance–voltage $(C–V)$ and conductance–voltage $(G–V)$ characteristics of capacitor memory, as shown in Figure 15.10, were obtained by sweeping the gate voltage (V_g) between inversion and accumulation regions at room temperature. A clockwise hysteresis in the depletion region was observed in both $C–V$ and $G–V$ characteristics. One also notices the presence of a conductance peak in either the forward or the reverse V_g sweep direction. The peak position is close to the flat-band voltage. Both the magnitude of hysteresis in $C–V$ and the shift of peak position in $G–V$ are 0.12 V. For a comparison, neither hysteresis in $C–V$ nor conductance peak in $G–V$ was observed in the sample without nc-Si dot deposition. Therefore, the hysteresis and the peak shift are attributed to electron trapping in the sandwiched nc-Si dots or at the interface with them but not to the defects in the oxide matrix or at the Si substrate/ tunneling oxide interface. Further measurements also indicated that the clockwise hysteresis or the conductance peak shift is independent of the scan direction and speed (5–500 mV/s). It is also worthwhile to figure out that hole trapping in the nc-Si dots is absent due to the higher energy barrier for holes at the valence band of Si/SiO_2

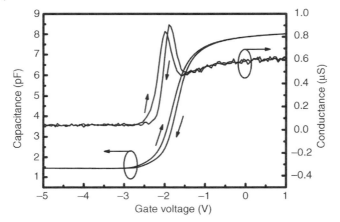

Figure 15.10 Typical *C–V* and *G–V* characteristics were obtained by sweeping V_g back and forth between −5 V and 1 V. The peak position in the *G–V* characteristics is around the flat-band voltage.

system, but possible at a relative large erasing voltage. Both electron and hole trapping in the nc-Si dots were identified by Nassiopoulou [65]. The hole trapping in the nc-Si dots may lead to negative shifts in *C–V* or current–voltage (*I–V*) curves like a positive shift by the electron accommodation. For simplicity, we restrict ourselves only to the electron trapping in the following discussions.

At an approximately large negative V_g, no electron resides in the nc-Si dots. At an approximately high positive V_g, a number of electrons may be stored in the nc-Si dots by the direct tunneling through the ultrathin oxide film. The storage results in a shift both in the capacitance and in the conductance. It should be emphasized that the occurrence of the conductance peak is associated with the AC loss due to the capture and emission of electrons by the nc-Si dots but not due to the Si substrate/tunneling oxide interfacial states (see later). Both *C–V* and *G–V* characteristics are neither dispersed nor distorted with different scan speeds, indicating a low interfacial state density. However, it is also worth to point out that the defect or interfacial states essentially depend on actual fabrication processes [37].

One can estimate the theoretical magnitude of ΔV_{FB}^T with the Eq. (15.1). The ΔV_{FB}^T is calculated to be 0.282 V for one electron per dot. The experimentally measured $\Delta V_{FB} = 0.12$ V suggests that less than one-half of the nc-Si dots are charged. By electrically detecting the capacitance differences at a reading voltage or detecting the ΔV_{FB} at a certain capacitance in the hysteresis region, one can obtain the stored information.

Using the SiO_2/nc-Si/SiO_2 capacitor structure as a floating gate, a plausible background charge-free MOSFET single/few electron(s) memory, as shown in Figure 15.11, was fabricated. Figure 15.12 shows the V_g dependences of channel current at 77 K. A clockwise hysteresis with a discrete V_g shift was demonstrated. By sensing the current at a certain V_g, for instance, one may distinguish the stored information by high current ("0") or low current ("1"). As the V_g was swept forward and reverse between −5 and 1 V (referred to as *Scan A*), a clear current drop in the *I–V*

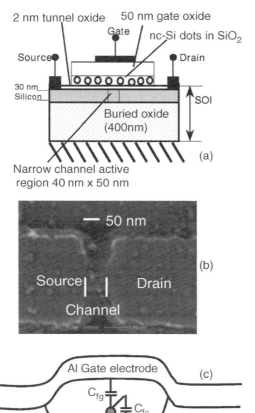

Figure 15.11   (a) Schematics of narrow/short-channel memory with randomly deposited nc-Si dots as a floating gate. (b) Scanning electron microscopy image of the short and narrow channel after the deposition of nc-Si dots; only several of them were located in the channel. (c) The equivalent circuit of memory with three mutual capacitances.

characteristics around -0.08 V was observed in the forward scan. The drop is a direct evidence of electron trapping in the nc-Si dot [18, 23]. When the programming voltage was raised from 1 to 5 V and the measurement was repeated (referred to as *Scan B*), the drop appeared at the same V_g in the corresponding forward scan.

Referring to Figure 15.12, it is worth noting that there is the appearance of a current jump at around -0.64 V in the reverse scan from 5 to -5 V. The jump is because of electron escaping from the nc-Si dot. The absence of such a feature in the Scan A apparently suggests that the stored electron is not erased from the nc-Si dot at such V_g. It is interesting to point out that (1) the reverse $I-V$ curve of Scan A coincides with the forward sweep after one-electron trapping, (2) the reverse $I-V$ curve of Scan B

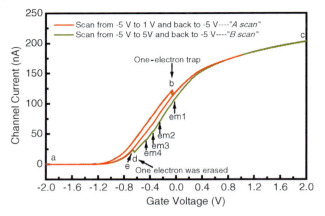

Figure 15.12 V_g dependences of channel current at 77 K. The symbol b indicates a trapping event by one electron; d indicates a detrapping event by one electron. The symbol *em1–em4* indicates some electron detrapping events by the nc-Si dots outside of the channel.

coincides with the reverse of Scan A after one-electron erasing, and (3) the ΔV_{th} corresponding to the programming voltage of 5 V is found to be nearly twice that at 1 V. These results imply that the sequential memory operations are from the same nc-Si dot located in the active channel region. The ΔV_{th}, caused by storing one electron in the floating gate, turns out to be approximately 92 mV for Scan A. Therefore, $\Delta V_{th} = 189$ mV for Scan B indicates two-electron storage in the nc-Si dot at the programming voltage of 5 V. Moreover, the reproducible small current steps are likely due to discharging of the nearest nc-Si dots out of the active region. Clearly, the contribution from the randomly distributed nc-Si dots is negligible with regard to the memory operations.

The nc-Si memory demonstrated a salient feature in ΔV_{th}. Though V_g varied continuously, V_{th} appeared to be a discrete shift with a certain increment. These phenomena derive from the self-limited charging of nc-Si dot determined by the Coulomb blockade effect. Several groups have also explored these self-limited charging phenomena in multiple nc-Si dots over a large area channel (Figure 15.3) or in a single poly-Si dot over a narrow channel (Figure 15.4). For the latter, a clear staircase of the threshold voltage shift was presented at room temperature [41]. However, such discrete ΔV_{th} becomes clear only at low temperature for the former [66]. The low temperature requirement could be due to a large percolation current flowing through the wide channel at high temperature. The discrete ΔV_{th} is potentially used in multiple bit storages [24].

15.4.2
Interfacial States

Although a lot of work has been devoted to fabricate nc-Si memory structures and to obtain reproducible memory characteristics [25, 29, 30], the interfacial states with

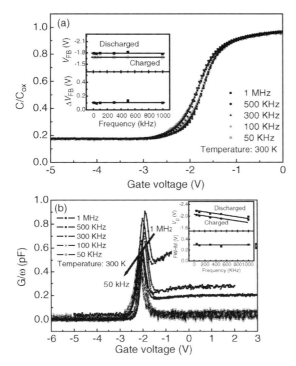

Figure 15.13 (a) C–V characteristics of nc-Si capacitor memory measured at 300 K for various frequencies; the inset shows the flat-band voltages and their shifts as a function of frequency. Both the flat-band voltage and the shift are independent of the frequency. (b) The G–V characteristics measured at identical conditions of (a); the inset shows that the conductance peak positions and the peak FWHMs are related to the frequency. However, the ΔV_p remains unchanged with the frequency. The same FWHM also implies that the electron charging and discharging have same origin.

nc-Si dots are still unclear. The interfacial traps, capturing and emitting electrons or holes with time constants in the range of picoseconds to hours, directly affect a number of device properties, such as leakage current, retention time, and noise voltage. In order to investigate the electron charging and discharging behaviors with the interfacial states in the SiO_2/nc-Si/SiO_2 sandwich structure, we use the temperature and frequency dependences of C–V and G–V measurements, which are sensitive to the interfacial states.

Frequency dependences of C–V and G–V characteristics were investigated at various temperatures, as shown in Figures 15.13 and 15.14. At room temperature, same clockwise C–V hysteresis (Figure 15.13a) and constant full width at half maximum (FWHM) of the conductance peak (Figure 15.13b) were found in the frequency range of 50 kHz–1 MHz. These suggest that the hysteresis and the conductance peak have the same origin [32]. Moreover, neither dispersion nor stretch-out was observed in the C–V characteristics at each frequency at 30 K. These results indicate that the C–V hysteresis and G–V peak are not due to interfacial traps,

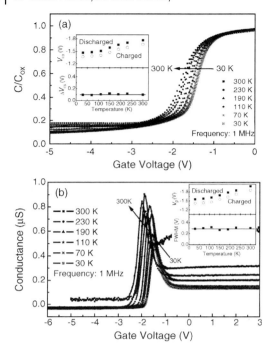

Figure 15.14 (a) C–V characteristics of nc-Si capacitor memory measured at 1 MHz for various temperatures; the inset shows the flat-band voltages and ΔV_{FB} versus the temperature. No change of ΔV_{FB} happens, while V_{FB} weakly relates to the temperature. (b) The G–V measurements made at identical conditions of (a); the inset shows the conductance peak position shift and the peak FWHM remain the same with changing temperature.

as they generally give rise to the time- or frequency-dependences of C–V and G–V characteristics [67]. However, small but noticeable shifts in the conductance peaks with frequency and temperature were observed in the G–V characteristics (Figures 15.13b and 15.14b). These may be explained by the fact that the conductance is directly related to the energy loss provided by the AC signal source during the capture and emission of carriers by the nc-Si dots and therefore is a more sensitive probe of the tunneling process than the capacitance. From the temperature-related measurements at a constant frequency (1 MHz), same C–V clockwise hysteresis and same conductance peak FWHMs were found in the temperature ranging from 30 to 300 K, as shown in Figure 15.14. A right-hand shift with regard to the V_g occurred in the C–V characteristics as temperature decreased. A corresponding nonnegligible right-hand shift in the peak position was also found in the G–V characteristics. These results will be compared with that of frequency related measurements in detail.

As noted in Figure 15.13a, the magnitude of hysteresis and the shift of V_{th} remained almost unchanged with changing frequency, whereas the conductance peaks shifted monotonically to the left-hand side with decreasing frequency

(Figure 15.13b). This may be attributed to the interfacial states. At a given frequency, all traps with time constants shorter than the reciprocal of frequency respond to the measuring signal and the flat-band voltage may be reached at a certain number. At a lower frequency, however, slower traps can respond as well. As a consequence, the previously trapped charges may be discharged. The flat-band voltage may reach at a lower number. The discharge of frozen charges in the interfacial traps gives rise to a stretch-out in the $C–V$ curves and a left-hand shift in the conductance peak in Figure 15.14a and b, respectively. However, the frozen charge in the interfacial states does not contribute to the total capacitance or to the V_g screening.

Another way to find the nature of interfacial states is to estimate the thermal activation energy [67]. Very weak temperature dependences of the conductance have been noted at 1 MHz and 50 kHz. The activation energies turn out to be smaller than 5 meV, which indicate that deep defects have a negligible contribution in charging and discharging with the nc-Si dots. It is likely that shallow interfacial states close to the edge of conductance band lead to the $C–V$ stretch-out and $G–V$ shift [68]. Therefore, it can be concluded that the memory performance depends on the nc-Si dots only.

15.4.3
Retention Characteristics with Electron Charge, Storage, and Discharge

A major requirement for the nc-Si flash memory is the data retention for not less than 10 years. Immediately after electrons are trapped, the stored electrons have a finite probability to tunnel back to the channel at the ground or a reading voltage. The random discharge causes a gradual shift of channel current or $SiO_2/nc-Si/SiO_2$ capacitance. The shift directly reflects the barrier height/width, internal electric field, defect, or interfacial state. The investigation into the time dependences of charging/discharging with the nc-Si dots may offer a better understanding of the retention characteristics.

In the $SiO_2/nc-Si/SiO_2$ capacitor memory, the retention time was found to exceed 5 h at room temperature, as shown in Figure 15.15. The time dependence of capacitance was measured at an initial flat-band voltage. A nominal decrease in the charge loss rate was found at low temperature. A logarithmic law can be found at the initial time (up to 10^4 s). This suggests that the tunneling probability varies with time, as a constant probability would give an exponential dependence [69]. That is to say that the electric field across the tunneling barrier is a function of time since the tunneling probability depends on the electric field [70]. The time dependence of electric field implies a varied tunneling barrier with time. The energy band diagram of the nc-Si memory is shown in Figure 15.16. Immediately after charging, the initial electric field across the tunneling oxide is zero, as shown by the solid line, at the flat-band voltage due to the stored charge. Then, some of the stored electrons may tunnel back to the substrate across the ultrathin tunnel barrier at the initial flat-band voltage. Accordingly, the increased downward band bending at the silicon channel surface happens, as shown by the dotted line. An electric field from the nc-Si dot layer to the silicon substrate is induced ($10^4–10^5$ V/cm around the nc-Si dot estimated by one

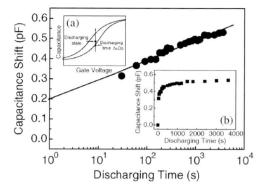

Figure 15.15 Time dependences of capacitance shift from the initial flat-band capacitance immediately after the charging of embedded nc-Si dots; a logarithmic discharging behavior is found. The gate voltage was kept at the initial flat-band voltage (-1.8 V) for the read. The insets show (a) the discharging history and (b) the capacitance shift as a function of the discharging time on linear scale.

electron releasing from the dot). The field may hinder discharging of the rest of the electrons. This "built-in" electric field changes with the stored charge. As a consequence, the tunneling probability also changes with the charge loss. The "built-in" field, created and controlled by the charge loss of nc-Si dots, could significantly improve the retention time and is speculated to be the reason for the long retention time and the logarithmic discharging.

It must be stated that the multiple nc-Si dots or the huge number of nc-Si dots contribute a gradual change in channel current or diode capacitance due to randomly tunneling events, though the single-electron tunneling takes place in each of the nc-Si

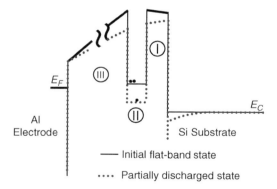

Figure 15.16 Energy band diagram of nc-Si dot memory (not drawn to scale); the flat-band state with electrons trapped in the nc-Si dots is denoted by the solid line. The loss of charge from the dots gives rise to the partially discharged state (dotted line). A "build-in" electric field was induced. The field direction is from the nc-dot to the substrate. The I, II, and III indicate tunneling oxide, nc-Si dot, and control gate oxide, respectively. The symbol • used inside II denotes trapped charges.

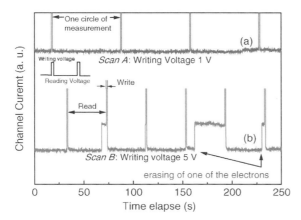

Figure 15.17 Time dependences of channel current immediately after the charging of nc-Si dot in the MOSFET single-electron memory; the read voltage was −0.46 V. The insert is the programming/reading voltage pulses. The measurement period is (a) 70 s for a programming voltage of 5 V and (b) 40 s for a programming voltage of 1 V. Apparently, one discrete current step in (b) shows one electron escaping from the nc-Si dot.

dots. With limiting the number of nc-Si dots, one may find that the gradual change evolves to the discrete. Ultimately, the discrete change of channel current can be observed in a one single nc-Si dot-based memory.

The time dependences of channel current in the MOSFET nc-Si floating gate memory (Figure 15.11) was measured at 77 K. Figure 15.17a and b depicts the discharging history of the memory at the V_g stress of 1 and 5 V, respectively. Clearly, no current jump, indicating the electron emission from the nc-Si dot, is found in each typical period (70 s) in Figure 15.17a. Even when the measurement period was raised to 200 s, no current jump was observed. However, one discrete current step was observed immediately after approximately 40 s, as shown in Figure 15.17b. Considering the fact that each current jump corresponds to the emission of one electron, one can conclude that one of the two electrons was erased from the nc-Si dot. Therefore, the retention time of one electron is more than 40 s in the two-electron storage. However, the retention time of one-electron storage is much longer, not less than 200 s. Hinds *et al.* [18] reported that charging with an nc-Si dot showed a polarization effect under a positive V_g by which the electron would be attracted to the top of the dot reducing the transparency. It is likely that the polarization effect and the suppressed thermal perturbation at 77 K prolong the lifetime of the electron stored in the nc-Si dot. However, when the second electron was stored in the dot, the equilibrium was destabilized due to the strong superposition of electron wavefunctions [35]. This resulted in the loss of one of the electrons. Therefore, one-electron emitting is exhibited in Scan B but not in Scan A.

Figure 15.17 is just a representative example of the memory retention time. There is a wide distribution of retention time for other pulses. To have an accurate measure of the retention time, one needs to use at least several hundred cycles. After 400 programming/erasing cycles, a large number of these retention times were collected and analyzed by statistical method [18]. The distribution of retention time can be fitted by the Poisson's distribution function, which indicates a binomial process (storage/emission) with a large number of events and a small occurrence possibility.

The retention time of nc-Si memory with a 2 nm tunneling oxide film and a huge number of nc-Si dots is found not more than 1 day. The storage of one single nc-Si dot provides several hundred seconds of retention time. Both of them do not match the nonvolatile requirement if the thickness of tunneling oxide is not increased. However, the thick tunneling oxide barrier, in turn, slows down the operation speed.

15.4.4
Characteristics Improvements of the nc-Si Flash Memory

In addition to the successful demonstrations of single-electron memory operations and the obvious advantages of nc-Si memory devices, the charge retention time is still not long enough for the nonvolatile memory ($\sim$10 years) [29, 30]. Since the ultrathin tunneling oxide film was involved for high-speed operations and the charge storage in a floating gate essentially raised its potential, the stored charge was apt to release from the memory nodes via direct tunneling. Therefore, nc-Si memory devices always encounter a trade-off between the operation speed and the retention time.

Ohba demonstrated [27] a self-aligned doubly stacked nc-Si dot memory device, as shown in Figure 15.18a. Since the charge leakage between the upper dots and the channel can be suppressed by the energy barrier due to the quantum confinement and Coulomb blockade in the lower dot, the charge retention time is improved exponentially by reducing lower dot size. Figure 15.18b clearly shows the difference of memory window between the doubly stacked and the single-layered nc-Si dot memory. The increase in retention time was obtained by two orders of magnitude. Zacharias extended the floating gate to multiple stacked nc-Si dots, as shown in Figure 15.19a and b. Charge storage in each nc-Si layer was clearly identified by the discrete ΔV_{FB}, as shown in Figure 15.19c. The multiple-layered nc-Si dots allow the multibit/transistor technology to be realized. Fully charging up with the multiple nc-Si layers also prolongs the retention time significantly, but programming/erasing the multiple nc-Si layers remains a challenge.

Alternatively, to improve charge retention time with consuming programming/erasing time, one may take advantages of trap-assisted charging/discharging that needs the thermal excitation process [71]. Horváth fabricated a Si_3N_4/nc-Si/Si_3N_4 capacitor memory, which gave rise to a huge memory window with charge storage in the upper Si_3N_4 film. However, the tunneling Si_3N_4 film was found providing leakage paths with the nc-Si dots to degrade the memory characteristics [72]. A more reliable structure is to use nc-Si dots capped with silicon nitride films as a floating gate and to keep a SiO_2 film as the tunneling barrier. The interfacial traps are intentionally introduced in terms of the ultrathin nitrides by N_2 plasma nitridations on the nc-Si dots, the so-called surface-nitrided nc-Si (SN-nc-Si) dots. The nitride system is commonly known to have a higher density of localized defect sites, but discharge properties are roughly similar to the Si/SiO_2 system [73]. Therefore, a nominal

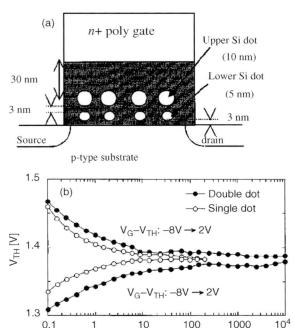

Figure 15.18 (a) Schematic diagram of Si self-aligned doubly stacked dot memory. The floating gates are self-aligned doubly stacked dots. Lower dot is smaller than upper one. (b) Retention characteristics of the self-aligned double dot memory and the single dot one. Reprinted with permission from Ref. [27]. Copyright (2002) by the IEEE.

number of trap sites on the dots may play their role in the charge storage but may not considerably degrade the device reliability or performance [74]. Figure 15.20 shows the schematics of the memory structure with the SN-nc-Si dots. Three samples are prepared, denoted by A (nc-Si dot deposition followed by nitridation), B (nc-Si dot deposition without nitridation), and C (nitridation without nc-Si dot deposition). Since the bottom of nc-Si dots is covered by the tunneling oxide layer, the introduced nitride traps are distributed only on the top or side of nc-Si dots. Based on the facts of the ultrathin nature of silicon nitrides and the defect density of approximately 10^{11}–10^{12} cm^{-2} [75], one may estimate that roughly one or two defects exist on one SN-nc-Si dot.

High frequency C–V measurements were performed at room temperature to investigate the memory operations, as shown in Figure 15.21. A clear hysteresis in the C–V curves is found with the ΔV_{FB} of 0.196 V in Sample B. No flat-band voltage shift is observed at the identical measurement in Sample C. The absence of shift may result from the negligible defects provided by the direct nitridation on the SiO$_2$ surface [76]. The hydrogen passivation effect on fabrication processes may be another reason for the absence of hysteresis [37, 77]. A significant increase in $\Delta V_{FB} = 0.372$ V can be found in Sample A. The ΔV_{FB} measured in Sample A is much greater than that

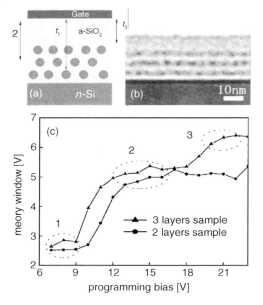

Figure 15.19 (a) Schematic image of the multilayer MOS structure; (b) TEM image of the realized structure with three layers of Si nanocrystals embedded in SiO$_2$; (c) the dependence of the memory window on programming voltage. There are three apparent stages observed for the three-layer sample and two stages for the two-layer sample. Each stage is spaced by approximately 7 V. Reprinted with permission from Ref. [34]. Copyright (2005) by the American Institute of Physics.

in Sample B and the calculation ($\Delta V_{FB}^T = 0.241$ V) based on the hypothesis that the trap sites is provided only by the nc-Si dots. Taking the structure difference between Samples A and B into account, one may attribute the extra ΔV_{FB} to the trap sites from the silicon nitride defects. An order estimate shows that the additional trap-site density is also consistent with the defect density of silicon nitrides.

To understand the contributions to charging and discharging with the nc-Si dots or with the traps of nitrides, one may investigate the time dependences of stored charge at room temperature. Immediately after the memory nodes were filled with electrons at a programming voltage of $+3$ V, the capacitance was measured at the initial flat-band voltages of -2.4 and -1.6 V for Samples A and B, respectively, as shown in Figure 15.22. The measured capacitance is converted to be the ratio of the stored charge over the initially stored charge. The Curves I and II correspond to discharging of Samples A and B, respectively. In general, the retention time is defined as the elapsed time that allows the memory window to become zero [27, 37]. Considering that the electron injection rate is usually smaller than the loss rate, one may suppose that the memory window may be zero, when the remaining electrons are 40%. Therefore, the retention time of Sample B turns out to be 16 h by extrapolating the Curve II to where 40% electrons remain. Accordingly, the retention time of Sample A is estimated to be 4580 h by three orders of magnitude longer than that of Sample B.

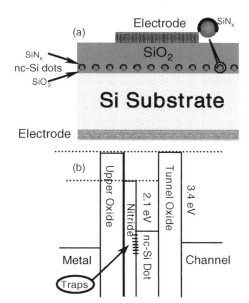

Figure 15.20 (a) Schematics of SN-nc-Si dot capacitor memory; the 7 nm nc-Si core was capped with 1 nm silicon nitride film. (b) Schematics of band diagram; the band bendings are not considered.

For a comparison, an increase by two orders of magnitude was observed in a doubly stacked nc-Si dot memory [27].

The steeper slope of Sample B indicates a larger charge loss rate. The difference in charge loss rates between Samples A and B could derive from the difference of charge

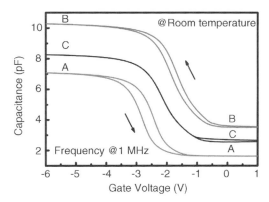

Figure 15.21 *C–V* characteristics of the samples with different modifications in the floating gates measured at the identical scan range and speed. The vertical shift of *C–V* curves is due to the different gate effective thickness. Samples are denoted by A (nc-Si dot deposition with nitridation), B (nc-Si dot deposition without nitridation), and C (nitridation without nc-Si dot deposition).

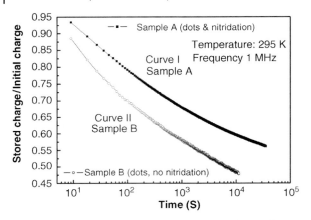

Figure 15.22 Time dependences of stored charge after the charging of floating gate. The zero point is the initial flat-band state immediately after the charging. The read voltage was kept at the initial flat-band voltage of −1.6 V for Sample A (solid square) and of −2.4 V for Sample B (open circle).

storage sites: nc-Si dots with traps for the former and nc-Si dots for the latter. It is notable that the Curve I can be fitted by two straight lines. The charge loss rate of the first part (left-hand side) is larger than that of the second part (right-hand side). The two charge loss rates imply different discharging mechanisms [78]. Empirically, the first discharge part of Sample A may result from the electrons stored and delocalized entirely over the nc-Si dots. The shorter distance apart from the channel and delocalized states may lead to a higher charge-loss rate, as what occurred in Sample B. The charge loss rate of Sample B is slightly greater than that of the first part (left-hand side) of Sample A. The discrepancy may result from less delocalized electrons in the SN-nc-Si dots. The second discharge part may result from the localized charges in the nitride traps, which is far away from the channel and localized.

The charge storage in the SN-nc-Si dots can be directly identified by means of the displacement current analysis. Since the displacement current is responsive to confined charges [79], the analysis allows to explore charge or potential variations in the embedded memory nodes. A group of typical *I–V* characteristics of Sample A at room temperature is shown in Figure 15.23. The V_g scan was performed at 10, 20, and 50 mV steps with the hold time of 10 ms and the delay time of 10 ms. Voltage step per such unit time is defined as the scan rate. A single current peak is observed in each forward and reverse scans. The inset demonstrates that the measured currents are consistent with each other after the normalization. Therefore, the observed currents are the displacement current rather than the leakage current. The former is not a true current across the space but due to charge displacements and is expressed as follows:

$$I_D = C_{MOS} \times \frac{dV_g}{dt} = C_{ox} \times \frac{dV_{ox}}{dt} \qquad (15.5)$$

where CMOS is the total capacitance of the dielectric layer and the surface space charge layer of the channel, V_{ox} the potential dropped on the floating gate, and *t* the scan time. If there is no change in charge taking place in the sandwich space, the

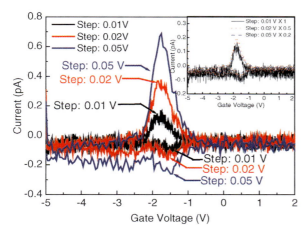

Figure 15.23 V_g scan-rate dependences of currents at room temperature. Inset shows the normalization of the scan rate in unit time.

displacement current must be consistent at an identical scan rate until a change in the total capacitance happens. Low-frequency MOS capacitances may manifest a valley at I–V spectroscopy around the depletion region due to the capacitance change in the space charge layer (C_s). On the contrary, once an electron tunnels directly into/out of an nc-Si dot, potential of the dot (V_{dot}) may rise/reduce some determined by the Coulomb blockade effect [23]. As a consequence, dV_{dot}/dt is no longer a constant at that moment and causes an additional peak in the corresponding displacement current spectroscopy. The height of current peak represents how fast the potential of memory nodes is varied. The smaller the variation rate, the lower the current height is. Therefore, the observed displacement current peaks in charging/discharging should be considered as a combination, comprising the current valley and peak from the total capacitance and the change in potential in the memory nodes.

In order to study the different contributions from the capacitance and from the memory nodes, one may compare the I–V characteristics of Samples A, B, and C, as shown in Figure 15.24. It is worthwhile to point out that charge trap centers (usually defects) could be introduced into the tunneling oxide area uncovered by the nc-Si dots after nitridation. Capturing and emitting electrons by defects may offer memory operations as well. However, Figure 15.24b indicates neither direction-dependent current nor current peaks in Sample C. Therefore, it can be concluded that (1) the change in total capacitance makes trivial contributions on the current and (2) the nitrided SiO₂ films are unable to perform as memory nodes. The former is probably derived from carrier generation and recombination at an equal rate in the space charge layer due to a very low frequency and a moderate doping level of the substrate. These give rise to a negligible shallow capacitance valley, which is too small to measure as happened in Sample C. The latter figures out that the nitrided oxide films in Sample A cannot impact the overall memory operation. Therefore, it is reasonable

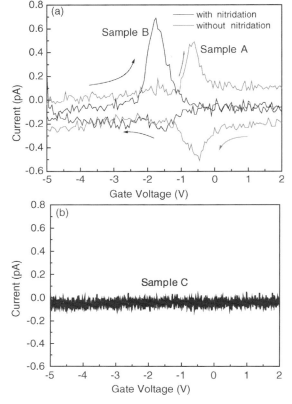

Figure 15.24 Room-temperature displacement current at a V_g step of 50 mV with hold time of 10 ms and delay time of 10 ms. (a) The comparison of Samples A and B; (b) The comparison of Sample C.

to attribute the observed current peaks to charging/discharging with the nc-Si dots and the SN-nc-Si dots.

As shown in Figure 15.24a, a pair of current peaks with same height and same FWHM is observed in each forward and reverse voltage scans in Sample B. The FWHM is commonly derived from the size variation of nc-Si dots and the thermal activation effect. The V_g at the center of current peak indicates the alignment between the Fermi level in the silicon substrate and an available energy level in the nc-Si dots. The peak positions located at −0.689 V for the forward scan and −0.460 V for the reverse scan show a shift of 0.229 V due to the stored charge. In Sample A, however, addition to a pair of current peaks, the height of the discharging peak in the reverse scan is smaller than that of the charging peak in the forward scan. Furthermore, the peak position shift in Sample A is 0.365 V, greater than that of Sample B. The shifts are analogous to memory windows in the C–V measurements. The enlarged peak position shift in Sample A, indicating more electrons stored in the SN-nc-Si dots, could be caused by a higher programming voltage due to a smaller flat-band voltage than that of Sample B.

The smaller discharging current height implies that the charge loss rate is smaller than the injection rate in the SN-nc-Si dots and the stored charge does not entirely release from them at that certain V_g. One may suppose that some of the charges remain and localize in the SN-nc-Si dots in Sample A, while all the stored charge is delocalize in the entire nc-Si dots in Sample B. The electron charging and discharging are a combined process by the dual memory nodes: nc-Si dot and silicon nitride film. Immediately after the programming, the stored electrons are polarized to the top of SN-nc-Si dot. Subsequently, some of them fall into defect states over there. The electrons in the defect states could be localized. The electron exchange between the defect trap and the nc-Si core slows down the charge loss rate and results in a reduced current peak height in discharging. Therefore, asymmetric pair current peaks of Sample A are the direct evidence that the stored charge are not only in the nc-Si core but also in the silicon nitride film. The integration under charging/discharging current peaks figures out that two-thirds of the stored charge fall into the defect states at room temperature. Erasing the trapped charges is indicated in the tilt current region after the discharging peak.

To evaluate different charging/discharging with the nc-Si core and the silicon nitride defects, one may investigate the temperature dependences of I–V characteristics. As shown in Figure 15.25, the heights of both charging and discharging current peaks are found proportionally increased with cooling samples. Correspondingly, the FWHMs become narrow as well. This can be attributed to direct tunneling into/out of the nc-Si cores because a low temperature causes a narrow distribution of electron energy and a lower phonon scattering. In charging, the electron relaxation from the nc-Si cores into the silicon nitride traps is manifested clearly in the transition that the charging current peak becomes asymmetric when temperature is below 190 K. Most notably, the height of charging current peak is remarkably raised and the discharging current peak evolves into two peaks located at -1.150 V (Peak I) and -2.075 V (Peak II) at 150 K. The Peaks I and II may correspond to the discharging from the nc-Si cores and from the silicon nitrides via the nc-Si cores, respectively. The integration of one charging peak and two discharging peaks shows that equal charge is involved in programming/erasing. The discharging Peak II replaces the tilted discharging curve in Figure 15.24 because the deep defects are fully occupied and cannot accept electrons any more at 150 K. Moreover, the stored charge in the silicon nitride defects turns out to be half the totally stored charge. The reduction from the two-thirds at room temperature could also result from the occupied deep defects. It seems that most of the electrons are trapped in the interface between the silicon nirtride film and the nc-Si core, but not in the silicon nitride bulk since an electric field is strong enough to manipulate the trapped electrons in the erasing even at 70 K. However, the defect states at the interface could be useful at a certain reading voltage (far smaller than the erasing voltage) and may improve the memory retention time.

The V_g spacing between these two neighboring discharging current peaks turns out to be a 117 meV potential drop between the nc-Si core and the nitride film. The calculation is based on the quantum effect corrected parallel plate-capacitor

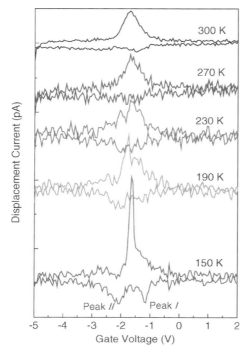

Figure 15.25 Temperature dependences of *I–V* characteristics at the temperature from 150 to 300 K.

model [80]. The potential drop can be comprehended as the energy difference between the confined state of nc-Si cores and the defect trap at the nc-Si/silicon-nitride interface. The number is consistent with the typical trap energy in silicon nitrides [81], taking the band offset of Si/Si_3N_4 system into account. The correlations between the nc-Si core and the defect traps are illustrated in Figure 15.26. In programming, electrons directly tunnel into the nc-Si core through the ultrathin SiO_2 barrier and are confined in the nc-Si core. Evidently, fast charging can be obtained. In the case of storage, some of the stored charges drop into the traps at the interface between the nc-Si core and the silicon nitride. Much longer charge storage is feasible. In erasing, the stored electrons in the nc-Si core first tunnel back to the substrate. Subsequently, the stored electrons at the nc-Si/silicon-nitride interface move back to the nc-Si core under a higher electric field and finally tunnel back to the substrate. Therefore, charging and discharging into/out of the SN-nc-Si dot are dominated by the nc-Si core and silicon nitride, respectively. These advantages allow the SN-nc-Si capable of nonvolatile memory applications with the fast programming. The erasing time may be improved using a much thinner silicon nitride film.

A narrow/short (40 nm/50 nm) channel MOSFET memory using a SN-nc-Si dot as a floating gate was fabricated on a p-type SOI substrate, as shown in the inset of Figure 15.27. The gate consists of control gate SiO_2 (40 nm), an nc-Si core (7 nm in

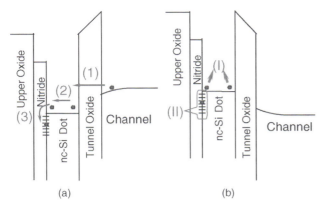

(a) (b)

Figure 15.26 Illustrations of (a) programming and (b) erasing processes in the SN-nc-Si dot memory. (1) Electron directly tunneling from the channel to nc-Si core. (2) Electron polarization to the top of core. (3) Electron drop into defect traps at the nc-Si/silicon–nitride interface. (I) The confinement state in the nc-Si core. (II) The localized state in the defect traps.

diameter) capped with silicon nitride (1 nm), and tunneling SiO_2 (1.5 nm). Only one SN-nc-Si dot is located in the active channel region. The typical V_g dependences of channel current, indicating a memory window, are shown in Figure 15.1. The V_g scan alters the channel into the accumulation, depletion, and inversion state. The hysteresis, which does not depend on the V_g sweeping speed and direction, is due to the electrons being stored in the floating SN-nc-Si dot. The ΔV_{th} is experimentally obtained as 180 mV. The mutual capacitances are estimated by the device geometry as follows: $C_{fg} = 1.82$ aF, $C_{fc} = 2.80$ aF, and $C_{cg} = 4.16$ aF. According to the differential conductance (dI/dV) in Figure 15.27, one may identify successive nine electrons charging into and three electrons escaping out of the SN-nc-Si dot in the forward scan

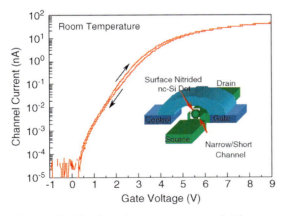

Figure 15.27 The channel current was measured at the source–drain voltage of 10 mV, when the V_g changed from −1 to 10 V and returned to −1 V. The inset shows the schematics of a SN-nc-Si dot floating gate memory.

(-1 to $10\,V$) and the reverse scan (10 to $-1\,V$), respectively. The calculated ΔV_{th} by Eq. (15.3) turns out to be $166\,mV$ by the nine-electron storage and is consistent with the experimental result. Evidently, losing three electrons from the SN-nc-Si dot reduces ΔV_{th} to $126\,mV$ when V_g is smaller than $2\,V$. In other words, the number of stored electrons decreases immediately after the programming, indicating a read degradation.

A reliable read operation should restrict such degradation and avoid the disappearance of the memory window within a desired time. The read stability of the device was examined by keeping read at $2\,V$ immediately after the programming ($10\,V$) or immediately after the erasing ($-4\,V$). Single-electron behaviors, derived from the electron charging/discharging into/out of the floating SN-nc-Si dot, are clearly observed at room temperature, as shown in Figure 15.28. Though first several of the stored electrons escape within initial $10^2\,s$, the memory window still survives after the consecutive read for $10^3\,s$. The time can be much longer because the remaining electrons in the interfacial traps between the nc-Si core and the silicon nitride film are quite stable [82].

A long-term retention more than 10 years is necessary for a typical flash memory. However, 10 years may not be absolutely needed for one who frequently updates data, for example, every week or month. Contrarily, fast programming, accessing, and erasing are much more desirable. However, 1 year retention may satisfy someone who needs a fast flash memory. To measure the retention time, we designed a series of sequent V_g pulses that allow programming/erasing the memory at $10/-4\,V$ and to read at $2\,V$ with grounding (power shutting down) separations, as shown in the inset of Figure 15.29. The duration for programming/erasing, reading, and grounding is 20, 2, and 40 s, respectively. The corresponding channel current was monitored, as

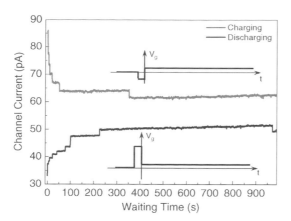

Figure 15.28 The channel current was monitored at $V_g = 2\,V$ as a function of waiting time immediately after the initial programming ($10\,V$) or erasing ($-4\,V$) voltage pulse ($20\,s$). The quantized current change is observed due to discrete emissions or injections of electrons from/out of the SN-nc-Si dot. The insets show the gate voltage transient patterns used in this measurement.

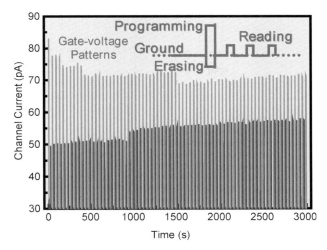

Figure 15.29 The channel current was monitored at $V_g = 2\,V$ for 2 s in every 40 s grounding immediately after the initial programming (10 V)/erasing (−4 V) voltage pulses (20 s). The inset shows the V_g transient patterns used in this measurement.

shown in Figure 15.29. It can be found that the memory turns out to serve with a superior long retention time by not less than 3×10^3 s. The slightly gradual increase in each current mesa may result from the impact of the SN-nc-Si dots out of the active channel. Although the electron injection and emission are random events, most of the injected/lost electrons happen within initial 10^2 s. The remained electrons allow the electron lifetime to become rather long due to the localized charges in the nc-Si core/Si nitride film interface. Experimentally, the memory state becomes much stable for 1 year.

15.4.5
Operation Speed and Device Reliability

Electron direct tunneling in the charging and discharging is a critical aspect of the nc-Si flash memory. Compared to the flash memory using the FN tunneling, the nc-Si flash memory provides faster programming/erasing by at least several orders of magnitude due to the direct tunneling. However, measuring the programming/erasing time is difficult because of the stochastic nature of the tunneling. Roughly, the programming/erasing time (t_t) can be estimated as $t_t \sim R_t C_{tt}$, where R_t is the resistance of the tunnel barrier and C_{tt} is the total capacitance between the dot and the channel. The total capacitance for a typical nc-Si memory device is on the order of 10^{-16} F and R_t is on the order of $10^8\,\Omega$ at the direct tunneling regime [83]. Therefore, the programming/erasing time is on order of 10^{-8} s. Tiwari *et al.* experimentally demonstrated that the programming time was less than 100s of a nanosecond at programming voltages below 2.5 V and the erasing time required milliseconds at 3 V [29]. Wasshuber figured out that one must wait for more than 5 ps for an electron to tunnel with an error probability lower than 10^{-20} [84].

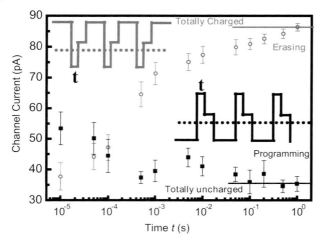

Figure 15.30 The channel current was measured at $V_g = 2$ V immediately after various programming (10 V) or erasing (-4 V) gate voltage pulse durations (t) to exploit the operation speed. The inset shows the V_g transient patterns used in this measurement.

As for the SN-nc-Si dot memory, the charging time should be the same as that of the nc-Si dot memory due to the silicon nitride film located on the top of nc-Si core. However, with regard to the erasing time, a longer time could be consumed to drive electrons out of the silicon-nitride traps and then to return them back to the channel. Apparently, the erasing time depends on the erasing voltage and the traps. Therefore, a trade-off between the erasing time and the retention time should be considered according to the specific requirements. The SN-nc-Si dot memory operation speeds were obtained by comparing the real-time channel current with the initial channel current at 2 V immediately after the programming (10 V)/erasing (-4 V) V_g pulse. The pulse duration varies from 10^{-5} to 100 s. The initial channel current refers to the totally charged/uncharged current when the SN-nc-Si dots were fully filled/unoccupied by electrons at $10/-4$ V. As shown in Figure 15.30, the totally uncharged state can be arrived from the totally charged state by the erasing voltage after not less than 10^{-3} s pulse duration. That indicates a minimum time of 10^{-3} s for a completely erasing. Meanwhile, no remarkable shift toward the totally uncharged state happens in the programming after not more than 10^{-5} s programming pulse duration. The result shows that a complete programming can be obtained within not more than 10^{-5} s. Judged by the disappearance of memory window, the acceptable memory operation needs an erasing pulse duration of not less than 10^{-4} s, which is actually faster than Tiwari's memory. Therefore, the programming/erasing time is consistent with the nc-Si dot memory, while the memory retention time is significantly extended by working with nine electrons in the SN-nc-Si dot memory. The fast operations could result from the trapped charge in the interface between the nc-Si core and the silicon nitride film.

The trapped charge is more sensitive with a modulated electric field than that in the bulk traps of silicon nitride film.

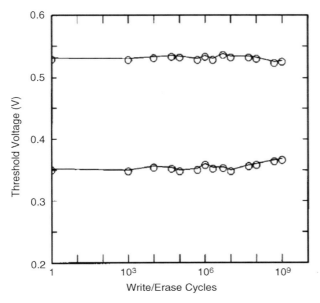

Figure 15.31 Endurance characteristics of a positive threshold voltage nc-Si MOSFET memory device due to ±2.5 V pulsing for writing (1 μs) and erasing (1 ms). Reprinted with permission from Ref. [26]. Copyright (1996) by the American Institute of Physics.

Direct tunneling in charging and discharging processes, on the other hand, is free of the hot carrier degradation. Therefore, the much better reliability of tunneling oxide film is expected in an nc-Si flash memory. As shown in Figure 15.31, Tiwari reported that the cyclability exceeded 10^{13} [26], which implies that the nc-Si floating gate with the ultrathin tunneling oxide film is a reliable structure for the high-endurance flash memory with the unlimited endurance.

Nitrogen atoms incorporated into a SiO_2 thin film is known to be the major reason of reliability degradation [85, 86]. However, in the SN-nc-Si dot memory, the utilized silicon nitride thin film is not the tunnel barrier but the memory node, which is located on the top of nc-Si core. Carriers do not transport across the nitride thin film so that the FN tunneling and the hot carrier effect are absent. Moreover, the $C–V$ measurements on Sample C indicated that the nitrided SiO_2 thin film do not contribute to the memory performance. As shown in Figure 15.32, the cycling endurance was examined over 10^6 times of the programming (10 V)/erasing (−4 V) voltage pulses with a duration of 1 ms. The cycling endurance is capable of extending to an unlimited extent like what a volatile memory exhibits due to the direct tunneling and the ultrathin silicon nitride film features. The result indicates a superior tolerance against the conventional Flash memory, which typically allows not more than 10^5 programming/erasing cycles due to the tunneling oxide degradation by the hot carrier effect. A high endurance was also reported based on silicon-nitride/nc-Si memory structure recently [71].

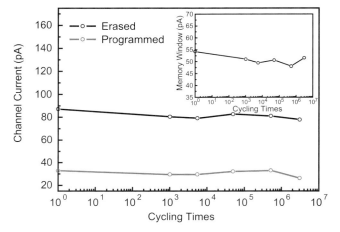

Figure 15.32 The channel current was measured at $V_g = 2$ V immediately after the memory was undertaken cycling of programming (10 V, 1 ms) and erasing (-4 V, 1 ms). The cycling endurance of more than 10^6 makes negligible change in the memory window.

15.4.6
Summary

Nanocrystalline silicon memory devices manifest low power consumption, fast programming/erasing speed, long retention time, and superior endurance with nondestructive read because (1) the Coulomb blockade and quantum confinement effect enable one single-electron transport. A few or even one single electron can guarantee a reliable memory state. (2) Electron charge/discharge through an ultra-thin tunneling SiO_2 film is dominated by the nc-Si dot. (3) A repulsive "built-in" electric field, created and controlled by the charge loss in the nc-Si dots, from nc-Si dots to silicon substrate may give rise to a long-term retention. (4) Programming and erasing in the direct tunnel regime and discrete nc-Si dots provide a high tolerance.

The improved retention time was demonstrated by consuming negligible erasing time in the multiple memory nodes. A smaller charge loss rate than that of single-layered nc-Si dots was experimentally demonstrated. Charge storage in a SN-nc-Si dot is identified by two states: delocalized states in the entire nc-Si dot and localized states in the defects at the nc-Si/silicon-nitride interface. The former provides fast programming and the latter enables long-term retention.

15.5
Comparisons of Emerging Nonvolatile Memory Devices

Providing single-cell electrical programming and fast simultaneous sector electrical erasing, flash memory architecture combines the best features of various memory devices. The flash memory has constantly improved over the past decade and is still expected to show further enhancements over the next decade. The nc-Si memory, featuring in distributed floating nc-Si QDs and an ultrathin tunneling oxide film, is considered one of the pioneering candidates for the next-generation flash memory. However, there are emerging competitors that are based on novel materials, technologies, or mechanisms.

15.5.1
Silicon Nanocrystals Flash Memory and Other Types of Nonvolatile Memory

Flash memory is a charge storage device, which may have the charge stored in the following forms: traps on the interface between the multilayer insulator gate structure (charge trapping) such as $SiO_2/Si_3N_4/SiO_2$ (ONO) or a conducting or semiconducting layer within insulators (floating gate). The embedded conducting or semiconducting layer can be of metallic materials [87] or bulk ploy semiconductors and semiconductor QDs. The floating gate MOSFET is the mainstream of the flash memory. One flash memory cell consists of only one transistor, the so-called two-bit-per-transistor technology. A sufficiently high V_g may drive electrons to tunnel from the channel Si conduction band to the embedded memory nodes, resulting in the negative charge accommodation to switch the threshold voltage. Changing the bit's state requires removing the stored charge, which demands a relatively large negative V_g to suck the electrons off the floating gate.

Of late, nonvolatile memory based on totally different mechanisms has been emerging, such as magnetic RAM (MRAM), ferroelectric RAM (FRAM), phase change memory (PRAM), and resistive RAM (ReRAM). They offer alternative advantages to match specific applications [88]. Here, we will not go into details on the theoretical and experimental work on them but give a brief summary. Reviews can be found in many literature [89–92].

MRAMs are based on the magnetoresistive effects in magnetic materials and structures that exhibit a resistance change when an external magnetic field is applied. Two types of MRAM are explored corresponding to giant magnetoresistance (GMR) and tunneling magnetoresistance (TMR), respectively. The magnetic devices have the advantages of nonvolatility, good operation speed, and low power consumption. However, the scalability of the MRAM is poor because (1) a transistor and tunneling junctions are needed, (2) the magnetic field causes a large fringing field when the memory cell is reduced, and (3) the current passing through the cell produces the magnetic field.

In an FRAM, the storage element is a ferroelectric crystal with a mobile atom in the center that can take either of the two stable positions. Applying an electric field across the face of crystal causes the atom to move in the direction of electric field. Once the atom is moved, it may remain in the position without the external field and is thus nonvolatile. The technology is completely different from the floating gate. Since no voltage-boost charge pump is needed, programming and erasing are fast. However, the access may destroy the storage state. Therefore, FRAM is not a real nonvolatile memory. Moreover, the FRAM is not as dense as the nc-Si flash memory because a ferroelectric crystal is incapable of storing multilevel data and an additional large capacitor is needed.

PRAM uses the unique behavior of chalcogenide glass, which can be switched between two states, crystalline with low resistance state and amorphous with high-resistance state, by the application of heat. The phase change is a thermally driven process rather than an electronic process. With appropriate activation energy for crystallization, it is possible to have high-speed crystallization under programming conditions while there is very slow crystallization under storage conditions. The greatest challenge for the PRAM has been the requirement of high programming current density ($>10^7$ A/cm^2) in the active volume. The phase change material itself is in sublithographic dimensions and has a cost disadvantage compared to flash memory.

ReRAM takes the advantage of voltage-induced resistance bistable switching effect based on materials with strong electron correlations, usually transition metal oxides. Fujitsu announced a breakthrough in which showed by wrapping titanium in nickel oxide, and by limiting the current flow from the transistor, the current needed to erase memory could be reduced down to 100 μA or less, while the whole operation can be completed in 5 ns [93].

In addition, molecular memory is another single-electron memory approach on a single-molecular scale [94]. The device exhibits electronically programmable and erasable memory bits compatible with the conventional flash memory. A molecular memory cell applicable to a random access has been demonstrated [95]. One of the most attractive features is that self-assembled circuits can be formed easily due to the self-growth mechanism [96]. Another example is a novel structure combining nc-Si dots with a nanoelectromechanical system (NEMS). The NEMS memory has a mechanically bistable floating gate beam made of nc-Si dots, which act as single-electron storages [97]. The memory is expected to have high speed in the GHz regime since the characteristic frequency increases with decreasing dimension [98].

15.5.2
Silicon Nanocrystals Memory Devices and Other Quantum Dot Memory Devices

Nanocrystalline silicon QDs are often used as memory nodes because of the perfect interface between Si and SiO$_2$. The interface is most important to the reliable and reproducible VLSI technology. However, since many other metal and semiconductor QDs manifest self-assembling growth by particular fabrication methods, additional advantages and functionalities can be provided [20, 99]. The single-electron charging/discharging effects observed in the materials imply that the advantages of the nc-Si memory can also be realized. Here, we restrict our discussions to representative semiconductor QDs.

The 4.16% mismatch between the Si and the Ge lattice constants allows nano-crystalline Ge (nc-Ge) dots self-assemble to incorporate into the Si CMOS technology [100]. The smaller bandgap and the light effective masses of carriers allow the nc-Ge dot to considerably enhance the tunneling current through an empty valence band. Accordingly, the programming/erasing time can be reduced down. King *et al.* demonstrated the superior characteristics of nc-Ge dot memory in terms of the operation speed and voltage [101]. *C–V* measurements indicated charge storage in the nc-Ge dots in many recent literatures [102–104]. Turan found that the hole loss rate was dramatically reduced by Coulomb blockade from the nc-Ge dots [105]. It has also been observed that very different characteristics were obtained from the nc-Ge dots synthesized with different technologies [106, 107]. The distortion and stretch-out of C–V curves, found in these works, might be related to deep defect traps or high interfacial states, which make the nc-Ge approach incapable of reliable application without a new dielectric material solution.

Self-assembled InAs QDs embedded in an InP matrix showed an unusually large confinement energy of approximately 400 meV for hole [108]. Medeiros-Ribeiro investigated the single-electron charging and Coulomb interaction in InAs QD arrays by a capacitance spectroscopy study [109]. The large dot areal density and the strong confinement of hole suggest that the InAs dot could be implemented in optical memory [110]. Imamura demonstrated a wavelength domain multiplication optical memory by burying InAs QDs into a Schottky-barrier diode. Instead of electrical gating, the memory relies on an optical signal by QD bleaching effects: two excited holes with up and down spins [111]. Lundstrom *et al.* recently reported the exciton storage in self-assembled InAs dots embedded in GaAs. The operation depends on the dissociation and separate storage of optically created excitons [112]. These achievements allow a possible application in the optical communication, but the charge holding time is extremely short at room temperature.

15.5.3
Summary

Comparisons of various types of emerging nonvolatile memory technologies show that the nc-Si memory is a typical hybrid memory with fast programming/erasing and superior endurance, long retention time with nondestructive read. Although there is a trade-off between the operation speed and the retention time, the challenge is to select the particular trade-off to match individual purposes.

The nc-Si memory, which is fully compatible with the well-established Si technology, shows a proven scalability record against the MRAM, FRAM, and PRAM. Many other semiconductor QDs can also be incorporated into memory devices in some specific application fields. However, the poor interfacial state and the low temperature requirement strongly limit the range of applicable areas and the mass production. The nc-Si memory is optimum in operational characteristics and fabrication technologies to match the requirements of next-generation flash memory.

15.6
Concluding Remarks and Prospects

The flash memory using electrically isolated nc-Si dots in an oxide film as a floating gate has been reviewed. They are operated in terms of the discrete threshold voltage shifts because charge storage in the dot is self-limited by the Coulomb blockade effect at one single level. The electrically isolated dots immunize the charge leakage through localized oxide defects. As a consequence, the performance reliability and the retention time have been improved. On the other hand, the much thinner tunneling oxide changes the tunnel current from the conventional FN tunneling to the direct tunneling in programming and erasing, hence allowing a higher operation speed and a better cycling endurance. Because the single-electron phenomena take place in the nanometer-scale dots rather than the other memory components, the nc-Si flash memory allows the ultradense scalability. A synergy of the top–down with bottom–up technologies could be a better way toward the practicable fabrications of nc-Si memory. The ultimate goal for the memory is few- or single-electron storage in one single dot, relying on the Coulomb blockade effect as a new transport principle.

Although there are very credible competitions from the emerging nonvolatile technologies using novel magnetic, ferroelectric, and phase change materials, the nc-Si memory provides the better scalability and the full compatibility with well-established VLSI technology against the MRAM, FRAM, and FRAM. Single-electron charging/discharging phenomena also take place in other QDs, such as Ge and InAs. However, when employing the materials in mass production, one may encounter the issues such as the poor interfacial states and low temperature requirement, which apparently limit the range of possible application areas. As a consequence, the nc-Si memory is considered optimum to match the requirements of next-generation flash memory.

Nevertheless, it is also clear that it will take considerable time to make the memory usable for consumers. To resolve fluctuations in threshold voltage, one may need the size-uniformed nc-Si dots for large-area assembling. Random background charges in the nanometer-scale memory cell may be still a problem, which hinders integration of technology on a large scale due to device-to-device variations. A peripheral verifying-circuitry in the memory chip may be required for acceptable error rates.

Dr. Gordon E. Moore had said that no exponential was forever, but we could delay "forever." For the flash memory technology, the nc-Si memory is one of the approaches to delay the virtual end of scaling down. Future memory structures may rely on totally different fabrication technologies and materials. Self-assembled molecular memory, for instance, might be an ultimate structure for building single-electron memory on a single-molecular level. However, there is a gap between the current submicrometer-scale memory and the molecular-scale memory technologies. The nc-Si memory is, therefore, helpful to bridge and realize a transition from the current VLSI technology to the molecular-scale technology. The well-established Si fabrication instructions are, therefore, still useful in the future.

References

1 Pavan, P., Bez, R., Olivo, P., and Zanoni, E. (1997) Flash memory cells: an overview. *Proc. IEEE*, **85**, 1271–1284.
2 Park, Y.-W., Choi, J.-D., Kang, C.-S., Lee, C.-H., Shin, Y.-C., Choi, B.-H., Kim, J.-H., Jeon, S.-H., Sel, J.-S., Park, J.-T., Choi, K.-H., Yoo, T.-H., Sim, J.-S., and Kim, K.-N. (2006) Highly manufacturable 32 Gb multi-level NAND flash memory with 0.0098 μm^2 cell size using TANOS

(Si-oxide-Al$_2$O$_3$-TaN) cell technology. International Electron Device Meeting Proceedings, pp. 29–32.

3 Noguchi, M., Yaegashi, T., Koyama, H., Morikado, M., Ishibashi, Y., Ishibashi, S., Ino, K., Sawamura, K., Aoi, T., Maruyama, T., Kajita, A., Ito, E., Kishida, M., Kanda, K., Hosono, K., Miyamoto, S., Ito, F., Hirata, Y., Hemink, G., Higashitani, M., Mak, A., Chan, J., Koyanagi, M., Ohshima, S., Shibata, H., Tsunoda, H., and Tanaka, S. (2007) A high-performance multi-level NAND flash memory with 43 nm-node floating-gate technology. International Electron Device Meeting Proceedings, pp. 445–448.

4 Meindl, J., Chen, Q., and Davis, J. (2001) Limits on silicon nanoelectronics for terascale integration. *Science*, **293**, 2044–2049.

5 Thompson, S., Packan, P., and Bohr, M. (1998) MOS scaling: transistor challenges for the 21st century. *Intel Technol. J.*, **Q3**, 1–19.

6 Hutchby, J., Bourianoff, G., Zhirnov, V., and Brewer, J. (2002) Extending the road beyond CMOS. *IEEE Circuits Devices Mag.*, **18**, 28–41.

7 Gautier, J. (1997) Beyond CMOS: quantum devices. *Microelectron. Eng.*, **39**, 263–272.

8 Agnello, P.D. (2002) Process requirements for continued scaling of CMOS: the need and prospects for atomic-level manipulation. *IBM J. Res. Dev.*, **46**, 317–338.

9 Bondyopadhyay, P.K. (1998) Moore's law governs the silicon revolution. *Proc. IEEE*, **86**, 78–81.

10 Davari, B., Dennard, R.H., and Shahidi, G.G. (1995) CMOS scaling for high performance and low power: the next ten years. *Proc. IEEE*, **83**, 595–606.

11 *The International Technology Roadmap for Semiconductors 2007.* (ITRS), 2007 update, International Sematech, Austin, TX, http://public.itrs.net.

12 http://www.intel.com/research/silicon/mooreslaw.htm.

13 Frank, D., Dennard, R., Nowak, E., Solomon, P., Taur, Y., and Wong, H. (2001) Device scaling limits of Si MOSFETs and their application dependencies. *Proc. IEEE*, **89**, 259–288.

14 Wong, H., Frank, D., Solomon, P., Wann, C., and Welser, J. (1999) Nanoscale CMOS. *Proc. IEEE*, **87**, 537–570.

15 Ferry, D., Grubin, H., Jacoboni, C., and Jauho, A. (1994) *Quantum Transport in Ultrasmall Devices*, Plenum Press, New York.

16 Salvo, B.D., Ghibaudo, G., Pananakakis, G., Masson, P., Baron, T., Buffet, N., Frnandes, A., and Guillaumot, B. (2001) Experimental and theoretical investigation of nano-crystal and nitride-trap memory devices. *IEEE Trans. Electron Devices*, **48**, 1789–1799.

17 White, M.H., Adams, D.A., and Bu, J. (2000) On the go with SONOS. *IEEE Circuits Devices Mag.*, **16**, 22–31.

18 Hinds, B.J., Yamanaka, T., and Oda, S. (2001) Emission lifetime of polarizable charge stored in nano-crystalline Si based single-electron memory. *J. Appl. Phys.*, **90**, 6402–6408.

19 Kirton, M.J., Uren, M.J., Collins, S., Schulz, M., Karmann, A., and Scheffer, K. (1989) Individual defects at the Si−SiO$_2$ interface. *Semicond. Sci., Technol*, **4**, 1116–1126.

20 Yoffe, A.D. (2001) Semiconductor quantum dots and related systems: electronic, optical, luminescence and related properties of low dimensional systems. *Adv. Phys.*, **50**, 1–208.

21 Trean, A. and Leburton, J.P. (2001) Geometry and strain effects on single-electron charging in silicon nano-crystals. *J. Appl. Phys.*, **90**, 6384–6390.

22 Williamson, A.J., Wang, L.W., and Zunger, A. (2000) Theoretical interpretation of the experimental electronic structure of lens-shaped self-assembled InAs/GaAs quantum dots. *Phys. Rev. B*, **62**, 12963–12977.

23 Yano, K., Ishii, T., Sano, T., Mine, T., Murai, F., Hashimoto, T., Kobayashi, T., Kure, T., and Seki, K. (1999) Single-electron memory for giga-to-tera bit storage. *Proc. IEEE*, **87**, 633–651.

24 Likharev, K.K. (1999) Single-electron devices and their applications. *Proc. IEEE*, **87**, 606–632.

25 Nakajima, A., Futatsugi, T., Kosemura, K., Fukano, T., and Yokoyama, N. (1997) Room temperature operation of Si single-electron memory with self-aligned floating dot gate. *Appl. Phys. Lett.*, **70**, 1742–1744.

26 Tiwari, S., Rana, F., Hanafi, H., Hartstein, A., Crabbé, E., and Chan, K. (1996) A silicon nanocrystals based memory. *Appl. Phys. Lett.*, **68**, 1377–1379.

27 Ohba, R., Sugiyama, N., Uchida, K., Koga, J., and Toriumi, A. (2002) Nonvolatile Si quantum memory with self-aligned doubly-stacked dots. *IEEE Trans. Electron Devices*, **49**, 1392–1398.

28 Huang, S., Arai, K., Usami, K., and Oda, S. (2004) Toward long-term retention-time single-electron-memory devices based on nitrided nanocrystalline silicon dots. *IEEE Trans. Nanotechnol.*, **3**, 210–214.

29 Tiwari, S., Rana, F., Chan, K., Shi, L., and Hanafi, H. (1996) Single charge and confinement effects in nano-crystal memories. *Appl. Phys. Lett.*, **69**, 1232–1234.

30 Guo, L.J., Leobandung, E., and Chou, S.Y. (1997) A silicon single-electron transistor memory operating at room temperature. *Science*, **275**, 649–651.

31 Ostraat, M.L., De Blauwe, J.W., Green, M.L., Bell, L.D., Atwater, H.A., and Flagan, R.C. (2001) Ultraclean two-stage aerosol reactor for production of oxide-passivated silicon nanoparticles for novel memory devices. *J. Electrochem. Soc.*, **148**, G265–G270.

32 Beenakker, C.W.J. (1991) Theory of Coulomb-blockade oscillations in the conductance of a quantum dot. *Phys. Rev. B*, **44**, 1646–1656.

33 Johansson, J. and Haviland, D.B. (2001) Random background charges and Coulomb blockade in one-dimensional tunnel junction arrays. *Phys. Rev. B*, **63**, 014201.1–014201.6.

34 Lu, T.Z., Alexe, M., Scholz, R., Talelaev, V., and Zacharias, M. (2005) Multilevel charge storage in silicon nanocrystal multilayers. *Appl. Phys. Lett.*, **87**, 202110.1–202110.3.

35 Sée, J., Dollfus, P., and Galdin, S. (2002) Comparison of a density functional

36 de Sousa, J.S., Thean, A.V., Leburton, J.P., and Freire, V.N. (2002) Three-dimensional self-consistent simulation of the charging time response in silicon nanocrystal flash memories. *J. Appl. Phys.*, **92**, 6182–6187.

37 Shi, Y., Saito, K., Ishikuro, H., and Hiramoto, T. (1998) Effects of traps on charge storage characteristics in metal–oxide–semiconductor memory structures based on silicon nanocrystals. *J. Appl. Phys.*, **84**, 2358–2360.

38 Thean, A. and Leburton, J.P. (2001) Geometry and strain effects on single-electron charging in silicon nano-crystals. *J. Appl. Phys.*, **90**, 6384–6390.

39 de Sousa, J.S., Leburton, J.P., Thean, A.V., Freire, V.N., and da Silva, E.F. (2003) Effects of crystallographic orientations on the charging time in silicon nanocrystal flash memories. *Appl. Phys. Lett.*, **82**, 2685–2687.

40 Molas, G., Salvo, B.D., Ghibaudo, G., Mariolle, D., Toffoli, A., Buffet, N., Puglisi, R., Lombardo, S., and Deleonibus, S. (2004) Single electron effects and structural effects in ultrascaled silicon nanocrystal floating-gate memories. *IEEE Trans. Nanotechnol.*, **3**, 42–48.

41 Guo, L.J., Leobandung, E., Zhuang, L., and Chou, S.Y. (1997) Fabrication and characterization of room temperature silicon single electron memory. *J. Vac. Sci. Technol. B*, **15**, 2840–2843.

42 Likharev, K.K. (1998) Layered tunnel barriers for nonvolatile memory devices. *Appl. Phys. Lett.*, **73**, 2137–2139.

43 Casperson, J.D., Bell, L.D., and Atwater, H.A. (2002) Materials issues for layered tunnel barrier structures. *J. Appl. Phys.*, **92**, 261–267.

44 Banerjee, S., Huang, S., Yamanaka, T., and Oda, S. (2002) Evidence of storing and erasing of electrons in a nanocrystalline-Si based memory device at 77 K. *J. Vac. Sci. Tech. B*, **20**, 1135–1138.

45 Yano, K., Ishii, T., Hashimoto, T., Kobayashi, T., Murai, F., and Seki, K. (1994) Room-temperature

theory and a Hartree treatment of silicon quantum dot. *J. Appl. Phys.*, **92**, 3141–3146.

single-electron memory. *IEEE Trans. Electron Devices*, **41**, 1628–1638.

46 Choi, B.H., Hwang, S.W., Kim, I.G., Shin, H.C., Kim, Y., and Kim, E.K. (1998) Fabrication and room-temperature characterization of a silicon self-assembled quantum-dot transistor. *Appl. Phys. Lett.*, **73**, 3129–3131.

47 Sheats, J.R. and Smith, B.W. (1998) *Microlithography: Science and Technology*, Marcel Dekker, Inc., New York.

48 Melngailis, J., Mondelli, A.A., Berry, I.L., and Mohondro, R. (1998) A review of ion projection lithography. *J. Vac. Sci. Technol. B*, **16**, 927–957.

49 Lercel, M.J., Craighead, H.G., Parikh, A.N., Seshadri, K., and Allara, D.L. (1996) Sub-10 nm lithography with self-assembled monolayers. *Appl. Phys. Lett.*, **68**, 1504–1506.

50 Li, Y., Maynor, B.W., and Liu, J. (2001) Electrochemical AFM "dip-pen" nanolithography. *J. Am. Chem. Soc.*, **123**, 2105–2106.

51 Wu, W., Gu, J., Ge, H., Keimel, C., and Chou, S.Y. (2003) Room-temperature Si single-electron memory fabricated by nanoimprint lithography. *Appl. Phys. Lett.*, **83**, 2268–2270.

52 Moore, G. (2003) No exponential is forever: but "forever" can be delayed! *IEEE Int. Solid-State Circuits Conf.*, **1**, 20–23.

53 Nakajima, A., Sugita, Y., Kawamura, K., Tomita, H., and Yokohama, N. (1996) Si quantum dot formation with low-pressure chemical vapor deposition. *Jpn. J. Appl. Phys.*, **35**, L189–L191.

54 Baron, T., Martin, F., Mur, P., Wyon, C., and Dupuy, M. (2000) Silicon quantum dot nucleation on Si_3N_4, SiO_2 and SiO_xN_y substrates for nanoelectronic devices. *J Cryst. Growth*, **209**, 1004–1008.

55 Fissel, A., Akhtariev, R., Kaiser, U., and Richter, W. (2001) MBE growth of Si on SiC(0001): from superstructures to islands. *J. Cryst. Growth*, **227**, 777–781.

56 Ziegler, J.F. (1988) *Ion Implantation: Science and Technology*, Academic Press, Boston.

57 González-Varona, O., Garrido, B., Cheylan, S., Pérez-Rodríguez, A., Guadras, A., and Morante, J.R. (2003) Control of tunnel oxide thickness in Si-nanocrystal array memories obtained by ion implantation and its impact in writing speed and volatility. *Appl. Phys. Lett.*, **82**, 2151–2153.

58 Maeda, T., Suzuki, E., Sakata, I., Yamanaka, M., and Ishii, K. (1999) Electrical properties of Si nanocrystals embedded in an ultrathin oxide. *Nanotechnol.*, **10**, 127–131.

59 Goncalves, C., Charvet, S., Zeinert, A., Clin, M., and Zellama, K. (2002) Nanocrystalline silicon thin films prepared by radiofrequency magnetron sputtering. *Thin Solid Films*, **403**, 91–96.

60 Ifuku, T., Otobe, M., Itoh, A., and Oda, S. (1997) Fabrication of nanocrystalline silicon with small spread of particle size by pulsed gas plasma. *Jpn. J. Appl. Phys.*, **36**, 4031–4034.

61 Oda, S. (1993) Frequency effects in processing plasmas of the VHF band. *Plasma Sources Sci. Technol.*, **2**, 26–29.

62 Arai, K. and Oda, S. (2003) Photoluminescence of surface nitrided nanocrystalline silicon dots. *Phys. Stat. Sol. C*, **0**, 1254–1257.

63 Baur, C., Bugacov, A., Koel, B.E., Madhukar, A., Montoya, N., Ramachandran, T.R., Requicha, A.A.G., Resch, R., and Will, P. (1998) Nanoparticle manipulation by mechanical pushing: underlying phenomena and real-time monitoring. *Nanotechnol.*, **9**, 360–364.

64 Guarini, K.W., Black, C.T., Zhang, Y., Babich, I.V., Sikorski, E.M., and Gignac, L.M. (2003) Low voltage, scalable nanocrystal flash memory fabricated by templated self assembly. *IEEE IEDM 2003 Technical Digest*, 22.2.1–22.2.4.

65 Kouvatsos, D.N., Ioannou-Sougleridis, V., and Nassiopoulou, A.G. (2003) Charging effects in silicon nanocrystals within SiO_2 layers, fabricated by chemical vapor deposition, oxidation, and annealing. *Appl. Phys. Lett.*, **82**, 397–399.

66 Tiwari, S., Wahl, J., Silva, H., Rana, F., and Welser, J. (2000) Small silicon memories: confinement, single-electron, and interface state considerations. *Appl. Phys. A*, **71**, 403–414.

67 Nicollian, E.H. and Brews, J.R. (1982) *MOS Physics and Technology*, John Wiley & Sons, New York.

68 Sze, S.M. (1981) *Physics of Semiconductor Devices*, 2nd Ed., John Wiley & Sons, New York, Chap. 1 and Appendix G therein.

69 Busseret, C., Souifi, A., Baron, T., and Guillot, G. (2000) Discharge mechanisms modeling in LPCVD silicon nanocrystals using $C–V$ and capacitance transient techniques. *Superlattices Microstruct.*, **28**, 493–500.

70 Fleischer, S., Lai, P.T., and Cheng, Y.C. (1992) Simplified closed-form trap-assisted tunneling model applied to nitrided oxide dielectric capacitors. *J. Appl. Phys.*, **72**, 5711–5715.

71 Steimle, R.F., Sadd, M., Muralidhar, R., Rao, R., Hradsky, B., Straub, S., and White, B.E. (2003) Hybrid silicon nanocrystal silicon nitride dynamic random access memory. *IEEE Trans. Nanotechnol.*, **2**, 335–340.

72 Basa, P., Horváth, Zs.J., Jászi, T., Pap, A.E., Dobos, L., Pécz, B., Tóth, L., and Szöllősi, P. (2007) Electrical and memory properties of silicon nitride structures with embedded Si nanocrystals. *Physcia E*, **38**, 71–75.

73 Ma, Y., Yasuda, T., and Lucovsky, G. (1994) Ultrathin device quality oxide–nitride–oxide heterostructure formed by remote plasma enhanced chemical vapor deposition. *Appl. Phys. Lett.*, **64**, 2226–2228.

74 Queisser, H.J. and Haller, E.E. (1998) Defects in semiconductors: some fatal, some vital. *Science*, **281**, 945–950.

75 Sekine, K., Saito, Y., Hirayama, M., and Ohmi, T. (2000) Highly robust ultrathin silicon nitride films grown at low-temperature by microwave-excitation high-density plasma for giga scale integration. *IEEE Trans. Electron Devices*, **47**, 1370–1374.

76 Green, M.L., Gusev, E.P., Degraeve, R., and Garfunkel, E.L. (2001) Ultrathin (<4 nm) SiO_2 and $Si–O–N$ gate dielectric layers for silicon microelectronics: understanding the processing, structure, and physical and electrical limits. *J. Appl. Phys.*, **90**, 2057–2121.

77 Lucovsky, G. (1999) Ultrathin nitride gate dielectrics: plasma processing, chemical characterization, performance, and reliability. *IBM J. Res. Dev.*, **43**, 301–326.

78 Miller, S.L., Mcwhorter, P.J., Dellin, T.A., and Zimmerman, G.T. (1990) Effect of temperature on data retention of silicon-oxide-nitride-oxide-semiconductor nonvolatile memory transistors. *J. Appl. Phys.*, **67**, 7115–7124.

79 Majima, Y., Oyama, Y., and Iwamoto, M. (2000) Measurement of semiconductor local carrier concentration from displacement current–voltage curves with a scanning vibrating probe. *Phys. Rev. B*, **62**, 1971–1977.

80 Huang, S., Banerjee, S., Tung, R.T., and Oda, S. (2003) Quantum confinement energy in nanocrystalline silicon dots from high-frequency conductance measurement. *J. Appl. Phys.*, **94**, 7261–7265.

81 Lusky, E., Shacham-Diamand, Y., Shappir, A., Bloom, I., and Eitan, B. (2004) Traps spectroscopy of the Si_3Ni_4 layer using localized charge-trapping nonvolatile memory device. *Appl. Phys. Lett.*, **85**, 669–671.

82 Huang, S. and Oda, S. (2005) Charge storage in nitrided nanocrystalline silicon dots. *Appl. Phys. Lett.*, **87**, 173107.1–173107.3.

83 Depas, M., Vermeire, B., Mertens, P.W., Vanmeirhaeghe, R.L., and Heyns, M.M. (1995) Determination of tunneling parameters in ultra-thin oxide layer poly-$Si/SiO_2/Si$ structures. *Solid State Electron.*, **38**, 1465–1471.

84 Wasshuber, C., Kosina, H., and Selberherr, S. (1998) A comparative study of single-electron memories. *IEEE Trans. Electron Devices*, **45**, 2365–2371.

85 Habermehl, S., Nasby, R.D., and Rightley, M.J. (1999) Cycling endurance of silicon-oxide-nitride-oxide-silicon nonvolatile memory stacks prepared with nitrided $SiO_2/Si(100)$ interfaces. *Appl. Phys. Lett.*, **75**, 1122–1124.

86 Fujieda, S., Miura, Y., Saitoh, M., Hasegawa, E., Koyama, S., and Ando, K. (2003) Interface defects responsible for negative-bias temperature instability in plasma-nitrided SiON/Si(100) systems. *Appl. Phys. Lett.*, **82**, 3677–3679.

87 Liu, Z.T., Lee, C., Narayanan, V., Pei, G., and Kan, E.C. (2002) Metal nanocrystal memories. I. Device design and fabrication. *IEEE Trans. Electron Devices*, **49**, 1606–1613.

88 Pinnow, C.U. and Mikolajick, T. (2004) Material aspects in emerging nonvolatile memories. *J. Electrochem. Soc.*, **151**, K13–K19.

89 Inomata, K. (2001) Present and future of magnetic RAM technology. *IEICE Trans. Electron.*, **E84C**, 740–746.

90 Tehrani, S., Slaughter, J.M., Chen, E., Durlam, M., Shi, J., and DeHerren, M. (1999) Progress and outlook for MRAM technology. *IEEE Trans. Magn.*, **35**, 2814–2819.

91 Kim, K. (2001) High density stand alone-FRAM: present and future. *Integr. Ferroelectr.*, **36**, 21–39.

92 Fox, G.R., Chu, F., and Davenport, T. (2001) Current and future ferroelectric nonvolatile memory technology. *J. Vac. Sci. Tech. B*, **19**, 1967–1971.

93 Tsunoda, K., Kinoshita, K., Noshiro, H., Yamazaki, Y., Iizuka, T., Ito, Y., Takahashi, A., Okano, A., Sato, Y., Fukano, T., Aoki, M., and Sugiyama, Y. (2007) Low power and high speed switching of Ti-doped NiO ReRAM under the unipolar voltage source of less than 3 V. International Electron Device Meeting Proceedings, pp. 767–770.

94 Reed, M.A. (1999) Molecular-scale electronics. *Proc. IEEE*, **87**, 652–658.

95 Reed, M.A., Chen, J., Rawlett, A.M., Price, D.W., and Tour, J.M. (2001) Molecular random access memory cell. *Appl. Phys. Lett.*, **78**, 3735–3737.

96 Mantooth, B.A. and Weiss, P.S. (2003) Fabrication, assembly, and characterization of molecular electronic components. *Proc. IEEE*, **91**, 1785–1802.

97 Tsuchiya, Y., Takai, K., Momo, N., Nagami, T., Mizuta, H., Oda, S., Yamaguchi, S., and Shimada, T. (2006) Nanoelectromechanical nonvolatile memory device incorporating nanocrystalline Si dots. *J. Appl. Phys.*, **94**, 094306.1–094306.6.

98 Huang, X.M.H., Zorman, C.A., Mehregany, M., and Roukes, M.L. (2003) Nanodevice motion at microwave frequencies. *Nature*, **421**, 496–496.

99 Wang, K.L. (2002) Issues of nano-electronics: a possible roadmap. *J. Nanosci. Nanotechnol.*, **2**, 235–266.

100 Wang, K.L., Liu, J.L., and Jin, G. (2002) Self-assembled Ge quantum dots on Si and their applications. *J. Cryst. Growth*, **237**, 1892–1897.

101 King, Y.C., King, T.J., and Hu, C.M. (2001) Charge-trap memory device fabricated by oxidation of $Si_{1-x}Ge_x$. *IEEE Trans. Electron Device*, **48**, 696–700.

102 Choi, W.K., Chim, W.K., Heng, C.L., and Teo, L.W., Ho F V. Ng, V., Antoniadis, D.A., Fitzgerald, E.A. (2002) Observation of memory effect in germanium nanocrystals embedded in an amorphous silicon oxide matrix of a metal–insulator–semiconductor structure. *Appl. Phys. Lett.*, **80**, 2014–2016.

103 Kim, J.K., Cheong, H.J., Kim, Y., Yi, J.Y., Bark, H.J., Bang, S.H., and Cho, J.H. (2003) Rapid-thermal-annealing effect on lateral charge loss in metal–oxide–semiconductor capacitors with Ge nanocrystals. *Appl. Phys. Lett.*, **82**, 2527–2529.

104 Kanoun, M., Souifi, A., Baron, T., and Mazen, F. (2004) Electrical study of Ge-nanocrystal-based metal–oxide–semiconductor structures for p-type nonvolatile memory applications. *Appl. Phys. Lett.*, **84**, 5079–5081.

105 Akca, I.B., Dâna, A., Aydinli, A., and Turan, R. (2008) Comparison of electron and hole charge–discharge dynamics in germanium nanocrystal flash memories. *Appl. Phys. Lett.*, **92**, 052103.1–052103.3.

106 Marstein, E.S., Gunnæs, A.E., Olsen, A., Finstad, T.G., Turan, R., and Serincan, U. (2004) Introduction of Si/SiO$_2$ interface states by annealing Ge-implanted films. *J. Appl. Phys.*, **96**, 4308–4312.

107 Hanafi, H.I., Tiwari, S., and Khan, I. (1996) Fast and long retention-time nano-crystal memory. *IEEE Trans. Electron Devices*, **43**, 1553–1558.

108 Pettersson, H., Pryor, C., Landin, L., Pistol, M.E., Carlsson, N., Seifert, W., and

Samuelson, L. (2000) Electrical and optical properties of self-assembled InAs quantum dots in InP studied by space-charge spectroscopy and photo-luminescence. *Phys. Rev. B*, **61**, 4795–4800.

109 Medeiros-Ribeiro, G., Pinheiro, M.V.B., Pimentel, V.L., and Marega, E. (2002) Spin splitting of the electron ground states of InAs quantum dots. *Appl. Phys. Lett.*, **80**, 4229–4231.

110 Pettersson, H., Bààth, L., Carlsson, N., Seifert, W., and Samuelson, L. (2001) Case study of an InAs quantum dot memory: Optical storing and deletion of charge. *Appl. Phys. Lett.*, **79**, 78–80.

111 Imamura, K., Sugiyama, Y., Nakata, Y., Muto, S., and Yokohama, N. (1995) New Optical memory structure using self-assembled InAs quantum dots. *Jpn. J. Appl. Phys.*, **34**, L1445–L1447.

112 Lundstrom, T., Schoenfeld, W.V., Mankad, T., Jaeger, A., Lee, H., and Petroff, P.M. (2000) Splitting and storing excitons in strained coupled self-assembled quantum dots. *Physica E*, **7**, 494–498.

16
Photonics Application of Silicon Nanocrystals

Aleksei Anopchenko, Nicola Daldosso, Romain Guider, Daniel Navarro-Urrios,
Alessandro Pitanti, Rita Spano, Zhizhong Yuan, and Lorenzo Pavesi

16.1
Introduction to Silicon Nanophotonics

For years the semiconductor industry has used larger and larger silicon wafers where smaller and smaller integrated circuits have been produced both to improve chip performances and manufacturing yields, and to accomplish the Moore's law. Nowadays, multicore processors where billions of transistors are integrated are facing the interconnection problem because signals at Tb/s rates have to be exchanged along fast, energy-efficient, and environment-friendly channels. The only option left is to use the enormous bandwidth of optics as in optical communications. However, optical interconnect is only one of the main application fields where silicon photonics is used [1, 2]. Indeed, silicon photonics is an enabling technology that could be applied to life science, medicine, sensing, automotive industry, and energetics. Apart from its good optical properties, its low cost and easy manufacturability, the main advantage of using silicon for photonics is the possibility to merge electronics and photonics in the same chip [3]. This results in a high level of functionality and versatility that can address a broad range of applications while keeping the cost low.

Silicon photonics has been pioneered by Richard Soref during the 1980s [4, 5]. Since silicon is a good optical material, but there is a lack of active devices, the demonstration of light emission from porous silicon at room temperature by Canham [6] boosted the research on silicon-based light sources. At the same time, the concept of silicon microphotonics or optoelectronics emerged impetuously [7–9]. But it was only from 2000 that the research field has really boomed (Figure 16.1).

Low-loss silicon stripe waveguides [10] and silicon photonic crystal optical wave-guides were realized as early as in 2000 [11]. Then, Si-based waveguides were shrunk resulting in silicon wires (0.1 μm × 0.2 μm) and photonic crystal-based waveguides with low losses (few dB/cm) fabricated both at IBM [12, 13] and at NTT [14]. Recently, a new concept, slot waveguide, was introduced at Cornell University [15] allowing mode confinement in a low index medium. This concept is getting very popular both for sensing and for nonlinear optical applications [16]. The problem of coupling

Silicon Nanocrystals: Fundamentals, Synthesis and Applications. Edited by Lorenzo Pavesi and Rasit Turan
Copyright © 2010 WILEY-VCH Verlag GmbH & Co. KGaA, Weinheim
ISBN: 978-3-527-32160-5

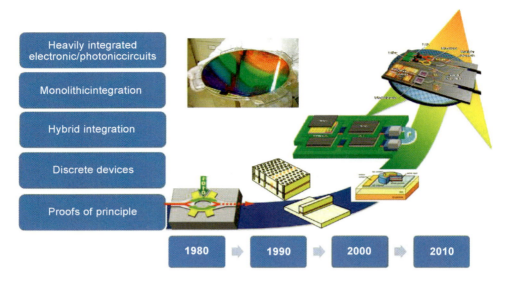

Figure 16.1 Schematic evolution of silicon photonics versus the years. The various diagrams in the insets are courtesy of Ghent University and INTEL.

between optical fiber and silicon waveguides has been solved, among the others, by the concept of inverted tapers [17, 18] and grating couplers [19].

Light receivers in silicon photonics are easily addressed by integration of transimpedance amplifier (TIA) and photodetectors. For these last components, the heterogrowth of germanium on silicon allows the development of high-speed CMOS-compatible optical receivers [20, 21]. Successful implementation of such a device has been shown; for example, the 30 GHz SiGe photodetector produced at IBM [22] and the SiGe photodetectors realized at Stuttgart [23] and at MIT [24].

Concerning the transmitters, different approaches to produce a light source have been attempted since 2000. Unfortunately, there are still no dominant technologies. Optical gain in silicon nanocrystals (Si-ncs) was demonstrated in Italy in 2000 [25]. In 2001–2004, several breakthroughs were achieved: optical gain in Er-doped Si-nc waveguide [26], stimulated blue emission in silicon nanoparticles [27], high efficient electroluminescence in Er-doped devices [28, 29], Raman gain in silicon wave-guides [30], and, finally, a pulsed Silicon Raman laser were demonstrated [31]. In 2005, bipolar electrical injection was shown in a Si-nc LED [32], while a CW silicon Raman laser was introduced by INTEL [33]. In 2006, the hybrid silicon evanescent laser was invented by UCSB and INTEL [34], later on developed into the mode-locked version [35], and an InP microdisk laser coupled to silicon waveguides was fabricated by IMEC and LETI [36].

Data encoding of the optical signal is not achieved by direct modulation of the light sources but by using an optical modulator. High-speed GHz modulators were fabricated at INTEL based on University of Surrey design [37, 38] and by Luxtera [39]. An ultracompact 1.5 Gb/s modulator based on ring resonators was fabricated by

Cornell [40], while an ultrafast design was proposed by MIT for a p–i–n [41] and by the University of Surrey for p–n structures [42]. Moreover, all optical switching in silicon was proposed [43] and demonstrated in Si-nc [16], the electro-optical effect in strained silicon was also demonstrated [44], and a quantum confinement Stark effect modulator based on SiGe was fabricated [45].

High-quality factor (high-Q) photonic crystal optical nanocavities were first demonstrated in Japan ($Q > 10^4$ [46] and $Q > 10^6$ [47]). Up to 16 Cascade ring add/drop filters were produced by IBM [48] by using ring resonators. The IBM team also demonstrated optical buffering of 10 bits at 20 Gbps in 100 cascaded ring resonators [49] and recently fast optical switching [50].

In 2008, the emphasis moved from research and demonstration to commercial products. Few companies emerged as front-runners in the silicon photonics market. The big microelectronic giants, such as Intel, IBM, SUN, HP, have aggressive research programs on silicon photonics. On the other hand, smaller companies and start-ups are going to market Si-photonic devices. Lightwire Inc. launched a high-speed interconnect platform optical application-specific integrated circuit (OASIC) based on its patented silicon photonics [51]. Luxtera Inc. introduced Blazar 40G cable connector: a monolithic optoelectronic optical active cable operating at 40 Gbps that contains four complete fiber optic transceivers per end, each operating at data rates from 1 to 10.5 Gbps and supporting a reach up to 300 m [52]. Kotura Inc. showed the first example of successful silicon photonics-based product [53]: the UltraVOA array that provides simple current-controlled optical attenuation (0–40 dB) and enables ultrafast (300 ns) power management in optical networks.

This short historical overview emphasizes the progress of silicon photonics, needless to say that we have mostly reviewed the impact on interconnects.

The possibility of low dimensional silicon to tune on one side its electronic properties and on the other side its dielectric properties allows new phenomena and device concepts [54, 55]. In this chapter, the exploitation of low dimensional silicon (i.e., Si-nc) to demonstrate various photonic building blocks for an all Si nanophotonics is reviewed. We will emphasize how Si-nc can serve silicon photonics by reviewing performances and possibilities of low dimensional silicon in guiding, modulating and, above all, generating and/or amplifying light. It is worth to note that the main advantage of using Si-nc is to integrate light sources and/or amplifiers within CMOS photonics platform. This is the biggest challenge facing silicon nanophotonics.

16.2
Nanosilicon Waveguides and Resonators

16.2.1
General Properties

In Figure 16.2, we show a schematic example of an optical planar waveguide.

It is composed of three dielectric layers: the core (with a refractive index n_2) and the cladding layers (with refractive indices of n_1 and n_3). Light confinement in the x–y

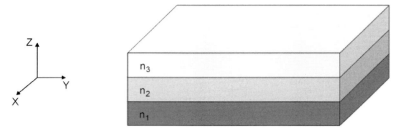

Figure 16.2 Schematic representation of an optical planar waveguide.

plane is achieved by the refractive index difference between the waveguide core and the cladding layers. Lateral patterning allows to produce light channels (Figure 16.3) where light is laterally confined too.

To obtain a single-mode channel waveguide with a high refractive index contrast, the waveguide cross section must be reduced to the order of $(\lambda/n_2)^2$, with λ the wavelength of the light. For very high contrasts, for example, silicon ($n_2 = 3.45$) to air ($n_3 = 1.0$) or silicon to silica ($n_3 = 1.45$), waveguides have widths smaller than 500 nm, with features that can be as small as 100 nm when operating at telecom wavelengths between 1.3 and 1.6 µm. Using high refractive index contrast implies not only that the geometrical features become very small (100–500 nm) but also that they have to be very accurately fabricated (few nm) to reduce the propagation losses in the waveguide. Different materials can be used to fabricate the waveguides. Silicon-on-insulators (SOIs) waveguides consist of a thin crystalline silicon layer on top of an oxide cladding layer (buried oxide or BOX) on a silicon wafer. SOI waveguides have a large refractive index contrast and are transparent at telecom wavelengths of 1.3 and 1.55 µm. For visible applications, other materials are needed such as Si oxynitride (SiON, n variable depending on the N content), Si nitride (Si$_3$N$_4$, $n = 2$) [56]. For nanosilicon photonics, we use Si nanocrystals in SiO$_2$ [57, 58].

A Si-nc waveguide is formed by a core layer of Si-rich Si oxide sandwiched between Si oxide cladding layers. Refractive indices ranging between 1.45 and 2.2 at 780 nm have been reported depending on the Si excess content [57]. In these waveguides, optical losses can have different origins, both intrinsic (absorption, excited carrier

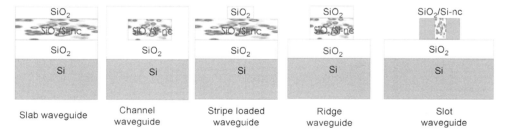

Figure 16.3 Different geometries to realize two-dimensional waveguides where Si-ncs are formed in the waveguide core.

absorption (CA), and Mie scattering) and extrinsic (scattering losses due to imperfections, side wall scattering, and radiation into the substrate). Optical losses of 120–160 dB/cm have been reported for Si-nc in the visible range [58]. Lower values (about 10 dB/cm) have been reported for thick-slab waveguides at 780 nm and about 3.5 dB/cm at 1000 nm, where Rayleigh scattering is decreased according to the well-known $1/\lambda^6$ law [57]. It is clear that the assessment of the losses strongly depends on the density and the size of the Si-nc. A detailed study as a function of the wavelength was recently performed [58]. We studied Si-rich silicon oxide and SiO_2 (SRSO)/SiO_2 multilayer samples grown by reactive magnetron sputtering and then annealed at high temperature to induce the formation of Si-nc with mean size of 3–4 nm and density of about $3.5 \times 10^{18}\,cm^{-3}$. Propagation losses decrease with increasing wavelength from about 73 dB/cm (at 785 nm) to 2 dB/cm (at 1630 nm). An analysis of the different contributions to the optical losses such as Mie scattering and scattering due to waveguide roughness has been done, allowing to isolate the contribution due to the absorption losses and, thus, to extract the absorption cross section at different wavelengths (Figure 16.4). Values of about $3.5 \times 10^{-18}\,cm^2$ have been found at 830 nm, increasing with decreasing wavelength.

In addition to the linear losses, other nonlinear losses exist in Si-nc waveguides. They are mainly due to two-photon absorption or due to excited carrier absorption. Two-photon absorption coefficient in Si-nc will be discussed in Section 16.7. Here, we discuss the excited carrier absorption. In bulk Si, this is due to free carrier and has been extensively studied [59], while only few works deal with Si-nc [60]. An extensive study on the carrier absorption mechanism in multilayered Si-nc rib waveguides has been performed [61]. A pump (532 nm) and probe (1535 nm) technique was used to assess the losses. When the waveguide is optically pumped, carriers are excited within the Si-nc and contribute to an additional loss term that is proportional to the number of excited carriers (N_{Carr}) and to their absorption cross section at the signal wavelength (σ_{CA}). Thus, the CA loss coefficient can be written as a function of the ratio between the transmitted signal when the waveguide is pumped to the

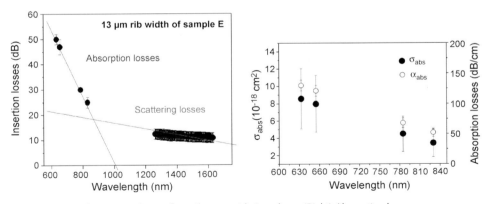

Figure 16.4 (Left) Insertion losses for a rib waveguide 3 cm long. (Right) Absorption losses α_{abs} (empty symbols) and absorption cross sections σ_{abs} (full symbols) as in Ref. [58].

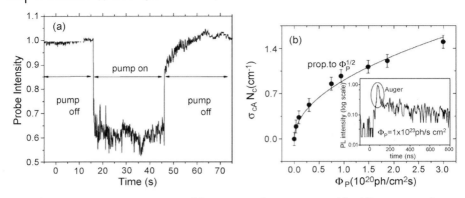

Figure 16.5 Direct measurement of the intensity of a 1535 nm signal for different pump photon fluxes: (a) full temporal dynamics and (b) carrier absorption losses at 1535 nm as a function of the photon flux. A square root fit to the experimental data is also shown (solid line). From Ref. [61].

transmitted signal and when the waveguide is not pumped (usually named signal enhancement, SE), in the following way:

$$\sigma_{CA} N_{Carr} = -\frac{\ln(SE)}{\Gamma L_{pump}}, \tag{16.1}$$

where Γ is the optical mode confinement factor and L_{pump} is the length of the waveguide that is actually excited by the pump. In Figure 16.5a, the transmitted signal is shown when the pump is switched on. A rapid decrease in transmission is observed. The dynamics of the decrease is characterized by two timescales, one fast (on the order of microseconds) and one slow (on the order of seconds). The slow one is due to thermal effects while the other is due to excited CA. Figure 16.5b shows the maximum of the excited CA losses as a function of the pump photon flux Φ_p. A square root dependence of $\sigma_{CA} N_{carr}$ is observed. CA losses increase with Φ_p, up to 6 dB/cm for 3×10^{20} ph/cm^2 s. Since σ_{CA} is independent of Φ_p, $N_{carr} \sim \Phi_p^{1/2}$. This is an indication of Auger-dominated recombination processes in the Si-nc, possibly between close Si-nc due to their particular close distribution in the multilayered samples. If we assume one excited carrier per Si-nc, from $\sigma_{CA} N_{carr} = 1.4$ cm^{-1}, we get $\sigma_{CA} = 4 \times 10^{-19}$ cm^2 at 1535 nm, when $N_{carr} = 3.5 \times 10^{18}$ cm^{-3}.

The excited carrier absorption has the same characteristic dynamics of the recombination of exciton luminescence in large Si-nc (see the inset in Figure 16.5). This indicates that the way to reduce the excited carrier absorption is to decrease the Si-nc size in the waveguide.

16.2.2
Si-nc Slot Waveguides

As Si-ncs within SiO$_2$ have a relatively low refraction index, their use in conventional channel waveguides would result in a large cross section and weak light confinement. So a new waveguide architecture where light propagates mostly in the low-index

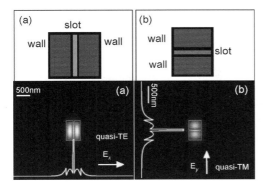

Figure 16.6 (a) Vertical slot waveguide field profile for the quasi-TE polarization. (b) Horizontal waveguide field profile for the quasi-TM polarization. The light propagates in the z-direction.

medium has been proposed: the slot waveguide (Figures 16.2 and 16.6) [15]. Their typical structure assumes a thin layer of a low refractive index material (n_S) sandwiched between two rectangular-shaped regions of a high-index material (n_H). The principle of operation is the discontinuity of the electric field at the interface between two dielectrics (Figure 16.6). The electric field component of the quasi-TE mode in the case of a vertical slot (quasi-TM mode in the case of horizontal slot) undergoes a discontinuity that is proportional to the square of the ratio between the refractive indices of the silicon and the low refractive index slot $(n_H/n_S)^2$ [15]. This discontinuity is such that the field is much more intense in the low refractive index slot region than in silicon. As the width of the slot is comparable to the decay length of the field, the electrical field remains high across the slot, so the power density in the slot part is much higher than in the silicon regions. For example, the fraction of power transmitted in a sub-100 nm wide slot can be higher than 40% of the total guided power. This waveguide geometry can be used in many photonic devices, as high-Q Fabry–Perot resonators [62] or ring resonators.

16.2.3
Ring Resonators Based on Slot Waveguides

Ring resonators are versatile building blocks widely used in photonics [49]. A way to excite the modes in a ring is to couple it with a waveguide so that the evanescent tails of the optical mode are confined in the waveguide and in the ring overlap. Under this condition, optical energy transfer to the ring and back to the waveguide may occur. Figure 16.7a represents schematically a ring resonator structure. To have a resonance, the optical path length in the ring should be a multiple of the wavelength or $m\lambda_m = 2\pi R\, n_{eff}$, where R is the ring radius, n_{eff} is the effective refractive index of the waveguide, λ_m is the resonance wavelength, and m is an arbitrary integer.

The gap width g drastically influences the coupling in the ring resonator. It can be controlled by lithographically adjusting it. We have tested in Ref. [63] several structures

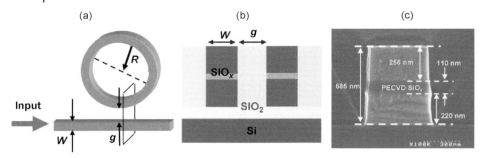

Figure 16.7 (a) Ring resonator layout, (b) cross-sectional view of the sandwich slot waveguide, and (c) SEM image of a horizontal slot waveguide structure [63]. R is the ring radius, g is the waveguide–ring gap distance, and W is the waveguide width.

with different gaps, ring radii, and waveguide widths. Figure 16.7b shows a cross-sectional view of the sandwich slot waveguide used to realize the ring resonator structures. The waveguides are fabricated on SOI wafers with 220 nm thick Si layer and 2 μm thick BOX. A SiO_x layer of 110 nm was grown by plasma-enhanced chemical vapor deposition (PECVD) and annealed at 1000 °C in order to promote Si-nc formation by Si/SiO_2 phase separation. 250 nm thick amorphous silicon was then deposited to produce the horizontal slot waveguide. A scanning electron microscopy (SEM) image of the slot structure is shown in Figure 16.7c.

The most important parameter of a resonator is its Q factor, which is defined as $Q = \lambda_m/\Delta\lambda$ with $\Delta\lambda$ the resonance mode linewidth. It is directly linked to the losses in the slot waveguide and to the single resonance width. Table 16.1 lists the results of a detailed test of slot waveguide-based ring resonators. Such Q-values are the largest ever reported for ring resonators fabricated by optical lithography and are of the same order of magnitude as the ones reported by [64] for similar structures fabricated by e-beam lithography.

The free spectra range (FSR), that is, the wavelength separation between two adjacent resonant peaks, may be tuned by properly choosing the ring radius R. The spectral responses of ring resonators as a function of the ring radius have been recorded in the wavelength range 1560–1620 nm. Figure 16.8 shows the results. From top to bottom, the radius of the ring R increases from 10 to 40 μm. As expected, a decrease in the FSR as R gets larger has been observed. We found FSR values of

Table 16.1 Experimental data of ring resonator structures fabricated by deep UV lithography.

W (μm)	g (μm)	R (μm)	λ_o (nm)	λ_{3dB} (nm)	ER (dB)	Q
0.5	0.25	10	1602.63	0.1418	− 15.2	11 300
		15	1608.43	0.1284	− 9.2	12 500
		20	1594.22	0.1630	− 8.2	9800

W denotes the waveguide width, g the gap, R the ring radius, λ_o resonance central wavelength, $\Delta\lambda_{3dB}$ the bandwidth, ER the extinction ratio, and Q the quality factor [63].

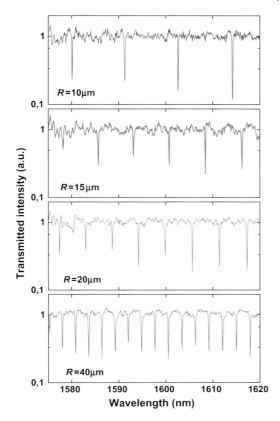

Figure 16.8 TM-like mode transmission spectra through a 500 nm waveguide coupled to rings with different radii of, from top to bottom, 10, 15, 20, and 40 μm, respectively.

about 11 nm in structures with a ring radius of 10 μm, regardless of the waveguide width and the gap distance. For some applications, in which a large FSR is required, small ring radii are needed. However, optical losses in such small rings, if fabricated with standard optical lithography, may drastically limit the Q factor to few hundreds [65].

16.3
Nanosilicon Cavity-Based Devices

Optical properties of nanosilicon-based materials, extensively described in the next sections of this chapter, can be further enhanced incorporating Si-nc in smart dielectric devices. In particular, dielectric microcavity-based devices increase the light–matter interaction slowing down the group velocity of light (slow-wave devices) or provide an optical feedback to allow lasing (Fabry–Perot, whispering gallery mode (WGM) active resonators).

16.3.1
Slow-Wave Devices

The decrease in the group velocity of light (i.e., the generation of slow light) inside a dielectric device allows to increase nonlinear effects [66], realize delay lines, or shift the carrier frequency of a pulse traveling inside the structure. Slow light technology is nowadays widely used in various devices such as optical fibers [67] and photonic crystals [68]. In waveguide technology, coupled resonator optical waveguides (CROWs) [69] in which the light propagates through coupled resonators have been realized, obtaining a delay as high as 500 ps [70]. The latest approach is the realization of slow-wave devices based on slot waveguide structures, in which the group velocity of light can be controlled and, at the same time, the electric field can be localized in the low-index slotted material. Structures based on photonic crystals waveguides (PhCWs) [71] and channel waveguide [72, 73] have been designed and realized.

Figure 16.9 shows the SEM picture and the photonic band structure of a slotted photonic crystal waveguide. It is clear that while PhCWs can achieve very low propagation losses of the propagating Bloch modes, their integration in optical circuits appears difficult due to their strong planar structure. A higher degree of compactness can be achieved by designing a photonic crystal structure along a

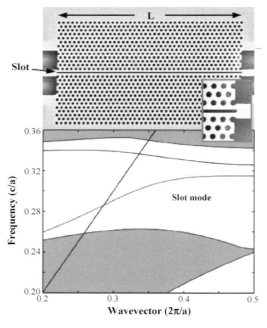

Figure 16.9 Top: an SEM image of a fabricated slotted photonic crystal waveguide SPhCW device and zoom-in image of the slot termination section. Bottom: band diagram in normalized frequency and wave vector units. Picture taken from Ref. [71].

channel waveguide. A careful design of a one-dimensional photonic crystal on a vertical slot waveguide structure (see Figure 16.7a) has been performed to realize Bragg mirrors with a photonic bandgap centered around 1.55 μm [72].

Figure 16.10 reports the bandgap maps for quasi-TE and quasi-TM polarizations for different realizations of the photonic crystal structure. The partial corrugation of the waveguide (Figure 16.10, panels b and c) lowers the propagation losses of the field inside the slot granting good guiding properties. A quasiflat band, few tens of nanometers wide, is obtained by careful engineering of a 5-coupled cavities structure (total length shorter than 10 μm). A particular care has been taken to the design of the first and last mirrors, relaxing the limitation induced by finite-size effect [74]. Moreover, a tapering of the first and last period in each mirror has been realized.

Since vertical slot waveguides are difficult to produce, a slow wave structure based on a completely etched Bragg mirror configuration in an horizontal slot waveguide has been designed and fabricated (Figure 16.11) [73].

The transmission spectrum of the slow wave structure is reported in Figure 16.12. It is possible to recognize the bandgap and the Bloch mode peak for the wavelength resonant with the cavity mode (Figure 16.12a). The spectral features simulated with finite-difference time-domain algorithm (Figure 16.12b) reproduce in the experimental data.

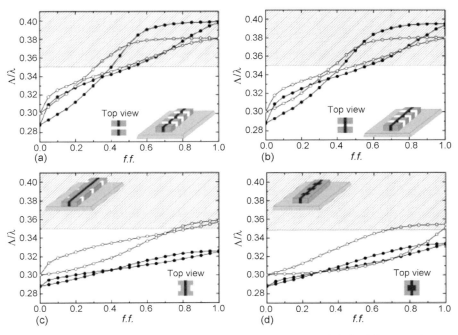

Figure 16.10 Bandgap maps for the quasi-TE (circle) and the quasi-TM (disk) for different comb-like structures realized on a Si-based vertical slot waveguide. Picture taken from Ref. [72].

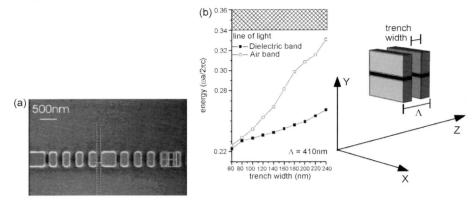

Figure 16.11 (a) SEM image of the photonic crystal structure processed on a horizontal slot waveguide (top view). (b) Quasi-TM bandgap in the high-symmetry X point for different trench etching widths.

16.3.2
Si-nc Active Microdisks

It is well known that light can propagate inside a dielectric structure by total reflection of light at the dielectric interface. Eventually, the reflected ray can interfere constructively leading to the formation of guided modes inside the structure. In a dielectric with azimuthal symmetry geometry (a disk, ring, or sphere), optical guided modes have to fulfill constructive interference after a round trip, which determines a mirrorless cavity. These optical modes are called whispering gallery modes, since

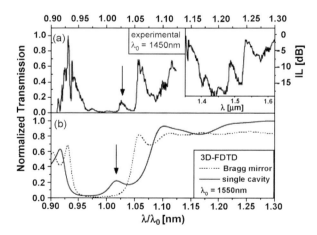

Figure 16.12 (a) Experimental measurement of the coupled resonators optical waveguide structure ($\lambda_0 = 1.45\,\mu m$) for quasi-TM-polarized light. The arrow shows the cavity peaks (inset, insertion losses of the device). (b) 3D FDTD simulation of the device with a single cavity and the Bragg mirror ($\lambda_0 = 1.55\,\mu m$).

they were first observed for acoustic wave in the dome of the St. Paul cathedral in London and mathematically described by Rayleigh [75]. In the 1960s, optical WGMs were observed for the first time in solid-state laser.

From the first studies on microspheres and microdroplets, advanced lithographic technique allowed the creation of micrometric size rings, disks, or even toroids [76]. The quality factors of this kind of cavity can be as high as 10^6 for passive SiO_2-based devices. Such high-quality cavities can be employed in a wide range of applications, such as frequency comb generators [77], optomechanics [78], or environmental sensors [79]. Lately, they are widely used as an experimental platform to study cavity quantum electrodynamic (CQED) fundamental physics [80].

Active based WGM cavities can be fabricated by using nano-Si. Si-ncs have a strong emission band (and a strong absorption cross section) in the visible: while the emission can be exploited to pump energy into the cavity without resorting to micrometric precision alignment systems, the material absorption losses lower the quality factor to 10^3–10^4 in the visible region for micrometric-sized devices [81, 82]. The most appealing feature of active microdisk devices is the possibility to obtain lasing in nano-Si, in analogy with the III–V material low-threshold WGM laser.

Active microdisks have been fabricated by using PECVD and optical lithography [82]. A typical spectrum from a single WGM microdisk cavity is shown in Figure 16.13.

In analogy with standard quantum mechanics, we can associate with each resonance an "optical quantum number": the radial number ϱ (number of field antinodes along radial direction inside the microdisk) and the azimuthal number m (number of field nodes along the circumference). In thin disk, it is possible to recognize the different WGM families, represented in Figure 16.13 with the notation (ϱ, m). The material characteristics influence the cavity behavior. For example, pump-induced losses as carrier absorption (see Section 16.2.1) degrade the cavity Q when the pump power is increased, as reported in Figure 16.14.

But nanosilicon-based materials have been widely employed in microcavities not only as luminescent center but also as thermo-optical switch [83] or as scattering center to perform fundamental physics measurement in the near-IR [84].

In the thermo-optical switch, a thin layer of Si-nc in silicon oxide matrix is coated around a silica microsphere. By pumping the thin layer of Si-nc with an Ar laser, it is possible to cause a geometric, heating-induced dilatation in the sphere, changing the resonance position, in such a way that the near-IR resonant signal can be transmitted through the tapered fiber coupled to the microsphere, as shown in Figure 16.15.

16.4
Si-nc-Based Visible Optical Amplifiers

Silicon nanocrystals provide a promising material for optically pumped (and even electrically, see Section 16.6) amplification both in the infrared (see Section 16.5) and in the visible wavelength regions. Hereafter, we focus on how Si-ncs are themselves

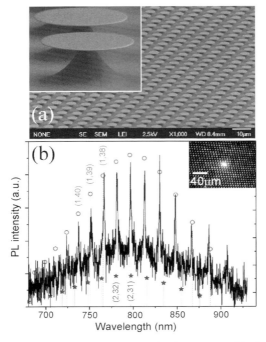

Figure 16.13 SEM picture (a) and WGM spectrum (b) of a single Si-nc-based microdisk resonator. The resonance positions have been calculated using an FDTD algorithm.

an optical active material at visible wavelengths, while in next section we review how they can efficiently sensitize Er ions for light amplification at 1.54 μm.

Optically pumped gain in Si-nc thin films has been reported by several research groups, notably by our laboratory [85–88]. We have shown amplified spontaneous

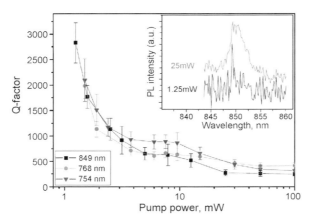

Figure 16.14 *Q*-factor dependence on pump power in a Si-nc-based microdisk resonator.

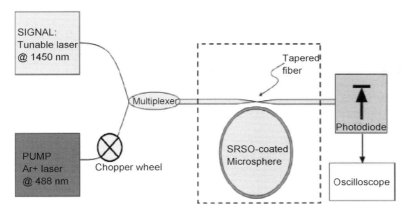

Figure 16.15 Experimental setup for the thermo-optical switch characterization. Picture taken from Ref. [83].

emission (ASE) from Si-nc grown by different techniques (ion implantation [25], PECVD [87, 88], electron evaporated superlattices [89], and magnetron sputtering) by means of VSL (variable stripe length) technique in the CW (continuous wavelength) and TR (time-resolved) regime, where the luminescence of Si-nc is used as a probe beam and one looks for enhancement as it propagates in an optically pumped waveguide. Figure 16.16 shows the most representative results on Si-nc PECVD samples. Loss or gain can be measured by VSL technique (a scheme of the method

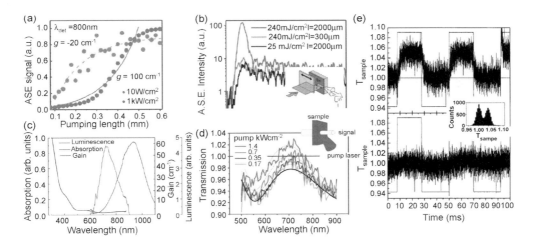

Figure 16.16 (a) Amplified spontaneous emission versus the pumping length for two pumping powers at 800 nm. (b) Time-resolved ASE for various pump powers and excited volumes. The inset shows a scheme of the VSL method. (c) Summary of the optical properties of Si-nc. (d) Transmitted intensity versus the wavelength for different power density by pump and probe measurements. Dark line refers to the transmission of the sample without pump. The inset shows the idea of the experiment. (e) Pump and probe experiments with chopped probe signal at 2 kW/cm^2 (top panel) and 50 W/cm^2 pump intensity (bottom panel).

is shown in the inset of panel b) depending on the pump power (Figure 16.16a). By modeling the system within a one-dimensional amplifier scheme the gain spectrum can be obtained. A summary of the emission, absorption, and gain spectra for a representative Si-nc sample is reported in Figure 16.16c. Absorption increases strongly at short wavelengths while emission (both spontaneous and stimulated) occurs at long wavelengths. At the same time, the gain and luminescence spectra peak at different wavelengths, which indicates that either only the small Si-ncs have strong gain or that gain and luminescence have a different origin. In the TR spectra obtained by VSL method, a fast recombination component appears in the decay dynamics (Figure 16.16b) that disappears when either the excitation length l is decreased at a fixed pump density power J_{pump} or when J_{pump} is decreased for a fixed l. These observations rule out the nonradiative Auger processes as the origin of the observed fast component since the intensity does not depend on the excitation length l, whereas the fast recombination peaks critically depend on the pumping length, keeping fixed the excitation conditions.

Gain has also been observed in signal amplification (i.e., pump and probe) experiments (Figure 16.16d and e) [87]. A red signal beam is transmitted through a thin (200 nm) layer of Si-nc on a quartz substrate and at the same time a blue pump beam is exciting the Si-nc. When the pump beam is weak, the transmission through the Si-nc is mostly unaffected by the presence of the probe beam. On the contrary, when the pump beam is strong, the transmission through the Si-nc gets larger than one: population inversion is reached and amplification of the signal observed.

Although a full theoretical model of the stimulated emission process in Si-nc is still lacking and the observed characteristics cannot be explained only on the basis of electron localization in the nanocrystals, a model to explain all these observations has been proposed [87]. The gain is associated with a four-level system. It is also worth to note that strong lattice relaxation (bond deformation) occurs when the Si-ncs are excited, and in a configurational coordinate diagram, this can be associated with a four-step process. Although this is still a phenomenological model, which does not refer to a developed theory of the optical properties of Si-nc in SiO_2 and of their interfaces, a physical interpretation of these levels has been suggested. It has been proposed that interface radiative states associated with oxygen atoms play a crucial role. They can be associated either with the formation of silicon dimers [90] or with Si=O bonds [91] at the interface between the Si-nc and the oxide or within the oxide matrix. X-ray measurements and *ab initio* calculations [92] showed the presence around the Si-nc of a strained SiO_2 region (about 1 nm) participating in the light emission process. One can speculate that this stressed SiO_2 shell enhances the formation of interface oxygen-related states on the surface of Si-nc or decreases the nonradiative Auger rate because of the resulting smoothing of the potential barriers, thus influencing the optical gain. Similar results have been obtained also by Monte Carlo simulations [93], and the role of the chemical passivation of Si-ncs has been pointed out in a recent experimental work [94], in which the coupling between surface vibrations and fundamental gap as well as the increase of interaction between them at the strong confinement regime are proposed to interpret light emission.

16.5
Nanosilicon-Based Infrared Optical Amplifiers

When a Si-nc-based waveguide is codoped with rare earth ions, an erbium-doped waveguide amplifier (EDWA) is within reach. EDWAs afford advantages with respect to erbium-doped fiber amplifiers (EDFAs), which are well established in long-haul transmission. Reducing their size and cost for widespread integration presents major difficulties: ion pair interactions, combined with the small excitation cross section of the Er^{3+} ion, necessitate the use of long lightly doped fibers. Moreover, high-power (and therefore expensive) laser diodes tuned to specific Er^{3+} transitions are required as pump sources. Clearly, a breakthrough would be a new gain medium that enables broadband optical or electrical excitation of rare earth ions, with a potential 100-fold reduction in pump costs, and provides order-of-magnitude enhancements in effective absorption cross sections. The new gain medium could be based on broadband sensitizers that absorb the excitation and transfer it to the rare earth ions. This relaxes the stringent conditions for the pump source and raises the efficiency of the optical amplifier. A good sensitizer has to have a high absorption cross section and has to efficiently transfer energy to Er^{3+}.

Si-ncs have typical absorption spectra that depend on the average size of the Si-nc but that usually start to be appreciable near 600 nm and grow toward shorter wavelengths (see Figure 16.16c). The absorption cross sections are in the order of 10^{-16} cm^2 around the 488 nm region that is five orders of magnitude higher than the absorption cross section of Er^{3+} in stoichiometric silica (1×10^{-22} cm^2 at 477 nm and 8×10^{-21} cm^2 at 488 nm). In addition, it has been demonstrated that Er^{3+}-doped silica containing Si-nc produced by cosputtering, plasma-enhanced chemical vapor deposition, or ion implantation exhibits a strong energy coupling between Si-nc and Er^{3+} [111]. Quantum efficiencies greater than 60% and fast Si-nc-to-Er^{3+} transfer rates have also been measured. In addition to the increase in effective excitation cross section (σ_{exc}) of the indirectly excited Er^{3+}, Si-ncs increase the average refractive index of the dielectrics, allowing good light confinement, and conduct electrical current that opens the route to electrically pumped optical amplifiers.

It has also been seen that the shape of the Er^{3+} photoluminescence (PL) spectra at low pumping powers when placed in a SiO_2 matrix is almost independent of the presence of the Si-nc on the matrix, which indicates that Er is surrounded by oxide. However, many aspects of the exact nature of the interaction between Si-nc and Er are still controversial. In particular, to engineer the system with the aim of achieving net optical gain in the amplifiers, the role of detrimental processes is to be figured out. Figure 16.17 summarizes the various mechanisms and defines the related cross sections for this system. Excitation of Er^{3+} occurs via an energy transfer from photoexcited e–h pairs that are excited in the Si-nc: the overall efficiency of light generation at 1.535 μm through direct absorption in the Si-nc is described by an effective Er^{3+} excitation cross section σ_{exc}. On the other hand, the direct absorption of Er^{3+} ions and the emission from Er ions are described by absorption (σ_{abs}) and emission (σ_{em}) cross sections, respectively. The typical radiative lifetime of Er^{3+} is of 9 ms, which is similar to the one of Er^{3+} in pure SiO_2. Several authors have suggested

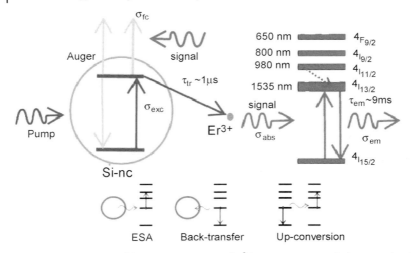

Figure 16.17 Diagram of the excitation process of Er^{3+} ions via a Si-nc, with the main related cross sections.

different channels for the quenching of the Er emission such as cooperative upconversion [98], excited state absorption [99], and Auger de-excitation [100].

Table 16.2 summarizes the various cross sections reported in the literature. It is important to note the five-order of magnitude increase in σ_{exc} and the fact that this value is conserved when electrical injection is used to excite the Si-nc. In addition, despite erroneous literature reports [101] on an enhanced σ_{em}, more reliable data show that its value is almost the same as that of Er in silica.

The main obstacles to achieve net optical amplification in Si-nc-based EDWA are carrier absorption losses [61] and the low number of Er ions coupled to Si-nc (few percentages). The reason for this low number is still unresolved [102, 103]. Recently, our group focused on eliminating the CA issue in silicon-rich silica oxide (SRSO) Er-doped layers [104]. The CA-induced losses are proportional to the exciton density in Si-nc; thus, to reduce the CA, a faster exciton recombination in small

Table 16.2 Summary of the various cross sections at 1.535 μm of Er^{3+} in various hosts.

Cross section	Er in SiO_2 (cm^2)	ER in Si (cm^2)	Er in Si-nc (cm^2)	References
Effective excitation cross section of luminescence by pumping at 488 nm	$1-8 \times 10^{-21}$	3×10^{-15}	$1.1-0.7 \times 10^{-16}$	[95]
Effective excitation cross section of electroluminescence		4×10^{-14}	1×10^{-14}	[96]
Emission cross section	6×10^{-21}		$5-10 \times 10^{-21}$	[97]
Absorption cross section	4×10^{-21}	2×10^{-20}	$5-10 \times 10^{-21}$	[97]

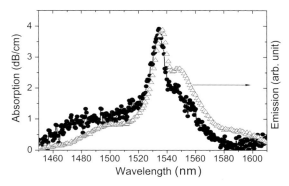

Figure 16.18 Absorption and emission (476 nm, 1.2×10^{18} ph/(cm^2 s)) spectrum of the rib waveguide sample (5 μm width) [105].

nanocrystals and/or a faster carrier population depletion (due to, for example, a transfer mechanism) is needed. In order to realize this, an intensive sample optimization has to be performed to achieve a high Er^{3+} photoluminescence signal with indirect excitation together with a long PL lifetime, maintaining a low Si excess in the sample to keep small Si-ncs. This ensures an efficient energy transfer and, in addition, a reduction in the CA within the Si-ncs.

The most general requirement for realizing an optical amplifier is to have an active material that can provide high enough net optical gain to compensate for the passive losses (propagation and coupling losses) of the waveguides. In our case, propagation losses (measured at 1600 nm) do not depend strongly on the channel widths and show an average value of about 3.5 dB/cm with the best value of 2.6 dB/cm. It is worth to note that material losses in a nonprocessed layer have been found as low as 1–2 dB/cm by SES (shift excitation spot) measurements [105].

Figure 16.18 reports the absorption and emission spectrum for 5 μm wide channel waveguide. From this result, it is possible to extract an absorption loss coefficient at 1535 nm of about 4 dB/cm. Hence, if the whole population is inverted and the propagation losses are of 3 dB/cm, net optical gain would be achieved.

Figure 16.19 shows pump and probe measurements performed for two different probe wavelengths: at the Er^{3+} emission spectrum peak (1535 nm) and almost outside the emission spectrum (1610 nm). Under these conditions, the ions are excited by pumping both nonresonant and resonant with Er^{3+} ions' internal transitions (respectively, $\lambda = 476$ nm and $\lambda = 488$ nm). The internal gain is almost zero within the error bars in a wide range of pumping powers. At higher powers, g shows a positive value for the Er-gain peak wavelength probe (1535 nm) while remaining zero for the probe wavelength outside the Er emission peak (1610 nm). In this particular case, the confinement factor of the waveguide, that is, the fraction of light traveling through the active region, is about 0.7. Therefore, at those photon fluxes a maximum internal gain of more than 1 dB/cm has been measured. The internal gain for the 1610 nm probe wavelength appears to be independent of the

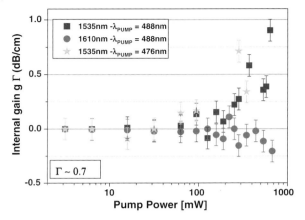

Figure 16.19 Pump and probe measurements on sample A for probe wavelengths of 1535 and 1600 nm.

pump photon flux, which means that the CA losses are negligible in this device. By comparing the measured signal enhancement (about 1 dB/cm) with the absorption coefficient at 1535 nm (about 4 dB/cm), the fraction of erbium coupled to Si-nc in our system can be estimated at 25% of the optically active Er ions. This represents by far the largest improvement from the few percentages reported in the literature so far.

To improve this result, our group looked into the possibility of an Auger back-transfer issue, performing fast (nanosecond) time-resolved IR spectroscopic measurements [106]. In fact, one interpretation of the transfer mechanism suggests an Auger-like transfer followed by a fast backtransfer between excited Er^{3+} and excitons within the Si-nc, limiting the active Er^{3+} to few percentages that is susceptible to be excited through indirect energy transfer [107]. Under this scenario, the dynamics of the process can be separated in two different contributions: (i) one fast (nanoseconds) transfer to the first excited state of Er^{3+}, through which it is possible to indirectly excite around 50% of the Er population, followed by a fast (nanoseconds) Auger backtransfer mechanism from excited Er^{3+} to excitons in Si-nc that acts as a nonradiative quenching; (ii) one slow (microseconds) transfer to higher Er^{3+} levels through which it is possible only to excite few percentages of the Er^{3+} population. In our samples also, there are two processes present in the 1.55 μm emission (inset of Figure 16.20): one fast (tens of nanoseconds, rise and decay) and one slow (microseconds, rise). However, spectral analysis showed a fast decaying signal even for wavelengths outside the Er^{3+} emission spectrum. Figure 16.20 shows the results of integrating the fast (first 200 ns) and slow (from 200 ns to 2 μs) decays separately as a function of the wavelength. Since it does not show the typical Er^{3+} spectral features, the fast mechanism is not associated with Er^{3+} ions emission. On the other hand, the slow part clearly reproduces the Er^{3+} emission spectra, showing both the 980 nm ($4I^{11/2} \rightarrow 4I^{15/2}$) and the 1.55 μm ($4I^{13/2} \rightarrow 4I^{15/2}$) transitions. It is worth noting that the intensity of the first transition appears higher than the second one due to the

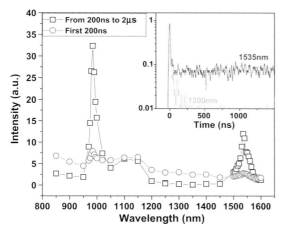

Figure 16.20 Time-integrated (fast, circle; slow, square) spectra for the sample A. The Er^{3+} spectral features are clearly visible for the slow process. Inset: decay signal inside (1535 nm) and outside (1300 nm) Er^{3+} emission spectrum.

much shorter lifetime (microseconds compared to milliseconds). Therefore, Er^{3+} ions are excited very fast (tens of nanoseconds), and once excited the level population dynamics is similar to that observed in a normal glass, without signs of energy backtransfer to the Si-nc. As a consistency check, the same measurements were performed in samples with the same Si-nc composition but without Er^{3+}, revealing only a fast component in the IR region. It is thus possible to conclude that Auger backtransfer is not relevant in high-quality samples.

16.6
Nanosilicon-Based LED and Solar Cells

16.6.1
Si-nc-Based LED

One of the more appealing properties of Si-nc is the possibility to get light by current injection into the Si-nc and thus enabling LEDs [108–111]. The main difficulty in obtaining efficient LEDs is to get efficient carrier injection. Interesting results have been obtained by the Stockholm group [112] in ion-implanted samples, showing maximum external quantum efficiency of about 3×10^{-5}. Similar data have been obtained in PECVD Si-nc [113]. Field effect-induced luminescence has been achieved by alternate tunnel injection of electrons and holes into Si nanocrystals with external quantum efficiencies of 0.03% [114]. Materials with optical properties different from SiO_2, such as Si nitride, have also been used as host matrices [115]. LEDs based on Si/SiO_2 superlattices were fabricated either by MBE [116] or by LPCVD [117, 118]: both PL and EL were observed. Lifetime tests of several LEDs showed stable continuous operation for over 1 year.

On the other hand, the electrical injection into Si-nc is a delicate task by itself. Indeed, in most of the reported devices, electroluminescence is produced either by blackbody radiation (the electrical power is converted into heat that raises the sample temperature which, in turn, radiates) or by impact excitation of electron–hole pairs in the Si-nc by energetic electrons that tunnel through the dielectric by a Fowler–Nordheim process (see Figure 16.21). Electron–hole pairs excited in this way recombine radiatively with an emission spectrum that is very similar to that obtained by photoluminescence. The problem with impact excitation is its inefficiency (maximum quantum efficiency of 0.1%) and the damage it induces in oxide due to the energetic electron flow. To get a high electroluminescence efficiency, one should try to get bipolar injection. However, bipolar injection is extremely difficult to achieve. In fact, the effective barrier for tunneling of electrons is much smaller than the one for holes (see Figure 16.21).

Despite some claims, most of the reported Si-nc LEDs are impact ionization devices: electron–hole pairs are generated by impact ionization by the energetic free carriers injected through the electrode. Another recent work reports on a FET structure where the gate dielectric is a thin oxide with a layer of Si-ncs [32, 119]. In this way, by changing the sign of the gate bias, separate injection of electrons and holes in the Si-nc is achieved. Luminescence is observed only when both electrons and holes are injected into the Si-nc. By using this pulsing bias technique, alternate charge carrier injection is achieved that leads to high efficiency in the emission of the LED. Electrical charge injection and charge trapping effects in Si-nc-based LEDs

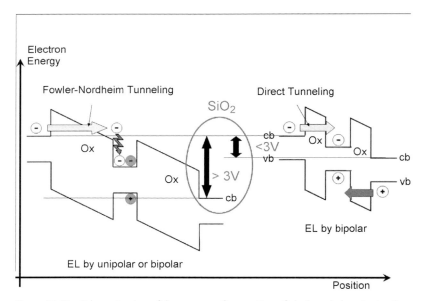

Figure 16.21 Schematic view of the process of generation of electron–hole pairs in silicon nanocrystals by impact excitation or direct tunneling: cb or vb refer to the conduction or valence band edges, while Ox refers to the silicon oxide barrier.

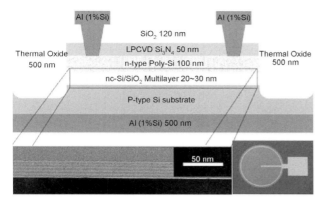

Figure 16.22 LED schematic cross section (top), top view of the LED (bottom right), and TEM image (bottom left) of the nc-Si/SiO$_2$ multilayer (annealed structure of SRO 4 nm/SiO$_2$ 2 nm, 5 periods) [127].

prepared by PECVD are examined in detail by current-voltage (I-V) capacitance-voltage (C-V) and impedance measurements [120, 121].

A typical Si-nc LED structure is a metal-oxide semiconductor (MOS) capacitor or transistor (Figure 16.22). Si-ncs are embedded in the SiO$_2$ layer of the MOS structure and have been formed from silicon-rich silicon oxide (SRO) or oxy-nitride by a high-temperature annealing. The Si-nc layer can be replaced with a multilayered Si-nc/SiO$_2$ oxide. Using multilayer structures in electroluminescence experiments has a relatively long history [122–126]. Nowadays, interest in multilayers has been revived [127–129], supported by several enhanced properties of the multilayer LED with respect to a single-layer nanocrystalline silicon LED: low turn-on voltages, low leakage currents, high nanocrystal density, and more uniform nanocrystal size. The main advantage of the multilayer structure is the possibility of exercising a better control over the nanocrystal formation through the confined silicon growth and, hence, the possibility of accurately designing the nanocrystalline silicon layer of the LED.

The conductivity of the multilayer Si-nc LED is controlled by direct tunneling of electrical charges between Si nanocrystals [130]. The I-V characteristic of such devices is strongly superlinear (Figure 16.23) with two distinct regions being observed in the log–log plot. The current density shows a weak increase in the low electric field region (for the multilayer devices, it is up to 0.5 MV/cm) and increases much faster in the region of high fields. At the low electric fields, the current density of a sample with a multilayer structure formed by SRO 3 nm/SiO$_2$ 2 nm is two orders of magnitude larger than that of a single-layer reference sample with the same average silicon content. With increasing electric field strength, the difference between the current densities of single layer and multilayer becomes smaller. The enhanced current at the low electric fields in the multilayer device originates from the direct charge tunneling from the substrate into the silicon nanocrystals and, then, into the gate. In the multilayer structure, high nanocrystal density is expected, that is, silicon nanocrystals are closer and the oxide thickness among them is reduced, which facilitates the charge transport.

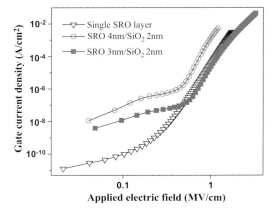

Figure 16.23 Gate current density as a function of applied electric field. Two distinct regions can be observed in this log–log plot. Multilayer devices show much higher current density in the low electric field region. The multilayer SRO 3 nm/SiO$_2$ 2 nm has the same average silicon content as the single layer and lower silicon content than the SRO 4 nm/SiO$_2$ 2 nm [127].

Direct tunneling is not only less destructive than the Fowler–Nordheim tunneling but also presents a more efficient way of injecting charges into the nanocrystals. This is evident from Figure 16.24, which compares two Si-nc LEDs with these two injection mechanisms: multilayer LED with the dominant direct tunneling and single-layer LED with the dominant Fowler–Nordheim tunneling. This figure also shows a typical dependence of electroluminescence emission on the injected current, which is a linear function in bilog coordinates. Another important property of the multilayer LED is the lower operating voltages than in the common Si-nc LEDs. In the

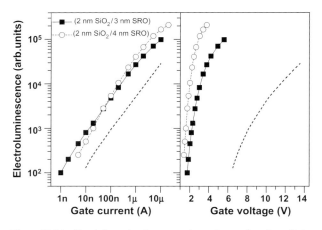

Figure 16.24 Total electroluminescence intensity as a function of injected current and gate voltage. The dotted line is the corresponding EL emission from a single layer LED approximately 50 nm thick with the same average silicon content as the multilayered LED, 2 nm SiO$_2$/3 nm SRO.

example shown in Figure 16.24, these voltages are below 5 V, which is of interest for Si-nc LED applications.

16.6.2
Si-nc-Based Solar Cells

For silicon solar cells, the efficiency limit is partly caused by an underuse of photons with energy higher than bandgap due to thermalization of hot carriers (thermalization losses) [131] and by the loss of subbandgap photons present in sunlight (IR losses). On the contrary, in the third-generation photovoltaics (PVs) [132–134], thermalization losses are reduced by using tandem cells [135], hot carrier cells [131], down converter cells [136, 137], and multiple exciton generation (MEG) solar cells [138–141]. IR losses might be overcome by intermediate band solar cells [142] where the subbandgap photons are absorbed through intermediate level/band present in the bandgap (known also as impurity photovoltaic effect [142]) or by coupling the cell with an upconverter material that absorbs two or more subbandgap photons and emits photons with energy larger than the bandgap of the cell material [136, 143]. Therefore, the theoretical efficiency of the third-generation solar cells can be well beyond the detailed balance limit of approximately 31% under direct AM1.5 sunlight for the so-called first-generation crystalline silicon solar cells [144]. In all these concepts, nanocrystalline materials and silicon, in particular, are supposed to play an important role [145, 146].

There are many applications of Si-ncs in the third-generation solar cells. One of them is all-silicon tandem cell [147]. It has stacks of subcells with different energy thresholds each absorbing a different band of the solar spectrum, usually connected together in a series. The active material of the subcell absorbing photons with higher energy than that of bulk silicon is Si-nc, which has larger and tunable bandgap compared to bulk silicon. This strategy can enhance the photocurrent of the solar cell. The other application is hot carrier solar cell [148], where photoexcited carriers with high energy (hot carrier) can be collected while they are still at elevated energies and thus allowing higher voltages to be achieved. Ideally, this collection would be isoentropic using monoenergetic contact, which has been attempted experimentally by a structure with single layer of Si-nc sandwiched by SiO_2 [149].

The Si-nc for all silicon tandem cells is mostly fabricated following the superlattice approach, where the phase separation is the main mechanism [150]. For solar cell application, the main challenge for this structure is to achieve sufficient carrier mobility and hence a reasonable conductivity. This generally requires formation of a true superlattice with overlap of the wave function for adjacent quantum wells or quantum dots, which in turn requires either close spacing between Si-ncs or low barrier height. That is to say that the inter-Si-nc distance is more important than Si-nc size [151]. However, for a given inter-Si-nc distance, the transport between two Si-ncs can also be modified by the host matrix in which the Si-ncs are embedded. It has been found that SiC and Si_3N_4 matrix give lower barrier heights [148] and also longer distance between Si-ncs for significant wavefunction overlapping [152] than those of SiO_2. The conductivity can also be improved by using a lateral multilayer Si-nc/SiO_2

structure [153]. This means that the carrier extraction takes place parallel to the Si/SiO$_2$ interfaces of two-dimensional Si-ncs while growth confinement is sustained in the vertical direction. It is shown that the developed lateral contact scheme is able to provide four orders of magnitude enhanced conductivity compared to Si-nc/SiO$_2$ multilayer with standard vertical contacts where the charge transport is limited by insulating SiO$_2$ barriers [154]. Another problem for this superlattice structure is how to precisely control the Si-nc size by the defined thickness of SRO. It has been found that in Si-nc/SiO$_2$ multilayer, crystallinity of approximately 5% and approximately 25% for the 2 and 5 nm thick SRO layers was obtained, respectively [155]. This is mainly influenced by stress, which depends on the periods of the multilayer, substrate, and annealing processes [156].

The Si-nc embedded in SiO$_2$ thin film is a complex system. There are some interesting photoresponse features in this system that were reported recently, which demonstrate the potential of Si-ncs for photovoltaic application.

Dark and illuminated (with 633 nm laser) *I–V* characteristics of a Si-nc device (MOS structure with a 50 nm thick Si-rich silicon oxynitride as gate oxide) along with the difference between the two currents (the so-called, photocurrent, $J_{ph} = J_L - J_D$) are shown in Figure 16.25 [157, 158]. The Si nanocrystals are formed in silicon oxynitride matrix with high silicon content by high-temperature annealing. This plot shows three distinct regions in the *I–V* characteristics. Under reverse bias, it shows a characteristic like a photodiode (PD). Under forward bias, the curve enters into the fourth quadrant showing the photovoltaic region with a large spectral response. Beyond the open-circuit voltage, current under illumination J_L has been found to be less than the dark current J_D, indicating the negative photoconducting region (–PC). Above a forward bias around 1.7 V, J_L becomes higher than J_D and the device acts as a photoconductor (PC).

An internal gain mechanism was discovered in the Si-nc [157], which brings about the PV effect shown in Figure 16.25. This has been demonstrated by a superlinear

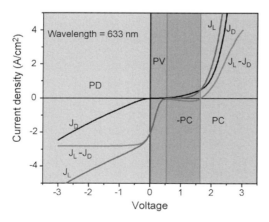

Figure 16.25 Dark current, current under illumination (633 nm, 700 mW/cm^2), and the difference of the two (photocurrent) as a function of the applied bias.

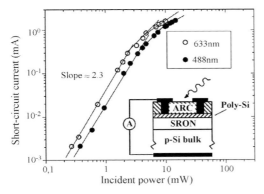

Figure 16.26 Variation in short-circuit current as a function of incident power intensity for two different wavelengths (488 nm, disks and 633 nm, circles). Inset: schematic cross section of the device [157].

variation of the short-circuit photocurrent with increasing incident optical power (Figure 16.26). The short-circuit current increases with the incident optical power as the power function of about 2 at low optical powers and saturates at very high optical power.

The internal gain mechanism is explained by secondary carrier generation from subbandgap states at the Si-nc interface, which is caused by energetic photogenerated electrons. The presence of the subbandgap states at the interface of Si-nc and silicon oxynitride matrix has been confirmed with spectral response (responsivity) measurements [158]. The spectral response of the device at a reverse bias of −5 V is shown in Figure 16.27. Figure 16.27a shows the responsivity at short wavelengths while Figure 16.27b depicts the same at longer wavelengths. The line shape of the responsivity in the 350–1200 nm range is very similar to that of crystalline silicon. On the other hand, a weak but significant responsivity is also measured below the optical bandgap of both crystalline silicon (1.1 eV) and of nanocrystalline silicon in

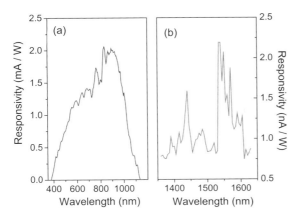

Figure 16.27 Spectral response of the device (a) in 350–1200 nm range and (b) in 1360–1630 nm range [158].

the SRO layer (1.6 eV in this case). This absorption is not found in a control silicon solar cell, which indicates that it is due to the presence of subbandgap energy levels in the SRO layer. The spectral peak at about 1550 nm in the responsivity curve can be tentatively associated with nitrogen-related mid-bandgap states formed at the Si-nc/silicon oxynitride matrix interfaces [159].

A simple phenomenological model was developed [157], which is illustrated with a scheme shown in Figure 16.28. When a 488 or 633 nm photon is absorbed by the active layer, generating an electron–hole pair inside a Si-nc (Figure 16.28a), the photogenerated hole is most likely trapped within the nanocrystal but the photogenerated electron either contributes to the short-circuit current (process labeled 1 in Figure 16.28a) or recombines with the trapped hole radiatively (process labeled 2) or gets trapped at the dielectric/Si-nc interface (process labeled 3). When second photon is absorbed by the same Si-nc (Figure 16.28b), fourth option is possible: since the energy of the absorbed photon is much larger than the nanocrystal bandgap, the photogenerated electron has extra kinetic energy that can be released by impact excitation of the trapped electron (Figure 16.28b). This mechanism generates the current of secondary carriers, which brings about the superlinear dependence of the short-circuit current shown in Figure 16.26.

These results indicate that a proper design of the Si-nc device may lead to a significant photovoltaic effect even with illumination with light where Si is transparent. Thus, Si-nc is a potential material for the third-generation photovoltaics not only for its application in MEG solar cells but also for intermediate band solar cells.

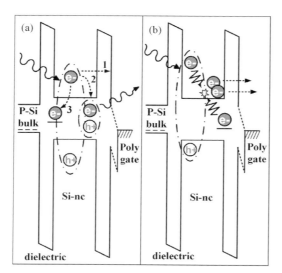

Figure 16.28 Schematic energy diagram of the active layer under short circuit bias. (a) Primary photoexcitation process where an electron–hole pair is generated and where the photoexcited electron either contributes to the short-circuit current (1), or recombines (2), or is trapped (3). (b) Secondary photoexcitation process where a photogenerated electron detraps a subbandgap electron and both contribute to the short-circuit current.

16.7
Nanosilicon Nonlinear Optical Properties and Devices

All optical devices based on nonlinear optical effects are very promising for meeting the need of high bandwidth in optical networking. In recent years, a lot of efforts were devoted to the research of the best material, with high nonlinear response but with low losses too. The first materials used for nonlinear applications were silicon and silica. The former assures a quite high nonlinear refractive index (as shown in Table 16.3) but, on the other hand, presents a high nonlinear absorption mainly due to free carrier effect. The latter has low nonlinear absorption but does not have high nonlinear refraction. Other candidates studied in recent times are the Chalcogenide Glasses [160] with nonlinear refractive index of the same order of magnitude of silicon, and small TPA coefficient. Moreover, large-scale diffusion of integrated devices requires materials with low production costs. Our main goal of investigating Si-nc nonlinear properties is to assess the adequacy of this material as a good candidate in all-optical nonlinear devices due to their high nonlinear response coupled with low nonlinear losses and their low production costs. In fact, a figure of merit for applications in optical nonlinear devices is given by $f = \frac{n_2}{\beta\lambda}$, where n_2 and β are the nonlinear refractive index and the nonlinear absorption coefficients, respectively. These are defined according to

$$n = n_0 + n_2 I$$
$$\alpha = \alpha_0 + \beta I, \tag{16.2}$$

where n_0 and α_0 are the linear refractive index and absorption coefficients, respectively, and I is the optical power density. On the one hand, an accurate characterization of nonlinear properties of Si-nc was conducted for second-order nonlinearities, such as second [161] and third [162] harmonic generation. Even the Kerr effect was investigated in the visible region, at 532 and 355 nm [163], or at 813 nm [164]. We extend these investigations at the relevant 1.55 μm wavelength.

Let us note that the optical nonlinearities in a semiconductor are a combination of bound electronic, free carrier, and thermal effects. Bound electronic response involves a distortion of the electronic cloud about an atom by the optical field. If the atom is highly polarizable, it can exhibit a significant electronic nonlinearity (n_{2be}). This kind of response is very fast since it instantaneously follows the field. However, other contributions are also present. Single- or two-photon absorption

Table 16.3 Comparison among different nonlinear materials.

Material	n_2 at 1550 nm (cm^2/W)	References
Silica	1.54×10^{-16}	[166]
Silicon	4.5×10^{-14}	[167]
Ga-As	1.59×10^{-13}	[168]
Si-nc	1–6×10^{-13}	[169]

processes can excite free carriers in a semiconductor. In turn, free carriers absorb the incident radiation: an effect that is related by Kramers–Kronig relation to a change in the refractive index. Thus, excitation via one- or two-photon absorption of a significant population of free carriers makes an additional contribution to the nonlinear refractive index (n_{2fr}). The induced free carrier refraction occurs on a timescale typical of carrier generation and recombination, a timescale of hundreds of nanoseconds to a few microseconds; this is why this effect can compromise the fast response of nonlinear devices. The thermalization of excited carriers via nonradiative recombination is responsible for the heating of the material and constitutes one of the sources of the thermal lensing effect. Thus, the nonlinear response n_2 of a semiconductor is the result of three terms: $n_2 = n_{2be} + n_{2fr} + n_{2th}$. To separate the various contributions, one method is to perform experiments at different timescales: with low repetition rate and fast pulses, one emphasizes the bound electronic contribution; with high repetition rate or long pulses, one emphasizes the others.

To measure n_2, we performed nonlinear transmission z-scan experiments [165]. Three laser sources were used in the measurements: (i) an optical parametric generator pumped by a Ti:sapphire laser, $\lambda = 1550$ nm, 1 kHz repetition rate with 100 fs pulse duration, peak intensity $I_0 \sim 10^{11}$–10^{12} W/cm^2; (ii) an optical parametric oscillator pumped by a Nd:YAG laser, $\lambda = 1550$ nm, 10 Hz repetition rate with a train pulse with 20 pulses of 20 ps duration, peak intensity $I_0 \sim 10^9$–10^{10} W/cm^2; and (iii) an optical parametric oscillator pumped by a Nd:YAG laser, $\lambda = 1550$ nm, 10 Hz repetition rate with 4 ns pulse duration, $I_0 = 10^8$–10^9 W/cm^2. To evaluate the nonlinear contributions, we have fitted the experimental transmittance $T(x)$ by the expressions

$$T(x) = 1 + 4 \times \Delta\varphi/[(x^2 + 9)(x^2 + 1)] - \beta I_0 d(3 - x^2)/[(x^2 + 9)(x^2 + 1)]$$

(16.3)

for closed aperture configuration and

$$T(x) = 1 + \beta I_0 d/(1 + x^2)$$

(16.4)

for open-aperture configuration. In Eqs (16.3) and (16.4), $x = z/z_0$ is defined as the reduced distance from the focal point, z being the absolute longitudinal distance, and z_0 the Rayleigh range of the beam. I_0 and d are the peak intensity at the focus position and the thickness of the film, respectively. Finally, the phase change $\Delta\varphi$ is related to the nonlinear refractive index n_2 by

$$n_2 = \Delta\varphi\lambda\alpha/[2\pi I_0(1 - e^{-\alpha d})],$$

(16.5)

where α is the linear absorption coefficient at the used wavelength λ. At 1550 nm, from the analysis of the z-scan traces considerably high nonlinear coefficients were obtained. Thus, nanosecond pulses generate a strong negative nonlinear response, on the order of $n_2 \sim -(10^{-9}$–$10^{-8})$ cm^2/W, which is due to thermal effects, and is responsible of a strong thermal lens effect; in this excitation regime, the nonlinear absorption coefficient β is found to be in the range of 10^{-7}–10^{-6} cm/W.

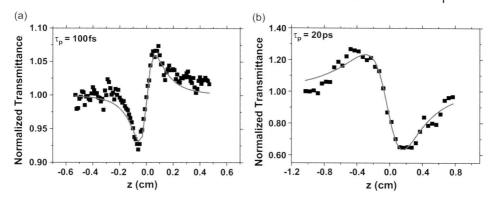

Figure 16.29 Comparison between z-scan measurements for high and low repetition rate exciting pulses (a) and for high repetition rate exciting pulses (b) on the sample with 21% of silicon excess annealed at 800 °C.

The z-scan traces change when femtoseconds with low repetition rate or picoseconds with high repetition pulses are used (Figure 16.29). Also, the magnitude of the nonlinear response changed. In fact, with femtosecond pulses, a positive nonlinear refractive index (valley–peak curve) on the order of $n_2 \sim 10^{-13}\,cm^2/W$ for a peak intensity in the range of $I_p = 10^{11}$–$10^{12}\,W/cm^2$ was measured. In this case, both the sign and the value show that the nonlinearity is due to bound electronic effects. While in the picosecond excitation regime, a strong negative nonlinear response (peak–valley curve) is found with a $n_2 \sim -(10^{-10})\,cm^2/W$ for a peak intensity of I_p 10^9–$10^{10}\,W/cm^2$. This negative nonlinear response is due to free carrier refractive effects and thermal cumulative effects due to the high repetition rate in each train pulse.

Table 16.4 shows a clear trend for n_{2be} both as a function of the annealing temperature and as a function of the Silicon excess. In particular, a strong nonlinearity is observed from the sample annealed at low temperature and with low silicon content. Low temperature and low silicon excess mean small Si-nc, which indicates an influence of quantum size effects on the nonlinearity values.

The situation is somehow different if the measurements are performed in resonance with Si-nc emission band at 800 nm. In order to obtain the electronic response of the SiO_x films, the real part of third-order nonlinear susceptibility has

Table 16.4 Nonlinear parameters for a few PECVD samples measured at fs excitation and at $\lambda = 1550\,nm$.

Si_{exc} (%)	T_{Ann} (°C)	n_{2be} (cm²/W)	β_0 (cm/W)
20	800	$(2.3 \pm 0.5) \times 10^{-13}$	$(4.0 \pm 0.3) \times 10^{-9}$
8	1100	$(1.3 \pm 0.7) \times 10^{-13}$	$(4.4 \pm 0.3) \times 10^{-9}$
8	800	$(4.8 \pm 0.6) \times 10^{-13}$	$(7.0 \pm 0.6) \times 10^{-9}$

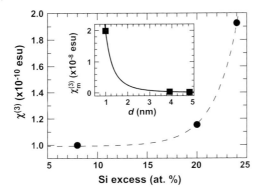

Figure 16.30 Electronic susceptibility versus Si excess at 1.5 eV; the dashed line is a guide to the eye. In the inset, the microscopic susceptibility at the same excitation energy has been represented as a function of the Si nanocrystal size; the solid line represents best fit to a $1/d^3$ dependence. From Ref. [171].

been calculated from $Re\chi^{(3)} = 2n^2\varepsilon_0 c n_2$, where ε_0 is the permittivity of free space and c is the velocity of light in vacuum. We have plotted the third-order nonlinear susceptibility as a function of the Si excess in Figure 16.30, where an increase is apparent with the amount of Si in the films. The observed $Re\chi^{(3)}$ accounts for an effective electronic response of the SiO_x material that at the same time includes both the density (or volume fraction) and the size of Si particles. Therefore, in order to single out the contribution of the Si-aggregates, we followed the approach of [170]. We considered the microscopic third-order nonlinear susceptibility $Re\chi_m^{(3)}$, which is related to the effective $Re\chi^{(3)}$ by experimentally measured linear parameters:

$$Re\chi_m^{(3)} = 2\pi/\lambda \times k_{si}/(f^2\sqrt{\varepsilon_0}) \times Re\chi^{(3)}/\alpha, \tag{16.6}$$

where f is the local field effect [$f = 3\varepsilon_0/(\varepsilon + 2\varepsilon_0)$], ε_0 and ε denote the dielectric constants of SiO_2 and Si, respectively, and k_{Si} is the extinction coefficient of Si. In this way, by looking at $Re\chi_m^{(3)}$, we have decorrelated the effects due to the variation in Si volume fraction between samples, and established the correct evolution of the susceptibility of the single Si-nc versus its size. The results are shown in the inset of Figure 16.30, where $Re\chi_m^{(3)}$ data quite nicely follow a $1/d^3$ dependence, typical of a contribution from intraband transitions induced by quantum confinement. The excellent agreement of our experimental data with the proposed trend shows that in the studied size range the main contribution to the electronic susceptibility of the Si-ncs comes from intraband transitions between discrete levels in the conduction band, and the quantum confinement strongly enhances the nonlinear response of the system, mostly for Si-nc with a diameter smaller than 2 nm.

The application of these optical nonlinearities in all-optical nonlinear devices has shown Gb/s switching rate with overall performances competitive with those achieved with more mature materials.

16.8
Conclusions

In this chapter, we have introduced various concepts on nanosilicon photonics. In particular, we have underlined how quantum size effects that occur in low-dimensional silicon together with the delicate interplay between the silicon and the embedding matrix allow to underpin new phenomena that in turn enable fabrication of new devices. Further progress can be foreseen in near future with the development and demonstration of a wide spectrum of new photonic devices.

Acknowledgments

This work is supported by EC through the LANCER (FP6-033 574), PHOLOGIC (FP6-017 158), POLYCERNET (MCRTN-019 601), WADIMOS (FP7-216 405), and HELIOS (FP7-224 312) projects, by PAT through the HCSC and NAOMI projects and by a grant from INTEL. We acknowledge the help of many coworkers from both the national and the international collaboration. They can be recognized in the cited literature. In particular, we would like to recognize the hard work of many present and past collaborators of the Nanoscience Laboratory without whom the research here reported would not have been performed. Last but not the least, the ongoing collaboration with the Center of Materials and Microsystems of the Bruno Kessler Foundation has permitted the achievement of most of the results shown in this chapter. Many friends and colleagues shared with us their expertise and experience.

References

1 Jalali, B. (2006) Silicon photonics. *J. Lightwave Technol.*, **24**, 4600.

2 Soref, R.A. (2006) The past, present, and future of silicon photonics. *J. Sel. Top. Quant. Electr.*, **12**, 1678.

3 Pavesi, L. and Lockwood, D. (2004) Silicon photonics, in *Topics in Applied Physics*, vol. 94, Springer-Verlag, Berlin.

4 Soref, R.A. and Lorenzo, J.P. (1985) Single-crystal silicon: a new material for 1.3 and 1.6 μm integrated-optical components. *Electron. Lett.*, **21**, 953–954.

5 Soref, R.A. and Lorenzo, J.P. (1986) All-silicon active and passive guided-wave components for $\lambda = 1.3$ and 1.6 μm. *IEEE J. Quant. Electron.*, **22**, 873.

6 Canham, L.T. (1990) Silicon quantum wire array fabrication by electrochemical and chemical dissolution of wafers. *Appl. Phys. Lett.*, **57**, 1046–1048.

7 Soref, R.A. (1993) Silicon-based optoelectronics. *Proc. IEEE*, **81**, 1687–1706.

8 Kimerling, L.C. (2000) Silicon microphotonics. *Appl. Surf. Sci.*, **159–160**, 8–13.

9 Bisi, O., Campisano, S.U., Pavesi, L., and Priolo, F. (eds) (1999) Silicon based microphotonics: from basics to applications, in *Proceedings of the International School of Physics E. Fermi: Course: CXLI*, IOS Press, Amsterdam.

10 Lee, K.K., Lim, D.R., Luan, H.-C., Agarwal, A., Foresi, J., and Kimerling, L.C. (2000) Effect of size and roughness on light transmission in a Si/SiO_2 waveguide: experiments and model. *Appl. Phys. Lett.*, **77**, 1617–1619.

11 Loncar, M., Doll, T., Vuckovic, J., and Scherer, A. (2000) Design and fabrication

of silicon photonic crystal optical waveguides. *J. Lightwave Technol.*, **18**, 1402–1411.

12 McNab, S., Moll, N., and Vlasov, Y. (2003) Ultra-low loss photonic integrated circuit with membrane-type photonic crystal waveguides. *Opt. Express*, **11**, 2927–2939.

13 Vlasov, Y.A. and McNab, S.J. (2004) Losses in single-mode silicon-on-insulator strip waveguides and bends. *Opt. Express*, **12**, 1622–1631.

14 Kuramochi, E., Notomi, M., Hughes, S., Shinya, A., Watanabe, T., and Ramunno, L. (2005) Disorder-induced scattering loss of line-defect waveguides in photonic crystal slabs. *Phys. Rev. B*, **72**, 161318.

15 Almeida, V., Xu, Q., Barrios, C., and Lipson, M. (2004) Guiding and confining light in void nanostructure. *Opt. Lett.*, **29**, 1209.

16 Marti, J. *et al. Nat. Photon.*, submitted for publication.

17 Shoji, T., Tsuchizawa, T., Watanabe, T., Yamada, K., and Morita, H. (2002) Low loss mode size converter from 0.3 μm square Si wire waveguides to singlemode fibers. *Electron. Lett.*, **38**, 1669–1670.

18 Almeida, V.R., Panepucci, R.R., and Lipson, M. (2003) Nano-taper for compact mode conversion. *Opt. Lett.*, **28**, 1302–1304.

19 Taillaert, D., Bogaerts, W., Bienstman, P., Krauss, T.F., Van Daele, P., Moerman, I., Verstuyft, S., De Mesel, K., and Baets, R. (2002) An out-of-plane grating coupler for efficient butt-coupling between compact planar waveguides and single-mode fibers. *IEEE J. Quant. Electron.*, **38**, 949–955.

20 Huang, F.Y., Sakamoto, K., Wang, K.L., Trinh, P., and Jalali, B. (1997) Epitaxial SiGeC waveguide photodetector grown on Si substrate with response in the 1.3–1.55 μm wavelength range. *IEEE Photon. Technol. Lett.*, **9**, 229–231.

21 Colace, L., Masini, G., and Assanto, G. (1999) Ge-on-Si approach to the detection of near-infrared light. *IEEE J. Quant. Electron.*, **35**, 1843–1852.

22 Koester, S.J., Schaub, J.D., Dehlinger, G., Chu, J.O., Ouyang, Q.C., and Grill, A. (2004) High-efficiency, Ge-on-SOI lateral PIN photodiodes with 29 GHz bandwidth. Device Research Conference.

23 Jutzi, M., Berroth, M., Wohl, G., Oehme, M., and Kasper, E. (2005) Ge-on-Si vertical incidence photodiodes with 39-GHz bandwidth. *IEEE Photon. Technol. Lett.*, **17**, 1510–1512.

24 Michel, J., Liu, J.F., Giziewicz, W., Pan, D., Wada, K., Cannon, D.D., Jongthammanurak, S., Danielson, D.T., Kimerling, L.C., Chen, J., Ilday, F.O., Kartner, F.X., and Yasaitis, J. (2005) High performance Ge p–i–n photodetectors on Si. Proceedings of the Group IV Photonics Conference, 2005, pp. 177–179.

25 Pavesi, L., Dal Negro, L., Mazzoleni, C., Franzo, G., and Priolo, F. (2000) Optical gain in Si nanocrystals. *Nature*, **408**, 440.

26 Han, H.-S., Seo, S.-Y., and Shin, J.H. (2001) Optical gain at 1.54 μm in erbium-doped silicon nanocluster sensitized waveguide. *J. Appl. Phys.*, **27**, 4568.

27 Nayfeh, M.H., Barry, N., Therrien, J., Akcakir, O., Gratton, E., and Belomoin, G. (2001) Stimulated blue emission in reconstituted films of ultrasmall silicon nanoparticles. *Appl. Phys. Lett.*, **78**, 1131.

28 Castagna, M.E., Coffa, S., Monaco, M., Muscara, A., Caristia, L., Lorenti, S., and Messina, A. (2003) *Mater. Res. Soc. Symp. Proc.*, **770**, I2.1.1.

29 Iacona, F., Pacifici, D., Irrera, A., Miritello, M., Franzò, G., Priolo, F., Sanfilippo, D., Di Stefano, G., and Fallica, P.G. (2002) Electroluminescence at 1.54 μm in Er-doped Si nanocluster-based devices. *Appl. Phys. Lett.*, **81**, 3242–3244.

30 Claps, R., Dimitropoulos, D., Raghunathan, V., Han, Y., and Jalali, B. (2003) Observation of stimulated Raman scattering in silicon waveguides. *Opt. Express*, **11**, 1731–1739.

31 Boyraz, O. and Jalali, B. (2004) Demonstration of a silicon Raman laser. *Opt. Express*, **12**, 5269–5273.

32 Walters, R.J., Bourianof, R.I., and Atwater, H. (2005) Field-effect electroluminescence in silicon nanocrystals. *Nat. Mater.*, **4**, 143.

33 Rong, H., Liu, A., Jones, R., Cohen, O., Hak, D., Nicolasecu, R., Fang, A., and

Paniccia, M. (2005) An all-silicon Raman laser. *Nature*, **435**, 292–294.

34 Fang, A.W., Park, H., Cohen, O., Jones, R., Paniccia, M., and Bowers, J. (2006) Electrically pumped hybrid AlGaInAs-silicon evanescent laser. *Opt. Express*, **14**, 9203–9210.

35 Koch, B.R., Fang, A.W., Cohen, O., and Bowers, J.E. (2007) Mode-locked silicon evanescent lasers. *Opt. Express*, **15**, 11225–11233.

36 Van Campenhout, P., Rojo Romeo, P., Regreny, P., Seassal, C., Van Thourhout, D., Verstuyft, S., Di Cioccio, L., Fedeli, J.-M., Lagahe, C., and Baets, R. (2007) Electrically pumped InP-based microdisk lasers integrated with a nanophotonic silicon-on-insulator waveguide circuit. *Opt. Express*, **15**, 6744–6749.

37 Liu, A., Jones, R., Liao, L., Samara Rubio, D., Rubin, D., Cohen, O., Nicolaescu, R., and Paniccia, M. (2004) A high-speed silicon optical modulator based on a metal-oxide-semiconductor capacitor. *Nature*, **427**, 615–618.

38 Liao, L., Samara-Rubio, D., Morse, M., Liu, A., Hodge, D., Rubin, D., Keil, U.D., and Franck, T. (2005) High speed silicon Mach-Zehnder. *Opt. Express*, **13**, 3129–3135.

39 Gunn, C. (2006) CMOS photonics for high-speed interconnects. *IEEE Micro*, **26**, 58–66.

40 Xu, Q., Schmidt, B., Pradhan, S., and Lipson, M. (2005) Micrometre-scale silicon electro-optic modulator. *Nature*, **435**, 325–327.

41 Gan, F. and Kartner, F.X. (2005) High-speed silicon electrooptic modulator design. *IEEE Photon. Technol. Lett.*, **7**, 1007–1009.

42 Gardes, F.Y., Reed, G.T., Emerson, N.G., and Png, C.E. (2005) A sub-micron depletion-type photonic modulator in silicon on insulator. *Opt. Express*, **13**, 8845.

43 Boyraz, O., Koonath, P., Raghunathan, V., and Jalali, B. (2004) All optical switching and continuum generation in silicon waveguides. *Opt. Express*, **12**, 4094–4102.

44 Fage-Pedersen, J., Frandsen, L.A., Lavrinenko, A., and Borel, P.I. (2006) A linear electrooptic effect in silicon, induced by use of strain. 3rd EEE/LEOS International Conference on Group IV Photonics, pp. 37–39.

45 Kuo, Y.-H., Lee, Y.-K., Ge, Y., Ren, S., Roth, J.E., Kamins, T.I., Miller, D.A.B., and Harris, J.S. (2005) Strong quantum-confined Stark effect in germanium quantum-well structures on silicon. *Nature*, **437**, 1334–1336.

46 Akahane, Y., Asano, T., Song, B.S., and Noda, S. (2003) High-Q photonic nanocavity in a two-dimensional photonic crystal. *Nature*, **425**, 944–947.

47 Song, B.S., Noda, S., Asano, T., and Akahane, Y. (2005) Ultra-high-Q photonic double-heterostructure nanocavity. *Nat. Mater.*, **4**, 207–210.

48 Xia, F., Sekaric, L., O'Boyle, M., and Vlasov, Y. (2006) Coupled resonator optical waveguides based on silicon-on-insulator photonic wires. *Appl. Phys. Lett.*, **89**, 041122.

49 Xia, F., Sekaric, L., and Vlasov, Y. (2007) Ultracompact optical buffers on a silicon chip. *Nat. Photon.*, **1**, 65.

50 Vlasov, Y., Green, W.M.J., and Xia, F. (2008) High-throughput silicon nanophotonic deflection switch for on-chip optical networks. *Nat. Photon.*, (April 1).

51 http://www.lightwire.com/.

52 http://www.luxtera.com/.

53 http://www.kotura.com/.

54 Gaburro, Z., Bettotti, P., Daldosso, N., Ghulinyan, M., Navarro, D., Melchiorri, M., Riboli, F., Saiani, M., Sbrana, F., and Pavesi, L. (eds) (2006) *Nanostructured Silicon for Photonics: From Materials to Devices*, vols. 27–28, Materials Science Foundation, TransTecH Publications Ltd.

55 Kumar, V. (ed.) (2007) *Nanosilicon*, Elsevier Ltd, Oxford.

56 Daldosso, N., Melchiorri, M., Riboli, F., Girardini, M., Pucker, G., Crivellari, M., Bellutti, P., Lui, A., and Pavesi, L. (2004) Design, fabrication, structural and optical characterization of thin Si_3N_4 waveguides. *IEEE J. Lightwave Technol.*, **22**, 1734.

57 Pellegrino, P., Garrido, B., Garcia, C., Arbiol, J., Morante, J.R., Melchiorri, M., Daldosso, N., Pavesi, L., Schedi, E., and

Sarrabayrouse, G. (2005) Low loss rib waveguides containing Si nanocrystals embedded in SiO$_2$. *J. Appl. Phys.*, **97**, 074312-1/8.

58 Daldosso, N., Melchiorri, M., Pavesi, L., Pucker, G., Gourbilleau, F., Chausserie, S., Belarouci, A., Portier, X., and Dufour, C. (2006) Optical losses and absorption cross section of silicon nanocrystals. *J. Lumin.*, **121**, 344–348.

59 Spitzer, W. and Fan, H.Y. (1957) *Phys. Rev.*, **108**, 268.

60 Forcales, M., Smith, N.J., and Elliman, R.G. (2006) *J. Appl. Phys.*, **100**, 014902.

61 Navarro-Urrios, D. *et al.* (2008) Quantification of the carrier absorption losses in Si-nanocrystal rich rib waveguides at 1.54 μm. *Appl. Phys. Lett.*, **92**, 051101.

62 Robinson, J.T., Manolatou, C., Chen, L., and Lipson, M. (2005) Ultrasmall mode volumes in dielectric optical micro-cavities. *Phys. Rev. Lett.*, **95**, 1439011–1439014.

63 Lebour, Y., Blasco, J., Guider, R., Pellegrino, P., Daldosso, N., Pavesi, L., Martinez, J.M., Martinex, A., Marti, J., Jordana, E., and Fedeli, J.M. and Garrido, B. Sandwich slot waveguides integrated ring resonators.

64 Sun, R., Dong, P., Feng, N.-N., Hong, C.-Y., Michel, J., Lipson, M., and Kimerling, L. (2007) Horizontal single and multiple slot waveguides: optical transmission at λ = 1550 nm. *Opt. Express*, **15**, 17967.

65 Little, B.E., Foresi, J.S., Steinmeyer, G., Thoen, E.R., Chu, S.T., Haus, H.A., Ippen, E.P., Kimerling, L.C., and Greene, W. (1998) Ultra-compact Si-SiO$_2$ microring resonator optical channel dropping filters. *IEEE Photon Technol. Lett.*, **10** (4), 549.

66 Soljacic, M., Johnson, S.G., Fan, S., Ibanescu, M., Ippen, E., and Joannopoulos, J.D. (2002) Photonic crystals slow-light enhancement of nonlinear phase sensitivity. *J. Opt. Soc. Am. B*, **19**, 2052.

67 Thévenaz, L. (2008) Slow and fast light in optical fibres. *Nat. Photon.*, **2**, 474.

68 Baba, T. (2008) Slow light in photonic crystals. *Nat. Photon.*, **2**, 465.

69 Yariv, A., Xu, Y., Lee, R.K., and Scherer, A. (1999) Coupled-resonator optical waveguide: a proposal and analysis. *Opt. Lett.*, **24**, 711.

70 Xia, F., Sekaric, L., and Vlasov, Y. (2006) Ultracompact optical buffers on a silicon chip. *Nat. Photon.*, **1**, 65.

71 Di Falco, A., O'Faolain, L., and Krauss, T.F. (2008) Dispersion control and slow light in slotted photonic crystal waveguides. *Appl. Phys. Lett.*, **92**, 083501.

72 Riboli, F., Bettotti, P., and Pavesi, L. (2007) Band gap characterization and slow light effects in one dimensional photonic crystals based on silicon slot-waveguides. *Opt. Express*, **19**, 11769.

73 Pitanti, A. *et al.* Coupled cavities in one-dimensional photonic crystal based on horizontal slot waveguide structure with Si-nc. Group IV Photonics, 2008 5th IEEE International Conference, 17–19 September 2008.

74 Ye, Y.-H., Ding, J., Jeong, D.-Y., Khoo, I.C., and Zhang, Q.M. (2004) Finite-size effect on one-dimensional coupled resonator optical waveguides. *Phys. Rev. E*, **69**, 0566041.

75 Rayleigh, L. (1914) *Philos. Mag.*, **27**, 100.

76 Vahala, K.J. (2003) Optical microcavities. *Nature (London)*, **424**, 839.

77 Del'Haye, P., Arcizet, O., Schliesser, A., Holzwarth, R., and Kippenberg, T.J. (2008) Full stabilization of a microresonator frequency comb. *Phys. Rev. Lett.*, **101**, 053903.

78 Schliesser, A., Anetsberger, G., Rivière, R., Arcizet, O., and Kippenberg, T.J. (2008) High-sensitivity monitoring of micromechanical vibration using optical whispering gallery mode resonators. *New J. Phys.*, accepted for publication.

79 Armani, A.M. and Vahala, K.J. (2006) Heavy water detection using ultra-high-Q microcavities. *Opt. Lett.*, **31** (12), 1896–1898.

80 Aoki, T. *et al.* (2006) Observation of strong coupling between one atom and a monolithic microresonator. *Nature*, **443**, 671.

81 Zhang, R.-J. (2006) Visible range whispering-gallery mode in microdisk array based on size-controlled Si nanocrystals. *Appl. Phys. Lett.*, **88**, 153120.

82 Ghulinyan, M., Navarro-Urrios, D., Pitanti, A., Lui, A., Pucker, G., and Pavesi, L. (2008) Whispering-gallery modes and light emission from a Si-nanocrystal-based single microdisk resonator. *Opt. Express*, **16**, 13218.

83 Tewary, A., Digonnet, M.J.F., Sung, J.-Y., Shin, J.H., and Brongersma, M.L. (2006) Silicon-nanocrystal-coated silica microsphere thermooptical switch. *IEEE J. Sel. Top. Quant. Electron.*, **6**, 1089.

84 Kippenberg, T.J., Tschebotareva, A., Kalkman, J., Polman, A., and Vahala, K.J., Purcell factor enhanced scattering efficiency in silicon nanocrystal doped micro-cavities. *Quantum Electronics and Laser Science Conference, 2005 QELS '05, May 22–27, 2005, Baltimore, MD*. The Optical Society of America, Washington, DC, pp. 62–64.

85 Pavesi, L., Dal Negro, L., Mazzoleni, C., Franzo, G., and Priolo, F. (2000) Optical gain in Si nanocrystals. *Nature*, **408**, 440.

86 Ruan, J., Fauchet, P.M., Dal Negro, L., Cazzanelli, M., and Pavesi, L. (2003) Stimulated emission in nanocrystalline silicon superlattices. *J. Appl. Phys.*, **83**, 5479.

87 Dal Negro, L., Cazzanelli, M., Danese, B., Pavesi, L., Iacona, F., Franzò, G., and Priolo, F. (2004) Light amplification in silicon nanocrystals by pump and probe transmission measurements. *J. Appl. Phys.*, **96**, 5747.

88 Dal Negro, L., Cazzanelli, M., Daldosso, N., Gaburro, Z., Pavesi, L., Priolo, F., Pacifici, D., Franzò, G., and Iacona, F. (2003) Stimulated emission in plasma enhanced chemical vapour deposited silicon nanocrystals. *Physica E*, **16**, 297.

89 Cazzanelli, M., Navarro-Urrios, D., Riboli, F., Daldosso, N., Pavesi, L., Heitmann, J., Yi, L.X., Scholz, R., Zacharias, M., and Rosele, U. (2004) Optical gain in monodispersed silicon nanocrystals. *J. Appl. Phys.*, **96**, 3164–3171.

90 Luppi, M. and Ossicini, S. (2003) Multiple $Si=O$ bonds at the silicon cluster surface. *J. Appl. Phys.*, **94**, 2130.

91 Wolkin, M.V., Jorne, J., Fauchet, P.M., Allan, G., and Delerue, C. (1999) Electronic states and luminescence in porous silicon quantum dots: the role of oxygen. *Phys. Rev. Lett.*, **82**, 197.

92 Daldosso, N., Luppi, M., Ossicini, S., Degoli, E., Magri, R., Dalba, G., Fornasini, P., Grisenti, R., Rocca, F., Pavesi, L., Boninelli, S., Priolo, F., Bongiorno, C., and Iacona, F. (2003) Role of the interface region on the optoelectronic properties of silicon nanocrystals embedded in SiO_2. *Phys. Rev. B*, **68**, 085327.

93 Hadjisavvas, G. and Kelires, P.C. (2004) Structure and energetics of Si nanocrystals embedded in a-SiO_2. *Phys. Rev. Lett.*, **93**, 226104.

94 Lioudakis, E., Antoniou, A., Othonos, A., Christofides, C., Nassiopoulou, A.G., Lioutas, Ch.B., and Frangis, N. (2007) The role of surface vibrations and quantum confinement effect to the optical properties of very thin nanocrystalline silicon films. *J. App. Phys.*, **102**, 083534.

95 Fujii, M., Yoshida, M., Kanzawa, Y., Hayashi, S., and Yamamoto, K. (1997) 1.54 μm photoluminescence of Er^{3+} doped into SiO_2 films containing Si nanocrystals: evidence for energy transfer from Si nanocrystals to Er^{3+}. *Appl. Phys. Lett.*, **71**, 1198–1200.

96 Pacifici, D., Irrera, A., Franzo, G., Miritello, M., Lacona, F., and Priolo, F. (2003) Erbium-doped Si nanocrystals: optical properties and electrolumine-scent devices. *Physica E*, **16**, 331–340.

97 Daldosso, N., Navarro-Urrios, D., Melchiorri, M., Pavesi, L., Sada, C., Gourbilleau, F., and Rizk, R. (2006) Refractive index dependence of the absorption and emission cross sections at 1.54 μm of Er^{3+} coupled to Si nanoclusters. *Appl. Phys. Lett.*, **88**, 161901.

98 Pacifici, D., Franzò, G., Priolo, F., Iacona, F., and Dal Negro, L. (2003) Modeling and perspectives of the Si nanocrystals–erbium interaction for optical ampli-fication. *Phys. Rev. B*, **67**, 245301.

99 Oton, C.J., Loh, W.H., and Kenyon, A.J. (2006) Er^{3+} excited state absorption and the low fraction of nanocluster-excitable Er^{3+} in SiO_x. *Appl. Phys. Lett.*, **89**, 031116.

100 Izeddin, I., Gregorkiewicz, T., and Fujii, M. (2007) *Physica E*, **38**, 144.

101 Han, H., Seo, S., and Shin, J. (2001) Optical gain at 1.54 µm in erbium-doped silicon nanocluster sensitized waveguide. *Appl. Phys. Lett.*, **79**, 4568.

102 (a) Wojdak, M., Klik, M., Forcales, M., Gusev, O.B., Gregorkiewicz, T., Pacifici, D., Franzò, G., Priolo, F., and Iacona, F. (2004) *Phys. Rev. B*, **69**, 233315; (b) Kik, P.G., Polman, A. (2000) *J. Appl. Phys.*, **88**, 1992.

103 Garrido, B., García, C., Seo, S.-Y., Pellegrino, P., Navarro-Urrios, D., Daldosso, N., Pavesi, L., Gourbilleau, F., and Rizk, R. (2007) Excitable Er fraction quenching phenomena in Er-doped SiO_2 layers containing Si nanoclusters. *Phys. Rev. B*, **76**, 245308.

104 Navarro-Urrios, D., Pitanti, A., Daldosso, N., Gourbilleau, F., Rizk, R., Pucker, G., and Pavesi, L. (2008) Quantification of the carrier absorption losses in Si-nanocrystal rich rib waveguides at 1.54 µm. *Appl. Phys. Lett.*, **92**, 051101.

105 Pitanti, A., Navarro-Urrios, D., Guider, R., Daldosso, N., Gourbilleau, F., Khomenkova, L., Rizk, R., and Pavesi, L. (2008) *Proc. SPIE*, **6996**, 699619.

106 Navarro-Urrios, D. *et al.* (2009) *Phys. Rev. B*, in press.

107 Izeddin, I., Moskalenko, A.S., Yassievich, I.N., Fujii, M., and Gregorkiewicz, T. (2006) *Phys. Rev. Lett.*, **97**, 207401.

108 Brus, L. (1998) Light emission in silicon: from physics to devices, in *Semiconductors and Semimetals*, vol. 49 (ed. D. Lockwood), Academic Press, p. 303.

109 DiMaria, D.J., Kirtley, J.R., Pakulis, E.J., Dong, D.W., Kuan, T.S., Pesavento, F.L., Theis, T.N., Cutro, J.A., and Brorson, S.D. (1984) Electroluminescence studies in silicon dioxide films containing tiny silicon islands. *J. Appl. Phys.*, **56**, 401–416.

110 Fiory, A.T. and Ravindra, N.M. (2003) Light emission from silicon: some perspectives and applications. *J. Electron. Mater.*, **32**, 1043–1051.

111 Ossicini, S., Pavesi, L., and Priolo, F. (2003) *Light Emitting Silicon for Micro-*

Photonics, vol. 194, Springer Tracts in Modern Physics, Springer-Verlag, Berlin.

112 Lalic, N. and Linnros, J. (1999) Light emitting diode structure based on Si nanocrystals formed by implantation into thermal oxide. *J. Lumin.*, **80**, 263.

113 Park, N.-M., Kim, T.-S., and Park, S.-J. (2001) Band gap engineering of amorphous silicon quantum dots for light-emitting diodes. *Appl. Phys. Lett.*, **78**, 2575.

114 Peralvarez, M., Garcia, C., Lopez, M., Garrido, B., Barreto, J., Dominguez, C., and Rodriguez, J.A. (2006) Field effect luminescence from Si nanocrystals obtained by plasma-enhanced chemical vapor deposition. *Appl. Phys. Lett.*, **89**, 051112.

115 Cho, K.S., Park, N.-M., Kim, T.-Y., Kim, K.-H., Sung, G.Y., and Shin, J.H. (2005) *Appl. Phys. Lett.*, **86**, 0713909.

116 Tsu, R., Filios, A., Lofgren, C., Dovidenko, K., and Wang, C.G. (1998) Silicon epitaxy on Si(100) with adsorbed oxygen. *Electrochem. Solid State Lett.*, **1**, 80.

117 Heikkila, L., Kuusela, T., and Hedman, H.P. (2001) Electroluminescence in Si/SiO_2 layer structure. *J. Appl. Phys.*, **89**, 2179.

118 Gaburro, Z., Pavesi, L., Pucker, G., and Bellutti, P. (2001) *MRS Proc.*, **638**, F18.5.1.

119 Walters, R.J., Carreras, J., Feng, T., Bell, L.D., and Atwater, H.A. (2006) Silicon nanocrystal field effect light-emitting devices. *IEEE J. Sel. Top. Quant.*, **12**, 1647–1656.

120 Anopchenko, A., Prezioso, S., Gaburro, Z., Ferraioli, L., Pucker, G., Bellutti, P., and Pavesi, L. (2007) Proceedings of Group IV Photonics 2007, vol. 1–3, p. 70.

121 Prezioso, S., Anopchenko, A., Gaburro, Z., Pavesi, L., Pucker, G., Vanzetti, L., and Bellutti, P. (2008) Electrical conduction and electroluminescence in nano-crystalline silicon-based light emitting devices. *J. Appl. Phys.*, **104**, 063103.

122 Lu, Z.H., Lockwood, D.J., and Baribeau, J.-M. (1995) Quantum confinement and light emission in SiO_2/Si superlattices. *Nature*, **378**, 258–260.

123 Wu, Wei, Huang, X.F., Chen, K.J., Xu, J.B., Gao, X., Xu, J., and Li, W. (1999) Room temperature visible electroluminescence in silicon nanostructures. *J. Vac. Sci. Technol*, **A 17**, 159–163.

124 Tsybeskov, L., Hirschman, K.D., Duttagupta, S.P., Zacharias, M., Fauchet, P.M., McCaffrey, J.P., and Lockwood, D.J. (1998) Nanocrystalline-silicon superlattice produced by controlled recrystallization. *Appl. Phys. Lett.*, **72**, 43–45.

125 Vinciguerra, V., Franzo, G., Priolo, F., Iacona, F., and Spinella, C. (2000) Quantum confinement and recombination dynamics in silicon nanocrystals embedded in Si/SiO_2 superlattices. *J. Appl. Phys.*, **87**, 8165–8173.

126 Gaburro, Z., Pucker, G., Bellutti, P., and Pavesi, L. (2000) Electroluminescence in MOS structures with Si/SiO_2 nanometric multilayers. *Solid State Commun.*, **114**, 33–37.

127 Wang, M., Anopchenko, A., Marconi, A., Moser, E., Prezioso, S., Pavesi, L., Pucker, G., Bellutti, P., and Vanzetti, L. (2008) Light emitting devices based on nanocrystalline-silicon multilayer structure. *Physica E*. doi: 10.1016/j.physe.2008.08.009

128 Huang, R., Dong, H., Wang, D., Chen, K., Ding, H., Wang, X., Li, W., Xu, J., and Ma, Z. (2008) Role of barrier layers in electroluminescence from SiN-based multilayer light-emitting devices. *Appl. Phys. Lett.*, **92**, 181106.

129 Scardera, G., Puzzer, T., Perez-Wurfl, I., and Conibeer, G. (2008) The effects of annealing temperature on the photoluminescence from silicon nitride multilayer structures. *J. Crystal Growth*, **310**, 3680–3684.

130 De Salvo, B., Ghibaudo, G., Luthereau, P., Baron, T., Guillaumot, B., and Reimbold, G. (2001) Transport mechanisms and charge trapping in thin dielectric/Si nano-crystals structures. *Solid State Electron.*, **45**, 1513–1519.

131 Ross, R.T. and Nozik, A.J. (1982) *J. Appl. Phys.*, **53**, 3813.

132 Ginley, D., Green, M.A., and Collins, R. (2008) Solar energy conversion toward 1 terawatt. *MRS Bull.*, **33**, 355–364.

133 Green, M.A. (2002) Third generation photovoltaics: solar cells for 2020 and beyond. *Physica E*, **14**, 65.

134 Green, M.A. (2001) Third generation photovoltaics: ultra-high conversion efficiency at low cost. *Prog. Photovolt. Res. Appl.*, **9**, 123.

135 Marti, A. and Araujo, G.L. (1996) *Sol. Energy Mater. Sol. Cells*, **43**, 203.

136 Strumpel, C., McCann, M., Beaucarne, G., Arkhipov, V., Slaoui, A., Svrcek, V., del Canizo, C., and Tobias, I. (2007) *Sol. Energy Mater. Sol. Cells*, **91**, 238.

137 Trupke, T., Green, M.A., and Wurfel, P. (2002) *J. Appl. Phys.*, **92**, 1668.

138 Landsberg, P.T., Nussbaumer, H., and Willeke, G. (1993) *J. Appl. Phys.*, **74**, 1451.

139 Beard, M.C., Knutsen, K.P., Yu, P., Luther, J.M., Song, Q., Metzger, W.K., Ellingson, R.J., and Nozik, A.J. (2007) Multiple exciton generation in colloidal silicon nanocrystals. *Nano Lett.*, **7**, 2506.

140 Wurfel, P. (1997) Solar energy conversion with hot electrons from impact ionization. *Sol. Energy Mater. Sol. Cells*, **46**, 43.

141 Timmerman, D., Izeddin, I., Stallinga, P., Yassievich, I.N., and Gregorkiewicz, T. (2008) Space-separated quantum cutting with silicon nanocrystals for photovoltaic applications. *Nat. Photon.*, **2**, 105.

142 Wolf, M. (1960) Limitations and possibilities for improvement of photovoltaic solar energy converters: Part I: considerations for Earth's surface operation. *Proc. IRE*, **48**, 1246.

143 Trupke, T., Green, M.A., and Wurfel, P. (2002) Improving solar cell efficiencies by up-conversion of sub-band-gap light. *J. Appl. Phys.*, **92**, 4117.

144 Shockley, W. and Queisser, H.J. (1961) Detailed balance limit of efficiency of p–n junction solar cells. *J. Appl. Phys.*, **32**, 510.

145 Morf, R.H. (2002) Unexplored opportunities for nanostructures in photovoltaics. *Physica E*, **14**, 78.

146 Nozik, A.J. (2002) Quantum dot solar cells. *Physica E*, **14**, 115.

147 Cho, E.C., Cho, Y.H., Trupke, T., Corkish, R., Conibeer, G., and Green, M.A. (2004)

Proceedings of the 19th European Photovoltaic Solar Energy Conference, Paris, France, pp. 235–238.

148 Conibeer, G., Green, M.A., Corkish, R., Cho, Y., Cho, E.C., Jiang, C., Fangsuwannarak, T., Pink, E., Huang, Y., Puzzer, T., Trupke, T., Richards, B., Shalav, A., and Lin, K. (2006) Silicon nanostructures for third generation photovoltaic solar cells. *Thin Solid Films*, **511–512**, 654–662.

149 Jiang, C., Cho, E.C., Conibeer, G., and Green, M.A. (2004) Proceedings of the 19th European Photovoltaic Solar Energy Conference, June 2004, WIP-Munich & ETA-Florence, Paris, pp. 80–83.

150 Zacharias, M., Heitmann, J., Scholz, R., Kahler, U., Schmidt, M., and Blasing, J. (2002) Size-controlled highly luminescent silicon nanocrystals: a SiO/SiO_2 superlattice approach. *Appl. Phys. Lett.*, **80**, 661–663.

151 Jiang, C.W. and Green, M.A. (2006) Silicon quantum dot superlattices: modeling of energy bands, densities of states, and mobilities for silicon tandem solar cell applications. *J. Appl. Phys.*, **99**, 114902.

152 Cho, E.C., Green, M.A., Conibeer, G., Song, D.Y., Cho, Y.H., Scardera, G., Huang, S.J., Park, S., Hao, X.J., Huang, Y., and Dao, L.V. (2007) Silicon quantum dots in a dielectric matrix for all-silicon tandem solar cells. *Adv. OptoElectr.*, **2007**, **Article ID 69578**.

153 Rolver, R., Berghoff, B., Batzner, D.L., Spangenberg, B., and Kurz, H. (2008) Lateral Si/SiO_2 quantum well solar cells. *Appl. Phys. Lett.*, **92**, 212108.

154 Rolver, R., Berghoff, B., Batzner, D., Spangenberg, B., Kurz, H., Schmidt, M., and Stegemann, B. (2008) Si/SiO_2 multiple quantum wells for all silicon tantem cells: conductivity and photocurrent measurements. *Thin Solid Films*, **516**, 6763–6766.

155 Arguirov, T., Mchedlidze, T., Kittler, M., Rolver, R., Berghoff, B., Forst, M., and Spangenberg, B. (2006) Residual stress in Si nanocrystals embedded in a SiO_2 matrix. *Appl. Phys. Lett.*, **89**, 053111.

156 Mchedlidze, T., Arguirov, T., Kouteva-Arguirova, S., Jia, G., Kittler, M., Rolver, R., Berghoff, B., Forst, M., Batzner, D.L., and Spangenberg, B. (2008) Influence of a substrate, structure and annealing procedures on crystalline and optical properties of Si/SiO_2 multiple quantum wells. *Thin Solid Films*, **516**, 6800–6803.

157 Prezioso, S., Hossain, S.M., Anopchenko, A., Pavesi, L., Wang, M., Pucker, G., and Bellutti, P. (2009) Super-linear photovoltaic effect in Si-nanocrystals based metal-insulator-semiconductor devices. *Appl. Phys. Lett.*, **94**, 062108.

158 Hossain, S.M., Anopchenko, A., Prezioso, S., Ferraioli, L., Pavesi, L., Pucker, G., Bellutti, P., Binetti, S., and Acciarri, M. (2008) Subband gap photoresponse of nanocrystalline silicon in a metal-oxide-semiconductor device. *J. Appl. Phys.*, **104**, 074917.

159 Ramos, L.E., Degoli, E., Cantele, G., Ossicini, S., Ninno, D., Furthmuller, J., and Bechstedt, F. (2007) *J. Phys.: Condens. Matter*, **19**, 466211.

160 Asobe, M. (1997) Nonlinear optical properties of chalcogenide glass fibers and their application to all-optical switching. *Opt. Fiber Technol.*, **3**, 142.

161 Figliozzi, P., Sun, L., Jiang, Y., Matlis, N., Mattern, B., Downer, M.C., Withrow, S.P., White, C.W., Mochan, W.L., and Mendoza, B.S. (2005) Single-beam and enhanced two-beam second-harmonic generation from silicon nanocrystals by use of spatially inhomogeneous femtosecond pulses. *Phys. Rev. Lett.*, **94**, 047401.

162 Golovan, L.A., Kuznetsova, L.P., Fedotov, A.B., Konorov, S.O., Sidorov-Biryukov, D.A., Timoshenko, V.Y., Zheltikov, A.M., and Kashkarov, P.K. (2003) Nanocrystal-size-sensitive third-harmonic generation in nanostructured silicon. *Appl. Phys. B*, **76**, 429.

163 Vijayalakshmi, S., Grebel, H., Yaglioglu, G., Pino, R., Dorsinville, R., and White, C.W. (2000) Nonlinear optical response of Si nanostructures in a silica matrix. *J. Appl. Phys.*, **88**, 6418.

164 Vijaya Prakash, G., Cazzanelli, M., Gaburro, Z., Pavesi, L., Iacana, F., Franzo, G., and Priolo, F. (2002) Linear and nonlinear optical properties of plasma enhanced chemical-vapour deposition

grown silicon nanocrystals. *J. Modern Opt.*, **49**, 719.

165 Sheik-Bahae, M., Said, A.A., Wei, T.-H. Hagan, D.A., and Van Stryland, E.W. (1990) Sensitive measurements of optical nonlinearities using a single beam. *IEEE J. Quant. Electron.*, **26**, 760–769.

166 Sutherland, R.L. (ed.) (2003) *Hand book of Nonlinear Optics*, Marcel Dekker, New York.

167 Adair, R., Chase, L.L., and Payne, S.A. (1989) Nonlinear refractive index of optical crystals. *Phys. Rev. B*, **39**, 3337.

168 Dinu, M. *et al.* (2003) *Appl. Phys. Lett.*, **82**, 2954.

169 Spano, R., Daldosso, N., Cazzanelli, M., Ferraioli, L., Tartara, L., Yu, J., Degiorgio, V., Jordana, E., Fedeli, J.M., and Pavesi, L. (2009) Bound electronic and free carrier nonlinearities in silicon nanocrystals at 1550 nm. *Opt. Express*, **17**, 3941–3950.

170 Ballesteros, J.M., Solís, J., Serna, R., and Afonso, C.N. (1999) Nanocrystal size dependence of the third-order nonlinear optical response of Cu:Al$_2$O$_3$ thin films. *App. Phys. Lett.*, **74**, 2791.

171 Hernández, S., Pellegrino, P., Martínez, A., Lebour, Y., Garrido, B., Spano, R., Cazzanelli, M., Daldosso, N., Pavesi, L., and Jordana, E. and Fedeli, J.M. (2008) Linear and non-linear optical properties of Si nanocrystals in SiO$_2$ deposited by PECVD. *J. Appl. Phys.*, **103**, 064309.

17
Lighting Applications of Rare Earth-Doped Silicon Oxides

Tyler Roschuk, Jing Li, Jacek Wojcik, Peter Mascher, and Iain D. Calder

Silicon nanoclusters (Si-ncs) are of substantial interest as they play a significant role in the development of Si-based light sources [1, 2]. One of the main application areas is in integrated photonics for telecommunications and computing; however, there is also a substantial market for the use of these materials in lighting and displays. Lighting accounts for approximately 20% of the $60 billion electricity market in the United States [3]. The potential to tap this market, combined with the inefficiency of incandescent bulbs, has driven research over the past decades in the area of solid-state lighting (SSL). The market for both incandescent and fluorescent light sources has been steadily dwindling with a shift toward the development of more efficient, cheaper, and environment friendly alternatives, such as SSL [4, 5]. In 2007, the market for LED lighting was estimated at $330 million (a 60% increase over the previous year) and is projected to grow to $1.4 billion by 2012 [6].

As with light sources for integrated photonics, the SSL market relies primarily on III–V, II–VI, or phosphor-based materials and, more recently, organic light emitting diodes (OLEDs) [5]. With the ability to attain efficient luminescence in Si-nc-based systems, however, the possibility of an all-Si approach has become attractive. Such an approach would allow manufacturers to use of the same large-scale production and processing facilities as used by the microelectronics and solar cell industries while maintaining low cost. There are two performance metrics generally of interest for solid-state lighting, the performance (in lumens) per dollar and the performance per watt (efficacy). Figure 17.1 illustrates these metrics for several common lighting sources. While LED and OLED technologies currently lead in terms of efficacy, recently explored RE-doped silicon-based materials (shown on the graph as "Rare Earth Films") have a clear advantage in terms of the performance per dollar (data courtesy of Group IV Semiconductor Inc.). Furthermore, integration with the necessary electronics for driving Si-based lighting cells or pixels would then be easily achieved due to the use of a Si-platform. In particular, the use of rare earth dopants to attain specific colored emissions for mixing to obtain white light presents a viable pathway to Si-based lighting. Such an approach borrows from that commonly used with color-converting phosphors, where rare earths are routinely used, while incorporating the rare earth elements into a silicon-based host for the advantages in fabrication.

Silicon Nanocrystals: Fundamentals, Synthesis and Applications. Edited by Lorenzo Pavesi and Rasit Turan
Copyright © 2010 WILEY-VCH Verlag GmbH & Co. KGaA, Weinheim
ISBN: 978-3-527-32160-5

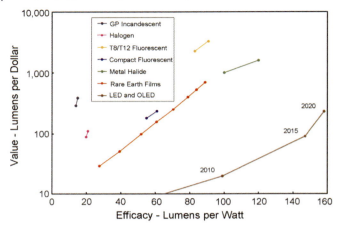

Figure 17.1 Performance per dollar versus performance per watt for several common forms of general illumination. Rare earth-doped Si-based devices are indicated on the plot as "Rare Earth Films"; the range shown represents expected improvements over time.

In this chapter, we discuss the general principles and operating parameters used for SSL, beginning with a discussion of the approaches to obtaining white light. This is followed by a discussion of rare earth doping of silicon-based materials to obtain specific colored emissions, which can be mixed to produce white light, including a discussion of the luminescence mechanisms of these materials. Finally, potential white light emitting device structures, which can be used for SSL, are considered. Although the focus of the discussion is on white light emission and SSL, it should be straightforward to see how the applications of these materials and devices could be extended to displays through proper cell/pixel design and red-green-blue (RGB) mixing.

17.1
Solid-State Lighting: A Basic Introduction

For both lighting and display applications, it is important to be able to precisely control the color of the emission from the emitting materials. Lighting applications in particular require a balance of the emission spectrum across a broad band of energies in order to best simulate the optical spectrum from natural light sources. Figure 17.2 shows the 1931 Commission Internationale D'Eclairage (CIE) chromaticity diagram [5]. The Planckian locus shown defines the color coordinates of a blackbody radiator for temperatures from 2500 to 10 000 K. The typical color balance sought for lighting applications is along the arc from 3000 to 6000 K. The exact color attained creates a "warm" (low color temperatures) or "cool" (high color temperatures) feel to certain light sources. A three-color-based system, such as RGB, is used to define three vertices of a triangle within the color diagram. In principle, the mixing of these three base colors allows any color within the bounds of the triangle to be produced. In practice, the nature of the output spectrum from the different color components in the system affects how accurately this can be done and how well white light is rendered.

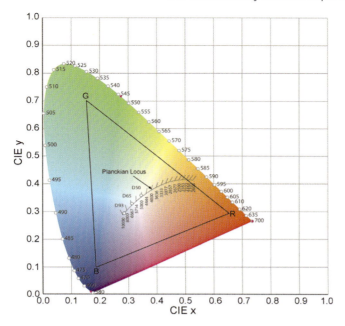

Figure 17.2 CIE chromaticity diagram. The planckian locus defines the color coordinates of a blackbody radiator. White light falls along the locus from 3000 to 9300 K. For general lighting applications, the target point lies typically on the arc from 3000 to 6000 K while for display applications the target point is near 6500 K, which is very close to the solar standard known as D65. Adapted from Ref. [7]. Three-color components such as RBG can be used for color mixing to obtain white light. The RGB points here are purely representative and do not necessarily correspond to the emissions of any particular material.

It is also important to note that the eye is most sensitive to certain colors, such as green or yellow-green. As such, the color balance necessary for a solid-state light cell to appear as white light does not have equal contributions from the different color components [8].

At present, there are a number of materials and integration schemes used to generate the necessary base set of colors to produce white light. RE dopants have a long-standing history in this area in order to achieve either specific emission colors from the RE ions themselves or as color-converting phosphors. Details of emission from the rare earths themselves will be discussed in the following section. Specific details of color-converting phosphors will be left to the literature; however, the integration schemes commonly used are similar to those being employed for RE-doped silicon-based materials that will be discussed later in this chapter.

17.2
Luminescence of Rare Earth-Doped Silicon-Based Materials

Although Si-ncs themselves exhibit tunability in their emission through the control of their size, it is difficult to obtain a precise control over the emission energy throughout the visible spectrum. This is because nanocluster surface states and defects limit the

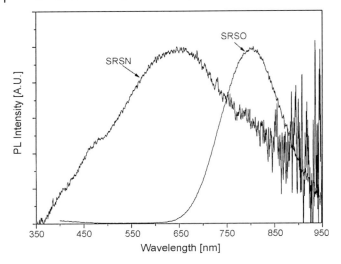

Figure 17.3 Normalized SRSN and SRSO PL emission spectra after UV excitation at 325 nm. Luminescence from these materials arises as a result of quantum confinement but is mediated by interface states at the Si-nc surface. Several interface states in the SRSN system enable a range of luminescence throughout the visible spectrum.

tunability range of these systems. In silicon-rich silicon oxides (SRSOs), oxygen-related defects, such as the SiO_2 double bond, limit the emission to the near-IR, while in silicon-rich silicon nitrides (SRSNs) defects related to both silicon and nitrogen bonds can limit the emissions to specific wavelength ranges from blue to red [9, 10]. Figure 17.3 shows representative PL spectra from the two material systems illustrating their broad nature. In SRSN materials, the detailed peak structure arises from contributions from several defects associated with the silicon nitride host matrix [10]. Because light emission from these materials strongly depends on composition and processing parameters, slight variations in the processing can lead to changes in the emission spectra and difficulties in color mixing. One method of obtaining specific emission energies is through the incorporation of RE dopants into these materials that leads to luminescence characteristic of the electronic energy levels of the RE.

As discussed in Section 17.1, the use of rare earth dopants to achieve specific color emissions for lighting has previously been applied to phosphor-based systems. Here, we are concerned only with the emissions arising from the RE ions themselves and not specifically in the use of RE ions as color converters, although such an application will be briefly considered in the following section. Much of the present work on RE-doped silicon-based materials for lighting involves efforts to characterize the luminescent properties of individually RE-doped thin films in order to optimize their performance, and full integration efforts are only now being explored. As such, it is worthwhile to first consider the details of luminescence from RE-doped thin films prior to considering some of the white light schemes that could be used in the final design of an SSL device.

Luminescence from RE ions typically occurs when they are in the optically active trivalent state formed by losing one 4f and both 6s electrons. Luminescence then arises from either intra-4f or 5d-to-4f transitions. The 5d-to-4f transitions strongly depend on details of the host matrix within which the RE ions are incorporated, as the 5d states interact with the surrounding environment. Intra-4f transitions, however, are relatively independent of the host matrix as the 4f orbitals are shielded from the surrounding environment by the 5d states. Intra-4f transitions are generally parity forbidden, particularly for the case of the free ions; however, the incorporation of the ion into the host matrix allows to mix opposite parity wavefunctions leading to radiative recombinations that have long lifetimes and low oscillator strengths [11, 12]. In contrast, 5d-to-4f transitions are parity permitted and have short luminescence lifetimes and high oscillator strengths. Three common rare earth ions used to obtain red, green, and blue emissions are Eu^{3+}, Tb^{3+}, and Ce^{3+}, respectively. It is also possible to obtain a blue-green emission from Eu^{2+}. The electronic energy levels and radiative emissions of the trivalent Eu, Tb, and Ce dopants, as well as Eu^{2+}, are shown in Figure 17.4. Specific details and features of the output spectra will be considered momentarily. The optimization of the luminescence from an RE-doped layer depends strongly on an understanding of the interactions of the RE dopant with the surrounding host matrix and of the possible RE excitation mechanisms, the details of which will vary depending on the type of RE ion.

To date, RE doping in silicon-based materials has focused primarily on Er due to its infrared emission at 1.54 μm, the transparency window for optical fibers, making it suitable for telecommunications [13–18]. In Er-doped Si-nc-embedded material, the Si-ncs act as sensitizers of the Er^{3+} ion, leading to a substantial increase in the Er^{3+} luminescence through an energy transfer from the Si-ncs to the RE ion [19–21]. It is not clear that the same energy transfer model applied to Er is suitable for all RE

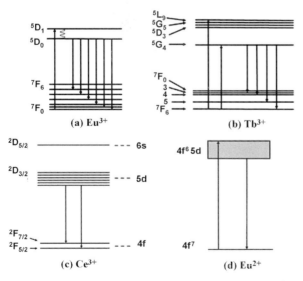

Figure 17.4 Electronic energy levels and radiative emissions of (a) Eu^{3+}, (b) Tb^{3+}, (c) Ce^{3+}, and (d) Eu^{2+} ions in SiO_2. Radiative recombination pathways for these dopants are indicated by the downward arrows.

dopants, particularly those with visible emissions such as Ce and Tb, although the model has been suitable for describing RE ions with near-infrared emissions, including Tm, Yb, and Nd [22]. Sensitization of the RE ions by Si-ncs is primarily observed in the case of optical excitation, as the optical absorption cross section of the Si-ncs is much larger than that of the RE ions. The use of Si-nc sensitizers is, therefore, particularly relevant for improving optically pumped devices such as Er-doped fiber amplifiers (EDFAs). For the development of electrically pumped devices, one must consider that the electronic cross section of the REs is larger than that of the Si-ncs. Electroluminescent (EL) devices, such as those being explored for SSL, then see no advantage in the use of sensitizers. In fact, as will be discussed, it is often advantageous to use stoichiometric SiO_2, or even oxygen-rich films, which results in more RE ions in optically active states.

The Ce^{3+} ion exhibits a blue emission due to transitions from the 5d to 4f levels. Figure 17.5 shows the PL spectra for a set of Ce-doped silicon oxides annealed for 60 min under flowing N_2 at the temperatures shown. The samples were grown using electron cyclotron resonance (ECR) plasma-enhanced chemical vapor deposition (PECVD) and were doped *in situ* during the growth process. For a Ce-doped SRSO, no emission from the Ce ions is observed at anneal temperatures below 1200 °C, rather, the PL spectra are dominated by an emission peak attributable to the presence of Si-ncs within the film. After annealing at 1200 °C, a weak emission band at approximately 450 nm, attributable to Ce^{3+}, is observed. In Ce-doped oxygen-rich silicon oxides (ORSO), however, a far brighter Ce^{3+} luminescence has been observed, as is shown in Figure 17.5b. PL from the sample increases for anneal temperatures up to 800 °C, decreases with annealing from 900 to 1000 °C, and then increases substantially with annealing up to 1200 °C [23]. A Gaussian deconvolution of the spectra has revealed the presence of three peaks centered at 400, 422, and 460 nm.

As with Er, the formation of a luminescent Ce state strongly depends on the coordination of Ce with its surrounding environment such that Ce is in the optically

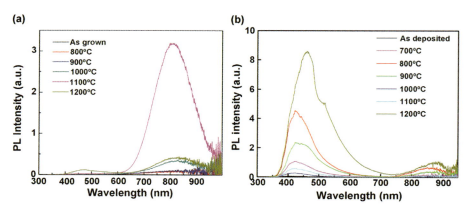

Figure 17.5 Luminescence from Ce-doped (a) SRSO and (b) ORSO samples. The samples have been annealed under N_2 at the temperatures shown. The SRSO sample contains 40% Si and 0.04% Ce, while the ORSO sample contains 32% Si and 1.0% Ce. Reproduced from Ref. [23].

active +3 state, as opposed to the inactive +4 state. For the case of Er, extended X-ray absorption fine structure (EXAFS) analysis has revealed that Er is in a luminescent state when coordinated with six oxygen atoms [24]. Likewise, the ability of Ce to coordinate with oxygen (Ce_6O_{11}) and to be in a luminescent state appears to readily occur in an ORSO environment. At higher anneal temperatures, clustering of the Ce initially reduces the PL; however, at 1100 and 1200 °C, evidence of a Ce-silicate phase (either $Ce_2Si_2O_7$ or $Ce_{4.667}(SiO_4)_3O$) has been observed leading to the increase in the total PL [25]. The exact details of the PL excitation from these samples remain under investigation. ORSO samples do, however, offer several advantages over their SRSO counterparts. The first is that excitation of Ce^{3+} by Si-ncs formed in SRSO is unlikely due to the higher energy of the Ce emission. Rather, the use of stoichiometric or oxygen-rich silicon oxides better serves to coordinate Ce such that it is in an optically active state. These films also allow to incorporate greater amounts of Ce, and through the formation of a Ce-silicate after high-temperature annealing, bright luminescence is obtained. It is important to use caution in the interpretation of PL, however, as Ce^{3+} may be resonantly excited by the 325 nm emission of the HeCd laser source used in these studies. Nevertheless, the results do provide good evidence of the ability to obtain optically active Ce within these films.

It has already been mentioned that the luminescence from 5d to 4f transitions strongly depends on details of the local environment due to the interactions of the 5d states with their environment (nephelauxetic effect) [26]. This effect is illustrated in Figure 17.6, which shows the emission from the Ce^{3+} ion in a series of silicon oxynitride host matrices with varying composition, as well as for an aluminum oxide host. The nephelauxetic effect leads to a broad emission band and spectral shifting of the emission peak. In cases of strong covalent bonding (such as for N in Si_3N_4), the effect is stronger and results in a redshift of the emission peak. In the silicon

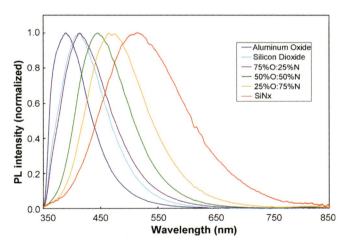

Figure 17.6 Nephelauxetic effect for Ce^{3+} embedded in the indicated host matrices. The normalized PL spectra arising from 5d–4f transitions in the RE ions strongly depend on the local environment due to covalent bonding of the outer 5d electrons to surrounding atoms in the host matrix.

oxynitrides this effect can be seen to be approximately linear with nitrogen content, resulting in a shift of approximately 100 nm in the emission peak between SiO_2 and Si_3N_4. It is, therefore, important to account for this effect in the design of an emitting layer in order to achieve the emission color necessary for SSL.

While the electrical properties of the Ce-ORSO and silicate thin films have not yet been explored, EL of Ce-doped SiO_2 MOSLEDs formed by implanting Ce into SiO_2 has been demonstrated [27–29]. As with ORSO samples, the authors report on the observation of luminescence from these samples in the absence of Si-ncs. Through the use of codoping with Gd^{3+}, which has an UV emission at 316 nm and can serve to sensitize Ce^{3+}, an increase in the external quantum efficiency of the luminescence of these materials from 0.5% for the case of Ce^{3+} doping alone to 1.8% and output powers up to 34 mW/cm^2 have been demonstrated [29]. EL spectra and the integrated EL power density as a function of current density for these MOSLEDs are shown in Figure 17.7. The EL behavior of the samples was explained through impact ionization due to hot carrier injection into the conduction band of SiO_2 or through excitation across the SiO_2 bandgap or of defects followed by a charge transfer to the Ce^{3+} (or Gd^{3+}) ions, with the latter being dominant.

The green photoluminescence of Tb^{3+} in SRSO and SiO_2 thin films has been studied by several groups [23, 30, 31]. Slightly different trends in behavior from that seen for Ce-doped samples have been observed for a set of ECR-PECVD-grown Tb-doped SRSO and ORSO samples, as is shown in Figure 17.8. For the SRSO sample, PL attributable to Si-ncs is observed at anneal temperatures up to 1100 °C. At this temperature, a slight peak attributable to Tb^{3+} appears, while at 1200 °C emission lines characteristic of the $^5D_4 - {}^4F_j$ ($j = 6, 5, 4, 3$) intra-4f transitions of the Tb^{3+} ion are seen at 487, 546, 588, and 620 nm.

In stark contrast to the above, a bright PL signal is observed in the as-grown Tb-doped ORSO sample. The luminescence increases with annealing at 800 °C, the lowest anneal temperature considered here and then decreases with higher anneal temperatures. X-ray absorption fine structure studies of Tb-implanted SiO_2 have

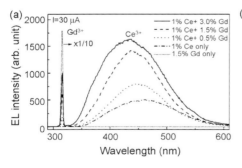
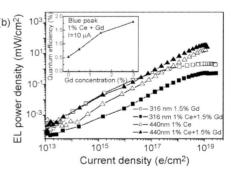
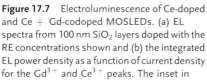

Figure 17.7 Electroluminescence of Ce-doped and Ce + Gd-codoped MOSLEDs. (a) EL spectra from 100 nm SiO_2 layers doped with the RE concentrations shown and (b) the integrated EL power density as a function of current density for the Gd^{3+} and Ce^{3+} peaks. The inset in (b) shows the increase in quantum efficiency of the Ce^{3+} emission as a function of Gd concentration. Reprinted with permission from Ref. [29]. Copyright (2006) by the American Institute of Physics.

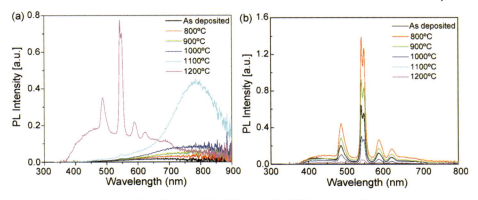

Figure 17.8 Luminescence from Tb-doped (a) SRSO and (b) ORSO samples. The samples have been annealed under N_2 at the temperatures shown. The SRSO sample contains 36% Si and 0.3% Tb, while the ORSO sample contains 32% Si and 0.4% Tb. Reproduced from Ref. [23].

demonstrated that as with Er^{3+}, Tb^{3+} is found to be in an optically active state when sixfold coordinated with surrounding oxygen atoms [32]. As with Ce, the formation of such an optically active configuration may be aided by the presence of excess oxygen in the host matrix, leading to the intense luminescence seen for the as-grown sample. With annealing at 800 °C, an increase in the number of Tb6O centers then leads to the increase in PL from these samples. At higher anneal temperatures, changes in the structure of the host matrix, as well as Tb clustering, may lead to a change in the bonding structure of the Tb ions to the optically inactive Tb-2O structure. In the SRSO sample, the simultaneous quenching of the Si-nc PL combined with an increase in the Tb^{3+} PL has been considered as evidence of an energy transfer occurring between the Si-ncs and the Tb^{3+} ions. Recent results from the analysis of both ORSO and SRSO samples through the study of their X-ray-excited optical luminescence and X-ray absorption near-edge structure at the Si K and $L_{3,2}$ edges and the O K-edge have revealed a strong correlation between excitation at O-related energy states and the Tb^{3+} luminescence, demonstrating that the host matrix and O-related energy states play an active role in the PL process for Tb-doped SiO_x [33]. The results do not, however, preclude possible contributions to the excitation through Si-ncs, particularly in the case of Si-ncs with diameters of approximately 2.5 nm or less, for which quantum confinement effects can lead to a widening of the Si-nc bandgap sufficient for the excitation of Tb^{3+}.

The electroluminescence of Tb-doped silicon oxides has been studied by several groups for both ion-implanted SiO_2 [34] and *in situ* doped SRSO thin films [35]. As for the case of Ce, MOSLED structures have been fabricated from Tb-implanted SiO_2 and the excitation mechanism is considered to be the same (impact ionization by hot electrons). The authors report external quantum efficiencies in excess of 16%. It has been further demonstrated that using a silicon oxynitride barrier layer in the MOSLED structure can improve the performance (increasing the breakdown field, injection current density, and operation lifetime) by protecting the structure from avalanche breakdown, a result that could be used to enhance device structures for SSL applications as well [36]. The EL of a Tb-doped SRSO (34.7% Si, 0.3% Tb) sample annealed at 1100 °C

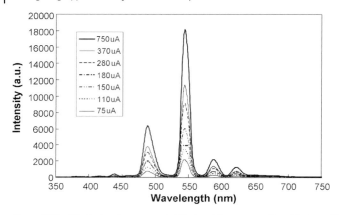

Figure 17.9 Electroluminescence at differing drive currents for a Tb-doped SRSO sample under AC excitation. The sample had a composition of 34.7% Si and 0.3% Tb and was annealed at 1100 °C. Reproduced from Ref. [35].

under AC excitation at 15 kHz with a sinusoidal input is shown in Figure 17.9. The behavior has been attributed to energy transfer from the Si-ncs to the Tb^{3+} ions. An AC drive current was used in order to try to reduce the effects of trapped oxide charges and the results have shown performance better than commercial green LEDs [35].

While O-coordination has been found to play an important role in the luminescence of Tb-doped silicon oxides, efficient luminescence has also been reported for Tb-doped silicon nitrides [37, 38] and oxynitrides [39]. In SRSN thin films, Tb was incorporated into the films after the deposition process via ion implantation. A far more intense Tb^{3+} photoluminescence was reported than for SRSO, as shown in Figure 17.10, a result attributed to the sensitization of the rare earths by recombination processes through the higher energy defect emission peaks present in SRSN thin films (as was shown in Figure 17.3). Bright luminescence from a nitride host matrix makes this an attractive alternative material for electroluminescent devices due to its reduced barrier to carrier injection, meaning that lower electric field strengths are required to excite the luminescence centers, which is in-line with the desire for reduced power consumption for SSL. Details of the bonding configuration of Tb^{3+} in SRSN materials are not known and are under investigation.

A similar trend of increasing PL with decreasing oxygen content has been reported for silicon oxynitride films doped with Tb through concurrent sputtering of Tb during the PECVD growth process [39]. In this case, however, the Tb^{3+} luminescence does not appear to arise from an energy transfer from defect or bandtail states, but rather through excitation across the reduced bandgap of the material as the films approach silicon nitride, followed by capture of the carriers at Tb^{3+} sites. In spite of the ongoing discussion of the origin of Tb^{3+} luminescence in silicon nitrides or silicon oxynitrides, and often conflicting results, they remain an attractive candidate material for the green emitting layer in SSL due to their reduced barrier to carrier injection in comparison to silicon oxides. One must keep in mind, however, that high defect densities in the Si_3N_4 or SiN_x host matrices may act as a barrier toward

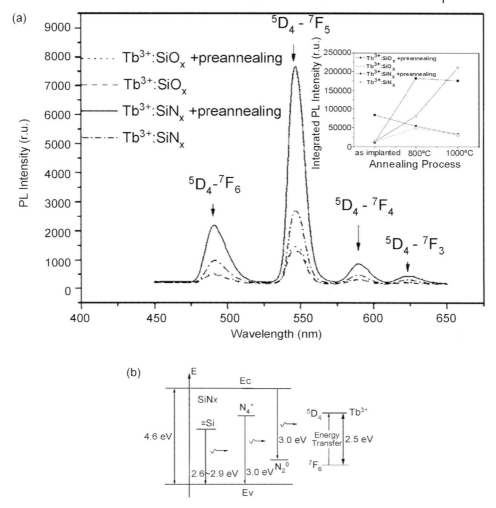

Figure 17.10 (a) Photoluminescence from Tb-doped SRSN and SRSO samples and (b) a model to explain the enhanced excitation of Tb^{3+} in the SRSN matrix through energy transfer from the recombination at defects in the SRSN matrix. In (a) some samples were preannealed prior to the Tb implant for 60 min at 1100 °C while all samples were annealed after the implant at 800 °C. The inset in (a) shows the integrated PL intensity of the $^5D_4 - {}^7F_5$ emission from the Tb^{3+} ion as a function of the annealing temperature. Reprinted with permission from Ref. [37]. Copyright (2006) by the American Institute of Physics.

the electrical excitation of Tb^{3+} ions in the nitride host matrix, which could pose a problem for their use in electroluminescent devices for SSL.

The final color component necessary for SSL is a red emission, obtainable from Eu, as mentioned earlier. Europium has two different optically active ionic states: Eu^{2+} and Eu^{3+}. Emission from the former involves the $4f^6 5d^1 - 4f^7$ transition and leads to two broad blue-green emission bands from 400–600 nm due to the fivefold degeneracy of Eu^{2+} in the crystal field of a surrounding lattice [40]. As the 5d states interact with the surrounding host matrix, the exact position of the emission bands will strongly depend on details of the host environment. For applications in SSL, it is the red emission from Eu^{3+} that is of primary interest. These emissions originate from the intra-4f transitions from $^5D_0$ to ^7F_j ($j=0$–6) states and give rise to a series of narrow emission lines from 570–700 nm. In samples with mixed Eu^{2+}/Eu^{3+} states, the output spectra will be dominated by the broad emission bands from the Eu^{2+} ions due to the higher oscillator strength and shorter lifetimes of the 4d–4f transitions.

The photoluminescence of Eu-implanted SiO_2 has been reported in Ref. [41]. In that work, the authors demonstrated the emission from both Eu^{2+} and Eu^{3+} from their samples after excitation with 377 and 464 nm sources, respectively. Care should be taken, however, in the interpretation of the red emission bands as originating from Eu^{3+}, as they were found to be quite broad, which is in contrast with the common behavior observed for intra-4f transitions. The Eu^{3+} emission was found to increase with annealing at temperatures up to 1100 °C (for 30 min under N_2) and then to decrease, while the Eu^{2+} emission increased with annealing up to 1200 °C, indicating a change in the ionic state of Eu within the films with high-temperature annealing.

The 325 nm excited PL from a SRSO sample grown by ECR-PECVD and doped *in situ* during the deposition process is shown in Figure 17.11. A Gaussian deconvolution of the spectra reveals the presence of three peaks in the emission spectrum centered within the ranges 442–486, 521–568, and 605–687 nm. The first two of these peaks arise from radiative transitions in Eu^{2+}. The lack of a fine line emission in the third peak indicates that it originates from Si-ncs formed within the SRSO matrix and not from the intra-4f transitions of the Eu^{3+} ion.

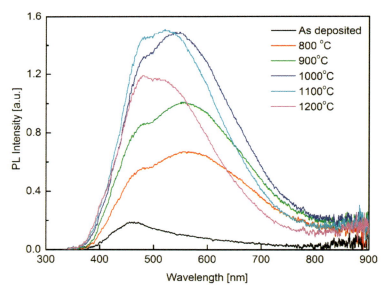

Figure 17.11 PL spectra for a Eu-doped SRSO sample (35% Si, 0.1% Eu) annealed under N_2 for 60 min at the temperatures shown. The PL spectra show luminescence corresponding only to the Eu^{2+} ion and Si-ncs. From Ref. [42].

The possibility that Eu^{3+} could excite Si-ncs through an energy transfer process has been studied through experiments on porous Si impregnated with a Eu-chloride solution [43]. The impregnation process directly introduces Eu^{3+} ions into the film, thus avoiding issues associated with deposition chemistry and Eu incorporation that will occur with a PECVD-based growth method. The results indeed show that even in a scenario where Eu is in the red emitting trivalent state its emission may be quenched by the presence of Si-ncs formed within the film after annealing. As a result, even if Eu is in the necessary trivalent state it may prove difficult to obtain a red luminescence from *in situ* Eu-doped SRSO thin films grown by ECR-PECVD. ORSO films, which, as discussed, have yielded some of the brightest emission from Ce- and Tb-doped samples, may provide a preferential host matrix to SRSO films and will be discussed later in this section.

As with Ce, Eu-silicates are an attractive option for incorporating high Eu concentrations in order to achieve bright luminescence, particular for silicates where Eu is in the desired red emitting Eu^{3+} state. Although the formation of pure Eu-silicates requires annealing at temperatures in excess of 1400 °C, mixed phase structures (containing Eu_2SiO_4, $EuSiO_3$, Eu_2O_3, and SiO_2) have been reported by reactive sputtering of Si and $EuSi_2$ in Ar or Ar/O_2 atmospheres followed by annealing at 900–1000 °C [44]. Unfortunately, in both of the silicates the authors demonstrate that Eu exists in the divalent rather than the trivalent state. The presence of Eu_2O_3 does, however, provide a by-product of the process with Eu in the trivalent state.

As with Ce and Tb, Eu-doped MOSLED structures have been fabricated through ion implantation of Eu into SiO_2 and characterized by the Institute of Ion Beam Physics and Materials Research in Dresden, Germany. The authors report on devices that exhibit both blue and red emissions (from the presence of Eu^{2+} and Eu^{3+}) [45] and on the ability to switch between emissions from the two ions by varying the excitation current [46]. Figure 17.12 shows the EL spectra for a series of Eu-doped

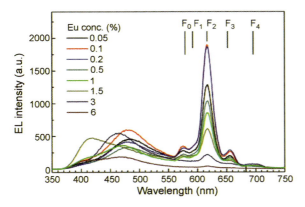

Figure 17.12 EL spectra for a set of Eu-implanted MOSLEDs with the dopant concentrations shown. The samples were subjected to rapid thermal annealing at 1000 °C for 6 s in nitrogen and the EL spectra were excited with an injection current of 5 μA. Reprinted with permission from Ref. [45]. Copyright (2008) by the American Institute of Physics.

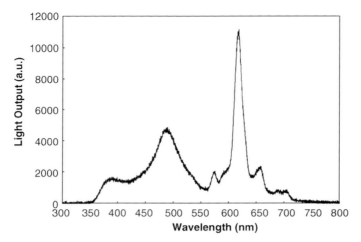

Figure 17.13 EL spectrum for an Eu-doped ORSO sample. The sample was grown with a composition close to SiO_2 to avoid Si-nc formation. The luminescence shows contributions from both Eu^{2+} and Eu^{3+} ions at 490 and 618 nm, respectively.

films with varying dopant concentrations. With increasing Eu concentration, the decreased red emission from the sample has been attributed to competition for oxygen between Eu and the SiO_2 host, which can lead to preferential formation of EuO structures (Eu^{2+}) rather than Eu^{3+}, similar to the trend observed in Ref. [41] after high-temperature annealing. Furthermore, the authors have also demonstrated the presence of Eu and Eu oxide clusters after high-temperature annealing in samples with higher Eu concentrations.

The results raise the interesting question of what would occur in the case of Eu-doped ORSO, as the increased oxygen coordination has been found to lead to favorable luminescent properties for Ce- and Tb-doped samples and plays an important role in determining the iconicity of Eu in the silicon oxide host matrix. An EL spectrum for a Eu-doped ORSO sample is shown in Figure 17.13. Excitation of peaks related to both Eu^{2+} and Eu^{3+} are readily apparent, as in the case of the Eu-implanted MOSLEDs discussed above, illustrating that the ORSO host does lead to favorable EL properties of the film, particularly a strong red Eu^{3+} emission, as required for SSL.

Furthermore, there is significant potential in the study of Eu-silicate formation, which could allow the incorporation of high Eu concentrations (with the goal to have Eu in the trivalent state), and in the study of Eu incorporated in other host matrices such as silicon nitrides and silicon oxynitrides, areas where few if any results exist.

The discussion above has been focused on RE-doped films formed through more conventional, CMOS compatible methods such as PECVD of, and/or ion implantation into, silicon oxide, nitride, or oxynitride hosts. Several reports exist on alternative methods of fabricating such materials, particularly Ce-doped silicon oxides, through sol-gel processes, but are not considered here [47–51]. In addition, reports exist on the synthesis and use of more "exotic" rare earth-doped silicon nitrides and oxynitrides such as $M_2Si_5N_8$ or $MSi_2O_{2-\delta}N_{2+2/3\delta}$ (M = Ca, Sr, Ba), primarily through sintering/curing of powders, for use in LEDs and as phosphor converters [52, 53]. While the results are both interesting and informative for understanding the RE excitation, it is more in line with the interests of this work to focus on RE-doped silicon-based materials fabricated through processes that can take advantage of conventional microelectronics fabrication technology since this serves as one of the motivating factors for the pursuit of solid-state lighting on a Si-platform.

17.3
White Light Emitting Si-Based Device Structures

Several groups have explored the fabrication of RE-doped LEDs, some of which were discussed in the previous section. Rather than reiterating details of their work, this section considers integration schemes that could be employed in order to obtain white light emitters based on a Si-platform.

There are three possible designs to a solid-state lighting cell incorporating red, green, and blue emitting layers (all of which borrow elements from designs already used for RE phosphors). The first involves the codoping of a single layer with several ions in order to obtain desired emission properties. White light emission is then achieved by balancing the numbers of specific color emitting species within the layer. The advantage of such a system is the simplicity of its fabrication. *In situ* doping during the PECVD growth process or ion implantation of specific doses can be used to accomplish this. Such a system, however, has several disadvantages. Although the system is conceptually simple to deposit, the solubility limits within the host matrix present a barrier to how many RE ions can be practically incorporated into the film. In order to increase the RE content of the films, thick layers must be grown, leading to issues with electrical pumping of the lighting cell. The second disadvantage lies in the complex energetics of the material. With several RE species within the sample, energy transfer can occur from the Si-ncs to the RE ions and between the RE ions themselves, leading to a competition between emission at different energies and often quenching of the total luminescence. As mentioned earlier, Gd^{3+} has been used as a sensitizer for Ce^{3+} emission. Similarly, the Ce^{3+} ion can serve as a sensitizer for both the Tb^{3+} and Eu^{3+} ions, leading to a loss of the blue emission component. It then becomes necessary to characterize the complex kinetics for the layer as a whole in order to determine how to optimize performance.

The second design involves the fabrication of a multilayered (superlattice) structure, with separate layers in the structure being used to produce separate colors, as is shown in Figure 17.14. In this case, white light emission may be tuned by balancing the number of layers for a specific color emission or by varying some of the layer thicknesses [54]. Through the appropriate design of the layer stacks, energy transfer or reabsorption effects between different RE ions can be minimized. Placing the high-energy blue emitting layer near the surface ensures that some fraction of the blue light will be emitted from the material without reabsorption. The development of such a system then relies on the characterization of individually doped layers and

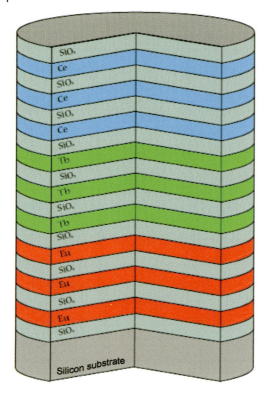

Figure 17.14 A superlattice structure for the generation of white light from Eu-, Tb-, and Ce-doped layers. The structure may make use of SiO_x layers, as shown, or SiN_x layers.

their optimization, a far simpler endeavor than dealing with a single multidoped layer. Such a structure is then better suited to the design of a solid-state lighting cell. It should be noted, however, that the emission from the layers is nondirectional and thus, emission from the blue or green layers toward the substrate can lead to reabsorption by the underlying layers resulting in increased sensitization. In the final structure, this detail must be taken into account in order to avoid distorting the color balance to the green or red. In general, this is not a large concern as the optical cross section for absorption is much smaller than the electronic cross section; however, Ce has been found to sensitize Tb, leading in some cases to a complete absence of Ce emission in the output spectrum. One must also be concerned with energy transfer between the RE ions themselves through a Förster process [55]. The use of individual, well-separated RE-doped layers mitigates such processes. The complexity of the design lies in the electrical excitation of the layers, as the thickness of the total structure and the use of ORSO films to achieve brighter Ce^{3+} or Tb^{3+} luminescence can lead to a structure where high fields are required in order to excite emission from all the layers.

A third possible design makes use of a luminescent, high-energy (near UV-blue or white) backing layer/substrate such as SiC or InGaN. The emission from this layer then serves as an optical pump for the RE-doped layer(s) that can then downconvert some of the light to perform the desired RGB mixing, as is commonly employed in RE-doped phosphors. The RE-doped layers could be configured with either of the previous two design geometries, keeping in mind their relative advantages and disadvantages. Alternatively, the order of layers in the superlattice structure shown in Figure 17.14 can be reversed. In such a scenario, one can focus on electrically exciting only the Ce^{3+} layer that would then serve to optically pump the Tb^{3+} or Eu^{3+} layers. As discussed earlier, although the emission from Ce can serve to optically pump Tb, a direct energy transfer process between the two is more efficient. In practice, better results can, therefore, be achieved by placing the rare earths in proximity to one another in order to increase this energy transfer process. This would avoid having to employ a series of thin layers ($\sim$100 nm) as in the aforementioned designs but rather thick layers with thicknesses on the order of microns.

Devices making use of the above integration schemes are the subject of research as the actual performance of these devices requires much further testing and optimization. In particular, the emission intensity, color balance, and device lifetime are key areas where extensive research is required in order to produce lighting cells that can compete with those in widespread use. In addition, it must be considered that the optimal annealing conditions differ from rare earth to rare earth, an issue that need not be addressed in the characterization of films doped with a single RE species. The determination of either a balance in the processing parameters or the determination of a suitable processing sequence remains to be done.

17.4
Fabrication of RE-Doped Silicon-Based Layers for SSL

In the previous section, potential device structures being developed for the fabrication of SSL lighting cells were considered. It is also worthwhile to consider those fabrication methods that are best suited to the implementation of such designs, keeping in mind that compatibility with current microelectronics fabrication is desired.

Ion implantation of rare earth dopants into SiO_2 has certainly proven a successful method of obtaining blue, green, and red electroluminescent devices, as discussed above. For a single-color emitting sample or for codoped layers, implantation is an ideal fabrication method, allowing precise control of the relative dopant concentrations. While implantation damage may be a concern in some instances, the majority of the films are subjected to a postdeposition anneal process that reduces this issue. For more complex superlattice-type structures, ion implantation may not be suitable, particularly in cases where the RE ions may be incorporated in layers with differing compositions (silicon oxide or silicon nitride) in order to enhance the emission of a respective layer. Although this could be accomplished through a series of film depositions followed by ion implantation into each layer, such a series of steps complicates the fabrication process.

In situ doping of thin-film layers, fabricated through either PECVD-based methods or sputtering, either via cosputtering of the RE species or introduction alongside the precursor gases, is the most attractive option for the fabrication of superlattice structures, as precise control of the layer structure, compositions, and interface can be achieved. Such systems also bring about a wide range of versatility in the growth process such that layers of alternating compositions can be easily grown in a series of sequential steps but during the same deposition run. The drawbacks of the method involve the sometimes complex deposition chemistry, particularly as it relates to incorporating the RE ions into the film, and requirements for extensive calibration of the system.

17.5
Conclusions and Future Outlook

As with Si-based light emitters in general, there is significant optimism toward the possibility of developing SSL lighting based on Si that offers significant advantages in terms of fabrication and cost. For integrated photonics, the primary competitors are III–V laser diodes that would require hybrid integration schemes. In contrast, OLEDs make a significant competitor in terms of performance for lighting applications. The present generation of OLEDs offer better performance with lower power demands. However, OLEDs are currently quite expensive and lag far behind Si in terms of performance per dollar. Both materials offer performance that is either better than or close to that of incandescent and fluorescent light sources and future improvements in the technology will likely allow Si-based emitters to surpass them.

Present research in the field has been primarily focused on optimizing the performance of individual thin films doped with a single type of RE. One of the particularly interesting aspects, particularly for the blue and green emitting, Ce^{3+}- and Tb^{3+}-doped materials, is that optimization of luminescent properties requires oxygen-rich (ORSO) materials. This enables the REs to better coordinate with oxygen atoms such that they are in the necessary trivalent luminescent state. Furthermore, the presence of oxygen-related states and the formation of RE-silicate phases have led to new considerations in optimizing the compositions of these materials to balance both the luminescent and electrical properties for better devices. Moreover, doping of SRSN materials with RE ions emitting in the visible regions remains little explored but the broad emission attainable with Si-nc-embedded SRSN films may prove suitable for the excitation of these dopants as has been seen for Tb-doped samples. Several reports have also considered slightly more exotic silicon oxynitride systems that have shown promising results and could be considered for use in device structures.

Although fully functional devices for generating white light have not yet been reported, several candidate designs have been discussed and are being explored. Significant challenges still remain in optimizing the materials for electrical, as opposed to optical, excitation, and in enhancing the lifetime of Si-based SSL cells such that they are comparable with or better than current lighting solutions. Nevertheless, the performance reported to date for RE-doped thin films and LEDs suggests that there is much to be optimistic about the potential of these materials for solid-state lighting.

Acknowledgments

The authors thank Howard Tweddle for providing market information related to the field of solid-state lighting and Carla Miner for supplying some recent results. Special thanks to O. H. Y. Zalloum and P. R. J. Wilson for valuable input and discussions on this research. The McMaster authors are grateful for financial support from the Center for Photonics of the Ontario Centers of Excellence Inc. (OCE), the Natural Sciences and Engineering Research Council of Canada (NSERC), and the Canadian Institute for Photonic Innovations (CIPI).

References

1 Ossicini, S., Pavesi, L., and Priolo, F. (2003) *Light Emitting Silicon for Microphotonics*, Springer, Berlin.

2 Pavesi, L. and Lockwood, D.J. (2004) *Silicon Photonics*, Springer, Berlin.

3 Bergh, A., Craford, G., Duggal, A., and Haitz, R. (2001) *Phys. Today*, **54**, 42.

4 U.S. Department of energy solid-state lighting webpage, http://www1.eere.energy.gov/buildings/ssl/.

5 Kitai, A. (2008) *Luminescent Materials and Applications*, John Wiley & Sons, Inc., Hoboken, NJ.

6 Anderson, S.G. (2008) *Laser Focus World*, **44**, 27.

7 Hoffmann, G. (1931) CIE chromaticity diagram. Available online at www.fho-emden.de/~hoffmann/ciesuper.txt.

8 Wyszecki, G. and Stiles, W.S. (2000) *Color Science: Concepts and Methods, Quantitative Data, and Formulae*, 2nd edn, John Wiley & Sons, Inc., New York.

9 Wolkin, M.V., Jorne, J., Fauchet, P.M., Allan, G., and Delerue, C. (1999) *Phys. Rev. Lett.*, **82**, 197.

10 Deshpande, S.V., Gulari, E., Brown, S.W., and Rand, S.C. (1995) *J. Appl. Phys.*, **77**, 6534.

11 Gaft, M., Reisfeld, R., and Panczer, G. (2005) *Modern Luminescence Spectroscopy of Minerals and Materials*, Springer, Berlin.

12 Kenyon, A.J. (2002) *Prog. Quantum Electron.*, **26**, 225.

13 Abram, R.A., Kenyon, A.J., Stoneham, A.M., and Dobson, P. (2003) *Philos. Trans. R. Soc. London, Ser. A*, **361**, 361.

14 Daldosso, N., Navarro-Urrios, D., Melchiorri, M., Pavesi, L., Gourbilleau, F., Carrada, M., Rizk, R., Garcia, C., Pellegrino, P., Garrido, B., and Cognolato, L. (2005) *Appl. Phys. Lett.*, **86**, 261103.

15 Iacona, F., Irrera, A., Franzo, G., Pacifici, D., Crupi, I., Miritello, M., Presti, C.D., and Priolo, F. (2006) *IEEE J. Sel. Top. Quantum Electron.*, **12**, 1596.

16 Kenyon, A.J., Loh, W.H., Oton, C.J., and Ahmad, I. (2006) *J. Lumin.*, **121**, 193.

17 Pacifici, D., Irrera, A., Franzo, G., Miritello, M., Iacona, F., and Priolo, F. (2003) *Physica E*, **16**, 331.

18 Zacharias, M., Richter, S., Fischer, P., Schmidt, M., and Wendler, E. (2000) *J. Non-Cryst. Solids*, **266**, 608.

19 Kenyon, A.J. (2003) *Philos. Trans. R. Soc. London, Ser. A*, **361**, 345.

20 Kik, P.G. and Polman, A. (2000) *J. Appl. Phys.*, **88**, 1992.

21 Polman, A. and van Veggel, F. (2004) *J. Opt. Soc. Am. B*, **21**, 871.

22 Franzo, G., Vinciguerra, V., and Priolo, F. (1999) *Appl. Phys. A*, **69**, 3.

23 Li, J., Zalloum, O.H.Y., Roschuk, T., Heng, C.L., Wojcik, J., and Mascher, P. (2008) *Adv. Opt. Technol.*, **2008**, 295601.

24 Adler, D.L., Jacobson, D.C., Eaglesham, D.J., Marcus, M.A., Benton, J.L.,

Poate, J.M., and Citrin, P.H. (1992) *Appl. Phys. Lett.*, **61**, 2181.

25 Li, J., Zalloum, O.H.Y., Roschuk, T., Heng, C.L., Wojcik, J., and Mascher, P. (2009) *Appl. Phys. Lett.*, **94**, 011112.

26 van Krevel, J.W.H. (2000) PhD Thesis, Technische Universiteit Eindhoven, the Netherlands.

27 Castagna, M.E., Coffa, S., Monaco, M., Muscara, A., Caristia, L., Lorenti, S., and Messina, A. (2003) *Mater. Res. Soc. Symp. Proc.*, **770**, I2.1.1.

28 Skorupa, W., Sun, J.M., Prucnal, S., Rebohle, L., Gebel, T., Nazarov, A.N., Osiyuk, I.N., Dekorsy, T., and Helm, M. (2005) *Mater. Res. Soc. Symp. Proc.*, **866**, V4.1.1/FF4.1.1.

29 Sun, J.M., Prucnal, S., Skorupa, W., Helm, M., Rebohle, L., and Gebel, T. (2006) *Appl. Phys. Lett.*, **89**, 091908.

30 Amekura, H., Eckau, A., Carius, R., and Buchal, C. (1998) *J. Appl. Phys.*, **84**, 3867.

31 Yoshihara, M., Sekiya, A., Morita, T., Ishii, K., Shimoto, S., Sakai, S., and Ohki, Y. (1997) *J. Phys. D*, **30**, 1908.

32 Ofuchi, H., Imaizumi, Y., Sugawara, H., Fujioka, H., Oshima, M., and Takeda, Y. (2003) *Nucl. Instrum. Methods Phys. Res. B*, **199**, 231.

33 Roschuk, T., Wilson, P.R.J., Li, J., Wojcik, J., and Mascher, P. (2009) Accepted for publication in Phy. Status Solidi B DoI: 10.1002/pssb. 2009–45531.

34 Sun, J.M., Skorupa, W., Dekorsy, T., Helm, M., Rebohle, L., and Gebel, T. (2005) *J. Appl. Phys.*, **97**, 123513.

35 MacElwee, T.W., Hill, S.E., Campbell, S., Ducharme, D., Ruoux, B.A., Calder, I.D., Flynn, M., Wojcik, J., Gujrathi, S., and Mascher, P. (2006) *Proceedings of the 3rd IEEE International Conference on Group IV Photonics*. IEEE, p. 216.

36 Sun, J.M., Rebohle, L., Prucnal, S., Helm, M., and Skorupa, W. (2008) *Appl. Phys. Lett.*, **92**, 071103.

37 Yuan, Z.Z., Li, D.S., Wang, M.H., Chen, P.L., Gong, D., Wang, L., and Yang, D. (2006) *J. Appl. Phys.*, **100**, 083106.

38 Yuan, Z.Z., Li, D.S., Wang, M.H., Gong, D.R., Cheng, P.H., Chen, P.L., and Yang, D.R. (2008) *Mater. Sci. Eng. B*, **146**, 126.

39 Jeong, H., Seo, S.-Y., and Shin, J.H. (2006) *Appl. Phys. Lett.*, **88**, 161910.

40 Russell, H.N., Albertson, W., and Davis, D.N. (1941) *Phys. Rev.*, **60**, 641.

41 Liu, F., Zhu, M., Wang, L., and Hou, Y. (2000) *J. Alloys Compd.*, **311**, 93.

42 Li, J. and Sc, M.A. (2008) Thesis, McMaster University, Hamilton, ON.

43 Moadhen, A., Elhouichet, H., Canut, B., Sandu, C.S., Oueslati, M., and Roger, J.A. (2003) *Mater. Sci. Eng. B*, **105**, 157.

44 Qi, J., Matsumoto, T., Tanaka, M., and Masumoto, Y. (2000) *J. Phys. D*, **33**, 2074.

45 Rebohle, L., Lehmann, J., Prucnal, S., Kanjilal, A., Nazarov, A., Tyagulskii, I., Skorupa, W., and Helm, M. (2008) *Appl. Phys. Lett.*, **93**, 071908.

46 Prucnal, S., Sun, J.M., Skorupa, W., and Helm, M. (2007) *Appl. Phys. Lett.*, **90**, 181121.

47 Vedda, A., Chiodini, N., Martino, D.D., Fasoli, M., Martini, M., Paleari, A., Spinolo, G., Nikl, M., Solovieva, N., Baraldi, A., and Capelletti, R. (2005) *Phys. Status Solidi C*, **2**, 620.

48 Malashkevich, G.E., Poddenezhny, E.N., Melnichenko, I.M., and Boiko, A.A. (1995) *J. Non-Cryst. Solids*, **188**, 107.

49 Otto, A.P., Brewer, K.S., and Silversmith, A.J. (2000) *J. Non-Cryst. Solids*, **265**, 176.

50 Boye, D.M., Ortiz, C.P., Silversmith, A.J., Nguyen, N.T.T., and Hoffman, K.R. (2008) *J. Lumin.*, **128**, 888.

51 Magyar, A.P., Silversmith, A.J., Brewer, K.S., and Boye, D.M. (2004) *J. Lumin.*, **108**, 49.

52 Li, Y.Q., van Steen, J.E.J., van Krevel, J.W.H., Botty, G., Delsing, A.C.A., DiSalvo, F.J., de With, G., and Hintzen, H.T. (2006) *J. Alloys Compd.*, **417**, 273.

53 Ru-Shi, L., Yu-Huan, L., Nitin, C.B., and Shu-Fen, H. (2007) *Appl. Phys. Lett.*, **91**, 061119.

54 Chik, G., MacElwee, T., Calder, I., and Hill, E.S. Filed – 2006 Published – 2008. Engineered structure for solid-state light emitters, U.S. Patent Application 20080093608.

55 Dexter, D.L. (1953) *J. Chem. Phys.*, **21**, 836.

18
Biomedical and Sensor Applications of Silicon Nanoparticles

Elisabetta Borsella, Mauro Falconieri, Nathalie Herlin, Victor Loschenov,
Guiseppe Miserocchi, Yaru Nie, Iparia Rivolta, Anastasia Ryabova, and Dayang Wang

18.1
Introduction

The rapid development of biomedical sciences demands new advanced techniques and instruments to investigate cells and cellular processes [1, 2]. Cellular components and processes are mostly visualized by attaching a fluorophore to a molecule (e.g., an antibody) that binds to the target via molecular recognition. Information on the exact location of the monitoring molecule can be achieved through fluorescence micros-copy, whereas the variation in the fluorescence parameters (such as intensity, emission spectrum, and lifetime) can be used to determine what is happening in the molecular environment [3].

Traditionally used fluorescent labels are organic dyes, fluorescent proteins, and lanthanide chelates that are highly water-soluble (even at high salt concentration) and easily bioconjugated [1]. However, widely employed organic dyes cannot be used in simultaneous measurement of several biological indicators (i.e., multicolor detec-tion) having narrow excitation spectra, which imposes the use of different excitation sources for different dyes, and broad emission spectra with a long tail at long wavelengths, which causes spectral cross-talk between different detection chan-nels [4]. Moreover, organic dyes can undergo either reversible transfer to "dark" states in which they do not emit photons, and consequently the fluorescence goes "on" and "off" (blinking), or irreversible photoinduced reactions such as photooxidation [1]. This last phenomenon, known as "photobleaching," is a major limiting factor for biological studies requiring either long observation times under a fluorescence microscope or temporally resolved imaging [1, 3]. In any case, wide-field optical microscopy does not allow to resolve structures smaller than a few hundred nanometers since the distance between fluorophores must be larger than the resolution limit. Biological structures can be resolved with better resolution by replacing fluorophores with metallic (especially gold) nanoparticles, which can be imaged by electron microscopy, but only in *in vitro* experiments.

Silicon Nanocrystals: Fundamentals, Synthesis and Applications. Edited by Lorenzo Pavesi and Rasit Turan
Copyright © 2010 WILEY-VCH Verlag GmbH & Co. KGaA, Weinheim
ISBN: 978-3-527-32160-5

Recently, a new class of fluorescent materials, the II–VI semiconductor quantum dots (QDs), has raised expectations in advanced biological research due to their peculiar optical properties. QDs are generally II–VI semiconductor structures with all the physical dimensions smaller than the exciton Bohr radius (typically below 10 nm) [5]. In this size range, QDs have size-dependent tunable optical properties that can be exploited for multiple applications in biomedical research, ranging from cell labeling to biosensing, animal imaging, and therapy [1–6].

Compared to organic fluorophores, inorganic QDs are rather resistant to photobleaching and can be observed in high-resolution electron microscopy. Moreover, QDs are ideal probes for multicolor experiments since nanocrystals with different sizes can be simultaneously excited by a single wavelength of light, at energies above the onset of their continuous excitation spectra, resulting in multicolor, symmetrical spectral emission without red tails [4]. As a result, many colors can be distinguished without spectral overlap and different cellular compartments/processes can be labeled simultaneously, each with a different color [3–6].

Another important feature of the QDs is their longer fluorescence lifetime (more than a few tens of nanoseconds) compared to the lifetime of the organic fluorophores (of a few nanoseconds), matching the fast decay time of the autofluorescence from many biological samples. As a consequence, QDs emit light at a rate too slow to eliminate most of the autofluorescence signal (through time-gated detection) but slow enough for time-resolved fluorescence bioimaging [7]. Finally, the reduced tendency to photobleaching is useful for long-term imaging studies such as fluorescent labeling of transport processes and protein tracking.

Early developed QDs of II–VI semiconductors (typically CdS, CdSe, CdTe, or ZnS) were uncapped nanoparticles (NPs), showing weak and unstable light emission [2]. Recently, QDs with a high and stable luminescence have been produced by passivation of their surfaces with different semiconductors with a wider electronic energy gap [2]. However, high-quality QDs are hydrophobic, which makes QDs insoluble in aqueous media, often not biocompatible, and short of reactive functional groups for bioconjugation with biomolecules. In 1998, the groups of Alivisatos [4] and Nie [8] succeeded in synthesizing QD bioconjugates via chemical exchange and demonstrated the potential of the application of QDs in biomedicine. So far, various approaches have been developed to render II–VI semiconductor QDs water-soluble and able to conjugate to biological molecules. The II–VI QDs have proven useful for cell labeling, tracking cell migration, flow cytometry, pathogen detection, genomic and proteomic detection, FRET sensors, and high-throughput screening of biomolecules [2]. However, their application in biology and medicine is hampered by their inherent chemical toxicity [1, 2].

Unlike bulk Si that is not a good light emitter, nanostructured Si can emit photons in the visible–near-IR range with a reasonable efficiency [9]. It follows that silicon nanoparticles (Si NPs) have the potential to overcome the inherent limitations in the biomedical use of QDs since silicon is inert, nontoxic, abundant, and economical. Moreover, the silicon surface is apt to chemical functionalization, thus allowing numerous stabilization and bioconjugation steps [10, 11].

To exploit this potential for biomedical applications, Si NPs should remain highly luminescent and well dispersed in water and biological fluids over a wide range of pH and salt concentration. However, preparing macroscopic quantities of single Si NPs with stable and intense photoluminescence (PL) emission and good dispersibility in water is still a challenging and difficult task that limits the use of Si NPs in biomedicine and is the objective of several studies and attempts that will be analyzed in Section 18.2.

Although Si is a nontoxic material, a major concern that might restrict the application of Si nanoparticles in medicine arises from the potential harm that could derive from their reduced size. In this respect, a critical review of recent studies dealing with cytotoxicity and inflammatory response of cells exposed to Si-based nanoparticles will be provided in Section 18.3. An attempt will also be made to analyze the published results in the more general context of the biological impact of nanoparticles, aiming at an assessment of short- and long-term risks for the health.

In Section 18.4, we will present an overview of the first, recently published applications of engineered Si NPs for bioimaging, as nucleotide carriers for gene therapy, and as sensors.

Finally, a critical insight into the possible future developments in this field will be given.

18.2
Synthesis and Surface Engineering of Si Nanoparticles for Bioapplications

18.2.1
Synthesis and Optical Properties of Si Nanoparticles

The origin of the PL emission from Si NPs has been intensely debated for almost two decades: whether the photons arise from highly localized defect states or the emission is ruled by the quantum confinement effect [12, 13]. There is now a consensus on the possibility that both effects play an important role, and the optical properties of nanostructured Si are strongly influenced by the presence of atoms/molecules/radicals or defects at or close to the surface. Thus, the challenge of preparing light emitting Si nanoparticles is to deliberately control both the particle size and morphology and the surface properties [14–16].

To date, a wide variety of techniques have been developed to synthesize Si NPs either in solution or in gas phase. Based on the wet chemistry, Si NPs can be produced in inverse micelles [17, 18] by synthesis in high-temperature supercritical solutions [19] and by reduction of alkylsilicon halides [20]. Gas-phase approaches are based on silane or disilane decomposition either via laser light irradiation, such as photolysis [21] and pyrolysis [22–25], or via electron impact [26]. In addition, Si NPs can be obtained via wet electrochemical etching of Si wafer in hydrofluoric acid, followed by ultrasound treatment [27–29]. In the last case, the as-prepared Si NPs are polydisperse in terms of shape and size, so post-size selection processes, such as centrifugation, selective precipitation, size exclusion chromatography, and capillary electrophoresis, are required to narrow the size distribution [27, 28, 30].

As expected [14–16], the optical properties of Si NPs significantly vary, depending on the preparation technique. Si NPs with an oxide surface passivation typically show a dipole-forbidden, moderately intense, PL emission ranging from yellow–orange to IR [22–24]. For example, in Figure 18.1 PL spectra measured under UV excitation on size-selected Si nanoparticles prepared by laser pyrolysis and exposed to air are shown [23]. The PL decay curves measured at different emission wavelengths, under laser excitation at 488 nm, are shown in Figure 18.2 (data are taken from Ref. [32]). Similar to porous Si, the decay of the PL emission of Si NPs can be fit by a stretched exponential function, a decay law often encountered in disordered systems, with average lifetime ranging from 10^{-4} to 10^{-3} s and increasing as the PL emission wavelength increases [31, 32].

Enhancement of luminescence emission from pyrolytic Si NPs was observed after thermal oxidation [25, 33] and wet chemical soft oxidation reactions [34, 35]. These results underline the importance of the quality of the oxide capping layer for the passivation of defect states that can otherwise quench the luminescence emission [35, 36]. For example, in Figure 18.3 PL emission spectra of pyrolytic Si NPs, as-

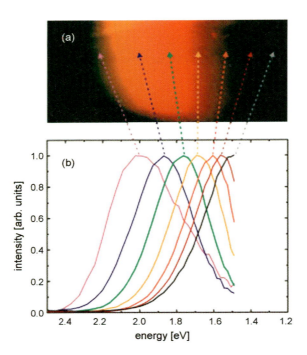

Figure 18.1 Photoluminescence emitted from size-separated Si NPs produced by cluster beam deposition [23]. Panel (a) shows a photo of the deposited nanoparticles (with size increasing from the left to the right side) under illumination with a UV lamp; panel (b) reports PL spectra recorded at the positions indicated by the arrows. Taken from Ref. [23].

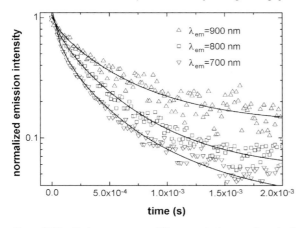

Figure 18.2 PL decay curves at different emission wavelengths from suspensions of wet oxidized pyrolytic Si NPs; fitting curves (solid lines) are based on the stretched exponential law (data are taken from Ref. [32]).

prepared and after oxidizing treatments in alcoholic and aqueous solvents (i.e., MeOH and the mixture H_2/MeOH 1/1), under laser excitation at 532 nm are shown (data are taken from Ref. [35]).

Several attempts were made to get bright green and blue light emission from Si NPs prepared by gas-phase methods with the intent of achieving multicolor cellular labeling [25, 36]. A significant reduction in the PL emission intensity was observed when HF etching followed by reoxidation was used to shift the PL emission wavelength by decreasing the Si NP size [36]. On the other hand, an intense and fast deep blue emission was observed after UV excitation of ultrasmall ($\leq$1 nm in

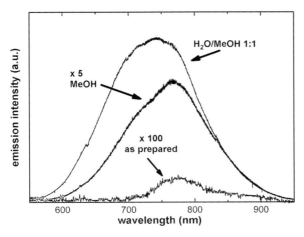

Figure 18.3 PL emission band from pyrolytic Si NPs as-prepared and after oxidizing treatments in alcoholic and aqueous solvents (data are taken from Ref. [34]).

size) Si NPs prepared by chemical etching of bulk crystalline silicon, followed by ultrasonication [36, 37].

It was proposed, on the basis of calculations reported in Ref. [16], that the intense and fast (a few nanoseconds) blue PL emission arises in H-passivated Si NPs having optically allowed, direct bandgap transitions, whereas the yellow–red, weaker, and slower PL emission originates from oxide-passivated Si NPs having optically forbidden, indirect gap-type transitions. More recently, the direct-like behavior of the fast blue emission was explained in terms of localization on radiative deep molecular-like Si–Si traps with size-dependent depth [38].

18.2.2
Surface Functionalization of Luminescent Si NPs for Bioapplications

For biomedical applications, the high hydrophilicity and colloidal stability in a biological environment are mandatory to prevent Si NP aggregation and precipitation. On the other hand, the surface termination, passivation, and functionalization can alter the internal electronic structure of Si NPs and, in turn, their PL properties and decay dynamics. Deliberate surface modification of Si NPs is of paramount importance for biological interaction.

18.2.2.1 Alkali–Acid Etching of Si NPs
Etching of as-prepared Si NPs is usually unavoidable either for reduction of the NP size to tune PL properties or for disaggregation of pyrolytic nanopowders. In fact, Si NPs produced by laser pyrolysis are agglomerated due to particle coagulation and coalescence during the growth process [22, 24, 39], with typical aggregate size of some tens of nanometers, as measured by dynamic light scattering on nanopowder suspensions. Etching of Si and SiO_2 is well established thus far. Silica can be effectively etched by both hydrogen fluoride [40–43] and alkaline hydroxide [44], while silicon can be etched only by alkali hydroxide [45], as shown in the reaction equations reported below:

1) HF etching:

$$SiO_2 + 6HF \rightarrow 2H^+ + SiF_6^{2-} + 2H_2O \quad [40\text{-}43]$$

2) Alkaline etching:

$$2OH^- + SiO_2 \rightarrow SiO_3^{2-} + H_2O \quad [44]$$

$$Si + 2OH^- + 2H_2O \rightarrow SiO_2(OH)_2^{2-} + 2H_2 \quad [45]$$

Since alkali etch silicon must faster than silica [46], it may not be a good choice to tune the size of Si NPs already coated with silica shell. In order to decrease the size of Si NPs in a properly controlled way, a mixture of HF and HNO_3 was mostly utilized [47]. Upon kicking off the SiO_2 layer by HF, the fresh Si core is immediately oxidized by HNO_3, thus reducing the core size. By controlling the HF/HNO_3 ratio

(a)

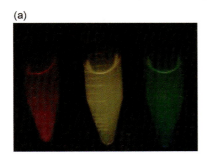

(b)

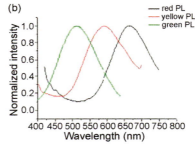

Figure 18.4 (a) Optical images of Si NPs dispersed in chloroform after etching by concentrated HF and HNO₃ with the HF/HNO₃ volume ratio of 10 : 1 in a mixture of methanol and water in the volume ratio of 1 : 2. The images were recorded under a UV lamp after 3, 9, and 12 min (from left to right). (b) Photoluminescence spectra of the corresponding samples under UV illumination.

and the etching time, Si—H terminated Si NPs with emission color varying from red to green were obtained (Figure 18.4).

The Si NPs after this treatment show unstable luminescence, bad dispersibility in most solvents, and susceptibility to oxidation in air, so surface modification is necessary to overcome these drawbacks. To date, Si—H terminated Si NPs have been modified by two ways: (1) silanization via Si—O bonding and (2) hydrosilylation through Si—C bonding.

18.2.2.2 Surface Modification of Si NPs by Silanization

Well-established silanization procedures allow one to coat Si NPs with organosilanes with various functional groups, thus providing flexibility for Si NP surface functionalization.

Swihart and coworkers [48] modified pyrolytic Si NPs, after oxidation by the piranha protocol or HNO₃, with octadecyltrimethoxysilane in toluene, thus creating a luminescent and clear dispersion of Si NPs [48]. The PL measurements demonstrated that surface treatments significantly stabilized the photoluminescent properties of the Si nanoparticles against degradation.

Kauzlarich and coworkers [49] synthesized blue luminescent Si—Cl terminated nanoparticles by sodium naphthalide reduction and modified their surfaces by silanization. The treated Si NPs exhibited photochemical stability in nonpolar organic solvents and did not show any significant photobleaching over 4000 s compared to fluorophores.

Ruckenstein and coworkers [50] capped luminescent Si nanoparticles with organosilane reaction with conductive polyaniline, for applications as optical emitters. This coating greatly stabilized the PL emission of the Si NPs against quenching and degradation, also in basic solutions.

Lee and coworkers [51] prepared water-dispersible core–shell 3 nm $Si@SiO_xH_y$ NPs and tuned their luminescence by oxidation using a mixture of EtOH and H_2O_2. The results suggested that the surface modification of Si nanoparticles by Si—O bonds did not affect the luminescence behavior of the Si cores.

Recently, we found that oxidation by heating under air can turn pyrolytic Si NPs luminescent in the red [24, 32]. However, this luminescence arises mainly from the dangling unsaturated bonds of Si–O on the NP surfaces rather than from the Si core [12, 13].

It has been demonstrated that the Si–O bonds dramatically affect the electronic distribution in the Si NPs and thus reduce the energy of indirect transitions [16, 52], which quenches the luminescence of Si NPs. Thus, better modification routes to avoid the formation of Si–O bonds on Si NPs are needed.

18.2.2.3 Surface Modification of Si NPs via Hydrosilylation

Si–H bonds terminated on Si NPs can be cleaved into silicon radicals at high temperature or via photoactivation. The highly active silicon radical can attach to C=C and C≡C groups and form Si–C bonds; this procedure is known as hydrosilylation.

Swihart and coworkers [53] first obtained Si–H dominated surface passivation on Si NPs by increasing the HF/HNO$_3$ ratio during etching and then using photo-activated hydrosilylation under argon they created Si NPs with stable luminescent emission varying from green to red [53].

Because various functional alkenes are commercially available, this hydrosilylation-based protocol is useful for further bioconjugation. Wang *et al.* [54] demonstrated the feasibility to couple DNA to luminescent Si NPs for biological labeling. Reipa and coworkers conjugated streptavidin to luminescent Si NPs through UV-initiated hydrosilylation [11].

Besides alkenes, alkynes react with silicon too. Mitchell and coworkers [55] reported the feasibility to couple alkynes and silicon dimers from blue luminescent Si NPs prepared by high-energy ball milling.

In order to render the Si NPs hydrophilic, acrylic acid was capped on the NPs via hydrosilylation, which was first reported by Ruckenstein and coworkers [56]. Warner *et al.* [17] prepared water-soluble Si quantum dots by covalently capping allylamine on the NPs via hydrosilylation, catalyzed by platinum (Figure 18.5). The so obtained highly blue luminescent Si NPs were used for bioimaging HeLa cells (see Section 18.4 for more details). The amine group on the particle surface allowed further modification with carboxy groups, epoxy groups, ketones, acyl chlorides, and aldehydes. However, this protocol may be disadvantageous for long wavelength emitting Si NPs because the amine group is known to quench the photoluminescence [17].

By conducting hydrosilylation of acrylic acid in the presence of HF, Swihart and coworkers [57] obtained stable aqueous dispersions of luminescent Si NPs with different emitting colors. The Si NP size and the corresponding PL emission colors, from yellow to green, were controlled by varying the etching time. Blue luminescent Si NPs were prepared by hydrosilylation followed by oxidation of yellow-emitting Si NPs, either using UV irradiation in solution or heating in air [58]. It was found, however, that the blue luminescence arises neither from quantum-confined states in smaller Si core nor from states in fully oxidized silica. The mechanism of emission from these particles remains an open question [58].

(a) **Hydrophobic**　　　　　　　　　　　**Hydrophilic**

(b)

Figure 18.5　(a) Schematic diagram of the procedure used in Ref. [17] to change the surface chemistry of the silicon quantum dots from hydrogen to allylamine. (b) Blue luminescence emitted by the allylamine capped silicon quantum dots [17]. Taken from Ref. [17].

Water-dispersible Si NPs/polyacrylic acid composite nanospheres were prepared by hydrosilylation of acrylic acid on the NPs and photopolymerization of the acrylic acid in the surrounding media under carefully tuned UV irradiation (Figure 18.6)[51, 59]. Despite the good solubility in water, however, the colloidal stability of the resulting composite spheres in biological fluids was not reached.

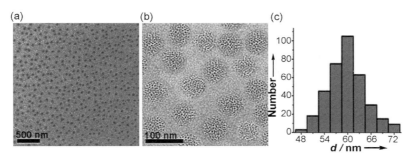

Figure 18.6　(a and b) TEM images of the Si NPs poly(acrylic acid) nanospheres with a size of approximately 60 nm and (c) the corresponding size distribution. Taken from Ref. [59]

Swihart's group [60] synthesized biocompatible luminescent Si NP composite vesicles for imaging cancer cells. First, luminescent Si NPs, obtained via HF/HNO$_3$ etching, were capped with styrene, octadecene, and ethyl undecylenate via hydrosilylation, which enabled the NPs to disperse in most nonpolar organic solvents. Then, the modified Si NPs were mixed with phospholipids functionalized by polyethylene glycol (PEG) in chloroform. After chloroform evaporation, the mixtures dispersed in water and formed luminescent composite vesicles of 50–120 nm in size. Their aqueous dispersion showed no precipitation after 2 months' storage at 4 °C. The luminescence intensity showed negligible decrease from 25 to 40 °C and remained constant in phosphate-buffered solution (PBS) [60]. Due to the nonimmunogenic and nonantigenic properties of PEG, the resulting composite vesicles showed low cytotoxicity.

Similar to the above-mentioned methods, in the framework of the EC Project BONSAI,[1)] we also were able to get luminescent Si NPs by oxidation or fluoride/nitric acid etching (Figure 18.3), and through suitable nonionic surfactant modification, we succeeded to get water-dispersible Si NPs. However, more efforts are still necessary in the scientific community in order to get truly monodispersed, stable, biocompatible, and luminescent Si NPs.

18.3
Biointeraction of Si-Based Nanoparticles

In the previous sections were described methodologies for preparing photoluminescent Si NPs with surfaces engineered for biomedical applications. Although silicon is a nontoxic material, potential toxicity induced by the nanostructure and engineered surface characteristics need to be accurately investigated before Si nanoparticles can be safely used for *in vivo* applications in bioimaging, cell staining, and drug delivery.

As previously discussed, the Si NPs are quickly oxidized in the presence of aqueous solvents. Thus, in case of uncapped Si NPs or biodamage of the capping material, the biological specimen may come in contact with silica first. Moreover, due to the medical relevance of silicosis, there is more abundant literature on this topic for silica than for Si NPs. Therefore, in the next sections, dealing with interaction with biomaterials, we will report on both silica and silicon nanoparticles.

The main problem concerning the evaluation of the potential toxicity of Si NPs (and other nanoparticles) is that there are not yet standard strategies and methods. On this premise, we will examine the main concerns related to exposure of cells to silica/silicon NPs in terms of biophysical issues, discussing the amount of particles that come in contact with cells and the key parameters one has to consider to discriminate toxicity.

In vitro studies have become an essential component in assessing NP toxicity since they are a relatively rapid and cost-effective way of screening [61]; moreover, they

---

1) FP6 EC-STREP Project BONSAI: "Bio-imaging with smart functional nanoparticles" (www.bonsai-project.eu).

restrict the animal use, as required by ethical issues. Although a large number of studies have been performed with *in vitro* systems, yet this model represents a serious challenge and in fact data on this topic obtained by several groups are often contradictory. Multiple factors are responsible for the differences found; some might be related to particles size, shape, and physicochemical surface properties, while other may pertain to the cell lines used, and finally one should consider differences in methodologies of the test itself.

In general, while studying nanomaterial–cell exposure, Teeguarden *et al.* [62] and Lison *et al.* [63] pointed out a challenging question, namely, the difference between nominal and effective dose in cytotoxicity assays. Since particles can diffuse, aggregate, and settle in media, the effective dose of NPs to which cells are exposed in an *in vitro* system has to be regarded in a somehow dynamic way. In the case of NPs, one might also consider the nature of both the particles and the solution in which they are dispersed (e.g., viscosity and the presence of proteins that interact with NPs). Therefore, the exposure to NPs cannot be equated with that of a soluble chemical, in which case the exposure can be described only by considering two "dimensions": chemical concentration and time of exposure.

Since the target of an *in vitro* study is the cell whose response relates specifically to the amount of material it comes in contact with, it seems appropriate to introduce the concept of "delivered cellular dose" instead of "media concentration."

Finally, one has to consider that in the presence of particle aggregation, like in the case of pyrolytic Si NPs, the relevant quantities are the aggregate dimension and morphology, which usually depend on both the material processing and the disaggregation procedures adopted; this clearly may affect the experimental conditions. Moreover, binding to proteins can also alter the hydrodynamic diameter of the particles and in turn affect the sedimentation velocity and diffusion rates.

One should also adjust for the effect due to the gravitational and diffusional delivery. In the case of NPs with diameter less than 50 nm, the gravitational effect is less important so that the "delivered cellular dose" reflects more the diffusional rather than the gravitational movement of NPs. In this case, the diffusional component is inversely proportional to the particles size.

Finally, another source of variability relates to the type of cells used for the studies.

All these factors, when not taken properly into account, hinder the attempt to integrate different results from different studies.

18.3.1
Cell Viability

With "cell viability" we intend a physiological condition under which cells grow at their regular doubling time (typically 24 h for a normal mammalian cell, see www.newton.dep.anl.gov) and the percentage of dead cells (typically around 8% of the cell population) is related only to physiological processes. Among these processes leading to cells death, we also include the "apoptosis" that is a form of programmed cell death. Apoptosis may originate as a consequence of a wide range of extra- or intracellular signals, for example, stress signals that may induce the cells to suicide. The same stress signals generate an acute response first inducing the cell to produce cytokines, a category of signal molecules, such as IL-1, IL-6, and TNF, which are synthesized in response to trauma or cell damage.

In order to characterize whether a compound may be toxic for a cell population, researchers investigate if parameters such as cell viability, cell death, or apoptosis (just to mention a few of them) are altered in the presence of the test substance compared to a control condition where the cells are growing in their regular medium.

The literature presents a wide variety of cell lines tested and assays to evaluate the geno- or cytotoxicity. The research groups performing *in vitro* testing mostly use lung epithelial cells or epithelial cells in general (HeLa, a cell line derived from a cervical cancer), endothelial cells, and macrophages. A cell line specifically used to evaluate the effects of NPs on the lung is the commercially available A549, while macrophages are commonly used to simulate a cell microenvironment and the clearance of NPs.

The most common tests to study the *in vitro* toxicity evaluate not only the cell morphology, apoptosis, and cell viability but also cytokine production as an indirect marker of inflammation and oxidative stress. There are different methods to study the cell vitality: some of them are luminescent assays to assess the metabolic activity of the cell population (MTT, WST-1 assay) or the plasma membrane integrity (Lactate Dehydrogenase, LDH, release). All these assays are based on the dosage of the activity of specific enzymes (mitochondrial, MTT, or WST-1 assay or cytoplasmic LDH assay) that should not be present outside their normal microenvironment (either mito-chondria or cytoplasm). If the enzyme activity is detected outside such microenvir-onments, this is an indication of ultrastructural damage. The enzyme activity is measured by means of a colorimetric assay that uses the red formazan, an artificial product obtained by reducing tetrazolium salts (MTT or WST-1) or lactate.

In terms of LDH release from cells exposed to silica NPs, some authors report no increase in the enzyme activity in the medium [64] while others found a silica NPs dose-dependent increase [61, 63, 65, 66].

Yang *et al.* [67] are the first to propose that luminescent tests may not be suitable to inspect the cytotoxicity of some categories of engineered silica NPs since the formation of the MTT (and also LDH) leads to production of formazan salt and this may alter the outcome of the test at high particle concentration. The authors came to this conclusion because the results of two tests done under the same experimental conditions (MTT, WST-1 assay) failed to agree.

18.3.2
Genotoxicity

Because of the potential medical applications, other groups tested the toxicity of NPs exposure on the genome by means of the COMET system that is the most widely used test able to detect both single- and double-stranded breaks of DNA chains. In fact, if any DNA is damaged, the cell would produce a comet tail whose length is propor-tional to the amount of DNA damaged.

Jin *et al.* in 2007 [68] investigated the genotoxicity of silica NPs by exposing A549 cells to different concentrations (1×10^{-4} to 0.50 mg/ml). They found that the particles enter the cells but remain outside the nuclei. In macrophages, the uptake rate was much higher. The DNA was not damaged within 72 h and no significant nucleotide base modification was induced by the NPs. These results are consistent with the observation that the silica NPs do not penetrate the nucleus. They found by Trypan blue exclusion test[2] that necrosis was dose dependent, but there was no sign of apoptosis. The conclusion of the authors is that silica nanoparticles at high concentration (>0.1 mg/ml) might attach to the cell surface and interfere with membrane function or metabolism.

18.3.3
Overall Cell Monolayer Electrical Resistance

Since NPs impact on physiological barriers, another important question is whether they interfere with the integrity of the whole monolayer of the cells when they come in contact with. Transepithelial/endothelial resistance (TER, Ω/cm^2) is a good parameter to estimate perturbations induced on the integrity of the monolayer since it is directly proportional to the integrity of the junctions between the cells.

Yacobi *et al.* report in their paper [64] a decrease of about 60% in TER of primary rat alveolar monolayer after exposure to ultrafine ambient particles. This result was interpreted as an increase in permeability of cell junctions and must also be taken into account in case of other NPs of the same size, including Si NPs.

18.3.4
Effect of NP Size

Toxicity may be related not only to the exposure dose of the NPs but also to their dimensions.

Lin *et al.* [65] found that it is very important to control the aggregation state in the cell culture medium. In fact, in their paper, they found no difference between the effect of 15 nm silica NPs and 46 nm NPs, the reason being that both nanoparticles reached a similar size due to aggregation in the culture medium. Therefore, NP aggregation state is a critical parameter that needs to be controlled.

In the same paper [65], the authors evaluated the oxidative stress by measuring the production of reactive oxygen species (ROS) and of glutathione (GSH, an antioxidant molecule produced under oxidative stress). The authors found that in A549 cells exposed to 10–100 μg/ml of NPs for time ranging from 24 to 72 h, ROS levels rose while GSH decreased, indicating an increase in oxidative stress and a decrease in antioxidant capacity.

In another study [69] on the same cell line incubated with Si nanoparticles ranging in size from 10 to 110 nm, up to 24–72 h of exposure, no change in mitochondrial activity (MTT assay) and no increase in GSH production was observed. Note that this

---

2) Trypan blue exclusion test: when incubated with the blue vital dye, cells with leaky membranes are stained in blue, while cells with intact membrane exclude the blue dye and are not stained.

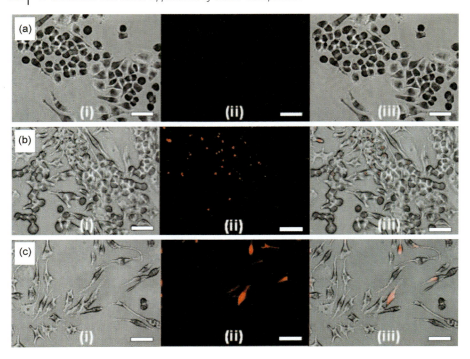

Figure 18.7 Phase (i) fluorescence (ii) and combined (iii) images of RAW 264.7 murine macrophages incubated (24 h) with red fluorescent ($\lambda_{em} = 640$ nm) silicon nanoparticles (<4 nm diameter). (a) Control (no Si NPs present); (b) Si NPs, 20 µg/ml; (c) Si NPs, 50 µg/ml; scale bar, 30 µm. Taken from Ref. [70].

last finding would indicate the lack of toxicity only if ROS production was unchanged, a point not addressed in the paper.

Other scientists [11, 70] have also compared the biological responses to nano- and micron-sized Si particles in RAW 264.7 cells, a mouse leukemia monocyte macrophage cell line. They monitored cell morphology, cytotoxicity, and cell viability, as well as cellular distress from the production of nitric oxide (NO), IL-6 and tumor necrosis factor-alpha (TNF-alpha). The authors found that with Si NP concentration of 1–200 µg/ml and exposure of 24–48 h (note that particles aggregated to a final diameter of 100–300 nm), the morphology of the cells did not change, although signs of latent toxicity (cells detached from the plate) were observed for Si NP concentration >20 µg/ml. Phase fluorescence and combined images of RAW 264.7 murine incubated with red fluorescent Si NPs are shown in Figure 18.7. Moreover, as shown in Figure 18.8, both the Trypan blue exclusion dye test and the MTT test were coherent in indicating nanosized Si NPs are more toxic than microsized NPs. The cells were treated with different concentrations (1–200 µg/ml) of nano- (SNs) and microparticles (SMs) for 24 and 48 h. At the end of exposure period, Trypan blue dye was added to an aliquot of cells to assess the cell live/dead ratio. MTT was introduced into wells containing cells incubated for 24 or 48 h with SNs or SMs. For the Trypan

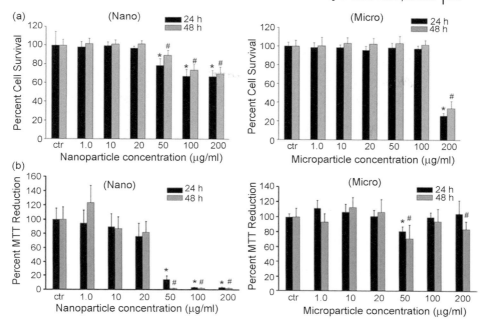

Figure 18.8 Effect of Si NPs and Si-MPs on cell survival in RAW 264.7 murine macrophages based on Trypan blue dye exclusion (a) and MTT (b) assays. Cells were treated with different concentrations (1–200 μg/ml) of nano (SNs) and micro (SMs) silicon particles for 24 and 48 h. Taken from Ref. [70].

blue dye exclusion cytotoxicity assay, data are expressed as "percent cell survival" and represent the ratio of number of dead cells divided by the number of total cells in a given sample. Cell viability for controls was >95%. For the MTT cell viability assay, data in Figure 18.8 are expressed as "percent MTT reduction" and were calculated by dividing the value of optical density (OD) of supernatants for each treatment well by the average OD from all corresponding supernatants from control wells (100×). Optical density for controls was 1.29 ± 0.19. Statistical difference compared to the control, $P < 0.05$.

Sayes *et al.* in their paper [61] found a dose dependency in the response to exposure of NPs in terms of production of inflammatory markers MIP-2 (macrophage inflammatory protein-2), TNF-alpha, and IL-6.

Another important issue in the study of the cellular impact of Si NPs was established by Alsharif and coworkers [71] who found that both alkyl-capped Si nanocrystals of 5 nm (suitable for diagnostic imaging) and all uncapped Si NPs are highly soluble only in nonpolar solvents such as tetrahydrofluorane, toluene, dimethyl sulfoxide (DMSO), and diethyl ether. All these solvents are very toxic for cells. They studied HeLa and SW1353 cells and found acute necrosis within 30 min from the exposure to the vehicle alone. But if the vehicle was DMSO at a fractional value of 0.2% and up to 24 h of exposure, there was no evidence of acute necrosis, apoptosis, or impact on the rate of cell proliferation.

It is often proposed that NP toxicity is related to NP cellular internalization rather than to physical damage to the cellular membrane. However, end results of toxicity are either apoptosis or necrosis that imply an opposite water imbalance, either a decrease or an increase in cell volume. It was found that relatively minor damages to cellular membrane may end in marked toxicity [72] likely due to changes in water fluxes at plasma membrane level. We therefore presume that the NPs–cell membrane interaction can result in local deformation that possibly affects membrane channel properties and water traffic. Bearing this in mind, in the framework of the BONSAI Project,[1] we tried to evaluate the initial phase of NPs–cell membrane interaction, addressing, in particular, the standard viability tests and their correlation with alteration in transendo-/epithelial resistance. Biocompatibility tests were performed on Si NPs prepared by laser pyrolysis [22, 24, 73].

In order to avoid alteration in results due to interference of the luminescence of Si NPs with luminescent formazan salt, the viability of cells exposed to NPs was investigated with the Trypan blue assay.[2] Tests were done on Si NPs dissolved in DMSO, a toxic solvent per se, at different concentrations (0.001–0.5 mg/ml) for up to 24 h of exposure on lung epithelial A30 cell line [74], not commercially available and very similar to A549.

It was determined that the maximum concentration of DMSO to avoid its toxicity was 0.01%. With this concentration and the corresponding NP concentration of 0.01 mg/ml, no sign of distress (e.g., detachment from the plate) was observed by comparing DMSO alone with NPs in DMSO to controls. In fact, the percentages of dead cells in the three conditions were 9.7, 6, and 5%, respectively, not significantly different (see Table 18.1).

Solubility in aqueous solution could be increased by a specific functionalization of Si NPs with PEG [60, 75] that is nonimmunogenic and nonantigenic. This should lower the toxicity of the encapsulated nanoparticles. However, exposure of PEGylated Si NPs led to a decrease in the cell number. This finding was ascribed to the sedimentation of the PEGylated nanoparticles, preventing the new cells coming from parental mitosis from attaching to the plate. Nevertheless, no sign of toxicity was found under these conditions. Our results are shown in Figure 18.9.

In a very recent paper [76], Fujioka *et al.* reported on the response of HeLa cells to luminescent oxidized Si NPs. The mitochondrial activity assays and LDH release

Table 18.1 Biocompatibility of Si NPs in DMSO tested in A30 cells.

Si NP concentration (mg/ml)	Percentage of dead cells in CTRL	Percentage of dead cells in DMSO alone	Percentage of dead cells in Si NPs in DMSO
0.5	6	100	100
0.1	6	18	24
0.01	4.9	9.7	6
0.01	10	11.6	7.2

Toxicity measured after 6 h of incubation.

(a)
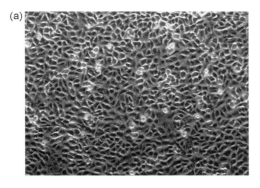

(b)
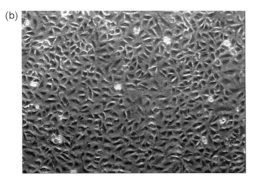

Figure 18.9 A30 cell exposed to PEGylated Si NPs. In panel (a) is shown a typical image, acquired with a light fluorescence microscope, of cells grown in control condition. Panel (b) represents A30 cell grown for 24 h in the presence of PEGylated Si NPs. Despite the seeded cell density was identical, the plate incubated with the particles after 24 h shows a minor population.

assays showed that the Si NPs are safer than CdSe quantum dots at high concentration, especially under UV exposure conditions. However, Si NPs may become slightly toxic to cultured human cells at concentrations higher than 112 μg/ml. As one of the toxicity mechanisms, the authors suggest that oxygen radicals from Si NPs can affect the plasma membrane of the cells [76].

18.3.5
In Vitro and *In Vivo* **Comparative Studies**

Comparative studies to extrapolate data to *in vivo* and to humans are very sparse. When comparing *in vivo* with *in vitro* studies, Sayes *et al.* [61] found that the rank of hazard potency from the *in vitro* assay did not correspond to the ranking results from the *in vivo* samples. Most of the samples they tested were not toxic in the *in vitro* assay but were very toxic after *in vivo* airway instillation in rats. Their conclusion was that the *in vitro* assay may provide mechanistic information specific for a cell type but does not allow extrapolation to the *in vivo* model. Therefore, *in vitro* assays have limitations in simulating the biological effects of inhaled or instilled particles in the whole body. The simplest limitations to list are again the "delivered cellular dose," the selection of cell types to create a microenvironment (e.g., single-cell culture system versus coculture system), the time course of the effects, and the end point reported.

The main conclusion from these studies is the need for standardized methodologies and experimental protocols to remove potential sources of irreproducibility when testing the toxicity of NPs and for more in-depth study of the mechanisms responsible for toxicity.

18.3.6
Exposure Risks

Exposure to Si NPs may occur in professional working conditions (e.g., during Si NPs collection after gas-phase synthesis processes) or for people undergoing a diagnostic screening test. In the first case, the risk is mostly related to the inhalation of nanoparticles; thus, the most likely route of entry is across the lung alveolar epithelium whose surface area averages $70\,m^2$. In the case of diagnostic imaging, nanoparticles are usually injected intravenously and Si NPs could impact on the endothelial wall. In both cases, outflow of NPs from the initial compartment requires translocation across biomembranes.

On the basis of the previous considerations, we can conclude that the concern deriving from "*in vitro* to *in vivo*" extrapolation and the need for standardized methodologies for toxicity tests represent the major hindrance to biomedical use of nanoparticles, including Si nanoparticles. In addition, it is at present still difficult to estimate whether subthreshold perturbations, not necessarily revealed by standard toxicity tests, might induce over time a pathological condition. In fact, a striking example of "latent toxicity" is that of inhaled ultrafine asbestos fibers as acute exposure only causes a modest pulmonary inflammatory reaction while a long-term exposure is a known cause of pleural mesothelioma. Yet, on the basis of available toxicity data, acute exposure to Si NPs for biomedical use might deserve a careful consideration of the risk/benefit ratio, accounting for the specific pathology, provided their concentration is not exceeding 0.01 mg/ml.

18.4
Applications of Si NPs in Biomedicine

The field of biomedical applications of Si NPs is still in its infancy as a consequence of the difficulties encountered in getting stable and intense PL emission, easy dispersibility in aqueous biological media, and biocompatibility [17].

To our knowledge, the preparation of multicolor, highly luminescent, single Si NPs well dispersed in biological systems is still problematic. In fact, as previously discussed, Si NPs that are not properly terminated are highly soluble only in nonpolar solvents that are toxic. However, recently, there was a surge of interest in this topic and several papers dealing with the interaction of Si NPs with biosystems have appeared in the literature [11, 59, 60, 68, 70]. Si NPs have in fact promising optical properties as labeling materials for bioimaging, for cell staining, for determining trace amounts of analytes, and as nucleotide carriers for gene therapy.

On the basis of the results reviewed in Section 18.2, two main classes of luminescent Si NPs can be identified, namely, orange–NIR emitting nanoparticles with long luminescence decay times (several tens of microseconds), usually produced by surface oxidation [22–25], and blue-emitting nanoparticles with fast (tens of nanoseconds) luminescence decay times [27, 29, 37]. These latter nanoparticles can have various surface terminations but are mostly terminated with hydrogen. An interface with biological system is then provided by further functional groups attached to the surface bonds (as detailed in Section 18.2).

Slow-emitting Si NPs are suitable for real-time tracking and monitoring of biological events at the cellular level. Moreover, PL emission in the red–IR is highly desirable for *in vivo* imaging because of the low tissue absorption and scattering effects in this spectral range [2]. In fact, the absorbance spectra of interfering biological molecules (i.e., hemoglobin and water) are minimal in the NIR region (Figure 18.10); thus, the ideal optical probe for *in vivo* imaging should exhibit excitation and emission wavelengths falling in this window (Figure 18.10). Oxide-covered Si NPs with size in the range of 2.5–8 nm emit in the range 560–850 nm [23]. Moreover, two-photon excited luminescence covering the same spectral interval was recently observed after NIR (900 nm) femtosecond laser excitation of Si NPs prepared by laser pyrolysis and oxidized by wet chemical procedures [77].

Another advantage offered by the use of slow-emitting Si NPs is the possibility to eliminate the fast (few nanoseconds) skin autofluorescence and the luminescence

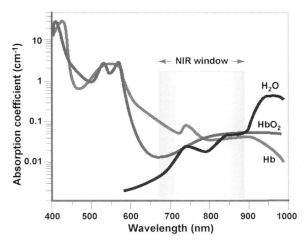

Figure 18.10 Near-IR *in vivo* imaging window. Taken from Ref. [2].

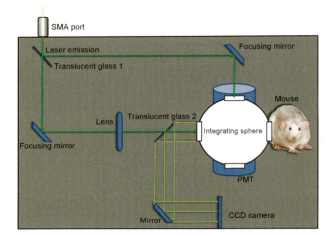

Figure 18.11 Experimental setup for *in vivo* bioimaging with an integrating sphere. To eliminate the fast (few nanoseconds) autofluorescence of biotissues, a time delay between the laser pulse for PL excitation and the PL signal detection/accumulation system is required. For using red-emitting silicon nanoparticles, the time delay between the laser pulse and the PL signal detection is 10 ns, the PL signal detection time is 1 ms.

emitted by most of the bioorganic materials by using time-resolved detection techniques [7, 78]. In fact, a delay of a few tens of nanoseconds between excitation and detection is sufficient to eliminate most of the parasitic light signals falling in the same wavelength range of the PL emission (Figure 18.11). The PL signal observed in continuous detection mode (1 min) after i.v. injection of a colloidal solution of pyrolytic Si NPs (2 mg/ml) in the tail vein of a mouse is shown in Figure 18.12. No

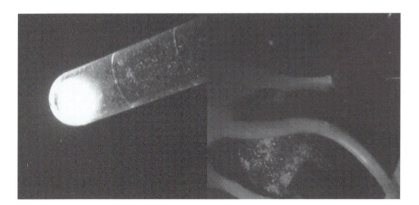

Figure 18.12 Photoluminescence detected in the tail vein of a mouse 1 min after the i.v. injection of a dose of 0.5 ml of silicon nanoparticles water colloid (2 mg/ml). No acute toxicity was observed. Experiments were performed at the General Physics Institute of Moscow (courtesy of V.B. Loschenov and A.V. Ryabova).

acute toxicity was observed. However, the signal intensity was not yet sufficient to evaluate the specific distribution of luminescent Si NPs inside the organs. Despite the current low PL intensity, these NPs show promise as an alternative for optical probes by taking advantage of their long lifetime to avoid background interference and autofluorescence by time-resolved spectroscopy. Thus, efforts are made to find new strategies for PL emission intensity enhancement.

The fast, intense blue PL emission from H-terminated Si NPs is exploited in a number of interesting applications in the biomedical field.

Wang et al. [54] at the Biotechnology Division of NIST conjugated 1–2 nm diameter blue luminescent Si NPs to a 5′-amino-modified oligonucleotide through a three-step procedure. The Si NPs were obtained with H-terminated surface through electro-chemical etching. The conjugation was implemented via two photoinduced reactions followed by a DNA labeling step through formation of a carboxamide bond. After conjugation with the oligonucleotides, the nanoparticles were water dispersible and showed blue stable luminescence for at least a week with a quantum efficiency equal to 8%. The ability of silicon NPs to carry sequences of nucleotides opens the possibility to interfere with cytoplasmic mRNA transcription phase. Si NPs, mono-dispersed in a stable suspension and functionalized with amino groups for DNA binding, might be used to develop nonviral vectors for in vitro and in vivo gene delivery [54]. The stability of the Si NPs–oligonucleotide conjugation makes this technique also valid in gene array.

Later on [11], the same group, in collaboration with the Department of Chemical and Biomolecular Engineering at the University of Maryland, reported on the conjugation of the blue luminescent Si NPs to streptavidin molecules by using a multistep photoassisted reaction and a bifunctional cross-linker. The characteristic blue photoluminescence was retained in the Si NP–protein complex. From the capillary electrophoresis analysis, it was determined that typically from four to five Si NPs were bound to the streptavidin molecule. Streptavidin molecules retain their capability of binding to biotin after the conjugation protocol. Covalent attachment of Si NPs to streptavidin provides a convenient pathway for biomolecule labeling in biotin–streptavidin affinity-based assays, thus opening the possibility to label a wide range of biomolecules by exploiting this interaction. Applications can be found in in vitro assays that rely on antibodies, cell receptors, and cell tracing probes [11].

As previously reported in Section 18.2, Warner et al. [17] prepared water-soluble Si quantum dots showing efficient and fast (tens of nanoseconds) blue luminescence by covalent bonding of allylamine on the surface of the Si NPs. The allylamine-capped Si NPs were then used to stain HeLa cells. The cells were incubated for 12 h with a 0.2 nM Si NP solution and then they were washed with a phosphate-buffered saline twice to remove the nonspecific binding nanoparticles. Luminescence images of the cells were captured using a digital camera coupled to a conventional wide-field microscope under UV lamp excitation at 365 nm, showing that the luminescence was emitted from the nanoparticles in the cytosol. Also, the blue luminescent hydrophilic particles were used to obtain luminescence images of Vero cells [82]. The nanoparticles were transfected into the cells by adding liposomal plasmid-transfection reagents. After incubation and washing, the cells were fixed and fluorescence images

(a)

(b)

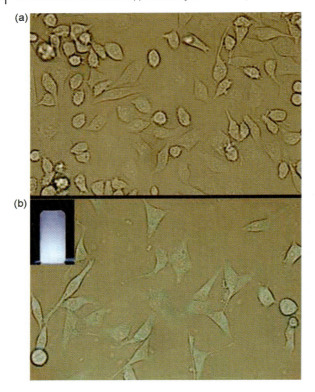

Figure 18.13 Overlay of the transmission and fluorescence microscope images of (a) HeLa cells in the absence of any quantum dots and (b) HeLa cells with silicon quantum dots incorporated inside the cytosol. Inset: fluorescence from a vial of allylamine-capped silicon. Taken from Ref. [17].

were captured on a cooled charge-coupled device (CCD) camera mounted on a fluorescence microscope using multi-bandpass filters and mercury lamp excitation luminescence. The suitability of allylamine-capped Si NPs as chromophores for biological imaging is demonstrated in Figure 18.13 (from Ref. [17]).

Very recently, Sudeep *et al.* [75] prepared and characterized photostable amphiphilic Si nanoparticles covered by a PEG corona. These particles are soluble in both organic solvents and aqueous environments and moreover maintained an absorbance spectrum characteristic of the Si NPs. Once again, these particles may turn useful for different biological applications *in vitro* [75].

He *et al.* [59] used the luminescent nanospheres (60–200 nm) formed by polyacrylic acid chains incorporating Si NPs to stain HEK293T human kidney cells, thus obtaining well-resolved and stable fluorescence microscopy images with a laser scanning confocal microscope. Luminescence from cells stained with Si NPs could be excited at 458 and 488 nm, in contrast with the case of cells stained with the commonly used dye FITC (fluorescein isothiocyanate) that could be excited only at 488 nm.

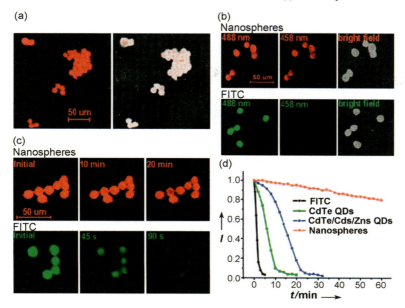

Figure 18.14 (a) Fluorescence microscopy images of HEK2-293T cells labeled with the as-prepared Si NPs loaded nanospheres excited at 488 nm (left) and superposition of fluorescence and transillumination images (right). (b) Comparison of fluorescent signals of HEK293T cells imaging with the nanospheres (top) and FITC (bottom) excited at different wavelengths. (c) Temporal evolution of fluorescence of the HEK293T cells labeled with the as-prepared nanospheres (top) and FITC (bottom). The nanospheres and FITC were both excited at 488 nm by argon laser with 8 ms dwell time and about 15 mW power. (d) Photostability comparison of fluorescent II/VI QDs (CdTe QDs and CdTe/CdS/ZnS core–shell–shell QDs) and the as-prepared nanospheres. All samples were continuously irradiated by a 450 W xenon lamp. Taken from Ref. [59].

Comparison with cells stained with FITC and with II–VI quantum dots showed that the Si NP-loaded nanospheres possess a significant property, a high resistance to photobleaching, for use as cellular labels in long-term, real-time imaging (see Figure 18.14).

In the medical field, diagnostic imaging is being used to visualize gross pathology [71]. Silicon NPs uptake seems higher in malignant (SW1353) than in normal (HeLa) cells with a maximum accumulation in the cytosol and, according to these authors [71], also in the nucleus (the discrepancy with the results reported by Jin [68] is to be noted). The NP uptake was found to be endocytosis mediated since more than 80% are blocked by endocytosis inhibitors.

Another group [66] made the point that silica NPs were more toxic for normal skin fibroblast cell lines (MRC-5, WS1, and CCD-966sk) compared to cancer cells (A549, HT-29, and MKN-28).

Erogbogbo *et al.* [60] reported on the use of luminescent Si NPs, encapsulated into phospholipidic micelles terminated with PEG groups, to image human pancreatic cancer (Panc-1) cells with a laser scanning confocal microscope. After functionalization with amine or transferrin groups, the micelle-encapsulated Si NPs at a concentration of 8 μg/ml were used to stain the cells by incubation for 2 h at 37 °C. After 2 h, the cells were washed thrice with PBS and directly imaged under the microscope using 405 nm laser excitation. Uptake of Si NPs from the cells could be clearly observed. Control experiments using PEG-terminated micelles, without amine or transferrin surface functionalization, showed no detectable uptake by the cells.

All these results show that Si NPs can be a valuable optical tool in biomedical diagnostic.

18.5
Si Nanoparticle-Based Sensors

Materials based on nanostructured silicon have a great potential to be compatible with current silicon technology, which is at the core of the modern electronics industry. In fact, it is expected that applications in several sectors such as medicine, textiles, cosmetics, and so on can be developed starting from new functionalities offered by incorporation of Si NPs into existing systems, leading to innovative tools.

A class of devices showing such useful cross-fertilization is the sensors. Use of Si NPs in sensors takes advantages from their optical and/or electronic transport properties; therefore, control of the nanoparticles surface chemistry and interface with the surrounding matrix is a key issue in developing such applications [79]. In addition, new functionalities due to collective behavior and interparticle interactions are expected if the nanoparticles could be organized in a controlled and regular pattern.

Building a sensor device often requires growing directly nanoparticles as a layer or depositing them in a film. Such materials are usually produced by PECVD (plasma-enhanced chemical vapor deposition), LRD (laser reactive deposition), or sputtering techniques in the form of thin films. These applications are not presented in this chapter, devoted to materials prepared from dispersed Si NPs.

Although several authors suggest possible applications as gas or biomolecule sensors, so far the only report on sensing applications of disperse Si NPs is the amperometric detection of glucose using an electrode formed by a layer of Si_{29} nanoclusters deposited on a heavily doped Si substrate [80], thus producing an all-silicon device. The sensor showed enhanced sensitivity with respect to the devices based on the glucose oxidase (GO) enzyme, together with improved selectivity and stability. Due to the highly efficient luminescence emission of the nanoparticles, the authors expected the possibility of dual amperometric/optical sensing for the detection of glucose. Furthermore, modulation of the sensor amperometric response by UV light was demonstrated, thus opening the possibility of phase-sensitive detection. It was concluded that this device can be used mainly in the amperometric mode and that it is suitable for miniaturization in view of further implantation and *in vivo* use.

In another paper [81], Si NPs prepared by electrochemical etching of bulk Si were used as substrates for oligonucleotide sensor chips, by immobilization of DNA fragments on the Si NPs surface. The binding capacity and the hybridization efficiency of the sensor depend upon the Si NPs particle size. The improved performances of the sensor chips are attributed to the large specific surface area of Si NPs compared to the existing conventional substrates (e.g., bulk Si). Moreover, the oligonucleotide array sensor showed high stability under test conditions and could be regenerated at least 12 times.

18.6
Conclusions

Luminescent Si NPs have a great potential for use as optical probes in biomedical diagnostics. Size-controllable Si NPs emitting photons in the yellow near-IR range (with lifetime of the order of 10^{-4} s) can be prepared by several gas- and solution-phase methods. A fast, deep blue luminescence emission is observed after UV excitation of ultrasmall (≤ 1 nm in size) Si NPs prepared by chemical etching of bulk crystalline silicon, followed by ultrasonication. However, surface engineering is necessary for enhancing and stabilizing the luminescence emission in liquids and to make water-soluble and biocompatible pristine Si NPs.

Modifications of freestanding Si NPs were realized either by forming Si−C covalent bond between unsaturated chemical compounds and Si−H or by reaction between organosilane and Si−X (X can be hydroxy group or halide). Both methods confer Si NPs stable luminescence and colloidal stability, either as small hydrophobic nanoparticles dispersible in organic solvents or as small quantum dots embedded in big water-dispersible nanospheres. Similar to the above-mentioned methods, in the framework of the EC Project BONSAI, we also were able to get luminescent Si NPs by oxidation or fluoride/nitric acid etching, and through suitable nonionic surfactant modification, we succeeded in getting water-dispersible Si NPs. However, more efforts are still necessary in order to get truly monodispersed, stable, biocompatible, and luminescent Si NPs.

An assessment of the biological impact of Si NPs is mandatory before their use in *in vivo* medical applications. The results reported in the current literature on the biointeraction of Si-based NPs were critically reviewed and we underlined how some concern derives from "*in vitro* to *in vivo*" extrapolation and the lack of standardized methodologies in toxicity tests. In addition, it is at present still difficult to estimate whether subthreshold perturbations, not necessarily revealed by standard toxicity tests, might induce over time a pathological condition. It was reported that Si NPs are less toxic than CdSe quantum dots at high concentration. Yet, based on the available toxicity data, acute exposure to Si NPs for biomedical use might deserve a careful consideration of the risk/benefit ratio, accounting for the specific pathology, provided their concentration is not exceeding 0.01 mg/ml.

Finally, we have reviewed the recent, pioneering applications of Si NPs as chromophores for *in vitro* cellular imaging and for the visualization of severe pathology, as a consequence of higher accumulation of Si NPs in malignant cells. In particular, it was demonstrated that Si-based nanospheres possess excellent photostability with respect to II–VI quantum dots and are suitable for multiwavelength excitation.

Other promising fields of applications for Si NPs are as luminescent labels to DNA and as glucose sensors.

In conclusion, a rapid expansion and development is expected in the field of biomedical applications of Si NPs as soon as the procedures for their surface modification are optimized and macroscopic quantities of highly luminescent, water dispersible, and biocompatible single Si NPs are easily available.

Acknowledgments

The authors thank the partners of the EC-BONSAI Project Professor F. Huisken, Professor G. Mattei, Dr. E. Trave, Dr. G. Sancini, and Dr. V. Pustovoy for their valuable contribution to getting a deeper insight into the topics treated in this chapter.

References

1 Wang, F., Tan, W.B., Zhang, Y., Fan, X., and Wang, M. (2006) Luminescent nanomaterials for biological labelling. *Nanotechnology*, **17**, R1–R13.

2 Klostranec, J.K. and Chen, W.C. (2006) Quantum dots in biological and biomedical research: recent progress and present challenges. *Adv. Mater.*, **18**, 1953–1964.

3 Parak, W.J., Pellegrino, T., and Plank, C. (2005) Labelling of cells with quantum dots. *Nanotechnology*, **16**, R9–R25.

4 Bruchez, M., Jr., Moronne, M., Gin, P., Weiss, S., and Alivisatos, A.P. (1998) Semiconductor nanocrystals as fluorescent biological labels. *Science*, **281**, 2013–2016.

5 Alivisatos, A.P. (1996) Semiconductor clusters, nanocrystals and quantum dots. *Science*, **271**, 933–937.

6 Chan, W.C.W. and Nie, S.M. (2004) Quantum dots in biology and medicine. *Physica E*, **25**, 1–12.

7 Dahan, M., Laurence, T., Pinaud, F., Chemla, D.S., Alivisatos, A.P., Sauer, M., and Weiss, S. (2001) Time-gated biological imaging by use of colloidal quantum dots. *Opt. Lett.*, **26**, 825–827.

8 Chan, W.C.W. and Nie, S.M. (1998) Quantum dots bioconjugates for

ultrasensitive nonisotopic detection. *Science*, **281**, 2016–2018.

9 Pavesi, L. and Lockwood, D.J. (eds) (2004) *Silicon Photonics*, Springer, Berlin.

10 Buriak, J.M. (2002) Organometallic chemistry on silicon and germanium surfaces. *Chem. Rev.*, **102**, 1271–1308.

11 Choi, J., Wang, N.S., and Reipa, V. (2008) Conjugation of the photoluminescent silicon nanoparticles to streptavidin. *Bioconj. Chem.*, **19**, 680–685.

12 Goesele, U. (2008) Shedding new light on silicon. *Nat. Nanotechnol.*, **3**, 134–135.

13 Godefroo, S., Hayne, M., Jivanescu, M., Stesmans, A., Zacharias, M., Lebedev, O.I., Van Tendeloo, G., and Moshchalkov, V.V. (2008) Classification and control of the origin of photoluminescence from Si nanocrystals. *Nat. Nanotechnol.*, **3**, 174–178.

14 Draeger, E., Grossman, J.C., Wiliamson, A., and Galli, G. (2004) Optical properties of passivated silicon nanoclusters: the role of synthesis. *J. Chem. Phys.*, **120**, 10807–10814.

15 Pudzer, A., Wiliamson, A.J., Grossman, J.C., and Galli, G. (2002) Surface control of optical properties of silicon nanoclusters. *J. Chem. Phys.*, **117**, 6721–6729.

16 Zhou, Z., Brus, L., and Friesner, R. (2003) Electronic structure and luminescence of 1.1- and 1.4-nm silicon nanocrystals: oxide shell versus hydrogen passivation. *Nano Lett.*, **3**, 163–167.

17 Warner, J.H., Hoshino, A., Yamamoto, K., and Tilley, R.D. (2005) Water soluble photoluminescent silicon quantum dots. *Angew. Chem., Int. Ed.*, **44**, 4550–4554.

18 Wilcoxon, J.P., Samara, G.A., and Provencio, P.N. (1999) Optical and electronic properties of silicon nanoclusters synthesized in inverse micelles. *Phys. Rev. B*, **60**, 2704–2714.

19 Ding, Z.F., Quinn, B.M., Haram, S.K., Pell, L.E., Korgel, B.A., and Bard, A.J. (2002) Electrochemistry and electrogenerated chemiluminescence from silicon nanocrystal quantum dots. *Science*, **296**, 1293–1297.

20 Baldwin, R.K., Pettigrew, K.A., Garno, J.C., Power, P.P., Liu, G.Y., and Kauzlarich, S.M. (2002) Room temperature solution synthesis of alkyl-capped tetrahedral shaped silicon nanocrystals. *J. Am. Chem. Soc.*, **124**, 1150–1151.

21 Batson, P.E. and Heath, J.R. (1993) Electron-energy-loss spectroscopy of single silicon nanocrystals: the conduction band. *Phys. Rev. Lett.*, **71**, 911–914.

22 Borsella, E., Botti, S., Cremona, M., Martelli, S., Montereali, R.M., and Nesterenko, A. (1997) Photoluminescence from oxidised Si nanoparticles produced by CW CO_2 laser synthesis in a continuous-flow reactor. *J. Mater. Sci. Lett.*, **16**, 221–223.

23 Huisken, F., Ledoux, G., Guillois, O., and Reynaud, C. (2002) Light emitting silicon nanocrystals from laser pyrolysis. *Adv. Mater.*, **14**, 1861–1864.

24 Herlin-Boime, N., Jursikova, K., Trave, E., Borsella, E., Guillois, O., Fabbri, F., Vicens, J., and Reynaud, C. (2004) Laser-grown silicon nanoparticles and photoluminescence properties. *Mater. Res. Soc. Symp. Proc.*, **818**, M13.4.1.

25 Li, X., He, Y., Talukdar, S.S., and Swihart, M.T. (2003) Process for preparing macroscopic quantities of brightly photoluminescent silicon nanoparticles with emission spanning the visible spectrum. *Langmuir*, **19**, 8490–8496.

26 Mangolini, L., Thimsen, E., and Kortshagen, U. (2005) High-yield plasma synthesis of luminescent silicon nanocrystals. *Nano Lett.*, **5**, 655–659.

27 Akcakir, O., Therrien, J., Belomoin, G., Barry, N., Muller, J.D., Gratton, E., and Nayfeh, M. (2000) Detection of luminescent single ultrasmall silicon nanoparticles using fluctuation correlation spectroscopy. *Appl. Phys. Lett.*, **76**, 1857–1859.

28 Choi, J., Wang, N.S., and Reipa, V. (2007) Photoassisted tuning of silicon nanocrystal photoluminescence. *Langmuir*, **23**, 3388–3394.

29 Yamani, Z., Asshab, S., Nayfeh, A., Thompson, W.H., and Nayfeh, M. (1998) Red to green rainbow luminescence from unoxidized silicon nanocrystallites. *J. Appl. Phys.*, **83**, 3929.

30 Wilson, W.L., Szajowskyi, P.F., and Brus, L.E. (1993) Quantum confinement in size-selected, surface oxidized silicon nanocrystals. *Science*, **262**, 1242–1244.

31 Guillois, O., Herlin-Boime, N., Reynaud, C., Ledoux, G., and Huisken, F. (2004) Photoluminescence decay dynamics of non interacting silicon nanoparticles. *J. Appl. Phys.*, **95**, 3677–3682.

32 Trave, E., Enrichi, F., Mattei, G., Bello, V., Borsella, E., Fabbri, F., Carpanese, M., Falconieri, M., Abate, C., Costa, F., Costa, L., and Gini, L. (2004) Investigation and application of size-dependent properties of silicon-based nanoparticles produced by laser pyrolysis, in *Advanced Laser Technologies* (eds I.A. Shcherbakov, A. Giardini, V.I. Konov, and V.I. Pustovoy), *Proc. SPIE*, **5850**, 25–32.

33 Trave, E., Bello, V., Enrichi, F., Mattei, G., Borsella, E., Carpanese, M., Falconieri, M., Abate, C., Herlin-Boime, N., Jursiokova, K., Costa, F., Costa, L., and Gini, L. (2005) Towards controllable optical properties of silicon based nanoparticles for applications in optoelectronics. *Opt. Mater.*, **27**, 1014–1019.

34 Trave, E., Bello, V., Mattei, G., Mattiazzi, M., Borsella, E., Carpanese, M., Fabbri, F., Falconieri, M., D'Amato, R., and Herlin-Boime, N. (2006) Surface control of optical properties in silicon nanocrystals produced by laser pyrolysis. *Appl. Surf. Sci.*, **252**, 4467–4471.

35 D'Amato, R., Falconieri, M., Carpanese, C., Fabbri, F., and Borsella, E. (2007) Strong luminescence emission enhancement by wet oxidation of pyrolytic silicon nanopowders. *Appl. Surf. Sci.*, **253**, 7879–7883.

36 Pi, X.D., Mangolini, L., Campbell, S.A., and Kortshagen, U. (2007) Room-temperature atmospheric oxidation of Si nanocrystals after HF etching. *Phys. Rev. B*, **75**, 085423-1.

37 Belomoin, G., Therrien, J., and Nayfeh, M. (2000) Oxide and hydrogen capped ultrasmall blue luminescent Si nanoparticles. *Appl. Phys. Lett.*, **77**, 779–781.

38 Smith, A., Yamani, Z.H., Roberts, N., Turner, J., Habbal, S.R., Granick, S., and Nayfeh, M.H. (2005) Observation of strong direct-like oscillator strength in the photoluminescence of Si nanoparticles. *Phys. Rev. B*, **72**, 205307-1.

39 Swihart, M.T. (2003) Vapor-phase synthesis of nanoparticles. *Curr. Opin. Colloid Interface Sci.*, **8**, 127–133.

40 Burg, A.B. (1950) in *Fluorine Chemistry*, vol. I (ed. J.H. Simons), Academic Press, New York, p. 150.

41 Blumberg, A.A. and Stavrinou, S.C. (1960) Tabulated functions for heterogeneous reaction rates: the attack of vitreous silica by hydrofluoric acid. *J. Phys. Chem.*, **64**, 1438–1442.

42 Berzelius, J.J. (1824) Untersuchungen über die Flußspathsäure und deren merkwürdigsten Verbindungen. *Poggendorffs Ann. Phys. Chem.*, **I**, 169–230.

43 Sidgwick, N.V. (1950) *The Chemical Elements and Their Compounds*, vol. I, Oxford University Press, Oxford, UK, p. 615.

44 Wiberg, E., Wiberg, N., and Holleman, A.F. (2001) *Inorganic Chemistry*, Elsevier, p. 858.

45 Palik, E.D., Bermudez, V.M., and Glembocki, O.J. (1985) Ellipsometric study of orientation-dependent etching of silicon in aqueous KOH. *J. Electrochem. Soc.*, **132**, 871–884.

46 Seidel, H., Csepregi, L., Heuberger, A., and Baumgaertel, H. (1990) Anisotropic etching of crystalline silicon in alkaline solutions. *J. Electrochem. Soc.*, **137**, 3612–3626.

47 Seraphin, A.A., Werwa, E., and Kolenbrander, K.D. (1997) Influence of nanostructure size on the luminescence behavior of silicon nanoparticle thin films. *J. Mater. Res.*, **12**, 3386–3392.

48 Li, X., He, Y., and Swihart, M.T. (2004) Surface functionalization of silicon nanoparticles produced by laser-driven pyrolysis of silane followed by $HF-HNO_3$ etching. *Langmuir*, **20**, 4720–4727.

49 Zou, J., Baldwin, R.K., Pettigrew, K.A., and Kauzlarich, S.M. (2004) Solution synthesis of ultrastable luminescent siloxane-coated silicon nanoparticles. *Nano Lett.*, **4**, 1181–1186.

50 Li, Z.F., Swihart, M.T., and Ruckenstein, E. (2004) Luminescent silicon nanoparticles capped by conductive polyaniline through the self-assembly method. *Langmuir*, **20**, 1963–1971.

51 Kang, Z., Liu, Y., Tsang, C.H.A., Ma, D.D.D., Fan, X., Wong, N.-B., and Lee, S.-T. (2009) Water soluble silicon quantum dots with wavelength-tunable photoluminescence. *Adv. Mater.*, **21**, 661–664.

52 Puzder, A., Williamson, A.J., Grossman, J.C., and Galli, G. (2003) Computational studies of the optical emission of silicon nanocrystals. *J. Am. Chem. Soc.*, **125**, 2786–2791.

53 Hua, F., Swihart, M.T., and Ruckenstein, E. (2005) Efficient surface grafting of luminescent silicon quantum dots by photoinitiated hydrosilylation. *Langmuir*, **21**, 6054–6062.

54 Wang, L., Reipa, V., and Blasic, J. (2004) Silicon nanoparticles as a luminescent label to DNA. *Bioconj. Chem.*, **15**, 409–412.

55 Heints, A.S., Fink, M.J., and Mitchell, B.S. (2007) Mechanochemical synthesis of blue luminescent alkyl/alkenyl-passivated silicon nanoparticles. *Adv. Mater.*, **19**, 3984–3988.

56 Li, Z.F. and Ruckenstein, E. (2004) Water-soluble poly(acrylic acid) grafted luminescent silicon nanoparticles and their use as fluorescent biological staining labels. *Nano Lett.*, **4**, 1463–1467.

57 Sato, S. and Swihart, M.T. (2006) Propionic-acid-terminated silicon nanoparticles: synthesis and optical characterization. *Chem. Mater.*, **18**, 4083–4088.

58 Hua, F., Erogbogbo, F., Swihart, M.T., and Ruckenstein, E. (2006) Organically capped silicon nanoparticles with blue photoluminescence prepared by hydrosylilation followed by oxidation. *Langmuir*, **22**, 4363–4370.

59 He, Y., Kang, Z.-H., Li, Q.-S., Tsang, C.H.A., Fan, C.-H., and Lee, S.-T. (2009) Ultrastable, highly fluorescent, and water-dispersed silicon-based nanospheres as cellular probes. *Angew. Chem., Int. Ed.*, **121**, 134–138.

60 Erogbogbo, F., Yong, K.-T., Roy, I., Xu, G., Prasad, P.N., and Swihart, M.T. (2008) Biocompatible luminescent silicon quantum dots for imaging of cancer cells. *ACS Nano*, **2**, 873–878.

61 Sayes, C.M., Reed, K.L., and Warheit, D.B. (2007) Assessing toxicity of fine and nanoparticles: comparing *in vitro* measurements to *in vivo* pulmonary toxicity profiles. *Toxicol. Sci.*, **97**, 163–180.

62 Teeguarden, J.G., Hinderliter, P.M., Thrall, B.D., and Pounds, J.G. (2007) Particokinetics *in vitro*: dosimetry consideration for *in vitro* nanoparticle toxicity assessments. *Toxicol. Sci.*, **95**, 300–312.

63 Lison, D., Thomassen, L.C.J., Rabolli, V., Gonzalez, L., Napierska, D., Seo, J.W., Kirsch-Volders, M., Hoet, P., Kirschhock, C.E.A., and Martens, L.A. (2008) Nominal and effective dosimetry of silica nanoparticles in cytotoxic assays. *Toxicol. Sci.*, **104**, 155–162.

64 Yacobi, N.R., Phuleira, H.C., Demaio, L., Liang, C.H., Peng, C., Sioutas, C., Borok, Z., Kim, K., and Crandall, E.D. (2007) Nanoparticle effects on rat alveolar epithelial cell monolayer barrier properties. *Toxicol. In Vitro*, **21**, 1373–1381.

65 Lin, W., Huang, Y., Zhou, X., and Ma, Y. (2006) *In vitro* toxicity of silica nanoparticles in human lung cancer cells. *Toxicol. Appl. Pharm.*, **217**, 252–259.

66 Chang, J.S., Chang, K.L.B., Hwang, D., and Kong, Z. (2007) *In vitro* cytotoxicity of silica nanoparticles at high concentrations strongly depends on the metabolic activity type of the cell line. *Environ. Sci. Technol.*, **41**, 2064–2068.

67 Yang, H., Liu, C., Yang, D., Zhang, H., and Xi, Z. (2009) Comparative study of cytotoxicity, oxidative stress and genotoxicity induced by four typical nanomaterials: the role of particle size, shape and composition. *J. Appl. Toxicology*, **29**, 69–78.

68 Jin, Y., Kannan, S., Wu, M., and Zhao, J.X. (2007) Toxicity of luminescent silica nanoparticles to living cells. *Chem. Res. Toxicol.*, **20**, 1126–1133.

69 Cha, K.E. and Myung, H. (2007) Cytotoxic effects of nanoparticles assassed *in vitro* and *in vivo*. *J. Microbiol. Biotechnol.*, **17**, 1573–1578.

70 Choi, J., Zhang, Q., Reipa, V., Wang, N.S., Stratmeyer, M.E., Hitchins, V.M., and Goering, P.L. (2008) Comparison of cytotoxic and inflammatory responses of photoluminescent silicon nanoparticles with silicon micron-sized particles in RAW264.7 macrophages. *J. Appl. Toxicology*, **29**, 52–60.

71 Alsharif, N.H., Berger, C.E.M., Varanasi, S.S., Chao, Y., Horrocks, B.R., and Datta, H.K. (2009) Alkyl capped silicon nanocrystal lack cytotoxicity and have enhanced intracellular accumulation in malignant cells via cholesterol-dependent endocytosis. *Small*, **5**, 221–228.

72 Beretta, E., Gualtieri, M., Botto, L., Palestini, P., Miserocchi, G., and Camatini, M. (2007) Organic extract of tire debris causes localized damage in the plasma membrane of human lung epithelial cells. *Toxicol. Lett.*, **173** (3), 191–200.

73 Lacour, F., Guillois, O., Portier, X., Perez, H., Herlin, N., and Reynaud, C. (2007) Laser pyrolysis synthesis and characterization of luminescent silicon nanocrystals. *Physica E*, **38**, 1–2.

74 Botto, L., Beretta, E., Bulbarelli, A., Rivolta, I., Lettiero, B., Leone, B.E., Miserocchi, G., and Palestini, P. (2008) Hypoxia-induced modifications in plasma membranes and lipid microdomains in A549 cells and primary human alveolar cells. *J. Cell. Biochem.*, **105**, 503–513.

75 Sudeep, P.K., Page, Z., and Emrick, T. (2008) PEGylated silicon nanoparticles: synthesis and characterization. *Chem. Commun.*, 6126–6127.

76 Fujioka, K., Hiruoka, M., Sato, K., Manabe, N., Myasaka, R., Hanada, S., Hoshimo, A., Tilley, R.D., Manome, Y., Hirakuri, K., and Yamamoto, K. (2008) Luminescent passive-oxidized silicon quantum dots as biological staining labels and their cytotoxicity effects at high concentration. *Nanotechnology*, **19**, 1–7.

77 Falconieri, M., D'Amato, R., Fabbri, F., Carpanese, M., and Borsella, E. (2008) Two-photon excitation of luminescence in pyrolytic silicon nanocrystals. *Physica E.* doi: 10.1016/j.physe.2008.08.055.

78 Stratonnikov, A.A., Meerovich, G.A., Ryabova, A.V., Savel'eva, T.A., and Loshchenov, V.B. (2006) Application of backward diffuse reflection spectroscopy for monitoring the state of tissues in photodynamic therapy. *Quantum Electron.*, **36** (12), 1103–1110.

79 Baraton, M.I. (2005) Chapter 10, in *Synthesis, Functionalization and Surface treatments of Nanoparticles*, American Scientific Publishers, Los Angeles, CA.

80 Wang, G., Yau, S., Mantey, K., and Nayfeh, M. (2008) Fluorescent Si nanoparticle-based electrode for sensing biomedical substances. *Opt. Commun.*, **281**, 1765–1770.

81 Zhu, Z., Zhu, B., Zhang, J., Zhu, J., and Fan, C. (2007) Nanocrystalline silicon-based oligonucleotide chips. *Biosens. Bioelectron.*, **22**, 2351–2355.

82 Tilley, R.D. and Yamamoto, K. (2006) The microemulsion synthesis of hydrophobic and hydrophilic silicon nanocrystals. *Adv. Mater.*, **18**, 2053–2056.

19
Nanosilicon-Based Explosives

Dominik Clément and Dimitry Kovalev

19.1
Introduction

Explosions are very fast and therefore very efficient chemical reactions. Their efficiency depends on the energy yield and the rate of the chemical reactions [1]. Most of the modern explosives are based on carbon. They usually combine in one molecule fuel (carbon atoms) and oxidizer (oxygen atoms) resulting in an enormous efficiency of the explosion when the detonation shock wave activates the reaction partners almost instantaneously. Bulk silicon (Si) was never considered as a basis for energy carriers or explosive materials. Contrary to carbon, its reaction product is solid and its oxidation rate is limited by the relatively slow diffusion of oxygen through the silicon dioxide layer at the Si surface [2]. However, the potential yield of energy of the exothermic reaction of silicon and oxygen is higher than that of most common carbon-based explosives.

Porous silicon (PSi), a nanoporous modification of bulk Si, can be prepared in different ways [3]. The most common procedure is the electrochemical or chemical etching of bulk Si wafers and Si powders. Etching results in the formation of a nanometer-sized sponge-like structure that retains the crystalline structure of the initial material. PSi has remarkable morphological properties leading to an enormous increase in the oxidation reaction rates that are crucial for explosive reactions. The PSi layers have enormous internal surface area (up to $10^3 \, \text{m}^2/\text{cm}^3$) [3] and therefore the spacing between Si and oxidizing atoms brought to the pores is at the atomic scale. For as-prepared samples, this internal surface is almost completely covered with hydrogen and the concentration of hydrogen atoms can be as high as $10^{22} \, \text{cm}^{-3}$. Other properties of PSi, such as porosity and pore sizes, can be easily adjusted to the required values, from nanometer to micron pore sizes, using different etching procedures.

Since 1992, PSi is known as a reactive material. The fast combustion of PSi immersed in nitric acid was discovered by McCord *et al.* [4]. In 2001, Kovalev *et al.* discovered the explosive reaction of PSi immersed in cryogenic oxygen [5]. This laboratory accident at the Technical University of Munich (TUM) drew the attention

Silicon Nanocrystals: Fundamentals, Synthesis and Applications. Edited by Lorenzo Pavesi and Rasit Turan
Copyright © 2010 WILEY-VCH Verlag GmbH & Co. KGaA, Weinheim
ISBN: 978-3-527-32160-5

of the pyrotechnic industry on PSi as a reactive material and led to a new boost in the development of nanosilicon-based energetic materials because this topic was advertised in several daily newspapers and public magazines [6–8]. Furthermore, several public science television programs showed the explosive reaction of PSi infiltrated with solid oxidizers [9–11]. In 2002, a true composite solid-state system based on PSi layers filled with gadolinium nitrate and operating at room temperature was demonstrated by Mikulec *et al.* [12]. The research on PSi-based energetic materials begun intensively after it was realized that a new highly energetic composite system was discovered. Up to date, this solid-state system has already been tested with a large number of oxidizers and several filling methods (see Ref. [13]). It has been demonstrated that the composite system of PSi layers with sodium perchlorate as an oxidizer is the most efficient one, and PSi layers are mechanically stable, applicable at room temperature, ignitable in a common and controllable way, and long-term stable. They show reproducible and predictable reaction rates and energy output. The first concrete industrial application was the content of the research project "SilAnz" (Si-based airbag initiator) in collaboration with industrial partners. In the framework of this project, the initial tests have demonstrated that $3 \times 3 \, mm^2$ PSi-based explosive elements can initiate the standard booster charge of an airbag [14].

In this chapter we are going to answer the question why nanosilicon is an interesting material as a fuel, give an overview of the production methods of PSi, present the experimental results obtained till date, and highlight its possible future applications.

19.2
Properties and Applications of Porous Silicon and Its Compounds as Energetic Materials

19.2.1
Production of Porous Silicon

Bulk Si was never considered as a promising fuel material for pyrotechnic applications since the chemical reactivity of bulk Si is very low. In addition, contrary to carbon, the reaction products are mainly solids, Si dioxide and silicates. Also, its oxidation rate is limited by the very slow diffusion of oxygen through the silicon dioxide layer at the Si surface. However, the potential yield of energy of the exothermic reaction of Si and oxygen is higher than that of the most common carbon-based explosives. The enthalpy of formation for Si dioxide is $\Delta_f H^0(SiO_2)$ $= 911 \, kJ/mol$ [15].

PSi can be produced in the form of layers via electrochemical etching of crystalline bulk Si wafers in a mixture of hydrofluoric acid (HF) and ethanol as wetting agent [3]. The growing rate of the layer depends on the doping type and level of bulk Si wafers, as well as current density and composition of the etching solution. The thickness of the layers is roughly proportional to the etching time. Normally, the current density is kept constant and stable layer thicknesses up to $150 \, \mu m$ can be achieved using this

Table 19.1 Preparation conditions and resulting parameters of PSi layers produced from different wafers using different current densities.

Type of wafer (doping level)	Current density (mA/cm^2)	Etching time (min)	Thickness (µm)	Growth velocity (µm/min)	Porosity (%)
p^{++} (1–15 mΩ cm)	43.5	30	70	2.32	59.1
	60.9	30	90	2.98	60.3
	87.0	30	112	3.73	71.0
	43.5	60	138	2.30	59.1
	60.9	60	170	2.82	65.8
p$^+$ (10–30 mΩ cm)	43.5	30	72	2.40	54.9
	60.9	30	94	3.13	58.8
	87.0	30	120	4.00	66.7
	43.5	60	132	2.20	59.0
	60.9	60	160	2.65	67.4
	20.0	40	57	1.43	49.3
p$^-$ (1–1000 Ω cm)	30.0	30	63	2.09	49.6
	43.5	20	58	2.90	50.3
	20.0	80	116	1.45	50.1
	30.0	60	125	2.09	51.0
	43.5	40	105	2.62	59.0

technique. We have tested different current densities, varied the doping levels of the bulk Si wafers, and changed the etching times because these parameters influence the layer thickness and the porosity (these are indicated in Table 19.1). If the etching current is kept constant, the porous layers become mechanically unstable after a certain time depending on the current density. The thickest mechanically stable layers we could achieve were about 150 µm thick. Even the mechanically most stable p^{++} 4-inch wafers with about 3.5-inch PSi layers, according to our observations, became unstable at a maximal thickness of about 200 µm.

To realize thicker stable PSi layers, we developed a special technique using varying current densities and breaks between etching steps. At the beginning we started with a higher current density (up to 120 mA/cm^2) to achieve larger pore diameters. This allowed the hydrogen gas generated during the etching process to escape from the pores more easily in the following etching stages. During the etching process, the current density was gradually decreased because the equilibration of the depletion of hydrogen at the etching front needs longer time. In addition, we inserted breaks in the etching process to allow the hydrogen to be released from the pores; otherwise, the mechanical stress due to the gas pressure cracks the porous structure. With this etching technique we were able to etch complete p^{++} 4-inch wafers up to thicknesses >500 µm. In Figure 19.1, an etched Si wafer, masked by silicon nitride mask stable in HF-based solutions protecting Si against etching, consisting of about 400 etched 3 × 3 mm^2 PSi elements (darker areas) and the remaining Si below silicon nitride mask (brighter areas) is shown. The structure of one of these single elements monitored using tomography is shown in Figure 19.2. Again the darker region is the porous layer and the brighter region is the surrounding remaining bulk Si. The upper

Figure 19.1 Fully etched 4-inch Si wafer with PSi areas (darker gray) and bulk Si bridges in between (light gray). The partial etching is realized by masking the wafer with silicon nitride (this image was taken by CiS Institute for Microsensorics, from the "SilAnz" report [14]).

picture shows the top view and the lower picture the cut-through of a single PSi element. The porous structure of the layers is continuous and can easily be filled with oxidizers dissolved in organic liquids from the top. Figure 19.3 shows the parallel structure of the pores, more open at the top, becoming smaller at the bottom and

Figure 19.2 Structure of a $3 \times 3\,mm^2$ single element observed with computer tomography, top view (top) and cut through the middle of the element (bottom). The porous region is less dense and therefore darker than the bulk Si region (this image was taken from the "SilAnz" report [14]). The round edges of PSi layer in the bottom picture are formed due to underetching of the silicon nitride mask.

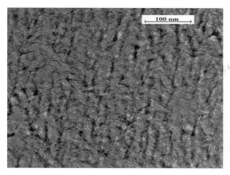 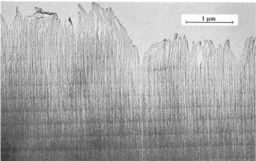

Figure 19.3 TEM image of the surface of an etched 3×3 mm^2 single element with the pores clearly seen (left side) and cut through a single element thinned out by etching where the parallel structure of the pores can be seen. The pores are wider at the top, where the sample surface is, and become smaller at the bottom due to the decrease in the current density. The horizontal structures come from the breaks during the etching process (this image was taken from the "SilAnz" report [14]).

perpendicular to the samples surface. To fulfill the stoichiometric conditions, the porosity of the PSi structure must be adjusted in a small range, depending on the type of oxidizer used. The highest energy yield of the reaction is assured when the oxidation reaction is complete. The stoichiometric ratio of SiX$_2$, where X is oxygen or sulfur, can be realized for PSi layers with porosities in the range of 70% for most of the oxidizers. For etched porous p^{++} Si wafers, the nanocrystals and the pore sizes are between 10 and 50 nm and all pores are interconnected, allowing an easy filling by liquid oxidizers or oxidizers dissolved in organic liquids.

Another method for producing PSi involves polycrystalline Si powder, which is simply etched chemically in a mixture of nitric acid and hydrofluoric acid. The nitric acid first oxidizes the impurities along Si grain boundaries and continues etching by oxidation of Si grains afterward, while the hydrofluoric acid dissolves the oxide. Therefore, the interconnected pores are etched in the polycrystalline Si powder, whereas the mean diameter of the Si powder grains remains unchanged. The rate of the chemical etching highly depends on the temperature of the etching solution. During etching of the Si powder, the etching solution heats up and the reaction becomes faster. Therefore, it is important to remove PSi powder from the etching bath quickly otherwise the material is dissolved completely. The correct end point of the etching is defined by the required porosity of the Si grains. Finally, it is possible to produce fully porous sponge-like particles without a remaining bulk core in their center.

The resulting PSi is low-density brown powder, exhibiting visible photoluminescence with a color depending on the size of the crystallites. Red photoluminescence of the PSi particles due to quantum confinement effects indicates crystal sizes <10 nm. The porosity of the Si powder is adjustable in the whole possible range from 0 to 100%. In Figure 19.4, SEM and HRTEM images demonstrate the morphology of PSi particles. The mean grain diameter remains nearly unchanged after the etching

Figure 19.4 The resulting PSi powder under different magnifications. Upper left side: SEM image of the raw material as used for stain etching. Upper right side: TEM image of the edge of a single porous grain. Lower left side: HR-TEM image showing the sponge-like structure of porous Si grains under larger magnification. Lower right side: HR-TEM image demonstrating that the PSi retains the diamond structure of bulk Si.

process (top left picture), but their inner cores are etched, which results in a sponge-like structure after the etching is completed (the other three images).

19.2.2
Stabilization of Porous Silicon Surface

After etching of Si, irrelevant whether layers or particles are produced, the highly reactive huge internal surface area of PSi has to be stabilized. The remaining dangling bonds have to be saturated with hydrogen to avoid spontaneous oxidation. However, the hydrogen-covered surface is not completely stable against natural aging. Natural oxidation introduces a monolayer of oxygen atoms back-bonded to the surface Si layer ($\cdots Si-Si-O-Si-H$), while the hydrogen atoms covering the

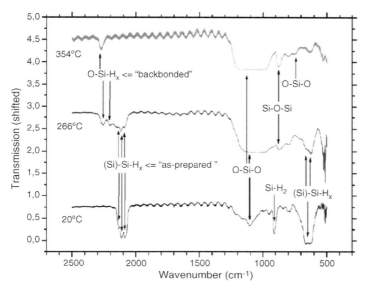

Figure 19.5 Stabilization of porous Si surface using an annealing procedure in air ambient. Specific surface bonds and annealing parameters are indicated (this image was prepared by D. Clément, Technical University of Munich, for the "SilAnz" report [14]).

surface Si atoms remain unaffected. This natural aging process takes several years for PSi. To achieve a structure with stable conditions, PSi can be artificially aged in ordinary ambience at elevated temperatures. The higher the temperature, the faster the PSi surface is oxidized. At a certain temperature, the hydrogen cover at the surface is affected. The H-atoms are thermally effused and the surface of the PSi suffers further oxidation. In Figure 19.5, infrared absorption spectra of PSi annealed at different temperatures in air are shown. From the features at $2100–2300\,\mathrm{cm}^{-1}$, the efficiency of hydrogen release from PSi surface can be estimated. The three features seen in the lower spectrum – from the sample kept at room temperature – around $2100\,\mathrm{cm}^{-1}$ are due to silicon–hydrogen bonds ($\mathrm{Si–Si–H}_x$, $x = 1, 2, 3$). If oxidation occurs, the frequency of this vibration shifts toward shorter wavelengths and therefore the introduction of back-bonded oxygen can be estimated using this shift. The middle spectrum has been measured for the sample annealed at $266\,°\mathrm{C}$ for 20 min, and the appearance of the features at about $2250\,\mathrm{cm}^{-1}$ is due to the shift of the vibrational frequency of the silicon–hydrogen bonds influenced by introduced oxygen. The integral area below these features is also useful, since it displays the total amount of hydrogen bonded to the surface Si atoms. Again in the middle spectrum, the integral area of Si−H related spectral features remains equal before and after annealing, since no hydrogen is effused from the surface. For the upper spectrum of the sample annealed at $354\,°\mathrm{C}$, a strong decrease in this spectral integral is observed, indicating almost complete effusion of hydrogen from the surface. In Figure 19.6, the integrated areas of Si−H bond-related spectral

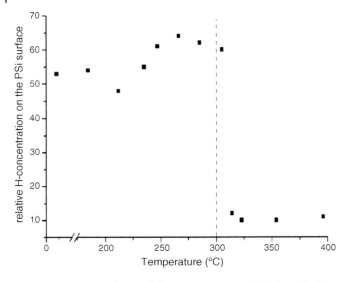

Figure 19.6 Integrated area of the silicon–hydrogen bond vibrations (representing a total number of surface hydrogen atoms) in the range from 2000 to 2300 cm^{-1}. A clear onset at 300 °C due to hydrogen effusion can be seen (this image was prepared by D. Clément, Technical University of Munich, for the "SilAnz" report [14]).

features measured for the samples annealed at different temperatures are compared. A step-like function is observed showing that the effusion of hydrogen occurs at about 300 °C. Therefore, in the case of artificial aging, temperatures below 300 °C should be chosen for the annealing process. The time of annealing is a much less critical factor than the temperature, as it has been shown in the framework of the SilAnz project [14].

We have demonstrated that stable conditions for the PSi surface can be achieved and at what temperature hydrogen is released from the surface. But how can a complete coverage of hydrogen atoms at the surface of Si be realized? To achieve this condition, the freshly prepared PSi has to be dipped in a solution of HF and ethanol directly after removal from the etching solution. This procedure will saturate the dangling bonds created during the etching process by hydrogen atoms and will also remove oxygen atoms being attached to the surface Si atoms in the short time during which the sample is exposed to air. The dangling bonds result from a depletion of hydrogen atoms at the etching front during the production process. These bonds, when being exposed to air, oxidize immediately. The refreshment procedure leads to the removal of this spontaneously built partial oxygen layer and simultaneously saturates the resulting dangling bonds with hydrogen (compare upper and lower spectra in Figure 19.7). Infrared absorption spectroscopy proves the stability of a fully hydrogen-covered PSi surface against oxidation for at least a month (see Figure 19.7). From older samples, it is known that the hydrogen cover of a crystalline Si sample is stable for several years and only submonolayer of back-bonded oxygen is formed.

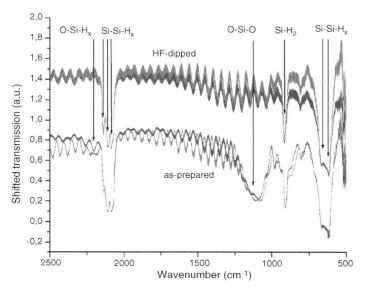

Figure 19.7 IR spectra of samples exposed to air: for 30 min (dark gray) and for 39 days (light gray). No significant change is observed. The HF-dipped sample shows no oxygen since the refreshing step removes the partial oxide layer and the resulting complete hydrogen cover is stable for at least a month (this image was prepared by D. Clément, Technical University of Munich, for the "SilAnz" report [14]).

19.3
Reaction Properties of Porous Silicon with Oxidizers

19.3.1
Infiltration of the Pores with Different Oxidizers

The physical properties of the PSi, such as porosity, can be adjusted from almost 0 to 100% and therefore a stoichiometric mixture with oxidizer can easily be achieved. The sponge-like structure of the PSi builds up the skeleton for the oxidizer and serves simultaneously as the fuel for explosives and as a packaging material. Ordinary pyrotechnical materials are usually mechanically mixed powders. In this case, mixtures of fuel and oxidizer on a molecular scale will never be achieved, resulting in relatively slow reaction rates. For binary PSi–oxidizer systems, however, this mixing is at the atomic scale (similar to traditional high explosives where fuel atoms (carbon) and oxidizing atoms (oxygen) are combined in one molecule) and the expected reaction rates should be extremely fast.

Standard oxidizers, for example, nitrates, perchlorates, sulfur, or fluorocarbon, were used as possible reaction partners for PSi. We calculated the standard enthalpies of reaction and the energy density using thermodynamic equations, and further the equilibrium reaction temperature by the NASA CEA code was calculated. The results

Table 19.2 Overview of some thermodynamic reaction parameters for several mixtures of Si with oxidizer.

Chemical reaction equations	ΔH_r^0 (kJ)	ΔH_r^0 (kJ/g)	T_{rct} (K)
$Si + O_2 \rightarrow SiO_2$	−911	−15.2	3131
$4Si + Ca(ClO_4)_2 \rightarrow 4SiO_2 + CaCl_2$	−3703	−10.5	3093
$2Si + NaClO_4 \rightarrow 2SiO_2 + NaCl$	−1850	−10.4	3057
$2Si + KClO_4 \rightarrow 2SiO_2 + KCl$	−1825	−9.4	3061
$5Si + 4NH_4ClO_4 \rightarrow 5SiO_2 + 2N_2 + 4HCl + 6H_2O$	−5453	−8.9	2917
$5Si + 4NaNO_3 \rightarrow 5SiO_2 + 2N_2 + 2Na_2O$	−3519	−7.3	2893
$5Si + 4KNO_3 \rightarrow 5SiO_2 + 2N_2 + 2K_2O$	−3305	−6.1	2980
$4Si + S_8 \rightarrow 4SiS_2$	−848	−2.3	1759
$nSi + (C_2F_4)_n \rightarrow nSiF_4 + 2nC$	−798	−6.2	3532

for several compositions are shown in Table 19.2. The highest energy density comes from the mixture of PSi with pure oxygen, which is not applicable. For a composite material, to create the largest contact areas of Si and oxidizers we need solvable solid oxidizers staying in the pores of the PSi after the evaporation of the solvent. One exception is sulfur, which can be infiltrated in the pores by melting. A list of chosen oxidizers tested together with PSi layers and powders is shown in Table 19.3. According to our observations, up to now the best oxidizers are

- **sodium perchlorate ($NaClO_4$)**: it has a high energy yield and very high solubility in methanol or comparable organic solvents;
- **sulfur (S_8)**: it has lower energy yield, but 100% filling of the pores can be achieved since it is melted and not solved;
- **fluorocarbon (e.g., Teflon $(C_2F_4)_n$)**: since very high temperatures can be reached.

The uniformity of filling of PSi single $3 \times 3\,mm^2$ elements was analyzed by tomography. In Figure 19.8, an empty single element (left side) is compared with an element filled with sodium perchlorate (middle image) and with a second one filled with sulfur (right side). The nearly 100% filling with sulfur is clearly seen and the filling of the pores with sodium perchlorate is quite uniform.

The calculated temperatures obtained for different mass ratios of Si to different oxidizers are shown in Figure 19.9. As already known from experiments, a porosity of 65–80% gives the highest temperatures since porosities for these sample mixtures with oxidizers are close to stoichiometry. The porous structures used for most of the experiments have porosities of about 72%.

Sodium perchlorate monohydrate ($NaClO_4 \cdot H_2O$) is one of the most powerful oxidizer with a high oxygen content (45.56%). Due to its very good solubility in organic solvents (about 180 g/100 g methanol), it fills the pores of PSi very efficiently. The chemical reaction equation is

$$5Si + NaClO_4 \cdot H_2O \rightarrow 5SiO_{2(s)} + 2NaCl_{(s)} + H_2 + 4020\,kJ, \ 954\,kJ\,g$$

leading to calculated reaction temperatures of about 3000 K.

Table 19.3 Comparison of chosen oxidizers and their solvents.

Oxidizer	Solvent (solubility)	Explosion		
		HP	**RH**	**SIB**
$Ca(ClO_4)_2 \cdot 4H_2O$	Me (237 g/100 g)	+	+	+
	Et (166 g/100 g)	+		
NH_4ClO_4	Me (6 g/100 g)	+		
	Et	+		
	Ac (>6 g/100 g)	+		
$LiClO_4 \cdot 3H_2O$	Me (182 g/100 g)	+	+	
	Et (152 g/100 g)	+		
	Ac (137 g/100 g)	+	+	
$NaClO_4 \cdot H_2O$	Me (~181 g/100 g)	+	+	+
	Et (<181 g/100 g)	+		
$Fe(ClO_4)_3 \cdot xH_2O$	Me	+		
	Et	+		
$KClO_4$	Me (<1.7 g/100 g)	−		
	Et (<1.7 g/100 g)	−		
	Ac (<1.7 g/100 g)	−		
$Ca(NO_3)_2 \cdot 4H_2O$	Me (>54 g/100 g)	+		
	Et (54 g/100 g)	+		
NH_4NO_3	Me (17 g/100 g)	+		
	Et (4 g/100 g)	−		
$LiNO_3 \cdot 3H_2O$	Me (good)	+		
	Et (good)	+		
Sulfur	CS_2 (good)	+	+	+
	Melting	+	+	+

HP: ignition on the hot plate; RH: resistive heating; SIB: standard ignition bridge; + indicates successful ignition; − indicates no successful ignition; empty boxes were not tried yet; Me: methanol; Et: ethanol; Ac: acetone. $Ca(ClO_4)_2$ was also successfully ignited with a laser.

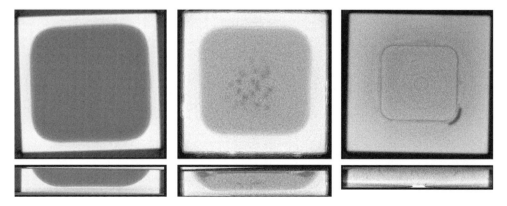

Figure 19.8 Computer tomograph images of PSi layers unfilled (left image), filled with sodium perchlorate (middle image), and filled with sulfur (right image) in the top view (top) and cut view through the middle of the element (bottom) (this image was prepared by D. Clément, Technical University of Munich, for the "SilAnz" report [14]).

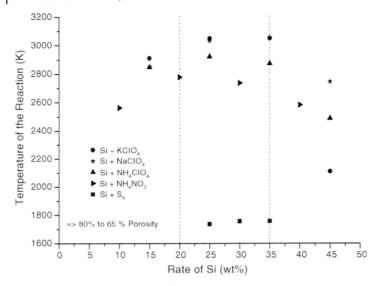

Figure 19.9 Thermodynamic equilibrium temperatures of the explosive reaction calculated for different mass ratios of Si to oxidizer and for the chosen promising oxidizers.

The samples were prepared by infiltrating etched wafers or infiltrating PSi powder by a saturated solution of $NaClO_4$ in methanol. To measure the speed of the explosive reaction and its lateral propagation rate, a 300 μm thick porous layer with the dimensions of 10–5 mm was filled with $NaClO_4$. This sample was ignited at one end with resistive heating and the spectrally integrated intensity of flash accompanying explosion was measured optically. The intensity of the flash shows the propagation of the reaction from the lowest intensities at the start to the highest intensities at the end, when the reaction front reaches the end of the 10 mm sample and the complete sample is exploded. The time of the reaction propagation over the 10 mm was measured to be ~500 ns. The recorded kinetics of the optical flash is shown in Figure 19.10. This reaction time corresponds to the reaction speeds parallel to the samples surface of about 10^4 m/s, which is expected for a very fast explosion.

Similar reaction propagation measurements were performed on longer sample layers and on PSi powders. The setup of the first experiment with layers is shown in Figure 19.11. The 50 mm long stripe of PSi was filled with sodium perchlorate and dried on a hot plate at about 80 °C throughout the experiment. The explosion was triggered by a short pulse of laser beam focused on one end of the sample, while reflection of the second CW laser from the other end was monitored. The propagation speed of the explosion was measured by measuring the time delay between the triggering of the explosion laser pulse and the disappearance of the reflected beam of CW laser. Using this technique, the propagation velocity was determined to be only about 2500 m/s. This value is lower than that for conventional high explosive due to completely unusual geometrical configuration. In high

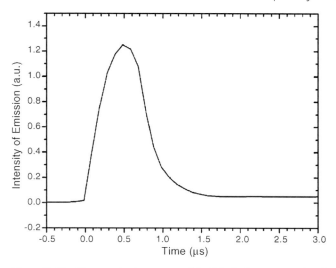

Figure 19.10 Time-resolved spectrum of the light flash accompanying the explosive reaction (this figure was prepared by D. Clément, Technical University of Munich, for the "SilAnz" report [14]).

explosives, a three-dimensional configuration is usually realized and the released energy is spent on sustaining the detonation wave propagation. For PSi stripe, however, significant fraction of the energy released was normal to the surface of the explosive layer, which reduced the velocity of explosion wave propagation.

The propagation times in a not fully dried sample were also measured by the experiments with PSi powder filled with sodium perchlorate. The experimental setup is shown in Figure 19.12. PSi powder was mixed with the sodium perchlorate

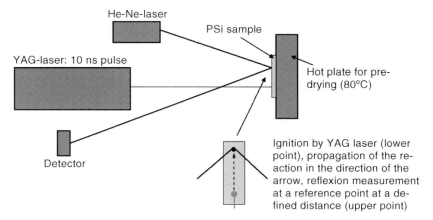

Figure 19.11 Experimental setup for the measurement of the reaction propagation velocity parallel to the porous layer surface. A porous layer filled with sodium perchlorate was used for these experiments (this image was prepared by D. Clément, Technical University of Munich, for the "SilAnz" report [14]).

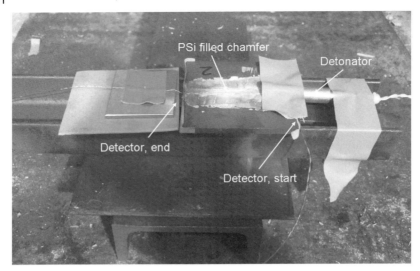

Figure 19.12 Experimental setup for the measurement of the reaction propagation velocity through porous Si powder filled with sodium perchlorate. The filled PSi powder was confined in an 80 mm long chamfer. The traces of the reaction are seen on the steel plate that has been bended by the reaction. The explosive reaction has been initiated by a detonator on one side of the chamber, while arrival of the detonation front has been measured by the detector on the other side.

solution in methanol in an 80 mm long chamfer. The samples were dried in vacuum in an oven at about 60 °C for about 3 h. The reaction was initiated by a detonator, and the start and end of the reaction were measured by short-circuit detection on two twisted pair cables. The propagation velocities found in these experiments were measured to be about 2500 m/s. We assume that these samples were not fully dried and the packing density of the material was relatively low.

Experimentally, the reaction temperature was estimated by measuring the spectral distribution of the flash accompanying the explosive reaction. We selected two samples: one was completely dried and the other one was still wet. For the dried sample, the explosive reaction was most efficient. Though explosive reactions are far from thermodynamic equilibrium, we estimated the temperature using the approach of a blackbody radiation. The measured spectra were fit with blackbody emission spectra and the temperature of explosive reaction was estimated to be $T = 3675$ K for the faster explosion of the dried sample (see Figure 19.13, upper spectrum). However, the temperature of the deflagration of the wet sample was similar and was estimated to be $T = 3585$ K (Figure 19.13, lower spectrum). In both cases, $NaClO_4 \cdot H_2O$ was used as an oxidizer. We expect hot spots with much higher temperatures since the blackbody temperature represents the equilibrium temperature. The presence of hot spots is proved by the observation of plasma lines of single or double ionized atoms in the spectral range from 550 to 800 nm in the emission spectrum of the explosion. We also performed spectroscopic measurements of some mixtures at the Fraunhofer

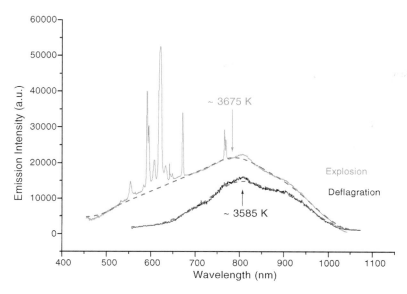

Figure 19.13 Emission spectra of flashes accompanying the explosions. The reaction temperatures are indicated and were estimated by the approach of a blackbody emission. They are very similar for the deflagration and the explosion. The spectrum of the explosion additionally shows the appearance of plasma lines of single or double ionized atoms demonstrating the presence of hot spots having temperatures much higher than the estimated ones (this image was prepared by D. Clément, Technical University of Munich, for the "SilAnz" report [14]).

ICT. From these measurements, temperatures of all the explosions were in a range from 3000 to 3400 K.

We used calorimetric bomb tests to define the energy yield of PSi samples filled with different oxidizers. Calorimetric bomb test results are listed in Table 19.4.

Table 19.4 Energy yields measured in a calorimetric bomb for PSi samples filled with different oxidizers.

Oxidizer	Mass (mg)			Amount (mmol)		Energy yield		
	m_{PSi}	m_{total}	m_{Ox}	Si	O	E_{mass} (kJ/g)	E_{total} (kJ)	$E_{relative}$ (kJ/g)
Ca(ClO$_4$)$_2$	252	660	408	9.0	10.5	7.3	4.8	8.6
	236	985	749	8.5	19.3	4.8	4.7	5.3
	259	1103	844	9.2	21.8	5.4	5.9	6.1
NaClO$_4$	233	649	416	8.3	11.9	8.0	5.2	8.9
	257	713	456	9.2	13.0	8.1	5.6	9.1
Sulfur	406	1879	1473	14.5	46.0	1.2	2.2	1.7

Figure 19.14 Time-integrated flame pictures of 3×3 mm PSi single elements filled with the indicated oxidizers (this image was prepared by D. Clément, Technical University of Munich, for the "SilAnz" report [14]).

We observed the highest energy yield, $E = 9.1$ kJ/g, for a sample filled with sodium perchlorate. It is important to mention that this yield is much higher than that measured for conventional high explosives; for example, the energy yield for trinitrotoluene is $E = 4.2$ kJ/g.

Figure 19.14 shows flame pictures of PSi single elements filled with different oxidizers. The amount of explosive mass is about 1 mg of Si plus 2 mg of oxidizer. The scale of one picture is about 40 cm in width. We found the most promising one to be sodium perchlorate, also showing the brightest flame and the biggest range of hot gases and particles.

19.3.2
Mechanical Mixing of Porous Silicon Powder with Oxidizers

The easiest way to produce pyrotechnic mixtures is mechanical mixing of fuel and oxidizer. Therefore, we performed experiments with PSi powder in mechanical mixtures with oxidizers to prove whether the extended surface of PSi has advantages for the reaction rate. The oxidizers used were different from the ones used for infiltration since they were solvable. We used potassium perchlorate (the nonhygroscopic perchlorate that is not solvable in common solvents in a sufficient amount), potassium nitrate, and lead oxide for our first experiments. For comparison, we used the same oxidizers in mixtures with polycrystalline Si powder with a mean diameter of 4 μm. The oxidizers were mechanically mixed with Si and a chamfer was filled with the mixture. The measured burning times of these mixtures are listed in Table 19.5. The fastest burning rate measured was expected to be and also found for PSi and potassium perchlorate to be about 100 cm/s. The comparison of both kinds of Si in mixtures with potassium nitrate showed that the mixture with PSi burned 125 times faster than that with the identical sizes of bulk polycrystalline Si grains. The only possible reason is the extended inner surface area of PSi grains.

Table 19.5 Overview of some burn rates measured on different mixtures of 4 μm large bulk Si powder and PSi with oxidizers.

Silicon	Oxidizer	Estimation of the burning rate
4 μm Si powder	Potassium perchlorate	No reaction
Porous Si	Potassium perchlorate	100 cm/s
4 μm Si powder	Potassium nitrate	0.4 cm/s
Porous Si	Potassium nitrate	50 cm/s
4 μm Si powder	Lead oxide	No reaction
Porous Si	Lead oxide	2 cm/s

19.4
Conclusions

The properties and applications of both micro- and nanomodifications of Si have been reported. We demonstrated that silicon-based composite explosives have high energy yields (up to 9.1 kJ/g), high reaction temperatures (above 3500 K), very short reaction times (500 ns), and a fast propagation of the shock wave (at least 2.5 km/s). The timescale of the reaction can be adjusted by varying the Si surface and/or the filling of the pores. PSi elements from electrochemical wafer etching show directed ejection of hot gases and particles normal to the surface. Due to its bulk, Si housing a single element has a long-term mechanical stability and due to its limited oxidation it has a very good chemical stability. Furthermore, no packaging is required since it is a self-confined system. It is important to note that in combination with the most efficient oxidizer, sodium perchlorate, only nonhazardous solids are formed. In pyrotechnic mixtures, the use of PSi powder leads to much faster reaction rates due to the high internal surface area. We assume that the reaction rate is directly coupled to the porosity and therefore the burning rate of a pyrotechnic mixture can be adjusted with the porosity of the PSi used. Due to the very fast reaction, the filled PSi material can be used as an igniter for several purposes as demonstrated in the framework of the SilAnz project for an airbag igniter.

Acknowledgments

D. Clément and D. Kovalev acknowledge the German Ministry of Education and Research for funding the SilAnz project (2003–2005). In the framework of this project, the authors acknowledge TRW Airbag Systems, CiS Institute of Microsensorics, and the Technical University of Munich for close cooperation. Furthermore, D. Clément acknowledges the BWB for funding the research at Diehl BGT Defence GmbH & Co. KG. The authors acknowledge Volker Weiser from the Fraunhofer ICT in Pfinztal for conducting the spectroscopic measurements on several mixtures.

References

1 Still, D.R. (1977) *Fundamentals of Fire and Explosion*, AIChE Monograph Series No. 10, vol. 73, American Institute of Chemical Engineers, New York, p. 100017.

2 Deal, B.E. and Grove, A.S. (1965) General relationship for the thermal oxidation of silicon. *J. Appl. Phys.*, **36**, 3770.

3 Cullis, A.G., Canham, L.T., and Calcott, P.D.J. (1997) The structural and luminescence properties of porous silicon. *J. Appl. Phys.*, **82**, 909.

4 McCord, P., Yau, S.-L., Bard, A.J. (1992) Chemiluminescence of anodized and etched silicon: evidence for a luminescent siloxene-like layer on porous silicon. *Science*, **257**, 68.

5 Kovalev, D., Timoshenko, V.Yu., Künzner, N., Gross, E., and Koch, F. (2001) Strong explosive interaction of hydrogenated porous silicon with oxygen at cryogenic temperatures. *Phys. Rev. Lett.*, **87**, 68301.

6 Sprengstoff aus Sand, Sueddeutsche Zeitung, March 13, 2001.

7 Laborunfall: Münchner Forscher stossen auf Super-Sprengstoff, Der Spiegel, August 2, 2001.

8 Biggest bang, New Scientist, 2001, 71, 15.

9 Licht und Silizium, Nano, 3sat, September 10, 2001.

10 Der große Knall – Supersprengstoff aus Silizium, Welt der Wunder, RTL2, August 7, 2005.

11 Explosiver Sandsturm, Kopfball, ARD, May 12, 2004.

12 Mikulec, F.V., Kirtland, J.D., Sailor, M.J. (2002) Explosive nanocrystalline porous silicon and its use in atomic emission spectroscopy. *Adv. Mater.*, **14** (1), 38.

13 Clément, D., Diener, J., Gross, E., Künzner, N., Timoshenko, V.Yu., and Kovalev, D. (2005) Highly explosive nanosilicon-based composite materials. *Phys. Stat. Solidi A*, **202** (8), 1357.

14 Reports of the SilAnz BMBF project (funded by the German Ministry of Education and Research, cooperation of TRW Airbag Systems (Aschau a. Inn), CiS Institute of Microsensorics (Erfurt), and the Technical University of Munich (München), 2003 and 2006).

15 Lide, D.R. (ed.) (1994) *Handbook of Chemistry and Physics*, special student edition, 75th edn, CRC Press Inc.

20
Applications of Si Nanocrystals in Photovoltaic Solar Cells

Gavin Conibeer

20.1
Introduction: Reasons for Application to Solar Cells

Photovoltaic cells absorb photons above their bandgap to create an electron–hole pair. This pair is separated in a built-in electric field generated by a p–n junction in the cell. Electrons are collected at an external contact on the n-type material and returned again to the cell at a lower chemical potential via the p-type material (i.e., holes are collected) – thus creating an electric current that can do useful work in the external circuit. The solar cell material and its quality are two of the most important factors determining the efficiency of the cell.

20.1.1
Limits of Single-Bandgap Cells

Photovoltaic cells can use a wide variety of semiconductor materials. The main parameter of interest is the electronic bandgap. This determines the range of the solar spectrum that is absorbed – the smaller the bandgap the more photons absorbed and hence the larger the current but at a relatively low voltage in the external circuit – as the voltage must always be less than the bandgap. The larger the bandgap the larger the voltage in the external circuit but the lower the current because fewer photons are absorbed. Therefore, the product of current and voltage, the power and hence also the efficiency, is a compromise value for the bandgap. For a standard solar spectrum, this compromise value is 1.3 eV. The efficiency of an ideal device with this optimum bandgap is 31% – the so-called Shockley–Queasier limit after the authors to first calculate it using detailed balance modeling [1]. Figure 20.1 shows this optimum bandgap energy against the solar spectrum [2]. It also shows the bandgaps of some semiconductors used for solar cells.

However, it is possible to improve on this and extract a much greater fraction of the available solar energy. The maximum possible efficiency for a solar cell occurs if the cell first absorbs all photons incident from the sun and second the energy from each of these photons (less those that are unavoidably emitted back out of the cell) is

Silicon Nanocrystals: Fundamentals, Synthesis and Applications. Edited by Lorenzo Pavesi and Rasit Turan
Copyright © 2010 WILEY-VCH Verlag GmbH & Co. KGaA, Weinheim
ISBN: 978-3-527-32160-5

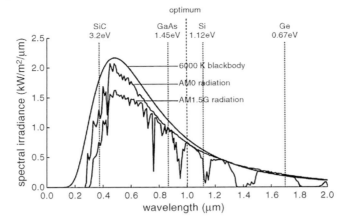

Figure 20.1 Solar spectrum both inside (AM1.5) and outside the Earth's atmosphere (AM0) as a function of wavelength (1.24/photon energy). (AM denotes the effective thickness of the atmosphere, or air mass, above the point of measurement.) Also shown is the spectrum from a blackbody at 6000 K, a good approximation for AM0 [2]. The figure also shows the bandgap energies of some materials used for solar cells. As the spectrum is peaked at about 2.6 eV (0.5 μm) there is an optimum bandgap at which the most power can be collected. Too small a bandgap results in high current and low voltage; too large a bandgap in small current and high voltage. The optimum bandgap is at about 1.3 eV (1 μm), close to that of Si.

converted to electricity using all of its energy (or chemical potential). This is theoretically possible if a large number of energy levels are exploited in a solar cell device and the theoretical maximum efficiency is then 68% [3].

These figures are for a cell illuminated at one sun concentration. If a concentrating lens is used to concentrate the light, the efficiency can be increased. To understand this, we must take a step back and note that a solar cell must at least emit light into the solid angle subtended by the sun or else it would not absorb light over this angle. But all light emitted in directions other than back toward the sun can in principle be reflected back into the cell and hence would not be lost. Under concentration such a condition is approached because the sun fills a greater proportion of the "sky" as far as the cell is concerned and a smaller proportion of the emitted light is directed into the hemisphere away from the sun, and hence less is wasted. Under maximum concentration conditions, that is, when the sun fills the whole "sky" for the cell, the efficiencies can be as high as 41% for a single bandgap at 1.1 eV [1] and 88% for a very large number of energy levels [3].

20.1.2
Solar Cells with Multiple Energy Levels

There are a number of suggested approaches for introducing such multiple energy levels in such a way as to give an overall advantage. The concept of third-generation photovoltaics seeks to achieve this while still using thin-film processes and abundant nontoxic materials. This seeming best of both worlds approach can be achieved by circumventing the Shockley–Queisser limit for single-bandgap devices, using multiple energy threshold approaches. Such an approach can be realized either by incorporating multiple energy levels in tandem or intermediate band devices; or by modifying the incident spectrum on a cell by converting either high-energy or low-energy photons to photons more suited to the cell bandgap; or by using an absorber that is heated by the solar photons with power extracted by a secondary structure. These methods have advantages and disadvantages and are at various stages of realization [3, 4].

Nanostructures and in particular quantum dot nanostructures feature in several of these approaches because of the extra degrees of freedom they give in the design of materials systems. Silicon quantum dots are particularly interesting because of the abundance of silicon and the wealth of experience in the material and the relative cheapness and low-energy usage of many of the thin-film deposition processes that can be employed.

There are several nanostructured cell approaches that aim at a higher efficiency than is possible using a single p–n junction. Among these ideas are quantum well [3], and quantum dot solar cells and intermediate band cells relying on a superlattice array of quantum dots. These are usually grown by mismatched epitaxial growth of III–V QDs [5]. However, conceptually the multijunction or "tandem" cell approach is simplest, with materials of increasing bandgap stacked on top of each other, each harvesting a different part of the solar spectrum at closer to the optimum energy of the absorbed photons. III–V epitaxially grown cells give the highest efficiencies but are also very expensive to fabricate [6]. The use of nanostructures and particularly QDs can facilitate engineering of the material bandgap and allow cheaper tandem cells to be made using thin-film processes.

20.2
Properties of Si Nanocrystals Relevant to Solar Cells

The primary property exploited in Si nanocrystals is the quantum confinement at small dimensions approaching the Bohr radius. In arrays of quantum dots with a significant overlap of the wave function, tunneling can give rise to a true superlattice and hence an effectively larger bandgap material. Methods to fabricate such Si QD arrays are hence also important for solar cell use.

20.2.1
Solid-Phase Formation of Si Nanocrystals

Si nanocrystals can be precipitated from a solid solution containing excess silicon in a range of materials. Most commonly, excess silicon in a dielectric is used. Deposition is by a thin-film process such as sputtering, plasma-assisted chemical vapor deposition (PECVD), or evaporation. After deposition, annealing of the samples leads to the formation of silicon nanocrystals by precipitation. If the temperature is high enough, a complete phase separation can occur [7]:

$$SiO_x \rightarrow \frac{x}{2} SiO_2 + \left(\frac{x}{2}\right) Si, \qquad where \quad 0 < x < 2.$$

The silicon nanocrystalline phase will precipitate only if either the temperature is high enough or the silicon excess is high enough. For a ratio of O:Si < 1 grown by PECVD, Si nanocrystals can be formed at temperatures between 800 and 950 °C [8]. For ion-implanted silicon in thick SiO_2 layers, Si nanocrystals form within a few seconds of the start of annealing [9, 10].

Upon annealing these porous as-deposited silicon-rich oxide (SRO) layers undergo significant densification. This can result in up to 15% reduction in layer thickness [11]. The assumption that all the excess Si precipitates to nanocrystals turns out to be an oversimplification. In fact, it has been observed that only half the excess Si clusters in these precipitates upon annealing at 1100 °C for 30 min, in material deposited by PECVD [12, 13], with a considerable amount of suboxide material forming in the matrix.

In Ref. [7], it was also found that the size of Si nanocrystals could be controlled by the layer thickness of the silicon-rich oxide. This has further been applied to the growth of Si quantum dots for bandgap engineering in Si QD-based solar cells [14, 15].

The quantum confinement that occurs in nanocrystals less than the Bohr radius of Si (5 nm) results in a blueshift in the optical bandgap. This has been observed for Si nanocrystals embedded in 500 nm thick thermally grown silicon oxide [16]. *Ab initio* calculations using density functional theory (DFT) indicate that this blueshift of the optical bandgap is expected to vary from 1.4 to 2.4 eV for a nanocrystal size of 8–2.5 nm [17]. However, further DFT calculations have found that in addition to quantum confinement effects in small nanocrystals, the matrix has a strong influence on the resultant energy levels [18]. With increasing polarity of the bonds between the nanocrystal and the matrix, there is an increasing dominance of the interface strain over quantum confinement. For a 2 nm diameter nanocrystal, this strain is such that the highest occupied molecular orbital (HOMO)–lowest unoccupied molecular orbital (LUMO) gap is dominated and significantly reduced by these effects in a polar SiO_2 matrix but not much affected in a less polar SiN_x or nonpolar SiC matrix [18, 19].

Raman spectroscopy is widely used for characterizing nanocrystalline materials. In general, an absorbed Raman photon is re-emitted at a lower energy due to the Stokes shift of an emitted Raman active phonon [20].[1] The small size of Si nanocrystals modifies this Stokes shift of emitted photon energies. In particular, the small crystal size sets up reflection of phonon modes from the nanocrystal interfaces such that they become standing waves and are optically, or Raman, active. This results in a shift to lower wave numbers for the Raman peak. Typically, a shift from ~520 to ~510 cm^{-1} is observed for nanoparticle sizes varying from ~12 to

---

1) An anti-Stokes shift due to absorption of a Raman-active phonon is also possible but critically depends on the occupancy of phonon modes. At room temperature, the occupancy of optical, or Raman-active, modes is very low so the Stokes shift of phonon emission is dominant.

~3 nm [21]. The increase in the range of phonon energies is only on the lower side of the original c-Si Stokes shift Raman energy and hence results in an asymmetric broadening of the Raman peak [22, 23]. This effect occurs whether confinement is in one, two, or three dimensions as in quantum wells, wires, or dots, respectively, and is characteristic of such phonon confinement at scales less than 10 nm. In principle, this asymmetric broadening should reflect the particular size of the nanocrystals, with formation of discrete peak doublets due to the particular allowed lower energy Raman-active phonon modes associated with this dimension. In practice, this is observable in high-quality epitaxial III–V multiple QW samples [24] but not in the much more variable size range of Si nanocrystals; the range of sizes blurring out these peaks. However, some particular peaks have been observed such as a Si-SiO$_2$ interface mode at 489 cm^{-1} [21]. Strain can also affect Raman energies because it alters the lattice spacing and hence the allowed phonon modes. Compressive microstrain associated particularly with nanocrystal interfaces can also give rise to small shifts to *higher* Raman wave numbers.

20.2.2
Quantum Confinement in Si QD Nanostructures

As discussed in this section, the effective mass approximation (EMA) is only of partial use in determining the absolute confined energy levels for small silicon nanocrystals. It does, however, correctly model the trend and relative increase in confined energy level as quantum dot size decreases.

The EMA solution of the Schrödinger equation for electrons confined in three dimensions in a quantum dot is similar to that for the case of a quantum well confined in only 1D. The increase in energy due to quantum confinement for the nth confined energy level is given by (e.g., [25, 26])

$$\Delta E_n = \hbar^2 k^2 / 2m^*, \qquad (20.1)$$

where k is the wave vector, $\hbar = h/2\pi$ is the reduced Planck's constant, and m^* is the effective mass of the particle.

For a very large confining potential and where "a" is the width of the QD:

$$ka \approx n\pi. \qquad (20.2)$$

Equation (20.1) then gives discrete solutions for the confined energy level with quantum numbers in each confined dimension given by n_1, n_2, and n_3.

$$\Delta E_n = \frac{\pi^2 \hbar^2}{m^* a^2}(n_1^2 + n_2^2 + n_3^2) = 3\frac{\pi^2 \hbar^2}{m^* a^2}n^2 \quad \text{for } n_1 = n_2 = n_3. \qquad (20.3)$$

This is a similar solution to that for the corresponding 2D quantum well case, except for the extra factor of 3 due to confinement in 3D instead of just 1D. The appropriate corresponding energy levels for the QW in the QD are those that are nondegenerate with the same quantum number. Hence, for n_{QW} in a QW the corresponding level occurs when $n_1 = n_2 = n_3 = n_{QW}$.

Therefore, for a given size of confinement, the QD has confined energy levels $3\times$ those of a QW. Or put another way, for a given confined energy level, a QD has a diameter that is $\sqrt{3}\times$ the width of the corresponding QW. (Strictly, this is for a cubic QD. For a spherical QD of diameter "a", the confinement is slightly greater and the factor slightly larger than $\sqrt{3}$, such that a factor of 2 is a good approximation.)

In this calculation, the isotropic conduction-band effective mass is taken as the most appropriate for Si, this being given by a weighted average of the longitudinal and transverse masses ($m_l^* = 0.92$ and $m_t^* = 0.18$). To give the equivalent isotropic effective mass of $m_e^* = 0.27m_0 a$. The closer to isotropic valence band gives a hole-effective mass, $m_h^* = 0.59m_0$ [27, 28]. Substituting these values in Eq. (20.3) gives, for the first quantized ground-state energy, E_1:

$$E_1 = E_g + \Delta E_1$$
$$\Delta E_1 = \Delta E_{1,n} + \Delta E_{1,p} = \frac{418}{a^2} + \frac{191}{a^2} = \frac{609}{a^2} \qquad \text{for cubic QDs,} \qquad (20.4)$$

where a is in Angstroms and E_1 and E_g (the bulk bandgap of Si) are in electron volts.

However, this calculation is for cubic quantum dots; hence, because of the slightly larger confinement in spherical dots of diameter equal to the side of a cubic dot discussed above, this value for ΔE_1 should be divided by $(\sqrt{3}/2)^2$.

$$E_1 \approx E_g + \frac{609}{(\sqrt{3}/2)^2 \cdot a^2} \qquad \text{for spherical QDs.} \qquad (20.5)$$

The results from this calculation are shown in Figure 20.2 [29] together with data from photoluminescence (PL) on QD nanostructures in SiO_2 and SiN_x from a number of authors. This shows a decreasingly inaccurate prediction of the confined energy level by the EMA as the QD size decreases. It also indicates that the trend of the

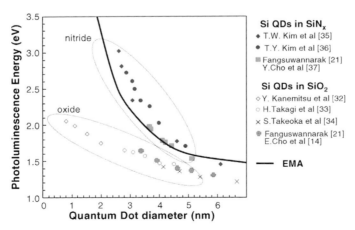

Figure 20.2 Measured photoluminescence energies for various authors' data for Si QDs in SiO_2 and SiN_x (300 K) as a function of QD size [14, 21, 32–37]. Also shown is the EMA calculation using Eq. (20.5) for spherical dots [29].

data is better predicted for QDs in nitride than in oxide. (The small discrepancy in actual values for nitrides is probably due to error in the measurement of QD diameter.)

Part of the explanation for this is that for a noninfinite confining barrier potential, that is, in a real dielectric matrix, the approximation of Eq. (20.2) is no longer valid. k is then given by the implicit equation:

$$ka = n\pi - \sin^{-1}(\hbar k/\sqrt{2m^* V_0}),$$

where V_0 is the corresponding barrier height. The value of E_n is reduced as given by [30], p. 578:

$$E_n \approx 3 \cdot \frac{\pi^2 \hbar^2}{m^* a^2} n^2 \Bigg/ \left[1 + \frac{\hbar}{a/2\sqrt{2m^* V_0}} \right]^2. \tag{20.6}$$

This is always less than E_n from Eq. (20.3). Hence, small confinement barriers will reduce the confinement energy. However, as discussed below PL data for confined energy levels do not match Eq. (20.6).

The PL results shown in Figure 20.2 for Si QDs from different authors are in good agreement where the matrix is the same but are quite different for QDs in oxide [14, 21, 31–34] compared to nitride [21, 35–37] particularly for small QDs. They are also qualitatively consistent with the results from *ab initio* modeling [18, 38, 39], which had been carried out for the confined energy levels in Si QD consisting of a few hundred atoms. This uses Gaussian03, a density functional–Hartree–Fock-based *ab initio* program. Calculations have been carried out on the gaps between the HOMO and the LUMO. These HMO–LUMO gaps are the ground state-confined energy levels for Si nanocrystals of various sizes terminated with either $-H$ (the closest to vacuum for a terminated surface); $-OH$ groups; and $-NH_2$ groups (using a nearest neighbor assumption, these two are assumed analogous to a SiO_2 and Si_3N_4 matrix).

The modeling not only shows the expected increasing confinement energy with decreasing QD size but also shows the reduction in energy on going from a QD effectively in vacuum to the one embedded in a dielectric [38]. It is also seen that the amino-terminated QDs (simulating nitride) have energies about 0.5 eV more than the hydroxyl-terminated ones (simulating oxide). This is qualitatively consistent with the PL results for QDs in oxide and nitride shown in Figure 20.2. However, both the PL data and the *ab initio* data are in *disagreement* with the modified EMA solution in Eq. (20.6), as the barrier height V_0 of nitride is *smaller* than oxide and hence Eq. (20.6) would give a *smaller* confined energy level not a larger one. This elucidates the underlying problem that the EMA breaks down for small QDs. Further Gaussian03 modeling has demonstrated this to be the case. This breakdown is not surprising as the Bloch assumption in the EMA of a carrier wave function varying only very gradually across the periodic potential of the lattice is no longer valid when the wave function is associated with a small spatial volume approaching the size of the lattice spacing, as in a small QD. It should be possible to calibrate a modified effective mass using the Gaussian modeling, in order to modify the EMA. This parameter will vary with the QD size and with the species at the interface, as for small QDs the lowest energy states dominate the confined energy levels [38, 39].

20.2.3
Carrier Tunneling Transport in Si QD Superlattices

Transport properties are expected to depend on the matrix in which the silicon quantum dots are embedded. As shown in Figure 20.3, different matrices produce different transport barriers between the Si dot and the matrix, with tunneling probability heavily depending on the height of this barrier. Si_3N_4 and SiC give lower barriers than SiO_2 allowing larger dot spacing for a given tunneling current.

The wave function of an electron confined to a spherical dot penetrates into the surrounding material, decreasing exponentially into the barrier. The slope of this exponential decay and hence the barrier to tunneling between quantum dots is reduced for a lower barrier height material. This is because – from transmission/ reflection probability – the tunneling probability T_e through a square potential well exponentially depends on three parameters, the barrier width $d =$ the spacing between quantum dots; the square root of the barrier height seen by the electron ($\Delta E^{1/2}$ the energy difference between the CB edge of the matrix and the confined energy level of the quantum dots $= (E_c - E_n)^{1/2}$); and the square root of the effective mass $(m^*)^{1/2}$ of the electron in the barrier. This gives the approximate relation (e.g., [30], p. 244):

$$T_e \approx 16 \exp\left\{ -d\sqrt{\frac{8m^*}{\hbar^2}} \Delta E \right\}. \tag{20.7}$$

Hence, the important parameter in determining the degree of interaction between quantum dots is $m^* \Delta E d^2$. As the barrier height decreases, the barrier thickness for a given probability increases, thus requiring a lower dot density for a given conductivity or higher conductivity for a given dot density. As the dot size decreases so does ΔE, thus increasing T_e and enhancing the effect further for smaller quantum dots [30]. The results suggest that dots in a SiO_2 matrix would have to be separated by no more than 1–2 nm of matrix, while they could be separated by more than 4 nm of SiC.

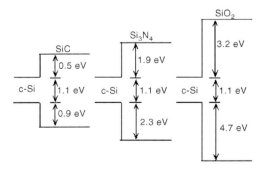

Figure 20.3 Bulk band alignments between silicon and its carbide, nitride, and oxide.

Fluctuations in spacing and size of the dots can be investigated using similar calculations. It is also found that the calculated Bloch mobilities do not depend strongly on variations in the dot spacing but do depend strongly on dot size within the QD material [40].

Fluctuations in spacing and size of the dots around their mean values can be investigated using similar calculations. Using this approach, it is also found that the calculated Bloch mobilities do not depend strongly on variability in the dot position around a mean position, Δd, but do depend strongly on variation in the dot size within the QD material [40]. This is an important result for engineering a real thin-film structure because although it is necessary to minimize the *mean* spacing between QDs, d, to give high mobilities for a given matrix, the variation around this mean value Δd is less critical.

Hence, transport between dots can be significantly increased by using alternative matrices with a lower barrier height ΔE with increasing spacing between QDs in oxide, nitride, and carbide to give the same effective transport. A similar deposition and quantum dot precipitation approach works for all three matrices.

20.3
The "All-Si" Tandem Cell: Si Nanostructure Tandem Cells

For photovoltaic applications, nanocrystal materials may allow the fabrication of higher bandgap solar cells that can be used as tandem cell elements on top of normal silicon cells. Silicon is of course a benign, readily available material, which is widely used for solar cell fabrication. Silicon also has a bandgap that is close to optimal not only for a standard, single p–n junction cell (a little too low, see Figure 20.1), but also for the bottom cell in a two-cell or even a three-cell tandem stack (a little too high). The radiative efficiency limit for a single-junction silicon cell is 29%. This increases to 42.5 and 47.5% for two-cell and three-cell tandem stacks, respectively. For an AM1.5 solar spectrum, the optimal bandgap of the top cell required to maximize conversion efficiency is <1.7–1.8 eV for a two-cell tandem with a Si bottom cell and 1.5 and 2.0 eV for the middle and upper cells for a three-cell tandem [41].

A cell based entirely on silicon and its dielectric compounds with other abundant elements (i.e., its oxides, nitrides, or carbides), fabricated with thin-film techniques, is therefore advantageous in terms of potential for large-scale manufacturability and in long-term availability of its constituents. Such thin-film implementation implies low-temperature deposition without melt processing; hence, it also involves imperfect crystallization with high defect densities. Hence, devices must be thin to limit recombination due to their short diffusion lengths, which in turn means they must have high absorption coefficients.

Quantum-confined nanostructures of silicon with barriers of SiO_2, Si_3N_4, or SiC can potentially fill these criteria and allow fabrication of a tandem cell. The quantum-confined energy levels resulting from a restriction in at least one dimension to close to the Bohr radius of bulk crystalline silicon (5 nm or less) will give an increase in the

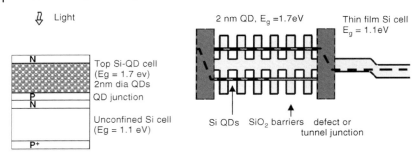

Figure 20.4 Schematic of a "all-silicon" tandem solar cell and its band diagram.

effective bandgap of the nanostructure compared to bulk Si. This will also result in increased absorption due to the pseudo direct bandgap of such localized quantum-confined systems [42–45], see Figure 20.4. To date, considerable work has been done on the growth and characterization of Si nanocrystals embedded in oxide [7, 31] and nitride [35, 37] dielectric matrices. However, little has been reported on the experimental properties of Si nanocrystals embedded in SiC matrix [46].

20.3.1
Alternative Matrices for Si QDs

As discussed in Section 20.2.3, alternative matrices with lower barrier heights than SiO_2 will give a greater tunneling probability between adjacent Si QDs and hence greater conductance (see Figure 20.3). Both SiN_x and SiC matrices have been investigated.

20.3.1.1 Si Quantum Dots in a Silicon Nitride Matrix
Si QDs can also be fabricated in a silicon nitride (SiN_x) matrix. Thick layers of silicon-rich nitride when annealed at above 1000 °C precipitate Si QD [47]. Multilayered structures also result in Si QD formation with a controlled size of the Si QDs [37]. The annealing temperature can also be used to modify the nitride matrix, with it being amorphous below 1150 °C but with crystalline nitride phases, in addition to the Si QDs, appearing at temperatures ranging from 1150 to 1200 °C [48]. Multilayered structures can be deposited by sputtering or by plasma-enhanced CVD with growth parameters and annealing conditions very similar to those for oxide giving good control of QD sizes. The main difference is the extra H incorporation with PECVD that requires an initial low-temperature anneal to drive off excess hydrogen and prevent bubble formation in the high-temperature anneal [37]. Si QDs in nitride can also be formed *in situ* during sputtering where they form in the gas phase during deposition. There is much less control over size and shape but no high-temperature anneal is required to form the Si QDs [49]. Multilayer growth using this *in situ* technique has also been attempted with irregular shaped but reasonably uniform-sized Si QDs resulting, see Figure 20.5 [45].

(a)

(b)

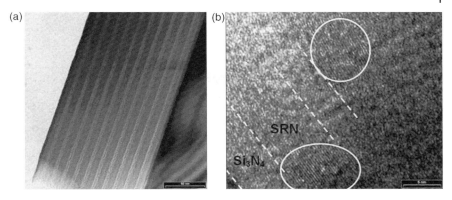

Figure 20.5 Gas phase *in situ* Si QDs dispersed in a Si_3N_4 matrix: (a) low magnification TEM and (b) high-resolution TEM [45].

The PL data in Figure 20.2 show that the PL energy for Si QDs in SiN_x increases with reduction in QD size at about the rate predicted by the EMA. This is because the relatively nonpolar Si—N bonds do not dramatically affect the quantum-confined energies. The other property in which we are interested is the increased transport that should be possible with the lower barrier height of the nitride matrix (see Section 20.2.3). Figure 20.6 shows the conductivity with temperature data from transfer length measurements of Si QD samples in nitride and oxide [21]. The nitride material shows a much greater conductivity, about six orders of magnitude at 300 K. This is indicative of the increased transport expected for the lower barrier height. However, Figure 20.6 also shows a much lower E_a for nitride compared to oxide. The oxide E_a at 0.76 eV at 300 K indicates an intrinsic material with a bandgap $= 2 \times E_a$ of about 1.5 eV (i.e., increased from that of bulk Si), whereas the E_a for nitride at 0.36 eV at 300 K indicates shallow levels as if the material were doped. This implies defect sites within the matrix that promote hopping transport. Nonetheless, this produces a higher conductivity material.

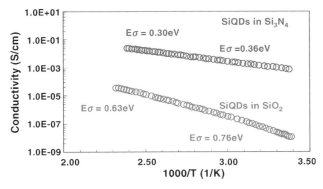

Figure 20.6 Conductivity against temperature for Si QDs in SiO_2 and SiN_x – also shown are activation energies, E_a [21].

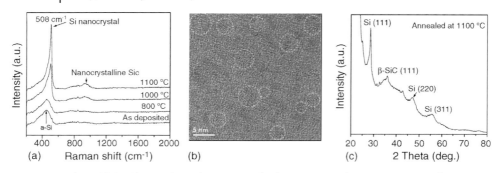

Figure 20.7 Silicon-rich SiC deposition and subsequent anneal: (a) Raman spectra for various annealing temperatures; (b) cross-sectional HRTEM image; and (c) X-ray diffraction both with a 1100 °C anneal.

20.3.1.2 Silicon QD Nanocrystals Embedded in Silicon Carbide Matrix

Si QDs in a SiC matrix offer an even lower barrier height and hence the promise of even better transport. However, the low barrier height also limits the minimum size of QDs to about 3 nm or else the quantum-confined levels are likely to rise above the level of the barrier, which should be around 2.3 eV for amorphous SiC. Si QDs in SiC matrix have been formed in a single thick layer by Si-rich carbide deposition followed by high-temperature annealing at between 800 and 1100 °C in a very similar process to that for oxide [50]. $Si_{1-x}C_x$/SiC multilayers have also been deposited by sputtering to give better control over the Si QD as with oxide and nitride matrices [50]. A feature that is seen in this materials system is that small SiC nanocrystals form in addition to the Si QDs as seen in the Raman and XRD data of Figure 20.7. The relatively uniform Si QDs are shown in the TEM image of Figure 20.7b. This Si QD in SiC material has also been doped both n- and p-type and fabricated into PV devices [51], as discussed in Section 20.3.3.2.

20.3.2
Doping of Si QD Arrays

A requirement for a tandem cell element is the presence of some form of junction for carrier separation. The impurities in bulk crystalline silicon play an important role in a semiconductor device. Dopants such as phosphorus and boron alter the conductivity of bulk Si by several orders of magnitude. There are several questions about the impurity doping in a low-dimensional structure [52]. Important questions arise as to whether the dopants will continue to play a role similar to that in bulk semiconductors or whether alternative methods of work function control will be required. It is not clear at present whether or not the doping of Si nanocrystals provides the generation of free charge carriers [53]. The junction can either be a grown or a diffused p–n junction or a p–i–n junction with the superlattice as the i-region. The latter requires careful control of the work functions (and therefore doping) of the p- and n-regions but also means that it is not essential for the QDs themselves to be doped.

Phosphorus and boron are excellent dopants in bulk Si as they have a high solid solubility at the annealing temperature. Hence, they are good initial choices to study the doping in Si QDs, although as discussed above there are reasons to suppose they will not dope them in the same way as bulk silicon. The number of atoms in a typical Si QD is in the range of 500–1000 atoms (for QDs with an approximate diameter of 5 nm). To translate a bulk doping density of 10^{18} cm^{-3} to such a nanostructure of 5 nm, spherical QDs would result in fewer than one dopant atom per QD. The doping used to date is with concentrations far higher than this such that no doubt much of the dopant is inactive. Nonetheless p- and n-type materials and rectifying junctions have been achieved [54–56].

20.3.2.1 Phosphorus Doping of Si Nanostructures

Phosphorus doping in the Si QD superlattices was achieved using P_2O_5 cosputtering during the deposition of silicon-rich oxide. The intention being that free carriers will be injected into the Si QDs that form on high-temperature postdeposition annealing. The P concentration in the SRO was controlled by varying the deposition rates from the three targets [54]. Transfer length measurement of the resistance was used to calculate both the resistivity at 300 K (see Figure 20.8) and the temperature dependence of the resistance R (see Figure 20.9). The values of the activation energy E_a calculated from the relation $R \approx \exp(E_a/kT)$ are also indicated in Figure 20.9. As the doping level increases from 0 at.% to 0.1 at.%, the activation energy decreases from 0.527 to 0.101 eV, together with a very significant decrease in resistance and resistivity of about seven orders of magnitude. The activation energy in a doped semiconductor is the energy difference between the conduction band (E_C) and Fermi level (E_F), $\Delta E = E_C - E_F$, which suggests effective n-type doping of the Si QDs. However, a further increase in P concentration to 0.35 at.% increases the activation energy to 0.149 eV and an increase again in resistance and resistivity by a factor of almost 1000. This small increase in activation energy may imply a reduction in the effective doping. The presence of P certainly has an effect on

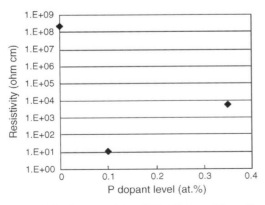

Figure 20.8 Dark resistivity of Si QD/SiO$_2$ multilayer films for various phosphorus dopant levels.

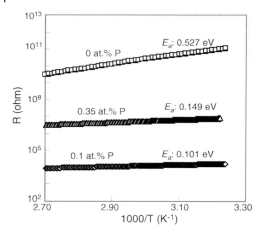

Figure 20.9 Temperature dependence of the resistance R of the Si QD films with various P concentrations.

nucleating crystallinity (see below), and it is probable that too much leads to a saturation of P in Si and/or degradation of the Si QD crystallinity leading to the increase again in resistivity.

TEM and XRD measurements indicate that the average Si nanocrystal size for P-doped material is increased compared to the undoped SRO under the same annealing conditions [54]. At 1100 °C, the average nanocrystal size in the 0.5 at.% P SRO film is almost double that of the undoped SRO film. In addition, P-doping in SRO films effectively reduces the annealing temperature required for the beginning of the Si crystallization. Based on these results, it can be concluded that the addition of P enhances the crystallization of the Si NCs through heterogeneous nucleation of nanocrystal growth [54].

20.3.2.2 Boron Doping of Si Nanostructures

Boron doping has been achieved by cosputtering from a boron Si and SiO_2 targets with subsequent annealing to form the Si nanocrystals [55]. X-ray photoelectron spectroscopy results indicate that the chemical environment of boron in both as-deposited and annealed boron-doped SRO films highly depends on the O/Si ratio of the SRO. A tendency for greater B—O bonding upon high-temperature annealing, indicating boron out-diffusion from B—B and/or B—Si to B—O, was found to be more pronounced in the high oxygen content SRO film. The results suggest a higher probability of effective boron doping in SRO films with low oxygen content.

Again using transfer length measurements, it was found that increasing the boron sputtering plasma power, and hence the B concentration, causes a dramatic decrease in resistivity, as shown in Figure 20.10, to a similar extent to that with P doping. This decrease in resistivity may be a consequence of an increase in carrier concentration due to more active dopants in the film and hence successful fabrication of p-type material [55].

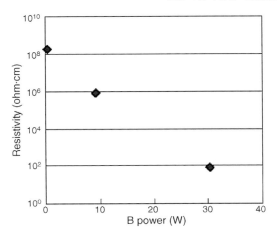

Figure 20.10 Dark resistivity of Si QD/SiO$_2$ multilayer films for various boron plasma power levels and hence concentrations [55].

As with P, B seems to have a strong effect on Si nanocrystal formation, but in this case with smaller Si nanocrystals forming compared to the undoped case [55]. TEM data indicate that the addition of boron slightly decreases the size of the Si nanocrystals. But the crystalline volume fraction was found to decrease with increasing boron concentration [55]. This suggests that boron suppresses Si crystallization perhaps due to local deformations induced by the impurity atoms.

20.3.2.3 Doping Mechanisms

In the fabrication of p- and n-type nanostructured material in this way, the doping mechanisms are not clear. Direct doping of the Si QDs is very unlikely due to the exclusion of impurities from the nanocrystals discussed in Sections 20.3.2. Hence, free carriers must be introduced by either doping of the matrix or doping of the interface between matrix and QDs. The interface is the most likely place to which these impurity atoms will migrate because of the relatively larger interstitial and defect sites at these locations. Also, the evidence of P and B modification of nanocrystal formation, discussed in Sections 20.3.2.1 and 20.3.2.2, strongly suggests that P or B atoms will be associated with nucleation of QD growth and hence will remain located close to the interface region.

The defects at the interface are also associated with multiple charge states. Hence, the formation of appropriately charged states such that free carriers are given up and captured by the QDs seems plausible. But these locations are also associated with many defects in the bandgap of the QDs and hence would be expected to dramatically enhance recombination. However, the passivation of these regions, which seems important for the device performance discussed in Section 20.3.3, presumably reduces the impact of these defects. Furthermore, the tendency of the strong local fields, associated with the interface regions, to sweep free carriers into the QDs, will somewhat mitigate the recombination.

Another possibility is that the enhancing effect on crystallization caused by the presence of P atoms, discussed in Section 20.3.2.1, could well also enhance transport and hence increase conductivity by reducing the average distance between QD nanocrystals. However, a similar mechanism does not explain enhanced conductivity in B-doped material as B suppresses rather than enhances crystallization, Section 20.3.2.2. Nonetheless, a combination of modification of crystallization behavior and the introduction of charged defects at interfaces could qualitatively explain increased conductivity and the two types of material produced.

20.3.3
Fabrication of Si QD PV Devices

20.3.3.1 Si QDs in SiO$_2$ Solar Cell

Sequential growth of multilayers of phosphorus-doped Si-rich oxide, followed by multilayers of undoped, and then boron-doped SRO layers, all interspersed with stoichiometric SiO$_2$ layers, has been carried out to fabricate rectifying p–n junction devices [57, 58]. The top B-doped bilayers were selectively etched to create isolated p-type mesas of about 0.1 cm^2, thus allowing access to the buried P-doped bilayers. Aluminum contacts were deposited by evaporation, patterned and sintered to create ohmic contacts on both p- and n-type layers.

I–V measurements in the dark and under 1-sun illumination indicate a good rectifying junction and generation of an open-circuit voltage, V_{OC}, up to 492 mV (see Figure 20.11). The high sheet resistance of the deposited layers, in conjunction with the insulating quartz substrates, causes an unavoidable high series resistance in the devices. The high resistance severely limits both the short-circuit current and the fill factor of the cells, particularly under illumination. It also makes it necessary to include effects due to in-plane current flow in the analysis of the measured electrical characteristics [58].

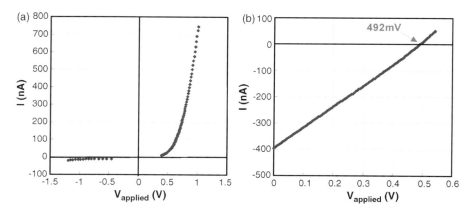

Figure 20.11 Electrical characteristics of p–i–n diodes with 4 nm SRO/2 nm SiO$_2$ bilayers with nominal O/Si = 0.8. (a) Dark and (b) illuminated *I–V* measurements showing V_{OC} = 492 mV [58].

Further evidence that this photovoltaic effect occurs in a material with an increased bandgap is given by temperature-dependent dark $I–V$ measurements, from which an electronic bandgap for the Si QD nanostructure materials can be extracted. A bandgap of 1.8 eV was extracted for a structure containing Si QDs with a nominal diameter of 4 nm. The extracted bandgap is larger than that of bulk silicon, highlighting the ability to alter the bandgap of a semiconductor using these bilayered nanostructures [58].

20.3.3.2 Si QDs in SiC Solar Cell

Homojunction Si QD devices have also been fabricated in a SiC matrix using the multiple level Si-rich carbide deposition method discussed in Section 20.3.1.2 [51]. Figure 20.12 is a schematic diagram of an n-type Si QD:SiC/p-type Si QD:SiC homojunction solar cell grown on a quartz substrate. The n-type Si QD emitter is approximately 200 nm thick and the p-type base layer is approximately 300 nm thick. The near-stoichiometric SiC layer was prepared by sputtering from the SiC target, while the Si-rich SiC layer was deposited by simultaneously sputtering from Si and SiC targets. Boron (for p-type Si QDs) and antimony (for n-type Si QDs) doping of Si QDs in the Si-rich layer was achieved by using a combination of Si, SiC, boron, or antimony targets.

This device gives an illuminated V_{OC} of 82 mV that is a promising initial value for a Si QDs in SiC solar cell on quartz substrate. Improvement of the device structure and optimization of dopant incorporation is expected to improve this value [51].

Current in both these SiO_2 matrix and the SiC matrix devices is very small, due principally to the very high lateral resistance and also because of the small amount of absorption in the approximately 200 nm of material used – but both these problems are being addressed. Hence, together with a further increase in the V_{OC}, this represents a promising approach to fabrication of a solar cell with an engineered bandgap. A full tandem cell would then require devices of different engineered bandgap (different QD sizes) to be grown on top of each other with a suitable connection between them. This connection would need to allow excited carriers

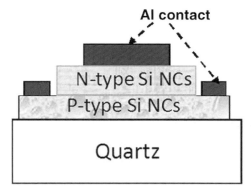

Figure 20.12 Schematic diagram of a (p)Si QD:SiC/(n)Si QD:SiC/quartz homojunction solar cell [51].

resulting from photogeneration at one wavelength, to cross into the adjacent cell, and be available to absorb another photon at another wavelength. This can be achieved using very thin layers with large defect densities that allow a recombination of electrons and holes but not a relaxation in their energy [58].

20.4
Intermediate Level Cells: Intermediate Band and Impurity Photovoltaic cell

The approach with these devices is to introduce one or more energy levels within the bandgap such that they absorb photons in parallel to the normal operation of a single bandgap cell. Hence, they also use multiple energy levels like the tandem cell, but the absorption of different energy photons is in parallel rather than in series. This semiparallel operation offers the potential to be much less spectrally sensitive but to still give high efficiencies.

Such a device with a single energy level in the bandgap has the same limiting efficiency as a three-level tandem – 63% under maximum concentration, 48% under 1 sun – because it has the same number of energy thresholds. However, this calculation does not take into account spectral sensitivity and assumes ideal properties such as ideal photon selection. This is a problem with the intermediate level concept, and the real devices would not be able to selectively absorb photons at the most appropriate energy levels, but the device will nonetheless collect photons that would not otherwise be absorbed. However, also note that the current now only has to be equal across the two lower energy levels while the main current across the bandgap is independent. This reduces the spectral sensitivity and compensates to some extent for the reduced photon selectivity.

These additional subbandgap absorbers can exist either as discrete energy levels in an impurity photovoltaic (IPV) cell or as a continuous band of levels, nonetheless, isolated from the conduction and valence bands – the intermediate band solar cell (IBSC) shown in Figure 20.13 [59].

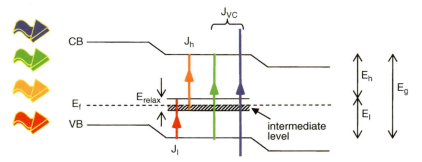

Figure 20.13 Intermediate level cell (IBSC): below-bandgap photons are absorbed by the two transitions to and from the intermediate level contributing to photocurrent, in parallel to normal operation of the cell.

Both devices can absorb two below-bandgap photons to create one electron–hole pair at the bandgap energy, but the IBSC has the advantage that the delocalization of carriers in its continuous band means that these photons do not necessarily have to be absorbed by the same electron. This gives a much longer lifetime to the intermediate level allowing much more time for absorption of the second photon. To maximize this advantage, the intermediate band should be half-filled with electrons – that is, it should have a Fermi level at half the band energy illustrated in Figure 20.13 – such that absorption of an electron from the valence band or emission of an electron to the conduction band is equally likely.

The formation of an intermediate band has been suggested in some III–V, II–VI, and chalcopyrite systems, usually alloyed with a transition metal (see, for example, [60–62]). But they have also been attempted experimentally using the confined energy levels of a InAs/GaAs QD superlattice [63]. These devices have demonstrated several of the indicators of true IBSC operation, although they have not yet achieved an efficiency advantage; nonetheless, this seems likely in the near future particularly if they are operated under concentration. If this is successful, it is possible that the technology could be transferred for the use of Si QDs, with the advantage of the use of much simpler and cheaper deposition techniques.

20.5
Multiple Carrier Excitation Using Si QDs

Carriers generated from high-energy photons (at least twice the bandgap energy) absorbed in a semiconductor can undergo "impact ionization" events that result in two or more carriers close to the bandgap energy. Thus, if this effect is incorporated in a solar cell the current resulting from absorption of high-energy photons can be increased, thus, boosting the overall efficiency. Impact ionization is extremely rare in bulk materials, but in quantum dots the localization of electrons and holes seems to dramatically increase the probability [64–66] (see Figure 20.14). The exact mechan-

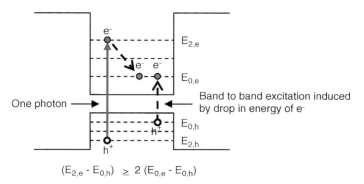

$$(E_{2,e} - E_{0,h}) \geq 2 (E_{0,e} - E_{0,h})$$

Figure 20.14 Multiple exciton generation in QDs: a high-energy photon is absorbed at a high confined energy level in the QD that then decays into two or more electron–hole pairs at the first confined energy level. Energy is conserved but momentum conservation in QDs is relaxed [66].

isms involved in this are not yet entirely clear but they are related to the reduced requirement for conservation of crystal momentum in the small spatial volume of a quantum dot. There is much experimental evidence showing production of up to seven electron–hole pairs for the absorption of a high-energy photon – that is, a quantum efficiency (QE) of 7. The QE must always be less than or equal to the ratio of the photon energy to the bandgap energy (E_{hv}/E_g) – hence small bandgap materials are preferable. This multiple exciton generation (MEG) was first seen in PbSe QDs but has now been seen in quite a wide range of II–VIs and other material QDs, including Si, which is significant for a possible future large-scale implementation. In fact, the effect is seen in quite large Si QDs of about 5 nm diameter, formed by a vapor phase method. Hence, a strong quantum confinement of energy is not important but rather the spatial localization that relaxes the momentum conservation requirement.

However, as yet the phenomenon has been observed only with absorption spectroscopy measurements and there is quite some controversy not only as to how it occurs but also whether it is repeatable. The most likely explanation seems to be that the mechanism is very sensitive to the exact surface states on the QDs [66]. The other problem is that so far no enhanced current has been observed at external contacts. This is probably related to the difficulty in transporting carriers through the QD array.

20.6
Hot Carrier Cells

The concept underlying the hot carrier solar cell is to slow the rate of photoexcited carrier cooling, caused by phonon interaction in the lattice, to allow time for the carriers to be collected while they are still at elevated energies ("hot") and thus allowing higher voltages to be achieved from the cell [3, 67, 68]. It thus tackles the major photovoltaic loss mechanism of thermalization of excess carrier energy down to the bandgap energy. In addition to an absorber material that slows the rate of carrier relaxation, a hot carrier cell must allow extraction of carriers from the device through contacts that accept only a very narrow range of energies – energy selective contacts (ESCs) (see Figure 20.15).

These requirements are very difficult, but the potential efficiency of the hot carrier cell is 65% at 1 sun (85% at maximum concentration), as it is close to an ideal multiple energy-level device. These very high limiting efficiencies mean that a real device, even if only partially successful, could significantly increase efficiencies. And furthermore, the device would conceptually be a fairly homogenous one without the very many multiple layers of different materials of other very high-efficiency concepts.

Energy-selective contacts can be fabricated using double-barrier resonant tunneling structures. These would use QDs or other discrete confined centers as the resonant centers to provide a discrete energy level between two insulating barriers. This would give conduction strongly peaked at the discrete energy level. The total energy filtering of a QD-based structure is required for a selective energy contact rather than 1D energy filtering because the 1D energy filtering in, for instance, a quantum well resonant tunneling device is effective only for carriers with momenta

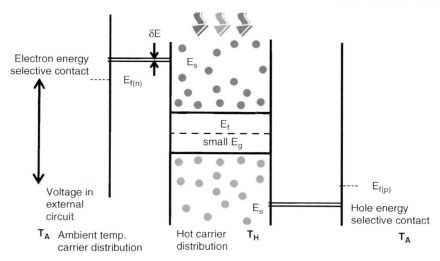

Figure 20.15 Band diagram of the hot carrier cell that requires slowed carrier cooling in the absorber and collection of carriers through ESCs.

entirely perpendicular to the plane of the well [69]. The required property for this energy selection is negative differential resistance (NDR). Such a filter should exhibit an NDR in all directions.

20.6.1
Photoluminescence of Si QD ESCs

Double-barrier tunneling structures consisting of SiO_2/Si QDs/SiO_2 layers have been fabricated by sputtering followed by annealing at $1100\,°C$ using the same Si-rich oxide technique described in Section 20.2.1 [70]. The control of the size of a single layer QDs in SiO_2 with the thickness of SRO layer has been studied with photoluminescence measurements. Figure 20.16 shows PL spectra for four samples with SRO thicknesses in the range of 2.4–6 nm. Thicknesses of the two oxide layers were about 6 nm for all the samples. A blueshift of the PL with the decrease in the SRO layer thickness is observed. Since PL energy is associated with the size of QDs, this demonstrates a correlation between QD size and SRO layer thickness. The broad PL peaks observed for all samples suggests that a fairly large distribution of QD sizes exist around the mean diameter. However, PL peaks from different samples with different SRO thicknesses are clearly separated [58].

A very significant improvement in the uniformity of the QDs has been obtained with samples fabricated with substrate heating. Figure 20.17 shows PL spectra from a single-layer SiO_2/Si QDs/SiO_2 structure [58]. The first SiO_2 layer was gown by thermal oxidation of a silicon substrate at $800\,°C$ in oxygen ambient, to give a good quality oxide. The other two layers, namely SRO and the top SiO_2, were deposited by sputtering. In this case, the substrate was heated at about $250\,°C$ during the film growth. The sample was annealed at $1100\,°C$ for 2 hours, followed by a forming gas

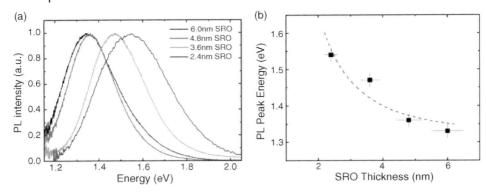

Figure 20.16 (a) Results of PL measurement on a single layer of Si QDs in SiO$_2$. Correlation of PL peak energy increase with SRO thickness decrease is evident. (b) PL peak energy, as derived from (a) as a function of SRO thickness. The dashed line is a least square fit of PL energy with inverse square of SRO thickness [58].

anneal. The two curves, grey and dashed, represent the measurements conducted over an interval of 3 weeks. For comparison, the PL spectrum (black curve) for a sample grown without substrate heating is also shown. It is evident that the PL for the samples grown with substrate heating is much sharper than for the samples grown without substrate heating. This demonstrates a much better QS size uniformity in these samples. In addition to the main PL peak, the spectra for the sample grown with substrate heating also show a shoulder peak at a slightly lower energy. This double peak feature is thought to be due to exciton recombination for the higher energy peak and free carrier recombination for the lower energy peak – these usually being masked by the breadth of QD sizes. In spatially confined QDs, the exciton binding energy is much greater than in bulk because of the lack of the opportunity for the carriers to delocalize. For the same reason, the exciton lifetime is also much longer

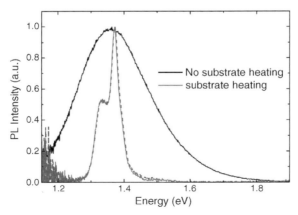

Figure 20.17 PL from a single layer of Si QDs in SiO$_2$ fabricated with (grey/dashed) and without (black) substrate heating [58].

than in bulk. Hence, if correct, this explanation would mean that the spacing between the double peak is the enhanced exciton binding energy – about 400 meV – and the ratio of the lower energy peak intensity to that of the higher energy peak will be the ratio of free carriers to excitons in the system and hence also related to the exciton lifetime.

20.6.2
Negative Differential Resistance in Si QD ESCs

For the measurement of resonant tunneling through the double-barrier structures, devices were fabricated as schematically shown in the inset of Figure 20.18. The devices consist of a single layer of Si QDs in SiO_2 grown on a heavily doped Si substrate. A Si capping layer is deposited from a highly doped target. The thickness of the SRO and SiO_2 layers is approximately 5 nm. The thickness of the silicon capping layer is about 50 nm. During the growth, a shadow mask is used to obtain isolated devices. Aluminum contacts are deposited by evaporation using a shadow mask.

Figure 20.18 shows results of room-temperature DC *I–V* measurements on a typical device. The two *I–V* profiles are measured on different mesas on the same sample. In both the *I–V* profiles, negative differential resistance can be observed, at about 1.4 and 1.5 V, respectively. NDR is a requisite characteristic of resonant tunneling and hence of ESCs. Similar results have been observed for a few other similar devices. Although the quality factor of the NDR resonance (ratio of peak height to peak width) is not good, such a result at *room temperature* is very encouraging as a proof of 1D energy selection. Subsequent measurement showed degradation and breakdown of the oxide; hence the quality of the oxide needs to be improved.

Current (A/cm²)

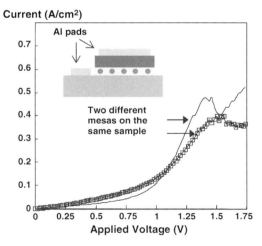

Figure 20.18 Result of *I–V* measurements on two mesas on a resonant tunneling device. NDR resonance is evident [70].

20.6.3
A Complete Hot Carrier Cell

Fabrication of an absorber material with slowed carrier cooling is very difficult. The effect has been observed at very high illumination intensities via a "phonon bottleneck" effect, in which carrier energy decay mechanisms are restricted. Compounds with large mass differences between their anions and cations have a gap in their allowed phonon modes that can slow down these decay mechanisms and enhance the bottleneck effect [71]. Examples are gallium nitride and indium nitride, with some experimental evidence for slowed cooling in the latter [72]. Theoretical work on replicating this effect by modifying the phononic band structures in QD nanostructure superlattices [73] is now being attempted experimentally.

As both absorbers and ESCs are likely to involve QDs, a working device may consist of a uniform array of QDs for the absorber that could include Si QDs – if an appropriate method can be found to fabricate the structure – and a resonant tunneling contact structure that is likely to involve single layers of Si QDs. Nonetheless, the hot carrier cell, while promising, is still a long way from demonstration.

20.7
Conclusions

Multiple energy thresholds in solar cells allow collection of solar photons in ways that facilitate a more efficient exploitation of their total energy. QD nanostructures allow incorporation of these multiple energy levels in materials that can be deposited with low-cost methods.

The principal property exploited in the use of Si nanocrystals, or quantum dots, for solar cells is the confined energy levels. Control of these by controlling the QD size allows a material with a larger bandgap than the bulk material to be engineered through the overlap of levels to create a miniband. This in turn allows the fabrication of tandem cells in which the bandgap can be optimized for the best extraction of energy from a particular part of the solar spectrum. Devices have been fabricated that demonstrate both increased effective bandgap and promising open-circuit voltages. Currents in these cells need to be increased and to create a full tandem cell devices with different bandgaps (QD size) need to be grown on top of each other with appropriate connection. But these two requirements should be met with further development.

Other devices exploit the localization of carriers in Si QDs. These include the multiple exciton generation approach in which this localization relaxes the requirement for conservation of momentum, allowing to generate multiple excitons from a single photon absorption.

The intermediate band cell exploits the discrete nature of the miniband formed by neighboring QDs, which allows a long lifetime of carriers generated by absorption of below-bandgap photons. Si QDs should also be applicable in such devices.

The ultimate efficiency for a solar cell can perhaps be achieved by preventing photogenerated carriers from losing their energies to the lattice as phonons. The phonon bandgap that can be engineered in QD superlattice arrays could be an effective way of preventing these decays. A uniform array of Si QDs could be used for such a superlattice. But this hot carrier solar cell device also requires narrow band energy selective contacts, and Si QDs have been demonstrated to allow such an effect in double-barrier resonant tunneling structures.

Quantum dots and particularly Si QD nanostructures are applicable to several types of solar cell, with the promise of higher efficiencies but in cells that can be made by cheap thin-film deposition processes. This is possible because the extra degrees of freedom given by the physics of quantum confinement allow multiple energy levels to be incorporated into semiconductor materials with independent control over device geometry.

Acknowledgments

The author acknowledges the support of the ARC Photovoltaics Center of Excellence, UNSW, during the collection of information for this chapter. The Center of Excellence is supported by the Australian Research Council (ARC) and the Global Climate and Energy Project (GCEP), the latter administered by Stanford University. The author would also like to thank his colleagues in the Center of Excellence for their work on and input to Third Generation PV concepts.

References

1 Shockley, W. and Queisser, H. (1961) Detailed balance limit of efficiency of p–n junction solar cells. *J. Appl. Phys.*, **32**, 510–519.

2 Wenham, S.R., Green, M.A., and Watt, M.E. *Applied Photovoltaics*, Centre for Photvoltaic Devices and Systems, Sydney.

3 Green, M.A. (2003) *Third Generation Photovoltaics: Advanced Solar Energy Conversion*, Springer-Verlag.

4 Conibeer, G. (2007) Third generation photovoltaics. *Mater. Today*, **10**, 42–50.

5 Cuadra, L., Marti, A., and Luque, A. (2002) Type II broken band heterostructure quantum dot to obtain a material for the intermediate band solar cell. *Physica E*, **14**, 162.

6 King, R.R. *et al.* (2007) 40% efficient metamorphic GaInP/GaInAs/Ge multijunction solar cells. *Appl. Phys. Lett.*, **90**, 183516.

7 Zacharias, M., Heitmann, J., Scholz, R., Kahler, U., Schmidt, M., and Bläsing, J. (2002) Size controlled highly luminescent silicon nanocrystals: a SiO/SiO superlattice approach. *Appl. Phys. Lett.*, **80**, 661.

8 Nesbit, L.A. (1985) Annealing characteristics of Si-rich SiO films. *Appl. Phys. Lett.*, **46**, 38.

9 Fernandez, B.G., Lopez, M., Garcia, C., Perez-Rodriguez, A., Morante, J.R., Bonafos, C., Carrada, M., and Claverie, A. (2002) Influence of average size and interface passivation on the spectral emission of Si nanocrystals embedded in SiO_2. *J. Appl. Phys.*, **91**, 798–807.

10 Sarikov, A., Litovchenko, V., Lisovskyy, I., Maidanchuk, I., and Zlobin, S. (2007) Role of oxygen migration in the kinetics of the phase separation of nonstoichiometric silicon oxide films during high-

temperature annealing. *Appl. Phys. Lett.*, **91**, 133109–133113.

11 Khomenkova, L., Bulakh, B., Korsunska, N., Stara, T., Goldstein, Y., Jedrzejewski, J., Savir, E., Sada, C., Bisello, D., Khomenkov, V., and Emirov, Y. (2007) Growth peculiarities of silicon nanoparticles in an oxide matrix prepared by magnetron sputtering. *Phys. Stat. Sol. C*, **4**, 3061–3065.

12 Spinella, C., Bongiorno, C., Nicotra, G., Rimini, E., Muscara, A., and Coffa, S. (2005) Quantitative determination of the clustered silicon concentration in substoichiometric silicon oxide layer. *Appl. Phys. Lett.*, **87**, 044102–044103.

13 Khomenkova, L., Korsunska, N., Baran, M., Bulakh, B., Stara, T., Kryshtab, T., Gómez Gasga, G., Goldstein, Y., Jedrzejewski, J., and Savir, E. (2009) Structural and light emission properties of silicon-based nanostructures with high excess silicon content. *Physica E*, **41** (6), 1015–1018.

14 Cho, E.-C. (2003) Optical transitions in SiO_2/crystalline Si/SiO_2 quantum wells and nanocrystalline silicon (nc-Si)/SiO_2 superlattice fabrication, PhD Thesis, University of New South Wales, Sydney.

15 Cho, E.-C., Park, S., Hao, X., Song, D., Conibeer, G., Park, S.-C., and Green, M.A. (2008) Silicon quantum dot/crystalline silicon solar cells. *Nanotechnology*, **19**, 245201.

16 Ding, L., Chen, T.P., Liu, Y., Ng, C.Y., and Fung, S. (2005) Optical properties of silicon nanocrystals embedded in a SiO_2 matrix. *Phys. Rev. B*, **72**, 125419.

17 Öğüt, S., Chelikowsky, J.R., and Louie, S.G. (1997) Quantum confinement and optical gaps in Si nanocrystals. *Phys. Rev. Lett.*, **79**, 1770.

18 König, D., Rudd, J., Green, M.A., and Conibeer, G. (2008) Role of the interface for the electronic structure of silicon quantum dots. *Phys. Rev. B*, **78**, 035339; selected for the *Virt. J. Nanosci. Technol.*, 2008, **18**/6.

19 König, D., Rudd, J., Green, M.A., and Conibeer, G. (2009) Impact of bridge and double-bonded oxygen on OH-terminated Si quantum dots: a DF–HF study. *Mater. Sci. Eng. B*, **159–160**, 117–121.

20 Droz, C. (2003) Thin film microcrystalline silicon layers and solar cells: microstructure and electrical performances. PhD Thesis, Université de Neuchâtel.

21 Fangsuwannarak, T. (2007) Electronic and optical characterisations of silicon quantum dots and its applications in solar cells, PhD Thesis, University of New South Wales, Australia.

22 Arora, A.K., Rajalakshmi, M., Ravindran, T.R., and Sivasubramanian, V. (2007) Raman spectroscopy of optical phonon confinement in nanostructured materials. *J. Raman Spectrosc.*, **38**, 604–617.

23 Iqbal, Z. and Veprek, S. (1982) Raman scattering from hydrogenated microcrystalline and amorphous silicon. *J. Phys. C Solid State*, **15**, 377–392.

24 Colvard, C., Gant, T.A., and Klein, M.V. (1985) Folded acoustic and quantized optic phonons in (GaAl)As superlattices. *Phys. Rev. B*, **31**, 2080.

25 Harrison, P. (2000) *Quantum Well, Wires and Dots: Theoretical and Computational Physics*, John Wiley & Sons Ltd, Chisester

26 Jiang, C.-W. (2006) Theoretical and experimental study of energy selective contacts for hot carrier solar cells and extensions to tandem cells, PhD Thesis, University of New South Wales, Australia.

27 Barber, H.D. (1967) Effective mass and intrinsic concentration in silicon. *Solid-State Electron.*, **10**, 1039.

28 Green, M.A. (1995) *Silicon Solar Cells: Advanced Principles and Practice*, UNSW.

29 Fangsuwannarak, T., Pink, E., Huang, Y.D., Cho, Y.-H., Conibeer, G., Puzzer, T., and Green, M.A. (2005) Effects of silicon nanocrystallite density on the Raman-scattering spectra of silicon quantum dot superlattices. *Proc. SPIE*, **6037**, 60370T-1.

30 Boer, K. (1990) *Survey of Semiconductor Physics*, Van Nostrand Reinhold.

31 Arguirov, T., Mchedlidze, T., Kittler, M., Rölver, R., Berghoff, B., Först, M., and Spangenberg, B. (2006) Residual stress in Si nanocrystals embedded in a SiO_2 matrix. *Appl. Phys. Lett.*, **89**, 053111.

32 Kanemitsu, Y. (1996) Light-emitting silicon materials. *J. Lumin.*, **70**, 333.

33 Takagi, H., Ogawa, H., Yamazaki, Y., Ishizaki, A., and Nakagiri, T. (1990) Quantum size effects on

photoluminescence in ultrafine Si particles. *Appl. Phys. Lett.*, **56**, 2379.

34 Takeoka, S., Fujii, M., and Hayashi, S. (2000) Size-dependent photoluminescence from surface-oxidized Si nanocrystals in a weak confinement regime. *Phys. Rev. B*, **62**, 16820.

35 Kim, T.-W., Cho, C.-H., Kim, B.-H., and Park, S.-J. (2006) Quantum confinement effect in crystalline silicon quantum dots in silicon nitride grown using SiH_4 and NH_3. *Appl. Phys. Lett.*, **88**, 123102.

36 Kim, T.Y., Park, N.M., Kim, K.H., Sung, G.Y., Ok, Y.W., Seong, T.Y., and Choi, C.J. (2004) Quantum confinement effect of silicon nanocrystals in situ grown in silicon nitride films. *Appl. Phys. Lett.*, **85**, 5355.

37 Cho, Y.-H., Cho, E.-C., Huang, Y., Jiang, C.-W., Conibeer, G., and Green, M.A. (2005) Proceeding of 20th European Photovoltaic Solar Energy Conference, June 6–10, Barcelona, Spain, p. 47.

38 König, D., Green, M.A., Conibeer, G., Takeda, Y., Ito, T., Motohiro, T., and Nagashima, T. (2006) Proceedings of the 21st European Photovoltaic Solar Energy Conference, Dresden, p. 164.

39 König, D., Rudd, J., Green, M.A., and Conibeer, G. (2008) Impact of interface on effective band gap of Si quantum dots. *Sol. Energy Mater. Sol. C*, electronic publication. doi: 10.1016/j.solmat.2008.09.026.

40 Jiang, C.-W. and Green, M.A. (2006) Silicon quantum dot superlattices: Modeling of energy bands, densities of states, and mobilities for silicon tandem solar cell applications. *J. Appl. Phys.*, **99**, 114902.

41 Meillaud, F., Shah, A., Droz, C., Vallat-Sauvain, E., and Miazza, C. (2006) Efficiency limits for single-junction and tandem solar cells. *Sol. Energy Mater. Sol. Cells*, **90**, 2952.

42 Green, M.A., Cho, E.-C., Cho, Y.-H., Pink, E., Trupke, T., Lin, K.-L., Fangsuwannarak, T., Puzzer, T., Conibeer, G., and Corkish, R. (2005) Proceeding of the 20th European Photovoltaic Solar Energy Conference, June 6–10, Barcelona, Spain, p. 3.

43 Green, M.A., Conibeer, G., Perez-Wurfl, I., Huang, S.J., Konig, D., Song, D., Gentle, A., Hao, X.J., Park, S.W., Gao, F., So, Y.H., and Huang, Y. (2008) Progress with silicon-based tandem cells using group IV quantum dots in a dielectric matrix. Proceedings of 23rd European Photovoltaic Solar Energy Conference, Valencia.

44 Conibeer, G., Green, M.A., Corkish, R., Cho, Y.-H., Cho, E.-C., Jiang, C.-W., Fangsuwannarak, T., Pink, E., Huang, Y., Puzzer, T., Trupke, T., Richards, B., Shalav, A., and Lin, K.-L. (2006) Silicon nanostructures for third generation photovoltaic solar cells. *Thin Solid Films*, **511/512**, 654.

45 Conibeer, G., Green, M.A., Cho, E.-C., König, D., Cho, Y.-H., Fangsuwannarak, T., Scardera, G., Pink, E., Huang, Y., Puzzer, T., Huang, S., Song, D., Flynn, C., Park, S., Hao, X., and Mansfield, D. (2008) Silicon quantum dot nanostructures for tandem photovoltaic cells. *Thin Solid Films*, **516**, 6748.

46 Kurokawa, Y., Miyajima, S., Yamada, A., and Konagai, M. (2006) Preparation of Nanocrystalline Silicon in Amorphous Silicon Carbide Matrix. *Jpn. J. Appl. Phys.*, **45**, L1064.

47 Kim, B.H., Cho, C., Kim, T., Park, N., Sung, G., and Park, S. (2005) Photoluminescence of silicon quantum dots in silicon nitride grown by NH_3 and SiH_4. *Appl. Phys. Lett.*, **86**, 091908.

48 Scardera, G., Puzzer, T., Perez-Wurfl, I., and Conibeer, G. (2008) The effects of annealing temperature on the photoluminescence from silicon nitride multilayer structures. *J. Crystal Growth*, **310**, 3680–3684.

49 Park, N.M., Kim, N.M, Park, T.S., and Park, S.J. (2001) Band gap engineering of amorphous silicon quantum dots for light-emitting diodes. *Appl. Phys. Lett.*, **78**, 2575.

50 Song, D., Cho, E.-C., Conibeer, G., Cho, Y.-H., Huang, Y., Huang, S., Flynn, C., and Green, M.A. (2007) Fabrication and characterization of Si nanocrystals in SiC matrix produced by magnetron co-sputtering. *J. Vac. Sci. Technol.*, **B25**, 1327.

51 Song, D., Cho, E.-C., Conibeer, G., and Green, M.A. Solar cells based on Si-NCs embedded in a SiC matrix. Technical Digest of the 18th International

Photovoltaic Science and Engineering Conference, Kolkata, India, 19–23 January 2009.

52 Ossicini, S., Iori, F., Degoli, E., Luppi, E., Magri, R., Poli, R., Cantele, G., Trani, F., and Ninno, D. (2006) Understanding Doping In Silicon Nanostructures. *IEEE J. Sel. Top. Quant. Electronics*, **12**, 1585.

53 Polisski, G., Kovalev, D., Dollinger, G., Sulima, T., and Koch, F. (1999) Boron in mesoporous Si – Where have all the carriers gone? *Physica B*, **273**, 951.

54 Hao, X.J., Cho, E.-C., Scardera, G., Shen, Y.S., Bellet-Amalric, E., Bellet, D., Conibeer, G., and Green, M.A. (2009) Phosphorus doped silicon quantum dots for all-silicon quantum dot tandem solar cells. *Sol. Energy Mater. Sol. Cells*, **93**, 1524.

55 Hao, X.J., Cho, E.-C., Flynn, C., Shen, Y.S., Park, S.C., Conibeer, G., and Green, M.A. (2009) Synthesis and characterization of boron-doped Si quantum dots for all-Si quantum dot tandem solar cells. *Sol. Energy Mater. Sol. Cells*, **93**, 273–279.

56 Park, S., Cho, E.-C., Song, D., Conibeer, G., and Green, M.A. (2009) n-Type silicon quantum dots and p-type crystalline silicon heteroface solar cells. *Sol. Energy Mater. Sol. Cells*, **93**, 684.

57 Hao, X., Perez-Wurfl, I., Conibeer, G., and Green, M.A. (2009) Study on properties of Si QDs junction in oxide matrix for "all-silicon" tandem solar cells. Proceedings of the PVSEC 19, Korea.

58 Annual Report of the Photovoltaics Centre of Excellence 2008, University of New South Wales, Sydney, Australia, June 2009, www.pv.unsw.edu.au/research/annualreports.asp.

59 Luque, A. and Martí, A. (1997) Increasing the Efficiency of Ideal Solar Cells by Photon Induced Transitions at Intermediate Levels. *Phys. Rev. Lett.*, **78**, 5014.

60 Palacios, P. *et al.* (2006) First-principles investigation of isolated band formation in half-metallic $Ti_xGa_{1-x}P$ (x = 0.3125–0.25). *Phys. Rev. B*, **73**, 085206.

61 Tablero, C. (2006) Electronic and magnetic properties of ZnS doped with Cr. *Phys. Rev. B*, **74**, 195203.

62 Yu, K.M., Walukiewicz, W., and Ager, J.W. (2006) Multiband GaNAsP quaternary alloys. *Appl. Phys. Lett.*, **88**, 092110.

63 Marti, A. *et al.* (2006) Production of Photocurrent due to Intermediate-to-Conduction-Band Transitions: A Demonstration of a Key Operating Principle of the Intermediate-Band Solar Cell. *Phys. Rev. Lett.*, **97**, 247701.

64 Schaller, R.D. and Klimov, V.I. (2004) High Efficiency Carrier Multiplication in PbSe Nanocrystals: Implications for Solar Energy Conversion. *Phys. Rev. Lett.*, **92**, 186601.

65 Hanna, M.C. and Nozik, A.J. (2006) Solar conversion efficiency of photovoltaic and photoelectrolysis cells with carrier multiplication absorbers. *J. Appl. Phys.*, **100**, 074510.

66 Nozik, A.J. (2008) Multiple exciton generation in semiconductor quantum dots. *Chem. Phys. Lett.*, **457**, 3–11.

67 Würfel, P. (1997) Solar energy conversion with hot electrons from impact ionisation. *Sol. Energy Mater. Sol. Cells*, **46**, 43.

68 Ross, R. and Nozik, A.J. (1982) Efficiency of hot-carrier solar energy converters. *J. Appl. Phys.*, **53**, 3813.

69 Conibeer, G., Jiang, C.-W., König, D., Shrestha, S., Walsh, T., and Green, M.A. (2008) Selective energy contacts for hot carrier solar cells. *Thin Solid Films*, **516**, 6968.

70 Jiang, C.-W. *et al.* (2006) 21st Euro PVSEC, Dresden, p. 168.

71 Conibeer, G. and Green, M.A. (2004) 19th Euro PVSEC, Paris, p. 270.

72 Chen, F. and Cartwright, A.N. (2003) Time-resolved spectroscopy of recombination and relaxation dynamics in InN. *Appl. Phys. Lett.*, **83**, 4984.

73 Conibeer, G.J., König, D., Green, M.A., and Guillemoles, J.F. (2008) Slowing of carrier cooling in hot carrier solar cells. *Thin Solid Films*, **516**, 6948.

21
Characterization of Si Nanocrystals

Selçuk Yerci, İlker Doğan, Ayşe Seyhan, Arife Gencer, and Rasit Turan

21.1
Introduction

In this chapter, we review and discuss some of the widely used analytical and optical methods for characterizing Si nanoclusters in various matrices. Universal diagnostic techniques such as transmission electron microscopy (TEM), X-ray photoelectron spectroscopy (XPS), photoluminescence (PL), atomic force microscopy (AFM), Raman spectroscopy (RS), Fourier transform infrared (FTIR) spectroscopy, and X-ray diffraction (XRD) are commonly used directly or indirectly to understand various features of nanostructures including Si. Large number of investigations and published reports are based on these measurements. Size determination and direct imaging of nanoclusters are very crucial in many applications as their functional properties are closely connected to their size and shape; therefore, imaging with TEM and AFM plays a central role in the nanocluster investigation. In addition, RS and XRD have been proven to be useful in size estimation. These methods can also be applied to estimate the stress on the nanoclusters and to monitor the amorphous (nanoclusters) to crystalline (nanocrystals) transition. XPS can be employed to investigate the formation of elemental Si ($Si-Si_4$) (i.e., nanoclusters) and suboxides ($Si-SiO_x$) (the shell around the nanoclusters). Thus, it is widely used in studies investigating the chemical environment of the Si nanoclusters. Recently, XPS is also used to understand the effect of quantum confinement effect (QCE) on the core-shell structures. Similarly, FTIR can be used to monitor the $Si-SiO_x$ states of nanoclusters and the matrix surrounding them. Although it does not probe $Si-Si$ bonds directly, it is useful to investigate the matrix between them, which is indeed very crucial for many aspects such as the origin of light emission from Si nanoclusters and the electronic transport between them. Since the light emission is one of the most significant motivations behind the Si nanocluster research, the origin of light generation and optical mechanisms need to be examined using various analytical and optical methods. PL spectroscopy has been used extensively to understand optical

Silicon Nanocrystals: Fundamentals, Synthesis and Applications. Edited by Lorenzo Pavesi and Rasit Turan
Copyright © 2010 WILEY-VCH Verlag GmbH & Co. KGaA, Weinheim
ISBN: 978-3-527-32160-5

properties that also shed light on chemical and structural properties. This chapter summarizes and discusses these characterization techniques with recently published research results.

21.2
Imaging

21.2.1
Si Nanocluster Imaging by Transmission Electron Microscopy Techniques

Si nanoclusters embedded in wide bandgap host matrices draw enormous attention due to their interesting physical properties apart from their bulk counterparts. There have been intensive studies on the optical and electrical properties of Si nanoclusters since they are promising candidates to construct novel microelectronic and optical devices monolithically integrated atop the widespread silicon platform. Since the optical and electrical properties of Si nanoclusters can be controlled by their dimensions, the most important achievement for constructing stable and reliable devices is to probe and control the size and shape of Si nanoclusters in host matrices. Till date, the most widely used imaging technique to monitor and track the nucleation of Si nanoclusters has been TEM. There are different methods for Si nanoclusters imaging closely related with TEM such as high-resolution TEM (HRTEM), energy filtered TEM (EFTEM), defocused bright-field imaging, dark-field imaging, and electron energy loss spectroscopy (EELS). Expectedly, these techniques are used for various aims on Si nanocluster imaging.

21.2.2
Transmission Electron Microscopy

TEM is a challenging task when studying Si nanoclusters in SiO_2 and Si_3N_4 matrices. In a Si/SiO_2 and Si/Si_3N_4 system, Si nanoclusters may be either in the amorphous or the crystalline phase. The difficulty lies in separating the host matrix and Si nanoclusters, especially when they are amorphous, since the difference in the atomic number and atomic density does not vary significantly. This results in a low contrast between the Si nanocluster and the SiO_2 or Si_3N_4 (SiO_2/Si_3N_4) matrix in TEM micrographs. Si nanoclusters in SiO_2/Si_3N_4 can be more securely imaged by TEM in two folds: cross section monitoring by (i) bright-field and (ii) HRTEM Dark-Field (DF) imaging. The distribution of Si nanoclusters can be monitored by these techniques if nanoclusters are in the crystalline phase (nanocrystals). Imaging process is orientation dependent and not all particles are monitored. The particles that are oriented in a well-defined direction with respect to the incident beam are the only ones to be monitored. Therefore, there is always a doubt in quantificatifying the density of Si nanoclusters in SiO_2/Si_3N_4 matrix. Moreover, this quantification based on TEM techniques assumes that the imaged sample is homogenous and the imaged area is a representative of the whole sample.

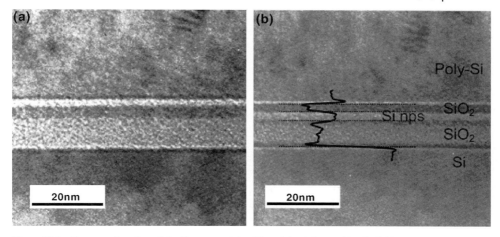

Figure 21.1 Cross-section view of the Si nanoparticles (Si-nps) by defocused bright-field imaging: (a) under focus where nanoclusters appeared as a thick black line between two bright regions, (b) over focus where nanoparticles appear as a thick white line between two dark regions.

Defocused bright-field imaging works on the principle of phase shifting of electron waves were propagated through Si nanoclusters in SiO_2 matrix. By assuming that the incident beam is coherent, the nanoclusters lead to a phase shift, which can be directly detected as Fresnel fringes seen as black/white or white/black contrasts between the interface of Si nanocluster-SiO_2 matrix. By this technique, the positions in the interface regions are detected from the cross section of 2D array of Si nanoclusters present in the SiO_2 matrix (Figure 21.1) [1].

On the other hand, HRTEM works on the principle of coherent superposition of incident and elastically scattered electromagnetic waves from Si nanoclusters. The success of this technique depends on the right orientation of the Si nanoclusters with respect to the incident beam and the thickness of the interested specimen (Figure 21.2) [2]. Disoriented Si nanoclusters cannot be monitored and if the specimen thickness is not comparable to the nanocrystal dimensions, no contrast is obtained. Due to these reasons, HRTEM is not quite effective when quantifying Si nanocluster density. Similarly, DF analysis works only on the right oriented Si nanoclusters. However, the thickness of the specimen in DF is not as crucial as in HRTEM technique. DF imaging allows lower magnifications with respect to HRTEM, resulting a better view of the nanocluster density. Since this technique also works only on Si nanocrystals, the exact number density of Si nanoclusters is generally underestimated.

21.2.3
Electron Energy Loss Spectroscopy

When an EELS is attached to a TEM, chemical mapping of the nanoclusters and the surrounding matrix can be performed. This technique does not observe the

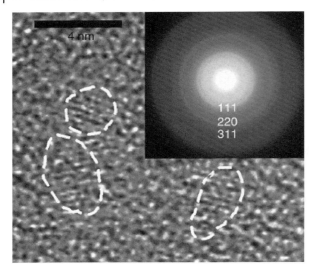

Figure 21.2 HRTEM micrograph of the Si nanocrystals implanted in SiO_2 matrix. Inset shows the electron diffraction pattern resulted from Si nanocrystals.

tailoring of atomic numbers and densities but the chemical phases of the elements existed (i.e., Si and SiO_2). The most important advantage of this technique is that any kind of morphology can be monitored, regardless of whether the nanocluster is amorphous or crystalline. The collective oscillations of the valence electrons are the main fingerprints in EELS for a chemical state, which are known as plasmons. By knowing the characteristic plasmon energies of Si and SiO_2 (17 and 26 eV, respectively) and by having an energy resolution narrow enough, Si and SiO_2 are easily distinguished, regardless of their phases.

There are two methods applicable for the imaging process: (i) monitoring a contrasted area by direct imaging of electrons (EFTEM) and (ii) indirect image forming after gathering the EELS spectra by extracting the distribution of the Si plasmon signal. This method refers to spectrum imaging in a scanning transmission electron microscope-parallel electron energy loss spectrometer (STEM-PEELS) [3].

In STEM-PEELS, the area of interest is scanned by using a probe and the superposition of various scanned trajectories has been done. These trajectories generally have signals both from Si and SiO_2, and the contribution from the SiO_2 signal should be subtracted in order to get a well-defined Si micrograph. For the subtraction of SiO_2, two methods can be used: linear interpolation (LI) and non-negative least square (NNLS) [1].

Although 2D image processing can be done by EELS, 3D modeling may also be possible by combining plasmon loss imaging with electron tomography [4]. This technique involves the reconstruction of 3D distribution of a Si nanocluster from a tilted series of micrographs [5] so that the shape of the nanocluster can be identified (Figure 21.3) [6].

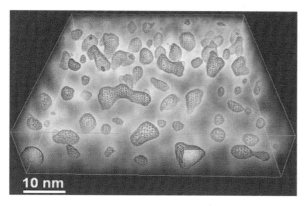

Figure 21.3 3D tilted tomographic reconstruction of Si nanoclusters in silica matrix. Tomographic image shows that most of the Si nanoclusters are nonspherical in shape [6].

21.2.4
Energy Filtered Transmission Electron Microscopy (EFTEM)

By using EFTEM, similar to HRTEM, spatial resolutions down to 1 nm can be achieved. However, unlike the HRTEM, EFTEM has no limitations related to the crystallinity of the Si nanocrystals and can detect both amorphous and crystalline nanoclusters. Moreover, high-energy resolution of EFTEM can make it possible to distinguish Si and SiO_2 plasmon energies.

The raw images obtained from EFTEM are less contrasted than the images obtained from STEM-PEELS. In order to improve the image quality, the contrast enhancement process, that is the subtraction of the contribution from SiO_2 background by yielding a well-defined Si nanocluster containing image should be applied. The contrast quality of EFTEM images after enhancement is well comparable to the STEM-PEELS images (Figure 21.4) [1].

In contrast to EFTEM, STEM-PEELS is more preferable for the detailed analysis of individual specimens. The quantitative analysis of the image is possible by LI and NNLS methods. By these methods, the contribution of Si nanoclusters can be singled out. On the other hand, EFTEM has an advantage when analyzing a number of specimens in a short time. Thus, EFTEM is more suitable for statistical analysis.

In general, it should be noted that the usage of TEM and its variations may induce surface and morphology changes, especially if the surface or the matrix is an insulator as in the case of Si nanoclusters SiO_2/Si_3N_4 matrix. There are various damage processes that may take place such as atomic displacement, ionization, electrostatic charging, sputtering by electron beam, and radiolysis [7]. For long exposure times, the irradiation may chemically reduce some local areas to Si from SiO_2, depending on the electron dose [1]. Similar morphological changes may occur at high-beam currents even if the exposure time is kept short.

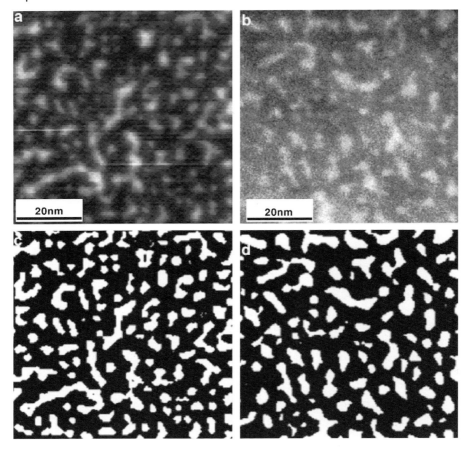

Figure 21.4 STEM-PEELS and EFTEM images of Si nanoclusters in SiO_2 before ((a) and (b), respectively) and after contrast enhancement ((c) and (d), respectively) [1].

21.2.5
Si Nanocluster Imaging by Atomic Force Microscopy

AFM is used for indirect surface processing of the samples. AFM is an alternative and in some cases, an easier method with respect to TEM, especially when probing Si nanoclusters. Si nanoclusters in SiO_2 matrix can be indirectly probed by AFM. Without understanding the shapes, positions, densities, and depth profiles of Si nanoclusters, it is difficult to draw a correct conclusion on the origin of light emission.

In the low saturation case of the Si nanoclusters, they can also be imaged by AFM as an alternative to TEM [8]. The most convenient way of "imaging" Si nanoclusters throughout the SiO_2 is depth profiling by etching [2, 8]. For Si implanted samples, the depth profiling process gives satisfactory data on the size distribution of the

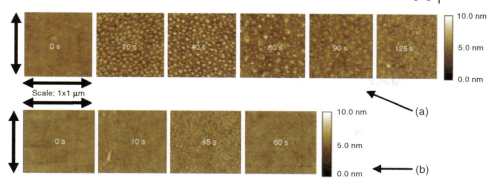

Scale; 1x1 µm

(a)

(b)

Figure 21.5 AFM tapping mode tomography of SiO$_2$ implanted SiO$_2$ samples with respect to etching time for (a) 1×10^{17} cm^{-2} Si ions and for (b) 5×10^{16} cm^{-2} Si ions.

nanoclusters with respect to the implantation projected range (Figure 21.5) [2]. Also, the height of the particle-like objects and nanoclusters can be identified by assuming that the layer on which the nanoclusters are attached is a flat surface.

Generally, the distribution of heights can be measured as an integral function of size distribution of Si nanoclusters. However, there are some difficulties when working on Si nanoclusters. When performing depth profiling, the nanoclusters in the etched layer may still remain attached on the surface. Since the measurements in AFM is always referenced to the main layer, the heights of the nanoclusters corresponding to that layer and the ones still attached on the surface from the upper layer are added up. The resulting image is "seen" as if higher nanoclusters are present in the film. This effect is referred as "memory effect" and may lead to misunderstandings on the height distribution of nanoclusters (Figure 21.6) [2].

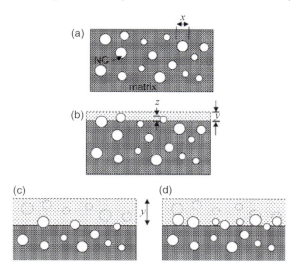

Figure 21.6 Etching illustration of the SiO$_2$ films containing Si nanoclusters. (a) The smooth surface before etching, (b) after etching a thickness of "y" and a nanocluster height of "z", (c) further etching with no memory effect, and (d) with memory effect [2].

There might be some limitations or problems when working on nanoclusters with AFM. The most important phenomenon that affects the reliability and reproducibility of AFM images is the tip convolution effects [9]. Since the AFM measurements rely on the physical interaction between the tip and the sample, the obtained image is definitely a convolution of both. Tip convolution effects are considerable when the size of the sample is comparable to the size of the tip as in the case of Si nanoclusters. There is an unfortunate probability that the image is highly distorted due to the finite size of the tip. If the tip geometry is highly distorted, a repeated pattern is seen as it is on the surface of the sample. In order to prevent these distortions, the tip of the AFM should be calibrated by "assorted" objects that have simple well-defined geometries or by a series of simulations and algorithms [9].

Consequently, in some cases AFM is a much more convenient, easier, and cheaper way for the investigation of the Si nanoclusters and their size variation with respect to TEM. For low dose Si containing SiO_2 films, it is hard to obtain a contrast in the TEM micrograph due to the small size or the amorphous phase of Si nanoclusters. However, a surface profile can be comprehensively done by AFM, and the Si nanoclusters can be detected even for the smallest sizes. Due to this reason, AFM is quite useful for studying Si nanoclusters especially when the partition of nanoclusters is low in the matrix or precipitates in the amorphous state.

21.3
Identification and Quantification of Nanocrystals

21.3.1
Raman Spectroscopy

When monochromatic light of wavenumber $\tilde{v}_0$ is incident on a material, scattering phenomena occurs. The frequency spectrum of scattered light reveals wavenumbers of light equal to $\tilde{v}_0$ and $\tilde{v}_0 \pm \tilde{v}_M$, where $\tilde{v}_M$ is a vibrational frequency of the molecule being investigated. While the former is called as Rayleigh scattering, the latter is called as Raman scattering—named after the discovery of C. V. Raman in 1928. Raman effect occurs due to the interaction of optical and vibrational oscillations and results in a change in vibrational energy of a molecule. A phonon is created (Stokes process) or annihilated (anti-Stokes process) during the Raman scattering [10, 11].

Raman spectroscopy as a fast and nondestructive method is frequently being used to characterize the Si nanocrystals. From the shape and the peak position of the first order Raman scattering band, the following properties of nanocrystals can be investigated.

1) The identification (i.e., determination of the composition of the nanocrystals)
2) The size variations (i.e., estimation of the size of the nanocrystals)
3) The evolution of the stress on nanocrystals (i.e., estimation of the stress on the nanocrystals)
4) The phase changes (i.e., estimation of amorphous to crystalline ratio)

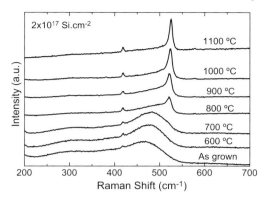

Figure 21.7 The Raman spectra of the samples implanted with a dose of 2×10^{17} Si/cm^2 and annealed at different temperatures [12]. The Raman spectrum of as-implanted sample is also shown for reference.

21.3.2
Identification of the Nanocrystals

Raman spectroscopy is very powerful to identify the presence of nanocrystals in different matrices. As shown in Figure 21.7, Raman spectra demonstrate the formation and the evolution of Si nanocrystals in Al_2O_3 matrix as a function of annealing temperature [12]. Amorphous Si clusters formed in the as-implanted sample transforms into crystalline Si nanostructures with a more intense and narrow Raman signal as the annealing temperature increases. While amorphous Si has a broad transverse optical (TO) band around $480\,cm^{-1}$, bulk Si has a sharp TO band with a natural linewidth of approximately $3\text{–}4\,cm^{-1}$ around $521\,cm^{-1}$ at room temperature. On the other hand, this band for nanocrystalline Si shows a broadening and a shift to lower wavenumbers due to the phonon confinement effect [12–17, 20]. A comparison of the Raman spectra shown in Figure 21.7 indicates the nanocrystal formation starts at temperatures between 700 and 800 °C for the sample with an implantation dose of 2×10^{17} Si/cm^2 [12].

21.3.3
Size Estimation of the Nanocrystals

Conservation of momentum in crystalline structures results in a narrow Raman line in spectrum. In other words, wave vectors of photons are much smaller than that of phonons. Therefore, phonons with wave vectors $k \approx 0$ can only participate in the Raman scattering. However, conservation rule does not apply to amorphous structures due to lack of long-range order. On the other hand, in nanocrystals, phonons are localized in small crystallites and the momentum is no longer well defined, which according to the uncertainty principle enables phonons with $q \neq 0$ to contribute to the Raman process. It is well known that the optical dispersion curves are flat for low k

values and gets smaller for larger ones. Thus, an asymmetric broadening and a redshift are observed in the first order Raman spectra of nanocrystals.

Various models are proposed in the literature for size calculation using Raman lineshape. A model, WRL, developed by Richter, Wang and Ley; modified by others has widely been used in the literature [13–17, 20]. This model stands on the multiplication of the wave function in an infinite crystal by a weighting function $W_D(r)$. Gaussian and Sinc weighting functions have been successfully applied and widely reported in the literature to find the dimensions of Si nanocrystals [13–17, 20]. However, the choice of the weighting function is arbitrary and results in quite different values at low dimensions of nanocrystals [15, 16]. The model has been modified and improved in recent years. For example, Paillard *et al.* [15] suggested maintaining the weighting function constant as Sinc function and adjusting the optical phonon dispersion relation by taking the anisotropy of the phonon dispersion curves. According to the modified model, the line shape of the Raman spectrum $I(\omega)$ is defined as [15]

$$I(\omega) \propto A \times [n(\omega) + 1] \times \int |C_D(q)|^2 \frac{\frac{\Gamma_0}{\pi}}{(\omega - \omega(q))^2 + \left(\frac{\Gamma_0}{2}\right)^2} dq \qquad (21.1)$$

where, $\omega(q)$ is the phonon dispersion of the bulk material, Γ_0 is the natural Raman line width of bulk material, $C(0,q)$ is the Fourier coefficient of the phonon confinement function, A is a constant, and $[n(\omega) + 1]$ is the Bose–Einstein factor. Furthermore, dispersion relation of optical phonons in nanocrystals $\omega(q)$ is assumed to be the same with bulk Si.

$$\omega^2(q) = 1.714 \times 10^5 + 1 \times 10^5 \cos\left(\frac{\pi q}{2}\right) \quad [20] \qquad (21.2)$$

or

$$\omega^2(q) = 522^2 - \frac{126,100q^2}{q + 0.53} \quad [15, 17] \qquad (21.3)$$

Here, q is expressed in units of $2\pi/a$, a is the lattice constant (0.543 nm) of Si, and Γ_0 is approximately $4\,\mathrm{cm}^{-1}$ and depends on the system configuration.

Furthermore, Islam and Kumar improved the RWL by including the size distribution of Si nanocrystals σ to find $I(\omega)$ [17].

$$I(\omega) \propto \int_q \left[1 + \frac{(\sigma q)^2}{\alpha}\right]^{-1/2} \frac{\exp\left(\frac{-q^2 L^2}{2\alpha^2}\right)}{(\omega - \omega(q))^2 + \left(\frac{\Gamma_0}{2}\right)^2} d^3 q \qquad (21.4)$$

where, a is a constant representing the degree of phonon confinement and L is the mean size of the nanocrystals.

In another study, Zi *et al.* correlated the Raman shift of the Si nanocrystals with their dimensions by a microscopic model, bond polarizability. The Raman shift $\Delta\omega$ is defined as [17]

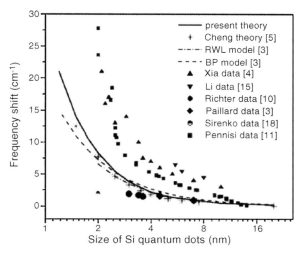

Figure 21.8 Plot of the Raman frequency redshift as a function of the QD size, according to the resent model. Also reported are the BP model and the RWL model. Experimental data from several authors are displayed for comparison [18].

$$\Delta\omega = -A \times \left(\frac{a}{L}\right)^{\gamma} \tag{21.5}$$

where A and γ are the fit parameters. The authors found that the latter one tends to be equal to 1.44 for Si spheres and 1.08 for Si columns from their fit.

A further modification in RWL model has been recently made by Faraci *et al.* using a specific spatial correlation function [18]. Figure 21.8 shows the study of Faraci *et al.* in which they calculated the Raman frequency redshifts due to quantum confinement for Si nanocrystal diameters between 1.2 and 100 nm using the models so-called bond polarizability [16], Cheng theory [19], RWL [13] and Faraci [18] (denoted as present theory in Figure 21.8). Moreover, they reprinted some of the data from the literature. While all the models show good agreement with the experiment at large sizes, they predict lower shift for small sizes, which can be related to the surface effects of nanocrystals such as stress [18].

21.3.4
Stress Estimation on Nanocrystals

It is now well established that the phonon confinement model predicts a redshift and a broadening in the first order Raman signal. However, in order to fully describe the Raman process in nanocrystals, the effects of stress on nanocrystals should be taken into account since the tensile and compressive stresses cause a redshift and blueshift, respectively.

In a stress-free diamond-type material such as Si and Ge, the optical phonons located at $k \approx 0$ are triply degenerated (F_{2g}) due to the cubic symmetry of the crystal. However, the introduction of stress breaks the cubic symmetry and splits the triplet.

The resultant stress can be estimated from the stress-induced wavenumber shift, $\Delta\omega_s$ [12, 21, 22], as

$$\sigma\,(\text{MPa}) \approx C \times \Delta\omega_s\,(\text{cm}^{-1}) \tag{21.6}$$

where C is a constant related with the elastic compliance and the deformation potential constants of the material [22, 23]. C is also stress direction dependent and approximately equal to 210 and 230 MPa for isotropic and biaxial stress, respectively.

21.3.5
Phase Determination of the Nanocrystals

It has been observed that the optical and electrical properties of amorphous Si nanoclusters and crystalline Si nanocrystals are quite different [24, 25]. On the other hand, as shown in Figure 21.7, transition from amorphous to crystalline form is gradual; therefore, quantification of crystalline volume fraction is significant to understand the properties of Si nanocrystals. The fraction can be calculated from the integrated intensities of the amorphous to crystalline peaks, I_a and I_c, respectively [26].

$$X_c = \frac{I_c}{I_c + \varrho I_a} \tag{21.7}$$

where ϱ is the ratio of the Raman efficiencies for crystalline to amorphous Si, and it can be described as

$$\varrho(L) = 0.1 + \exp\left(\frac{-L}{250}\right) \tag{21.8}$$

where L (in Å units) is the nanocrystal size. However, this relation is only valid when $L > 3$ nm [27].

21.3.6
X-Ray Diffraction Analysis of Nanocrystals

XRD in classical sense can be defined as elastic scattering of X-rays by the electrons of atoms. In other words, an electron in the path of incoming X-ray is excited to periodic vibrations by the changing field. Therefore, it behaves as a source of electromagnetic waves of the same frequency. In crystals, the scattered X-ray can be considered as centers for series of spherically spreading waves, which forms zero-, first-, second-, or higher order diffracted beams in certain directions. The theory and mathematical representation of XRD by a simple lattice was studied in detail by M. V. Laue [28]. Later, W. L. Bragg [29] developed the theory and found that the diffracting plane is a lattice plane, and stated the formula, which is known with his name.

$$n\lambda = 2d\sin\Theta \tag{21.9}$$

where n is an integer representing the orders of the diffraction, λ is the wavelength of the X-ray, Θ is the half of the deviation of the diffracted beam, and d is the interplanar spacing for a plane that is given by

$$d = \frac{a}{\sqrt{h^2 + k^2 + l^2}} \qquad (21.10)$$

where a is the constant spacing between the atoms, and h, k and, l are the corresponding Miller indices.

Scherrer showed that the coherence length, L, which represents the mean size of the crystallites forming a powder, is related to the pure X-ray diffraction broadening, β, by the equation [30].

$$L = \frac{\varkappa\lambda}{\beta\cos\Theta} \qquad (21.11)$$

where $\varkappa$ is Scherrer constant and approximately equal to unity, Θ is the Bragg angle, and λ is the wavelength of X-ray.

This formula has become more popular with the advances in micro- and nanotechnology. Recently, many studies have focused on the calculation of the mean dimension of nanocrystals using Scherrer's formula [12, 31–34]. For example, the X-ray diffractograms for Si_yO_{1-y} ($y = 0.45$) samples annealed at 900, 1000, and 1100 °C are demonstrated in Figure 21.9 [32]. The peaks located at approximately 22.0°, 28.3°, 47.5°, 55.9° are attributed to amorphous SiO_2 and Si (111), 220, and (311), respectively [32]. Comedi $et\ al.$ calculated the Si nanocrystals sizes from these diffractograms and obtained the activation energies using a diffusion model and a

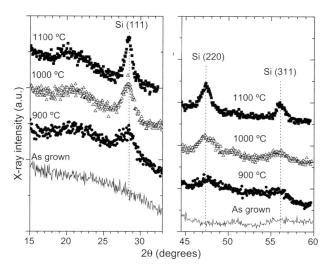

Figure 21.9 X-ray diffractograms for Si_yO_{1-y} ($y = 0.45$) samples fabricated by IC-PECVD and annealed at 900, 1000, and 1100 °C for 2 h. The expected positions of Bragg peaks for crystalline Si (111), (220), and (311) are indicated with dotted lines [32].

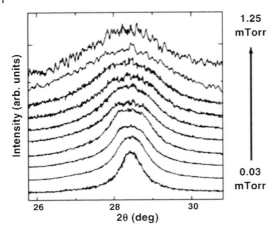

Figure 21.10 Normalized Si (111) Bragg peak for Si nanocrystals in SiO$_2$ matrix fabricated by reactive laser deposition varying the oxygen pressure between 0.03 and 1.25 mTorr [33].

model based on Oswald ripening theory [32]. In another study, XRD measurements were performed on SiO$_2$ films containing Si nanocrystals fabricated by reactive pulsed laser deposition under low oxygen pressures and subsequent annealing. Normalized Si (111) X-ray diffraction peak as a function of oxygen pressure is given in Figure 21.10. The dimensions of the formed nanocrystals calculated using Scherrer's formula altering between 2.6 and 13.3 nm [33].

Size calculation using Scherrer's formula has advantages of being simple in instrumentation, nondestructive, and easy in data manipulation; however, it has some physical and mathematical limitations. In the Scherrer's formula, λ and Θ are two constants that can be measured precisely. However, the determination of $\varkappa$ and the measurement of β values are not simple. The $\varkappa$ value, which depends on several factors (e.g., nanocrystal shape and size), differs in several publications [12, 32]. The widely accepted value of $\varkappa$ for cubic crystals is 0.94 [35]. The $\varkappa$ value is strongly size dependent for nanocrystals with dimensions less than 3 nm. Therefore, the validity of the formula is questionable below this size. Moreover, the formula becomes more questionable when complicated nanocrystal shapes, bimodal, and/or size distributions are present [35]. On the other hand, experimentally measured β value is a convolution of the physical broadening due to the finite grain size and lattice distortion as well as the instrumental broadening due to the response of the experimental setup being used. Therefore, a deconvolution procedure should be followed [35, 37].

FW$\frac{1}{5}$/$\frac{4}{5}$M method is proposed as an extension of the Scherrer's formula to find the grain size and the related mean size distribution from a single measurement [36]. However, it is hard to apply this method to Si nanocrystals embedded in thin films dielectrics where the signal is generally weak. The small number of the data point at the top and high noise level at the bottom of the peak increases the error bar of the estimation.

It is known that the XRD lines of the stressed materials shift or exhibit asymmetrical and broadened line profiles depending on the magnitude of the stress [31, 32, 34]. Therefore, misinterpretation may arouse from the fact that both stress and the decrease in coherence length can cause broadening. Method of integral breadths [35] and Warren–Averbach [37] can be used to calculate the effect of both finite grain size and lattice distortion at the same time since the broadening due to the lattice distortion is diffraction order dependent unlike the grain size. Both the methods require the precise measurement of several diffraction lines. In the case of Si nanocrystals embedded in dielectric matrices Si (111), (220), and (311), Bragg peaks are commonly observed [12, 31–34]. However, the signal due to the Bragg peaks of Si (220) and (311) are generally weak to perform these methods accurately.

21.3.7
Method of Integral Breadths

Bragg line broadening β_S generated by finite grain size can be calculated from Scherrer's equation as described above [30, 35].

$$\beta_S(2\Theta) = \frac{\varkappa\lambda}{L\cos\Theta} \quad \text{or} \quad \beta_S(s) = \frac{\varkappa}{L} \tag{21.12}$$

where $\beta_S(2\Theta)$ and $\beta_S(s)$ are the broadening of Bragg peak in 2Θ and s-axis due to the finite size, respectively, where $s = 2\lambda\sin\Theta$ On the other hand, the Bragg line broadening originated only from lattice distortion can be expressed as [38]

$$\beta_D(2\Theta) = 4e\tan\Theta \quad \text{or} \quad \beta_D(s) = 2es \tag{21.13}$$

where $e \approx (\Delta d/\bar{d})$, which is approximately the upper limit of the lattice microstrain and related with the root mean square strain $1.25\langle\varepsilon^2\rangle^{1/2}$ [39]

When the size and strain broadening are simultaneously present, integral breadths method can be performed to calculate the broadening due to each mechanism. The assumption, both effects have bell-shaped profiles can be described either Gaussian or Cauchy, used to calculate the instrumental broadening can be applied.

$$\beta = \beta_S + \beta_D = \frac{\varkappa\lambda}{L\cos\Theta} + 4e\tan\Theta \quad \text{C–C} \tag{21.14}$$

$$\beta^2 = \beta_S^2 + \beta_D^2 = \left(\frac{\varkappa\lambda}{L\cos\Theta}\right)^2 + (4e\tan\Theta)^2 \quad \text{G–G} \tag{21.15}$$

$$\frac{\varkappa\lambda}{\beta L\cos\Theta} + \left(\frac{4e\tan\Theta}{\beta}\right) = 1 \text{ or } \frac{\beta^2}{\tan^2\Theta} = \frac{\varkappa\lambda}{L}\left(\frac{\beta}{\tan\Theta\sin\Theta}\right) + 16e^2 \quad \text{C–G} \tag{21.16}$$

Broadening is better described by Gaussian. A comprehensive discussion and comparison on the validity of these methods can be found in Ref. [35].

21.3.8
Method of Warren–Abervach

It is based on the Fourier transformation of the physically broadened profile. XRD line profile can be expressed a Fourier transformation of crystalline structure in the reciprocal space [37, 39].

$$f(s) \propto \sum_{L=-\infty}^{\infty} A(L)\cos\left[2\pi(s-s_0)L\right] + B(L)\sin\left[2\pi(s-s_0)L\right] \tag{21.17}$$

where $A(L)$ and $B(L)$ are cosine and sine Fourier coefficients, respectively. L is related with the column length of the unit cells perpendicular to the diffracting planes corresponding to the Bragg peak. The $s = 2\sin\Theta/\lambda$ is the reciprocal x-axis, and the XRD peak is centered at s_0.

Warren showed that cosine coefficient A_L is a product of a size coefficient and a lattice distortion coefficient.

$$A_L = A_L^D \times A_L^S \tag{21.18}$$

where size coefficient A_L^S is independent of the order of s whereas, strain coefficient A_L^D is dependent. Since

$$A_L^D = \left\langle \cos\frac{2\pi L \varepsilon_L}{a} h_0 \right\rangle \approx \left(1 - \frac{2\pi^2 L^2 < \varepsilon_L^2 >}{a^2} h_0^2\right) \tag{21.19}$$

Equation (21.18) can be written as

$$\ln(L, h_0) = \ln A^s(L) - \frac{2\pi^2 L^2 < \varepsilon_L^2 >}{a^2} h_0^2 \tag{21.20}$$

a is the lattice constant, n is the Fourier order, $h_0^2 = h^2 + k^2 + l^2$ where h, k, and l are Miller indices, and $\langle \varepsilon_L^2 \rangle$ is the mean square strain perpendicular to the reflection planes. $L = n\lambda/2(\sin(\Theta_2 - \Theta_1))$, $\Theta_2 - \Theta_1$ is the Bragg angle interval.

From the plots of $A(L, h_0)$ versus L^2, the size Fourier coefficients are obtained from the intercepts, whereas the strain Fourier coefficients can be obtained from the slopes. For example, Kohli *et al.* applied the Warren–Averbach method to calculate the Si nanocrystals size and root mean square of the strain on samples fabricated by thermal oxidation of amorphous Si: H films [40].

21.3.9
X-Ray Photoelectron Spectroscopy

In XPS, X-ray is used to eject the electrons of an atom from its core shell and the emitted electrons are analyzed according to their kinetic energies. The method is one of the most widely used surface analysis technique, and it supplies information about the atomic composition of the surface and chemical environments (i.e., binding states, oxidation states) of the elements, except Hydrogen and Helium [41]. When the incident photon interacts with an electron with a binding energy of E_B in a core shell, the photoelectric effect occurs if the energy of the photons is greater than the binding energy. The energy of the photon is partly used to remove the electron from the core shell, and the rest is transferred to electron as kinetic energy (E_{kin}). Since the Fermi level, by definition, corresponds to zero binding energy, some more energy (work function, Φ) must be spent to extract to electron to the vacuum level. The kinetic energy of the emitted electron can be written as [42]

$$E_{kin} = h\upsilon - E_B - \Phi \qquad (21.21)$$

It should be noted that the determination of atomic percentage and chemical environment requires a successive quantification study (i.e., background subtraction and peak fitting). Shirley [43] and Tougaard [44] developed two methods that are widely used for background subtraction. Moreover, Voigt function is commonly applied to fit the XPS lines [45–47].

Si 2p and Auger KLL bands have been studied to understand the evolution, oxidization, charging, and quantum size effect of Si nanocrystals. Si 2p peak is a doublet and $2p_{1/2}$ is located 0.6 eV higher energies than $2p_{3/2}$, whereas its intensity approximately equal to the half of $2p_{3/2}$. Si 2p band can be decomposed in 5 peaks as Si^{+n} ($n = 1, 2, 3, 4$ and 5) representing Si, Si_2O, SiO, Si_2O_3, and SiO_2 chemical structures, respectively. Peak position for Si—Si bonds is located around 99.5 eV [48, 56]. The peak positions of oxidization states Si_2O, SiO, Si_2O_3, SiO_2 shift from this value approximately by 0.93, 1.73, 2.56, 3.88 eV, respectively, and their FWHM increase with the same order for thin film of SiO_2 formed on Si [45, 46, 48]. According to an outstanding research from Renault *et al.* [46], although the peak positions of Si^{+n} stated does not change noticeably, their FWHMs for Si nanocrystals on Al_2O_3 are larger than thin films of SiO_2 on Si as shown in Figure 21.11. They attributed this broadening to the stress due to the variations of the Si—O bond lengths and Si—O—Si bond angles in oxides on Si nanocrystals. On the other hand, they observed an increase in Si° signal in the spectrum taken 320 eV photons due to the higher penetration depth of photoelectrons. Moreover, the shift of -0.5 eV in binding energy is observed. This shift was explained by pinning of the Fermi level close to the conduction band maximum of Si and can be observed with photons of 160 eV thanks to enhanced surface sensitivity [46].

XPS is employed in nanostructure analysis to find the relative concentrations of elements, to monitor the Si—SiO_2 phase separation of Si nanocrystals with SiO_2 host matrix, and to investigate the chemical states of their interfaces [45–47, 50]. In 1996, Min *et al.* [50] investigated the effect of Si concentration, annealing temperature, and time on PL spectra of Si nanocrystals. They used XPS and TEM cooperatively to prove the existence of Si nanocrystals. Increase in the Si—Si 2p was correlated to the increase in the Si—Si bonds, which can be associated with the Si nanocrystals density. They observed an increase in PL intensity with Si—Si bond density, which increases with implantation dose, annealing temperature and time [50]. It is also possible to quantify the oxidization states by using a successive fitting procedure. A decrease in the SiO_x peaks and increase in the SiO_2 and Si related peaks are accepted as the formation of Si nanocrystals in SiO_2 matrix [7, 10, 11]. Liu *et al.* [51] implanted 1 keV energetic Si ions into 30 nm SiO_2 films grown onto Si (100) and subsequently annealed at 1000 °C for 20 min. XPS spectra of the as-implanted sample and the annealed sample are shown

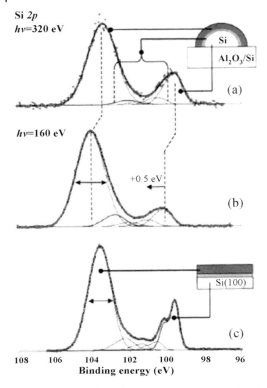

Si *2p*
hv=320 eV

Si
Al$_2$O$_3$/Si
(a)

hv=160 eV

+0.5 eV
(b)

Si(100)
(c)

108   106   104   102   100   98   96
Binding energy (eV)

Figure 21.11 High-resolution Si 2p XPS spectra of oxidized Si nanocrystals grown onto alumina taken at a photon energy of (a) 320 eV and (b) 160 eV; (c) oxidized bulk Si substrate at a photon energy of 160 eV. The different chemical regions of the samples with their corresponding contribution to the spectrum are given in the inset schematics. Pure Si, pure SiO$_2$, and suboxides SiO$_x$ are represented by white, dark gray, and gray, respectively [46].

in Figure 21.12. Although a phase separation was monitored, the decrease in Si–Si peak is attributed to the oxidization of Si nanocrystals [51].

The oxidization of different sizes of Si nanocrystals in O$_2$ and NO ambient as a function of annealing time, temperature was monitored using XPS and TEM in controlled experiment by Scheer *et al.* [52]. They concluded that oxidization in O$_2$ atmosphere is self limited below 850 °C and at all temperatures in NO. The reason for the former one is the compressive stress normal to the Si/SiO$_x$ interface. Moreover, the latter one is due to the incorporation of N in the interface layer between SiO$_2$ and air or in the shell oxide of the Si nanocrystals [52].

Recently, Gillet *et al.* have published an interesting study on size nanocharacterization of the core shell nanoclusters using XPS [53]. Assuming that a nanocrystal has a spherical core and surrounded by a uniform spherical shell and using an analytical model, they propose a general equation for XPS intensity of a nanocrystal with a diameter smaller than or equal to the XPS probing depth ~10 nm [53].

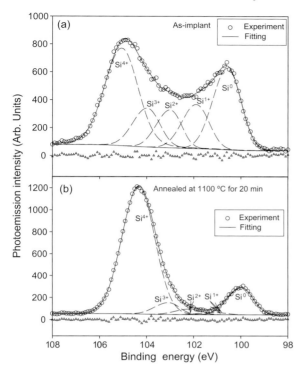

Figure 21.12 The deconvolution of the Si 2p peak for (a) as-implanted sample and (b) the sample annealed at 1000 °C for 20 min [51].

Si nanocrystals are usually formed in thick ∼100–1000 nm host materials such as SiO_2, Al_2O_3, and Si_3N_4 while the penetration depth of the photoelectrons is ∼10 nm. Therefore, it is important to investigate the depth profile of the films, which are inhomogeneous in depth such as multilayers. Wet or dry etching (traditionally Ar sputtering) has been used alternatively [47, 54, 55]. Although sputtering used to remove the material from the surface creates some artifacts on the surface such as atomic mixing, knock on implantation, bond breaking, and preferential sputtering, reliable depth profiling can be obtained with this technique. Use of low energy ions and ion species with higher masses (i.e., Xe, Kr) can minimize some of these artifacts [54]. XPS depth profile of Si nanocrystals embedded in Al_2O_3 matrix to observe both the formation and charging mechanism of Si nanocrystals was performed by Yerci *et al.* as shown in Figure 21.13 [47]. The Gaussian-like depth profile is due to the implantation of 100 keV energetic Si ions and matches with the secondary ion mass spectrometry profile. As seen in Figure 21.13, when nanocrystals formed in Al_2O_3 similar to SiO_2, suboxide peaks decrease and Si° peak increases. However, unlike SiO_2, Si^{4+} peak decreases, as well [47]. This decrease suggests that Si-nc shell is mainly formed from suboxides and all the detectable SiO_2 bonds are replaced to form the shell of Si nanocrystals and Al_2O_3 matrix [47]. In a similar study, Liu *et al.* performed a depth

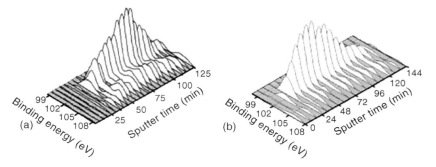

Figure 21.13 XPS depth profile of Si 2p signals for (a) as-implanted sample and (b) the sample annealed at 1000 °C for 2 h [47].

profile of SiO_2 layer with Si nanocrystals to investigate the charging effect of the nanocrystals [56]. Both Yerci *et al.* and Liu *et al.* observed less charging at the depth where dense Si nanocrystals layer is located [47, 56].

Chen *et al.* measured a decrease of 0.6 eV in the binding energy of the core electrons of nanocrystalline Si° with respect to crystalline Si°, which is in the opposite direction than expected from the quantum confinement effect [57]. They used the C 1s signal exists due to the carbon contamination on the surface of the films as reference. The same amount of decrease was also reported by Aykutlu *et al.* using the final state Auger parameter, which is free from charging effects [58]. In this study, the shift in the Si° binding energy of Si nanoclusters was attributed to the relaxation energy differences. Recently, Thogersen *et al.* have shown that the amount of the shift is different for amorphous and crystalline phases, as well as varies with the size of the Si nanoclusters. They analyzed a set of samples with various fabrication and post-annealing treatments using initial and final state Auger parameters, and Wagner diagram [49]. The authors designated that the shift is ruled by initial state rather than final state. In another recent study, Kim *et al.* performed synchrotron radiation XPS and measured an increase in the binding energy of Si° signal [45]. In order to explain the shift, they obtained the quantum confinement effect from PL spectra for different sizes of Si nanocrystals, and calculated the theoretical shifts based on the screening and the strain effects. They believe that the tensile strains of the embedded Si nanocrystals are the most probable reason for the shift [45].

21.4
Looking at the Nanocluster's Surrounding

21.4.1
Fourier Transform Infrared Spectroscopy

FTIR spectroscopy that trades with the absorption of infrared light in the vibrational states of the investigated species is widely used to characterization of the semiconductor materials. FTIR absorption of molecules basically depends on the change in the dipole moment. Therefore, FTIR is very sensitive to the chemical compositions and the structural variations of the species. The structural variations of SiO_x matrix during the formation of Si nanocrystals can be characterized using FTIR [101]. In a recent study, the Si−O−Si asymmetric stretching mode has been deconvoluted to monitor the evolution of SiO_x matrix during the annealing process. The integrated area and the shift in the SiO_x peak positions are found to be well correlated with the change in the film stoichiometry and nanocrystal formation. It has been shown that the nonstoichiometric SiO_x matrix turns into stoichiometric SiO_2 as the excess Si atoms precipitate to form nanocrystals.

The samples containing excess Si ions were prepared by implantating ^{28}Si ions into 250 nm thick SiO_2 films grown thermally on both sides of a double polished Si substrate. The implantation dose and the implantation energy of Si ions was 1.5×10^{17} ions/cm^2 and 100 keV, respectively. The implanted Si ions have a Gaussian distribution with a peak position of 140 nm from the surface calculated using SRIM program. The samples were annealed at 800–1200 °C for 2 under N_2 atmosphere in quartz furnace to induce the nucleation of Si nanocrystals.

Three main well-known transverse optical bands can be observed in FTIR spectrum of SiO_2 located at around 459 cm^{-1} (TO_1), 815 cm^{-1} (TO_2), and 1080 cm^{-1} (TO_3) corresponding to the rocking, bending, and asymmetric stretching (AS) modes of Si−O−Si, respectively [59, 60]. In addition to these peaks, longitudinal optical (LO) vibrational modes at 1253 (LO_3), 1162 (LO_4), and TO mode at 1205 (TO_4) cm^{-1} were recorded at an oblique incident angle of IR light of 40° [61].

Figure 21.14 shows the variation of the FTIR spectra with annealing temperature for Si implanted SiO_2 films. It is known that AS vibration of SiO_2 located at 1080 cm^{-1} in the IR spectrum is very sensitive to the structural variation of the matrix [61, 62]. Its intensity, peak position, and the full width at half maximum (FWHM) strongly depend on the stoichiometry of the matrix [62]. It is seen from Figure 21.14 that a

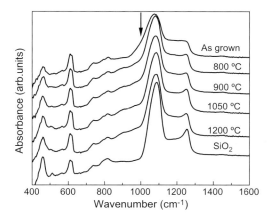

Figure 21.14 FTIR spectra of Si implanted SiO_2 films on Si substrate annealed at various temperatures [101]. The FTIR spectrum of as-implanted SiO_2 is shown as reference. The arrow shows the Si−O−Si AS band for nonstoichiometric silicon oxide.

shoulder (highlighted with an arrow) appears at the lower wavenumber side of the Si−O−Si AS band belongs to SiO$_2$ in the as-implanted sample and in the samples annealed at low temperatures. This shoulder is commonly attributed to the Si−O−Si AS band in nonstoichiometric silicon oxide (i.e., SiO$_x$ ($x < 2$)) [64, 68, 72]. The peak shape and the position of the Si−O−Si AS vibration of SiO$_2$ implanted with Si gradually approach to those of pure SiO$_2$ with the annealing temperature.

In order to understand the evolution of the matrix with annealing, a deconvolution procedure was applied to the samples in the range between 900 and 1400 cm^{-1}. In this process, the position of Si−O−Si asymmetric stretching band of stoichiometric SiO$_2$ was fixed to 1080 cm^{-1} assuming that this band is present in all annealing temperatures due to the presence of unimplanted SiO$_2$ at the other side of the Si substrate. Thus, one can observe the variation in the Si−O−Si AS band of SiO$_x$ with $x < 2$ with respect to the fixed stoichiometric oxide peak. The peak position of thermally grown stoichiometric SiO$_2$ (1080 cm^{-1}) is well known from many studies reported previously [65, 66] and verified using the reference sample whose FTIR spectrum, and its deconvolution is given in Figure 21.15a. Other three peaks obtained from the deconvolution process were allowed to vary in peak position and width as the samples represent a dynamical system varying with the annealing process.

FTIR spectra and deconvoluted peaks for the as-implanted sample with Si at a dose of 1.5 × 10^{17} ions/cm^2 are shown in Figure 21.15b. The Si−O−Si AS band related with nonstoichiometric SiO$_x$ is distinguishable from the one related with the stoichiometric SiO$_2$ and positioned at 1017 cm^{-1}. Formation of the SiO$_x$ with $x < 2$ is expected as a result of excess Si implanted and the recoil of Si and O atoms during the implantation process. The presence of excess Si in the oxide may also generate new chemical bonds such as Si−SiO$_3$, Si−Si$_2$O$_2$, and Si−Si$_3$O and some individual dangling bonds [64, 67–69].

The variation of the SiO$_x$ peak resolved from the deconvolution process is analyzed by monitoring its peak position and integrated area. As shown in

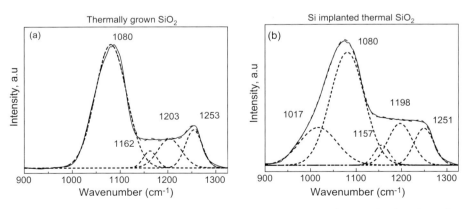

Figure 21.15 The deconvolution of the Si−O−Si asymmetric stretching mode of (a) thermally grown SiO$_2$, and (b) ^{28}Si implanted thermally grown SiO$_2$ at a dose of 1.5 × 10^{17} ions/cm^2 [101].

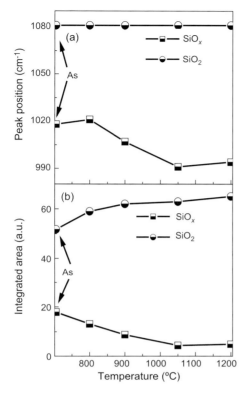

Figure 21.16 (a) The shift in the peak position and (b) the variation of integrated area of Si−O−Si asymmetric stretching band of SiO_2 and SiO_x ($x < 2$) with annealing of SiO_2 sample implanted with 1.5×10^{17} Si/cm^2 [101].

Figure 21.16a, the integrated area of the peak corresponding to the nonstoichiometric SiO_x ($x < 2$) decreases while that arise from the Si−O−Si AS band of stoichiometric SiO_2 increases with annealing temperature. The latter one is clearly related with the recovery of the SiO_2 matrix with annealing. On the other hand, the former one can be related with the nucleation of additional atoms to form nanoclusters since Si−Si bonds are FTIR inactive [63]. This argument can be further justified since the peak corresponding to the nonstoichiometric SiO_x ($x < 2$) shifts to lower wavenumbers with annealing temperature as it is shown in Figure 21.16a). It is known that the peak position of the Si−O−Si AS mode of SiO_x band shifts to lower wavenumbers with increase in excess Si atoms, and reaches $965\ cm^{-1}$ for $x = 1$ (SiO) [64, 66]. Therefore, the shift in the peak position of the Si−O−Si AS mode of SiO_x toward lower values is directly related to the increase in the Si content of the SiO_x matrix [71]. As a result, one can argue that Si-rich regions get more Si rich and forms Si nanoclusters while oxygen-rich areas become more oxygen rich and form stoichiometric SiO_2 matrix between the nanoclusters [67, 70, 72, 73].

21.5
Optical Techniques

21.5.1
Photoluminescence Spectroscopy

Photoluminescence spectroscopy is a powerful tool used for the characterization of semiconductors. It is a simple, versatile, contactless, and nondestructive method of probing the electronic structure of materials. In a PL spectroscopy experiment, excitation is provided by laser light whose energy is typically larger than the optical bandgap. The photoexcited carriers are created, which relax toward their respective band edges or radiative centers and return back to a lower energy level.

For two decades, characterization of nanostructures with PL spectroscopy has become a very active field of science. Particularly, characterization of Si nanostructures with PL spectroscopy has attracted huge attention due to their potential applications in optoelectronics and photonics. The first observation of strong visible PL from porous Si [74] and from Si nanocrystals embedded in SiO_2 [75–77] initiated remarkable studies on low dimensional Si materials. When the size of the Si nanocrystals decreases, bandgap widens and consequently the separation of the energy levels increases. This phenomenon is called as "quantum confinement effect." PL peak position can then be altered with the size of the Si nanocrystal. Many studies showed that QCE plays an active role in efficient visible PL from porous Si and Si nanocrystals [78–82, 88]. Although the quantum confinement of the carriers in Si nanocrystals is the general elucidation for the strong PL, there are still some controversial approaches. A model states that PL originates from the localized defects at the Si/SiO_2 interface [78, 83–85], while the other assigns that the red band is due to quantum confinement supplemented by surface states. The blue band is generally connected to the presence of oxide related defects and the infrared band is related with dangling bonds or bandgap luminescence in large crystallites [86]. Recently, the general agreement about the origin of PL from Si nanocrystals is that both localized defects at the interface and the quantum confinement of excitons play important roles but it is difficult to explain the mechanism of the luminescence in detail [84, 87].

Figure 21.17 shows PL spectra of samples implanted with different doses of ^{28}Si into SiO_2 substrate. [88]. It can be seen that PL spectra is blueshifted as the implantation dose decreases and PL peaks are broadened due to size distribution. The authors explained the shift of the PL peak position by QCE.

PL dynamics of Si nanocrystals strongly depend on the temperature. While the temperature decreases, PL peak shifts to shorter wavelengths and PL intensity shows a bell shape with its peak position at approximately 100 K. The temperature dependence of the PL peak position and its intensity imply that the mechanisms of PL are different at low and high temperatures [89, 90]. In order to explain temperature dependence of the PL intensity and the peak position, Y. Kanemitsu *et al.* proposed three region models [78]. According to the model, active interface state is formed between c-Si and the SiO_2 layer. The photo

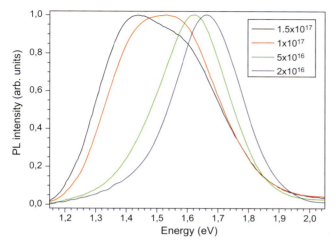

Figure 21.17 The PL spectra of samples implanted with different doses of ^{28}Si into SiO$_2$ substrate. Samples are annealed at 1050 °C Sample labels (Si ion dose) are A (1.5×10^{17} cm^{-2}), B (1×10^{17} cm^{-2}), C (5×10^{16} cm^{-2}), and D (2×10^{16} cm^{-2}) [88].

generation of the excitons mostly occurs in the c-Si core. Some of the excitons in the core are localized in the localized states on the nanocrystal surface. The radiative recombination of excitons confined in the interfacial layer results in strong PL [78, 90, 91].

In addition to continuous wave PL, time-resolved PL spectroscopy is employed to explore the dynamical properties of optical emission from Si nanocrystals. Numerous studies have been reported on the PL decay characteristic of porous Si [92–94] and Si nanocrystals [95, 96]. The time-resolved experiments of Si nanocrystals indicate that the PL decay is not single exponential and instead can be explained with a stretched exponential function, which is used to describe dispersive process in disordered systems [97]. J. Linnros *et al.* [98] have presented time-resolved PL decays and the results from fitting of stretched exponentials for porous Si and Si nanocrystals embedded in SiO$_2$ as shown in Figure 21.18 [98].

The temperature dependence of the lifetime of Si nanocrystals provides significant information about the dynamics of the PL mechanism. Calcott *et al.*, 1993 proposed a two level model to explain the increase in the PL lifetime with decreasing temperatures. According to this model, the exciton level is split into two levels; the lower level is the triplet state and the upper level is the singlet state. At high temperatures, both states are populated. In contrast, at low temperatures only the triplet state is occupied. Since the lifetime of triplet state (approximately millisecond range) is longer than the singlet state (approximately microsecond range), the longer lifetime at cryogenic temperatures can be explained with QCE in the Si nanocrystals. Later, M. Dovrat *et al.* [99] proposed that the environment of the nanocrystals affects the lifetime and the dispersion exponent of the lower triplet state. Therefore, the lifetime is dominated by the nonradiative processes in the matrix [99] and PL intensity can dramatically be enhanced by the exclusion of nonradiative channels.

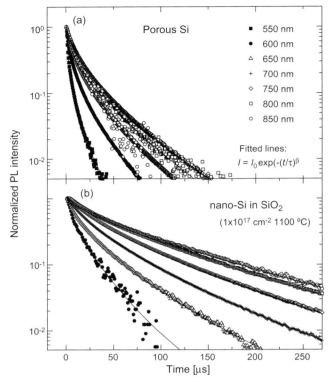

Figure 21.18 Normalized time resolved PL decay of porous Si (a) and Si implanted SiO$_2$ (b) (λ_{exc} = 355 nm). Solid lines are stretched exponential fit whose formula is given in (b) [98].

Recently, S. Godefroo *et al.* performed electron spin resonance (ESR) and magneto-PL experiments to determine the origin of the PL from Si nanocrystals embedded in SiO$_2$ [100]. ESR analysis showed that numerous defects that are nonradiative recombination centers placed between the Si nanocrystals and the surrounding SiO$_2$ exist. The density calculation of nanocrystals and defects demonstrated that every 1.4 nanocrystals have 1 P$_b$ defect, which means at least 30% of the Si nanocrystals are P$_b$ defect free, and potentially optically active. P$_b$ defects are deep nonradiative recombination centers, probably located at Si/SiO$_2$ interface and known to quench the PL arising from Si nanocrystals. Magneto-PL measurements were performed in pulsed magnetic field up to 50 T since free exciton of nanocrystals are expected to show blueshift unlike the localized ones in high magnetic fields. The authors observed a very small shift and concluded that the PL is dominated by highly localized defect states. After annealing the sample under hydrogen atmosphere, ESR analysis evidenced that the all the paramagnetic defects were removed from the matrix. After the H passivation, the increase in the PL signal is solely attributed to QCE, since a parabolic shift of ~1.5 meV was also observed in magneto-PL experiment. Using UV illumination from an Ar + laser, the effect of the passivation was reversed and the defects were reintroduced. As a result, it is shown that the origin of PL can be classified using high magnetic fields as from either QCE or defects and the mechanism of PL can be altered to QCE by defect passivation and to localized defects by UV illumination.

References

1 Schamm, S., Bonafos, C., Coffin, H., Cherkashin, N., Carrada, M., Ben Assayag, G., Claverie, A., Tence, M., and Colliex, C. (2008) *Ultramicroscopy*, **108**, 346–357.

2 Mayandi, J., Finstad, T.G., Thogersen, A., Foss, S., Serincan, U., and Turan, R. (2007) *Thin Solid Films*, **515**, 6375–6380.

3 Jenguillaume, C. and Colliex, C. (1989) *Ultramicroscopy*, **28**, 252–257.

4 Gass, M.H., Koziol, K.K., Windle, A.H., and Midgley, P.A. (2006) *Nano Lett.*, **6**, 376.

5 Frank, J. (1992) *Electron Tomography: Three Dimensional Imaging with the Transmission Electron Microscope*, Plenum Press, New York.

6 Yurtsever, A., Weyland, M., and Muller, D.A. (2006) *Appl. Phys. Lett.*, **89**, 151920.

7 Egerton, R.F., Li, P., and Malac, M. (2004) *Micron*, **35**, 399–409.

8 Mayandi, J., Finstad, T.G., Foss, S., Thorgesen, A., Serincan, U., and Turan, R. (2006) *Phys. Scr.*, **T126**, 77–80.

9 Markiewicz, P. and Goh, M.C. (1995) *J. Vac. Sci. Technol. B*, **13**, 1115.

10 Long, D.A. (1997) *Raman Spectroscopy*, McGraw-Hill Inc.

11 Ferraro, J.R. and Nakamoto, K. (1994) *Introductory Raman Spectroscopy*, Academic Press Limited.

12 Yerci, S., Serincan, U., Dogan, I., Tokay, S., Genisel, M., Aydinli, A., and Turan, R. (2006) *J. Appl. Phys.*, **100**, 074301.

13 Richter, H., Wang, Z.P., and Ley, L. (1981) *Solid State Commun.*, **39**, 625.

14 Campbell, I.H. and Fauchet, P.M. (1986) *Solid State Commun.*, **58**, 739.

15 Paillard, V., Puech, P., Laguna, M.A., Carles, R., Kohn, B., and Huisken, F. (1999) *J. Appl. Phys.*, **86**, 1921.

16 Zi, J., Buscher, H., Falter, C., Lugwig, W., Zhang, K., and Xie, X. (1996) *Appl. Phys. Lett.*, **69**, 200.

17 Islam, M.N. and Kumar, S. (2001) *Appl. Phys. Lett.*, **78**, 715.

18 Faraci, G., Gibilisco, S., Russo, P., and Penniisi, A. (2006) *Phys. Rev. B*, **73**, 033307.

19 Cheng, W. and Reng, S.-F. (2002) *Rhys. Rev. B*, **65**, 205305.

20 Mishra, P. and Jain, K.P. (2002) *Mater. Sci. Eng. B*, **95**, 202.

21 Macia, J., Martin, E., Perez-Rodriquez, A., Jimenez, J., Morante, J.M., Aspar, B., and Margial, J. (1997) *J. Appl. Phys.*, **82**, 3730.

22 Yoshikawa, M., Murakami, M., Matsuda, K., Matsunobe, T., Sugie, S., Okada, K., and Ishada, H. (2005) *J. Appl. Phys.*, **98**, 063531.

23 Cerdeira, F., Buchenauer, C., Pollak, F.H., and Cardona, M. (1972) *Phys. Rev. B*, **5**, 580.

24 Iacona, F., Franzò, G., Ceretta Moreira, E., and Priolo, F. (2001) *J. Appl. Phys.*, **89**, 8354.

25 Franzò, G., Irrera, A., Ceretta Moreira, E., Miritello, M., Iacona, F., Sanfilippo, D., Di Stefano, G., Fallica, P.G., and Priolo, F. (2002) *Appl. Phys. A*, **74**, 1.

26 Viera, G., Huet, S., and Boufendi, L. (2001) *J. Appl. Phys.*, **90**, 4175.

27 Bustarret, E., Hachicha, M.A., and Brunel, M. (1988) *Appl. Phys. Lett.*, **52**, 1675.

28 Laue, M.v. (1913) *Ann. Phys.*, **41**, 971.

29 Bragg, W.L. (1913) *Proc. Cambridge Philos. Soc.*, **17**, 43.

30 Scherrer, P. (1918) *Gottinger Nachrichten*, **2**, 98; an extensive discussion on Scherrer's formula and the value of K can be found in Ref. [9].

31 Morales, M., Leconte, Y., Rizk, R., and Chateigner, D. (2004) *J. Appl. Phys.*, **97**, 034307.

32 Comedi, D., Zalloum, O.H., Irving, E.A., Wojcik, J., Roschuk, T., Flynn, M.J., and

Mascher, P. (2006) *J. Appl. Phys.*, **99**, 023518.

33 Riabinina, D., Durand, C., Margot, J., Chaker, M., Botton, G.A., and Rosei, F. (2006) *Phys. Rev. B*, **74**, 075334.

34 Dinh, L.N., Chase, L.L., Balooch, M., Siekhaus, W.J., and Wooten, F. (1996) *Phys. Rev. B*, **54**, 5029.

35 Klug, H.R. and Alexander, L.E. (1974) *X-Ray Diffraction Procedures for Polycrystalline and Amorphous Materials*, John Wiley & Sons, Inc., New York.

36 Pielaszek, R. (2004) *J. Alloys Compd.*, **382**, 128–132.

37 Warren, B.E. (1969) *X-Ray Diffraction*, Addison-Wesley, New York, pp. 251–314.

38 Stokes, A.R. and Wilson, A.J.C. (1944) *Proc. Phys. Soc. London*, **56**, 174.

39 Zhang, Z. and Lavernia, E.J. (2003) *Metall. Mater. Trans. A*, **34A**, 1349.

40 Kohli, S., Theil, J.D., Snyder, R.D., Rithner, C.D., and Dorhout, P.K. (2003) *J. Vac. Sci. Technol. B*, **21**, 719.

41 Bubert, H. and Janett, H. (eds) (2002) *Surface and Interface Analysis*, Wiley-VCH Verlag GmbH.

42 Hufner, S. (2003) *Photoelectron Spectroscopy Principles and Applications*, Springer-Verlag, Berlin.

43 Shirley, D.A. (1972) *Phys. Rev.*, **55**, 4709.

44 Tougaard, S. (1988) *Surf. Interface Anal.*, **11**, 453.

45 Kim, S., Kim, M.C., Choi, S.-H., Kim, K.J., Hwang, H.N., and Hwang, C.C. (2007) *Appl. Phys. Lett.*, **91**, 103113.

46 Renault, O., Marlier, R., Gely, M., De Salvo, B., Baron, T., Hansson, H., and Barrett, N.T. (2005) *Appl. Phys. Lett.*, **87**, 163119.

47 Yerci, S., Yildiz, I., Kulakci, M., Serincan, U., Barozzi, M., Bersani, M., and Turan, R. (2007) *J. Appl. Phys.*, **102**, 024309.

48 Dreiner, S., Schurmann, M., Krause, M., Berges, U., and Westphal, C. (2005) *J. Electron. Spectrosc. Relat. Phenom.*, **405**, 144–147.

49 Thogersen, A., Diplas, S., Mayandi, J., Finstad, T., Olsen, A., Watts, J.F., Mitome, M., and Bando, Y. (2008) *J. Appl. Phys.*, **103**, 024308.

50 Min, S.K., Shcheglov, K.V., Yang, C.M., Atwater, H., Brongersma, M.L., and Polman, A. (1996) *Appl. Phys. Lett.*, **69**, 2033.

51 Liu, Y., Chen, T.P., Ding, L., Zhang, S., Fu, Y.Q., and Fung, S. (2006) *J. Appl. Phys.*, **100**, 096111.

52 Scheer, K.C., Rao, R.A., Muralidhar, R., Bagchi, S., Conner, J., Lozano, L., Perez, C., Sadd, M., and White, B.E., Jr. (2003) *J. Appl. Phys.*, **93**, 5637.

53 Gillet, J.-N. and Meunier, M. (2005) *J. Phys. Chem. B*, **109**, 8733.

54 Oswald, S. and Reiche, R. (2001) *Appl. Phys. Sci.*, **179**, 307.

55 Mulloni, V., Bellutti, P., and Vanzetti, L. (2005) *Surf. Sci.*, **585**, 137.

56 Liu, Y., Chen, T.P., Ng, C.Y., Ding, L., Zhang, S., Fu, Y.Q., and Fung, S. (2006) *J. Phys. Chem. B*, **110**, 16499.

57 Chen, T.P., Lui, Y., Sun, C.Q., Tse, M.S., Hsieh, J.H., Fu, Y.Q., Lin, Y.C., and Fund, S. (2004) *J. Phys. Chem. B*, **108**, 16609.

58 Dane, A., Demirok, U.K., Aydinli, A., and Suzer, S. (2006) *J. Phys. Chem. B*, **110**, 1137.

59 Yi, L.X., Heitmann, J., Scholz, R., and Zacharias, M. (2002) *Appl. Phys. Lett.*, **81** (22), 4248.

60 Berreman, D.W. (1963) *Phys Rev.*, **130** (6), 2193.

61 Lange, P. (1988) *J. Appl. Phys.*, **66** (1), 201.

62 Montero, I., Galan, L., Najimi, O., and Albella, J.M. (1994) *Phys. Rev. B*, **50** (7), 4881.

63 Liu, Y., Chen, T.P., Fu, Y.Q., Tse, M.S., Hsieh, J.H., Ho, P.F., and Liu, Y.C. (2003) *J. Phys. D: Appl. Phys.*, **36**, L97–L100.

64 Tsu, D.V., Lucovsky, G., and Davidson, B.N. (1989) *Phys. Rev. B*, **40** (3), 1795.

65 Kirk, C.T. (1988) *Phys. Rev. B*, **38**, 1255–1273.

66 Nesheva, D., Bineva, I., Levi, Z., Areneva, Z., Merdzahanova, Ts., and Pivin, J.C. (2003) *Vacuum*, **68**, 1.

67 Turan, R. and Finstand, T.G. (1992) *Semicond. Sci. Technol.*, **7**, 75.

68 Drinek, V., Pola, J., Bastl, Z., and Subrt, J. (2001) *J. Non-Cryst. Solids*, **288**, 30.

69 Yamamoto, M., Koshikawa, T., Yasue, T., Harima, H., and Kajiyama, K. (2000) *Thin Solid Films*, **369**, 100.

70 Serincan, U., Yerci, S., Kulakci, M., and Turan, R. (2005) *Nucl. Instrum. Methods Phys. Res. B*, **239**, 419.

71 Trshova, M., Zerrek, J., and Jurek, K. (1997) *J. Appl. Phys.*, **82** (7), 3519.

72 Swart, H.C., Van Hattum, E.D., Arnoldbik, W.M., and Habraken, F.H.P.M. (2004) *Phys. Status Solidi C*, **1**, 2286.

73 Serincan, U., Kartopu, G., Guennes, A., Finstad, T.G., Turan, R., Ekinci, Y., and Bayliss, S.C. (2004) *Semicond. Sci. Technol.*, **19**, 247.

74 Canham, L.T. (1990) *Appl. Phys. Lett.*, **57**, 1046.

75 Wang, X.X., Zhang, J.G., Ding, L., Cheng, B.W., Ge, W.K., Yu, J.Z., and Wang, Q.M. (2005) *Phys. Rev. B*, **72**, 195313.

76 Iacona, F., Guillois, O., Porterat, D., Reynaud, C., Huisken, F., Kohn, B., and Paillard, V. (2000) *J. Appl. Phys*, **87**, 1295.

77 Heitmann, J., Müller, F., Zacharias, M., and Gösele, U. (2005) *Adv. Mater.*, **17**, 795.

78 Kanemitsu, Y. and Ogawa, T. Shiraishi, K. and Takeda, K. (1993) *Phys. Rev. B*, **48**, 4883.

79 Takagahara, H. and Takeda, K. (1992) *Phys. Rev. B*, **46**, 15578.

80 Proot, J.P., Delerue, C., and Allan, G. (1992) *Appl. Phys. Lett.*, **61**, 1948.

81 Ledoux, G., Guillois, O., Porterat, D., Reynaud, C., Huisken, F., Kohn, B., and Paillard, V. (2000) *Phys. Rev. B*, **62**, 15942.

82 Ögüt, S., Chelikowsky, J.R., and Louie, S.G. (1997) *Phys. Rev. Lett*, **79**, 1770.

83 Wolkin, M.V., Jorne, J., Fauchet, P.M., Allan, G., and Delerue, C. (1999) *Phys. Rev. Lett.*, **82**, 197.

84 Hadjisavvas, G. and Kelires, P.C. (2004) *Phys. Rev. Lett.*, **93**, 226104.

85 Kobitski, A.Y., Zhuravlev, K.S., Wagner, H.P., and Zahn, D.R.T. (2001) *Phys. Rev. B*, **63**, 115423.

86 Fauchet, P.M. *et al.* (1996) *J. Lumin*, **70**, 294.

87 Averboukh, B., Huber, R., Cheah, K.W., Shen, Y.R., Qin, G.G., Ma, Z.C., and Zong, W.H. (2002) *J. Appl. Phys.*, **92**, 3564.

88 Wang, J. *et al.* (2008) *Solid State Commun.*, **147**, 461.

89 Vial, J.C., Bsiesy, A., Gaspard, F., Herino, R., Ligeon, M., Muller, F., Romestain, R., and Macfarlane, R.M. (1992) *Phys. Rev. B*, **45**, 14171.

90 Kanemitsu, Y. (1996) *J. Lumin.*, **70**, 333–342.

91 Kanemitsu, Y. (1994) *Phys. Rev. B*, **49**, 16845.

92 Lockwood, D.J. (ed.) (1998) *Light Emission in Silicon: From Physics to Device*, Academic Press, New York.

93 Calcott, P.D., Nash, K.J., Canham, L.T., Kane, M.J., and Brumhead, D. (1993) *J. Phys. Condens. Matter*, **5**, L91.

94 Pavesi, L. and Ceschini, M. (1993) *Phys. Rev. B*, **48**, 17625.

95 Brongersma, M.L., Kik, P.G., Polman, A., Min, K.S., and Atwater, H.A. (2000) *Appl. Phys. Lett.*, **76**, 351.

96 Linnros, J., Galeckas, A., Lalic, N., and Grivickas, V. (1997) *Thin Solid Films*, **297**, 167.

97 Kakalios, R., Kakalios, J., Street, R.A., and Jackson, W.B. (1997) *Phys. Rev. Lett.*, **59**, 1037.

98 Linnros, J., Lalic, N., Galeckas, A., and Grivickas, V. (1999) *J. Appl. Phys.*, **86**, 6128.

99 Dovrat, M., Goshen, Y., Jedrzejewski, J., Balberg, I., and Sa'ar, A. (2004) *Phys. Rev. B*, **69**, 155311.

100 Godefroo, S., Hayne, M., Jivanescu, M., Stesmans, A., Zacharias, M., Lebedev, O., Van Tendeloo, G., and Moshchalkov, V.V. (2008) *Nat. Nanotechnol.*, **3**, 174.

101 Gencer Imer, A., Yerci, S., Alagoz, A.S., Kulakci, M., Serincan, U., Finstad, T.G., and Turan, R. (2010) *J. Nanosci. Nanotechnol.*, **10**, 525.

Index

Silicon Nanocrystals: Fundamentals, Synthesis and Applications. Edited by Lorenzo Pavesi and Rasit Turan
Copyright © 2010 WILEY-VCH Verlag GmbH & Co. KGaA, Weinheim
ISBN: 978-3-527-32160-5